Alpha Herpesviruses

Molecular and Cellular Biology

Edited by Rozanne M. Sandri-Goldin

Caister Academic Press

Caister Academic Press
32 Hewitts Lane
Wymondham
Norfolk NR18 0JA
UK

www.caister.com

British Library Cataloguing-in-Publication Data

A catalogue record for this book is available from the British Library
ISBN10: 1-904455-09-3
ISBN13: 978-1-904455-09-7

Printed and bound in Great Britain

Contents

Contributors

Ann M. Arvin
Stanford University School of Medicine
Stanford, CA
USA

aarvin@stanford.edu

Joel D. Baines
Cornell University
Ithaca, NY
USA

jdb11@cornell.edu

Bruce W. Banfield
Department of Microbiology
University of Colorado Health Sciences Center at Fitzsimmons
Aurora, CO
USA

John A. Blaho
Department of Microbiology
Mount Sinai School of Medicine
New York, NY
USA

john.blaho@mssm.edu

David C. Bloom
Department of Molecular Genetics and Microbiology
University of Florida College of Medicine
Gainesville
USA

dbloom@mgm.ufl.edu

Gabriella Campadelli-Fiume
Department of Experimental Pathology
Alma Mater Studiorum University of Bologna
Bologna
Italy

gabriella.campadelli@unibo.it

Randall J. Cohrs
Department of Neurology
University of Colorado Health Sciences Center
Denver
Colorado
USA

Donald M. Coen
Department of Biological Chemistry & Molecular Pharmacology
Harvard Medical School
Boston, MA
USA

donald_coen@hms.harvard.edu

Steven Deitch
Department of Neurology
University of Colorado Health Sciences Center
Denver, CO
USA

Carol Duffy
Cornell University
Ithaca, NY
USA

Roger D. Everett
MRC Virology Unit
Church Street
Glasgow
Scotland
UK

r.everett@vir.gla.ac.uk

Kathryn A. Fraser
Department of Microbiology
University of Minnesota
Minneapolis, MN
USA

Tatiana Gianni
Department of Experimental Pathology
Alma Mater Studiorum University of Bologna
Bologna
Italy

Dr. Donald H. Gilden
Department of Neurology
University of Colorado Health Sciences Center
Denver, CO
USA

don.gilden@uchsc.edu

Charles Grose
University of Iowa Hospital
Iowa City, IA
USA

charles-grose@uiowa.edu

Jeremy O. Jones
Stanford University School of Medicine
Stanford, CA
USA

Chia Chi Ku
Stanford University School of Medicine
Stanford, CA
USA

Gary D. Luker
Department of Radiology
University of Michigan Medical School
Ann Arbor, MI
USA

gluker@umich.edu

Lucie Maresova
Virology Laboratory
Department of Pediatrics
University of Iowa
Iowa City, IA
USA

Ravi Mahalingam
Department of Neurology
University of Colorado Health Sciences Center
Denver, CO
USA

Guruprasad Medigeshi
Vollum Institute
Oregon Health & Science University
Portland, OR
USA

Ian Mohr
New York University School of Medicine
Dept. of Microbiology
New York, NY
USA

ian.mohr@med.nyu.edu

Karen L. Mossman
Department of Pathology and Molecular Medicine
McMaster University
Hamilton, ON
Canada

mossk@mcmaster.ca

Christopher D. Pack
Department of Pathology
Emory University School of Medicine
Atlanta, GA
USA

btr@utk.edu

Stephen A. Rice
Department of Microbiology
University of Minnesota
Minneapolis, MN
USA

ricex019@umn.edu

Dr. Bernard Roizman
Department of Microbiology
The Marjorie B. Kovler Viral Oncology Laboratories
The University of Chicago
Chicago, IL
USA

bernard.roizman@bsd.uchicago.edu

Barry T. Rouse
Department of Microbiology
The University of Tennessee
Knoxville, TN
USA

William T. Ruyechan
Department of Microbiology and Immunology and Witebsky Center for Microbial Pathogenesis and Immunology
University at Buffalo
State University of New York
Buffalo, NY
USA

ruyechan@buffalo.edu

Rozanne M. Sandri-Goldin
Department of Microbiology and Molecular Genetics
School of Medicine
University of California
Irvine, CA
USA

rmsandri@uci.edu

Christine M. Sanfilippo
Department of Microbiology
Mount Sinai School of Medicine
New York, NY
USA

Gregory K. Scott
Vollum Institute
Oregon Health & Science University
Portland, OR
USA

Anne C. Schaap
Stanford University School of Medicine
Stanford, CA
USA

Gregory A. Smith
Department of Microbiology-Immunology
Northwestern University Feinberg School of Medicine
Ward Building
Chicago. IL
USA

g-smith3@northwestern.edu

Marvin Sommer
Stanford University School of Medicine
Stanford, CA
USA

Brunella Taddeo
Department of Microbiology
The Marjorie B. Kovler Viral Oncology Laboratories
The University of Chicago
Chicago, IL
USA

Gary Thomas
Vollum Institute
Oregon Health & Science University
Portland, OR
USA

Edward K. Wagner
Department of Molecular Biology and Biochemistry
University of California
Irvine, CA
USA

Dr Wagner died in January 2006

Sandra K. Weller
Professor and Chair of Molecular Microbial and Structural Biology
University of Connecticut Health Center
Farmington, CT
USA

weller@nso2.uchc.edu

Leigh Zerboni
Stanford University School of Medicine
Stanford, CA
USA

Preface

In recent years, the study of the pathogenesis, molecular biology, and control of infection of the alpha-herpesviruses has made great strides. The chapters in this volume summarize and highlight some of the more provocative and exciting findings, each with a view to probing or defining the mechanisms of viral infection. For this reason, not all aspects of the biology of herpesvirus infection will be covered here. But instead, the emphasis is on recent advances. Each chapter herein serves as a review of the area but concentrates more on the latest developments rather than on a more historical perspective. As a consequence, transcriptional regulation during herpes simplex virus type 1 (HSV-1) infection is not covered here because the roles of VP16 and ICP4 have been reviewed extensively in the past. Instead, transcriptional control during varicella zoster virus (VZV) and the roles of VZV immediate early proteins have been reviewed (Chapter 1, Ruyechan, W.T.) because the contributions of the IE proteins to VZV infection are still being delineated. However, the modification of the host cellular RNA polymerase II during HSV-1 infection has been described with an eye to recent findings on the differences in phosphorylated species compared with uninfected cells and the potential implications of these modifications to viral and host transcription (Chapter 2, Rice, S.A. and Fraser, K.A.). The intensely studied HSV-1 IE proteins, ICP0 (Chapter 3, Everett, R.D.) and ICP27 (Chapter 4, Sandri-Goldin, R.M.), are covered in detail because each protein has been shown to be multifunctional and to be involved in a number of diverse viral regulatory events and virus–cell interactions. Furthermore, advances in our understanding of DNA replication in HSV-1 have been detailed here (Chapter 5, Weller, S.K.) because our conception of this process has been moved forward with new insights into the roles and properties of the proteins in the replication complex. Similarly, new information on the regulation of HSV-1 gene expression at the level of translational control has been detailed here (Chapter 6, Mohr, I.) because this is an emerging story. Further, the emergence of new technologies and their use in probing HSV-1 gene expression has been represented with a chapter on the development and use of microarrays (Chapter 7, Wagner, E.K.).

The important topics of viral entry and egress and the role of glycoproteins in these processes have been covered here for both HSV-1 (Chapter 8, Campadelli-Fiume, G. and Gianni, T.) and VZV (Chapter 9, Grose, C. *et al.*) because of some striking differences between the two viruses. In contrast, more of the process of assembly into virions and envelopment has been uncovered for HSV-1 and, therefore, the summary of what is known of these events is limited to HSV-1 (Chapter

10, Baines, J.D. and Duffy, C.). Recent technological advances in the development of fluorescent probes that can be introduced into the viral genome have made it possible to track virus particles during infection in living cells. The development and use of alphaherpesviruses expressing fluorescent proteins is described herein (Chapter 11, Smith, G.A. and Banfield, B.W.).

The subversion of host cell barriers to viral replication has been approached from several different perspectives. The induction of apoptosis and the HSV-1 gene products that interfere with cell death (Chapter 12, Sanfilippo, C.M. and Blaho, J.A.), and the mechanism by which HSV-1 disrupts interferon responses (Chapter 13, Mossman, K.L.), demonstrate how effective the virus is in replicating efficiently without actually killing its host. In Chapter 14, intriguingly titled "HSV-1 and the host cell: a story of global conspiracies, plots and counterplots" (Roizman, B. and Taddeo, B.), this point is lucidly illustrated in the detailed review of the interactions of just a few HSV-1 multifunctional proteins with host cellular pathways. The success of alphaherpesviruses in replicating *in vivo* is not, however, without consequence to the host. The discussion of molecular mechanisms of VZV pathogenesis through the development and use of the SCID-hu-mouse model (Chapter 15, Arvin, A.M. *et al.*) and a summary of what is known of VZV pathogenesis in humans, including serious neurological complications (Chapter 16, Gilden, D.H. *et al.*), clearly illustrate that herpesviruses cause disease. A striking feature of all herpesviruses is their ability to become latent in the host, and thus persist for the lifetime of the individual with the ability to reactivate and cause symptoms anew. What is known of the molecular basis of HSV-1 latency and the roles of the LATs (Chapter 17, Bloom, D.C.) as well as our current understanding of VZV latency (Chapter 16, Gilden, D.H. *et al.*) is presented here.

The host is not a passive player during infection with herpesviruses—an immune response is mounted and is reviewed here (Chapter 18, Pack, C.D. and Rouse, B.T.). However, despite considerable efforts to develop vaccines against HSV, efficacious vaccines are not yet available. In contrast, there is an effective vaccine against VZV. While vaccine development has not yet been successful for HSV, drug therapies have been much more successful beginning with acyclovir, which is one of the most successful antiviral agents. Details of drug design are outlined here with an emphasis on how genetic, biochemical, and structural analyses have revealed details on the mechanisms of action of antiherpetic drugs (Chapter 19, Coen, D.M.). The final entry in this compendium describes exciting new approaches in imaging HSV in living animals to gain insights about the effects of host immunity or therapeutic interventions on HSV pathogenesis (Chapter 20, Luker, G.D.). I hope it will become clear to the reader that, while much has been learned about herpesvirus infection, much still remains to be learned. However, new approaches and technologies combined with more standard tools should allow us to continue to unravel the complex interplay of the virus and the host.

Rozanne M. Sandri-Goldin

Varicella Zoster Virus Transcriptional Regulation and the Roles of VZV IE Proteins

1

William T. Ruyechan

Abstract

Varicella zoster virus (VZV) is the causative agent of varicella (chickenpox) upon primary infection. The virus is spread from its initial sites of replication through T cell viremias to the internal organs and skin. VZV establishes latency in dorsal root ganglia and upon reactivation causes zoster (shingles). During productive infection the entire coding capacity of the VZV genome (some 69 unique open reading frames) is believed to be expressed. In contrast, during latent infection, a small number of lytic phase genes are expressed while the remainder of the genome is silent. Thus, the biology of VZV infection indicates that control of expression of the viral genome is essential to the ability of the virus to replicate productively in a variety of cell types with minimal complications for the host, and to the establishment and maintenance of latency in neurons. In recent years, our understanding of some of the mechanisms of VZV gene expression and of the proteins which regulate those processes has significantly increased. This includes information on the roles of individual viral proteins and the involvement of ubiquitous cellular transcription factors.

Introduction

Varicella zoster virus (VZV) is a member of the Herpesviridae and one of eight herpesviruses which infect humans as their natural host. VZV is the causative agent of varicella (chickenpox) during primary infection and zoster (shingles) after reactivation from latency. The site of latent VZV infection lies primarily in the dorsal root ganglia (DRG) and based on this, in conjunction with the structure of the viral genome, VZV is designated as an alpha neurotropic herpesvirus as are herpes simplex virus types 1 and 2 (HSV-1 and HSV 2) (Arvin, 2001).

As will be described in more detail below, the biology of VZV infection involves numerous organs and cell types of the infected host. In each of these cell types there is a careful balance struck between the need for viral replication and propagation and the ability of the infected cell and the human host to survive the infection. Thus, regulation of expression of the viral genome is of paramount importance. In the case of VZV this regulation is carried out by a small number of viral genes in concert with the cellular transcription apparatus (Cohen and Straus, 2001). This chapter will describe our current understanding of the roles of the viral and cellular proteins known to be involved in expression of the VZV genome.

Structure of the VZV virion and viral genome

The virion

The VZV virion, like those of all herpesviruses is enveloped and approximately spherical, with a diameter of 100–300 nm. The viral double-stranded DNA genome is contained in an icosahedral nucleocapsid 100 nm in diameter that is surrounded by or embedded within a structure known as the tegument (Fig. 1.1). The tegument is composed of virus-encoded proteins, several of which are involved in expression of the viral genome. The thickness of the tegument can vary, even within a single virion. The lipid envelope of the virus, derived from cellular membranes, surrounds the tegument and contains a minimum of nine viral glycoproteins (Cohen and Straus, 2001).

The viral genome

The VZV genome is a linear, double-stranded DNA molecule comprised of approximately 125 kbp nucleotides with a G+C content of 46%. The genome consists of two covalently joined segments, the unique long (U_L) and unique short (U_S) regions, each bounded by a terminal repeat and an internal inverted repeat (TR_L/IR_L and TR_S/IR_S, respectively) (Davison and Scott, 1986). The repeats allow structural rearrangements of the unique regions to occur during viral DNA replication and, as a result, the encapsidated genomes exist as a mixture of four isomers. Two isomers are found predominantly in viral particles and are termed P (prototype) and I_S (inverted S). Two other isomers, which contain the L region inverted with respect to the S region in the P and I_S configurations, are designated I_L and I_{SL}. These represent 2–5% of the virion DNA. The preponderance of only two isomers is believed to reflect differences in recombination efficiency due, in part, to the significant difference in size of the inverted sequences bounding the U_S region (7.3 kbp) as compared with the U_L region (88.5 bp). A small amount of

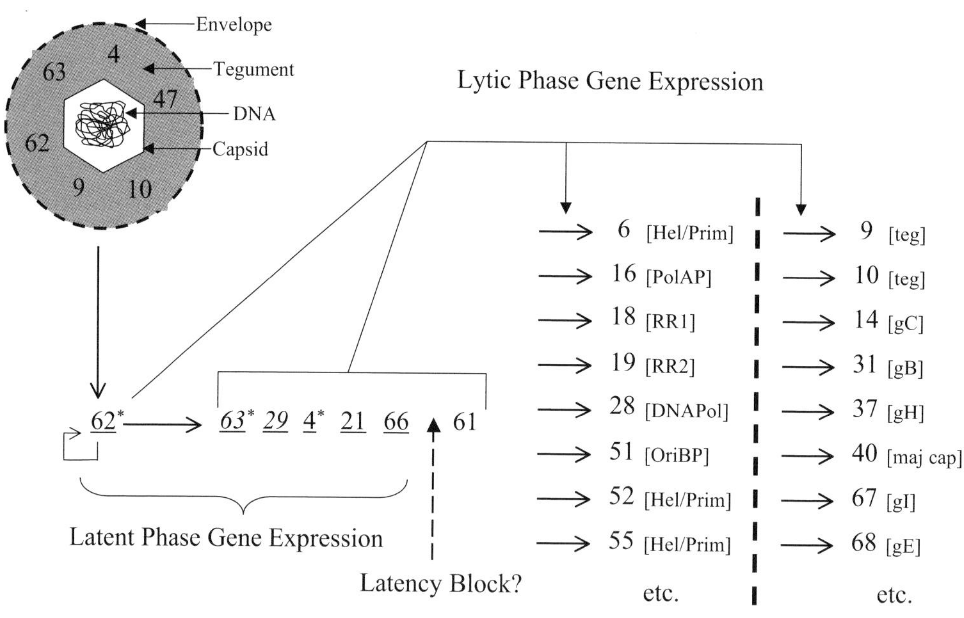

Figure 1.1 Schematic of the VZV virion and the current data on lytic and latent phase gene expression. Numbers refer to the open reading frames (ORFs) within the VZV genome. Genes whose expression has been detected in latently infected tissue are underlined. All genes indicated to the right of the latency block are not expressed during latency. The two ORFs most abundantly expressed in latently infected tissues are indicated in italics. The VZV IE genes are denoted by asterisks.

full length closed circular DNA (representing 0.1–5% of total) has also been isolated from virions. The function of these molecules in relation to the viral life cycle remains unknown (Ruyechan and Hay, 2000).

Analysis of the complete sequence of the VZV Dumas strain (Davison and Scott, 1986) revealed the coding potential for some 71 genes which were named numerically according to the location of their open reading frame (ORF) within the DNA structure relative to the 5′ end of the genome. The viral genome contains at least 69 unique ORFs, the first of which (ORF1) is located at the 5′ end of the genome with ORF68 the last single-copy sequence at the 3′ end of the U_S segment. ORF62 to ORF64 are duplicated within the TR_S and IR_S as ORFs 69 to 71, respectively. These ORFs encode proteins involved in regulation of expression of viral genes, replication of the viral DNA and regulation of nucleotide pools during infection, and structural proteins including those required for DNA packaging, capsid assembly, as well as the viral glycoproteins (Cohen and Straus, 2001).

Productive infection

Productive VZV infection begins with fusion of the virus envelope with the membrane of a susceptible cell. This occurs via interactions between one or more of the VZV glycoproteins and as yet unidentified receptors on the cell surface. Upon entry and uncoating, the nucleocapsid is translocated to the nuclear membrane and the viral genome enters the nucleus presumably through nuclear pores. Viral tegument proteins also enter the nucleus including several which regulate viral gene expression (see below). Transcription and viral DNA replication are then believed to occur in a regulated temporal cascade typical of all herpesviruses. Both viral and cellular proteins are involved in these processes. Concatameric progeny viral DNA molecules, presumably generated via a rolling circle mechanism, are cleaved and packaged into preassembled nucleocapsids. Tegument assembly is believed to at least in part occur in the nucleus. The nascent virion buds through the nuclear membrane to obtain an envelope. This enveloped particle is then excreted from the nucleus by an as yet undiscovered mechanism or buds into the cytoplasm, losing the initial envelope, and then buds again into the Golgi apparatus to regain an envelope into which the viral glycoproteins are inserted yielding infectious virions. The virus grows slowly in tissue culture, is highly cell associated, and purified virions exhibit low levels of infectivity making genetic and molecular characterization of the virus difficult. In contrast viral replication appears to be quite efficient in infected individuals resulting in high concentrations of infectious virus within the vesicular lesions typical of both varicella and zoster (Arvin, 2001, Wang *et al.*, 2001).

Pathogenesis of primary VZV Infection

VZV is transmitted primarily through the respiratory route via virus in air-borne water droplets or through close contact. The site of initial infection is most likely the conjunctivae or the nasal mucosa. The location of the primary cycles of replication remains unknown but is believed to include the regional cervical lymphatic tissue and the lungs. In the current model of VZV pathogenesis, replication at the local site is followed by primary viremia and distribution of the virus throughout the body via T cell infection, which is defined as the first phase of pathogenesis. The major internal sites of viral replication are not known with certainty. However, histologic data collected during autopsies of children who succumbed to neonatal chickenpox indicate that the liver and spleen are involved. After replication at internal sites, a secondary viremia occurs during which VZV is again released into the bloodstream and the virus quickly invades the cutaneous tissues and results in chickenpox (varicella). The viremia is detectable from 5 days before appearance of rash up to 3 days after the appearance of the varicella rash, which are within the second phase of pathogenesis. VZV tropism for $CD4^+$ and $CD8^+$ T lymphocytes which carry homing markers for skin has been demonstrated and it is these cells which are believed to be responsible for the spread of the virus during the viremic phases of infection (Arvin, 2001, Ku *et al.*, 2005).

Latency and reactivation

The primary infection model described above suggests two mechanisms by which VZV can access sensory nerve tissue. The first mechanism entails direct hematogenous spread to sensory ganglia; the second possible route to the ganglia is by retrograde axonal transport from infected epidermal and dermal tissues. Once VZV has infected neural tissues, the virus establishes a latent infection where the majority of the viral genome is transcriptionally silent and no infectious virus is produced. Latent VZV infection is maintained in neurons (primarily) and satellite cells (infrequently) in dorsal root ganglia. The mechanisms by which establishment and maintenance of the latent state occur are not known but appear to be different than those involving HSV. VZV does not encode the latency associated transcript (LAT) sequence, expression of which is associated with HSV latency. In contrast expression of several lytic phase genes at the levels of both RNA and protein has been detected in dorsal root ganglia from human cadavers and in animal model studies (Kennedy, 2002; Mitchell *et al.*, 2003).

Upon reactivation it is assumed that the molecular mechanisms permitting VZV to persist latently in neural tissues are compromised and a switch to productive virus replication occurs. The full set of viral genes is expressed, new virions are assembled, and transported by anterograde axonal flow to nerve terminals in the dermis. There, progeny virus is released; and spreads to adjacent epithelial cells and the infection is manifested clinically as zoster or shingles, mainly in the thoracic and cranial areas, and involving a single dermatome (Silverstein and Straus, 2000). Based on these descriptions of the pathogenesis of varicella and zoster infection, it is clear that the virus has adapted to replication and therefore expression of its genome in a variety of cell and tissue types in order to spread from cell to cell and from individual to individual. Potential mechanisms by which VZV expresses its genome during productive infection will be the subject of the remainder of this chapter.

Overview of varicella zoster virus gene regulation

Upon its entry into the infected cell nucleus, the VZV genome is transcribed by the host cell RNA polymerase II (RNA Pol II) and the cell's general transcription apparatus. The ORFs are believed to be expressed as three distinct kinetic classes during viral infection: immediate early (IE), early (E) and late (L) by analogy with other alpha-herpesviruses. Clear evidence for this temporal cascade, however, has been difficult to obtain due to the low titers and cell associated nature of VZV grown in tissue culture. Only three proteins have been shown to be expressed under IE conditions. These include IE4, IE63 and IE62, the last two of which are expressed from ORFs, 62 and 71 and 63 and 70, respectively (Kinchington and Cohen, 2000).

VZV promoters have, in general, been shown to be essentially quiescent in the absence of virus encoded transcriptional activators (Perera *et al.*, 1992, Kantakamalakul *et al.*, 1995). Several VZV-encoded proteins have been shown to be capable of transactivating and/or transrepressing specific viral promoters, promoters from heterologous virus families and cellular promoters in transient transfection assays. These include the VZV IE62, IE4, IE63, ORF10, and ORF61 proteins. The IE62 protein, however, is considered to be the primary viral transactivator during the lytic phase of VZV replication. It is known to activate expression of genes from all three putative kinetic classes of VZV genes and can act independently of all other viral factors (Kinchington and Cohen, 2000). When other viral factors are present, they have been shown to be capable of modulating the level of IE62 activation, even, in some cases where the promoter in question appeared responsive only to IE62 (Perera *et al.*, 1992, Lynch *et al.*, 2002).

VZV promoter structure

To date, a relatively small number of VZV promoters have been analyzed in detail. However, the results from these analyses as well as predictions of the positions of putative promoters within the VZV genome, indicate they appear to be relatively typical RNA Pol II promoters (Smale and Kadonaga, 2003) with TATA-like elements and binding sites for ubiquitous cellular transcription factors upstream of the TATA elements. Owing to the extremely compact nature of the VZV genome many promoters are predicted to lie completely or partially within the coding sequences of genes immediately upstream. As noted above, a characteristic of virtually all VZV promoters thus far tested is the fact that they show very low levels of activity in the absence of VZV transcriptional regulatory proteins, but the vast majority show a significant increase in activity in the presence of the IE62 protein. VZV promoters are also characterized by the fact that the TATA-elements required and/or predicted to be positioned correctly for expression are frequently atypical and therefore less likely to be bound with high affinity by the TATA-binding protein (TBP) (Davison and Scott, 1986; Ruyechan *et al.*, 2003). Thus far, no TATA-less VZV promoters have been identified and no promoters showing dependence on the presence of an initiator element (INR) at or just downstream of the start site of transcription are known. This most likely reflects the fact that only a few VZV promoters have been analyzed at the functional level, rather than any specific absence of these two possibilities in VZV gene regulatory mechanisms.

Potential VZV promoters can be separated into three categories, A, B, and C (Ruyechan, 2004). It must be said from the outset that this classification is preliminary and incomplete as it is based on what is currently a meager amount of functional data. A-type promoters contain only a TATA element. Thus far, only the VZV ORF14 (glycoprotein C, gC) and ORF 21 promoters are known to fall into this category. The putative gC promoter contains an atypical TATA element and is refractory to activation by IE62. In contrast, the ORF21 promoter contains a TATA element similar to the consensus TATA sequence and is responsive to IE62 (Kantakamalakul *et al.*, 1995; Cohrs *et al.*, 1998). B-type promoters contain a TATA-element and at least one functional upstream binding site for a ubiquitous cellular transcription factor (UTFBS) with which IE62 can interact physically and/or functionally. An example of such a promoter is the VZV ORF4 (IE4) minimal promoter which contains a consensus binding site for the ubiquitous cellular factor USF (Michael *et al.*, 1998). The requirement of this site for activation in conjunction with the presence of viral factors was demonstrated via site-specific mutagenesis. The functional validation of such sites via site specific mutagenesis and protein–DNA interaction assays is vital, since computer predictions of VZV promoters identify numerous potential transcription factor binding sites, only a few of which may be required for, or involved in, IE62 activation (Ruyechan *et al.*, 2003). B-type promoters can also contain a TATA element and two UTFBSs for cellular factors with which IE62 can interact. Examples of such promoters are the VZV ORF 67 (glycoprotein I, gI) promoter and the regulatory element controlling the divergent expression of VZV ORF28 (DNA polymerase) and ORF29 (major DNA-binding protein) (Meier *et al.*, 1994; Peng *et al.*, 2003; Yang *et al.*, 2004). In both of these cases, two cellular transcription factors appear to be required for full IE62-mediated activation of the promoter, one being required for synergistic IE62 activation, and the second acting to enhance the activation.

C-type promoters contain a TATA element, UTFBSs for factors with which IE62 can interact, and UTFBSs for factors that function to activate the promoter independently of IE62. This arrangement would allow for control of expression from such a promoter via both IE62-dependent and IE62-independent mechanisms. Currently, the only known example of this type of promoter is the promoter driving expression of the IE62 gene (Narayanan *et al.*, 2005). However, others may exist since recombinant viruses lacking one of the two genes encoding IE62 and producing low levels of the protein were still capable of expressing the IE4 gene at wild type levels which was not the case for the glycoprotein E (gE) gene whose promoter is also responsive to IE62 (Sato *et al.*, 2003a). Thus, the IE4 promoter may contain both IE62-responsive and non-IE62 responsive elements allowing expression of IE4 under a variety of conditions. This situation underlines the requirement for a full and extensive analysis of a significant number of VZV promoter elements before broad conclusions concerning structure and activation mechanisms can be made.

The VZV IE proteins

The three proteins expressed under IE conditions during VZV infection are commonly designated IE4, IE63 and IE62. They are orthologues of the HSV ICP27, ICP22 and ICP4 proteins, respectively (Kinchington and Cohen, 2000). Like their HSV counterparts, all three VZV IE proteins have been shown to be involved in transcription of the viral genome. However, the VZV IE proteins all exhibit distinct structural and functional features as compared with their HSV orthologues. One of these features lies in the fact that all three VZV IE proteins are present in the virion tegument in readily detectable quantities (Kinchington *et al.*, 1992; Kinchington *et al.*, 1995a) suggesting that they may play an important role in viral replication prior to the production of any viral gene products via transcription of the viral genome.

IE62

The IE62 protein is the most extensively studied of the VZV IE proteins due to its central role in VZV gene regulation and its readily assayable trans-activating activity. The VZV IE62 protein contains 1310 amino acids and is predicted to have a molecular weight of 140 kDa. The observed molecular weight of IE62 by SDS-PAGE, however, is ~175 kDa and the difference between the predicted and experimentally determined values is believed to be due to post-translational modification of the polypeptide primarily, if not exclusively, via phosphorylation by viral and cellular kinases (Kinchington and Cohen, 2000; Kenyon *et al.*, 2003). IE62 under native conditions is most likely an elongated dimer by analogy with the HSV ICP4 protein and based on the fact that the IE62 DNA-binding domain when expressed as a recombinant protein in bacteria dimerizes (Tyler and Everett, 1994). Recent work from this laboratory (Peng and Ruyechan, unpublished data) has shown that the sedimentation coefficient of IE62 is similar to that of ICP4, further supporting the existence of a native IE62 dimer.

IE62 structure

The primary sequence of IE62 can be divided into five domains designated I-V (Fig. 1.2) based on comparisons with other herpesvirus IE transactivators (Cheung, 1989). Of these domains, II and IV are the most conserved at the amino acid level, whereas, domains I, III, and V show considerable variability. Numerous functions have been mapped to the various regions of IE62. Region I (aa 1–467) contains an acidic activation domain essential for the function of IE62 and sequences which have been shown to be involved in interaction with viral (ORF 9, IE4) and cellular (Sp1) factors (Spengler *et al.*, 2000; Peng *et al.*, 2003). Region I also contains one of two SEAC (serine/acidic) domains within the IE62 sequence. Similar SEAC domains are present in ICP4, and the pseudorabies and equine herpesvirus IE transactivators. These domains were identified based on homology with the activation domain of the cellular PC4 transcriptional regulatory protein (Kretschmar *et al.*, 1994). Region II (aa 468–640) contains a highly conserved DNA-binding and dimerization domain (see below). This region, along with region III and the carboxy-terminal portion of Region I has been implicated in the direct binding of the viral IE63 protein and the cellular factors TBP and TFIIB (Perera *et al.*, 2000; Lynch *et al.*, 2002).

Region III (aa 641–734) contains a nuclear localization signal and also falls into the carboxy-terminal portion of IE62 involved in binding of IE63 and TBP/TFIIB. Region IV (aa735–1147) has been implicated in IE62 gene regulatory activity since mutations in this region result in alteration of IE62 transactivation (Baudoux *et al.*, 1995). Region V (aa 1148–1310) contains the second SEAC domain found in IE62. No additional functions have been ascribed to the region. However, it

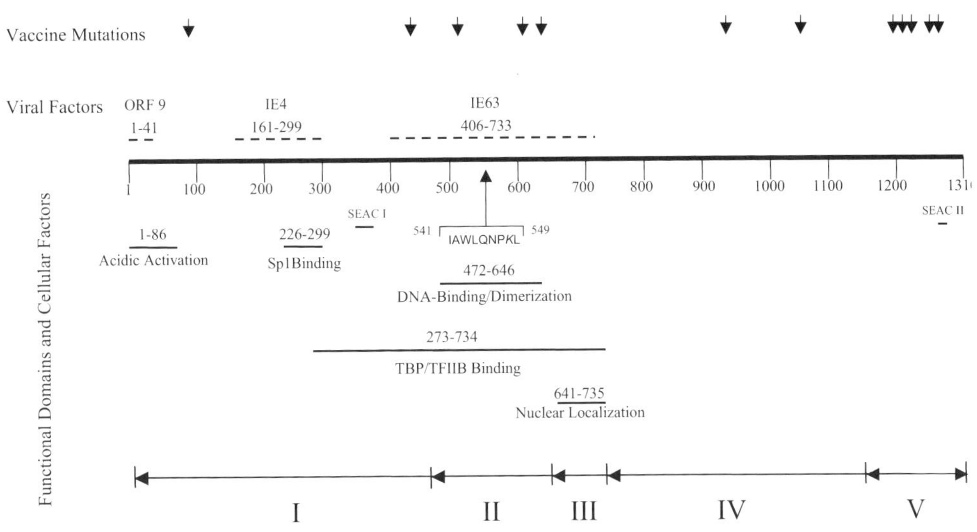

Figure 1.2 Functional map of the varicella zoster virus IE62 protein. The Roman numerals refer to the five regions of IE62 that are common to all alpha-herpesvirus transactivators. The Arabic numerals refer to amino acid positions within the predicted primary structure of IE62. Positions of functional domains and regions involved in interactions with viral and cellular factors are indicated as horizontal bars. Positions of mutations identified in IE62 in vaccine strains are indicated at the top as arrowheads.

is the site of numerous mutations in the IE62 genes present in attenuated vaccine lots of VZV and insertion of these mutations into the wild type protein has shown that some decrease the transactivating capabilities of IE62 in transient transfection assays (Gomi *et al.*, 2002).

IE62 DNA-binding activity

IE62 like all members of the alpha-herpesvirus major transacvtivator family has been demonstrated to be a DNA-binding protein. Work by Wu and Wilcox (1991) and Tyler and Everett (1993, 1994) showed that bacterially expressed fragments of IE62 containing amino acids which span region II of the protein primary structure are sufficient for binding of IE62 to a variety of VZV and non-VZV promoter elements. The ability of IE62 to interact directly with DNA has been shown to be an important aspect of the mechanism of IE62 activation since mutations in IE62 which ablate DNA binding also abrogate transactivation (Tyler *et al.*, 1994). However, it is unclear if IE62 is required to bind to a specific sequence. Wu and Wilcox (1991) identified a consensus sequence (–ATCGT–) to which the IE62 bound tightly. This sequence is identical to the first five bases present in the bipartite recognition sequence identified for the HSV major transactivator ICP4 (ATCGTCnnnnPyCGPuC) where n represents any nucleotide and Py and Pu designate pyrimidines and purines, respectively (Tyler and Everett, 1993). Other studies, however, indicated that IE62 is capable of binding numerous sequences within promoters in addition to ATCGT (Tyler and Everett, 1993). Finally, an extensive study by Perera (2000) showed that IE62 is capable of transactivation of minimal promoters containing only a TATA-box and lacking all known or permuted IE62 binding sites. Thus, the requirement for site specific IE62 binding either in the context of transactivation or transrepression remains an open question.

One additional and potentially significant point regarding the interaction of IE62 with DNA lies in the fact that in experiments using a recombinant protein containing the VZV DNA-binding domain, no bending of oligonucleotides containing the consensus binding sequence was observed. In contrast, the ICP4 DBD was shown to significantly bend DNA containing its cognate site (Tyler *et al.*, 1994). Further, the IE62 DBD produced greater DNAse hypersensitivity at the 5′ end of protected regions than did the ICP4 DBD in DNAse protection assays. Thus, the nature of the IE62-DNA complex is predicted to show significant differences as compared with that of HSV ICP4 and has been predicted to result in broad-ranging alterations in the structure of the DNA to which it binds. Such alterations could influence the binding of cellular transcription factors to their cognate sites in viral and cellular promoters. In addition, IE62 has been shown to strongly activate a variety of cellular and heterologous viral promoters (reviewed in Kinchington and Cohen, 2000). IE62 has also been reported to have self-regulatory properties, both in terms of auto-stimulation and auto-repression of its promoter depending on cell type (Baudoux *et al.*, 1995; Narayanan *et al.*, 2005).

Transcriptional regulatory domains

The major functional domain with respect to VZV transactivation is its N-terminal acidic activation domain (AD). This region while compositionally similar to other acidic activation domains found in HSV VP16 and pseudorabiesvirus, shows little homology to those domains at the individual amino acid level. The IE62 AD encompasses the first 86–90 amino acids of IE62 as determined using Gal-4-fusion constructs containing specific regions of IE62. Deletion of the N-terminal 86–90 amino acids resulted in a protein with very low levels of transactivating activity (1–5%) in comparison to the full length IE62 indicating that this region is vital to IE62 transcriptional function. Further analysis suggested that the first 38 amino acids of IE62 are the most important for activation in the context of the entire protein since IE62 proteins deleted for amino acids 39–90 retained 40% of wild type transactivating activity (reviewed in Kinchington and Cohen, 2000, and Ruyechan, 2004).

While the presence of the N-terminal acidic activation domain appears to be a necessary structural feature of IE62 transactivation, several additional regions of the protein may also be involved. Two of these regions are the SEAC domains. As indicated above, the SEAC domains map in regions I (aa 345–71) and V (aa 1285–1295) of the IE62 primary structure and were identified in sequence comparisons with the cellular transcriptional regulatory protein, PC4. PC4 is a 127 amino acid protein found in mammalian cells that mediates class II gene transcription (Kretschmar *et al.*, 1994). The SEAC domain in region I of IE62 shows homology as high as 80% homology with portions of the SEAC domain in PC4 including amino acids N-terminal to the serine/acidic stretch. The homology of the SEAC domain in region V of IE62 to the PC4 activation domain is limited to the serine/acidic stretch only. The role of these SEAC domains in IE62 is unknown. It is possible that they may substitute for or augment PC4 activity in infected cells. Both of the 62 SEAC domains are predicted to be substrates of the VZV ORF 47 kinase as well as casein kinase II based on the consensus sites for those two kinases (Kenyon *et al.*, 2003). Thus, phosphorylation of these regions by one or the other of these kinases may influence the transcriptional regulatory activity of IE62. Other portions of IE62 which appear to show some involvement in transactivation are Region IV and all of Region V. In the latter, as indicated above, mutations arising in vaccine strains exhibit consistent but limited attenuation of IE62 transcriptional activity.

IE62 protein–protein interactions

IE62 has been demonstrated to interact physically and/or functionally with numerous viral and cellular proteins. Cellular proteins include components of the general transcription apparatus (TBP and TFIIB) and the cellular transcriptional activators Sp1 and USF. All of these proteins have been shown to be capable of interacting directly with IE62 in the absence of other viral and cellular proteins and the regions of IE62 with which they interact have been mapped to varying levels of precision. As shown in Fig. 1.2 and documented above, these interaction sites map to regions I, II, and III of IE62. Viral proteins known to interact with IE62 include IE4, ORF9 and IE63 and the two viral kinases encoded by ORF47 and ORF66. As was the case with the cellular transcription factors, IE4, ORF9 and IE63 have all been shown to be capable of direct physical interaction with IE62 in the absence of other proteins. Their interaction sites map to regions I, II and III and in some cases overlap with those regions required for binding of the cellular factors. Specifically, the IE4 binding region overlaps with the region required for Sp1 binding and the IE63 binding region overlaps with the region shown to be capable of binding TBP and TFIIB. It is currently unknown if the binding of any or all of these viral and cellular factors is mutually exclusive, neutral or cooperative.

The two viral kinases encoded by ORF47 and ORF66 have been shown to be capable of phosphorylating IE62. Thus, both obviously can interact with IE62 in the context of an enzymatic reaction. The ORF47 kinase forms a stable complex with IE62 and the region of the kinase involved in this complex is distinct from the enzyme's catalytic domain (Besser *et al.*, 2003). The sites within IE62 which are phosphorylated by the ORF47 kinase are unknown, but at least nine have been predicted based on their similarity to the consensus ORF47 phosphorylation motif (Kenyon *et al.*, 2003). The ORF66 kinase phosphorylates IE62 at late times during infection, and this phosphorylation event is required for the redistribution/exclusion of IE62 from the nucleus to the cytoplasm and subsequent incorporation into the viral particle (Kinchington *et al.*, 2000).

IE4

The predicted molecular weight of the VZV IE4 protein is 51.4 kD (Davision and Scott, 1986). However, the native form of the protein is homodimeric and IE4 is phosphorylated although the specific kinases are unknown (Baudoux *et al.*, 2000). Deletion mutants of IE4 are not viable, hence the protein is required for viral replication both *in situ* and *in vivo* (Sato *et al.*, 2003b). It is also important for the establishment of latency in a rodent model (Cohen *et al.*, 2005b). IE4 acts at the level of transcription to activate gene expression both alone and in concert with IE62. IE4 has been shown to activate promoters representing all three putative kinetic classes of VZV genes including those regulating expression of IE62, ORF 61, ORF 29 and gI as well as its own promoter. This protein can also activate heterologous viral promoters including the HSV ICP4, ICP0, thymidine kinase (TK), and LAT promoters, the HCMV IE promoter, the SV40 early promoter, and the HIV LTR (reviewed in Kinchington and Cohen, 2000). Unlike IE62, however, the IE4 protein cannot activate promoters containing only a TATA element (Defechereux-Thibaut de Maiseres *et al.*, 1998; Baudoux *et al.*, 2000). Thus, IE4 exhibits the properties of a co-activator (Smale and Kadonaga, 2003) which stimulates activator dependent transcription whereas IE62 can act as both an activator and a co-activator. Analysis of the HIV LTR and the SV40 E promoter has shown that key players in the IE4 co-activator role are the cellular transcription factors NFκB and Sp1. A direct physical interaction between IE4 and NFκB has been demonstrated and involves the N-terminal portion of IE4 (Baudoux *et al.*, 2000).

IE4 also stimulates IE62-dependent promoter activation. This stimulation appears to be synergistic rather than additive (Perera *et al.*, 1992) and suggests that in the context of IE62 activation, IE4 acts as a bridging molecule with elements of the general transcription machinery (Baudoux *et al.*, 2000). Relevant to this, IE4 has been shown to bind to TBP and TFIIB. IE4 also interacts directly with IE62 as noted above (Spengler *et al.*, 2000) and the region of IE62 required for IE4 binding has been mapped to aa 161–299 (Fig. 1.2). This region overlaps that identified as being important for the interaction of IE62 with Sp1 and leaves open the possibility that IE4 may be able to substitute for or synergize with Sp1 in the context of IE62 activation. There is currently no evidence that IE4 binds directly to nucleic acids although the protein's primary structure contains KH-like motifs which are involved in RNA binding. It is unknown whether IE4, like its HSV orthologue ICP27 is involved in post-transcriptional events. There appears to be very little functional complementation between IE4 and ICP27 (Kinchington and Cohen, 2000).

IE63

The VZV IE63 protein, encoded by ORFs 63 and 70 is initially expressed at immediately early times during VZV infection and its expression continues at high levels throughout the remainder of the virus life cycle. There are currently conflicting data as to whether IE63 is absolutely required for VZV replication in tissue culture (Baiker *et al.*, 2004; Cohen *et al.*, 2005a). The protein is predominantly localized in the nuclei of infected cells even at late times in infection and it has been shown to interact with the host cell RNA Pol II and general transcription factor TFIIH (Lynch *et al.*, 2002; Di Valentin *et al.*, 2005). IE63 is phosphorylated by the VZV ORF 47 kinase and cellular kinases including casein kinase I and II and the Cyclin-dependent kinases CDK1 and CDK5 (Bontems *et al.*, 2002; Lynch *et al.*, 2002; Baiker *et al.*, 2004; Habran *et al.*, 2005). IE63 is also the most frequently detected and most abundant of the viral proteins identified in latently infected ganglia. During latency the protein is localized to the cytoplasm in contrast to its predominantly nuclear localization during lytic infection. The function or functions of IE63 during latent infection are currently unknown (Kennedy, 2002; Mitchell *et al.*, 2003), although a recent study indicates that IE63 is critical for efficient establishment of latency in the cotton rat model (Cohen *et al.*, 2005a).

The functions of IE63 during lytic infection also remain largely unknown and the data obtained to date from several laboratories do not present a consistent story. Some studies have shown that IE63 inhibits both basal level and IE62-mediated transcription (Bontems *et al.*, 2002; Di Valentin *et al.*, 2005). Others have shown that IE63 has little or no effect on basal level transcription, but is capable of upmodulating IE62-mediated activation of a specific promoter (Lynch *et al.*, 2002). A much earlier study indicated that IE63 showed no ability to affect basal level or IE62-mediated transactivation (Kost *et al.*, 1995). IE63 has also recently been shown to affect expression of cellular promoters in the absence of other viral proteins, up regulating the EF-1α promoter (Zuranski *et al.*, 2005) and down regulating the human interleukin 8 (IL8) promoter (Di Valentin *et al.*, 2005). While the reported data appear at odds with one another, these studies were performed with different cell types and different promoters, thus making direct comparisons difficult. As they currently stand, these findings suggest that the effects of IE63 are pleiotropic with regard to transcription in general and IE62 specifically.

IE63 can be co-immunoprecipitated with IE62 from infected cell extracts and the two proteins interact directly based on studies with purified recombinant polypeptides. The region of IE62 with which IE63 interacts encompasses amino acids 406–733. This includes all of domains II and III and the carboxy-terminus of domain I thus overlapping regions involved in DNA-binding, TBP/TFIIB binding and nuclear localization. The region of IE63 required for direct physical interaction of IE62 has been mapped to the amino-terminal half (aa 1–142) of IE63. This lies upstream of the major viral and cellular phosphorylation sites identified in the IE63 molecule. Recent fine mapping indicates that the interaction site is centered between aa 55–67 with amino acids R59 and L60 playing a critical role, since deletion of these two amino acids results in loss of IE62/IE63 complexes based on protein pulldown assays (Lynch *et al.*, 2002; Baiker *et al.*, 2004). It may be that one or more of the functions of IE63 relevant to IE62 function will be identified in assays using IE63 containing this mutation.

Non-IE proteins involved in VZV gene expression

ORF29

The VZV ORF29 protein is one of seven VZV gene products believed to be required for VZV DNA replication. The ORF29 protein binds to single-stranded DNA and accumulates within replication compartments in a manner analogous to its HSV orthologue ICP8 (Ruyechan and Hay, 2000). The ORF29 protein is one of the VZV proteins reported to be expressed during latency. Its relative frequency of expression and abundance are nearly as high as that observed with the IE63 protein (Kennedy, 2002; Mitchell *et al.*, 2003). The ORF29 protein has been shown to be capable of modulation of IE62 activation in a promoter and cell type dependent manner (Boucaud *et al.*, 1998). Its mechanism of action with regard to positive IE62 activation has been investigated in the context of the VZV gI promoter and involves a 40 base pair sequence termed the 29 responsive element (29RE) which is located some 200 bp upstream of the minimal gI cis-acting elements and falls partially within the coding sequence of the upstream ORF66 gene. Mutations introduced in the 29RE show some attenuation of virus growth and in one case yielded no virus progeny using a VZV recombinant cosmid system (Ito *et al.*, 2003). Details of the mechanism of ORF29 action in the context of the 29RE and the existence of other 29RE elements within the VZV genome are currently unknown.

ORF61

The VZV ORF61 protein is a 62–66 kDa phosphoprotein which is capable of activation of VZV immediate early and early promoters in transient transfection assays. The ORF61 protein also modulates the transactivation of promoters by IE62 and IE4 in a promoter and cell-type dependent manner. ORF61 contains a zinc RING finger domain in its amino terminus and this domain is required for its transactivating activities. ORF61 exhibits low levels of homology to HSV ICP0 but the two proteins can substitute for each other in transient transfection assays. However, unlike ICP0, the ORF61 protein is not expressed under immediate early conditions. The ORF61 protein is not absolutely required for growth in tissue culture, but deletion mutants are impaired for growth compared with wild type virus due at least in part to lower than normal levels of expression of VZV gE (reviewed in Kinchington and Cohen, 2000).

ORF10

The VZV ORF10 protein is a 47 kDa polypeptide that is a major component of the virion tegument. It is the orthologue of the HSV VP16 protein but lacks that protein's acidic activation domain and is dispensable for VZV replication whereas VP16 is essential for HSV replication. Nevertheless, the ORF10 protein transactivates expression of the VZV IE62 promoter as well as the promoters of HSV ICP4 and ICP0 but not those of the VZV IE4 and IE63 genes (Kinchington and Cohen, 2000). The ability of the ORF10 protein to activate specific promoters stems from the fact that it, like VP16, contains a binding motif for the cellular transcription factor HCF-1 (Freiman and Herr, 1997). This allows the formation of a tripartite complex of ORF10/HCF-1/Oct1 on promoters containing TAATGARAT elements which are required for the binding of the cellular factor Oct-1. Since the VZV IE4 and IE63 promoters lack such a sequence only IE62 expression is activated by ORF10 (Moriuchi *et al.*, 1995). Thus, the VP16 mechanism which drives expression of all HSV IE genes may be an alternate pathway for expression of only one IE gene in VZV (see below).

ORF47

The VZV ORF47 protein is a serine/threonine protein kinase which exhibits homology to cellular casein kinase II (CKII) and to the family of kinases encoded by all herpesvirus genomes thus far sequenced, whose prototype is the HSV UL13 kinase (Kinchington and Cohen, 2000). It is dispensable for VZV replication in tissue culture but essential for VZV infection of differentiated human T-cell and skin xenografts in a SCID mouse model of VZV pathogenesis implying that ORF47 is required for viremia and cutaneous infection, both of which are hallmarks of primary infection with VZV (Ku *et al.*, 2005).

It is also critical for VZV replication in dendritic cells and for spread of the virus from those cells to other cell types (Hu and Cohen, 2005). The specific role or roles played by ORF47 in VZV gene expression are not fully understood. Several studies have shown that IE62 is phosphorylated by ORF47 (reviewed in Kenyon *et al.*, 2003) thus phosphorylation by this protein may play a key role in tissues or cells where cellular kinases are less active, or where specific phosphorylation of ORF47 kinase sites leads to functional IE62 interactions with tissue-specific cellular factors. A stable interaction between IE62 and ORF47 has been demonstrated. The region of ORF47 involved in this interaction mapped to the N-terminal half of the protein rather than the C-terminal half which contains the kinase activity (Besser *et al.*, 2003). Surprisingly, ablation of the kinase activity still allowed some viral replication to occur in skin xenografts *in vivo* provided the ability of the two proteins to interact was preserved. Thus, it appears that the direct interaction between these two proteins rather than phosphorylation of IE62 is the critical element required for VZV replication in skin. The specific functional significance of this requirement for IE62 and ORF47 interaction is unknown and represents a novel and unlooked for finding.

ORF66

ORF66 is also a serine/threonine kinase. Its herpesvirus orthologues, typified by the HSV U_S3 kinase, are found only within the alphaherpesviruses. ORF66 is required for VZV growth in T lymphocytes but not skin (Ku *et al.*, 2005). ORF66 plays a unique and important role in IE62 function in infected cells. Specifically, ORF66 phosphorylates IE62 and inhibits nuclear localization of IE62 (Kinchington *et al.*, 2000). This has two effects: it *de facto* decreases the transcriptional activity of IE62 by decreasing the concentration of IE62 in the nucleus late in infection, and it allows IE62 to be incorporated into the virion tegument. VZV strains lacking the ORF66 kinase show low levels of IE62 incorporation into viral particles.

Expression of VZV IE genes

Examination of the putative promoter regions of the three VZV IE genes reveals, as noted above, that only the promoters for ORFs 62 and 71, which encode IE62, contain binding sites for the cellular factor Oct 1 (Moriuchi *et al.*, 1995). The presence of the Oct-1 recognition sequence and the existence of ORF 10, indicate that IE62 can be expressed via the same mechanism as are the HSV IE genes, whereas IE4 and IE63 could be expressed via mechanisms possibly novel to VZV. By analogy with the HSV IE expression mechanism, VZV ORF10 would interact with Oct-1 bound to the promoter and the cellular transcription factor HCF-1 to activate IE62 expression (Fig. 1.3). This does appear to be one mechanism by which IE62 is expressed since the existence of Oct-1/ORF10/HCF-1 complexes interacting with the IE62 promoter has been demonstrated (Moriuchi *et al*, 1995) and a recent study has shown that depletion of HCF-1 by SiRNA very significantly reduces the ORF10-mediated transactivation of the IE62 core promoter (Narayanan *et al.*, 2005).

IE62, however, has also been shown to regulate expression of its own promoter and, unlike HSV ICP4, IE62 is capable of positive as well as negative regulation. Thus, an alternate pathway for initial expression of the three VZV IE genes would involve their expression via the direct action of IE62 that enters the infected cell as part of the virion tegument and/or is synthesized immediately after infection. This then leads to the question as to whether specific cellular factors are involved in IE62 mediated expression of all three IE genes. Surprisingly, the study by Narayanan *et al.* (2005) showed that IE62 activation of its own promoter was also abrogated by the

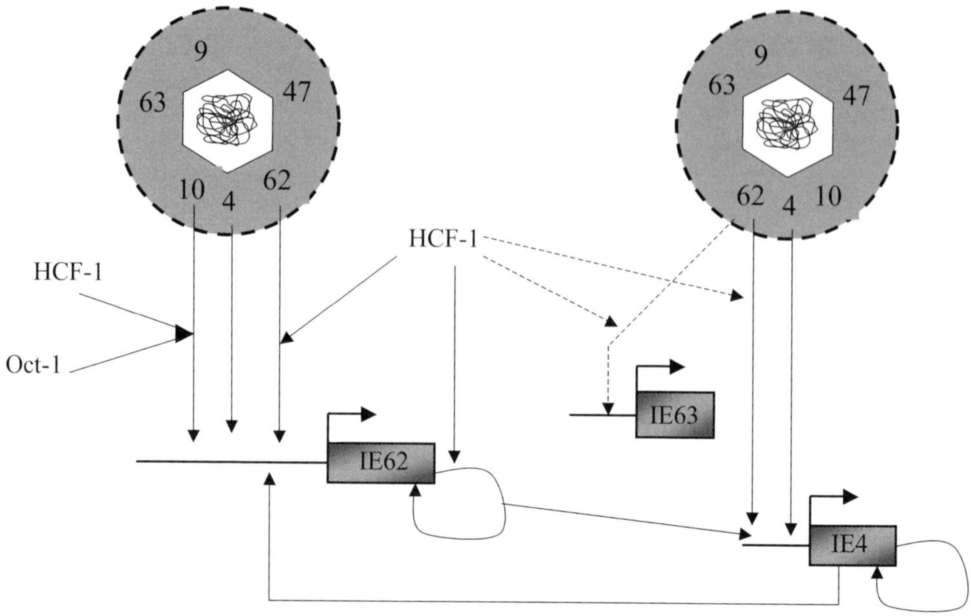

Figure 1.3 Possible model for activation of expression of VZV IE genes. Activation may occur via virion delivered proteins representing classic IE activation in the absence of *de novo* protein synthesis or via expression of IE62 in the infected cell. This second pathway is important in VZV infection since the IE proteins are expressed throughout the infectious cycle and incorporated into the virion. Solid lines indicate known involvement of proteins in expression of the respective genes. Dashed lines indicate potential involvement. No data are currently available regarding involvement of IE63 in VZV IE gene expression.

depletion of HCF-1. Thus, IE62 may interact directly with HCF-1 or the interaction may be bridged by another cellular factor. This latter possibility seems more likely since inspection of the IE62 sequence indicates that it lacks the four amino acid consensus motif (D/E-H-X-Y) required for interaction with HCF-1 which is present in both VP16 and the VZV ORF10 protein (Freiman and Herr, 1997). Both IE62 and HCF-1, however, have been shown to be able to utilize Sp1 as a partner in their activation mechanisms and the promoters for all three VZV IE genes contain predicted Sp1 binding sites (Ruyechan *et al*, 2003). Thus, Sp1 and HCF-1 would be leading candidates as mechanistic partners of IE62 in expression of the IE genes at our present level of understanding of this process.

Finally, as indicated above, IE4 alone has also been shown to be capable of stimulation of expression of both the ORF62 and ORF4 promoters. Thus, the IE4 protein present in the VZV virion can also be postulated be involved in IE gene expression either by directly stimulating IE promoters or in concert with IE62. It is not known if IE4 stimulation involves HCF-1. Like IE62, IE4 also lacks the HCF-1 binding motif. However, the IE4 protein can transactivate promoters in concert with Sp1(Baudoux *et al.*, 2000) which, in turn, has been shown to influence HCF-1 activity (Narayanan *et al.*, 2005). It is unknown at this time whether the IE63 protein has any influence either positive or negative, on the IE62-mediated activation of IE promoters. A model summarizing the potential scenarios for IE promoter activation based on currently published data is presented in Fig. 1.3.

Expression of non-IE VZV genes

The expression of the majority of VZV genes homologous to HSV early and late genes is believed to be mediated primarily through the action of the IE62 protein in conjunction with viral (IE4 and IE63) and cellular factors (Cohen and Straus, 2001). Thus far, the overall information regarding expression of these genes is limited due to the small number of promoters which have been analyzed. It is currently believed that IE62 is capable of transactivating the great majority of VZV promoters in the absence of other viral gene products. This rule, however, is not hard and fast since the responsiveness to IE62 of promoters expressing structural proteins such as VZV gI appears to be much less than that observed with IE promoters and the promoters controlling expression of the VZV DNA polymerase and major DNA binding protein (Perera *et al.*, 1992, Yang *et al.*, 2004). Further, based on their ability to interact directly with IE62 and the frequent presence of their binding sites in VZV promoters, the ubiquitous cellular factors USF and Sp1 appear to play a significant role in IE62 mediated expression of VZV genes (Ruyechan *et al.*, 2003). They are believed to stabilize TBP binding, and both have been shown to interact directly with IE62 (Peng *et al.*, 2003; Rahaus *et al.*, 2003).

Much of our understanding concerning the roles of Sp1 and USF come from recent analyses of the mechanism of IE62 activation of the promoter driving expression of the VZV gI gene and the regulatory elements controlling divergent expression of the VZV ORF28 and ORF29 genes. The gI promoter is a weak promoter and requires the presence of IE62 for activation above the levels seen with promoterless reporter constructs. It consists of only two elements, an atypical TATA-box and a sequence designated the activating upstream sequence (AUS) located 12–24 bases 5′ to the TATA element (Kantakamalakul *et al.*, 1995). The AUS sequence contains overlapping binding sites for three transcription factors, AP1, Sp1 and USF. Transient transfection and site specific mutagenesis showed that the Sp1 site is required for IE62 activation whereas the USF site, while dispensable, is required for full activation. The AP1 site did not appear to have a functional role in activation of this promoter. The roles of these sites were validated in the SCIDhu mouse model of VZV pathogenesis where viruses carrying either the Sp1 or both Sp1 and USF mutations failed to grow in skin xenografts which requires expression of the gI protein. Further, Sp1 and IE62 exhibited mutual stabilization of their binding to the promoter and the presence of IE62 stabilized TBP binding over and above any stabilization afforded by Sp1 (Ito *et al.*, 2003, Peng *et al.*, 2003).

Several studies have examined the functional architecture of the ORF28/ORF29 intergenic region. Work by Meier *et al.* (1994) identified a binding site for the ubiquitous cellular factor USF roughly midway within this region which was essential for IE62 mediated activation of reporter genes placed in both the ORF28 and 29 genes. A more recent analysis of this regulatory region by Yang *et al.* (2004) has shown that it is made up of two overlapping unidirectional promoters which share the USF site as a required element for IE62 activation. This study further showed that expression of the ORF28 gene requires the presence of a single TATA element directing expression in the ORF28 direction whereas two TATA elements direct ORF29 gene expression and are alternatively and differentially utilized for transcription initiation. An Sp1 site localized proximal to the ORF28 gene was also identified and functions as a nonrequired activator element for expression in both directions. Thus, the picture emerging relative to Sp1 and USF, predicted sites for which are found in at least 50% of putative VZV promoters (Ruyechan *et al.*, 2003), is that each, depending on the promoter, can be required for IE62 activation while the other can play a secondary role. Both of these factors contain their own activation domains thus one of the questions to be answered in the future is what the roles of their activation domains and the IE62 activation domain are in mechanisms of expression from viral promoters.

These results, however, represent only the beginnings of our understanding of the full spectrum of VZV gene regulation. As indicated above, IE4 and the ORF61 protein can also activate some VZV promoters in the absence of IE62. Further, Sato *et al.* (2003a) showed that in the case of low levels of IE62 in the infected cell, expression of the VZV ORF47 kinase as well as IE4 remained at essentially wild type levels in contrast to other known IE62-responsive genes suggesting that there may be alternative non-IE62 mediated mechanisms for expression of some VZV genes. Other cellular transcription factors are also likely to be involved in VZV gene expression including those in the CREB family (Rahaus and Wolff, 2003). Finally, it is currently unknown if HCF-1 is required for activation at all IE62-responsive promoters, or if this factor is important only for IE gene expression. All of these avenues remain to be explored.

Expression of VZV genes during latency

As was indicated above, there is evidence that during VZV latency, a small number of VZV lytic genes are expressed at both the RNA and protein levels. These include ORFs 4, 21, 29, 62 and 66 (Kennedy, 2002; Mitchell *et al.*, 2003; Cohrs *et al.*, 2003). The role or roles these proteins play during latency are currently unknown and are made even more intriguing in that the expressed proteins are localized primarily in the cytoplasm rather than in the nucleus. The relative frequencies of detection of expression of these genes varies considerably with the two most frequently detected being the products of ORF63 (IE63) and ORF29 (the major DNA-binding protein) (Kennedy, 2002). Therefore, the promoters regulating expression of these two genes appear to be the best candidates in which to find clues regarding mechanisms of latent gene expression.

The putative IE63 promoter borders on the origin of VZV DNA replication contained within the inverted repeat regions flanking the unique short segment of VZV DNA. Little published work has appeared concerning the promoter regulating expression of IE63. A putative TATA element has been identified 28 nucleotides upstream of the transcriptional start site by computer analysis as have several sites for Sp1 and CAAT (Kinchington *et al.*, 1995b) however none of these elements has, as yet been authenticated by site specific mutation. In contrast, considerably more information is available concerning expression of the ORF29 gene. The unidirectional ORF29 promoter, as described above, lies within an intergenic element situated between the divergently oriented ORF28 and ORF29 genes and shares a required USF site with the unidirectional ORF28 promoter. The results of Yang *et al.* (2004) indicate that the ORF28 and ORF29 genes can be expressed either coordinately or independently and that the observed expression of only the ORF29 gene during VZV latency may involve neuron specific cellular factors that recognize sequences within the ORF29 promoter and lead to its activation. Examination of the putative promoter elements of these two proposed VZV latency genes, as well as the others listed above, however, reveals no obvious similarities at the level of DNA sequence (Yang and Ruyechan, unpublished observations).

An alternative mechanism could involve variations in chromatin structure associated with the promoter regions of the latently expressed genes. The VZV genome has, like the HSV genome, been shown to be episomal during latency and is, therefore, likely to be bound by histones (reviewed in Mitchell *et al.*, 2003). This raises the possibility that during latency, the structure of the viral DNA could be such that the promoters and coding regions of the putative latency associated genes, particularly ORF63/70 and ORF29 could be histone free and nonrepressed. Precedence for this possibility comes from a recent report by Kubat *et al.* (2004) which showed that histone hypermethylation is a determinant of HSV-1 latent gene expression. Perhaps even more relevant are the findings of Schuttengruber *et al.* (2003) who showed that alternate activation of the mouse thymidine kinase and kynurenin formamidase genes which are divergently transcribed from a bidirectional regulatory element is also linked to histone methylation levels. This would imply that the regions of the VZV genome encoding ORFs 63 and 29 as well as 62, 21, 4 and 66 are all at least sometimes in a euchromatic structure whereas the remainder of the genome is not. How and why these specific regulatory elements would be singled out remains to be established.

HCF-1 may also play a role in VZV latency either mechanistically with regard to VZV gene expression, or by sharing the same mechanism involving nuclear exclusion with the VZV latency associated gene products. As indicated above, VZV proteins reported to be expressed during latency have all been detected only in the cytoplasm of latently infected neurons (reviewed in Kennedy, 2002; Mitchell *et al.*, 2003). Kristie *et al.* (1999) have shown that, unlike in other tissues, HCF-1 is not detected in the nuclei of sensory ganglia but rather in the cytoplasm. Experimental conditions that induce the reactivation of herpes simplex, however, result in rapid nuclear localization of HCF-1. Thus, the redistribution of HCF-1 may trigger the redistribution of latent VZV proteins and allow transcription of the VZV genome to begin via the function of this factor through expression of the IE62 promoter as described above. It is not known if the mechanism for sequestration of HCF-1 and the putative VZV latency expressed proteins is the same.

Future directions

Despite nearly two decades of work on VZV gene expression our knowledge of the mechanisms involved in this process are just beginning to emerge. Thus, the future holds considerable promise for important new information to come to light regarding this fundamental process. The following paragraphs describe several questions that could be considered for future experimentation in addition to, or re-enforcing, those already raised in previous sections.

- Additional VZV promoters representing all three putative kinetic classes of VZV genes should be analyzed in order to either confirm or allow re-structuring of the current models and generalities concerning expression of the VZV genome. The relevance of the promoter elements and protein–protein interactions identified and characterized *in vitro* and *in situ* must be authenticated in the context of viral infection.
- Since IE62, Sp1 and USF all contain potent transcriptional activation domains, the question arises as to whether both activation domains are required at promoters whose activation involves both IE62 and one of these factors. If the answer is yes, then what role does each of these domains play in the mechanism of activation of specific VZV promoters? Is there, perhaps another cellular factor such as the mammalian Mediator com-

plex (Bjorklund and Gustafsson, 2004) whose binding is bridged or stabilized by IE62 and/or the cellular transcription factor?

- What is the role of IE62 in the attenuation of the live varicella zoster virus vaccine strains? IE62 molecules carrying some or all of the vaccine mutations have been shown to be somewhat impaired in transactivation. Which of these mutations is most relevant to attenuation and is the attenuation phenotype manifested in a specific cell or tissue type as compared with the results obtained in cultured cells?

Answers to any or all of these questions will add to our understanding of the unique biology of VZV infection.

Acknowledgment

The work from this laboratory described here was supported by grant number AI18449 from the National Institute for Allergy and Infectious Diseases.

References

Arvin, A.M. (2001) Varicella zoster virus. In Virology B.N. Fields, D.M. Knipe, and P.M. Howley, P.M., eds. (Philadelphia, US: Lippincott-Raven) pp. 2731–2767.

Baudoux, L., Defechereux, P. Schoonbroodt, S., Merville, M.P., Rentier, B., and Piette, J. (1995). Mutational analysis of the varicella-zoster virus major immediate early-protein IE62. Nucleic Acids Res. 23, 1341–1349.

Baudoux, L., Defechereux, P., Rentier, B., and Piette, J. Gene activation by varicella-zoster virus IE4 protein requires its dimerization and involves both the arginine-rich sequence, the central part, and the carboxyl-terminal cysteine-rich region. (2000) J. Biol. Chem. 275, 32822–32831.

Baiker, A., Bagowski, C., Ito, H., Sommer, M., Fabel, K., Hay, J., Ruyechan, W., and Arvin, A. (2004). The immediate early 63 protein of varicella zoster virus: analysis of functional domains required for replication *in vitro* and for T-cell and skin tropism in the SCIDhu model *in vivo*. J. Virol. 78, 1181–1194.

Besser, J. Sommer, M.H., Zerboni, L., Bagowski, C.P., Ito, H., Moffat, J., Ku, C-C. and Arvin, A.M. (2003). Differentiation of varicella zoster virus ORF47 protein kinase and IE62 protein binding domains and their contributions to replication in human skin xenografts in the SCID-hu mouse. J. Virol. 77, 5964–5974.

Bjorklund, S. and Gustafsson, C.M. (2004). The mediator complex. Adv. Prot. Chem. 67, 43–65.

Bontems, S. Valentin, D., Baudoux, E.L., Rentier, B., Sadzot-Delvaux, C., and Piette, J. (2002). Phosphorylation of varicella zoster virus IE63 protein by casein kinases influences its cellular localization and gene regulation activity. J. Biol. Chem. 277, 21050–21060.

Boucaud, D., Yoshitake, H., Hay, J. and Ruyechan, W.T. (1998). The VZV ORF 29 protein acts as a modulator of a late VZV gene promoter. J. Infect. Dis., *178(Suppl 1)*, S34-S38.

Cheung, A.K. (1989). DNA sequence analysis of the immediate early gene of pseudorabies virus. Nucelic Acids Res. 17, 4637–4646.

Cohen, J.I. and Straus, S.E. (2001). Varicella zoster virus and its replication. In Virology B.N. Fields, D.M. Knipe, and P.M. Howley, P.M., eds. (Philadelphia, US: Lippincott-Raven) pp. 2707–2730.

Cohen, J.I., Krogmann, T., Bontems, S., Sadzot-Delvaux, C., and Pesinicak, L. (2005a). Regions of the varicella-zostervirus open reading frame 63 latency-associated protein important for replication *in vitro* are also critical for establishment of latency. J. Virol. 79, 5069–5077.

Cohen, J.I., Krogmann, T., Ross, J.P., Pesnicak, L., and Prikhod'ko, E.A. (2005b). Varicella-zoster virus ORF4 latency associated protein is important for establishment of latency. J. Virol. 79, 6969–6075.

Cohrs, R.J., Barbour, M., and Gilden, D.H. (1998). Varicella-zoster virus gene 21: transcriptional start site and promoter region. J. Virol. 72, 42–47.

Cohrs. R. J., Gilden, D.H., Kinchington, P.R., Grinfeld, E., and Kennedy, P.G. (2003). Varicella-zoster virus gene 66 transcription and translation in latently infected human ganglia. J. Virol. 77, 6660–6665.

Davison, A.J. and J. Scott. (1986). The complete sequence of varicella-zoster virus. J. Gen. Virol. 67, 1715–1816.

Defechereux-Thibaut de Maisieres, P., Baudoux-Tebache, L., Merville, M-P., Rentier, B., Bours, V., and Piette, J. (1998) Activation of the human immunodeficiency virus long terminal repeat by varicella-zoster virus IE4 protein requires nuclear factor-κB and involves both the amino-terminal and the carboxyl-terminal cysteine rich region. J. Biol. Chem. 273, 13636–13644.

Di Valentin, E., Bontems, S., Habran, L., Jolois, O., Markine-Goriaynoff, N., Vanderplasschen, A., Sadzot-Delvaux, C., and Piette, J. (2005). Varicella-zoster virus IE63 protein represses the basal transcription machinery by disorganizing the pre-initiation complex. Biol. Chem. *386*, 255–267.

Freiman, R.N. and Herr, W. (1997). Viral mimicry: common mode of association with HCF by VP16 and the cellular protein LZIP. Genes Dev. *11*, 3122–3127.

Gomi, Y., Sunamachi, H., Mori, Y., Nagaike, K., Takahashi, M., and Yamanishi, K. (2002). Comparison of the complete DNA sequences of the varicella vaccine and its parental virus. J. Virol. *76*, 11447–11459.

Habran, L., Bontems, S., Di Valentin, E., Sadzot-Delvaux, C., and Piette, J. (2005). Varicella-zoster virus IE63 protein phosphorylation by roscovitine-sensitive Cyclin-dependent kinases modulates its cellular localization and activity. J. Biol. Chem. *280*, 29135–29143.

Hu, H. and Cohen, J.I. (2005). Varicella-zoster virus open reading frame 47 (ORF47) protein is critical for virus replication in dendritic cells and for spread to other cells. Virology *337*, 304–311.

Ito, H., Sommer, M.H., Zerboni, L., He, H., Boucaud, D., Hay, J., Ruyechan, W. and Arvin, A.M. (2003). Promoter sequences of varicella-zoster virus glycoprotein I targeted by cellular transcription factors Sp1 and USF determine virulence in skin and T cells in SCIDhu mice *in vivo*. J. Virol. *77*, 489–498.

Kantakamalakul, W., Ruyechan, W.T., and Hay, J. 1995.Analysis of varicella-zoster virus promoter sequences. Neurology *45*, S28-S29.

Kennedy, P G E. (2002).Varicella zoster virus in human ganglia. Rev. Med. Virol. *12*, 327–334.

Kenyon, T.K., Homan, E., Storlie, J., Ikoma, M., and Grose, C. (2003).Comparison of varicella zoster virus ORF47 protein kinase and casein kinaseII and their substrates. J. Med. Virol. *70*, 95–102.

Kinchington, P.R., Hougland, J.K., Arvin, A.M., Ruyechan, W.T., and Hay, J. (1992). The varicella-zoster virus immediate early protein IE62 is a major component of virus particles. J. Virol. *66*, 359–366.

Kinchington, P.R., Bookey, D., and Turse, S.E. (1995a) The transcriptional regulatory proteins encoded by varicella zoster virus open reading frames (ORFs) 4 and 63, but not ORF 61are associated with purified virus particles. J. Virol. *69*, 4274–4284.

Kinchington, P.R., Vergnes, J., and Turse, S.E. (1995b) Transcriptional mapping of varicella zoster virus regulatory proteins. Neurology *45*, S33-S35.

Kinchington, P.R. and Cohen, J.I. (2000). Viral proteins. In Varicella Zoster Virus: Virology and Clinical Management, A.M. Arvin and A.A. Gershon, eds. (Cambridge, UK: Cambridge University Press) pp. 74–104.

Kinchington, P.R., Fite, K., and Turse, S.E. (2000). Nuclear accumulation of IE62, the varicella zoster virus (VZV) major transcriptional regulatory protein, is inhibited by phosphorylation mediated by the VZV open reading frame 66 protein kinase. J. Virol., *74*, 2265–2277.

Kost, R.G., Kupinsky, H. and Straus, S.E. (1995). Varicella-zoster virus gene 63: transcript mapping and regulatory activity. Virology *209*, 218–224.

Kretzschmar, M., Kaiser, K., Lottspeich, F., and Meisterernst, M. (1994). A novel mediator of class II gene transcription with homology to viral immediate-early transcriptional regulators. Cell *78*, 525–534.

Kristie, T.M., Vogel, J.L., and Sears, A.E. (1999). Nuclear localization of of the C1 factor in sensory neurons correlates with initiation of reactivation of HSV from latency. Proc. Natl. Acad. Sci. USA *96*, 1229–1233.

Ku, C-C., Besser, J., Abendroth, A., Grose, C., and Arvin, A.M. (2005). Varicella zoster virus pathogenesis and immunobiology: new concepts emerging from investigations with the SCIDhu mouse model. J. Virol. *79*, 2651–2658.

Kubat, N.J., Tran, R., McAnany, P., and Bloom, D.C. (2004). Specific histone tail modification and not DNA methylation is a determinant of herpes simplex virus type 1 latent gene expression. J. Virol. *78*, 1139–1149.

Lynch, J.M., Kenyon, T.K., Grose, C. Hay, J., and Ruyechan, W.T. (2002). Physical and functional interaction between the varicella zoster virus IE63 and IE62 proteins. Virology *302*, 71–82.

Meier, J.L., Luo, X., Sawadogo, M. and Straus, S.E. (1994). The cellular transcription factor USF cooperates with varicella-zoster virus immediate-early protein 62 to symmetrically activate a bi-directional viral promoter. Mol. Cell. Biol. *14*, 6896–6906.

Michael, E., Kuck, K., and Kinchington, P.R. (1998). Anatomy of the varicella zoster virus open reading frame 4 promoter. J. Infect. Dis. *178*, S27-S33.

Mitchell, B.M., Bloom, D.C., Cohrs, R.J., Gilden, D.H. and Kennedy, P.G.H. (2003) Herpes simplex virus-1 and varicella-zoster virus latency in ganglia. J. Neurol.Virol. *9*, 194–204.

Moriuchi, H., Moriuchi, M., and Cohen, J.I. (1995) Proteins and cis-acting elements associated with transactivation of the varicella zoster virus immediate-early gene 62 promoter by VZV open reading frame 10 protein. J. Virol. *69*, 4693–4701,.

Narayanan, A., Nogueira, M.L., Ruyechan, W.T., and Kristie, T.M. (2005) Combinatorial transcription of the HSV and VZV IE genes is strictly determined by the cellular coactivator HCF-1. J. Biol. Chem. *280*, 1369–1375.

Peng, H., He, H., Hay, J., and Ruyechan, W.T. (2003) Interaction between the varicella zoster virus IE62 major transactivator and cellular transcription factor Sp1. J. Biol. Chem. *278*, 38068–38075.

Perera, L.P. (2000). The TATA motif specifies the differential activation of minimal promoters by varicella zoster virus immediate-early regulatory protein IE62. J. Biol. Chem. *275*, 487–496.

Perera, L.P., Mosca, J., Ruyechan, W.T. and Hay, J. (1992). Regulation of varicella zoster virus gene expression in human T-Lymphocytes. J. Virol. *66*, 2468–2477.

Rahaus, M. and Wolff, M.H. (2003). Reciprocal effects of varicella zoster virus (VZV) and Ap1: activation of Jun, Fos, and ATF-2 after VZV infection and their importance for the regulation of viral genes. Virus Res. 92, 9–21.

Rahaus, M., Desloges, N., Yang, M., Ruyechan, W.T., and Wolff, M H. (2003). Transcription factor USF, expressed during the entire phase of VZV infection, interacts physically with the major viral transactivator IE62 and plays a significant role in virus replication. J. Gen. Virol. *84*, 2957–2967.

Ruyechan, W.T. (2004). Mechanism(s) of activation of varicella zoster virus promoters by the VZV IE62 protein. Recent Res. Devel. Virol. *6*, 145–172.

Ruyechan, W.T. and Hay, J. (2000). DNA Replication. In Varicella Zoster Virus: Virology and Clinical Management, A.M. Arvin and A.A. Gershon, eds. (Cambridge, UK: Cambridge University Press) pp. 51–73.

Ruyechan, W.T., Peng, H., Yang, M. and Hay, J. (2003). Cellular factors and IE62 activation of VZV promoters. J. Med. Virol. *70*, S90-S94.

Sato, B., Ito, H., Hinchliffe, S., Sommer, M.H., Zerboni, L., and Arvin, A.M. (2003a). Mutational analysis of open reading frames 62 and 71, encoding the varicella zoster virus immediate early transactivating protein, IE62, and effects on replication *in vitro* and in skin xenografts in the SCID-hu mouse *in vivo*. J. Virol. *77*, 5607–5620.

Sato, B., Sommer, M., Ito, H., and Arvin, A.M. (2003b). Requirement of varicella zoster virus immediate-early 4 protein for viral replication. J. Virol. *77*,12369–12372.

Schuttengruber, G., Doeltzlhoher, A. Kroboth, A., Wintersberger, E., and Seiser, C. (2003). Alternate activation of two divergently transcribed mouse genes fro a bi-directional promoter is linked to changes in histone modification. 2003. J. Biol. Chem. *278*, 1784–1793.

Silverstein, S. and Straus, S.E. (2000). Pathogenesis of latency and reactivation. In Varicella Zoster Virus: Virology and Clinical Management, A.M. Arvin and A.A. Gershon, eds. (Cambridge, UK: Cambridge University Press) pp. 123–141.

Smale, S.T., and Kadonaga, J.T. (2003). The RNA polymerase II core promoter. Annu. Rev. Biochem. *72*, 449–479.

Spengler, M.L., Ruyechan, W.T., and Hay, J. (2000). Physical interaction between two varicella zoster gene regulatory proteins, IE4 and IE62. Virology *272*, 375–381.

Tyler, J.K., and Everett, R.D. (1993). The DNA-binding domain of the varicella zoster virus gene 62 protein interacts with multiple sequences which are similar to the binding site of the related protein of herpes simplex virus type 1. Nucleic Acids Res. *21*, 513–522.

Tyler, J.K., and Everett, R.D. (1994). The DNA-binding domains of the varicella zoster virus gene 62 and herpes simplex virus type 1 ICP4 heterodimerize and bind to DNA. Nucleic Acids Res. 22, 711–721.

Tyler, J.K., Allen, K.E., and Everett, R.D. (1994). Mutation of a single lysine residue severely impairs the DNA recognition and regulatory functions of the VZV gene 62 transactivator protein. Nucleic Acids Res. *22*, 27–278.

Wang, Z.H., Gershon, M.D., Lungu, O., Zhu, Z., Mallory, S., Arvin, A.M., and Gershon, A.A. (2001). Esential role played by the C-terminal domain of glycoprotein I in envelopment of varicella-zoster virus in the trans-Golgi network: interactions of glycoproteins with tegument. J. Virol. *75*, 323–340.

Wu, C.-L. and Wilcox, K.W. (1991). The conserved DNA-binding domains encoded by the herpes simplex virus TYP1 ICP4, Pseudorabies virus IE180, and varicella zoster virus IE62 genes recognize similar sites in the corresponding promoters. J. Virol. *65*, 1149–1159.

Yang, M., Hay, J., and Ruyechan, W.T. (2004). The DNA element controlling expression of the varicella zoster virus ORF 28 and ORF 29 genes consists of two divergent unidirectional promoters which share a common USF site. J. Virol. *78*, 10939–10952.

Zuranski, T., Nawar, H., Czechowski, D., Lynch, J.M., Arvin, A.M., Hay, J., and Ruyechan, W.T. (2005). Cell-type-dependent activation of the cellular EF-1α promoter by the varicella-zoster virus IE63 protein. Virology 338, 35–42.

The Modification of Cellular RNA Polymerase II During HSV-1 Infection

2

Stephen A. Rice and Kathryn A. Fraser

Abstract

Like other nuclear-replicating DNA viruses, herpes simplex virus type 1 (HSV-1) utilizes the host cell RNA polymerase II (Pol II) to transcribe its genome during its productive infection. Moreover, HSV-1 has evolved mechanisms that allow it to effectively compete with the host cell for the use of Pol II. This chapter will focus on the ability of HSV-1 to alter the Pol II enzyme itself. Specifically, HSV-1 alters the localization of Pol II, recruiting it into virus-induced replication compartments, where viral DNA replication and transcription take place. In addition, infection alters the normal phosphorylation of the Pol II carboxyl-terminal domain (CTD), creating a novel form of the enzyme known as Pol II_I. Current evidence suggests that the HSV-1-induced changes to the CTD involve two separate pathways, both of which require viral immediate-early proteins. The biological ramifications of these viral changes to Pol II are discussed.

Introduction

Subversion of RNA polymerase II transcription by HSV-1

Herpes simplex virus type 1 (HSV-1) is a widespread human pathogen, affecting a majority of the world's population (Whitley and Roizman, 2001). Typically, primary infection occurs in epithelial cells of the oral mucosa. After a period of productive growth, HSV-1 invades neighboring sensory neurons where it establishes a lifelong latent infection. These neuronal sites are the source of the recurrent oral infections which typify HSV-1 disease. The closely related virus herpes simplex virus type 2 (HSV-2) usually infects the genital tract and has a very similar biology. Despite its predilection for latency during its natural infection, HSV-1 usually grows in its productive mode in cultured cells, often quite robustly. This efficient *in vitro* growth allows biochemical, genetic, and cell biological techniques to be readily applied to HSV-1. As a result, HSV-1 has become a valuable experimental model system to investigate fundamental aspects of alpha-herpesviral replication.

One area of HSV-1 biology that has received considerable attention is the regulation of viral gene expression. Similar to other nuclear-replicating DNA viruses, HSV-1 utilizes Pol II for the transcription of its genes. During its productive infection, expression of the approximately 80 viral genes is tightly regulated (Roizman and Knipe, 2001). Accordingly, viral genes have been categorized into three groups based in part on when they are active during infection: immediate-early [IE; approximately 2–5 hours post infection (hpi)], delayed-early (DE; approximately

4–8 hpi), and late (L; 6 hpi and onwards). To a first approximation, viral genes resemble cellular Pol II-transcribed genes in that they possess TATA box-containing promoters and typical downstream polyadenylation signals. However, as infection ensues, the level of viral gene transcription far surpasses that of cellular genes. The mechanisms by which HSV-1 commandeers Pol II and its associated machinery are only partially understood.

IE proteins play a key role in viral gene regulation. By definition, IE genes do not require viral protein synthesis for their expression. They are transcribed by Pol II immediately after infection, with the help of a viral tegument protein, VP16 (reviewed in Flint and Shenk, 1997). In contrast, the expression of DE and L genes is highly dependent on synthesis of IE proteins (Honess and Roizman, 1974). Four of the five IE proteins (ICP0, ICP4, ICP22, and ICP27) are nuclear-localized and regulate viral gene activity (Roizman and Knipe, 2001). The fifth (ICP47) is a cytoplasmic protein that acts to inhibit MHC class I antigen presentation (Hill *et al.*, 1995). Of the IE proteins, ICP4 has the most critical role in stimulating viral transcription: in its absence, no DE/L mRNAs are synthesized (Watson and Clements, 1980; Godowski and Knipe, 1986). The mechanism by which ICP4 activates viral genes is only partially understood (for recent discussion, see Zabierowski and DeLuca, 2004). It is a sequence-specific DNA-binding protein that physically interacts with Pol II general transcription factors (GTFs) or their components including the TATA-binding protein (TBP) and TAF250. Through these interactions, ICP4 stimulates the formation of Pol II preinitiation complexes on DE and L gene promoters. The exact role of DNA-binding by ICP4 is unclear. Although ICP4's DNA-binding activity is critical for its ability to turn on viral genes, specific ICP4 DNA-binding sites are not found in all DE/L promoters, and where present, they can be mutated with little or no effect (Smiley *et al.*, 1992). ICP0, ICP22, and ICP27 have also been implicated in the transcriptional activation of DE/L genes (Rice *et al.*, 1995; Jordan and Schaffer, 1997; Jean *et al.*, 2001). However, the molecular mechanisms by which these IE proteins enhance transcription are not well defined.

Host shutoff at the level of primary transcription

Activation of HSV-1 genes is accompanied by a corresponding decrease in cellular Pol II-transcribed gene expression, a phenomenon known as host shutoff. To date, two viral proteins have been clearly implicated in host shutoff. The first is the vhs (for *v*irion *h*ost *s*hutoff) protein, the product of the UL41 gene (reviewed in Smiley, 2004). The vhs polypeptide, introduced into the cell as a component of the virion's tegument layer, is an RNase that is believed to trigger global mRNA destabilization in infected cells. Viral vhs mutants demonstrate defective host shutoff. The second viral protein implicated in host shutoff is the multifunctional ICP27 protein, which acts by inhibiting pre-mRNA splicing (reviewed in Sandri-Goldin, 1998). Since nearly all HSV-1 DE/L genes are unspliced, splicing inhibition by ICP27 serves as an effective viral strategy to specifically repress cellular genes.

There is evidence for yet a third mechanism of host shutoff: the inhibition of primary Pol II transcription on cellular genes. Early work measuring ^{3}H-uridine incorporation into mRNA showed that the transcription of integrated adenoviral or polyomaviral sequences in transformed cells was repressed by HSV-1 infection (Pizer and Beard, 1976; Stenberg and Pizer, 1982). Nuclear run-on transcription assays (Groudine *et al.*, 1981) have more recently become the gold-standard for measuring *in vivo* transcription rates, and these have been used to measure transcription of cellular genes after infection. These studies have shown that HSV-1 infection decreases transcription rates on most cellular genes (Mayman and Nishioka, 1985; Kemp and Latchman, 1988; Smibert and Smiley, 1990; Spencer *et al.*, 1997). For example, it was found that HSV-1 infection of HeLa cells caused transcription of the c-myc, GAPDH, and histone H2b genes to decline to < 10% of the uninfected rate by nine hpi (Spencer *et al.*, 1997).

Spencer *et al.* addressed the question of which viral factor or factors are responsible for host transcriptional shutoff (Spencer *et al.*, 1997). Virion proteins are insufficient, since UV-inactivated HSV-1 failed to inhibit host cell transcription. Moreover, DE/L gene expression is not required, since transcription shutoff occurred in ICP4 mutant-infected cells. Thus, IE proteins appear to be the key players in the inhibition of host transcription. Studies by Latchman's laboratory implicated ICP22 in host transcriptional shutoff (Kemp and Latchman, 1988). In contrast, Spencer *et al.* found that no one IE protein (amongst ICP0, ICP4, ICP22, and ICP27) is absolutely required for significant inhibition of cellular gene transcription (Spencer *et al.*, 1997). Rather, ICP4, ICP0 and especially ICP27 all contributed to transcription shutoff, whereas ICP22 had no effect. In addition, it was found that individual cellular genes differed in which IE proteins effected their shutoff. For example, ICP27 inhibited the GAPDH gene, whereas both ICP4 and ICP27 inhibited the histone H2B gene.

Goals of this chapter

It is likely that HSV-1 has evolved multiple mechanisms that allow it to efficiently recruit Pol II to its genes. One possible mechanism is that the virus directly modifies Pol II to create a form which is specialized for viral transcription. It is thus intriguing that published work indicates that HSV-1 infection induces striking changes to both the localization and phosphorylation of Pol II. The purpose of this chapter is to (i) summarize the literature on HSV-1-induced changes to Pol II, and (ii) discuss if and how these changes might contribute to the dramatic alteration of transcription specificity that is observed following viral infection.

Recruitment of Pol II into viral replication compartments

Within 4–6 hours of HSV-1 infection of cultured cells, there is a dramatic rearrangement of Pol II within the host cell nucleus (Rice *et al.*, 1994). This can be seen in the immunofluorescence experiment shown in Fig. 2.1, in which two different anti-Pol II monoclonal antibodies have been used to localize the enzyme. In uninfected Vero cells (panels a and e), Pol II localizes in the non-nucleolar areas of the nucleus in a somewhat diffuse pattern, with some foci of more intense accumulation. By 6 hpi with HSV-1, however, Pol II becomes concentrated in large globular areas (c, g). Co-staining for the viral ICP8 single-stranded DNA-binding protein (d, h) reveals that these areas correspond to HSV-1-induced structures known as replication compartments (RCs), for which ICP8 is a marker (Quinlan and Knipe, 1985). RCs are sites of viral DNA synthesis and accumulation, and also appear to be sites of viral genome transcription at late times (de Bruyn Kops and Knipe, 1988; Phelan *et al.*, 1997). Consistent with this, several cellular GTFs, including TBP, TFIIE, and TFIIH localize to RCs following HSV-1 infection (Jenkins and Spencer, 2001), as do ICP4 (Knipe *et al.*, 1987), ICP27 (de Bruyn Kops *et al.*, 1998), and ICP22 (Leopardi *et al.*, 1997). Interestingly, host cell chromatin is excluded from RCs (Randall and Dinwoodie, 1986; Monier

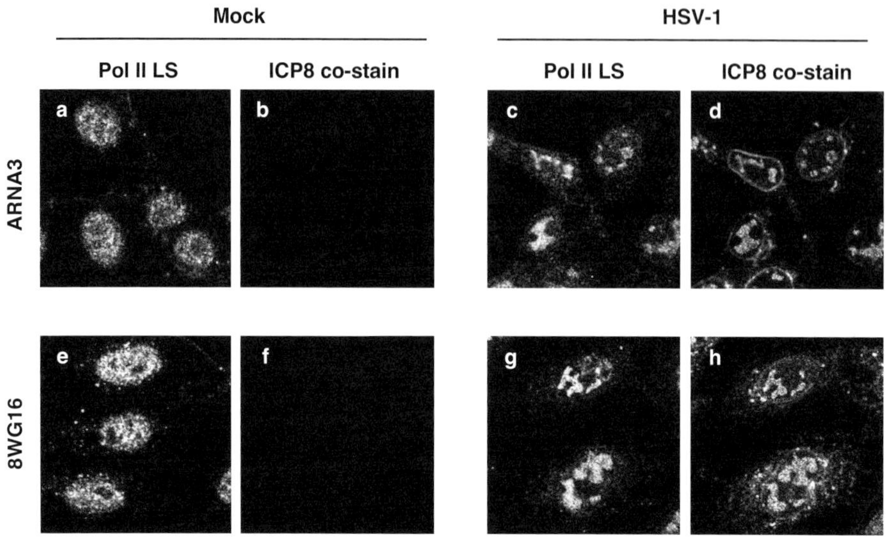

Figure 2.1 Recruitment of Pol II into HSV-1 replication compartments. Vero cells were mock-infected (a, b, e, f) or infected with HSV-1, strain KOS1.1 (c, d, g, h). At 6 hpi, the cells were fixed, permeabilized, and processed for immunofluorescence using mouse monoclonal antibody ARNA3, specific for the body of the Pol II LS (a, c), or 8WG16, specific for the CTD (e, g). Cells were co-stained using a rabbit antisera specific for HSV-1 ICP8, a marker for HSV-1 replication compartments (b, d, f, h).

et al., 2000), consistent with the long-standing observation that cellular chromatin becomes marginated to the nuclear periphery during HSV-1 infection (Schwartz and Roizman, 1969). Because of this segregation of viral and cellular chromatin, and the recruitment of the transcription machinery into RCs, the viral genome may not have to compete with the host genome for the use of Pol II at late times after infection.

The mechanism by which Pol II is recruited into RCs is unknown. ICP22, which has a clear role in altering Pol II phosphorylation (see below), is not required (Rice *et al.*, 1995). The involvement of other viral proteins is unknown. It is worth noting that several viral proteins (ICP0, ICP4, ICP27 and ICP8) have been reported to physically associate with Pol II or Pol II complexes in infected cells (Jenkins and Spencer, 2001; Zhou and Knipe, 2002), and thus could aid in recruitment.

Modification of the Pol II CTD by HSV-1 infection

Regulation of Pol II by CTD phosphorylation

Before discussing the biochemical effects of HSV-1 infection on Pol II phosphorylation, a brief review of Pol II transcription is in order, including its regulation by phosphorylation. Pol II is a large multi-subunit enzyme composed of 12 polypeptides (reviewed in Lee and Young, 2000). Transcription of a given gene can be broken into several discrete phases: preinitiation, initiation, promoter-clearance, elongation, and termination. Preinitiation complex formation and initiation can be readily studied *in vitro*, and many essential factors and mechanistic details are known for these early steps. In contrast, later steps, such as elongation and termination, are also very important in terms of regulation, but are less well understood and far fewer factors have been identified.

The Pol II transcription cycle is regulated by the phosphorylation of its largest subunit, commonly referred to as the large subunit or LS (reviewed in Kobor and Greenblatt, 2002; Lin *et al.*, 2003). The target for phosphorylation on the LS is an unusual sequence at its C-terminus known as the C-terminal domain, or CTD. The CTD consists of multiple repeats (52 in humans) of the heptapeptide consensus sequence YSPTSPS. Serine is the predominate residue phosphorylated, although trace amounts of phosphothreonine and phosphotyrosine have also been reported. The CTD is conserved throughout eukaryotes, and is essential for cell viability. In rapidly growing mammalian cells, Pol II exists in two states which differ in the extent of CTD phosphorylation. Pol II_A is the hypophosphorylated form, whereas Pol II_O is hyperphosphorylated. Phosphorylation on Pol II_O is extensive; it has been estimated that most of the 52 CTD repeats contain at least one phosphate moiety. Because of this, the LS species from Pol II_A and Pol II_O (known as IIa and IIo, respectively) migrate as distinct bands on SDS-PAGE and can be readily distinguished by immunoblotting.

Numerous studies demonstrate that CTD phosphorylation, regulated by cellular CTD kinases and phosphatases, modulates the function of Pol II during transcription (reviewed in Kobor and Greenblatt, 2002; Lin *et al.*, 2003; Palancade and Bensaude, 2003; Sims *et al.*, 2004). The CTD is thought to regulate transcription in at least three ways. First, contacts made by the unphosphorylated CTD with transcription initiation factors help to recruit Pol II to the promoter to begin transcription. Second, transcriptional elongation is stimulated by CTD phosphorylation. Third, phosphorylation of the CTD helps recruit pre-mRNA processing factors to Pol II, through protein–protein interactions. Such recruitment allows transcription to be appropriately coupled to essential mRNA processing reactions including 5′ capping, splicing, and 3′ end formation.

Although several potential cellular CTD kinases have been identified, two are most clearly of physiological importance (reviewed in Prelich, 2002). These are the cyclin-dependent kinase (cdk7) subunit of TFIIH, and the cdk9 subunit of transcriptional elongation factor P-TEFb. These kinases are involved in distinct steps in transcription and target different residues on the CTD consensus repeat. cdk7 preferentially modifies serine 5 (Ser-5) during initiation/promoter clearance, whereas cdk9 targets serine 2 (Ser-2) during elongation. To date, only one CTD phosphatase, FCP1 (Kong *et al.*, 2004), has been well characterized, although others have been identified (Yeo *et al.*, 2003) and are likely to function *in vivo* to regulate transcription.

The current literature suggests a model for how the Pol II transcription cycle is regulated by CTD phosphorylation (Kobor and Greenblatt, 2002; Lin *et al.*, 2003; Palancade and Bensaude, 2003; Sims *et al.*, 2004). First, hypophosphorylated Pol II_A is recruited to the promoter via contacts between the unphosphorylated CTD and transcription initiation factors (Fig. 2.2A). Coincident with or shortly after initiation, the cdk7 subunit of TFIIH phosphorylates the CTD on Ser-5, disrupting the association of Pol II with initiation factors and allowing it to clear the promoter region (Fig. 2.2B). Ser-5 phosphorylation also aids in recruitment of capping factors needed for the appropriate modification of the 5′ end of the pre-mRNA. As transcription proceeds, Pol II often becomes stalled, due to inherent signals in the gene and/or the action of negative transcriptional elongation factors. Resumption of transcription is associated with phosphorylation of the CTD on Ser-2 by the cdk9 subunit of P-TEFb (Fig. 2.2C). Ser-2 modification also helps to recruit poly-

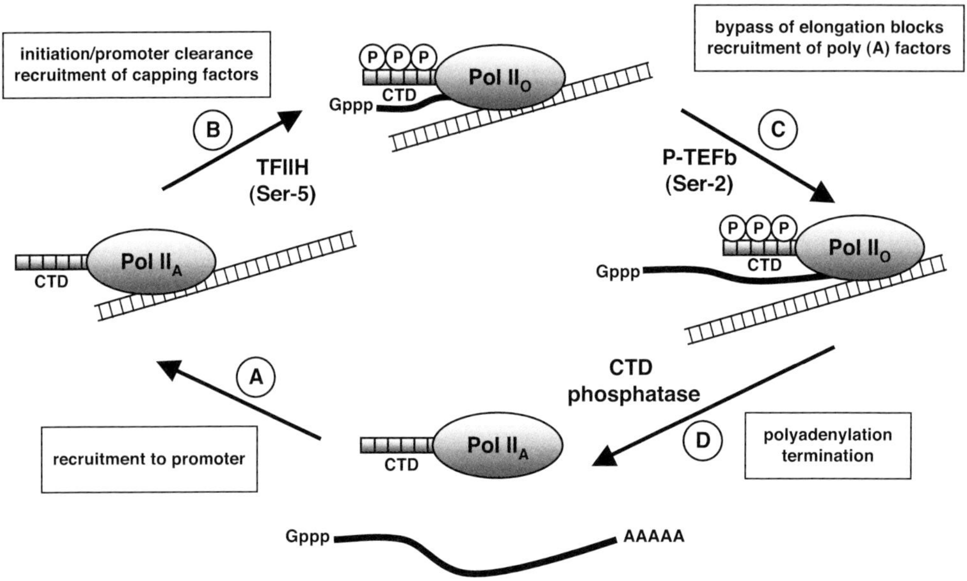

Figure 2.2 The regulation of transcription by Pol II CTD phosphorylation. See text for details.

adenylation and possibly splicing factors to Pol II. Upon cleavage/polyadenylation of the primary transcript, termination of transcription occurs, and CTD phosphatases must dephosphorylate Pol II prior to the beginning of a new round of transcription (Fig. 2.2D). CTD phosphatases are also likely active during transcriptional elongation. In fact, the CTD phosphatase FCP1, which is associated with the GTF TFIIF, likely travels with Pol II during transcription.

Modification of CTD phosphorylation by HSV-1

Rice *et al.* originally showed that HSV-1 infection dramatically alters the phosphorylation of Pol II (Rice *et al.*, 1994). These changes are illustrated in the experiment shown in Fig. 2.3A, in which an anti-LS monoclonal antibody, ARNA3, has been used to assess LS species after infection. ARNA3 binds to the body of the LS, and thus recognizes all forms equally, irrespective of their phosphorylation. In mock-infected Vero cells (lane 1), there are approximately equal levels of IIa and IIo, indicating comparable amounts of Pol II$_A$ and Pol II$_O$ in the uninfected cell. Soon after HSV-1 infection (lanes 2–7), however, IIo levels rapidly decline, and this form is replaced by a new set of polydisperse species that have electrophoretic mobilities intermediate between that of IIa and IIo. These forms were collectively designated IIi (for *i*ntermediately migrating), and the form of Pol II having them was designated Pol II$_I$. IIi is a phosphorylation variant of the LS, since *in vitro* treatment with calf intestinal phosphatase converts it to IIa. Based on its intermediate electrophoretic mobility, it is highly likely that IIi is partially phosphorylated compared with IIo. The HSV-1-induced changes to Pol II have been observed in numerous other cell lines including HeLa and human embryonic lung (HEL) cells (Rice *et al.*, 1994; Rice *et al.*, 1995). Moreover, similar changes occur in HSV-2-infected cells (Fig. 2.3C, lane 3).

To determine whether Pol II$_I$ is the transcriptionally engaged form of the en-

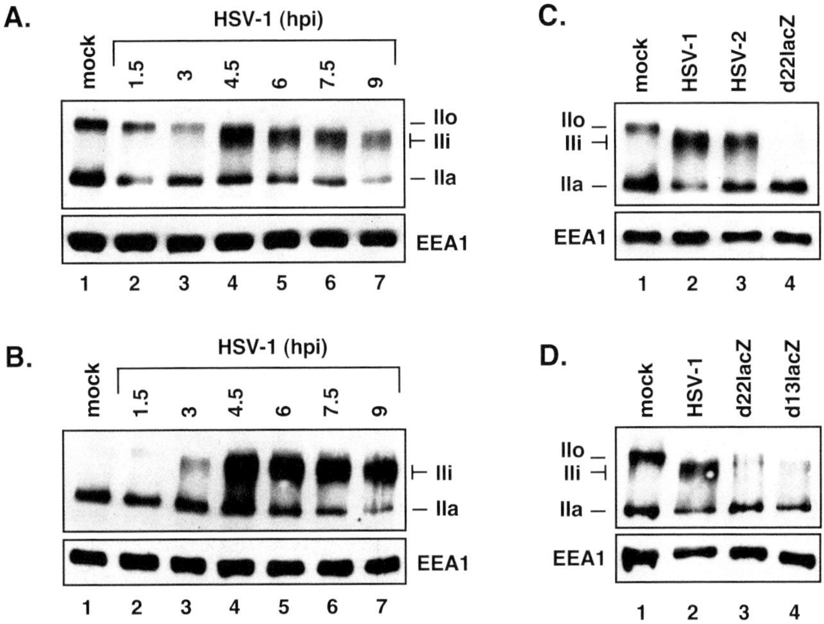

Figure 2.3 Altered phosphorylation of Pol II after HSV-1 infection. (A and B) Time course of Pol II alterations. Vero cells were mock-infected or infected with HSV-1. At 1.5 hour intervals after infection, protein extracts were prepared and analyzed by immunoblotting using anti-LS monoclonal antibodies ARNA3 (A) or 8WG16 (B). The various LS forms are indicated. As a loading control, blots were also probed for EEA1, a cellular endosomal antigen which does not vary in abundance during HSV-1 infection. (C) Altered phosphorylation in HSV-2- and ICP22-mutant infected cells. Vero cells were mock-infected, or infected with HSV-1, HSV-2 (strain G), or the HSV-1 ICP22 deletion mutant d22lacZ (Long *et al*., 1999). Total cell proteins were prepared at 6 hpi and analyzed by immunoblotting using ARNA3. (D) Altered phosphorylation in UL13 mutant infected cells. Vero cells were mock-infected or infected with HSV-1, d22lacZ, or the UL13 mutant d13lacZ (Long *et al*., 1999). Analysis was with monoclonal antibody ARNA 3, as in (C).

zyme in HSV-1-infected cells (as might be expected from its hyperphosphorylated status), Spencer *et al.* (Spencer *et al.*, 1997) used a photoaffinity-labeling technique that was originally used to show that Pol II_O is the transcriptionally active form in normal cells (Cadena and Dahmus, 1987). In this method, nuclear run-on transcription in isolated nuclei is carried out in the presence of ^{32}P-CTP and 4-thio-UTP, the latter being a photoactivatable UTP analog. Upon UV irradiation, nascent ^{32}P-labeled mRNA is crosslinked to transcriptionally active LS species, thus tagging them for subsequent identification on SDS-PAGE. When the technique is applied to HSV-1-infected cells, Pol II_I is identified as the transcriptionally active form. Since transcription of most host genes is repressed following HSV-1 infection, it is very likely that Pol II_I is primarily engaged in transcription of the viral genome.

Incompletely phosphorylated forms of Pol II are not unique to HSV-1 infected cells. For example, such a form is observed during mammalian embryogenesis (Bellier *et al.*, 1997). The early stages of mammalian embryogenesis are characterized by a period of transcriptional inactivity. The resumption of transcription after several cell divisions is known as zygotic genome activation (ZGA), and is preceded by the transient appearance of an intermediately phosphorylated LS form termed IIe (for embryonic). A similar form has been described during *Xenopus laevis* embryogenesis (Palancade *et al.*, 2001). An intermediately phosphorylated form of the LS subunit, termed IIm, is also generated in mammalian somatic cells upon signaling by ERK-type MAP kinases (Bonnet *et al.*, 1999). However, the corresponding form of Pol II may be transcriptionally inactive, as IIm does not appear to be bound to chromatin and its appearance is insensitive to transcription inhibitors. The relationship between the HSV-1-induced Pol II_I form and these other alternately phosphorylated forms is unknown.

Involvement of ICP22 and UL13 in Pol II modifications

Early studies addressed the question of which events in HSV-1 infection, and which viral proteins, are required for the observed changes to Pol II. It was demonstrated that IE protein synthesis is required, since the alterations were not observed after infection with UV-inactivated virus, or when cycloheximide was included in the media (Rice *et al.*, 1994). Furthermore, IE protein expression is sufficient, since the wild type HSV-1 Pol II pattern (i.e. loss of IIo and induction of IIi) was seen in cells infected with an ICP4 null mutant, which is unable to progress past the IE stage (Rice *et al.*, 1995). These experiments implicated the IE proteins as the key players in Pol II modification. Experiments using virus strains mutated in individual IE genes showed that, along with ICP4, neither ICP0 or ICP27 are required for Pol II alterations (Rice *et al.*, 1995). However, it was found that infection with an ICP22 mutant fails to efficiently induce Pol II_I, indicating that ICP22 plays an important role in modifying Pol II. This can be seen in the experiment shown in Fig. 2.3C, wherein cells have been infected with the ICP22 null mutant d22lacZ (lane 4). Note that loss of IIo still occurs in ICP22 mutant-infected cells. This suggests that loss of Pol II_O is mechanistically distinct from the induction of Pol II_I.

The observation that ICP22 affects CTD phosphorylation provided an intriguing clue as to the molecular function of this relatively poorly characterized IE protein. ICP22 homologs are conserved among alpha-herpesviruses, but are not found in beta-herpesviruses or gamma-herpesviruses (Roizman and Pellett, 2001). In cell culture, ICP22 is required for efficient HSV-1 growth in certain cell lines, including HEL cells and several rodent cells (Sears *et al.*, 1985; Poffenberger *et al.*, 1993). However, it is not required in other lines, including Vero cells (Post and Roizman, 1981). In those cell lines where ICP22 is needed, its primary function appears to be to stimulate expression of a subset of L genes (Sears *et al.*, 1985; Poffenberger *et al.*, 1993). Nuclear run-on assays indicate that this occurs, at least in part, at the level of transcription (Rice *et al.*, 1995). An inspection of ICP22's polypeptide sequence provides few clues as to its function, and no nucleic acid-binding properties or other biochemical activities have been described. ICP22 interacts *in vivo* with at least two cellular proteins, p60 and p78 (Bruni and Roizman, 1998; Bruni *et al.*, 1999), the functions of which are obscure.

ICP22's function in viral gene activation is dependent on UL13, a virus-encoded protein kinase (Purves *et al.*, 1993). UL13 is a L protein which is packaged into the tegument of the virion, and hence is present in infected cells from the very onset of infection. ICP22 is hypophosphorylated in UL13 mutant-infected cells, suggesting that it is phosphorylated by UL13 (Purves and Roizman, 1992). Moreover, HSV-1 UL13 mutants show a cell type-specific growth defect which is very similar to that of ICP22 mutants (Purves *et al.*,

1993). Given the functional linkages between ICP22 and UL13, it is perhaps not surprising that UL13 was also found to be required for HSV-1's induction of Pol II_I (Long *et al.*, 1999). As can be seen in Fig. 2.3D, the Pol II phenotype of a UL13 mutant (lane 4) is very similar to that of an ICP22 mutant (lane 3): IIo disappears, but there is very little induction of IIi. One noteworthy point is that Pol II_I is efficiently induced in an ICP4 mutant infection (Rice *et al.*, 1995). This suggests that virion UL13 is capable of carrying out the Pol II modification function. However, Roizman's group found that the modification of ICP22 by UL13 requires new expression of the UL13 gene (Purves *et al.*, 1993). Thus, the function of UL13 in CTD alteration may be distinct from its ability to phosphorylate ICP22.

Pol II_I: a specialized form for viral transcription?

As discussed above, photoaffinity experiments indicate that Pol II_I is the transcriptionally active form of Pol II in HSV-1-infected cells. However, ICP22 and UL13 mutants replicate efficiently in some cell lines, such as Vero cells, in the absence of high levels of Pol II_I. Therefore, abundant Pol II_I is clearly not essential for productive infection, at least under some conditions. Given the vast literature on CTD phosphorylation, it is in fact quite surprising that efficient transcription occurs in ICP22- and UL13 mutant-infected cells in the apparent absence of CTD phosphorylation. To address this puzzling issue, the photoaffinity crosslinking method described earlier was used to identify the transcriptionally active form of Pol II in ICP22 mutant-infected cells (Spencer *et al.*, 1997). It was found that the actively transcribing Pol II is indeed hyperphosphorylated, although it was not possible from this study to determine whether it corresponded to Pol II_I or Pol II_O. Thus, CTD phosphorylation occurs in ICP22/UL13 mutant-infected cells, but the steady-state levels of hyperphosphorylated forms are significantly reduced, such that they are not readily visible on immunoblots.

In contrast to the situation in Vero cells, ICP22 and UL13 mutants grow poorly in some cultured cells, such as HEL cells, and in these cases, defects in viral L gene transcription are seen (Rice *et al.*, 1995; Long *et al.*, 1999). Moreover, ICP22 mutants are avirulent in both mouse and guinea pig models of HSV-1 disease (Poffenberger *et al.*, 1994). Thus, the efficient alteration of Pol II phosphorylation by ICP22/UL13 may be required for high-level transcription in some cell lines, or during natural infections. It is possible that, in Vero and some other cells, host cell-specific factors compensate for the absence of ICP22/UL13.

How does the specific phosphorylation pattern of Pol II_I form differ from that of Pol II_O? This question is difficult to answer given the potential complexity of CTD phosphorylation. However, the pattern of serine phosphorylation can be probed to some extent through the use of phosphospecific anti-CTD monoclonal antibodies (reviewed in Palancade and Bensaude, 2003). One such antibody is 8WG16, which does not efficiently recognize the CTD when Ser-2 is phosphorylated (Patturajan *et al.*, 1998; Cho *et al.*, 2001). Consistent with this, 8WG16 does not recognize IIo in mock-infected Vero cells (Fig. 2.3B, lane 1). However, it does efficiently react with the HSV-1-induced IIi form (lanes 2–7) (Rice *et al.*, 1994). These results suggest that Pol II_I is not significantly phosphorylated on Ser-2. This conclusion has recently been confirmed using H5, an antibody specific for the Ser-2 phosphorylated CTD (K. Fraser and S. Rice, unpublished). Moreover, another antibody, H14, which is specific for Ser-5 phosphorylated CTD, reacts strongly with IIi (K. Fraser and S. Rice, unpublished). Thus, it appears that the HSV-1 specific Pol II_I is phosphorylated on Ser-5 of the CTD, but lacks significant phosphorylation on Ser-2.

The fact that the actively transcribing form of Pol II in HSV-1-infected cells is not significantly phosphorylated on Ser-2 is intriguing, since this modification is essential for transcription in normal cells (Chao and Price, 2001; Shim *et al.*, 2002). As reviewed earlier, the two major roles of Ser-2 phosphorylation are to promote transcriptional elongation, via the activity of P-TEFb, and to recruit polyadenylation and possibly splicing factors to Pol II. How could HSV-1 transcription proceed without these two important functions of the CTD? In regards to elongation, there are several reasons to think that HSV-1 genes may be less subject to transcriptional elongation blocks than cellular genes, and thus may be less dependent on P-TEFb. Nearly all viral DE/L genes are intronless, and are thus significantly shorter than cell genes. In addition, their chromatin structure is likely less compact, and possibly non-nucleosomal (Leinbach and Summers, 1980; Muggeridge and Fraser, 1986). In regards to recruitment of pre-mRNA processing factors, the intronless viral genes could reasonably dispense with recruiting splicing factors. However, viral genes have typical polyadenylation signals and certainly need to utilize the cellular polyadenylation machinery. It is interesting that ICP27, which physically interacts with Pol II complexes (Jenkins and Spencer, 2001; Zhou and Knipe, 2002), can enhance mRNA 3′ end formation both *in vivo* and *in vitro* (McLauchlan *et al.*, 1992; Sandri-Goldin and Mendoza, 1992; McGregor *et al.*, 1996). It is thus conceivable that ICP27 compensates for loss of Ser-2 phosphorylation on the CTD by itself recruiting polyadenylation factors to the sites of viral transcription.

It is not known how ICP22 and UL13 lead to the formation of Pol II_I. The possibility that UL13 itself is a viral CTD kinase, possibly in association with ICP22, cannot be excluded. In this regard, it is interesting that UL13 appears to target SP motifs (Kawaguchi *et al.*, 2003), making both Ser-2 and Ser-5 of the CTD potential targets. It is also possible that cellular CTD kinases are induced or modified by ICP22/UL13. One prime candidate is the cdk7 subunit of TFIIH, since it is the major Ser-5-specific CTD kinase active during transcription. It is noteworthy that the essential GTF TFIIE, which positively regulates the CTD kinase activity of TFIIH (Lu *et al.*, 1992; Okamoto *et al.*, 1998), has been shown to be absent from the Pol II holoenzyme in HSV-1 infected cells (Jenkins and Spencer, 2001). It is thus possible that ICP22/UL13 compensates for the loss of TFIIE's TFIIH-stimulating activity, and in so doing alters the pattern of CTD phosphorylation. Another cellular protein kinase that could be involved is cdc2, a known CTD kinase (Cisek and Corden, 1989; Gebara *et al.*, 1997). Its activity has been shown to be stimulated during HSV-1 infection in an ICP22/UL13-dependent manner (Advani *et al.*, 2000; Advani *et al.*, 2003).

HSV-1-induced loss of Pol II_O: a role in host transcriptional shutoff?

As mentioned earlier, Pol II_O loss during HSV-1 infection can be uncoupled from Pol II_I induction, since mutations in ICP22 or UL13 prevent Pol II_I induction but have no effect on Pol II_O loss. Although IE protein expression is required and sufficient for Pol II_O loss, experiments using viral null mutants indicate that ICP0, ICP4, ICP22 and ICP27 are not individually required (Rice *et al.*, 1995). It is interesting that similar findings apply to the shutoff of host genome transcription by HSV-1 (Spencer *et al.*, 1997). This correlation strongly suggests that the physical loss of Pol II_O is mechanistically associated with host transcriptional shutoff.

If IE protein expression is required for Pol II_O loss, but ICP0, ICP4, ICP22 and ICP27 are each dispensable, then what IE function is responsible? One possibility that needs to be investigated is whether the fifth IE protein, ICP47, could play a role. Another possibility is that the known IE nuclear regulatory proteins are redundant in their ability to cause IIo loss, thus explaining why mutants singly disrupted for these genes retain the ability to mediate this effect. This hypothesis could be tested by using HSV-1 mutants disrupted in multiple IE genes. It would also be interesting to express the IE proteins in uninfected cells to see whether they can individually affect Pol II_O and/or inhibit host transcription.

IE proteins could trigger Pol II_O loss by various mechanisms, including virus-induced inhibition of an important CTD kinase (such as P-TEFb), or activation of a CTD phosphatase. Another possible mechanism is that IE proteins could induce the specific proteolysis of Pol II_O. This possibility is supported by a recent study which suggests that the proteasome tracks with Pol II during transcription (Gillette *et al.*, 2004). Moreover, under certain circumstances such as DNA damage, Pol II is ubiquitinated and degraded via the proteasomal pathway (Lee *et al.*, 2002; Woudstra *et al.*, 2002; Inukai *et al.*, 2004). It should be noted that a proteolysis mechanism is not mutually exclusive with other mechanisms such as inhibition of a CTD kinase. The proteasome may function generally to remove Pol II complexes that have become arrested during transcription for a variety of reasons (Svejstrup, 2003). Thus, if one or more HSV-1 IE proteins caused transcribing Pol II to stall (e.g. by inhibiting P-TEFb), this could lead to proteolysis of the arrested complexes.

Modification of the Pol II CTD by other viruses

Although both HSV-1 and HSV-2 modify Pol II phosphorylation, it is not yet known how general this phenomenon is among the alpha-herpesviruses. However, it was recently shown that a beta-herpesvirus, human cytomegalovirus (HCMV), induces an intermediately migrating form of Pol II that resembles Pol II_I (Baek *et al.*, 2004). This study also showed that the purified UL13 homolog of HCMV, UL97, is able to phosphorylate the CTD *in vitro*, predominantly on Ser-5. However, the role of UL97 *in vivo* is uncertain, since the levels of intermediately phosphorylated Pol II were only minimally affected by a UL97 null mutation. It was also found that the levels of Pol II_O are unaffected by HCMV infection. As HCMV does not shut off host gene expression (Mocarski and Courcelle, 2001), this observation is consistent with the notion that HSV-1-induced Pol II_O loss is associated with the inhibition of host cell gene transcription.

Several non-herpesviruses have been shown to alter the levels or phosphorylation of Pol II, or to modify the activity of cellular CTD kinases. One of the best-studied examples is human immunodeficiency virus (HIV) and its regulatory protein Tat (reviewed in Taube *et al.*, 1999). When bound to nascent HIV mRNA, Tat recruits P-TEFb to the HIV-1 transcription complex, thus enhancing transcriptional elongation on HIV genes. Interestingly, the interaction of Tat with P-TEFb alters its CTD kinase specificity, allowing it to phosphorylate Ser-5 in addition to Ser-2 (Zhou *et al.*, 2000). Alteration of Pol II phosphorylation is not restricted to viruses which utilize Pol II for their own transcription, as several RNA viruses encoding their own transcriptases also modify Pol II or cellular CTD kinases. The nonstructural protein NSs of Rift Valley fever virus, a bunyavirus, interacts with the p44 subunit of TFIIH, and in so doing lowers TFIIH levels (Le May *et al.*, 2004). This activity correlates with an inhibition of host cell transcription. Another bunyavirus, Bunyamwera virus, encodes a nonstructural protein that prevents Pol II CTD phosphorylation on Ser-2 (Thomas *et al.*, 2004). Finally, poliovirus infection has been reported to lead to a general loss of Pol II_O, an effect that is associated with the inhibition of host transcription (Rangel *et al.*, 1987).

Summary and model

As described above, HSV-1 dramatically alters Pol II during its productive infection. First, there is a massive redistribution of Pol II into virus-specific compartments, believed to be sites of viral late transcription. Second, there are striking changes to Pol II phosphorylation. This includes the loss of the normal hyperphosphorylated Pol II_O form, as well as induction of a novel, intermediately phosphorylated Pol II_I form. Some evidence suggests that loss of Pol II_O is associated with host gene transcription shutoff, whereas generation of Pol II_I may be important for viral gene transcription under some conditions.

It is not yet clear how the HSV-1 induced changes to Pol II localization relate to changes in its phosphorylation. Recruitment of Pol II into viral RCs is clearly not dependent on formation of Pol II_I, since Pol II is effectively recruited to RCs in ICP22 mutant infections (Rice *et al.*, 1995). However, it is not known whether alteration of CTD phosphorylation is dependent on viral compartment formation, particularly if this process is viewed broadly. Recent work indicates that compartmentalization begins early in infection, when viral genomes localize to ND10 subnuclear domains (Ishov and Maul, 1996) and associate with newly expressed ICP4 (Everett *et al.*, 2003; Everett *et al.*, 2004). At least some of these complexes mature into RCs (Everett *et al.*, 2003) through a series of defined steps (Taylor *et al.*, 2003; Wilkinson and Weller, 2004). Thus, it seems quite possible that viral and cellular chromatin are segregated into distinct compartments from the very earliest stages of infection.

We propose a model for the alteration of Pol II phosphorylation by HSV-1 which attempts to take into account the separate compartmentalization of viral and cellular chromatin. The model proposes that two distinct HSV-1-induced pathways alter Pol

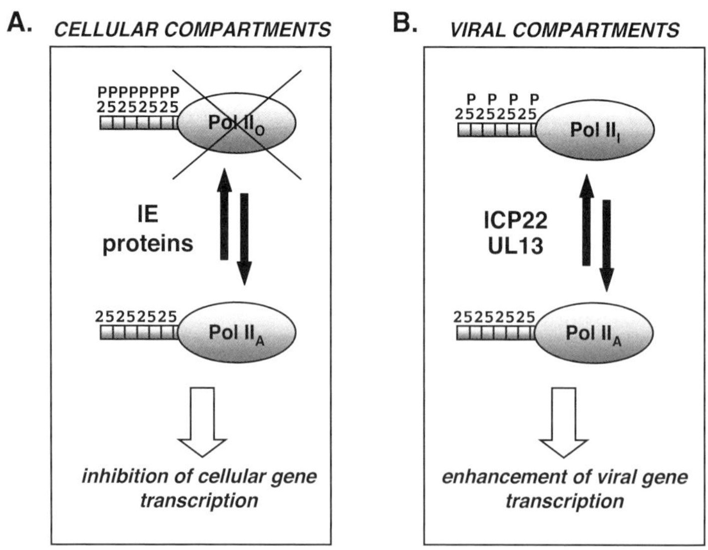

Figure 2.4 Model for modification of Pol II by HSV-1 infection. See text for details.

II phosphorylation, and that these are differentially active in the two chromatin compartments. One pathway (Fig. 2.4A), dependent on viral IE proteins but independent of ICP22/UL13, acts in the cellular chromatin compartment to cause the loss of the normal hyperphosphoryated Pol II_O form and the inhibition of host genome transcription. One scenario that is consistent with much of the existing data is that one or more viral IE proteins cause elongating Pol II to become arrested on most cellular genes. A cellular rescue pathway, perhaps similar to that which is operative when Pol II is stalled by DNA damage, leads to the removal of Pol II in a process that involves its dephosphorylation and/or degradation. A second pathway (Fig. 2.4B), involving ICP22 and UL13, operates in viral chromatin compartments to create Pol II_I, a form of the enzyme specialized for viral transcription. Pol II_I differs from the normal hyperphosphorylated form in that it is phosphorylated primarily on Ser-5, rather than on both Ser-2 and Ser-5. This unusual pattern of phosphorylation likely reflects inherent differences in the manner in which transcription occurs on viral genomes. Although high levels of Pol II_I are dispensable in some cell lines, this specialized form may be crucial for efficient viral transcription in other cell lines or during natural infections.

Obviously, numerous questions remain concerning the alteration of Pol II by HSV-1. Foremost among them is whether the HSV-1-induced modifications to Pol II phosphorylation are the causes or the effects of the dramatic changes in transcription specificity which occur during the productive infection. That is, does HSV-1 directly affect CTD phosphorylation, and in this way modulate transcription? Alternatively, HSV-1 could affect transcription at steps that do not involve the CTD, and only indirectly affect the modification of this structure. In either case, the mechanisms involved are certain to be important and interesting, and can likely be productively pursued by the continued study of Pol II in HSV-1-infected cells.

Acknowledgments

We thank Charlotte Spencer for past and current collaborations in this area, and for many helpful discussions and insights over the years. We also thank Rebecca Duerst and Lucia Zacchi for helpful comments on this chapter. The work carried out in authors' laboratory was supported by NIH grant RO1-AI50127.

References

Advani, S.J., Weichselbaum, R.R., and Roizman, B. (2000). The role of cdc2 in the expression of herpes simplex virus genes. Proc. Natl. Acad. Sci. USA *97*, 10996–11001.

Advani, S.J., Weichselbaum, R.R., and Roizman, B. (2003). Herpes simplex virus 1 activates cdc2 to recruit topoisomerase IIalpha for post-DNA synthesis expression of late genes. Proc. Natl. Acad. Sci. USA *100*, 4825–4830.

Baek, M.C., Krosky, P.M., Pearson, A., and Coen, D.M. (2004). Phosphorylation of the RNA polymerase II carboxyl-terminal domain in human cytomegalovirus-infected cells and *in vitro* by the viral UL97 protein kinase. Virology *324*, 184–193.

Bellier, S., Chastant, S., Adenot, P., Vincent, M., Renard, J.P., and Bensaude, O. (1997). Nuclear translocation and carboxyl-terminal domain phosphorylation of RNA polymerase II delineate the two phases of zygotic gene activation in mammalian embryos. EMBO J. *16*, 6250–6262.

Bonnet, F., Vigneron, M., Bensaude, O., and Dubois, M.F. (1999). Transcription-independent phosphorylation of the RNA polymerase II C-terminal domain (CTD) involves ERK kinases (MEK1/2). Nucleic Acids Res. *27*, 4399–4404.

Bruni, R., and Roizman, B. (1998). Herpes simplex virus 1 regulatory protein ICP22 interacts with a new cell cycle-regulated factor and accumulates in a cell cycle-dependent fashion in infected cells. J. Virol. *72*, 8525–8531.

Bruni, R., Fineschi, B., Ogle, W.O., and Roizman, B. (1999). A novel cellular protein, p60, interacting with both herpes simplex virus 1 regulatory proteins ICP22 and ICP0 is modified in a cell-type specific manner and Is recruited to the nucleus after infection. J. Virol. *73*, 3810–3817.

Cadena, D.L., and Dahmus, M.E. (1987). Messenger RNA synthesis in mammalian cells is catalyzed by the phosphorylated form of RNA polymerase II. J. Biol. Chem. *262*, 12468–12474.

Chao, S.H., and Price, D.H. (2001). Flavopiridol inactivates P-TEFb and blocks most RNA polymerase II transcription *in vivo*. J. Biol. Chem. *276*, 31793–31799.

Cho, E.J., Kobor, M.S., Kim, M., Greenblatt, J., and Buratowski, S. (2001). Opposing effects of Ctk1 kinase and Fcp1 phosphatase at Ser 2 of the RNA polymerase II C-terminal domain. Genes Dev. *15*, 3319–3329.

Cisek, L.J., and Corden, J.L. (1989). Phosphorylation of RNA polymerase by the murine homologue of the cell-cycle control protein cdc2. Nature *339*, 679–684.

de Bruyn Kops, A., and Knipe, D.M. (1988). Formation of DNA replication structures in herpes virus-infected cells requires a viral DNA binding protein. Cell 55, 857–868.

de Bruyn Kops, A., Uprichard, S.L., Chen, M., and Knipe, D.M. (1998). Comparison of the intranuclear distributions of herpes simplex virus proteins involved in various viral functions. Virology *252*, 162–178.

Everett, R.D., Sourvinos, G., and Orr, A. (2003). Recruitment of herpes simplex virus type 1 transcriptional regulatory protein ICP4 into foci juxtaposed to ND10 in live, infected cells. J. Virol. *77*, 3680–3689.

Everett, R.D., Sourvinos, G., Leiper, C., Clements, J.B., and Orr, A. (2004). Formation of nuclear foci of the herpes simplex virus type 1 regulatory protein ICP4 at early times of infection: localization, dynamics, recruitment of ICP27, and evidence for the de novo induction of ND10-like complexes. J. Virol. *78*, 1903–1917.

Flint, J., and Shenk, T. (1997). Viral transactivating proteins. Annu. Rev. Genet. *31*, 177–212.

Gebara, M.M., Sayre, M.H., and Corden, J.L. (1997). Phosphorylation of the carboxy-terminal repeat domain in RNA polymerase II by cyclin-dependent kinases is sufficient to inhibit transcription. J. Cell Biochem. *64*, 390–402.

Gillette, T.G., Gonzalez, F., Delahodde, A., Johnston, S.A., and Kodadek, T. (2004). Physical and functional association of RNA polymerase II and the proteasome. Proc. Natl. Acad. Sci. USA *101*, 5904–5909.

Godowski, P.J., and Knipe, D.M. (1986). Transcriptional control of herpesvirus gene expression: gene functions required for positive and negative regulation. Proc. Natl. Acad. Sci. USA *83*, 256–260.

Groudine, M., Peretz, M., and Weintraub, H. (1981). Transcriptional regulation of hemoglobin switching in chicken embryos. Mol. Cell. Biol. *1*, 281–288.

Hill, A., Jugovic, P., York, I., Russ, G., Bennink, J., Yewdell, J., Ploegh, H., and Johnson, D. (1995). Herpes simplex virus turns off the TAP to evade host immunity. Nature *375*, 411–415.

Honess, R.W., and Roizman, B. (1974). Regulation of herpesvirus macromolecular synthesis. I. Cascade regulation of the synthesis of three groups of viral proteins. J. Virol. *14*, 8–19.

Inukai, N., Yamaguchi, Y., Kuraoka, I., Yamada, T., Kamijo, S., Kato, J., Tanaka, K., and Handa, H. (2004). A novel hydrogen peroxide-induced phosphorylation and ubiquitination pathway leading to RNA polymerase II proteolysis. J. Biol. Chem. *279*, 8190–8195.

Ishov, A.M., and Maul, G.G. (1996). The periphery of nuclear domain 10 (ND10) as site of DNA virus deposition. J. Cell Biol. *134*, 815–826.

Jean, S., LeVan, K.M., Song, B., Levine, M., and Knipe, D.M. (2001). Herpes simplex virus 1 icp27 is required for transcription of two viral late (gamma2) genes in infected cells. Virology *283*, 273–284.

Jenkins, H.L., and Spencer, C.A. (2001). RNA polymerase II holoenzyme modifications accompany transcription reprogramming in herpes simplex virus type 1-infected cells. J. Virol. *75*, 9872–9884.

Jordan, R., and Schaffer, P.A. (1997). Activation of gene expression by herpes simplex virus type 1 ICP0 occurs at the level of mRNA synthesis. J. Virol. *71*, 6850–6862.

Kawaguchi, Y., Kato, K., Tanaka, M., Kanamori, M., Nishiyama, Y., and Yamanashi, Y. (2003). Conserved protein kinases encoded by herpesviruses and cellular protein kinase cdc2 target the same phosphorylation site in eukaryotic elongation factor 1delta. J. Virol. *77*, 2359–2368.

Kemp, L.M., and Latchman, D.S. (1988). Induction and repression of cellular gene transcription during herpes simplex virus infection are mediated by different viral immediate-early gene products. Eur. J. Biochem. *174*, 443–449.

Knipe, D.M., Senechek, D., Rice, S.A., and Smith, J.L. (1987). Stages in the nuclear association of the herpes simplex virus transcriptional activator protein ICP4. J. Virol. *61*, 276–284.

Kobor, M.S., and Greenblatt, J. (2002). Regulation of transcription elongation by phosphorylation. Biochim. Biophys. Acta *1577*, 261–275.

Kong, S.E., Kobor, M.S., Krogan, N.J., Somesh, B.P., Soegaard, T.M., Greenblatt, J.F., and Svejstrup, J.Q. (2004). Fcp1 phosphatase: Interaction with elongating RNA polymerase II holoenzyme, enzymatic mechanism of action, and genetic interaction with elongator. J. Biol. Chem.

Le May, N., Dubaele, S., Proietti De Santis, L., Billecocq, A., Bouloy, M., and Egly, J.M. (2004). TFIIH transcription factor, a target for the Rift Valley hemorrhagic fever virus. Cell *116*, 541–550.

Lee, K.B., Wang, D., Lippard, S.J., and Sharp, P.A. (2002). Transcription-coupled and DNA damage-dependent ubiquitination of RNA polymerase II *in vitro*. Proc. Natl. Acad. Sci. USA *99*, 4239–4244.

Lee, T.I., and Young, R.A. (2000). Transcription of eukaryotic protein-coding genes. Annu. Rev. Genet. *34*, 77–137.

Leinbach, S.S., and Summers, W.C. (1980). The structure of herpes simplex virus type 1 DNA as probed by micrococcal nuclease digestion. J. Gen. Virol. *51*, 45–59.

Leopardi, R., Ward, P.L., Ogle, W.O., and Roizman, B. (1997). Association of herpes simplex virus regulatory protein ICP22 with transcriptional complexes containing EAP, ICP4, RNA polymerase II, and viral DNA requires posttranslational modification by the U(L)13 proteinkinase. J. Virol. *71*, 1133–1139.

Lin, P.S., Tremeau-Bravard, A., and Dahmus, M.E. (2003). The repetitive C-terminal domain of RNA polymerase II: multiple conformational states drive the transcription cycle. Chem. Rec. *3*, 235–245.

Long, M.C., Leong, V., Schaffer, P.A., Spencer, C.A., and Rice, S.A. (1999). ICP22 and the UL13 protein kinase are both required for herpes simplex virus-induced modification of the large subunit of RNA polymerase II. J. Virol. *73*, 5593–5604.

Lu, H., Zawel, L., Fisher, L., Egly, J.M., and Reinberg, D. (1992). Human general transcription factor IIH phosphorylates the C-terminal domain of RNA polymerase II. Nature *358*, 641–645.

Mayman, B.A., and Nishioka, Y. (1985). Differential stability of host mRNAs in Friend erythroleukemia cells infected with herpes simplex virus type 1. J. Virol. *53*, 1–6.

McGregor, F., Phelan, A., Dunlop, J., and Clements, J.B. (1996). Regulation of herpes simplex virus poly (A) site usage and the action of immediate-early protein IE63 in the early-late switch. J. Virol. *70*, 1931–1940.

McLauchlan, J., Phelan, A., Loney, C., Sandri-Goldin, R.M., and Clements, J.B. (1992). Herpes simplex virus IE63 acts at the posttranscriptional level to stimulate viral mRNA 3′ processing. J. Virol. *66*, 6939–6945.

Mocarski, E.S., Jr., and Courcelle, C.T. (2001). Cytomegaloviruses and their replication. In Fields Virology, 4th edn, D.M. Knipe, and P. M. Howley, eds. (Philadelphia, PA, Lippincott, Williams & Wilkins), pp. 2629–2673.

Monier, K., Armas, J.C., Etteldorf, S., Ghazal, P., and Sullivan, K.F. (2000). Annexation of the interchromosomal space during viral infection. Nat. Cell Biol. *2*, 661–665.

Muggeridge, M.I., and Fraser, N.W. (1986). Chromosomal organization of the herpes simplex virus genome during acute infection of the mouse central nervous system. J. Virol. *59*, 764–767.

Okamoto, T., Yamamoto, S., Watanabe, Y., Ohta, T., Hanaoka, F., Roeder, R.G., and Ohkuma, Y. (1998). Analysis of the role of TFIIE in transcriptional regulation through structure-function studies of the TFIIEbeta subunit. J. Biol. Chem. *273*, 19866–19876.

Palancade, B., and Bensaude, O. (2003). Investigating RNA polymerase II carboxyl-terminal domain (CTD) phosphorylation. Eur. J. Biochem. *270*, 3859–3870.

Palancade, B., Bellier, S., Almouzni, G., and Bensaude, O. (2001). Incomplete RNA polymerase II phosphorylation in Xenopus laevis early embryos. J. Cell Sci. *114*, 2483–2489.

Patturajan, M., Schulte, R.J., Sefton, B.M., Berezney, R., Vincent, M., Bensaude, O., Warren, S.L., and Corden, J.L. (1998). Growth-related changes in phosphorylation of yeast RNA polymerase II. J. Biol. Chem. *273*, 4689–4694.

Phelan, A., Dunlop, J., Patel, A.H., Stow, N.D., and Clements, J.B. (1997). Nuclear sites of herpes simplex virus type 1 DNA replication and transcription colocalize at early times postinfection and are largely distinct from RNA processing factors. J. Virol. *71*, 1124–1132.

Pizer, L.I., and Beard, P. (1976). The effect of herpes virus infection on mRNA in polyoma virus transformed cells. Virology *75*, 477–480.

Poffenberger, K.L., Raichlen, P.E., and Herman, R.C. (1993). *In vitro* characterization of a herpes simplex virus type 1 ICP22 deletion mutant. Virus Genes *7*, 171–186.

Poffenberger, K.L., Idowu, A.D., Fraser-Smith, E.B., Raichlen, P.E., and Herman, R.C. (1994). A herpes simplex virus type 1 ICP22 deletion mutant is altered for virulence and latency *in vivo*. Arch. Virol. *139*, 111–119.

Post, L.E., and Roizman, B. (1981). A generalized technique for deletion of specific genes in large genomes: alpha gene 22 of herpes simplex virus 1 is not essential for growth. Cell *25*, 227–232.

Prelich, G. (2002). RNA polymerase II carboxy-terminal domain kinases: emerging clues to their function. Eukaryot. Cell *1*, 153–162.

Purves, F.C., and Roizman, B. (1992). The UL13 gene of herpes simplex virus 1 encodes the functions for posttranslational processing associated with phosphorylation of the regulatory protein alpha 22. Proc. Natl. Acad. Sci. USA *89*, 7310–7314.

Purves, F.C., Ogle, W.O., and Roizman, B. (1993). Processing of the herpes simplex virus regulatory protein alpha 22 mediated by the UL13 protein kinase determines the accumulation of a subset of alpha and gamma mRNAs and proteins in infected cells. Proc. Natl. Acad. Sci. USA *90*, 6701–6705.

Quinlan, M.P., and Knipe, D.M. (1985). Stimulation of expression of a herpes simplex virus DNA-binding protein by two viral functions. Mol. Cell. Biol. *5*, 957–963.

Randall, R.E., and Dinwoodie, N. (1986). Intranuclear localization of herpes simplex virus immediate-early and delayed-early proteins: evidence that ICP 4 is associated with progeny virus DNA. J. Gen. Virol. *67 (Pt 10)*, 2163–2177.

Rangel, L.M., Fernandez-Tomas, C., Dahmus, M.E., and Gariglio, P. (1987). Modification of RNA polymerase IIO subspecies after poliovirus infection. J. Virol. *61*, 1002–1006.

Rice, S.A., Long, M.C., Lam, V., and Spencer, C.A. (1994). RNA polymerase II is aberrantly phosphorylated and localized to viral replication compartments following herpes simplex virus infection. J. Virol. *68*, 988–1001.

Rice, S.A., Long, M.C, Lam, V., Schaffer, P.A., and Spencer, C.A. (1995). Herpes simplex virus immediate-early protein ICP22 is required for viral modification of host RNA polymerase II and establishment of the normal viral transcription program. J. Virol. *69*, 5550–5559.

Roizman, B., and Knipe, D.M. (2001). Herpes simplex virus and their replication. In Fields Virology, 4th edn, D.M. Knipe, and P. M. Howley, eds. (Philadelphia, PA, Lippincott, Williams & Wilkins), pp. 2399–2460.

Roizman, B., and Pellett, P.E. (2001). The family Herpesviridae: a brief introduction. In Fields Virology, 4th edn, D.M. Knipe, and P. M. Howley, eds. (Philadelphia, PA, Lippincott, Williams & Wilkins), pp. 2381–2398.

Sandri-Goldin, R.M. (1998). Interactions between a herpes simplex virus regulatory protein and cellular mRNA processing pathways. Methods *16*, 95–104.

Sandri-Goldin, R.M., and Mendoza, G.E. (1992). A herpesvirus regulatory protein appears to act post-transcriptionally by affecting mRNA processing. Genes Dev. *6*, 848–863.

Schwartz, J., and Roizman, B. (1969). Concerning the egress of herpes simplex virus from infected cells: electron and light microscope observations. Virology 38, 42–49.

Sears, A.E., Halliburton, I.W., Meignier, B., Silver, S., and Roizman, B. (1985). Herpes simplex virus 1 mutant deleted in the alpha 22 gene: growth and gene expression in permissive and restrictive cells and establishment of latency in mice. J. Virol. *55*, 338–346.

Shim, E.Y., Walker, A.K., Shi, Y., and Blackwell, T.K. (2002). CDK-9/cyclin T (P-TEFb) is required in two postinitiation pathways for transcription in the C. elegans embryo. Genes Dev. *16*, 2135–2146.

Sims, R.J., 3rd, Belotserkovskaya, R., and Reinberg, D. (2004). Elongation by RNA polymerase II: the short and long of it. Genes Dev. *18*, 2437–2468.

Smibert, C.A., and Smiley, J.R. (1990). Differential regulation of endogenous and transduced beta-globin genes during infection of erythroid cells with a herpes simplex virus type 1 recombinant. J. Virol. *64*, 3882–3894.

Smiley, J.R. (2004). Herpes simplex virus virion host shutoff protein: immune evasion mediated by a viral RNase? J. Virol. *78*, 1063–1068.

Smiley, J.R., Johnson, D.C., Pizer, L.I., and Everett, R.D. (1992). The ICP4 binding sites in the herpes simplex virus type 1 glycoprotein D (gD) promoter are not essential for efficient gD transcription during virus infection. J. Virol. *66*, 623–631.

Spencer, C.A., Dahmus, M.E., and Rice, S.A. (1997). Repression of host RNA polymerase II transcription by herpes simplex virus type 1. J. Virol. *71*, 2031–2040.

Stenberg, R.M., and Pizer, L.I. (1982). Herpes simplex virus-induced changes in cellular and adenovirus RNA metabolism in an adenovirus type 5-transformed human cell line. J. Virol. *42*, 474–487.

Svejstrup, J.Q. (2003). Rescue of arrested RNA polymerase II complexes. J. Cell Sci. *116*, 447–451.

Taube, R., Fujinaga, K., Wimmer, J., Barboric, M., and Peterlin, B.M. (1999). Tat transactivation: a model for the regulation of eukaryotic transcriptional elongation. Virology *264*, 245–253.

Taylor, T.J., McNamee, E.E., Day, C., and Knipe, D.M. (2003). Herpes simplex virus replication compartments can form by coalescence of smaller compartments. Virology *309*, 232–247.

Thomas, D., Blakqori, G., Wagner, V., Banholzer, M., Kessler, N., Elliott, R.M., Haller, O., and Weber, F. (2004). Inhibition of RNA polymerase II phosphorylation by a viral interferon antagonist. J. Biol. Chem. *279*, 31471–31477.

Watson, R.J., and Clements, J.B. (1980). A herpes simplex virus type 1 function continuously required for early and late virus RNA synthesis. Nature *285*, 329–330.

Whitley, R.J., and Roizman, B. (2001). Herpes simplex virus infections. Lancet *357*, 1513–1518.

Wilkinson, D.E., and Weller, S.K. (2004). Recruitment of cellular recombination and repair proteins to sites of herpes simplex virus type 1

DNA replication is dependent on the composition of viral proteins within prereplicative sites and correlates with the induction of the DNA damage response. J. Virol. *78*, 4783–4796.

Woudstra, E.C., Gilbert, C., Fellows, J., Jansen, L., Brouwer, J., Erdjument-Bromage, H., Tempst, P., and Svejstrup, J.Q. (2002). A Rad26-Def1 complex coordinates repair and RNA pol II proteolysis in response to DNA damage. Nature *415*, 929–933.

Yeo, M., Lin, P.S., Dahmus, M.E., and Gill, G.N. (2003). A novel RNA polymerase II C-terminal domain phosphatase that preferentially dephosphorylates serine 5. J. Biol. Chem. *278*, 26078–26085.

Zabierowski, S., and DeLuca, N.A. (2004). Differential cellular requirements for activation of herpes simplex virus type 1 early (tk) and late (gC) promoters by ICP4. J. Virol. *78*, 6162–6170.

Zhou, C., and Knipe, D.M. (2002). Association of Herpes Simplex Virus Type 1 ICP8 and ICP27 Proteins with Cellular RNA Polymerase II Holoenzyme. J. Virol. *76*, 5893–5904.

Zhou, M., Halanski, M.A., Radonovich, M.F., Kashanchi, F., Peng, J., Price, D.H., and Brady, J.N. (2000). Tat modifies the activity of CDK9 to phosphorylate serine 5 of the RNA polymerase II carboxyl-terminal domain during human immunodeficiency virus type 1 transcription. Mol. Cell. Biol. *20*, 5077–5086.

The Roles of ICP0 During HSV-1 Infection

3

Roger D. Everett

Abstract

ICP0, one of the five immediate-early proteins encoded by herpes simplex virus type 1 (HSV-1), plays a central role in regulating whether the virus progresses to lytic or latent infection. ICP0 is a member of the family of E3 ubiquitin ligase enzymes that have a so-called RING finger zinc-binding domain. This domain confers on ICP0 the ability to induce the proteasome-dependent degradation of a number of cellular proteins. This process results in multiple consequences, including the disruption of cellular nuclear substructures known as ND10 or PML nuclear bodies. Interestingly, prior to the disruption of ND10, parental HSV-1 genomes become closely associated with these structures during the early stages of HSV-1 infection. The E3 ubiquitin ligase activity of ICP0 correlates very well with its role in the stimulation lytic virus infection and induction of reactivation of latent or quiescent viral genomes. However, despite intensive investigation, the mechanism of the connection between the E3 ubiquitin ligase activities of ICP0 and its roles during HSV-1 infection remain poorly understood. This chapter summarizes recent work on the biochemical and biological activities of ICP0, and discusses the possible mechanisms that might explain how ICP0 regulates HSV-1 infection.

Introduction

Because of its roles in lytic and latent infection, and its profound effects on the host cell, ICP0 is one of the most interesting and intensely studied of the proteins encoded by HSV-1. The first part of this chapter will deal with the physical characteristics of ICP0, its ability to activate gene expression, its biochemical activities and interactions with cellular proteins, and the consequences of these activities for cells in which ICP0 is expressed. The second part will deal with the phenotype of HSV-1 viruses that fail to express ICP0, the ability of ICP0 to induce reactivation of gene expression from quiescent HSV-1 genomes in cultured cells, and the role of ICP0 in HSV-1 latency. Finally, the issue of the possible connections between functions of ICP0 and the potential role of cellular nuclear substructures known as ND10 or PML nuclear bodies during HSV-1 infection will be considered. Some of these issues are controversial, and for alternative views of these topics and more detailed description of some of the earlier work on ICP0 the reader can consult other recent reviews (Everett, 2000b; Hagglund and Roizman, 2004) and other chapters in this volume.

Although other members of the alphaherpesvirinae subfamily express proteins that are related to ICP0, for reasons of space they will not be discussed in this chapter. However, the available literature suggests that in many but not all respects, members of the ICP0 protein family have similar activities and perform similar roles during their respective virus infections.

Primary sequence of ICP0 and its post-translational modification

The IE-1 gene encoding ICP0

ICP0 is the product of gene IE-1 (also known as α0) that lies in the repeated sequences bounding the long unique segment of the HSV-1 genome. The primary IE-1 transcript is one of the few produced by HSV-1 that is spliced, and it encodes a protein product that contains 775 amino acids in HSV-1 strain 17 (Fig. 3.1) (Perry *et al.*, 1986). This protein has been referred to as Vmw110 or IE110 in many older publications, but the name ICP0 has now been almost universally adopted.

Sequence motifs within ICP0

The most notable sequence motif within ICP0 is a zinc-binding RING finger domain that is located near the N-terminus of the protein (Fig. 3.1). The RING finger domains in the ICP0 family of proteins are of the C3HC4 type, which binds two atoms of zinc in a cross-braced structure in which the polypeptide chain zig-zags between the two zinc atoms (Barlow *et al.*, 1994). The presence of a RING finger domain is characteristic of the members of the ICP0 family of proteins expressed by the alpha-herpesviruses, and indeed it is the only motif that is strongly conserved within all members of the family. Although the core ICP0 RING domain is compact (the first and last cysteine residues are at amino acids 116 and 156 respectively), insertion and deletion mutations in a region extending at least as far as residue 212 give phenotypes related to those in the core RING structure (Everett, 1987; Everett, 1988; Chen *et al.*, 1991; Everett, 2000a), suggesting that the functional RING domain extends for a further 50–60 residues on the C-terminal side of the core.

Downstream of the RING domain is a relatively poorly characterised region of the protein that includes three segments that have been mapped to contain major phosphorylation sites (Davido *et al.*, 2005), the third of which lies close to a functional nuclear local-

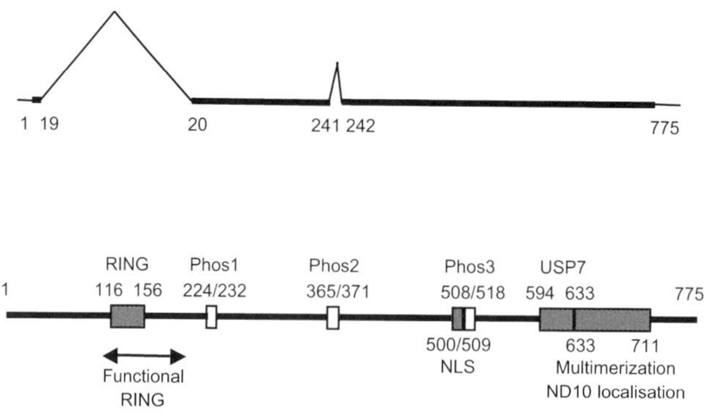

Figure 3.1 The upper part indicates the primary IE-1 transcript, with the introns shown as chevrons and the coding region as a thicker line. The numbers refer to the positions of the first and last codons of the entire protein and the codons at the boundaries of the introns. The lower part summarizes the major regions of the ICP0 amino acid sequence that have defined functions. The boundaries are not intended to be precise. Mutations in other parts of the protein may reduce ICP0 function in reporter gene expression and viral replication assays.

ization sequence in the vicinity of residue 500 (Everett, 1988; Mullen *et al.*, 1994). ICP0 is localised in the nucleus at the early stages of HSV-1 infection of most cell types, but after replication compartments develop it is found in the cytoplasm in increasing quantities (Lopez *et al.*, 2001).

The other well-characterised domains within ICP0 include a short region that is required for interaction with the cellular ubiquitin specific protease enzyme USP7, and downstream of this motif is a longer region that includes sequences required for self-multimerization of ICP0 (Ciufo *et al.*, 1994; Mullen *et al.*, 1994; Meredith *et al.*, 1995) and its localization to ND10 (Everett and Maul, 1994; Maul and Everett, 1994). The interactions of ICP0 with USP7 and ND10 will be described in detail below.

Post-translational modification of ICP0

ICP0 undergoes extensive post-translational modification, including phosphorylation by viral and cellular kinases (Ackermann *et al.*, 1984; Ogle *et al.*, 1997; Advani *et al.*, 2001; Advani *et al.*, 2000), nucleotidylation (Blaho *et al.*, 1993) and ubiquitination (Canning *et al.*, 2004). As will be described in detail below, ubiquitination of ICP0 and its degradation by the ubiquitin–proteasome pathway reflect a fundamental aspect of ICP0 activity.

The phosphorylation status of ICP0 alters as infection progresses, suggesting that the change in ICP0 localization from nuclear to predominantly cytoplasmic at later times of infection may in part be regulated by phosphorylation (Advani *et al.*, 2001). Phosphorylation also plays a role in the intracellular localization of ICP0 in primary rat trigeminal neurones and in differentiated PC12 neuronal cells. In these cases, ICP0 is found predominantly in the cytoplasm, but nuclear accumulation of the protein can be induced by inhibition of the phosphatase calcineurin (Chen *et al.*, 2000). In that the cyclin kinase inhibitor roscovitine reduces phosphorylation of ICP0 and inhibits ICP0-mediated activation of HSV-1 gene expression (Davido *et al.*, 2002), it is likely that phosphorylation of ICP0 has an important effect on its activities. Functional analysis of the phenotype of ICP0 mutants with alterations in major phosphorylated residues is consistent with this hypothesis (Davido *et al.*, 2005).

ICP0 as an activator of gene expression

ICP0 is a promiscuous activator of gene expression in transfection reporter assays

Initial work on the functions of ICP0 demonstrated that it activates gene expression from reporter plasmids in transfected cells (Everett, 1984; Gelman and Silverstein, 1985; O'Hare and Hayward, 1985a; Quinlan and Knipe, 1985). This activity is not restricted to any particular type of promoter or gene, as HSV-1 promoters of all three temporal classes (O'Hare and Hayward, 1985b; Cai and Schaffer, 1992), and cellular and heterologous viral promoters (Ostrove *et al.*, 1987; Nabel *et al.*, 1988; Gius and Laimins, 1989; Golden *et al.*, 1992; Margolis *et al.*, 1992; Cheung *et al.*, 1997; Kwun *et al.*, 2002) are all responsive to ICP0.

The diversity of ICP0-responsive promoters suggests that no particular specific promoter sequence or transcription factor is essential for ICP0 to activate gene expression, and where this has been analysed (with one possible exception; Jang *et al.*, 1991) this has turned out to be the case. Because ICP0 activity results in an increase in transcription (Jordan and Schaffer, 1997), it appears that ICP0 activates gene expression by a mechanism that is common to or acts prior to all transcription factor mediated transcriptional stimulation. What this mechanism is remains unknown, but the most commonly held hypothesis is that ICP0 relieves a repression mechanism that assembles HSV-1 genomes into a repressed chromatin structure. If such a mechanism also applies to transfected plasmid DNA, it would explain how ICP0 could activate a wide range of transfected plasmid reporter constructs.

Domains of ICP0 required for gene expression activation

Analysis of the domains of ICP0 that are involved in its ability to activate gene expression has revealed that the ND10 localization/self multimerization, USP7 binding and especially the RING finger domain all contribute to ICP0 activity in transfection reporter assays (Everett, 1987; Everett, 1988; Cai and Schaffer, 1989; Chen *et al.*, 1991; Everett *et al.*, 1999b). However, it is the RING finger domain that is the most important since disruption of this domain drastically reduces ICP0 activity in all its aspects so far analysed. Because of this, the role of the RING finger has been the subject of many studies.

ICP0 as an E3 ubiquitin ligase

The ubiquitin–proteasome pathway

Ubiquitination of proteins, involving either mono-ubiquitination or the sequential addition of ubiquitin molecules onto a substrate protein to form covalently linked polyubiquitin chains, plays an essential role in the cell. Mono-ubiquitination frequently acts as a marker for downstream events, while polyubiquitination generally leads to recognition by the 26S proteasome and subsequent substrate degradation. These processes are tightly regulated, and are involved in controlling the levels of intracellular proteins (particularly short-lived regulatory molecules), and in DNA repair, transcription, signalling pathways, antigen presentation, endocytosis of receptor molecules, and indeed probably in almost all intracellular processes. There is an increasing number of examples of the ubiquitin pathway being required for events during virus infection (Banks *et al.*, 2003).

The ubiquitin pathway starts with the expression of ubiquitin fusion protein precursors from which free ubiquitin is generated by ubiquitin specific protease (USP) enzymes (Fig. 3.2). Monomeric ubiquitin is then activated in an ATP-dependent reaction to form a thiolester linkage with the active site cysteine residue of the E1 ubiquitin activating enzyme. The ubiquitin is then transferred to another thiolester linkage with the active site cysteine of one of a large number of members of the E2 ubiquitin conjugating enzyme family. In conjunction with an E3 ubiquitin ligase enzyme, the ubiquitin is then transferred to form an isopeptide linkage with a lysine side chain in target substrate molecules. The specificity of the reaction is imparted by the characteristics of the substrate (and frequently its phosphorylation status) and by the E3 ubiquitin ligases, which may be single protein molecules or large multicomponent complexes. Ubiquitination may end with the addition of a single ubiquitin moiety (mono-ubiquitination), or further ubiquitin molecules can be added by isopeptide linkage to a lysine residue in the preceding ubiquitin molecule, thereby forming a polyubiquitin chain. Once the chain length exceeds four ubiquitin monomers, the substrate is then recognised by the 26S proteasome and degraded. The polyubiquitin chains are removed from the substrate and ubiquitin monomers are regenerated by the activities of USP enzymes.

Given that there are hundreds of proteins in the human proteome with sequence motifs characteristic of E3 ubiquitin ligases (Robinson and Ardley, 2004), and given that well over 50 potential human USP enzymes have been identified (Chung and Baek, 1999; Balakirev *et al.*, 2003; Quesada *et al.*, 2004;

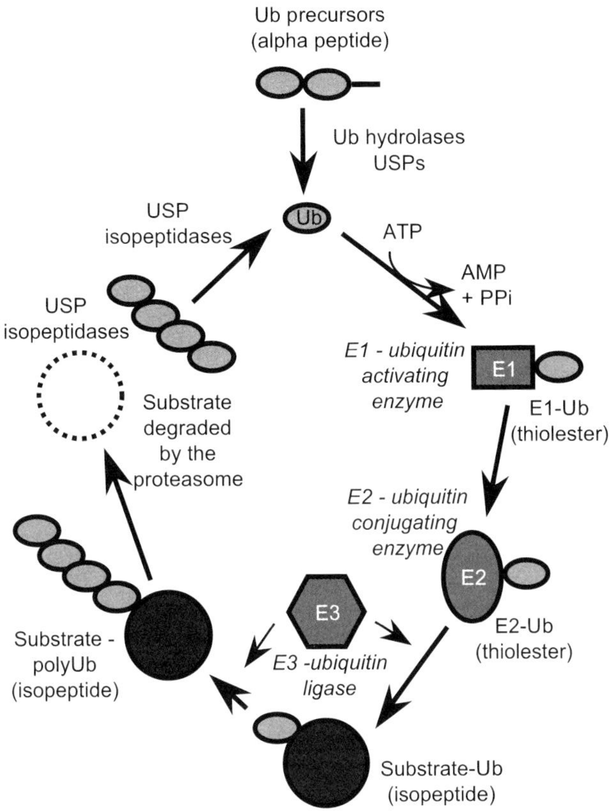

Figure 3.2. A basic summary of the core enzymes involved in the ubiquitin conjugation and processing pathway.

Soboleva and Baker, 2004), the complexity of the ubiquitin pathway and its potential for specific regulation rivals that of the kinases and phosphorylation.

ICP0 is a RING finger E3 ubiquitin ligase

A significant advance in ICP0 research occurred after it was established that many proteins of the RING finger class have E3 ubiquitin ligase activity (Joazeiro and Weissman, 2000). This observation provided a likely explanation for the well documented disruption of ND10 and centromeres by ICP0 through the proteasome-dependent degradation of PML and Sp100 (major components of ND10), and centromere proteins CENP-C and CENP-A (Everett *et al.*, 1998a; Chelbi-Alix and de The, 1999; Everett *et al.*, 1999a; Parkinson and Everett, 2000; Lomonte *et al.*, 2001). ICP0 also induces the degradation of the catalytic subunit of DNA protein kinase (DNA-PKcs) in a proteasome- and RING finger-dependent manner (Lees-Miller *et al.*, 1996; Parkinson *et al.*, 1999). It was then found that ICP0 induces the formation of colocalizing conjugated ubiquitin at both ND10 and centromeres, and at other locations where ICP0 accumulates (Everett, 2000a).

Consistent with these functional observations, it has been demonstrated that both full-length ICP0 and its bacterially expressed isolated RING finger domain have potent E3 ubiquitin ligase activity *in vitro*, producing high molecular weight conjugated ubiquitin chains in conjunction with the E2 ubiquitin conjugating enzymes UbcH5a and UbcH6 (Boutell *et al.*, 2002; Hagglund *et al.*, 2002). The involvement of UbcH5a in the ICP0-induced degradation of PML and Sp100 during infection has been confirmed using a dominant negative approach (Gu and Roizman, 2003).

The RING finger mediated E3 ubiquitin ligase activity of ICP0 makes a major contribution to the activity of ICP0 in gene expression and virus replication assays. Indeed, it has been clearly demonstrated that the activities of ICP0 in viral lytic cycle gene expression and in the reactivation of expression from quiescent HSV-1 genomes are dependent on the ubiquitin–proteasome pathway (Everett *et al.*, 1998b). Mutational analysis of the ICP0 RING finger domain and its environs demonstrated an excellent correlation between the degree of loss of ICP0 function in gene expression and virus replication assays, and the E3 ubiquitin ligase activity of the mutant protein both *in vitro* and in terms of producing co-localizing conjugated ubiquitin in transfected and infected cells (O'Rourke *et al.*, 1998; Everett, 2000a; Boutell *et al.*, 2002). Such studies also demonstrated that the similarity of the phenotypes of mutations downstream of the core RING domain to those within the core itself (Everett, 1987; Everett, 1988; Chen *et al.*, 1991) was largely consistent with the effects of the mutations on the RING finger E3 ubiquitin ligase activity of ICP0 (Everett, 2000a; Boutell *et al.*, 2002).

An important question is whether the proteins that are degraded in an ICP0-dependent manner during HSV-1 infection are direct substrates for the E3 ubiquitin ligase activity of ICP0, or whether they are degraded as an indirect response to the degradation of some other target of ICP0. For example, although ICP0 induces the formation of increased amounts of ubiquitinated PML in transfected cells, *in vitro* experiments using purified ICP0 and purified core enzymes of the ubiquitination pathway did not replicate the ICP0-mediated ubiquitination of PML that is seen in cultured cells (Boutell *et al.*, 2003). Similarly, ICP0 does not ubiquitinate Sp100 directly *in vitro*. Therefore, either PML and Sp100 are not direct substrates of ICP0 and their degradation is a consequence of the ubiquitination and loss of some other protein, or another as yet unidentified component is required to allow ICP0 to ubiquitinate these substrates.

The first protein to be identified that serves as a direct substrate of ICP0 *in vitro* was p53 (Boutell and Everett, 2003). Ubiquitination of p53 *in vitro* by ICP0 occurs with an efficiency similar to that performed by mdm2 in parallel *in vitro* reactions (Boutell *et al.*, 2005); mdm2 is the cellular E3 ubiquitin ligase that plays a central role in the regulation of p53 levels *in vivo*. Although p53 can also be ubiquitinated by ICP0 in transfected cells, this activity is very inefficient compared with that of mdm2 in similar assays (Boutell and Everett, 2003). It is possible that only a subset of p53 molecules, perhaps defined by post-translational modification or subcellular localization, are sensitive to ICP0-mediated ubiquitination in cultured cells. This issue is further complicated by the observation that during HSV-1 infection p53 becomes stabilised by multiple mechanisms, and this stabilization is dominant over any ICP0-induced ubiquitination of p53, at least in human fibroblast cells (Boutell and Everett, 2004). Therefore, the biological significance of the potential of ICP0 to ubiquitinate p53 remains unclear, although ICP0 can protect osteosarcoma U2OS cells from apoptosis induced by ultraviolet irradiation, presumably through a p53-dependent mechanism (Boutell and Everett, 2003).

The other direct substrates of the ICP0 RING finger E3 ubiquitin ligase activity that have been identified are ICP0 itself (Canning *et al.*, 2004) and ubiquitin specific protease USP7 (Boutell *et al.*, 2005). As described below, these activities have clear significance for the function of ICP0 and the biology of HSV-1.

Does ICP0 include two biologically functional E3 ubiquitin ligase domains?

Experiments using GST-fusion proteins that include a segment of ICP0 in the C-terminal third of the protein have suggested that ICP0 may contain a second E3 ubiquitin ligase activity (named HUL1) distinct from that of the RING finger domain (Hagglund and Roizman, 2002; Hagglund *et al.*, 2002; Van Sant *et al.*, 2001a). In this case, the activity gave rise to increased auto-ubiquitination of the E2 ubiquitin conjugating enzyme cdc34 (UbcH3) *in vitro*. It has been suggested that HUL1 activity leads to degradation of cdc34 during HSV-1 infection, and since cdc34 is the E3 ubiquitin ligase involved in the normal regulation of cyclins D1 and D3, this would explain the reported ICP0-dependent stabilization of these cyclins during HSV-1 infection (Kawaguchi *et al.*, 1997b; Van Sant *et al.*, 2001a; Van Sant *et al.*, 2001b; Hagglund and Roizman, 2002; Hagglund *et al.*, 2002; Hagglund and Roizman, 2003). However, unlike the HUL1 GST fusion protein, full-length ICP0 does not stimulate cdc34 *in vitro* (Boutell *et al.*, 2002), and it appears that stimulation of cdc34 *in vitro* can be induced by GST fusion proteins that lack authentic E3 activity (Deffenbaugh *et al.*, 2003). Furthermore, cdc34 is stable during high multiplicity HSV-1 infection of a wide variety of cell lines (Everett, 2004), and the reported apparent ICP0-mediated stabilization of cyclins D1 and D3 occurred in the original studies because the ICP0 null mutant virus was used at doses much greater than that of the wild type (wt) virus (Everett, 2004). As described below in the section on the phenotype of ICP0-null mutant viruses, this problem occurred because of experimental complications that arise from the properties of ICP0-negative HSV-1 (Everett *et al.*, 2004a). Therefore, it remains controversial whether or not ICP0 includes a second E3 ubiquitin ligase domain that is active in the full-length protein and that has biological significance during HSV-1 infection.

ICP0 and its effects on SUMO modified proteins

The SUMO modification pathway

ND10 proteins PML and Sp100 are covalently modified by conjugation of the ubiquitin-like protein SUMO-1 to specific lysine residues within their sequences (Sternsdorf *et al.*, 1997). SUMO-1 and the related proteins SUMO-2 and SUMO-3 can be conjugated to a growing repertoire of cellular proteins (Johnson, 2004; Seeler and Dejean, 2003), and in the case of PML, this modification has been reported to be important for the maintenance of ND10 structural integrity (Lallemand-Breitenbach *et al.*, 2001). The SUMO modification pathway is similar to that of ubiquitin conjugation, except that enzymes specific to the SUMO pathway are used and the end result is not degradation by the proteasome. The role of SUMO modification is likely to be varied according to the substrate protein in question, but it can include regulation of activity or localization. In the cases of PCNA (Hoege *et al.*, 2002) and IkBα (Desterro *et al.*, 1998), SUMO modification occurs at the same lysine residues that are used for ubiquitination and thus protects these proteins from degradation. In an increasing number of examples, SUMO modification leads to the repression of the activities of transcription factors and other proteins involved in the regulation of gene expression (Schmidt and Muller, 2003; Verger *et al.*, 2003; Yang *et al.*, 2003; Gill, 2004).

ICP0 induces the loss of SUMO-modified forms of PML

In the initial studies on the effects of ICP0 on PML, it was observed that the SUMO-modified forms of PML were particularly sensitive to loss during HSV-1 infection (Everett *et al.*, 1998a) and indeed in transfected cells ICP0 causes the ablation of SUMO-1 modification of PML (Muller and Dejean, 1999). Similarly, it is the SUMO-modified forms of Sp100 that are more sensitive than the unmodified protein to ICP0-induced degradation (Parkinson and Everett, 2000). These observations raise the possibility that ICP0, in addition to or related to its E3 ubiquitin ligase activity, might be working through SUMO modification pathways. Given that ICP0-induced loss of the SUMO-modified forms of PML and Sp100 is inhibited by the proteasome inhibitor MG132 (Everett *et al.*, 1998a; Parkinson and Everett, 2000) it is possible that ICP0 preferentially targets SUMO-modified proteins for degradation. However, CENP-C, CENP-A and DNA-PKcs (the other well-characterised proteins that are degraded in the presence of ICP0) have not been demonstrated to be modified by SUMO-1, at least *in vivo*. On the other hand, ICP0 co-localizes with SUMO isopeptidase SENP-1 in defined foci at the early stages of HSV-1 infection (Bailey and O'Hare, 2002), and although no direct interaction between ICP0 and SENP-1 has been reported, it is possible that SENP-1 or other proteins in the SUMO isopeptidase family could be involved in the activity of ICP0.

Interactions of ICP0 with viral and cellular proteins

Given the substantial effects of ICP0 on the cellular proteins described above, it would be expected that ICP0 might interact directly with these or other proteins. Although ICP0 does not interact directly with either PML or Sp100 *in vitro* (Boutell *et al.*, 2003), a number of proteins have been reported to interact with ICP0 *in vitro* and/or in yeast 2 hybrid assays. These include the major HSV-1 transcriptional activator ICP4 (Yao and Schaffer, 1994), cyclin D3, translation elongation factor EIF-1δ, transcription factor BMAL1 (Kawaguchi *et al.*, 1997a; Kawaguchi *et al.*, 1997b; Kawaguchi *et al.*, 2001), an otherwise uncharacterised protein known as p60 (Bruni *et al.*, 1999), and p53 (Boutell and Everett, 2003). ICP0 also interacts with cdc34 *in vitro* (Van Sant *et al.*, 2001a; Hagglund and Roizman, 2002). Whether these *in vitro* interactions contribute to roles of ICP0 during the early stages of HSV-1 infection remains to be substantiated.

Given its strong effects on the E2 ubiquitin conjugation enzymes UbcH5a and UbcH6 *in vitro*, it is perhaps surprising that no strong interaction between ICP0 and these proteins *in vitro* has been detected, at least under the assay conditions so far explored (Boutell *et al.*, 2002; Hagglund *et al.*, 2002). In transfected and infected cells however, these proteins are clearly recruited into ICP0 foci (Boutell *et al.*, 2002).

ICP0 has been shown to recruit histone deacetylase (HDAC) enzymes of both class I and class II types into foci in transfected cells, and to co-localise with the class II enzyme HDAC4 during infection (Lomonte *et al.*, 2004). ICP0 also interacts directly with class II HDAC enzymes *in vitro*, via a domain in the latter that has been implicated in the regulation of MEF-2, a transcription factor involved in muscle development and neuronal survival (Lomonte *et al.*, 2004). Furthermore, ICP0 overcomes the HDAC5-mediated repression of MEF-2 in transfection reporter assays (Lomonte *et al.*, 2004). Given that relief of HSV-1 genomes from repression is widely thought to be the consequence of ICP0 activity, the concept that ICP0 might impinge on the activities of the HDAC class of enzymes is extremely attractive because their role in the formation of a repressed chromatin structure. This hypothesis has been strengthened by the finding that HDAC1 becomes dissociated from the CoREST/REST repression complex during HSV-1 infection, and that ICP0 was required for this process (Gu *et al.*, 2005). Taken together with observations that histone deacetylase inhibitors can partially complement ICP0-null mutant viruses (see below), these data suggest that ICP0 might impede the assembly of a repressed chromatin structure on viral genomes by sequestering or affecting the properties of HDAC enzymes.

Perhaps the best characterised interaction between ICP0 and a cellular protein is that with ubiquitin specific protease USP7 (Meredith *et al.*, 1994; Meredith *et al.*, 1995; Everett *et al.*, 1997). USP7 is a member of a large family of enzymes involved in generation of free ubiquitin from precursor molecules, recycling of ubiquitin from polyubiquitin chains, or release of ubiquitin from ubiquitinated proteins (Fig. 3.3). The interaction between ICP0 and USP7 is particularly strong and has been authenticated by a variety of *in vivo* and *in vitro* assays. The region of ICP0 that is required for the interaction has been mapped and single amino acid substitution mutations that disrupt the interaction have been constructed (Everett *et al.*, 1999b). These studies have shown that interaction with USP7 contributes to the functions of ICP0 in transfection reporter and virus replication assays (Meredith *et al.*, 1995; Everett *et al.*, 1999b; Hsu and Everett, 2001).

The rationale behind the interaction between ICP0 and USP7 is now understood in greater detail (Fig. 3.4). The RING finger E3 ubiquitin ligase activity of ICP0 results in efficient auto-ubiquitination *in vitro*, and during infection this can lead to instability of the ICP0 protein. If ICP0 is unable to bind to USP7, or if USP7 levels are decreased by siRNA treatment, the stability of ICP0 is greatly decreased (Canning *et al.*, 2004), particularly at early times of low multiplicity infections (Boutell *et al.*, 2005). Consequently, viruses expressing forms of ICP0 that are unable to bind to USP7 express ICP0 and other viral proteins at reduced rates, and in plaque assays this correlates with a reduced probability that the initial infection will lead to plaque development (Boutell *et al.*, 2005). It is important to note that the extent of these defects varies in a cell type dependent manner, and they can be overcome by using a high multiplicity of infection.

Reciprocal activities of USP7 and ICP0

The stabilization of ICP0 by protection from auto-ubiquitination through its interaction with USP7 is complicated by the ability of ICP0 to ubiquitinate USP7 *in vitro* and induce the degradation of USP7 during HSV-1 infection (Boutell *et al.*, 2005). ICP0 induced degradation of USP7 is dependent on the RING finger domain of ICP0 and the ability of the two proteins to interact. The reciprocal activities of the two proteins, where each serves as a substrate for the other, poses the question of the relative importance of the two apparently conflicting processes. The available evidence indicates that it is the stabilization of ICP0 through its interaction with USP7 that is particularly important during the early stages of a low multiplicity infection. It remains to be determined whether USP7 is degraded by ICP0 simply as a consequence of the interaction between the two proteins, or whether ICP0-mediated loss of USP7 affects the cell in a way that is beneficial for virus infection (Boutell *et al.*, 2005).

Consequences of the RING finger E3 ligase activity

The multiple interactions of ICP0 with cellular proteins and its ability to induce the degradation of such a variety of proteins suggest that ICP0 has substantial effects on the

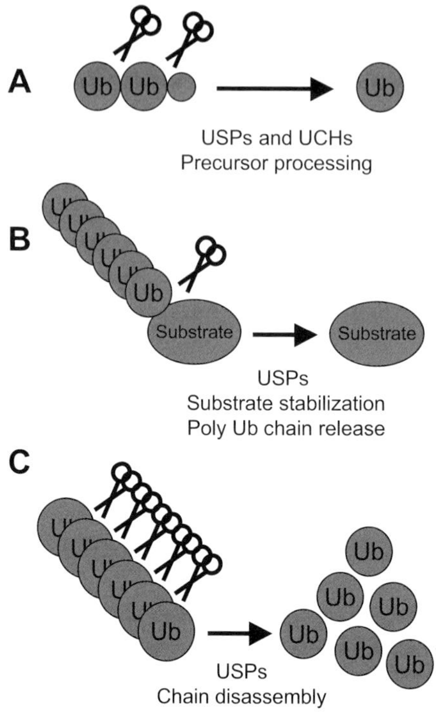

Figure 3.3. A summary of the basic roles of enzymes that cleave ubiquitin precursors, poly-ubiquitin chains and ubiquitinated proteins. A. Cleavage of the alpha-peptide bonds linking tandemly repeated ubiquitin moieties and/or C-terminal extensions following the di-glycine motif at the C-terminus of ubiquitin. Enzymes having this activity may have active sites that are characteristic of either the ubiquitin C-terminal hydrolase (UCH) or the ubiquitin specific protease (USP) families of proteins. B. Isopeptide-linked polyubiquitin chains may be cleaved from substrate proteins to protect them from proteasome-dependent degradation, or from peptide remnants after the substrates have been degraded. C. Monomeric ubiquitin is regenerated by the isopeptidase activity of USP enzymes.

intracellular environment. Indeed, it has long been known that ICP0 accumulates in ND10 nuclear substructures then brings about their disruption in a RING finger dependent manner (Maul *et al.*, 1993; Everett and Maul, 1994; Maul and Everett, 1994). This activity clearly occurs because of the ICP0-induced degradation of PML and Sp100 described above. Similarly, the ICP0-mediated degradation of centromere proteins CENP-C and CENP-A during HSV-1 infection leads to disruption of centromeres, so that if an infected cell is in the G2 phase of the cell cycle and attempts to go through mitosis, the process becomes stalled at a stage known as pseudo-prometaphase, in which correct alignment of the chromosomes at the metaphase plate cannot be achieved. Therefore, ICP0 induces a G2-M block to the cell cycle (Everett *et al.*, 1999a; Hobbs and DeLuca, 1999; Lomonte *et al.*, 2001).

ICP0-induced degradation of DNA-PKcs (Lees-Miller *et al.*, 1996; Parkinson *et al.*, 1999) would be expected to cause a defect in nonhomologous end-joining DNA repair and recombination reactions, while degradation of USP7 would affect the stabilities of all the natural substrates of this isopeptidase. Since these include p53 and mdm2 (Li *et al.*, 2002; Cummins *et al.*, 2004; Li *et al.*, 2004),

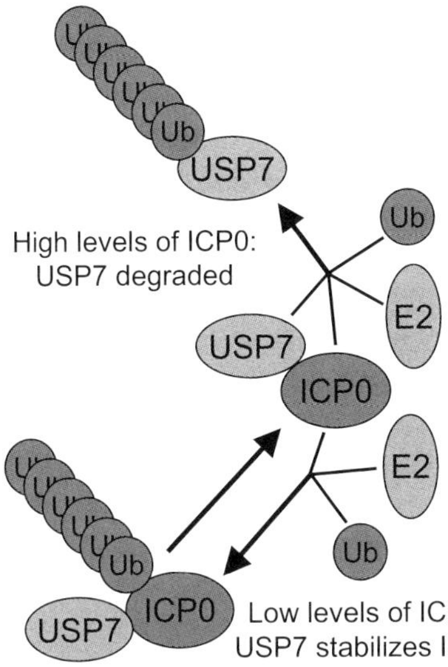

Figure 3.4. The consequences of the ICP0-USP7 interaction. When ICP0 is present at low levels, the dominant feature appears to be stabilization of ICP0 by USP7 through protection from ICP0 auto-ubiquitination. When ICP0 accumulates to higher levels, the degree of ICP0-mediated ubiquitination and degradation of USP7 increases. The consequent relative loss of ICP0 stabilization may be offset by higher rates of ICP0 synthesis (due to increased multiplicity or increased viral genome copy number) and perhaps by limiting availability of the relevant E2 ubiquitin conjugating enzymes.

this would be expected to have severe detrimental effects on the cell in the long term. In uninfected cells, reductions in USP7 levels lead to increased mdm2 auto-ubiquitination and therefore a rise in p53 levels that could cause a block in cell cycle progression from G1 to S phase (Cummins *et al.*, 2004; Li *et al.*, 2004). It is interesting that ICP0 also induces a G1 to S phase block in the cell cycle (Hobbs and DeLuca, 1999; Lomonte and Everett, 1999). However, the metabolism of p53 during HSV-1 infection is extremely complicated, and as yet no role for USP7 binding by ICP0 in these events has been detected (Boutell and Everett, 2003; Boutell and Everett, 2004). Nonetheless, the combination of all the potential effects of ICP0 on the cell (summarised in Fig. 3.5) doubtless explains why it has proved impossible to establish cell lines that constitutively express measurable levels of ICP0 protein.

The powerful events that ICP0 induces in the cell present an experimental problem in differentiating between the direct effects of ICP0 itself, and the indirect consequences that result from prolonged expression of ICP0 within a cell. This is particularly evident in plasmid transfection assays, in which ICP0 will be expressed in greater amounts and for longer times than the critical period at early times of infection when its biological function is most important. As these issues become better understood, it should become the norm that any phenotype that is attributed to ICP0 in transfected cells should be corroborated by studies of cells in the early stages of a low multiplicity HSV-1 infection.

The potential significance of these effects of ICP0 on cellular proteins and structures remains controversial. However, the important point is that the RING finger-dependent effects of ICP0 correlate extremely well with its roles during lytic virus infection and reactivation of quiescent HSV-1 genomes.

Phenotype of HSV-1 mutants lacking ICP0

ICP0-null viruses have a high particle to PFU ratio

To understand how the biochemical activities of ICP0 and its effects on cellular proteins contribute to the roles of ICP0 during HSV-1 infection it is essential to have a firm understanding of the phenotype of ICP0-negative virus mutants. This is more complicated than might appear at first sight. Studies using the first ICP0-null mutant viruses to be isolated showed that they could be propagated in cultured cells, but that they had a high particle to plaque forming unit (PFU) ratio (Stow and Stow, 1986; Sacks and Schaffer, 1987). This does not reflect a high proportion of damaged or inactive particles in the virus population, since the extent of the defect varies with cell type, being particularly marked in limited passage human fibroblasts, intermediate in BHK and Vero cells, and absent in U2OS cells (Stow and Stow, 1986; Sacks and Schaffer, 1987; Stow and Stow, 1989; Yao and Schaffer, 1995). The fact that ICP0-null mutant viruses are as infectious as wt strains in U2OS cells shows that ICP0 negative virus particles are as intrinsically infectious as the wt, but that in restrictive cell lines such as human fibroblasts, there is a greatly reduced probability that a particle of an ICP0-defective virus will induce the formation of a plaque in a low multiplicity infection. There is evidence that the extent of this defect varies at different stages of the cell cycle (Cai and Schaffer, 1991). Cells that receive ICP0-null viral genomes and fail to initiate lytic infection harbour the viral DNA in a quiescent state, and these genomes can be reactivated by later provision of ICP0.

The multiplicity dependence of the ICP0-null phenotype

The next important features of infection with ICP0-null viruses are that the defect can be overcome by increasing the viral genome load in high multiplicity infections, and that once lytic infection has been initiated, viral gene expression and yield of progeny particles are similar to those of wt HSV-1 infections (Stow and Stow, 1986; Everett, 1989; Cai and Schaffer, 1992). Thus, the defect is cell type-dependent, multiplicity-dependent and occurs during the early stages of infection. The multiplicity dependence of the ICP0-null mutant phenotype may reflect the failure of a cellular repression mechanism to act sufficiently rapidly on a high dose of viral genomes, thereby allowing sufficient viral gene expression to commit the cell into the lytic cycle.

The ICP0-null phenotype is nonlinear with respect to virus dose

A fundamental aspect of the phenotype of ICP0-null viruses is the concept of threshold. This underlies the observation that although the particle/PFU ratio of an ICP0-null mutant is about 500 times greater in human fibroblasts than in U2OS cells, it does not take a 500-fold greater dose of the mutant virus in the former cell type to establish an infection that is as active as that in the latter. For example, if human fibroblast cells are infected in parallel with wt and ICP0-null mutant HSV-1 at equal doses of virus particles that equate to a multiplicity of 10 PFU per cell (based on titration of both viruses in U2OS cells), the difference between the mutant and wt viruses in terms of viral gene expression and progeny production is not pronounced (Everett *et al.*, 2004a). This is despite the fact that this amount of the ICP0-null mutant corresponds to a multiplicity of only 0.02 PFU per cell based on titration in human fibroblast cells. On the other hand, if the multiplicity of both viruses is reduced to 1 PFU per cell (based on U2OS cell titres), viral gene expression and progeny virus production proceed much more efficiently in the wt than in the mutant virus infection (Everett *et al.*, 2004a). Thus, the multiplicity-dependent defect of ICP0 negative viruses is not linear and there is a broad threshold of input virus dose above which the infection becomes ICP0-independent. This threshold varies with cell type, such that it is about 10-fold less in Vero and BHK cells than in human fibroblasts (Everett *et al.*, 2004a). If infections are conducted below the threshold, then the true consequence of the ICP0 defect can be studied. If however infections are conducted above the threshold, the infection becomes ICP0-independent and the rate of viral gene expression is proportional to the input viral genome copy number.

Experimental complications concerning the use of ICP0-null viruses

The behaviour of ICP0-null mutant viruses outlined above has extremely important practical consequences. If conducting studies in Vero cells, for example, it might at first sight appear reasonable to use wt and ICP0-mutant viruses on the basis of their titres in Vero cells. In fact, this leads to severe experimental problems. The particle/PFU ratio of ICP0-null mutants is about 30-fold greater than that of wt strains in Vero cells (Cai and Schaffer, 1992; Everett *et al.*, 2004a), thus infection of Vero cells at multiplicities determined on Vero cell titres leads to a 30-fold greater dose of mutant virus genomes. This increased dose is highly likely to be above the threshold for ICP0-dependence in most cell types, so the mutant virus infection will actually proceed faster than that of the wt virus (Everett *et al.*, 2004a). This situation explains why the apparent ICP0 dependent stabilization of cyclins D1 and D3 (Kawaguchi *et al.*, 1997b; Van Sant *et al.*, 2001b) is not in fact due to a specific property of ICP0. The infections in these studies were conducted in Vero cells on the basis of Vero cell titres, and under these conditions the ICP0-null mutant infection progresses more rapidly than the wt because of the higher viral genome load (Everett *et al.*, 2004a). It turns out that these two cyclins are degraded during HSV-1 infection at a rate that is proportional to the speed of the developing infection and is independent of whether or not ICP0 itself is present (Everett, 2004). This example serves to illustrate a problem that could influence the interpretation of many studies that use viruses with lesions in ICP0, particularly in some older studies in which infections were not normalised according to genome numbers or viral titres on cells in which ICP0-mutant viruses initiate plaque formation as readily as the wt virus.

The genomes of ICP0-null mutant viruses have a high probability of becoming repressed when infection is conducted under ICP0-dependent conditions

What then of the phenotype of ICP0-null mutant viruses in restrictive cell lines when the infections are conducted at low multiplicity under ICP0-dependent conditions? In this instance, there is a low probability that a cell receiving a single or a low number of viral genomes will progress to productive lytic infection. Instead, these cells establish a quiescent infection in which viral gene expression eventually becomes undetectable. Thus, instead of being active, the viral genome becomes repressed. However, this repression is by no means immediate and all-embracing. It had been noted that infection of Vero cells at low multiplicity with an ICP0-null mutant virus gave a far higher number of cells that were positive for ICP4 expression than would be expected from the number of cells that initiate plaque formation (Cai and Schaffer, 1992). Even in the most restrictive human fibroblast cells, the number of ICP0-null mutant virus infected cells expressing ICP4 at six hours after infection exceeds the number of expected plaque-forming events by almost two orders of magnitude (Everett *et al.*, 2004a). Furthermore, the number of cells with an active IE-1 transcription unit in such infections is equivalent to that expected of a wt virus infection conducted at the same viral genome number input (Everett *et al.*, 2004a). Thus, the repression of the viral genome that occurs during HSV-1 infection in the absence of ICP0 is far from complete or very rapid, and it is surprising that in the absence of ICP0, infected cells can express and maintain readily detectable levels of ICP4 for extended periods yet fail to activate the bulk of the viral Early gene transcription units.

Stochastic viral gene expression in populations of cells infected at low multiplicity with ICP0-null mutant HSV-1

A final characteristic of infection with ICP0-null mutant virus under ICP0-dependent conditions is that the outcomes of the infection of individual cells are not uniform. There is a low probability that the cells will progress to plaque formation, and a very high probability that they will initially engage in IE-1 gene transcription. In between these extremes, individual cells may be positive for ICP4 and/or ICP27 and, at much reduced probability, even for Early gene products such as ICP8 or UL42 (Cai and Schaffer, 1992; Everett *et al.*, 2004a). This viral gene expression is stochastic, in that the cells that express these viral proteins do so in levels that are similar to those expressed in the early stages of a productive infection (Everett *et al.*, 2004a). If a cell has expressed viral proteins in the initial stages of what becomes a nonproductive or repressed infection, readily detectable levels of these proteins can remain for at least 24 hours. It appears that the repression that eventually takes over the viral genome does so gradually or unevenly, so that some transcription units in some cells escape and achieve significant levels of expression.

The complications concerning the phenotype of ICP0-null mutant viruses should be borne in mind when using such viruses in any experimental situation. Furthermore, definition of the phenotype of particular mutations within ICP0 will be subject to the same complications, and their true phenotype may only be revealed by conducting experiments under very carefully controlled conditions. It is also pertinent that only small amounts of ICP0 are required to stimulate efficient HSV-1 infection (Hobbs *et al.*, 2001).

ICP0 induces reactivation of quiescent HSV-1 genomes

Establishment of quiescent HSV-1 infection in cultured cells

The eventual outcome of infection of most restrictive cells at low multiplicity with ICP0-defective HSV-1 is that viral gene expression is repressed below the level of detection and the cells harbour quiescent viral genomes for extended periods (Preston, 2000). Experimental systems of this type have frequently utilised viruses with additional lesions in viral regulatory molecules such as VP16 (Mossman and Smiley, 1999), VP16 and ICP4 (Preston and Nicholl, 1997), or several other viral IE proteins (Samaniego *et al.*, 1998). Such multiply mutant HSV-1 viruses are highly defective and allow the establishment of cell populations in which the majority of cells harbour a quiescent HSV-1 genome.

ICP0 induces the reactivation of gene expression from repressed HSV-1 genomes

Using cells with quiescent infections established by other methods, early studies had shown that subsequent expression of ICP0 had the very special property of being able to reactivate gene expression from the repressed HSV-1 or HSV-2 genomes (Russell *et al.*, 1987; Harris *et al.*, 1989; Zhu *et al.*, 1990; Harris and Preston, 1991). The ICP0-induced reactivation is dependent on the RING finger domain of ICP0, but does not require sequences involved in its localization to ND10 and self-multimerization (Harris *et al.*, 1989). ICP0 expressed from an adenovirus vector also stimulated reactivation of HSV-1 from latently infected mouse primary trigeminal ganglion cells (Halford *et al.*, 2001). More recent studies using multiply defective HSV-1 viruses in Vero and human fibroblast cells have established that although limited reactivation can be achieved by the expression of VP16, ICP4, or the histone deacetylase inhibitor TSA in the first few days after establishment of the quiescent infection, after that initial period it is only ICP0 that can achieve efficient reactivation (Preston and Nicholl, 1997; Samaniego *et al.*, 1998).

However, in contrast to the situation in quiescently infected human fibroblast cells, reactivation of latent HSV-1 from primary mouse trigeminal ganglion cultures can be stimulated by adenoviruses expressing ICP0, ICP4 or VP16 (Halford *et al.*, 2001). It is possible that despite HSV-1 gene expression being repressed below the level of detection in latently or quiescently infected cells, there may be very low levels of residual gene expression that vary with cell type or between individual cells. While ICP0 appears to be able to reactivate repressed HSV-1 genomes in all cells and experimental systems, variations in the absolute degree of viral genome repression could allow reactivation by other mechanisms in a proportion of latently infected cells.

ICP0 reactivates gene expression from integrated, repressed AAV genomes

Similar ICP0-dependent reactivation of integrated quiescent adeno-associated virus (AAV) genomes in a cultured cell system has been observed. HeLa cells containing multiple copies of AAV integrated into chromosome 19 have been established. Normally, the AAV Rep proteins are not expressed in these cells, but expression can be induced by infection with either adenovirus or HSV-1. In the case of the latter, this reactivation is dependent on ICP0 (Geoffroy *et al.*, 2004). Reactivation of AAV Rep protein expression can also be induced by transfection of plasmids expressing ICP0, and this approach has identified the ICP0 RING finger domain as being essential for this activity. As in the case of reactivation of quiescent herpesvirus genomes (Harris *et al.*, 1989), sequences in the ND10 localization/self-multimerization domain of ICP0 were not required for AAV reactivation, but interestingly the region of ICP0 that is required for interaction with USP7 was important for efficient AAV reactivation (Geoffroy *et al.*, 2004). Therefore, the mechanism by which ICP0 brings about the reactivation of gene expression from quiescent episomal HSV-1 genomes may be very similar to that which induces reactivation of chromosomally integrated AAV genomes.

ICP0 and its role in latency

ICP0-null viruses reactivate poorly from latently infected mouse ganglia

Herpes simplex virus latency is an extremely complicated subject that is covered in more detail in the chapter by David Bloom in this volume. Therefore, the discussion here will be restricted to the information on the role of ICP0. Consistent with the phenotype of ICP0-null mutant viruses during lytic infection and in the reactivation of quiescent genomes from cultured cells, ICP0 negative mutants reactivate poorly from the explanted neurones of latently infected mice (Clements and Stow, 1989; Leib *et al.*, 1989). These early studies showed slight inconsistencies between each other, perhaps as a result of the type of problem regarding ICP0 mutant virus dose and titre explained above. Once it became clear that ICP0-null mutant virus replication in animals was strongly influenced by the host immune response (Leib *et al.*, 1999), treatments that delayed the kinetics of the innate immune response to HSV-1 (such as reduced viral inoculum and transient immuno-suppression) were used to establish equal numbers of wt and ICP0-null mutant viral genomes in latently infected mouse ganglia (Halford and Schaffer, 2000). With this technical hurdle cleared, comparisons of the reactivation efficiency of wt and ICP0-null mutant genomes definitively established that ICP0 was indeed very important for efficient reactivation of latent HSV-1 (Halford and Schaffer, 2001). As yet, studies of this type have not been extended to investigate whether the domains of ICP0 that contribute to its functions during lytic infection are the same as those required for efficient reactivation from latency.

Factors that could influence ICP0 during reactivation of latent HSV-1

If ICP0 expression is indeed a dominant factor in reactivation, the important question arises concerning the factors that could initially lead to ICP0 expression. It has been found that the IE-1 promoter is activated in neuronal cells following treatments that are known to induce HSV-1 reactivation (Minagawa *et al.*, 1994; Loiacono *et al.*, 2003). The type of factors that could be involved include neuronal-specific transcription factors that interact with the IE-1 promoter (Devireddy and Jones, 2000), transport into the nucleus of the cellular factor C1 (also known as HCF) that is involved in IE gene transcription (Kristie *et al.*, 1999), and the presence of multiple binding sites for the YY1 transcriptional repressor in the IE-1 intron (Gu *et al.*, 1995). However, initial functional studies have found no evidence for sequences in the IE-1 introns being involved in latency or reactivation (Natarajan *et al.*, 1991), while studies on the IE-1 promoter have yet to reveal specific elements that are required for HSV-1 reactivation (Davido and Leib, 1996; Thompson *et al.*, 2003). The intracellular localization and phosphorylation status of ICP0 itself might also be involved in the reactivation process (Chen *et al.*, 2000). Given the role of USP7 in stabilizing ICP0 at early times of a low multiplicity infection (Canning *et al.*, 2004) it is tempting to suggest that USP7 may play an important role in reactivation of HSV-1 from latency.

ICP0 and inhibition of the interferon response

A further important property of ICP0 is its role in counteracting the host interferon response (described in more detail in the chapter by Karen Mossman in this volume). HSV-1 infection induces expression of interferon-responsive genes by a mechanism that is inhibited by viral protein expression (Nicholl *et al.*, 2000; Mossman *et al.*, 2001). Because ICP0-negative viruses are hypersensitive to interferon treatment in cultured cells (Mossman *et al.*, 2000; Harle *et al.*, 2002) and ICP0 negative mutants replicate more efficiently in mice lacking intact interferon response pathways (Leib *et al.*, 1999), ICP0 is a good candidate for the viral protein that interferes with the interferon response. However, it is important to recognise that sensitivity to interferon and inhibition of the activation of the interferon response may not be precisely the same phenomena. Since interferon treatment results in a general reduction in the efficiency of IE gene transcription (Nicholl and Preston, 1996), it is possible that the interferon sensitivity of ICP0-negative viruses reflects the additive effects of interferon inhibition of IE gene transcription that in ICP0-negative mutant viruses is already inefficient because of the lack of ICP0.

The HSV-1 induced activation of interferon-responsive genes has been shown at least in part to be due to activation of interferon response factor 3 (IRF-3) (Preston *et al.*, 2001). ICP0 has been implicated in the inhibition of the interferon response (Eidson *et al.*, 2002) and more particularly inhibition of IRF-3 function (Lin *et al.*, 2004). These effects of ICP0 are also dependent on its RING finger domain. It has been suggested that ICP0 inhibits the interferon response through its degradation of PML (Chee *et al.*, 2003). The mechanism by which this might occur is not clear—although PML expression is stimulated by interferon, PML itself has not been identified as a component in the mechanism of the interferon response pathway. The situation is further complicated by the observation that HSV-1 induces mechanisms that inhibit the interferon beta response that are additional to those invoked by ICP0 (Melroe *et al.*, 2004).

ICP0, ND10, PML and the initiation of lytic virus infection

Despite the wealth of evidence concerning the phenotype of ICP0 negative mutant viruses, the biochemical functions of ICP0, its interactions with cellular proteins and its effects on cellular proteins and substructures, the mechanisms by which ICP0 stimulates virus infection and reactivation from latency or quiescence remain poorly understood. The diversity of the established and apparent targets of the RING finger E3 ubiquitin ligase activity of ICP0, and the requirement for the RING finger domain in all functional assays of ICP0 (Fig. 3.5), has made it impossible as yet to determine which if any of these documented effects are critical for ICP0 activity. In other words, the information to date has firmly established that RING finger dependent activities are essential for ICP0 function, but the diversity of the targets of ICP0-induced degradation has thus far ruled out confirmation that the degradation of any one of them is directly involved in the efficiency of HSV-1 infection. It is likely that the process by which ICP0 targets its truly relevant biological substrates also causes degradation of other proteins that happen to share a property that determines ICP0 susceptibility—some of the established targets of ICP0 may be innocent victims.

The controversy surrounding these questions is well illustrated by the potential role of ND10 in HSV-1 infection. As described above, ICP0 first localizes to ND10 during the early stages of infection then disrupts them through its E3 ubiquitin ligase activity. Given that ICP0-negative viral genomes become repressed in low multiplicity infections, that they are hypersensitive to the effects of interferon, and that PML and Sp100 are both interferon-induced proteins, an obvious hypothesis is that ND10 have a repressive effect on viral gene expression. In this model, disruption of ND10 by ICP0 relieves this repression. Support for this hypothesis comes from evidence that PML, Sp100 and hDaxx (another major component of ND10) have all been implicated in repression of gene expression in reporter assays (for reviews, references and views of these issues, see: Everett, 2001; Everett *et al.*, 2004a; Everett *et al.*, 2004b; Hagglund and Roizman, 2004).

Also consistent with the repression hypothesis is the important observation that parental HSV-1 genomes become juxtaposed to ND10 during the early stages of infection (Ishov and Maul, 1996; Maul *et al.*, 1996), and that these locations serve as "immediate-early transcript environments" at which viral transcription takes place and from which viral replication compartments develop (reviewed in Everett, 2001; Maul, 1998). More recent work has established that the association of ND10 structures with parental viral genomes occurs because ND10 proteins are deposited

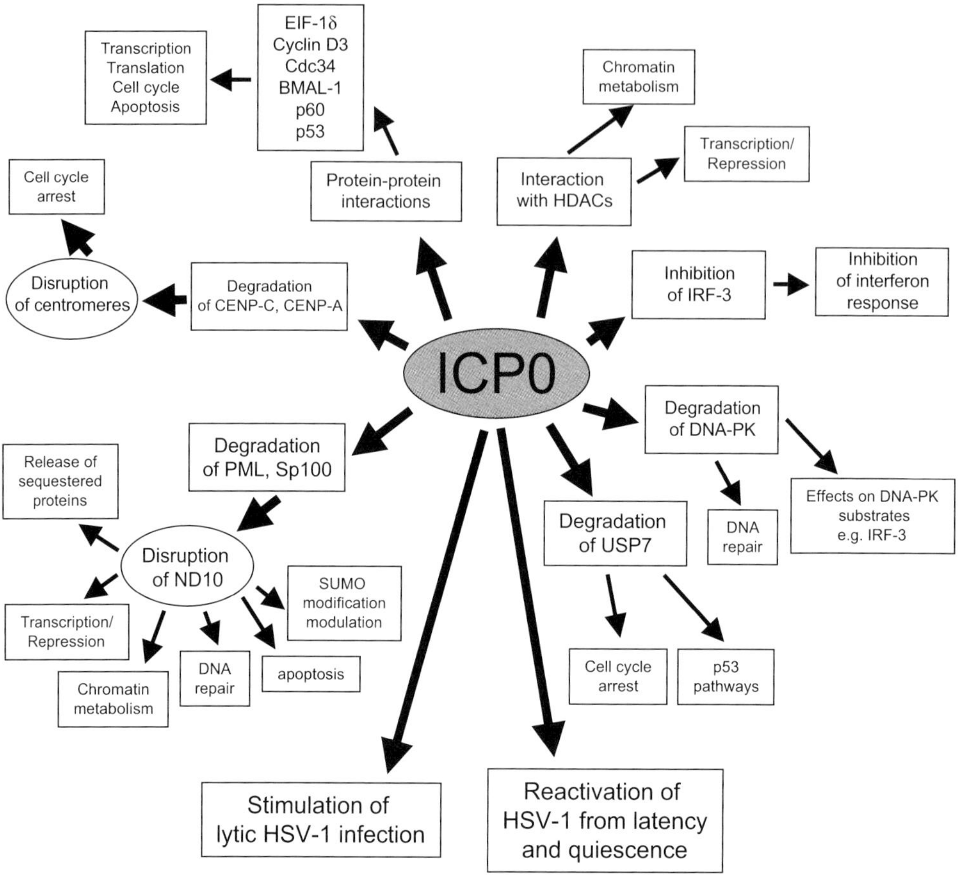

Figure 3.5. A summary of the published biochemical activities and interactions of ICP0, and the complex potential consequences of these functions for the cell and virus infection. Some of the activities and interactions have yet to be confirmed. The consequences that are noted here are in many cases hypothetical, but they are based on the known functions of the target proteins and structures. For example, PML and ND10 structures have been implicated in a plethora of cellular events and pathways, but the consequences for these phenomena of the ICP0-induced loss of PML and ND10 has not been determined directly. No attempt has been made to distinguish between those possible effects that have no direct experimental evidence to support them, and those that have been clearly validated. The connections between the effects of ICP0 on the cell and ICP0-mediated stimulation of viral lytic infection and reactivation from latency have not been established, except that, with the exception of some of the protein–protein interactions, all are dependent on the ICP0 RING finger domain.

at the sites juxtaposed to the incoming viral genomes, thus identifying this association as a cellular response to virus infection (Everett *et al.*, 2004b; Everett and Murray, 2005). This response is impeded by the activities of ICP0 that cause the degradation of PML and Sp100.

Despite these indications that can be interpreted in terms of ICP0 counteracting the repressive effect of a cellular response that involves ND10 proteins and their juxtaposition to viral nucleoprotein loci, there are a number of conflicting observations. These include that high-level expression of PML does not inhibit HSV-1 infection (Chelbi-Alix and de The, 1999; Lopez *et al.*, 2002), that an ICP0-null mutant virus does not replicate more efficiently in PML null compared with normal mouse fibroblasts (Chee *et al.*, 2003), and that HSV-1 infection proceeds normally without apparent disruption of ND10 structures in cells expressing very high levels of exogenous PML (Lopez *et al.*, 2002).

While the first two of the above observations have been confirmed independently in other laboratories, live cell microscopy has revealed that although exogenous PML expressed at very high levels is not completely degraded during wt HSV-1 infection, ND10 structures in such cells still undergo substantial disruption and rearrangement during infection (Everett and Zafiropoulos, 2004). Furthermore, in cells in which ICP0-null mutant viruses progress to DNA replication and productive infection (high multiplicity conditions), despite the failure to degrade PML and other ICP0 substrates, substantial modification of ND10 structures and repositioning of ND10 proteins occurs (Everett *et al.*, 2004b). On the other hand, in low multiplicity ICP0-mutant virus infections which fail to progress to lytic infection, reorganization of ND10 components and structures does not appear to occur, although subtle effects may have escaped detection (Everett *et al.*, 2004b). Therefore, while high-level expression of PML is not detrimental to HSV-1 infection, and lack of PML does not overcome the defect in HSV-1 replication caused by the absence of ICP0 (at least in mouse cells), disruption or reorganization of ND10 occurs in all cells that become productively infected with HSV-1.

Because viral replication compartments develop from genomes that are juxtaposed to ND10, another view of the ND10 association is that rather than being a repressive response of the cell, ND10 components may actually aid virus infection. This interpretation is related to the concept that ND10 serve as nuclear depots (Negorev and Maul, 2001). In this hypothesis, the disruption of ND10 by ICP0 releases active factors that are then used by the virus to stimulate viral gene expression and replication. Thus, ND10 may have elements of both positive and negative regulation; negative in the sequestering of factors that could aid virus replication, positive in that release of such factors could stimulate viral gene expression. An intriguing possibility couples this hypothesis with the increasing number of reports of transcription factors that are repressed by SUMO modification. If ICP0-mediated disruption of ND10 and degradation of PML, Sp100 and their SUMO-modified forms impinges on the status of SUMO-modified transcription factors, increased rates of viral gene transcription may ensue.

These are complicated and controversial issues that will be very difficult to resolve. It may even be that ICP0-mediated stimulation of lytic viral gene expression, relief from cellular repression, and reactivation of expression from quiescent genomes involve distinguishable pathways that are related by the requirement for the RING finger E3 ubiquitin ligase activity of ICP0. It is likely that ICP0 will continue to be the subject of much experimentation and debate.

Acknowledgments

The work in the author's laboratory is supported by the Medical Research Council. I am grateful for the constructive comments on the text of Chris Boutell, Chris Preston, Amanda Sykes, David Davido, Priscilla Schaffer, Karen Mossman and Bill Halford.

References

Ackermann, M., Braun, D.K., Pereira, L., and Roizman, B. (1984). Characterization of herpes simplex virus 1 alpha proteins 0, 4, and 27 with monoclonal antibodies. J. Virol. 52, 108–118.

Advani, S.J., Weichselbaum, R.R., and Roizman, B. (2000). The role of cdc2 in the expression of herpes simplex virus genes. Proc. Natl. Acad. Sci. USA *97*, 10996–11001.

Advani, S.J., Hagglund, R., Weichselbaum, R.R., and Roizman, B. (2001). Posttranslational processing of infected cell proteins 0 and 4 of herpes simplex virus 1 is sequential and reflects the subcellular compartment in which the proteins localize. J. Virol. *75*, 7904–7912.

Bailey, D., and O'Hare, P. (2002). Herpes simplex virus 1 ICP0 co-localizes with a SUMO-specific protease. J. Gen. Virol. *83*, 2951–2964.

Balakirev, M.Y., Tcherniuk, S.O., Jaquinod, M., and Chroboczek, J. (2003). Otubains: a new family of cysteine proteases in the ubiquitin pathway. EMBO Rep. *4*, 517–522.

Banks, L., Pim, D., and Thomas, M. (2003). Viruses and the 26S proteasome: hacking into destruction. Trends Biochem. Sci. *28*, 452–459.

Barlow, P.N., Luisi, B., Milner, A., Elliott, M., and Everett, R. (1994). Structure of the C3HC4 domain by 1H-nuclear magnetic resonance spectroscopy. A new structural class of zinc-finger. J. Mol. Biol. *237*, 201–211.

Blaho, J.A., Mitchell, C., and Roizman, B. (1993). Guanylylation and adenylylation of the alpha regulatory proteins of herpes simplex virus require a viral beta or gamma function. J. Virol. *67*, 3891–3900.

Boutell, C., and Everett, R.D. (2003). The herpes simplex virus type 1 (HSV-1) regulatory protein ICP0 interacts with and Ubiquitinates p53. J. Biol. Chem. *278*, 36596–36602.

Boutell, C., and Everett, R.D. (2004). Herpes simplex virus type 1 infection induces the stabilization of p53 in a USP7- and ATM-independent manner. J. Virol. *78*, 8068–8077.

Boutell, C., Sadis, S., and Everett, R.D. (2002). Herpes simplex virus type 1 immediate-early protein ICP0 and is isolated RING finger domain act as ubiquitin E3 ligases *in vitro*. J. Virol. *76*, 841–850.

Boutell, C., Orr, A., and Everett, R.D. (2003). PML residue lysine 160 is required for the degradation of PML induced by herpes simplex virus type 1 regulatory protein ICP0. J. Virol. *77*, 8686–8694.

Boutell, C., Canning, M., Orr, A., and Everett, R.D. (2005). Reciprocal activities between herpes simplex virus type 1 regulatory protein ICP0, a ubiquitin E3 ligase, and ubiquitin-specific protease USP7. J. Virol. *79*, 12342–12354.

Bruni, R., Fineschi, B., Ogle, W.O., and Roizman, B. (1999). A novel cellular protein, p60, interacting with both herpes simplex virus 1 regulatory proteins ICP22 and ICP0 is modified in a cell-type-specific manner and Is recruited to the nucleus after infection. J. Virol. *73*, 3810–3817.

Cai, W.Z., and Schaffer, P.A. (1989). Herpes simplex virus type 1 ICP0 plays a critical role in the de novo synthesis of infectious virus following transfection of viral DNA. J. Virol. *63*, 4579–4589.

Cai, W., and Schaffer, P.A. (1991). A cellular function can enhance gene expression and plating efficiency of a mutant defective in the gene for ICP0, a transactivating protein of herpes simplex virus type 1. J. Virol. *65*, 4078–4090.

Cai, W., and Schaffer, P.A. (1992). Herpes simplex virus type 1 ICP0 regulates expression of immediate-early, early, and late genes in productively infected cells. J. Virol. *66*, 2904–2915.

Canning, M., Boutell, C., Parkinson, J., and Everett, R.D. (2004). A RING finger ubiquitin ligase is protected from autocatalyzed ubiquitination and degradation by binding to ubiquitin-specific protease USP7. J. Biol. Chem. *279*, 38160–38168.

Chee, A.V., Lopez, P., Pandolfi, P.P., and Roizman, B. (2003). Promyelocytic leukemia protein mediates interferon-based anti-herpes simplex virus 1 effects. J. Virol. *77*, 7101–7105.

Chelbi-Alix, M.K., and de The, H. (1999). Herpes virus induced proteasome-dependent degradation of the nuclear bodies-associated PML and Sp100 proteins. Oncogene *18*, 935–941.

Chen, J.X., Zhu, X.X., and Silverstein, S. (1991). Mutational analysis of the sequence encoding ICP0 from herpes simplex virus type 1. Virology *180*, 207–220.

Chen, X., Li, J., Mata, M., Goss, J., Wolfe, D., Glorioso, J.C., and Fink, D.J. (2000). Herpes simplex virus type 1 ICP0 protein does not accumulate in the nucleus of primary neurons in culture. J. Virol. *74*, 10132–10141.

Cheung, P., Panning, B., and Smiley, J.R. (1997). Herpes simplex virus immediate-early proteins ICP0 and ICP4 activate the endogenous human alpha-globin gene in nonerythroid cells. J. Virol. *71*, 1784–1793.

Chung, C.H., and Baek, S.H. (1999). Deubiquitinating enzymes: their diversity and emerging roles. Biochem. Biophys. Res. Commun. *266*, 633–640.

Ciufo, D.M., Mullen, M.A., and Hayward, G.S. (1994). Identification of a dimerization domain in the C-terminal segment of the IE110 transactivator protein from herpes simplex virus. J. Virol. *68*, 3267–3282.

Clements, G.B., and Stow, N.D. (1989). A herpes simplex virus type 1 mutant containing a deletion within immediate early gene 1 is latency-competent in mice. J. Gen. Virol. *70*, 2501–2506.

Cummins, J.M., Rago, C., Kohli, M., Kinzler, K.W., Lengauer, C., and Vogelstein, B. (2004). Tumour

suppression: disruption of HAUSP gene stabilizes p53. Nature *428*, 487.
Davido, D.J., and Leib, D.A. (1996). Role of cis-acting sequences of the ICPO promoter of herpes simplex virus type 1 in viral pathogenesis, latency and reactivation. J. Gen. Virol. *77*, 1853–1863.
Davido, D.J., Leib, D.A., and Schaffer, P.A. (2002). The cyclin-dependent kinase inhibitor roscovitine inhibits the transactivating activity and alters the posttranslational modification of herpes simplex virus type 1 ICP0. J. Virol. *76*, 1077–1088.
Davido, D.J., von Zagorski, W.F., Lane, W.S., and Schaffer, P.A. (2005). Phosphorylation site mutations affect herpes simplex virus type 1 ICP0 function. J. Virol. *79*, 1232–1243.
Deffenbaugh, A.E., Scaglione, K.M., Zhang, L., Moore, J.M., Buranda, T., Sklar, L.A., and Skowyra, D. (2003). Release of ubiquitin-charged Cdc34-S-Ub from the RING domain is essential for ubiquitination of the SCF(Cdc4)-bound substrate Sic1. Cell *114*, 611–622.
Desterro, J.M., Rodriguez, M.S., and Hay, R.T. (1998). SUMO-1 modification of IkappaBalpha inhibits NF-kappaB activation. Mol. Cell. *2*, 233–239.
Devireddy, L.R., and Jones, C.J. (2000). Olf-1, a neuron-specific transcription factor, can activate the herpes simplex virus type 1-infected cell protein 0 promoter. J. Biol. Chem. *275*, 77–81.
Eidson, K.M., Hobbs, W.E., Manning, B.J., Carlson, P., and DeLuca, N.A. (2002). Expression of herpes simplex virus ICP0 inhibits the induction of interferon-stimulated genes by viral infection. J. Virol. *76*, 2180–2191.
Everett, R.D. (1984). Trans activation of transcription by herpes virus products: requirement for two HSV-1 immediate-early polypeptides for maximum activity. EMBO J. *3*, 3135–3141.
Everett, R.D. (1987). A detailed mutational analysis of Vmw110, a trans-acting transcriptional activator encoded by herpes simplex virus type 1. EMBO J. *6*, 2069–2076.
Everett, R.D. (1988). Analysis of the functional domains of herpes simplex virus type 1 immediate-early polypeptide Vmw110. J. Mol. Biol. *202*, 87–96.
Everett, R.D. (1989). Construction and characterization of herpes simplex virus type 1 mutants with defined lesions in immediate early gene 1. J. Gen. Virol. *70*, 1185–1202.
Everett, R.D. (2000a). ICP0 induces the accumulation of colocalizing conjugated ubiquitin. J. Virol. *74*, 9994–10005.
Everett, R.D. (2000b). ICP0, a regulator of herpes simplex virus during lytic and latent infection. Bioessays *22*, 761–770.
Everett, R.D. (2001). DNA viruses and viral proteins that interact with PML nuclear bodies. Oncogene *20*, 7266–7273.
Everett, R.D. (2004). Herpes simplex virus type 1 regulatory protein ICP0 does not protect cyclins D1 and D3 from degradation during infection. J. Virol. *78*, 9599–9604.
Everett, R.D., and Maul, G.G. (1994). HSV-1 IE protein Vmw110 causes redistribution of PML. EMBO J. *13*, 5062–5069.
Everett, R.D., and Murray, J. (2005). ND10 components relocate to sites associated with herpes simplex virus type 1 nucleoprotein complexes during virus infection. J. Virol. *79*, 5078–5089.
Everett, R.D., and Zafiropoulos, A. (2004). Visualization by live-cell microscopy of disruption of ND10 during herpes simplex virus type 1 infection. J. Virol. *78*, 11411–11415.
Everett, R.D., Meredith, M., Orr, A., Cross, A., Kathoria, M., and Parkinson, J. (1997). A novel ubiquitin-specific protease is dynamically associated with the PML nuclear domain and binds to a herpesvirus regulatory protein. EMBO J. *16*, 1519–1530.
Everett, R.D., Freemont, P., Saitoh, H., Dasso, M., Orr, A., Kathoria, M., and Parkinson, J. (1998a). The disruption of ND10 during herpes simplex virus infection correlates with the Vmw110- and proteasome-dependent loss of several PML isoforms. J. Virol. *72*, 6581–6591.
Everett, R.D., Orr, A., and Preston, C.M. (1998b). A viral activator of gene expression functions via the ubiquitin–proteasome pathway. EMBO J. *17*, 7161–7169.
Everett, R.D., Earnshaw, W.C., Findlay, J., and Lomonte, P. (1999a). Specific destruction of kinetochore protein CENP-C and disruption of cell division by herpes simplex virus immediate-early protein Vmw110. EMBO J. *18*, 1526–1538.
Everett, R.D., Meredith, M., and Orr, A. (1999b). The ability of herpes simplex virus type 1 immediate-early protein Vmw110 to bind to a ubiquitin-specific protease contributes to its roles in the activation of gene expression and stimulation of virus replication. J. Virol. *73*, 417–426.
Everett, R.D., Boutell, C., and Orr, A. (2004a). Phenotype of a herpes simplex virus type 1 mutant that fails to express immediate-early regulatory protein ICP0. J. Virol. *78*, 1763–1774.
Everett, R.D., Sourvinos, G., Leiper, C., Clements, J.B., and Orr, A. (2004b). Formation of nuclear foci of the herpes simplex virus type 1 regulatory protein ICP4 at early times of infection: localization, dynamics, recruitment of ICP27, and evidence for the de novo induction of ND10-like complexes. J. Virol. *78*, 1903–1917.

Gelman, I.H., and Silverstein, S. (1985). Identification of immediate early genes from herpes simplex virus that transactivate the virus thymidine kinase gene. Proc. Natl. Acad. Sci. USA *82*, 5265–5269.

Geoffroy, M.C., Epstein, A.L., Toublanc, E., Moullier, P., and Salvetti, A. (2004). Herpes simplex virus type 1 ICP0 protein mediates activation of adeno-associated virus type 2 rep gene expression from a latent integrated form. J. Virol. *78*, 10977–10986.

Gill, G. (2004). SUMO and ubiquitin in the nucleus: different functions, similar mechanisms? Genes Dev. *18*, 2046–2059.

Gius, D., and Laimins, L.A. (1989). Activation of human papillomavirus type 18 gene expression by herpes simplex virus type 1 viral transactivators and a phorbol ester. J. Virol. *63*, 555–563.

Golden, M.P., Kim, S., Hammer, S.M., Ladd, E.A., Schaffer, P.A., DeLuca, N., and Albrecht, M.A. (1992). Activation of human immunodeficiency virus by herpes simplex virus. J. Infect. Dis. *166*, 494–499.

Gu, H., and Roizman, B. (2003). The degradation of promyelocytic leukemia and Sp100 proteins by herpes simplex virus 1 is mediated by the ubiquitin-conjugating enzyme UbcH5a. Proc. Natl. Acad. Sci. USA *100*, 8963–8968.

Gu, W., Huang, Q., and Hayward, G.S. (1995). Multiple tandemly repeated binding sites for the YY1 repressor and transcription factors AP-1 and SP-1 are clustered within intron-1 of the gene encoding the IE110 transactivator of herpes simplex virus type 1. J. Biomed. Sci. *2*, 203–226.

Gu, H., Liang, Y., Mandel, G., and Roizman, B. (2005). Components of the REST/CoREST/histone deacetylase repressor complex are disrupted, modified, and translocated in HSV-1-infected cells. Proc. Natl. Acad. Sci. USA *102*, 7571–7576.

Hagglund, R., and Roizman, B. (2002). Characterization of the novel E3 ubiquitin ligase encoded in exon 3 of herpes simplex virus-1-infected cell protein 0. Proc. Natl. Acad. Sci. USA *99*, 7889–7894.

Hagglund, R., and Roizman, B. (2003). Herpes simplex virus 1 mutant in which the ICP0 HUL-1 E3 ubiquitin ligase site is disrupted stabilizes cdc34 but degrades D-type cyclins and exhibits diminished neurotoxicity. J. Virol. *77*, 13194–13202.

Hagglund, R., and Roizman, B. (2004). Role of ICP0 in the strategy of conquest of the host cell by herpes simplex virus 1. J. Virol. *78*, 2169–2178.

Hagglund, R., Van Sant, C., Lopez, P., and Roizman, B. (2002). Herpes simplex virus 1-infected cell protein 0 contains two E3 ubiquitin ligase sites specific for different E2 ubiquitin-conjugating enzymes. Proc. Natl. Acad. Sci. USA *99*, 631–636.

Halford, W.P., and Schaffer, P.A. (2000). Optimized viral dose and transient immunosuppression enable herpes simplex virus ICP0-null mutants To establish wild-type levels of latency *in vivo*. J. Virol. *74*, 5957–5967.

Halford, W.P., and Schaffer, P.A. (2001). ICP0 is required for efficient reactivation of herpes simplex virus type 1 from neuronal latency. J. Virol. *75*, 3240–3249.

Halford, W.P., Kemp, C.D., Isler, J.A., Davido, D.J., and Schaffer, P.A. (2001). ICP0, ICP4, or VP16 expressed from adenovirus vectors induces reactivation of latent herpes simplex virus type 1 in primary cultures of latently infected trigeminal ganglion cells. J. Virol. *75*, 6143–6153.

Harle, P., Sainz, B., Jr., Carr, D.J., and Halford, W.P. (2002). The immediate-early protein, ICP0, is essential for the resistance of herpes simplex virus to interferon-alpha/beta. Virology *293*, 295–304.

Harris, R.A., and Preston, C.M. (1991). Establishment of latency *in vitro* by the herpes simplex virus type 1 mutant in1814. J. Gen. Virol. *72*, 907–913.

Harris, R.A., Everett, R.D., Zhu, X.X., Silverstein, S., and Preston, C.M. (1989). Herpes simplex virus type 1 immediate-early protein Vmw110 reactivates latent herpes simplex virus type 2 in an *in vitro* latency system. J. Virol. *63*, 3513–3515.

Hobbs, W.E., 2nd, and DeLuca, N.A. (1999). Perturbation of cell cycle progression and cellular gene expression as a function of herpes simplex virus ICP0. J. Virol. *73*, 8245–8255.

Hobbs, W.E., Brough, D.E., Kovesdi, I., and DeLuca, N.A. (2001). Efficient activation of viral genomes by levels of herpes simplex virus ICP0 insufficient to affect cellular gene expression or cell survival. J. Virol. *75*, 3391–3403.

Hoege, C., Pfander, B., Moldovan, G.L., Pyrowolakis, G., and Jentsch, S. (2002). RAD6-dependent DNA repair is linked to modification of PCNA by ubiquitin and SUMO. Nature *419*, 135–141.

Hsu, W.L., and Everett, R.D. (2001). Human neuron-committed teratocarcinoma NT2 cell line has abnormal ND10 structures and is poorly infected by herpes simplex virus type 1. J. Virol. *75*, 3819–3831.

Ishov, A.M., and Maul, G.G. (1996). The periphery of nuclear domain 10 (ND10) as site of DNA virus deposition. J. Cell Biol. *134*, 815–826.

Jang, K.L., Pulverer, B., Woodgett, J.R., and Latchman, D.S. (1991). Activation of the cellular transcription factor AP-1 in herpes simplex virus infected cells is dependent on the viral immediate-early protein ICPO. Nucleic Acids Res. *19*, 4879–4883.

Joazeiro, C.A., and Weissman, A.M. (2000). RING finger proteins: mediators of ubiquitin ligase activity. Cell *102*, 549–552.

Johnson, E.S. (2004). Protein modification by sumo. Annu. Rev. Biochem. *73*, 355–382.

Jordan, R., and Schaffer, P.A. (1997). Activation of gene expression by herpes simplex virus type 1 ICP0 occurs at the level of mRNA synthesis. J. Virol. *71*, 6850–6862.

Kawaguchi, Y., Bruni, R., and Roizman, B. (1997a). Interaction of herpes simplex virus 1 alpha regulatory protein ICP0 with elongation factor 1delta: ICP0 affects translational machinery. J. Virol. *71*, 1019–1024.

Kawaguchi, Y., Van Sant, C., and Roizman, B. (1997b). Herpes simplex virus 1 alpha regulatory protein ICP0 interacts with and stabilizes the cell cycle regulator cyclin D3. J. Virol. *71*, 7328–7336.

Kawaguchi, Y., Tanaka, M., Yokoymama, A., Matsuda, G., Kato, K., Kagawa, H., Hirai, K., and Roizman, B. (2001). Herpes simplex virus 1 alpha regulatory protein ICP0 functionally interacts with cellular transcription factor BMAL1. Proc. Natl. Acad. Sci. USA *98*, 1877–1882.

Kristie, T.M., Vogel, J.L., and Sears, A.E. (1999). Nuclear localization of the C1 factor (host cell factor) in sensory neurons correlates with reactivation of herpes simplex virus from latency. Proc. Natl. Acad. Sci. USA *96*, 1229–1233.

Kwun, H.J., Han, H.J., Lee, W.J., Kim, H.S., and Jang, K.L. (2002). Transactivation of the human endogenous retrovirus K long terminal repeat by herpes simplex virus type 1 immediate early protein 0. Virus Res. *86*, 93–100.

Lallemand-Breitenbach, V., Zhu, J., Puvion, F., Koken, M., Honore, N., Doubeikovsky, A., Duprez, E., Pandolfi, P.P., Puvion, E., Freemont, P., and de The, H. (2001). Role of promyelocytic leukemia (PML) sumolation in nuclear body formation, 11S proteasome recruitment, and As2O3-induced PML or PML/retinoic acid receptor alpha degradation. J. Exp. Med. *193*, 1361–1371.

Lees-Miller, S.P., Long, M.C., Kilvert, M.A., Lam, V., Rice, S.A., and Spencer, C.A. (1996). Attenuation of DNA-dependent protein kinase activity and its catalytic subunit by the herpes simplex virus type 1 transactivator ICP0. J. Virol. *70*, 7471–7477.

Leib, D.A., Coen, D.M., Bogard, C.L., Hicks, K.A., Yager, D.R., Knipe, D.M., Tyler, K.L., and Schaffer, P.A. (1989). Immediate-early regulatory gene mutants define different stages in the establishment and reactivation of herpes simplex virus latency. J. Virol. *63*, 759–768.

Leib, D.A., Harrison, T.E., Laslo, K.M., Machalek, M.A., Moorman, N.J., and Virgin, H.W. (1999). Interferons regulate the phenotype of wild-type and mutant herpes simplex viruses *in vivo*. J. Exp. Med. *189*, 663–672.

Li, M., Chen, D., Shiloh, A., Luo, J., Nikolaev, A.Y., Qin, J., and Gu, W. (2002). Deubiquitination of p53 by HAUSP is an important pathway for p53 stabilization. Nature *416*, 648–653.

Li, M., Brooks, C.L., Kon, N., and Gu, W. (2004). A dynamic role of HAUSP in the p53-Mdm2 pathway. Mol. Cell. *13*, 879–886.

Lin, R., Noyce, R.S., Collins, S.E., Everett, R.D., and Mossman, K.L. (2004). The herpes simplex virus ICP0 RING finger domain inhibits IRF3- and IRF7-mediated activation of interferon-stimulated genes. J. Virol. *78*, 1675–1684.

Loiacono, C.M., Taus, N.S., and Mitchell, W.J. (2003). The herpes simplex virus type 1 ICP0 promoter is activated by viral reactivation stimuli in trigeminal ganglia neurons of transgenic mice. J. Neurovirol. *9*, 336–345.

Lomonte, P., and Everett, R.D. (1999). Herpes simplex virus type 1 immediate-early protein Vmw110 inhibits progression of cells through mitosis and from G(1) into S phase of the cell cycle. J. Virol. *73*, 9456–9467.

Lomonte, P., Sullivan, K.F., and Everett, R.D. (2001). Degradation of nucleosome-associated centromeric histone H3-like protein CENP-A induced by herpes simplex virus type 1 protein ICP0. J. Biol. Chem. *276*, 5829–5835.

Lomonte, P., Thomas, J., Texier, P., Caron, C., Khochbin, S., and Epstein, A.L. (2004). Functional interaction between class II histone deacetylases and ICP0 of herpes simplex virus type 1. J. Virol. *78*, 6744–6757.

Lopez, P., Van Sant, C., and Roizman, B. (2001). Requirements for the nuclear-cytoplasmic translocation of infected-cell protein 0 of herpes simplex virus 1. J. Virol. *75*, 3832–3840.

Lopez, P., Jacob, R.J., and Roizman, B. (2002). Overexpression of promyelocytic leukemia protein precludes the dispersal of ND10 structures and has no effect on accumulation of infectious herpes simplex virus 1 or its proteins. J. Virol. *76*, 9355–9367.

Margolis, D.M., Rabson, A.B., Straus, S.E., and Ostrove, J.M. (1992). Transactivation of the HIV-1 LTR by HSV-1 immediate-early genes. Virology *186*, 788–791.

Maul, G.G. (1998). Nuclear domain 10, the site of DNA virus transcription and replication. Bioessays *20*, 660–667.

Maul, G.G., and Everett, R.D. (1994). The nuclear location of PML, a cellular member of the C3HC4 zinc-binding domain protein family, is rearranged during herpes simplex virus infection by the C3HC4 viral protein ICP0. J. Gen. Virol. *75*, 1223–1233.

Maul, G.G., Guldner, H.H., and Spivack, J.G. (1993). Modification of discrete nuclear domains induced by herpes simplex virus type 1 immediate early gene 1 product (ICP0). J. Gen. Virol. *74*, 2679–2690.

Maul, G.G., Ishov, A.M., and Everett, R.D. (1996). Nuclear domain 10 as preexisting potential replication start sites of herpes simplex virus type-1. Virology *217*, 67–75.

Melroe, G.T., DeLuca, N.A., and Knipe, D.M. (2004). Herpes simplex virus 1 has multiple mechanisms for blocking virus-induced interferon production. J. Virol. *78*, 8411–8420.

Meredith, M., Orr, A., and Everett, R. (1994). Herpes simplex virus type 1 immediate-early protein Vmw110 binds strongly and specifically to a 135-kDa cellular protein. Virology *200*, 457–469.

Meredith, M., Orr, A., Elliott, M., and Everett, R. (1995). Separation of sequence requirements for HSV-1 Vmw110 multimerisation and interaction with a 135-kDa cellular protein. Virology *209*, 174–187.

Minagawa, H., Tanaka, S., Toh, Y., and Mori, R. (1994). Detection of herpes simplex virus type 1-encoded RNA by polymerase chain reaction: different pattern of viral RNA detection in latently infected murine trigeminal ganglia following *in vitro* or *in vivo* reactivation. J. Gen. Virol. *75*, 647–650.

Mossman, K.L., and Smiley, J.R. (1999). Truncation of the C-terminal acidic transcriptional activation domain of herpes simplex virus VP16 renders expression of the immediate-early genes almost entirely dependent on ICP0. J. Virol. *73*, 9726–9733.

Mossman, K.L., Saffran, H.A., and Smiley, J.R. (2000). Herpes simplex virus ICP0 mutants are hypersensitive to interferon. J. Virol. *74*, 2052–2056.

Mossman, K.L., Macgregor, P.F., Rozmus, J.J., Goryachev, A.B., Edwards, A.M., and Smiley, J.R. (2001). Herpes simplex virus triggers and then disarms a host Antiviral response. J. Virol. *75*, 750–758.

Mullen, M.A., Ciufo, D.M., and Hayward, G.S. (1994). Mapping of intracellular localization domains and evidence for colocalization interactions between the IE110 and IE175 nuclear transactivator proteins of herpes simplex virus. J. Virol. *68*, 3250–3266.

Muller, S., and Dejean, A. (1999). Viral immediate-early proteins abrogate the modification by SUMO-1 of PML and Sp100 proteins, correlating with nuclear body disruption. J. Virol. *73*, 5137–5143.

Nabel, G.J., Rice, S.A., Knipe, D.M., and Baltimore, D. (1988). Alternative mechanisms for activation of human immunodeficiency virus enhancer in T cells. Science *239*, 1299–1302.

Natarajan, R., Deshmane, S., Valyi-Nagy, T., Everett, R., and Fraser, N.W. (1991). A herpes simplex virus type 1 mutant lacking the ICP0 introns reactivates with normal efficiency. J. Virol. *65*, 5569–5573.

Negorev, D., and Maul, G.G. (2001). Cellular proteins localized at and interacting within ND10/PML nuclear bodies/PODs suggest functions of a nuclear depot. Oncogene *20*, 7234–7242.

Nicholl, M.J., and Preston, C.M. (1996). Inhibition of herpes simplex virus type 1 immediate-early gene expression by alpha interferon is not VP16 specific. J. Virol. *70*, 6336–6339.

Nicholl, M.J., Robinson, L.H., and Preston, C.M. (2000). Activation of cellular interferon-responsive genes after infection of human cells with herpes simplex virus type 1. J. Gen. Virol. *81*, 2215–2218.

Ogle, W.O., Ng, T.I., Carter, K.L., and Roizman, B. (1997). The UL13 protein kinase and the infected cell type are determinants of posttranslational modification of ICP0. Virology *235*, 406–413.

O'Hare, P., and Hayward, G.S. (1985a). Evidence for a direct role for both the 175,000- and 110,000-molecular-weight immediate-early proteins of herpes simplex virus in the transactivation of delayed-early promoters. J. Virol. *53*, 751–760.

O'Hare, P., and Hayward, G.S. (1985b). Three trans-acting regulatory proteins of herpes simplex virus modulate immediate-early gene expression in a pathway involving positive and negative feedback regulation. J. Virol. *56*, 723–733.

O'Rourke, D., Elliott, G., Papworth, M., Everett, R., and O'Hare, P. (1998). Examination of determinants for intranuclear localization and transactivation within the RING finger of herpes simplex virus type 1 IE110k protein. J. Gen. Virol. *79*, 537–548.

Ostrove, J.M., Leonard, J., Weck, K.E., Rabson, A.B., and Gendelman, H.E. (1987). Activation of the human immunodeficiency virus by herpes simplex virus type 1. J. Virol. *61*, 3726–3732.

Parkinson, J., and Everett, R.D. (2000). Alphaherpesvirus proteins related to herpes simplex virus type 1 ICP0 affect cellular structures and proteins. J. Virol. *74*, 10006–10017.

Parkinson, J., Lees-Miller, S.P., and Everett, R.D. (1999). Herpes simplex virus type 1 immediate-early protein vmw110 induces the proteasome-dependent degradation of the catalytic subunit of DNA-dependent protein kinase. J. Virol. *73*, 650–657.

Perry, L.J., Rixon, F.J., Everett, R.D., Frame, M.C., and McGeoch, D.J. (1986). Characterization of the IE110 gene of herpes simplex virus type 1. J. Gen. Virol. *6*, 2365–2380.

Preston, C.M. (2000). Repression of viral transcription during herpes simplex virus latency. J. Gen. Virol. *81*, 1–19.

Preston, C.M., and Nicholl, M.J. (1997). Repression of gene expression upon infection of cells with herpes simplex virus type 1 mutants impaired for immediate-early protein synthesis. J. Virol. *71*, 7807–7813.

Preston, C.M., Harman, A.N., and Nicholl, M.J. (2001). Activation of interferon response factor-3 in human cells infected with herpes simplex virus type 1 or human cytomegalovirus. J. Virol. *75*, 8909–8916.

Quesada, V., Diaz-Perales, A., Gutierrez-Fernandez, A., Garabaya, C., Cal, S., and Lopez-Otin, C. (2004). Cloning and enzymatic analysis of 22 novel human ubiquitin-specific proteases. Biochem. Biophys. Res. Commun. *314*, 54–62.

Quinlan, M.P., and Knipe, D.M. (1985). Stimulation of expression of a herpes simplex virus DNA-binding protein by two viral functions. Mol. Cell. Biol 5, 957–963.

Robinson, P.A., and Ardley, H.C. (2004). Ubiquitin-protein ligases. J. Cell Sci. *117*, 5191–5194.

Russell, J., Stow, N.D., Stow, E.C., and Preston, C.M. (1987). Herpes simplex virus genes involved in latency *in vitro*. J. Gen. Virol. *68*, 3009–3018.

Sacks, W.R., and Schaffer, P.A. (1987). Deletion mutants in the gene encoding the herpes simplex virus type 1 immediate-early protein ICP0 exhibit impaired growth in cell culture. J. Virol. *61*, 829–839.

Samaniego, L.A., Neiderhiser, L., and DeLuca, N.A. (1998). Persistence and expression of the herpes simplex virus genome in the absence of immediate-early proteins. J. Virol. *72*, 3307–3320.

Schmidt, D., and Muller, S. (2003). PIAS/SUMO: new partners in transcriptional regulation. Cell. Mol. Life Sci. *60*, 2561–2574.

Seeler, J.S., and Dejean, A. (2003). Nuclear and unclear functions of SUMO. Nat. Rev. Mol. Cell. Biol. *4*, 690–699.

Soboleva, T.A., and Baker, R.T. (2004). Deubiquitinating enzymes: their functions and substrate specificity. Curr. Protein Pept. Sci. *5*, 191–200.

Sternsdorf, T., Jensen, K., and Will, H. (1997). Evidence for covalent modification of the nuclear dot-associated proteins PML and Sp100 by PIC1/SUMO-1. J. Cell Biol. *139*, 1621–1634.

Stow, N.D., and Stow, E.C. (1986). Isolation and characterization of a herpes simplex virus type 1 mutant containing a deletion within the gene encoding the immediate early polypeptide Vmw110. J. Gen. Virol. *67*, 2571–2585.

Stow, E.C., and Stow, N.D. (1989). Complementation of a herpes simplex virus type 1 Vmw110 deletion mutant by human cytomegalovirus. J. Gen. Virol. *70*, 695–704.

Thompson, R.L., Shieh, M.T., and Sawtell, N.M. (2003). Analysis of herpes simplex virus ICP0 promoter function in sensory neurons during acute infection, establishment of latency, and reactivation *in vivo*. J. Virol. *77*, 12319–12330.

Van Sant, C., Hagglund, R., Lopez, P., and Roizman, B. (2001a). The infected cell protein 0 of herpes simplex virus 1 dynamically interacts with proteasomes, binds and activates the cdc34 E2 ubiquitin-conjugating enzyme, and possesses *in vitro* E3 ubiquitin ligase activity. Proc. Natl. Acad. Sci. USA *98*, 8815–8820.

Van Sant, C., Lopez, P., Advani, S.J., and Roizman, B. (2001b). Role of cyclin D3 in the biology of herpes simplex virus 1 ICPO. J. Virol. *75*, 1888–1898.

Verger, A., Perdomo, J., and Crossley, M. (2003). Modification with SUMO. A role in transcriptional regulation. EMBO Rep. *4*, 137–142.

Yang, S.H., Jaffray, E., Senthinathan, B., Hay, R.T., and Sharrocks, A.D. (2003). SUMO and transcriptional repression: dynamic interactions between the MAP kinase and SUMO pathways. Cell Cycle *2*, 528–530.

Yao, F., and Schaffer, P.A. (1994). Physical interaction between the herpes simplex virus type 1 immediate-early regulatory proteins ICP0 and ICP4. J. Virol. *68*, 8158–8168.

Yao, F., and Schaffer, P.A. (1995). An activity specified by the osteosarcoma line U2OS can substitute functionally for ICP0, a major regulatory protein of herpes simplex virus type 1. J. Virol. *69*, 6249–6258.

Zhu, X.X., Chen, J.X., Young, C.S., and Silverstein, S. (1990). Reactivation of latent herpes simplex virus by adenovirus recombinants encoding mutant IE-0 gene products. J. Virol. *64*, 4489–4498.

The Functions and Activities of Herpes Simplex Virus Protein ICP27, a Multifunctional Regulator of Gene Expression

4

Rozanne M. Sandri-Goldin

Abstract

ICP27 is a multifunctional regulator of gene expression that assumes different roles throughout the course of infection. At early times, ICP27 is an inhibitor of cellular splicing, whereas, at later times it is involved in recruiting RNA polymerase II to viral sites of replication and in facilitating viral RNA export. Evidence also suggests that, at late times, it may be involved in the stimulation of translation of viral transcripts. ICP27 performs its activities by interacting with RNA and with a myriad of proteins. ICP27 binds viral RNAs to facilitate their export. Further, ICP27 interacts with itself to form multimers, and with HSV-1 proteins ICP4 and ICP8, as well as with an increasing number of cellular proteins. The regions of ICP27 involved in these interactions have been broadly mapped to both the N-terminal and C-terminal halves of the molecule. Several protein motifs have been identified based upon sequence comparisons; however, no structural information has yet been generated. Although much has been learned about the mechanisms by which ICP27 performs its roles, little is known about how its varied activities are regulated. The roles and activities of ICP27 are detailed in this chapter.

Introduction

ICP27 is an immediate-early (IE) protein that is required for viral replication (Sacks *et al.*, 1985). It is a 512-amino-acid phosphoprotein that performs a number of functions during infection including contributing to the shut off of host protein synthesis (Hardwicke and Sandri-Goldin, 1994), stimulating viral early gene expression and DNA synthesis (Rice and Knipe, 1990; Uprichard and Knipe, 1996), and inducing late gene expression (Hann *et al.*, 1998; Jean *et al.*, 2001). ICP27 has also been implicated in the repression of host gene transcription after infection (Spencer *et al.*, 1997), in blocking the cell cycle in the G1 phase (Song *et al.*, 2001) and in the prevention of apoptosis (Aubert and Blaho, 1999; Aubert *et al.*, 2001). Early studies on the regulatory functions of ICP27 focused on its ability to upregulate or down regulate target gene expression in transfection experiments (Hardwicke *et al.*, 1989; Rice *et al.*, 1989). These studies demonstrated that ICP27 can both positively and negatively affect the expression of different target genes. Later studies on the mechanisms by which ICP27 performs its functions have demonstrated that ICP27 acts predominantly at the post-transcriptional level, affecting RNA processing and export, although it has also been shown to be required for the stimulation of late gene transcription (Jean *et al.*, 2001). ICP27 interacts with a myriad of proteins, and its various activities appear to be regulated temporally throughout the course of infection. That is, at early times it associates with splicing complexes whereas at later times, it shuttles between the nucleus and cytoplasm. It has not yet been revealed how ICP27 switches among its various activities, although the protein is post-translationally modified by phosphorylation and arginine methylation. Both modifications have been demonstrated to affect protein–protein interactions (Bedford and Richard, 2005; Hsu *et al.*, 2005). In this chapter, the mechanisms by which ICP27 performs its various roles will be detailed along with recent insights into how its various activities may be regulated.

Post-transcriptional activities of ICP27

ICP27 mediates the inhibition of host cell pre-mRNA splicing

Infection with some viruses can alter cellular mRNA processing to favor viral gene expression. ICP27 inhibits host cell splicing (Hardwicke and Sandri-Goldin, 1994; Hardy and Sandri-Goldin, 1994; Bryant *et al.*, 2001; Linberg and Kreivi, 2002), which results in decreased levels of cytoplasmic cellular spliced mRNAs, thus contributing to the shut off of host protein synthesis (Hardwicke and Sandri-Goldin, 1994; Soliman *et al.*, 1997). This does not cause problems for HSV-1 gene expression because most HSV-1 transcripts are intronless. Because spliced mRNAs are more efficiently exported than intronless transcripts (Luo and Reed, 1999), splicing inhibition may also result in alteration of cellular splicing-dependent export pathways. This is fact does occur and will be described in detail in the next section. Early studies showed that ICP27 caused a redistribution of splicing factors (Phelan *et al.*, 1993; Sandri-Goldin *et al.*, 1995), co-immunoprecipitated with anti-Sm antisera and altered the phosphorylation of two co-precipitating splicing proteins (Sandri-Goldin and Hibbard, 1996). Furthermore, ICP27 was found to interact with splicing protein SAP145 in a yeast two hybrid analysis (Bryant *et al.*, 2001). These findings suggested that ICP27 mediates inhibition by interacting with splicing factors.

In studies to address the mechanism of splicing inhibition by ICP27, we reported that ICP27 interacted with SRp20 in a yeast two-hybrid screen (Sciabica *et al.*, 2003). SRp20 is a member of a highly conserved family of serine-arginine-rich essential splicing factors termed SR proteins with roles in both constitutive and alternative splicing (Graveley, 2000). Furthermore, SR proteins have been shown to be required for spliceosome assembly (Roscigno and Garcia-Blanco, 1995). Co-immunoprecipitation and co-localization studies showed that ICP27 also interacts with other members of the SR protein family and the C-terminus of ICP27 was required for these interactions. Significantly, ICP27 expression resulted in hypophosphorylation of SR proteins (Sciabica *et al.*, 2003). Proper phosphorylation is required for SR proteins to perform their roles in splicing (Roscigno and Garcia-Blanco, 1995; Cao *et al.*, 1997; Prasad *et al.*, 1999) and consequently, it was found that spliceosome complex formation is stalled in wild type HSV-1 infected cells at the prespliceosomal complex A stage before the first transesterification reaction (Bryant *et al.*, 2001; Sciabica *et al.*, 2003). Furthermore, ICP27 was shown to interact with SR protein kinase 1 (SRPK1), a member of a family of conserved kinases that are highly specific for arginine/serine dipeptides (Gui *et al.*, 1994a,b). Interaction with ICP27 resulted in a compartmental redistribution of SRPK1 *in vivo*, in that this predominantly cytoplasmic kinase was relocalized to the nucleus (Sciabica *et al.*, 2003). The RGG box RNA binding motif region of ICP27 was required to be intact for its interaction with SRPK1. Further, the interaction of ICP27 with SRPK1 altered the activity of the kinase *in vitro*. SR protein substrates purified from HeLa cells or expressed as recombinant proteins in bacteria were less

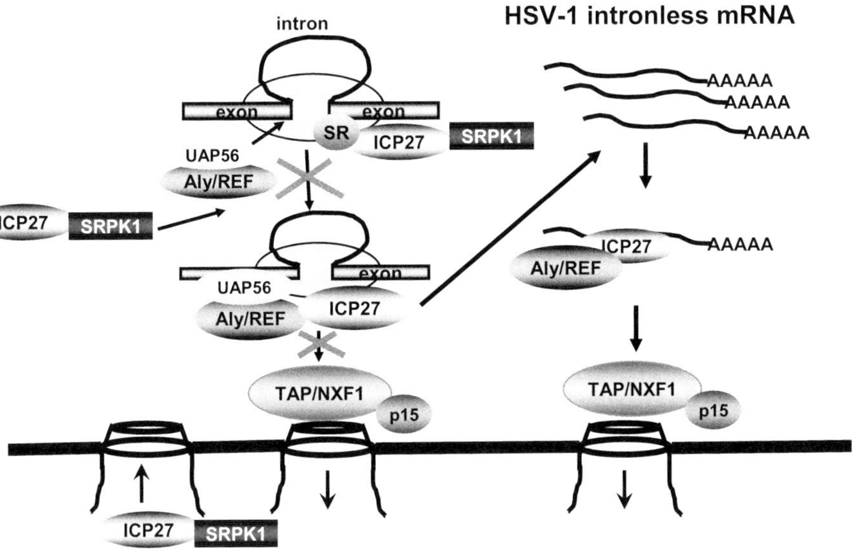

Figure 4.1 Schematic depiction of splicing inhibition during HSV-1 infection and the pathway for export of viral transcripts. Early in infection, ICP27 recruits SRPK1 to the nucleus. ICP27 interacts splicing factors, including members of the SR protein family of splicing factors. SR proteins are involved in spliceosome assembly and their activity is regulated by phosphorylation. Hypophosphorylation, mediated by the interaction of ICP27 with SRPK1, results in stalled spliceosomal complexes and the inhibition of host cell splicing (Sciabica *et al.*, 2003). Cellular mRNAs are exported through the nuclear export receptor TAP/NFX1 through nuclear export adaptor proteins, including Aly/REF and UAP56, which are part of the exon junction complex. ICP27 encounters Aly/REF in stalled spliceosomal complexes and recruits Aly/REF to sites of HSV-1 transcription (Chen *et al.*, 2005). Cellular pre-mRNAs remain incompletely spliced and are retained in the nucleus. Viral mRNAs are bound by ICP27 and the ICP27-RNA-Aly/REF complex interacts with TAP/NXF1 and its heterodimeric partner p15 for export through the nuclear pore complex.

phosphorylated in *in vitro* kinase assays in the presence of ICP27 compared with its absence (Sciabica *et al.*, 2003). These findings indicate that pre-mRNA splicing inhibition by ICP27 may result from its interaction with SRPK1, thus triggering aberrant SR protein phosphorylation. A model depicting the inhibition of cellular pre-mRNA splicing by ICP27 is shown in Fig. 4.1.

ICP27 facilitates the export of viral transcripts to the cytoplasm

Recent advances have led to an understanding of how eukaryotic mRNAs are exported from the nucleus to the cytoplasm. This process involves an elaborate machinery that is highly conserved from yeast to humans, and which is coupled to upstream events in RNA metabolism. Eukaryotic pre-mRNAs are processed after synthesis in the nucleus by capping at the 5′ end, cleavage and polyadenylation to form the 3′ end, and splicing to remove intervening sequences. Following processing, mRNAs must be exported through the nuclear pore complex (NPC) to the cytoplasm for translation, which requires recognition by export factors to direct the mRNAs to nuclear export receptors for translocation through the nuclear pore complex (NPC) (Reed and Magni, 2001; Zenklusen and Stutz, 2001; Cullen, 2003). A large body of evidence shows that TAP/NXF1, in conjunction with its heterodimeric partner, p15/NXT is the nuclear export receptor for mRNAs in metazoans (Gruter *et al.*, 1998; Kang and Cullen, 1999; Braun *et al.*, 2001; Fribourg *et al.*, 2001; Weigand *et al.*, 2002; Katahira *et al.*, 2002). The yeast homolog of TAP/NXF1, termed Mex67p, has been shown to function as the mRNA export receptor in yeast (Segref *et al.*, 1997; Santos-Rosa *et al.*, 1998; Strasser *et al.*, 2000; Hurt *et al.*, 2000). TAP/NXF1 has been shown to shuttle between the nucleus and cytoplasm, cross-link to poly(A^+) RNA, localize at the nuclear pores and interact directly with nucleoporins (Segref *et al.*, 1997; Santos-Rosa *et al.*, 1998; Kang and Cullen, 1999; Herold *et al.*, 2000; Braun *et al.*, 2001). Further, overexpression of TAP/NXF1 stimulates mRNA export (Herold *et al.*, 2000; Braun *et al.*, 2001) and knock down of TAP/NXF1 by RNA interference blocked nuclear export of poly(A+) RNA indicating a direct role in mRNA export (Tan *et al.*, 2000; Herold *et al.*, 2003).

In metazoans, nuclear export of mRNAs has been closely linked to splicing (Luo and Reed, 1999). The basis of this connection was uncovered with the discovery of a protein complex that is deposited on pre-mRNAs undergoing splicing at a specific position just upstream of exon junctions (LeHir *et al.*, 2000a; LeHir *et al.*, 2000b; LeHir *et al.*, 2001). This exon junction complex or EJC consists of at least six proteins, which have been shown to function in splicing, nuclear export and mRNA surveillance (Reichert *et al.*, 2002; Wagner and Lykke-Andersen, 2002). One of these proteins, termed Aly/REF, the metazoan homolog of the yeast export factor Yra1p (Straber and Hurt, 2000; Zenklusen *et al.*, 2001) interacts directly with TAP/NXF1(Stutz *et al.*, 2000), and is recruited to pre-mRNA sites near exon junctions (LeHir *et al.*, 2000a; LeHir *et al.*, 2001) by a DEAD-box helicase termed UAP56, which functions in spliceosome assembly (Luo *et al.*, 2001) and which also appears to have role in mRNA export (Gatfield and Izaurralde, 2002). Thus, one or more proteins in the EJC mark the mRNA for export through the TAP/NXF1 pathway (Cullen, 2000).

Most cellular mRNAs are multiply spliced, whereas the majority of HSV-1 transcripts are intronless and as described above, ICP27 inhibits cellular pre-mRNA splicing at early times of infection. This presents an interesting paradox for HSV-1 because of the intimate coupling of export to splicing (Luo and Reed, 1999). Following infection with HSV-1, the cell is presented with thousands of viral intronless transcripts that would compete poorly with cellular transcripts for export. ICP27 shifts the balance by shutting down the cellular splicing machinery (Hardy and Sandri-Goldin, 1994), thus cellular pre-mRNAs are not fully processed and exported (Hardwicke and Sandri-Goldin, 1994; Sciabica *et al.*, 2003). ICP27 interacts with SR proteins and other spliceosomal components (Bryant *et al.*, 2001; Sciabica *et al.*, 2003), and most likely then encounters Aly/REF, which is part of the EJC (Koffa *et al.*, 2001; Chen *et al.*, 2002). We have reported that ICP27 co-immunoprecipitates and colocalizes with Aly/REF in HSV-1 infected cells, and that ICP27 recruits Aly/REF from spliceosomes to sites of HSV-1 transcription (Chen *et al.*, 2002; Chen *et al.*, 2005). ICP27 binds to HSV-1 intronless transcripts through its RGG box RNA binding motif (Sandri-Goldin, 1998). The RNA-ICP27-Aly/REF complex binds to TAP/NFX1 through the interaction of ICP27 with Aly/REF (Koffa *et al.*, 2001; Chen *et al.*, 2002), and in addition, ICP27 binds directly to TAP/NXF1 in the absence of Aly/REF (Chen *et al.*, 2005). Thus, HSV-1 transcripts are exported through the nuclear pore complex. A model showing viral mRNA export mediated by ICP27 is shown in Fig. 4.1.

Two regions of ICP27 have been shown to be required to be intact for interaction with TAP/NXF1(Chen *et al.*, 2005), which are the leucine-rich N-terminus and the C-terminus, which encodes a predicted KH domain and a zinc-finger-like motif. Immunofluorescence studies and *in vitro* export assays have demonstrated that mutations in either the N-terminus or the C-terminus confine ICP27 to the nucleus (Sandri-Goldin, 1998; Mears and Rice, 1998; Chen *et al.*, 2002; Chen *et al.*, 2005). To determine the fate of viral mRNA in infections with ICP27 mutants that are defective in export, two approaches have been taken. In the first approach, a combination *in vitro* export assay and *in situ* hybridization were performed (Johnson, L. A., Devi-Rao, G., Wagner, E.K. and Sandri-Goldin, R.M., unpublished results). HeLa cells were infected with wild type HSV-1 or with ICP27 mutants. At 8 hours after infection, when viral transcription is robust and cellular mRNA processing has been halted, infected cells were permeabilized and *in vitro* export assays were performed as described previously (Chen *et al.*, 2005). At various intervals, the cells were fixed and *in situ* hybridization was performed with an oligo-dT probe to monitor poly(A+) mRNA. Samples were then stained with a monoclonal antibody to ICP27. These studies showed that by 30 minutes after the start of the *in vitro* export assay, greater than 90% of ICP27 had left the nucleus and there was a concomitant decrease in nuclear poly(A+) RNA. In contrast, in a mutant with a deletion of the leucine-rich N-terminus, which cannot interact with TAP/NXF, ICP27 was confined to the nucleus during the course of the assay, and so was poly(A+) RNA, confirming that ICP27's interaction with TAP/NXF is crucial for export of both ICP27 protein and mRNA. Accordingly, in assays with an ICP27 mutant with a deletion of the RGG box RNA binding motif, ICP27 was exported as quickly as wild type ICP27 because the RGG box resides outside the regions required for interaction with TAP/NXF. However, poly(A+) RNA mostly remained in the nucleus, reaffirming that ICP27 RNA binding is required for proficient viral RNA export (Johnson, L. A., Devi-Rao, G., Wagner, E.K. and Sandri-Goldin, R.M., unpublished results).

In the second approach, microarray analysis was performed on nuclear and cytoplasmic RNA fractions from cells infected with wild type HSV-1 or with ICP27 mutants. The microarrays consisted of DNA oligonucleotide probes representing the complete repertoire of HSV-1 transcripts (Sun *et al.*, 2004). The interaction of ICP27 with TAP/NXF1 was required for efficient export of viral transcripts because HSV-1 RNAs were retained in the nucleus of cells infected with ICP27 mutants in the TAP/NXF interaction region. Further, ICP27 must bind viral transcripts for them to be exported efficiently because there was a nuclear retention seen in the samples from infections with the RGG box deletion mutant. On the other hand, HSV-1 RNA was exported efficiently by an ICP27 mutant that cannot interact with Aly/REF (Johnson, L. A., Devi-Rao, G., Wagner, E.K. and Sandri-Goldin, R.M., unpublished results). There were some differences among the HSV-1 transcripts in the extent of export in the ICP27 mutant infections, suggesting that some viral transcripts can use other export pathways or export adaptors. This will be pursued in future research.

The role of ICP27 in 3′ end formation

Studies on the mechanisms by which ICP27 performs its functions have demonstrated that ICP27 acts predominantly at the post-transcriptional level, thus the induction of late gene expression by ICP27 was proposed to stem in part from its proposed role in mRNA 3′ processing. Specifically, *in vitro* studies showed that some viral polyadenylation sites, present primarily on late transcripts, were utilized inefficiently by cellular polyadenylation factors, and these sites were operationally defined as weak (McLauchlan *et al.*, 1989). ICP27 stimulated the use of these sites (Sandri-Goldin and Mendoza, 1992; McLauchlan *et al.*, 1992). Further analysis suggested that ICP27 recruits the cellular cleavage stimulation factor, Cst64, which is required for poly(A) site recognition, to weak HSV-1 poly(A) sites (McGregor *et al.*, 1996). However, a study investigating differential polyadenylation of two transcripts that arise from the UL24 gene by the use of two different poly(A) sites, one weak and one strong, found that the accumulation of transcripts during infection from the weak poly(A) site did not require ICP27 (Hann *et al.*, 1998). In addition, a study using various HSV-1 plasmid constructs of the late glycoprotein C gene showed that the stimulatory effect of ICP27 occurred post-transcriptionally but was independent of the inserted polyadenylation site (Perkins *et al.*, 2003). Thus, the role of ICP27 in enhancing the use of weak poly(A) sites has not been firmly established.

A role for ICP27 in translation?

A recent report on proteomic studies involving immunoprecipitation of ICP27 and mass spectrometric identification of coprecipitated proteins showed an association of ICP27 with the cellular translation initiation factors poly A binding protein (PABP), eukaryotic initiation factor 3 (eIF3), and eukaryotic initiation factor 4G (eIF4G) (Fontaine-Rodriquez *et al.*, 2004). Co-immunoprecipitation studies confirmed these associations *in vivo*, and purified MBP-tagged ICP27 (MBP-27) interacted with eIF3 subunits p47 and p116 *in vitro*. These results suggest that ICP27 may also play a role in stimulating translation of certain viral and host mRNAs and/or in inhibiting host mRNA translation. A feasible hypothesis for the proposed role of ICP27 in translation is that ICP27 binds to nascent transcripts in its role as an RNA export factor. ICP27 may remain bound to viral transcripts upon translocation through the nuclear pore, and its interaction with translation initiation factors may result in recruiting these factors to viral mRNAs to facilitate their translation in the cytoplasm (see Fig. 4.2). One situation where this may have already been demonstrated is in the case of the HSV-1 UL24 gene short transcript, which is expressed in the absence of ICP27 but not translated efficiently into UL24 protein (Pearson *et al.*, 2004). In another example, Ellison *et al.* (2005) examined how ICP27 influences the expression of the HSV-1 protein VP16, which is both a tegument protein and transactivator of immediate-early gene expression. The effects of ICP27 were monitored on the levels, nuclear export, and polyribosomal association of VP16 mRNA and on the amount and stability of VP16 protein. Deletion of ICP27 reduced the levels of VP16 mRNA without altering its nuclear export or the stability of the encoded protein. However, the translational yield of the VP16 mRNA produced in the absence of ICP27 was reduced up to 80-fold relative to that for wild type infection, suggesting a defect in translation. In the absence of ICP27, the majority of cytoplasmic VP16 mRNA was not associated with actively translating polyribosomes but instead co-sedimented with 40S ribosomal subunits, indicating that the translational defect is likely at the level of initiation. These effects were mRNA specific, as polyribosomal analysis of two cellular transcripts and two early HSV transcripts indicated that ICP27 was not required for efficient translation of these mRNAs (Ellison *et al.*, 2005). Thus, it is possible that ICP27 may regulate translation of specific viral transcripts.

Currently, the RNA binding specificity of ICP27 has not been well defined. The only published study to date that attempted to define the RNA binding specificity of ICP27 used a yeast three-hybrid approach with DNA

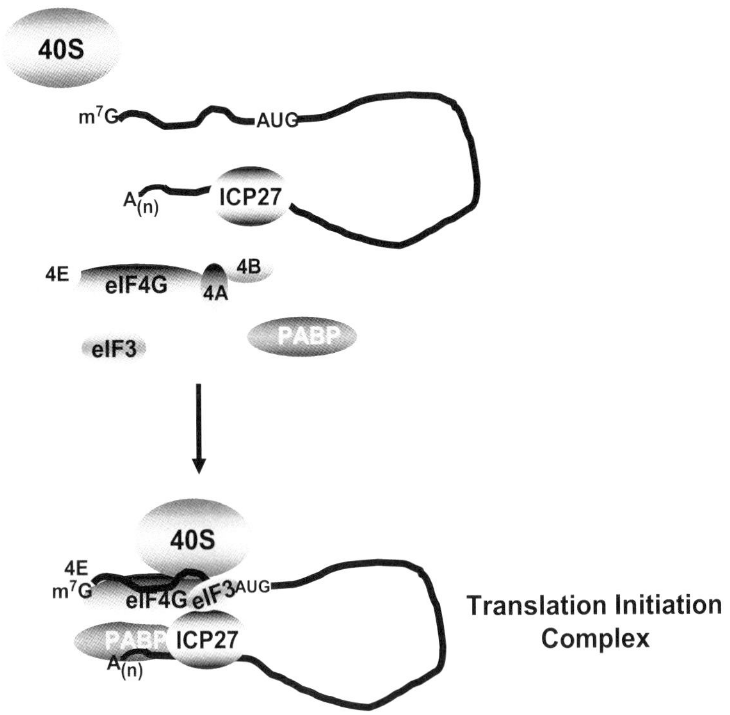

Figure 4.2 Model for the recruitment of translation initiation factors to viral transcripts by ICP27. ICP27 binds to viral RNAs to facilitate their export to the cytoplasm. In the cytoplasm, ICP27 remains bound to the exported RNA where it interacts with PABP (Fontaine-Rodriquez *et al*., 2004), which binds to the poly(A) tail of mRNA in the cytoplasm. The eIF4 complex binds to the m^{7}G-cap through the eIF4E component. ICP27 also interacts with eIF4G, a large scaffolding protein that interacts with eIF3 and with PABP. The eIF4G/PABP interaction brings about the circularization of the RNA, which stimulates translation (for review see Kahvejian *et al*., 2005).

fragments ranging in size from 33 nucleotides to 230 nucleotides from the entire HSV-1 genome (Sokolowski *et al*., 2003). In this study 31 sense RNAs were identified that mapped to 28 different ORFs, however, no consensus binding sequence was revealed through analysis of these RNAs. Interestingly, an RNA sequence within the VP16 gene was identified in the three hybrid screen, which would correlate with ICP27 binding to the VP16 transcript to stimulate its translation (Ellison *et al*., 2005); however, no sequences within UL24 were identified, although ICP27 is also required for efficient translation of the short transcript emanating from this gene (Pearson *et al*., 2004). Further, in the three hybrid screen ICP27 bound an RNA sequence within the UL29 or ICP8 gene, and it was shown that ICP27 had no effect on the translation of ICP8 (Ellison *et al*., 2005). Thus, it will be necessary to perform additional studies to determine to which transcripts ICP27 binds and what are the functional consequences of this binding.

Transcriptional stimulation of HSV-1 expression by ICP27

ICP27 associates with RNA polymerase II

ICP27 has also been shown to stimulate expression of some early genes and transcription of some late viral genes and the accumulation of early and late gene products is severely reduced in ICP27 viral mutant infections (Rice and Knipe, 1990; Rice *et al.*, 1993; Rice and Lam, 1994; Jean *et al.*, 2001; Lengyel *et al.*, 2002). Furthermore, ICP27 has been shown to associate with cellular RNA polymerase II (RNAP II) holoenzyme, and this interaction was found to be independent of DNA and RNA (Zhou and Knipe, 2001). A number of cellular proteins that are involved in RNA processing, including 5′ capping, splicing, and 3′ cleavage and polyadenylation, have been found to bind to the C-terminal domain (CTD) of RNAP II, which acts a platform to bring these factors to sites on the nascent transcript where their action is needed (for review see Hirose and Manley, 2000; Bentley, 2005). Therefore, we asked if ICP27 interacted directly with RNAP II CTD, in accord with its post-transcriptional activities. *In vitro* binding assays demonstrated that ICP27 does interact with the CTD, and further both the N-terminus and C-terminus of ICP27 were required to be intact for this interaction to occur. In virus infected cells, ICP27 N-terminal and C-terminal mutants did not coimmunoprecipitate or colocalize with RNAP II (Dai-Ju *et al.*, 2006).

Some cellular RNA processing factors that associate with RNAP II CTD interact predominantly with a specific phosphorylated form of the CTD, whereas, others do not show a marked preference when binding (Kim *et al.*, 1997; Cho *et al.*, 1997; Komarnitsky *et al.*, 2000; Ahn *et al.*, 2004). The RNAP II CTD in all eukaryotes is highly conserved and consists of tandem repeats of a heptapeptide, YSPTSPS, which is repeated 52 times in humans. The CTD is reversibly phosphorylated on serine-2 and serine-5 in the heptapeptide repeat (Dahmus, 1996). Unphosphorylated RNAP II is recruited to the promoter, where a preinitiation complex forms. At initiation of transcription, serine-5 is phosphorylated primarily by cdk7, which is associated with general transcription factor TFIIH (Dahmus, 1996; Komarnitsky *et al.*, 2000). Thus, RNAP II found at the promoters of genes is mostly phosphorylated on serine-5 (for review see Kober and Greenblatt, 2002). This form is also referred to as hypophosphorylated or RNAP IIA. During elongation, another kinase P-TEFb, which consists of the cyclin-dependent kinase cdk9 and one of several cyclin T isoforms, phosphorylates serine-2 (for review see Price, 2000). The elongating form of RNAP II is hyperphosphorylated and is called the RNAP IIO form, which has a slower migration (Zhang and Corden, 1991). In HSV-1 infected cells, an intermediately phosphorylated form has been found beginning around 5 hours after infection (Rice *et al.*, 1994). HSV-1 protein ICP22 and kinase UL13 have been shown to mediate the phosphorylation of RNAP II resulting in the RNAP IIi form (Rice *et al.*, 1995; Long *et al.*, 1999). Further, it has been shown that ICP22 interacts with cdk9 in a complex and that the HSV-1 kinase Us3 may also be involved in the intermediate phosphorylation of RNAP II (Durand *et al.*, 2005). The exact nature of the modifications and role of the RNAP IIi form during viral transcription have not been elucidated.

To determine if ICP27 associates with the RNAP II initiating or elongating transcription complex, we used monoclonal antibodies that are specific for different phosphorylated forms of the RNAP II CTD in colocalization studies. Immunofluorescence analysis was performed at early times after infection, and ICP27 was seen to colocalize in part with RNAP II stained by antibody 8WG16, which recognizes unphosphorylated and phospho-serine-5 hypophosphorylated RNAP IIA (Patturajan *et al.*, 1998). Unphosphorylated RNAP II CTD is the form that is recruited to promoters and hypophosphorylated RNAP IIA is the form that is found in the preinitiation and initiation complexes, wherein the

CTD is phosphorylated by TFIIH/CDK7 on serine-5 in the heptad repeat.(Dahmus, 1996) This colocalization was also seen in staining with antibody ARNA3, which recognizes an epitope in the N-terminus of the large subunit of RNAP II, thus all forms, phosphorylated and unphosphorylated are recognized. We were particularly interested in determining if ICP27 colocalized with the hyperphosphorylated phosphoserine-2 form of RNAP II, which is found predominantly in elongating transcription complexes, and with which factors required for splicing and 3′ end formation have been found to associate. (Kim *et al.*, 1997; Ahn *et al.*, 2004; Adamson *et al.*, 2005). Antibody H5 has been reported to specifically recognize serine-2 phosphorylation in the CTD (Patturajan *et al.*, 1998). At early times after infection with wild type HSV-1, H5 staining was seen to colocalize with ICP27. Thus, these data suggest that ICP27 binds to the CTD of RNAP II in a phosphorylation independent manner because it associates with both phosphorylated and unphosphorylated forms (Dai-Ju *et al.*, 2006). This is an important finding because this indicates that ICP27 does not just associate with the elongating CTD to use it as a platform to reach RNA processing sites, but instead that ICP27 could also play a role in the recruitment of unphosphorylated RNAP II to HSV-1 transcription sites, and namely to HSV-1 promoters. This would suggest a transcriptional role for ICP27.

ICP27 helps to recruit RNA polymerase II to viral replication sites

It has been shown that RNAP II is redistributed in HSV-1 infected cells to sites of viral transcription, and further that ICP27 is one of the immediate early gene products that is required for this to occur (Rice *et al.*, 1994). Therefore, we looked at the distribution of RNAP II in wild type and ICP27 mutant virus infected cells. In particular, we looked at viral transcription-replication compartments that contain the HSV-1 transcriptional activator ICP4. In wild type infected cells, RNAP II recognized by 8WG16 and ARNA3 was redistributed into ICP4-containing viral transcription-replication compartments by 6 hours after infection. Redistribution was also seen in cells infected with ICP27 mutants that lie outside of the N-terminal and C-terminal regions of ICP27 that are required for interaction with RNAP II. These mutants had been shown to shown to interact with RNAP II in infected cells in immunoprecipitation studies (Dai-Ju *et al.*, 2006). In contrast, in cells infected with mutants in the N-terminus and C-terminus that fail to coimmunoprecipitate with or colocalize with RNAP II, staining of RNAP II remained diffusely distributed throughout the nucleus, and further, ICP4-containing transcription-replication compartments were poorly formed (Dai-Ju *et al.*, 2006). These data suggest that ICP27 is required for the efficient recruitment of RNAP II to viral transcription sites, and further that formation of transcription-replication compartments requires RNAP II redistribution. Thus, stimulation of transcription of viral genes by ICP27 may stem from its interaction with RNAP II CTD and its role in recruiting RNAP II to viral replication sites.

A model for the transcriptional stimulation of HSV-1 early and late gene expression by ICP27 is as follows. During viral infection, ICP27 interacts with the CTD of RNAP II, which acts as a platform for RNA processing factors. ICP27 also interacts with ICP8 (Olesky *et al.*, 2005), a protein that is required for viral DNA replication, and ICP8 has also been shown to associate with RNAP II (Zhou and Knipe, 2001). Further, evidence has been presented that ICP27 interacts with ICP4 (Panagiotidis *et al.*, 1997), the major transcriptional transactivator of HSV-1. Both ICP8 and ICP4 are part of HSV-1 replication compartments, and ICP27 recruits unphosphorylated RNAP II to these sites such that transcription initiation can commence on viral genes. ICP27 mutants that cannot interact with RNAP II show reductions in viral transcription and this may result from the inability of these mutants to recruit sufficient amounts of RNAP II to viral transcription-replication sites, which in turn do not assemble into full-fledged compartments (Fig. 4.3).

How are the activities of ICP27 regulated throughout the course of infection?

Post-translation modifications: arginine methylation

Protein arginine methylation by type I methyltransferases, which add one or two methyl groups to arginine, has been shown to affect a number of eukaryotic processes including protein transport, transcription and cell signaling (Bedford and Richard, 2005). ICP27 was shown to be methylated *in vivo* and the arginine-glycine-rich RGG box motif, which is required for binding to RNA (Mears and Rice, 1996; Sandri-Goldin, 1998), was shown to be one of the targets for methylation (Mears and Rice, 1996). Because ICP27 is multifunctional and it appears to alter its activities throughout the course of infection, we postulate that arginine methylation could modify its affinity for any of the plethora of interacting partners that it encounters during its various roles, and thus act as a switching control. We probed the arginine methylation of ICP27. We found that ICP27 was methylated *in vivo* and a deletion of the RGG box reduced but did not eliminate arginine methylation (Souki, S.K. and Sandri-Goldin, R.M. unpublished results). To precisely map the methylated residues, we performed MALDI-TOF Mass Spectrometry. Two arginines within the RGG box motif (residues 138 to 152) were found to be methylated in ICP27 purified from wild type HSV-1 infected cells. These were R138 and R150. In addition, arginine methylation was found on two other arginine residues outside of this motif, indicating that some arginine residues outside of the RGG box can also be methylated in ICP27. The MALDI-TOF MS results also showed that nuclear ICP27 was more methylated than cytoplasmic ICP27 and that di-methylated arginine residues occurred more frequently in nuclear fractions compared with cytoplasmic fractions (Souki, S.K. and R.M. Sandri-Goldin, unpublished results). There are at least eight characterized protein-arginine-methyltransferases in mammalian cells, and five of these are mostly nuclear. Therefore, arginine methylation may act as a switch for ICP27 to leave splice sites and to begin its role as an RNA export protein. That is, as with other export proteins, arginine methylation may modulate its export activity. Arginine methylation can affect both nucleocytoplasmic transport and protein–protein interactions of RNA binding proteins (McBride *et al.*, 2005). In future studies, it will be important to determine which ICP27 protein–protein interactions may be affected or modulated by arginine methylation through mutational analysis of the targeted arginines.

Post-translation modifications: phosphorylation

Another post-translational modification that has been demonstrated to regulate protein function is phosphorylation. ICP27 has been demonstrated to be phosphorylated on serine residues during infection (Zhi and Sandri-Goldin, 1999). Phospho-peptide mapping of ICP27 purified from HSV-1 infected cells showed that there are two serine residues at positions 16 and 18 within the leucine-rich N-terminal region required for export that are phosphorylated by CK2, and a serine at position 114 within the nuclear localization signal

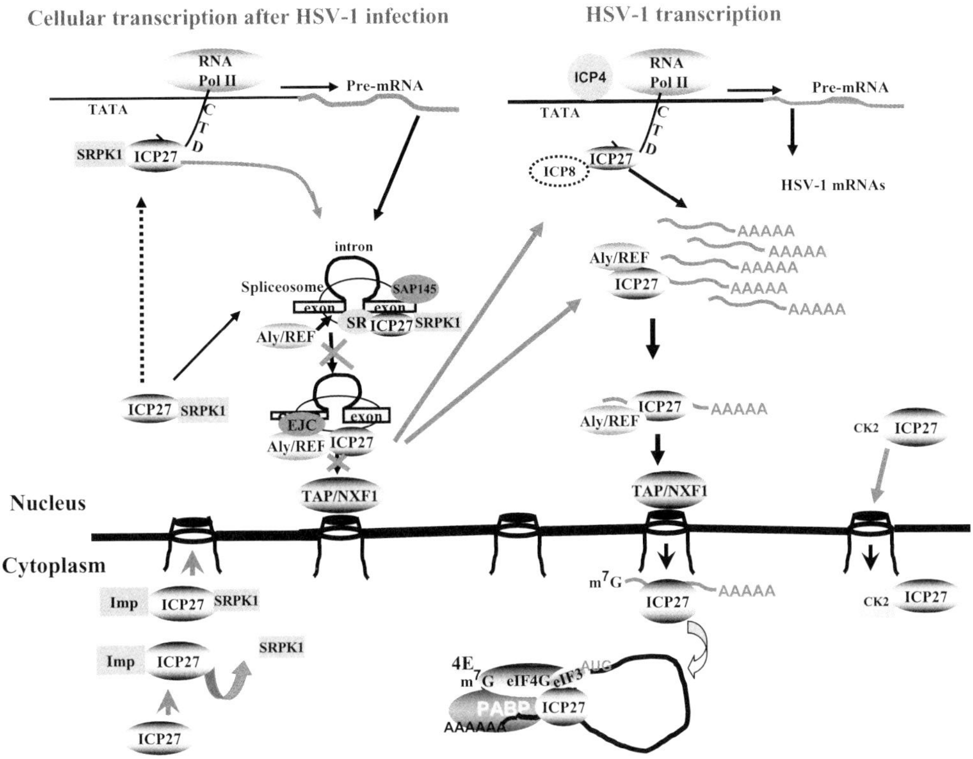

Figure 4.3 Schematic depiction showing the myriad roles and activities of ICP27. At early times after infection, ICP27 enters the nucleus and brings with it the SR protein specific kinase SRPK1. ICP27 interacts with the CTD of RNAP II, although it has not been worked out when this occurs. ICP27 interacts with spliceosomal complexes through its interaction with SR proteins. Aberrant phosphorylation of SR proteins mediated by ICP27's interaction with SRPK1 results in stalled prespliceosomal complexes and the inhibition of cellular splicing. ICP27 encounters Aly/REF, an exon junction protein that functions as an RNA export adaptor, and recruits Aly/REF away from spliceosomes to HSV-1 transcription-replication sites. ICP27 accesses these sites by binding to the CTD of RNAP II and further, is involved in the efficient recruitment of RNAP II to HSV-1 genome replication sites. It is not known if interactions with ICP4 and/or ICP8 are also required. ICP27 binds viral RNA and the ICP27-Aly/REF-RNA complexes are directed to TAP/NXF, the cellular mRNA export receptor. ICP27 interacts with TAP/NXF and the RNP complex is exported to the cytoplasm. Evidence has been presented that ICP27 remains bound to the RNA in the cytoplasm to recruit translation initiation factors.

(NLS) that is phosphorylated by PKA (Zhi and Sandri-Goldin, 1999). Mutation of the serine at 114 to alanine diminished import of ICP27 to the nucleus (Zhi and Sandri-Goldin, 1999), suggesting that phosphorylation by PKA may regulate the ability of ICP27 to interact with importins, perhaps by unmasking or altering the conformation of the NLS. The CK2 sites are within one of the regions required for ICP27 to interact with TAP/NXF1 (Chen *et al.*, 2005), and thus these sites may modulate ICP27's export activity. In addition, it has been reported that ICP27 interacts with CK2 (Wadd *et al.*, 1999) and that HSV-1 infection stimulates CK2 activity (Koffa *et al.*, 2003). CK2 stimulation correlates with redistribution of the holoenzyme from the nucleus to the cytoplasm. Both CK2 stimulation and redistribution to the cytoplasm required expression and cytoplasmic accumulation of ICP27 (Koffa *et al.*, 2003). These findings suggest that CK2 may be involved in regulating the trafficking of ICP27.

Interestingly, ICP27 is also responsible for the redistribution of another kinase, SRPK1, but in this case ICP27 recruits SRPK1 from the cytoplasm to the nucleus at early times after infection in its role as an inhibitor of splicing (Sciabica *et al.*, 2003). Further, ICP27 was phosphorylated by SRPK1 *in vitro* but the sites of phosphorylation have not been mapped (Sciabica, K.S. and Sandri-Goldin, R.M., unpublished results). To determine the functional importance of phosphorylation in modulating ICP27's protein–protein and perhaps, protein–RNA interactions, it will be necessary to identify all phosphorylation sites through mass spectrometric analysis and then to mutate these sites individually and in combination and to perform functional analysis during infection with viruses harboring these mutations.

The multiple interactions of ICP27 and the domains required

ICP27 binds viral RNAs to facilitate their export (Sandri-Goldin, 1998). Further, ICP27 interacts with itself to form multimers (Zhi *et al.*, 1999), and with HSV-1 proteins ICP4 (Panagiotidis *et al.*, 1997) and ICP8 (Olesky *et al.*, 2005), as well as with cellular proteins including: SR splicing proteins (Sciabica *et al.*, 2003), SRPK1 (Sciabica *et al.*, 2003), spliceosomal protein SAP145 (Bryant *et al.*, 2001), splicing associated factor p32 (Bryant *et al.*, 2000), hnRNP K (Wadd *et al.*, 1999), CK2 (Wadd *et al.*, 1999), RNA export protein Aly/REF (Chen *et al.*, 2002), mRNA export receptor TAP/NXF1 (Chen *et al.*, 2002) and RNA polymerase II (Zhou and Knipe, 2001). The regions of ICP27 that are involved in its interactions with proteins and RNA have been mapped by the analysis of mutants, and have been broadly mapped to protein motifs in both the N-terminal and C-terminal halves of the molecule and are shown in Fig. 4.4. Although several protein motifs were identified based upon sequence comparisons, no structural information on ICP27 has been generated to date.

The predicted protein structural motifs in the interacting regions of ICP27 (Fig. 4.4) were based upon sequence comparisons with proteins whose structures have been determined. Modular motifs within multidomain proteins can specify interacting partners and the identification of these interactions helps to define biochemical functions and the cellular pathways in which they function. However, identification of modular domain type through sequence homology studies alone can lead to ambiguous and sometimes conflicting results. For example, the N-terminus of ICP27 contains a leucine-rich hydrophobic stretch of amino acids that strongly resembles a nuclear export sequence (NES) of the type that interacts with the exportin family receptor, CRM1 (Fornerod *et al.*, 1997). This sequence can in fact confer export activity on a heterologous reporter protein (Sandri-Goldin, 1998), and deletion of this region decreases the efficiency of ICP27 export to the cytoplasm (Sandri-Goldin, 1998; Chen *et al.*, 2002; Lengyel *et al.*, 2002). However, ICP27 was not found to interact with CRM1 in yeast two-hybrid assays, co-immunoprecipitation experiments or co-localization studies (Chen *et al.*, 2002), and despite reports that ICP27 export was sensitive to the CRM1 inhibitor, leptomycin B (Soliman and Silverstein, 2000a; Soliman and Silverstein, 2000b), we (Chen *et al.*, 2002) and others (Koffa *et al.*, 2001) did not observe any effect on the export of ICP27 in the presence of LMB in virus-infected cells or micro-injected oocytes. On the contrary, we found that the N-terminal leucine-rich region is required for efficient interaction with the

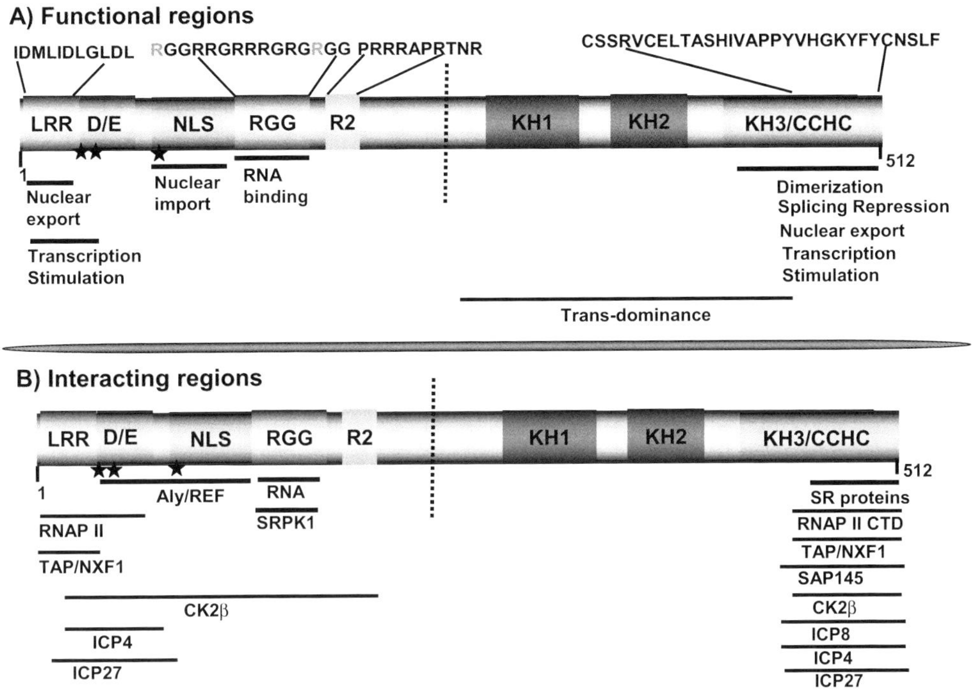

Figure 4.4 Schematic depiction of ICP27 showing known functional regions and structural motifs based on sequence homology to other proteins. The motifs include: leucine-rich sequence (LRR); acidic region (D/E); nuclear localization signal (NLS); RGG-box RNA binding motif (RGG); arginine-rich R2 region (R2); KH1, KH2 and KH3 RNA binding domains, and a zinc-finger-like domain (CCHC). The stars mark phosphorylation sites. A) Functional regions as mapped by mutant analysis. The amino acid sequence of selected motifs is shown. B) Regions of ICP27 shown to interact with cellular and viral proteins and with RNA are indicated.

major mRNA export receptor TAP/NXF1 (Chen *et al.*, 2005). Thus, it possible that the N-terminal NES-like region is important for ICP27 structural folding and deletion or alteration of this region disrupts its interaction with TAP/NXF1.

Another example of ambiguous results from the assignment of putative structural domains through sequence comparison occurs within the C-terminus of ICP27. A predicted KH RNA binding domain (KH3) was proposed to encompass the C-terminal 70 or so amino acids (Soliman and Silverstein, 2000b), although a role in RNA binding for this region has not been confirmed. In fact, we found that an ICP27 mutant with a substitution within the KH3 domain bound RNA as efficiently as the wild type protein (Chen *et al.*, 2005). On the other hand, it has been reported that a KH domain was involved in the self interaction of the cellular protein, PCBP2, which is similar to ICP27 in that it encodes three predicted KH domains (Bedard *et al.*, 2004). The region spanned by the predicted KH3 domain of ICP27 has been implicated in many of its protein–protein interactions (Fig. 4.4). Further, the C-terminus is rich in cysteine-histidine residues around a phenylalanine core, which resembles zinc-finger domains and we showed that ICP27 binds zinc *in vitro* through its C-terminus (Vaughan *et al.*, 1992). Zinc-finger domains have also been shown to mediate protein–protein interactions (Webster and Riccardi, 1991; Fridell *et al.*, 1995; Singh *et al.*, 1999). The putative KH3 motif and zinc-finger-binding domain overlap, yet the structures of KH domains and zinc-binding domains are different. Thus, structural data on ICP27 will be required to resolve the true nature of this critical protein interaction domain of ICP27.

Another interesting dichotomy in the structure of ICP27 is that for several of its protein–protein interactions, mutation of either the N-terminal region or the C-terminal KH3 and zinc-finger-like region reduces or abrogates the interaction. This has been shown for the interaction of ICP27 with itself to form multimers (Zhi *et al.*, 1999), as well as for the interaction with RNAP II CTD (Dai-Ju *et al.*, 2006), TAP/NXF1 (Chen *et al.*, 2005), CK2 (Wadd *et al.*, 1999) and ICP4 (Panagiotidis *et al.*, 1997). These findings strongly suggest that alterations in the N-terminus and C-terminus of ICP27 alter or disrupt correct folding of the protein, which in turn prevents interaction with other proteins. Again, structural data will be required to resolve these issues.

Future directions

ICP27 is a key player in HSV-1 infection. ICP27 is multifunctional and has been implicated in all aspects of gene expression from transcription through RNA processing and export to translation, and in cell cycle control and the prevention of apoptosis, although the latter two functions were not addressed in this chapter. Key questions remain including how ICP27 switches among its various activities during the course of infection; how its export and import activities may be controlled by post-translational modifications; how ICP27 recognizes viral transcripts by determining the RNA recognition sequence/structure, and how and when mRNA is released from ICP27 in the cytoplasm. Future studies will be focused on charting the dynamics of ICP27's intermolecular interactions. In the model of HSV-1 gene expression showing the known functional activities and interactions of ICP27 (Fig. 4.3), it is currently unclear which of these activities and interactions are essential for HSV-1 infection. Further, which activities attributed to ICP27 are actually directly mediated by ICP27 and which are mediated indirectly, for example by products whose expression requires ICP27? Structural information on ICP27 will be required to design targeted mutations that will disrupt specific interactions without disrupting the structural integrity of the protein. It will be necessary to dissect each of these interactions and activities to determine their overall importance to viral infection.

Acknowledgments

The studies on ICP27 from my laboratory were supported by Public Health Service grants from the National Institute of Allergy and Infectious Diseases.

References

Adamson, T.E., Shutt, D.C., and Price, D.H. (2005). Functional coupling of cleavage and polyadenylation with transcription of mRNA. J. Biol. Chem. *280*, 32262–32271.

Ahn, S.H., Kim, M., and Buratowski, S. (2004). Phosphorylation of serine 2 within the RNA polymerase II C-terminal domain couples transcription and 3′ end processing. Mol. Cell *13*, 67–76.

Aubert, M. and Blaho, J.A. (1999). The herpes simplex virus type 1 protein ICP27 is required for the prevention of apoptosis in infected human cells. J. Virol. *73*, 2803–2813.

Aubert, M., Rice, S.A., and Blaho, J.A. (2001). Accumulation of herpes simplex virus type 1 early and leaky-late proteins correlates with apoptosis prevention in infected human HEp-2 cells. J. Virol. *75*, 1013–1030.

Bedard, K.M., Walter, B.L., and Semler, B.L. (2004). Multimerization of poly(rC) binding protein 2 is required for translation initiation mediated by a viral IRES. RNA *10*, 1266–1276.

Bedford, M.T. and Richard, S. (2005). Arginine methylation: an emerging regulator of protein function. Mol. Cell *18*, 263–272.

Bentley, D.L. (2005). Rules of engagement: co-transcriptional recruitment of pre-mRNA processing factors. Curr. Opin. Cell Biol. *17*, 251–256.

Braun, I.C., Herold, A., Rode, M., Conti, E., and Izaurralde, E. (2001). Overexpression of TAP/p15 heterodimers bypasses nuclear retention and stimulates nuclear mRNA export. J. Biol. Chem. *276*, 20536–20543.

Bryant, H.E., Matthews, D.A., Wadd, S., Scott, J.E., Kean, J., Graham, S., Russell, W.C., and Clements, J.B. (2000). Interaction between herpes simplex virus type 1 IE63 protein and cellular protein p32. J. Virol. *74*, 11322–11328.

Bryant, H.E., Wadd, S., Lamond, A.I., Silverstein, S.J., and Clements, J.B. (2001). Herpes simplex virus IE63 (ICP27) protein interacts with spliceosome-associated protein 145 and inhibits splicing prior to the first catalytic step. J. Virol. *75*, 4376–4385.

Cao, W., Jamison, S.F., and Garcia-Blanco, M.A. (1997). Both phosphorylation and dephosphorylation of ASF/SF2 are required for pre-mRNA splicing *in vitro*. RNA *3*, 1456–1467.

Chen, I.B., Sciabica, K.S., and Sandri-Goldin, R.M. (2002). ICP27 interacts with the export factor Aly/REF to direct herpes simplex virus 1 intronless RNAs to the TAP export pathway. J. Virol. *76*, 12877–12889.

Chen, I.B., Li, L., Silva, L., and Sandri-Goldin, R.M. (2005). ICP27 recruits Aly/REF but not TAP/NXF1 to herpes simplex virus type 1 transcription sites although TAP/NXF1 is required for ICP27 export. J. Virol. *79*, 3949–3961.

Cho, E.-J., Takagi, T., Moore, C.R., and Buratowski, S. (1997). mRNA capping enzyme is recruited to the transcription complex by phosphorylation of the RNA polymerase II carboxy-terminal domain. Genes Dev. *11*, 3319–3326.

Cullen, B.R. (2000). Connections between the processing and nuclear export of mRNA: evidence for an export license? Proc. Natl. Acad. Sci. USA *97*, 4–6.

Cullen, B.R. (2003). Nuclear RNA export. J. Cell Sci. *116*, 587–597.

Dahmus, M.E. (1996). Reversible phosphorylation of the C-terminal domain of RNA polymerase II. J. Biol. Chem. *271*, 19009–19012.

Dai-Ju, J.Q., Li, L., Johnson, L.A., and Sandri-Goldin, R.M. (2006). ICP27 interacts with the C-terminal domain of RNA polymerase II and facilitates its recruitment to herpes simplex virus-1 transcription sites, where it undergoes proteasomal degradation during infection. J. Virol. *80*, 3567–3581.

Durand, L.O., Advani, S.J., Poon, A.P.W., and Roizman, B. (2005). The carboxyl-terminal domain of RNA polymerase II is phosphorylated by a complex containing cdk9 and infected-cell protein 22 of herpes simplex virus 1. J. Virol. *79*, 6757–6762.

Ellison, K.S., Maranchuk, R.A., Mottet, K.L., and Smiley, J.R. (2005). Control of VP16 translation by the herpes simplex virus type 1 immediate-early protein ICP27. J. Virol. *79*, 4120–4131.

Fontaine-Rodriquez, E.C., Taylor, T.J., Olesky, M., and Knipe, D.M. (2004). Proteomics of herpes simplex virus infected cell protein 27: association with translation initiation factors. Virology *330*, 487–492.

Fornerod, M., Ohno, M., Yoshida, M., and Mattaj, I.W. (1997). CRM1 is an export receptor for leucine-rich nuclear export signals. Cell *90*, 1051–1060.

Fribourg, S., Braun, I.C., Izaurralde, E., and Conti, E. (2001). Structural basis for the recognition of a nucleoporin FG repat by the NTF2-like domain of the TAP/p15 mRNA nuclear export factor. Mol. Cell *8*, 645–656.

Fridell, R.A., Harding, L.S., Bogerd, H.P., and Cullen, B.R. (1995). Identification of a novel human zinc finger protein that specifically interacts with the activation domain of lentiviral Tat proteins. Virology *209*, 347–357.

Gatfield, D. and Izaurralde, E. (2002). REF/Aly and the additional exon junction complex proteins are dispensable for nuclear mRNA export. J. Cell Biol. *159*, 579–588.

Graveley, B.R. (2000). Sorting out the complexity of SR protein functions. RNA *6*, 1197–1211.

Gruter, P., Tabernero, C., von Kobbe, C., Schmitt, C., Saavedra, C., Bacchi, A., Wilm, M., Felber, B.K., and Izaurralde, E. (1998). TAP, the human homologue of Mex67p, mediates CTE-dependent RNA export from the nucleus. Mol. Cell *1*, 649–659.

Gui, J.-F., Lane, W. S., and Fu, X.-D. (1994a). A serine kinase regulates intracellular localization of splicing factors in the cell cycle. Nature *369*, 678–682.

Gui, J-F., Tronchere, H., Chandler, S.D., and Fu, X.-D. (1994b). Purification and characterization of a kinase specific for the serine- and arginine-rich pre-mRNA splicing factors. Proc. Natl. Acad. Sci. USA *91*, 10824–10828.

Hann, L.E., Cook, W.J., Uprichard, S.L., Knipe, D.M., and Coen, D.M. (1998). The role of herpes simplex virus ICP27 in the regulation of UL24 gene expression by differential polyadenylation. J. Virol. *72*, 7709–7714.

Hardwicke, M.A. and Sandri-Goldin, R.M. (1994). The herpes simplex virus regulatory protein ICP27 can cause a decrease in cellular mRNA levels during infection. J. Virol. *68*, 4797–4810.

Hardwicke, M.A., Vaughan, P.J., Sekulovich, R.E., O'Conner, R., and Sandri-Goldin, R.M. (1989). The regions important for the activator and repressor functions of the HSV-1 alpha protein ICP27 map to the C-terminal half of the molecule. J. Virol. *63*, 4590–4602.

Hardy, W.R. and Sandri-Goldin, R.M. (1994). Herpes simplex virus inhibits host cell splicing, and regulatory protein ICP27 is required for this effect. J. Virol. *68*, 7790–7799.

Herold, A., Suyama, M., Rodrigues, J.P., Braun, I.C., Kutay, U., Carmo-Fonseca, M., Bork, P., and Izaurralde, E. (2000). TAP (NXF1) belongs to a multigene family of putative RNA export factors with a conserved modular architecture. Mol. Cell. Biol. *20*, 8996–9008.

Herold, A., Teixeira, L., and Izaurralde, E. (2003). Genome-wide analysis of nuclear mRNA export pathways in *Drosophila*. EMBO J. *22*, 2472–2483.

Hirose, Y. and Manley, J.L. (2000). RNA polymerase II and the integration of nuclear events. Genes Dev. *14*, 1415–1429.

Hsu, I., Hsu, M., Li, C., Chuang, T., Lin, R., and Tarn, W.-Y. (2005). Phosphorylation of Y14 modulates its interaction with proteins involved in mRNA metabolism and influences methylation. J. Biol. Chem. *280*, 34507–34512.

Hurt, E., Straber, K., Segref, A., Bailer, S., Schlaich, N., Presutti, C., Tollervey, D., and Jansen, R. (2000). Mex67p mediates nuclear export of a variety of RNA polymerase II transcripts. J. Biol. Chem. *275*, 8361–8368.

Jean, S., LeVan, K.M., Song, B., Levine, M., and Knipe, D.M. (2001). Herpes simplex virus 1 ICP27 is required for transcription of two viral late (γ2) genes in infected cells. Virology *283*, 273–284.

Kahvejian, A., Svitkin, Y.V., Sukarieh, R., M'Boutchou, M.N., and Sonenberg, N. (2005). Mammalian poly(A)-binding protein is a eukaryotic translation initiation factor, which acts via multiple mechanisms. Genes Dev. *19*, 104–113.

Kang, Y. and Cullen, B.R. (1999). The human Tap protein is a nuclear mRNA export factor that contains novel RNA-binding and nucleocytoplasmic transport sequences. Genes Dev. *13*, 1126–1139.

Katahira, J., Straesser, K., Saiwaki, T., Yoneda, Y., and Hurt, E. (2002). Complex formation between Tap and p15 affects binding to FG-repeat nucleoporins and nucleocytoplasmic shuttling. J. Biol. Chem. *277*, 9242–9246.

Kim, E., Du, L., Bregman, D.B., and Warren, S.L. (1997). Splicing factors associate with hyperphosphorylated RNA polymerase II in the absence of pre-mRNA. J. Cell Biol. *136*, 19–28.

Kober, M.S. and Greenblatt, J. (2002). Regulation of transcription elongation by phosphorylation. Biochim. Biophys. Acta *1577*, 261–275.

Koffa, M.D., Clements, J.B., Izaurralde, E., Wadd, S., Wilson, S.A., Mattaj, I.W., and Kuersten, S. (2001). Herpes simplex virus ICP27 protein provides viral mRNAs with access to the cellular mRNA export pathway. EMBO J. *20*, 5769–5778.

Koffa, M.D., Kean, J., Zachos, G., Rice, S.A., and Clements, J.B. (2003). CK2 protein kinase is stimulated and redistributed by functional herpes simplex virus ICP27 protein. J. Virol. *77*, 4315–4325.

Komarnitsky, P., Cho, E.-J., and Buratowski, S. (2000). Different phosphorylated forms of RNA polymerase II and associated mRNA processing factors during transcription. Genes Dev. *14*, 2452–2460.

LeHir, H., Izaurralde, E., Maquat, L.E., and Moore, M.J. (2000a). The spliceosome deposits multiple proteins 20–24 nucleotides upstream of mRNA exon-exon junctions. EMBO J. *19*, 6860–6869.

LeHir, H., Moore, M.J., and Maquat, L.E. (2000b). Pre-mRNA splicing alters mRNP composition: evidence for stable association of proteins at exon-exon junctions. Genes Dev. *14*, 1098–1108.

LeHir, H., Gatfield, D., Izaurralde, E., and Moore, M.J. (2001). The exon-exon junction complex provides a binding platform for factors involved in mRNA export and nonsense-mediated mRNA decay. EMBO J. *20*, 4987–4997.

Lengyel, J., Guy, C., Leong, V., Borge, S., and Rice, S.A. (2002). Mapping of functional regions in the amino-terminal portion of the herpes simplex virus ICP27 regulatory protein: importance of the leucine-rich nuclear export signal and RGG box RNA-binding domain. J. Virol. *76*, 11866–11879.

Long, M.C., Leong, V., Schaffer, P.A., Spencer, C.A., and Rice, S.A. (1999). ICP22 and the UL13 protein kinase are both required for herpes simplex virus-induced modification of the large subunit of RNA polymerase II. J. Virol. *73*, 5593–5604.

Luo, M.J. and Reed, R. (1999). Splicing is required for rapid and efficient mRNA export in metazoans. Proc. Natl. Acad. Sci. USA *96*, 14937–14942.

Luo, M.J., Zhou, Z., Magni, K., Christoforides, C., Rappsilber, J., Mann, M., and Reed, R. (2001). Pre-mRNA splicing and mRNA export linked by direct interactions between UAP56 and Aly. Nature *413*, 644–647.

McBride, A.E., Cook, J.T., Stemmler, E.A., Ruteledge, K.L., McGrath, K.A., and Rubens, J.A. (2005). Arginine methylation of yeast mRNA-binding protein Npl3 directly affects its function, nuclear export, and intranuclear protein interactions. J. Biol. Chem. *280*, 30888–30896.

McGregor, F., Phelan, A., Dunlop, J., and Clements, J.B. (1996). Regulation of herpes simplex virus poly(A) site usage and the action of immediate-early protein IE63 in the early-late switch. J. Virol. *70*, 1931–1940.

McLauchlan, J., Simpson, S., and Clements, J.B. (1989). Herpes simplex virus induces a processing factor that stimulates poly(A) site usage. Cell *59*, 1093–1105.

McLauchlan, J., Phelan, A., Loney, C., Sandri-Goldin, R.M., and Clements, J.B. (1992). Herpes simplex virus IE63 acts at the posttranscriptional level to stimulate viral mRNA 3′ processing. J. Virol. *66*, 6939–6945.

Mears, W.E. and Rice, S.A. (1996). The RGG box motif of the herpes simplex virus ICP27 protein mediates an RNA-binding activity and determines *in vivo* methylation. J. Virol. *70*, 7445–7453.

Mears, W.E. and Rice, S.A. (1998). The herpes simplex virus immediate-early protein ICP27 shuttles between the nucleus and cytoplasm. Virology *242*, 128–137.

Olesky, M., McNamee, E.E., Zhou, C., Taylor, T.J., and Knipe, D.M. (2005). Evidence for a direct interaction between HSV-1 ICP27 and ICP8 proteins. Virology *331*, 94–105.

Panagiotidis, C.A., Lium, E.K., and Silverstein, S. (1997). Physical and functional interactions between herpes simplex virus immediate-early proteins ICP4 and ICP27. J. Virol. *71*, 1547–1557.

Patturajan, M., Schulte, R.J., Sefton, B.M., Berezney, R., Vincent, M., Bensaude, O., Warren, S.L., and Corden, J.L. (1998). Growth-related changes in phosphorylation of yeast RNA polymerase II. J. Biol. Chem. *273*, 4689–4694.

Pearson, A., Knipe, D.M., and Coen, D.M. (2004). ICP27 selectively regulates the cytoplasmic localization of a subset of viral transcripts in herpes simplex virus type 1-infected cells. J. Virol. *78*, 23–32.

Perkins, K.D., Gregonis, J., Borge, S., and Rice, S.A. (2003). Transactivation of a viral target gene by herpes simplex virus ICP27 is posttranscriptional and does not require the endogenous promoter or polyadeylation site. J. Virol. *77*, 9872–9884.

Prasad, J., Colwill, K., Pawson, T., and Manley, J.L. (1999). The protein kinase Clk/Sty directly modulates SR protein activity: Both hyper- and hypophosphorylation inhibit splicing. Mol. Cell. Biol. *19*, 6991–7000.

Price, D.H. (2000). P-TEFb, a cyclin-dependent kinase controlling elongation by RNA polymerase II. Mol. Cell. Biol. *20*, 2629–2634.

Reed, R. and Magni, K. (2001). A new view of mRNA export: separating the wheat from the chaff. Nature Cell Biol. 3, 201–204.

Reichert, V.L., LeHir, H., Jurica, M.S., and Moore, M.J. (2002). 5′ exon interactions within the human spliceosome establish a framework for exon junction complex strucutre and assembly. Genes Dev. *16*, 2778–2791.

Rice, S.A. and Knipe, D.M. (1990). Genetic evidence for two distinct transactivation functions of the herpes simplex virus alpha protein ICP27. J. Virol. *64*, 1704–1715.

Rice, S.A. and Lam, V. (1994). Amino acid substitution mutations in the herpes simplex virus ICP27 protein define an essential gene regulation function. J. Virol. *68*, 823–833.

Rice, S.A., Su, L., and Knipe, D.M. (1989). Herpes simplex virus alpha protein ICP27 possesses separable positive and negative regulatory activities. J. Virol. *63*, 3399–3407.

Rice, S.A., Lam, V., and Knipe, D.M. (1993). The acidic amino-terminal region of herpes simplex virus type 1 alpha protein ICP27 is required for an essential lytic function. J. Virol. *67*, 1778–1787.

Rice, S.A., Long, M.C., Lam, V., and Spencer, C.A. (1994). RNA polymerase II is aberrantly phosphorylated and localized to viral replication compartments following herpes simplex virus infection. J. Virol. *68*, 988–1001.

Rice, S.A., Long, M.C., Lam, V., Schaffer, P.A., and Spencer, C.A. (1995). Herpes simplex virus immediate-early protein ICP22 is required for viral modification of host RNA polymerase II and establishment of the normal viral transcription program. J. Virol. *69*, 5550–5559.

Roscigno, R.F. and Garcia-Blanco, M.A. (1995). SR proteins escort the U4/U6-U5 tri-snRNP to the spliceosome. RNA *1*, 692–706.

Sacks, W.R., Greene, C.C., Ashman, D.P., and Schaffer, P.A. (1985). Herpes simplex virus type 1 ICP27 is an essential regulatory protein. J. Virol. 55, 796–805.

Sandri-Goldin, R.M. (1998). ICP27 mediates herpes simplex virus RNA export by shuttling through a leucine-rich nuclear export signal and binding viral intronless RNAs through an RGG motif. Genes Dev. *12*, 868–879.

Sandri-Goldin, R.M. and Mendoza, G.E. (1992). A herpes virus regulatory protein appears to act posttranscriptionally by affecting mRNA processing. Genes Dev. 6, 848–863.

Sandri-Goldin, R.M., and Hibbard, M.K. (1996). The herpes simplex virus type 1 regulatory protein ICP27 coimmunoprecipitates with anti-Sm antiserum and the C-terminus appears to be required for this interacion. J. Virol. *70*, 108–118.

Sandri-Goldin, R.M, Hibbard, M.K., and Hardwicke, M.A. (1996). The C-terminal repressor region of HSV-1 ICP27 is required for the redistribution of small nuclear ribonucleoprotein particles and splicing factor SC35; however, these alterations are not sufficient to inhibit host-cell splicing. J.Virol. *69*, 6063–6076.

Santos-Rosa, H., Moreno, H., Simos, G., Segref, A., Fahrenkrog, B., Pante, N., and Hurt, E. (1998). Nuclear mRNA export requires complex formation between Mex67p and Mtr2p at the nuclear pores. Mol. Cell. Biol. *18*, 6826–6838.

Sciabica, K.S., Dai, Q.J., and Sandri-Goldin, R.M. (2003). ICP27 interacts with SRPK1 to mediate HSV-1 inhibtion of pre-mRNA splicing by altering SR protein phosphorylation. EMBO J. *22*, 1608–1619.

Segref, A., Sharma, K., Doye, V., Hellwig, A., Huber, J., Luhrmann, R., and Hurt, E. (1997). Mex67p, a novel factor for nuclear mRNA export, binds to both poly(A)$^+$ RNA and nuclear pores. EMBO J. *16*, 3256–3271.

Singh, B.B., Patel, H.H., Roepman, R., Schick, D., and Ferreira, P.A. (1999). The zinc-finger cluster domain of RanBP2 is a specific docking site for the nuclear export factor, exportin-1. J. Biol. Chem. *274*, 37370–37378.

Sokolowski, M., Scott, J.E., Heaney, R.P., Patel, A.H., and Clements, J.B. (2003). Identification of herpes simplex virus RNAs that interact specifically with regulatory protein ICP27 *in vivo*. J. Biol. Chem. *278*, 33540–33549.

Soliman, T.M. and Silverstein, S.J. (2000a). Herpesvirus mRNAs are sorted for export via CRM1-dependent and -independent pathways. J. Virol. *74*, 2814–2825.

Soliman, T.M. and Silverstein, S.J. (2000b). Identification of an export control sequence and a requirement for the KH domains in ICP27 from herpes simplex virus type 1. J. Virol. *74*, 7600–7609.

Soliman, T.M., Sandri-Goldin, R.M., and Silverstein, S.J. (1997). Shuttling of the herpes simplex virus type 1 regulatory protein ICP27 between the nucleus and cytoplasm mediates the expression of late proteins. J. Virol. *71*, 9188–9197.

Song, B., Yeh, K.-C., and Knipe, D.M. (2001). Herpes simplex virus gene products required for expression of G1-phase functions. Virology *290*, 320–328.

Spencer, C.A., Dahmus, M.E., and Rice, S.A. (1997). Repression of host RNA polymerase II transcription by herpes simplex virus type 1. J. Virol. *71*, 2031–2040.

Straber, K. and Hurt, E. (2000). Yra1p, a conserved nuclear RNA-binding protein, interacts directly with Mex67p and is required for mRNA export. EMBO J. *19*, 410–420.

Strasser, K., Bassler, J., and Hurt, E. (2000). Binding of the Mex67p/Mtr2p heterodimer to FXFG, GLFG and FG repeat nucleoporins is essential for nuclear mRNA export. J. Cell Biol. *150*, 695–706.

Stutz, F., Bachi, A., Doerks, T., Braun, I.C., Seraphin, B., Wilm, M., Bork, P., and Izaurralde, E. (2000). REF, an evolutionary conserved family of hnRNP-like proteins, interacts with TAP/Mex67p and participates in mRNA nuclear export. RNA *6*, 638–650.

Sun, A., Devi-Rao, G.V., Rice, M.K., Gary, L.W., Bloom, D.C., Sandri-Goldin, R.M., Ghazal, P., and Wagner, E.K. (2004). Immediate-early expression of the herpes simplex virus type 1 ICP27 transcript is not critical for efficient replication *in vitro* or *in vivo*. J. Virol. *78*, 10470–10478.

Tan, W., Zolotukhin, A.S., Bear, J., Patenaude, D.J., and Felber, B. (2000). The mRNA export of *Caenorhabditis elegans* is mediated by Ce-NFX-1, an ortholog of human TAP/NFX and *Saccharomyces cerevisiae* Mex67p. RNA *6*, 1762–1772.

Uprichard, S.L. and Knipe, D.M. (1996). Herpes simplex ICP27 mutant viruses exhibit reduced expression of specific DNA replication genes. J. Virol. *70*, 1969–1980.

Vaughan, P.J., Thibault, K.J., Hardwicke, M.A., and Sandri-Goldin, R.M. (1992). The herpes simpex virus type 1 immediate early protein ICP27 encodes a potential metal binding domain and is able to bind to zinc. Virology *189*, 377–384.

Wadd, S., Bryant, H., Filhol, O., Scott, J.E., Hsieh, T.-Y., Everett, R.D., and Clements, J.B. (1999). The multifunctional herpes simplex virus IE63 protein interacts with heterogeneous ribonucleoprotein K and with casein kinase 2. J. Biol. Chem. *274*, 28991–28998.

Wagner, E. and Lykke-Andersen, J. (2002). mRNA surveillance: the perfect persist. J. Cell Sci. *115*, 3033–3038.

Webster, L.C. and Riccardi, R.P. (1991). trans-dominant mutants of E1A provide genetic evidence that the zinc-finger of the trans-activating domain binds a transcription factor. Mol. Cell. Biol. *11*, 4287–4296.

Weigand, H.L., Coburn, G.A., Zeng, Y., Kang, Y., Bogerd, H.P., and Cullen, B.R. (2002). Formation of Tap/NXT1 heterodimers activiates Tap-dependent nuclear mRNA export by enhancing recruitment to nuclear pore complexes. Mol. Cell. Biol. *22*, 245–256.

Zenklusen, D. and Stutz, F. (2001). Nuclear export of mRNA. FEBS Lett. *498*, 150–156.

Zenklusen, D., Vinciguerra, P., Strahm, Y., and Stutz, F. (2001). The yeast hnRNP-like proteins Yra1p and Tra2p participate in mRNA export through interaction with Mex67p. Mol. Cell. Biol. *21*, 4219–4232.

Zhang, J. and Corden, J.L. (1991). Phosphorylation causes a conformational change in the carboxyl-terminal domain of the mouse RNA polymerase II largest subunit. J. Biol. Chem. *266*, 2297–2302.

Zhi, Y. and Sandri-Goldin, R.M. (1999). Analysis of the phosphorylation sites of the herpes simplex virus type 1 regulatory protein ICP27. J. Virol. *73*, 3246–3257.

Zhi, Y., Sciabica, K.S., and Sandri-Goldin, R.M. (1999). Self interaction of the herpes simplex virus type 1 regulatory protein ICP27. Virology *257*, 341–351.

Zhou, C. and Knipe, D.M. (2001). Association of herpes simplex virus 1 ICP8 and ICP27 with cellular RNA polymerase II holoenzyme. J. Virol. *76*, 5893–5904.

HSV-1 DNA Replication

5

Sandra K. Weller

Abstract
Herpes simplex virus (HSV) DNA replication exhibits a complex relationship between DNA replication and recombination. In this respect HSV is similar to the large DNA bacteriophages, lambda and T4 which replicate via concatemers of tandemly repeated monomeric units using rolling circle replication as well as multiple pathways to carry out homologous recombination. The seven replication proteins of HSV may work in combination with both viral and cellular recombination proteins during this process, the details of which remain unclear. It appears that HSV DNA replication occurs in two stages. The first stage involves an origin-UL9 dependent step and in the second stage DNA replication may proceed in an origin-independent manner via a recombination-driven and/or rolling circle mechanism.

Introduction

Replication of DNA genomes is a highly coordinated process that guarantees accurate and efficient inheritance of the genetic information. The molecular mechanisms involved in viral DNA replication and its regulation are the subjects of intense research for several reasons. Viral enzymes involved in replication have provided a rich store of useful targets for antiviral therapy against herpesviruses as well as many other viral agents. Furthermore, viruses provide important models for the study of eukaryotic DNA replication because of their relative simplicity and ease of study. In fact, much of what we know about eukaryotic DNA replication has come from the study of viral systems. The herpesviruses encode a large number of replication proteins, many of which represent functional analogs of the eukaryotic DNA replication machinery. Since the HSV system is amenable to both genetic and biochemical manipulation, characterization of HSV DNA replication is expected to facilitate our understanding of analogous host processes. In addition, the herpesviruses have evolved an unusual strategy of DNA replication which utilizes viral and host recombination and repair machinery. Thus, the study of HSV DNA replication is also important to the understanding of human DNA replication, recombination and repair. Finally, the propensity of herpesviruses to usurp host proteins and subvert host mechanisms presents a remarkable opportunity to probe these cellular mechanisms from a new direction.

General features of herpes simplex virus DNA replication

Herpes simplex virus type I (HSV-1) genomes are linear double stranded molecules which replicate through longer-than-unit length concatemers. Monomeric genomic units are packaged by cleavage from concatemers in a reaction which occurs in conjunction with the uptake of genomes into preassembled capsids. In this chapter I will review the cis- and trans-acting viral elements needed for HSV DNA replication and describe what is known about the mechanism of viral DNA synthesis. The relationship between HSV DNA replication and recombination and repair will be described as well as other virus-host interactions relevant to the viral DNA replication process. Of particular interest is the manner in which the virus commandeers cellular factors to its own advantage during lytic infection for its own purposes. For instance, the virus reorganizes and redirects many host proteins within infected cells, recruiting desirable proteins to replication centers and sequestering others into separate domains. Material for this chapter draws primarily from papers published in the last three years; older material is cited primarily by reference to published reviews.

HSV genome

The HSV genome is large (152 kbp) and structurally complex, consisting of two unique regions (U_L and U_S) flanked by inverted repeat sequences (reviewed in Weller, 1995). The *a* sequence is repeated at both termini and an inverted copy is present at the U_L-U_S junction; the *b* sequences flank UL and the *c* sequences flank the US segment (Fig. 5.1A). The HSV genome contains three origins of replication: OriS is present twice in the viral genome in the c repeat region and OriL is present in the unique long region. OriS contains a 45-bp palindrome in which an A/T rich region is flanked by two recognition sites for the origin binding protein (box I and box II). A third weaker binding site (box III) is located adjacent to box I (Fig. 5.1B). OriL contains a much larger near perfect palindrome of 144 bp containing four recognition sites for UL9 (Fig. 5.1C). Both oriL and oriS are located in the promoter-regulatory regions of divergently transcribed genes.

Viral replication proteins

The HSV-1 genome encodes seven essential replication proteins and several nonessential replication proteins (reviewed in Knipe, 1989; Weller, 1995; Lehman and Boehmer, 1999; Marintcheva and Weller, 2001). Six HSV replication proteins can be considered as core functions: a single strand DNA binding protein (known as ICP8 or UL29), a two subunit DNA polymerase (UL30 and UL42), and a three subunit helicase/primase complex (UL5, UL8 and UL52). In addition to the 6 core replication proteins which are shared among all human and animal herpesviruses, HSV and the other alpha-herpesviruses also encode an origin binding protein (UL9). It is likely that the overall strategy of lytic viral DNA replication is generally conserved in all herpesviruses; however, the mechanism and regulation of initiation of viral DNA synthesis is probably shared only by the alpha-herpesviruses. The origins of replication in beta- and gamma-herpesviruses are more complex and no clear UL9 homologs have been identified.

Figure 5.1 (opposite) HSV-1 genome and origins of replication. A. The HSV-1 genome consists of two unique regions UL and US flanked by repeated sequences. UL is flanked by *ab* and *b'a'*, whereas US is flanked by *ac* and *c'a'*. During infection the two unique regions invert relative to each other (Hayward *et al.*, 1975). The arrows reflect possible orientations of the UL and ULS segments as a result of genomic inversion. B. OriS is positioned between two divergently transcribed mRNAs. Recognition sites for UL9 are boxed in thin lines (Boxes I, II and III and the AT-rich regions is boxed in black. C. OriL is positioned between two divergently transcribed mRNAs. Recognition sites for UL9 are boxed in grey and the AT-rich regions are boxed in heavy lines. Since OriL is a perfect palindrome, the recognition sequences are designated I and III on each side of the palindrome.

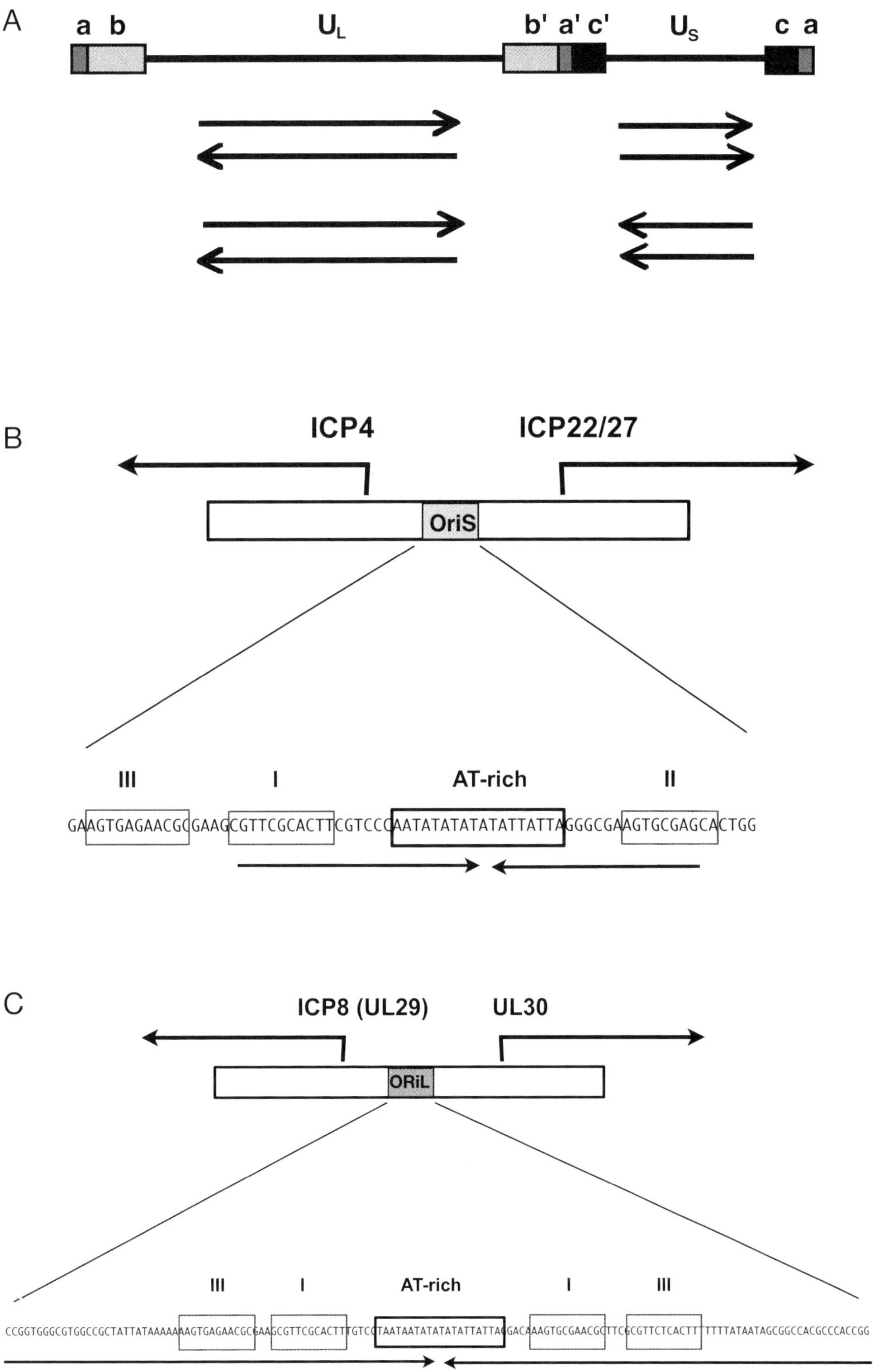
A
a b
U_L
b' a' c'
U_S
c a
B
ICP4
ICP22/27
OriS
III
I
AT-rich
II
GAAGTGAGAACGCGAAGCGTTCGCACTTCGTCCCAATATATATATATTATTAGGGCGAAGTGCGAGCACTGG
C
ICP8 (UL29)
UL30
ORiL
III
I
AT-rich
I
III
CCGGTGGGCGTGGCCGCTATTATAAAAAAAGTGAGAACGCGAAGCGTTCGCACTTTGTCCTAATAATATATATATTATTAGGACAAAGTGCGAACGCTTCGCGTTCTCACTTTTTTTATAATAGCGGCCACGCCCACCGG

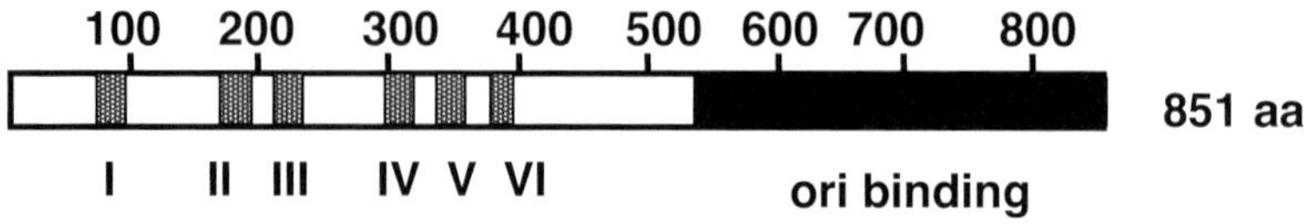

Figure 5.2 Domain structure of UL9. The UL9 gene is shown with seven stippled boxes depicting each of the motifs shared in SF 2 helicases. The N- terminus also contains sequences required for dimerization, cooperative binding to the origin, UL8 and UL42 binding (not shown). The DNA binding C- terminal domain is depicted by a black box. The very C terminus contains regions involved in nuclear localization and binding to ICP8 (not shown).

UL9

UL9 is a 94 kDa protein possessing the following activities: DNA-dependent nucleoside triphosphatase, DNA helicase on partially double-stranded substrates, ability to form dimers in solution, and cooperative origin-specific DNA binding (reviewed in Weller, 1995; Marintcheva and Weller, 2001). Genetic mapping experiments have revealed a complex domain structure for UL9 (Fig. 5.2). The origin-specific DNA binding domain has been localized to the carboxy-terminal one-third (residues 534 to 832) of UL9. The N terminal domain (residues 1–534) contains seven conserved helicase motifs, characteristic of the SF2 family of helicases. Mutagenic analysis of these helicase motifs indicates that they are essential for UL9 function. Although the DNA binding domain resides in the C-terminal region, residues in the N-terminal domain have been shown to be required for cooperative binding of UL9 to the origin. The cooperative binding likely reflects the ability of UL9 to form dimers and other higher order structures, and the residues responsible appear to map to the N-terminus (Chattopadhyay and Weller, 2006). The N-terminal region of UL9 also contains two conserved domains which may be important for function, a putative leucine zipper motif between residues 147 and 171 and a putative PEST sequence between 265 and 282. The functional significance of these motifs is not clear but may be involved in protein–protein interactions, post-translational modifications and protein stability.

Interaction of UL9 with other viral and cellular proteins

In addition to its ability to form homodimers, UL9 has been shown to interact with several other viral proteins including ICP8, UL8 protein, a component of the trimeric helicase/primase, and UL42 protein, the DNA polymerase accessory protein (reviewed in Marintcheva and Weller, 2001). ICP8 appears to act in conjunction with UL9 at least *in vitro* to increase the efficiency with which substrate binding and ATP hydrolysis are coupled to translocation/DNA unwinding (Arana *et al.*, 2001; Aslani *et al.*, 2002; Makhov *et al.*, 2003). Although the mechanism for the stimulation is not entirely understood, Arana *et al.* have proposed that ICP8 may act as a positive regulator of UL9 by neutralizing an inhibitory region of UL9 (Arana *et al.*, 2001). UL42 has also been reported to enhance the 3′-to-5′ helicase activity of UL9 (Trego and Parris, 2003). The region required for interaction with ICP8 lies within the C terminal 27 amino acids (Boehmer and Lehman, 1993). UL9 interactions with UL8 and UL42 have been localized to the N-terminal two-thirds of the molecule. Interactions with other cellular proteins have been reported including the large subunit of DNA pol alpha and human neural F-box 42 kDa protein (NFB42 or FBX2) (Lee *et al.*, 1995; Eom *et al.*, 2004). Furthermore, the *in vitro* activity of UL9 has been reported to be modulated by the addition of cellular heat shock proteins (Eom and Lehman, 2002; Tanguy Le Gac and Boehmer, 2002).

UL30/UL42

The catalytic polymerase activity resides on the 140 kDa UL30 subunit, while UL42 encodes a 53 kDa processivity factor. Since no viral DNA synthesis can be detected in the presence of drugs which inhibit the viral polymerase or in cells infected with mutants defective for viral polymerase, it is believed that UL30/UL42 is the sole polymerase needed during viral DNA replication and that it functions as both the leading and lagging strand polymerase. The HSV polymerase (Pol) has been studied extensively and is the primary target for antiviral drugs commonly in use today (reviewed in Coen, 1996; Coen and Schaffer, 2003). The crystal structure of a complex of the C-terminal 36 residues of Pol bound to residues 1–319 of UL42 has been solved and indicates that there are structural similarities between UL42 and PCNA, the processivity factor of eukaryotic polymerase (Zuccola *et al.*, 2000). This structural similarity is surprising considering that PCNA is a trimer which forms a ring that can move along DNA; whereas, UL42 functions as a single monomer in a 1:1 complex with the HSV DNA polymerase (Randell and Coen, 2004). UL42 has high intrinsic affinity for DNA and may slide along DNA by a very different mechanism than that of PCNA. It has been proposed that UL42 utilizes electrostatic interactions to slide on DNA, similar to the reported properties of thioredoxin which acts as a processivity factor for the T7 DNA polymerase (Randell and Coen, 2001; Randell and Coen, 2004). The interaction between UL30 and UL42 is currently being exploited in the development of novel anti herpesviral drugs (Coen and Schaffer, 2003; Pilger *et al.*, 2004).

UL5/UL8/UL52

The HSV-1 helicase/primase is a heterotrimer consisting of the products of the UL5, UL8 and UL52 genes (reviewed in Marintcheva and Weller, 2001; Weller, 1995). All three genes are essential for viral DNA replication in cell culture. The UL5/8/52 complex exhibits DNA-dependent ATPase, primase and helicase activities. A subcomplex consisting of UL5/UL52 displays similar activities. The precise function of UL8 within the heterotrimeric complex is not clear. *In vitro*, the helicase and primase activities do not depend on UL8; however, the ATPase and primase activities of the UL5/UL52 subcomplex can be stimulated *in vitro* by the addition of UL8. UL8 may also play a role in the proper localization of the UL5/UL8/UL52 complex to the nucleus (Barnard *et al.*, 1997). Furthermore, since UL8 can interact with other members of the replication machinery including UL9, Pol and ICP8, it appears that UL8 may act to coordinate complex protein–protein interactions which occur at the replication fork. UL52 contains a two aspartate motif conserved in many primases which, when altered, abolishes primase but not ATPase and helicase activity of the complex (Klinedinst and Challberg, 1994; Dracheva *et al.*, 1995). This suggests that the UL52 subunit is likely the primase of the complex. Sequence analysis indicates that UL5 contains seven motifs found in a large superfamily of known and putative helicases, Superfamily I (SF1). Thus, UL5 is likely to contain the helicase active site.

Although sequence analysis indicates that UL5 and UL52 are likely to comprise the helicase and primase subunit, respectively, recent work from our laboratory indicates a more complex interaction between these two subunits. The C-terminus of HSV-1 UL52 contains a Cys-His-Cys-Cys type of putative zinc binding motif. A similar, highly conserved zinc binding domain is present among herpesviruses primases and other prokaryotic and eukaryotic primases. The zinc finger of the T7 primase has been shown to play a role in DNA binding specificity. In addition, this region is involved in delivery of the primed DNA template to the DNA polymerase (Kato *et al.*, 2004). We previously showed that a mutation in the putative zinc finger at the C-terminus of UL52 abrogates not only primase activity but also ATPase, helicase and DNA binding activities of a UL5/UL52 subcomplex (Biswas and Weller, 1999). Although we have shown that both UL5 and UL52 can be cross linked to a single stranded DNA oligonucleotide (Biswas and Weller, 2001), the DNA binding and helicase defects of the zinc-binding region mutant indicate a complex interdependence between the two subunits for DNA binding and helicase activity. Four additional zinc-binding region mutations have now been analyzed and show greatly reduced primase activity as well as severe defects in ATPase, helicase and DNA binding activities (Chen *et al.*, 2005). The observation that UL52 zinc binding region mutants were defective in helicase, ATPase and DNA binding activities indicates that UL52 binding to DNA via the zinc finger may be necessary for loadingUL5 onto DNA. Alternatively, UL5 and UL52 may share a DNA binding site created by interaction between the two subunits. We conclude that the UL52 Zn finger motif is important for the activities of the helicase/primase complex.

Although most of the antiviral agents developed to combat herpesvirus infections have targeted the viral polymerase, two related classes of compounds have now been described that target the helicase/primase (Crute *et al.*, 2002; Kleymann *et al.*, 2002; and reviewed in Coen and Schaffer, 2003). These agents exhibit potent *in vitro* antiherpes activity, low resistance rates and excellent efficacy against HSV in animal models. Resistant mutations to these new agents have been mapped to both the UL5 and UL52 subunits, suggesting that these inhibitors may bind to the helicase-primase complex at the interface shared by the two subunits (Betz *et al.*, 2002; Kleymann *et al.*, 2002; Liuzzi *et al.*, 2004).

UL29 or ICP8

ICP8 encoded by the UL29 gene is a multifunctional 130 kDa zinc metalloprotein also known as the HSV single strand DNA binding protein (SSB). ICP8 preferentially binds ssDNA in a nonsequence specific and cooperative manner. ICP8 interacts with many viral and cellular proteins and plays an essential role in viral DNA replication. ICP8 is known to interact with other DNA replication proteins such as UL9, UL8 and Pol and is believed to be essential for both the initiation and elongation of viral DNA synthesis. ICP8 functionally interacts with the viral DNA polymerase and several other viral proteins by enhancing their activities. For instance it can stimulate helicase activity of both UL9 and the helicase-primase complex (Hamatake *et al.*, 1997; Boehmer, 1998; Arana *et al.*, 2001).

ICP8 is believed to play a major role in the reorganization of the cell prior to the commencement of viral DNA synthesis. ICP8 is required for the formation of prereplicative sites, discrete foci of viral proteins which lead to the formation of replication compartments, large globular domains in which viral gene expression and DNA replication are known to occur (reviewed in Knipe, 1989). ICP8 is also believed to play a role in the high level of recombination exhibited during HSV DNA replication. It can promote DNA strand transfer and exhibits helix-destabilizing activity (Nimonkar and Boehmer, 2002; Reuven *et al.*, 2003; Reuven *et al.*, 2004b; and reviewed in Wilkinson and Weller, 2003). Using immunoprecipitation and mass spectrometry, ICP8 was shown to interact either directly or indirectly with over 50 cellular and viral proteins (Taylor and Knipe, 2004). Many of the proteins identified in this study play roles in important cellular processes such as DNA replication, repair and recombination and chromatin remodeling, although it is not clear which of these interactions are functionally significant. The possible involvement of cellular proteins in viral DNA replication is consistent with previous suggestions that HSV may utilize cellular replication and recombination machinery during replication (Boehmer and Nimonkar, 2003; Boehmer and Villani, 2003; Wilkinson and Weller, 2003, 2004). ICP8 interacts specifically with the viral 5′ to 3′ exonuclease UL12. This two subunit complex may constitute a simple recombinase analogous to the red recombinase system encoded by bacteriophage lambda (Reuven *et al.*, 2003; Reuven *et al.*, 2004a; Reuven *et al.*, 2004b).

In addition to its role in DNA synthesis and recombination, ICP8 has also been shown to regulate viral gene expression by repressing transcription from the parental genome and stimulating late gene expression from progeny genomes (Chen and Knipe, 1996; McNamee *et al.*, 2000; reviewed in Knipe, 1989). A recently described interaction between ICP8 and the immediate-early protein ICP27 may play a role in the stimulation of late gene transcription (Zhou and Knipe, 2002; Olesky *et al.*, 2005).

The 3.0 Å resolution crystal structure of a C-terminal deletion construct of ICP8 was recently published which identifies a large N-terminal domain (residues 9–1038) and a small C-terminal domain (residues 1049–1129) (Mapelli *et al.*, 2004). Predictions about the site of ssDNA binding and the basis for co-operative binding were made which will need to be tested through genetic analysis.

Nonessential viral gene products

In addition to the seven required viral replication proteins, HSV also encodes several nonessential auxiliary proteins such as thymidine kinase, deoxyuridine triphosphatase, ribonucleotide reductase, uracil N-glycosylase and alkaline nuclease (reviewed in Weller, 1995). Thymidine kinase and ribonucleotide reductase participate in nucleotide biosynthesis. Ribonucleotide reductase consists of two subunits, RR1 and RR2, and both are needed for enzymatic activity. Interestingly, however, not all herpesviruses encode an active RR activity; beta-herpesviruses encode an RR1 subunit but not RR2, suggesting that RR1 may play a role in the viral life cycle in addition to nucleotide biosynthesis. Recent reports that the HSV RR1 has chaperone and antiapoptotic activities are of interest (Langelier *et al.*, 2002; Perkins *et al.*, 2002; Chabaud *et al.*, 2003; Perkins *et al.*, 2003). Thymidine kinase is well conserved in the alpha and gamma-herpesviruses. Anti-HSV drug therapy with acyclovir depends on the phosphorylation of the drug by the HSV tk. Interestingly, the beta-herpesviruses do not encode a thymidine kinase, and they have been shown to use a different enzyme for phosphorylation of gancyclovir, a related nucleoside analog, dUTPase and uracil-N-glycosylase may be important in preventing misincorporation of uracil residues into the viral genome. In cytomegalovirus, the uracil-N-glycosylase is required for replication in quiescent cells and may be involved in the introduction of strand breaks during DNA replication (Courcelle *et al.*, 2001). The alkaline nuclease is a 5′ to 3′ exonuclease which may play a role in recombination-dependent DNA replication (see below).

Overall model for HSV DNA replication

Despite the identification of viral cis- and trans-acting elements required for viral DNA synthesis, the overall mechanism of HSV DNA replication is not understood. It has long been recognized that the products of DNA replication are larger-than-unit length concatemers consisting of tandem repeats of the viral genome. It was previously thought that linear virion DNA circularized in infected cells and that concatemers were generated by rolling circle replication; however, this model has never been directly confirmed experimentally. Several aspects of this model including the initial circularization and the rolling circle mechanism have recently been challenged experimentally.

Conformation of incoming DNA

Most of the incoming linear viral DNA molecules are thought to lose their free ends shortly after infection by a process that does not require *de novo* protein synthesis. The simplest interpretation of these results is that the viral genome loses its free ends through the formation of a covalently closed circular molecule by a recombination process known as non-homologous end joining (NHEJ). It appears, however, that during lytic infection with wild type virus, endless viral DNA does not arise as a result of simple end-to-end joining of viral termini (Jackson and DeLuca, 2003). These authors demonstrated that HSV-1 genome circularization by NHEJ is actually inhibited during lytic infection of Vero cells by the action of the immediate early protein, ICP0 (Jackson and DeLuca, 2003). It is possible that circularization is prevented via inhibition of the NHEJ pathway through ICP0-mediated degradation of DNA-PKcs, which is observed in some cell types (Lees-Miller *et al.*, 1996; Parkinson *et al.*, 1999). We and others have demonstrated, however, that DNA-PKcs degradation is cell type specific and that in some cells, significant levels of degradation is not seen (Taylor and Knipe, 2004; Wilkinson and Weller, 2004). We have observed, however, that Ku86, a DNA end-binding protein that participates in NHEJ, is not recruited to the earliest sites of viral replication suggesting that, in some cell types HSV-1 inhibits NHEJ via spatial segregation, either by excluding or failing to recruit components of this pathway to viral foci. Additional evidence that NHEJ does not play a role in lytic HSV infection comes from the observation that HSV replicates more efficiently in cells which are genetically deficient in one or more components of the NHEJ pathway (Taylor and Knipe, 2004). The observation that linear genomes do not circularize by the NHEJ pathway during lytic infection raises the question of exactly what is the fate of the incoming genomes. It is possible that at the earliest stages of infection, the viral genome adopts an endless configuration by intra- or inter-molecular homologous recombination events (discussed below).

Initiation

HSV DNA synthesis is believed to initiate at one of the three viral origins of replication. As described above, both oriL and oriS contain two high affinity binding sites for UL9 on either side of an A/T rich region. These binding sites are themselves bipartite, consisting of two inverted pentanucleotide repeats. UL9 exists as a dimer, and one subunit of the dimer likely contacts the pentanucleotide repeat in the major groove of one strand, while the other subunit is thought to contact the partially overlapping pentanucleotide site on the other strand (Aslani *et al.*, 2002; Macao *et al.*, 2004). After the first dimer binds at site I, a second dimer of UL9 has been proposed to bind in a cooperative fashion to site II. Binding to sites I and II is believed to alter the conformation of the origin region and distort the A/T-rich region, in turn facilitating DNA unwinding via the ATP-dependent helicase activity of UL9. Cooperative binding requires not only the appropriate spacing of the two binding sites flanking the A/T rich region but also sequences within the N-terminus of the UL9 protein. The N-terminus appears to contain residues required for dimer and higher order multimer formation that contributes to the ability to cooperatively bind to the origin region.

Many aspects of the function of UL9 during initiation of viral DNA replication remain unresolved. For instance, although UL9 exhibits ATP-dependent DNA helicase activity *in vitro* on substrates which are partially single stranded, it has not been shown to unwind duplex DNA containing an origin of replication. This is in contrast to SV40 large T antigen which is able to unwind duplex DNA in an ATP dependent fashion. Elias and colleagues have suggested that UL9 and ICP8 together can convert double-stranded oriS to a novel conformation, oriS*, which can then be stably bound by UL9 (Aslani *et al.*, 2002; Macao *et al.*, 2004). OriS* contains a stable hairpin formed by complementary base pairing between box I and box III in oriS. Consistent with this model, Macao *et al.* have shown that ICP8 and UL9 can assemble more efficiently on negatively supercoiled DNA than on linear DNA (Macao *et al.*, 2004). It is possible that negative supercoiling may facilitate the opening of the A/T-rich element in oriS, allowing the formation of the oriS* formation which promotes the stable binding of UL9 and ICP8.

The inability to accomplish origin-dependent unwinding *in vitro* has severely hampered attempts to reconstitute HSV DNA replication in the test tube. It is possible that other viral or cellular factors or even structural components within the nucleus may be necessary to recapitulate events which occur during the initiation of viral DNA synthesis. No information is available on how UL9 functions *in vivo*. Host and viral factors could possibly promote some of the conformational changes in oriS that have been observed *in vitro*. As mentioned above, the UL9 protein interacts with ICP8, UL8, and UL42; in addition, sequences flanking the origins of replication contain binding sites for transcription factors (Nguyen-Huynh and Schaffer, 1998) and other cellular proteins such as OF-1 (Baker *et al.*, 2000; Murata *et al.*, 2004). It is also possible that *in vivo*, UL9 does not function as a helicase in the traditional sense. In conjunction with other viral or cellular proteins, it is possible that cooperative binding of two dimers of UL9 to the origin results in changes in the superhelicity of the region leading to distortion. According to this model, UL9 may act more like one of the chromatin remodeling proteins such as swi/snf family of proteins (Havas *et al.*, 2000; Owen-Hughes and Bruno, 2004). This class of proteins contains conserved helicase and ATPase motifs; however, they do not act to unwind DNA. Instead they act as ATP-dependent DNA translocases and can alter the conformation of DNA and chromatin. The molecular mechanism surrounding initiation of viral DNA replication remains one of the most intriguing mysteries in the HSV replication field.

Elongation

By analogy with other replication systems, it is anticipated that the six core HSV replication proteins function at the HSV DNA replication fork as part of the viral replisome. The replisome consisting of HSV DNA polymerase holoenzyme, the helicase/primase heterotrimer, and single-stranded DNA-binding protein is thought to mediate coordinated replication (Fig. 5.3). The elongation helicase (UL5) would be expected to unwind duplex DNA to create a single-stranded template for the polymerase and the primase (UL52) to synthesize short RNA primers on DNA that serve as start sites for DNA replication. The DNA binding and unwinding activities of the UL5/UL52/UL8 complex have been studied, and we have found that the complex does not bind or unwind substrates which are totally double stranded, regardless of whether they are linear or cruciform. On the other hand substrates with single strand regions, either containing two partially homologous oligo-

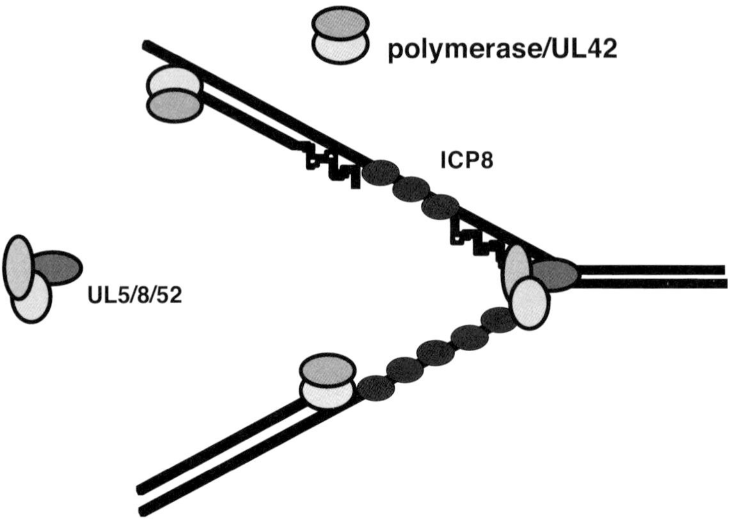

Figure 5.3 Model for HSV DNA replication. An HSV-1 replication fork would be expected to contain the helicase-primase complex (UL5/UL52/UL8) at the fork: UL5 is predicted to unwind duplex DNA ahead of the fork and U52 is thought to lay down RNA primers which could then be extended by the two-subunit DNA polymerase (UL30/UL42). The HSV-1 Pol is also thought to carry out leading strand synthesis. ICP8 (UL29, SSB) appears to bind to ssDNA generated during HSV DNA synthesis.

nucleotides or three way forked substrates are preferentially bound and unwound by the UL5/UL52/UL8 complex (Biswas and Weller, 2001; Chen and Weller, unpublished results). These results are consistent with the notion that the UL5/UL52/UL8 complex must be loaded on a partially unwound or distorted substrate during DNA replication.

The primase activity of the complex prefers templates containing a deoxyguanylate-pyrimidine-pyrimidine (3′-G-pyr-pyr-5′) triplet (Ramirez-Aguilar *et al.*, 2002). Flanking sequences are also important for optimal primase activity. Interestingly, the UL52 has recently been show to exhibit a remarkably low level of fidelity and is apparently one of the least faithful nucleotide polymerizing enzymes known (Ramirez-Aguilar and Kuchta, 2004). ICP8 is believed to stabilize the template strands while the polymerase holoenzyme (UL30/UL42) catalyzes DNA synthesis and proofreads the new DNA strand with its 3′ to 5′ exonuclease. As mentioned above, UL42, the accessory subunit of HSV polymerase, binds tightly to DNA and is believed to slide along the template via electrostatic interactions thereby contributing to the processivity of Pol (Randell and Coen, 2001; Randell and Coen, 2004). For the HSV replisome, it is not clear how or even if leading and lagging strand DNA syntheses are coordinated.

Once elongation has begun, leading and lagging strand DNA synthesis would be expected to proceed until the replication apparatus either reaches a break in the DNA molecule or receives some other signal to disengage. We thus must come back to the question of what is the conformation of the HSV genome during DNA replication. As described above, Jackson and Deluca have shown that incoming genomes do not form the covalently closed circles which would be expected from a nonhomologous end joining event (Jackson and DeLuca, 2003). However the Gardella gels used in these experiments would not distinguish between linear genomes and molecules which have undergone a circularization by homologous recombination in which the ends were held together by proteins (i.e. a noncovalently closed circle). Another complication lies in the observation that the HSV genome as packaged in virions has been shown to contain nicks and gaps (reviewed in Wilkinson and Weller, 2003). It is possible that nicks and gaps are repaired before DNA replication begins, possibly by cellular enzymes; however, if left unrepaired, double strand breaks would be expected to occur when a replication fork reaches and tries to replicate through a nick or a gap (Fig. 5.4) (Kowalczykowski, 2000). Since double strand breaks are known to be highly recombinogenic in all other systems studied, it is likely that double strand breaks generated during HSV-1 replication would stimulate recombination. Double strand breaks occur frequently within replicating DNA in both bacterial and eukaryotic systems, and the repair and recombination machinery is often called in to "restart" these replication forks (reviewed in Haber, 2000). A number of mammalian enzymes and overlapping pathways have been identified which play a role in repair of double strand breaks including NHEJ and homologous recombination (HR) (Hoeijmakers, 2001). The potential role of recombination in HSV DNA replication is discussed below.

Formation of replication compartments, sites of viral DNA synthesis

HSV carries out gene expression, DNA replication and DNA encapsidation in globular nuclear domains designated replication compartments (reviewed in Knipe, 1989; Weller, 1995). The formation of replication compartments is mediated in part by association with nuclear structures called ND10 (nuclear domains 10), also called promyelocytic leukemia (PML) bodies or PODs (Maul, 1998). HSV infection results in the disruption of ND10 and the degradation of some isoforms of

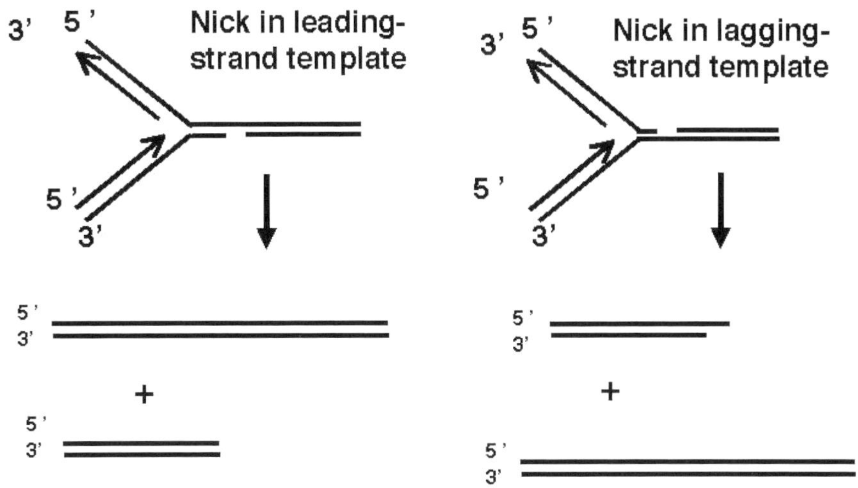

Figure 5.4 Generation of double stranded DNA breaks during replication through nicks at the replication fork. Double-stranded DNA breaks can be created by nicks or gaps at the replication fork. When DNA polymerase encounters a nick in either the leading-strand or lagging strand template, double-strand breaks would be expected (Kowalczykowski, 2000). This figure adapted from (Kowalczykowski, 2000).

PML (Everett *et al.*, 1998; Everett and Zafiropoulos, 2004).

The formation of replication compartments has been examined and it appears that their formation follows an ordered assembly process. Based on the staining patterns of ICP8 and PML, four stages of a wild type HSV-1 infection have been identified (Fig. 5.5). In stage I, PML (ND10) foci are intact and no ICP8 staining is visible. In stage II, the ND10 foci have been disrupted, and diffuse ICP8 staining is observed in the nucleus, indicating that immediate early and early protein synthesis has occurred. Stage III is defined by the formation of a limited number of ICP8-containing foci. The formation of stage III foci was shown to be dependent on the presence of UL5, UL8, UL9 and UL52, as cells infected with null mutants in these proteins never progressed past stage II. Stage III can be divided into two-stage IIIa foci contain the five viral proteins ICP8, UL5, UL8, UL9 and UL52; whereas, stage IIIb foci contain these five proteins along with HSV Pol, UL42 and PML. The recruitment of the polymerase holoenzyme to the five protein scaffold requires the presence of an active primase subunit (Carrington-Lawrence and Weller, 2003). In addition to the HSV polymerase holoenzyme and certain PML isoforms (Burkham *et al.*, 1998; Burkham *et al.*, 2001), stage IIIb foci appear to recruit several other cellular recombination proteins such as RPA, Rad51 and p95 (Wilkinson and Weller, 2004). These PML-associated foci likely represent relevant intermediates in replication compartment formation. We have shown that the formation of stage IIIb foci correlates with the induction of a DNA damage response (Wilkinson and Weller, 2004), and this may cause the recruitment of additional DNA damage response proteins to these foci. If replication is allowed to proceed, replication compartments, which stain with both the ICP8 and PML antibodies, are observed (stage IV).

In summary, HSV infection proceeds through many stages in the formation of the globular domains in which DNA replication and gene expression occur. Once the origin is distorted by UL9 and ICP8 as described above, the heterotrimeric helicase-primase complex can be recruited to form a five-protein scaffold (stage IIIa). Pol recruitment to replication foci requires primer synthesis and results in stage IIIb foci which also are able to recruit cellular proteins which may play

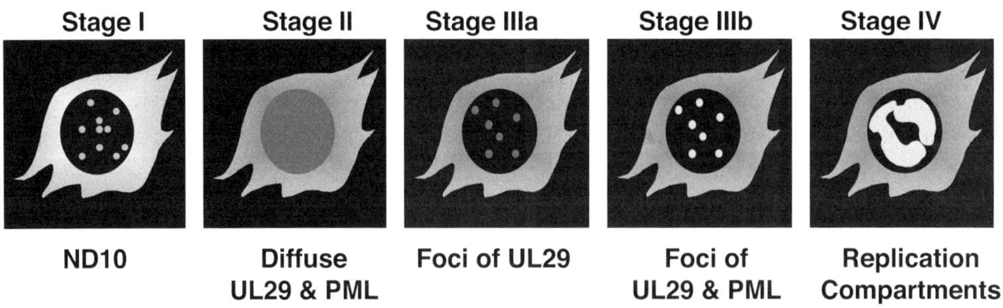

Figure 5.5 Five stages of infection as defined by localization of ICP8. In stage I, PML (ND10) foci are intact and no ICP8 staining is visible. In stage II the ND10 foci become disrupted and PML staining is diffuse throughout the nucleus and the cytoplasm of the infected cell. Diffuse ICP8 staining is also present in the nucleus, indicating that immediate early and early protein synthesis has occurred. Stage IIIa foci are dependent on the presence of ICP8, UL5, UL8, UL9 and UL52. If the UL52 in stage IIIa foci is active, Pol and UL42 can be recruited resulting in the formation of stage IIIb foci. PML, RPA and Rad 51 s are also found in stage IIIb but not stage IIIa foci (Wilkinson and Weller, 2004). If viral DNA synthesis is allowed to proceed, replication compartments are formed (stage IV).

important roles in viral DNA synthesis (see below).

Role of recombination in HSV DNA replication

Although rolling circle replication was originally proposed as the mechanism to generate concatemers, this mechanism alone is not consistent with the observation that replicating DNA is not simply concatenated, but instead is highly complex, possibly branched. Furthermore, newly replicated DNA has also undergone inversion. It has been known for some time that recombination is a frequent event within the HSV-1 genome as well as between infecting genomes and that this recombination is stimulated by the HSV DNA replication machinery but not the SV40 DNA replication machinery. High rates of homologous recombination in HSV have also been demonstrated *in vivo* using animal models (Kintner *et al.*,1995; Lingen *et al.*, 1997). A new study in human populations presents evidence of widespread recombination and concludes that recombination plays a major role in generating HSV-1 diversity in natural populations (Bowden *et al.*, 2004). Recent studies using amplicon vectors also confirm the importance of homologous recombination in HSV replication (Fu *et al.*, 2002). This recombination was dependent on HSV replication and it was not seen when plasmids were replicated by the SV40 replication machinery. Interestingly, the majority of the homologous recombination events resulted in sequence replacement or targeted gene repair (Fu *et al.*, 2002). Taken together, these observations are not consistent with the previously proposed rolling circle mechanism and instead suggest a role for recombination mediated replication, although it is possible that rolling circle replication also occurs (reviewed in (Weller, 1995; Boehmer and Lehman, 1997; Boehmer and Villani, 2003; Wilkinson and Weller, 2003). We have suggested that HSV DNA replication occurs in two stages: in the first stage replication may initiate at an origin of replication and require UL9, and in the second stage of DNA replication may proceed in an origin-independent manner via a recombination-driven and/or rolling circle mechanism (Fig. 5.6). This second stage of DNA replication is not expected to require UL9 (Blumel and Matz, 1995; Blumel *et al.*, 2000); in fact, UL9 appears to inhibit HSV late in infection (Malik *et al.*, 1992; Malik *et al.*, 1996; Marintcheva and Weller, 2003). If as we and others propose, HSV DNA replication involves DNA recombination events, it will be very important to determine which viral and/or cellular pathways are involved. It appears that the NHEJ pathway does not play a role in lytic DNA replication (as discussed above); however, several lines of evidence suggest that the HR pathway may play a role. The HR pathway requires the products of the *RAD52* epistasis group which includes RAD51, RAD52, RAD54 and the MRN complex (consisting of Mre11/Rad50/NBS1). As described above, ICP8 has been shown to directly or indirectly interact with cellular repair and recombination proteins including Rad50, Ku 86, RPA, Mre11 and PARP (Taylor and Knipe, 2004). In addition, host recombination proteins such as RPA, RAD51, and NBS1 are recruited to the earliest prereplicative sites observed in infected cells (Wilkinson and Weller, 2004). Furthermore, NBS1 is phosphorylated during productive HSV infection which indicates that the MRN pathway has been activated (Wilkinson and Weller, 2004). Based on these results we propose that the MRN complex and other components of the HR pathway play an active role in HSV DNA synthesis either alone or in conjunction with viral enzymes (see below).

Although recombination in HSV-1 likely involves cellular recombination pathways, viral proteins may also play a role in recombination-dependent replication. It has

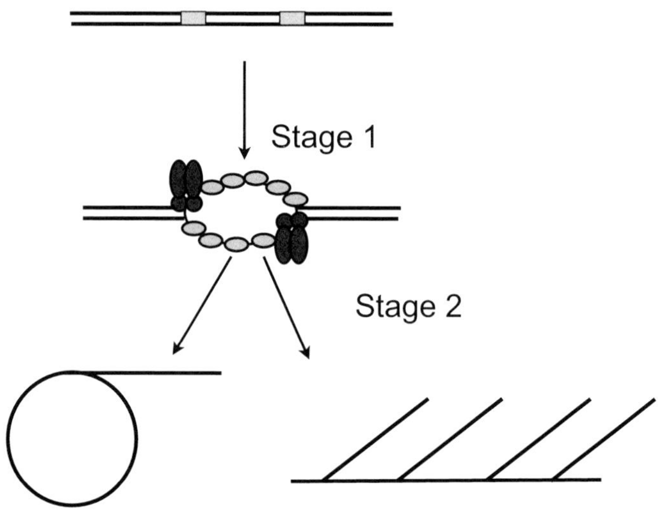

Figure 5.6 Model for UL9-dependent and UL9-independend HSV DNA replication. According to this model, the earliest stages of DNA replication after origin unwinding would be dependent on UL9 (stage 1) followed by either recombination-dependent or rolling circle replication (or both) (stage 2).

been recognized for some time that the viral ss DNA binding protein, ICP8, can participate in strand annealing (Bortner *et al.*, 1993; Nimonkar and Boehmer, 2003; Reuven *et al.*, 2004b). In addition, as mentioned above, we have recently demonstrated that HSV-1 alkaline nuclease and ICP8, mediate *in vitro* strand exchange in a manner that is analogous to the Redαβ recombinase system of lambda and the RecE/T recombinase system of *Eschericia coli* (Reuven *et al.*, 2003; Reuven *et al.*, 2004a; Reuven *et al.*, 2004b).

Thus, HSV appears to exhibit a complex relationship between DNA replication and recombination, sharing similarities with large DNA bacteriophages, T4 and lambda which replicate via concatemers of tandemly repeated monomeric units using rolling circle replication as well as multiple pathways to carry out homologous recombination. It should be kept in mind that if HSV is similar to lambda and T4, there may be several overlapping pathways which can participate in the process. The seven replication proteins of HSV may work in combination with both viral and cellular recombination proteins. Interestingly, other animal DNA viruses including SV40 and adenovirus also exhibit complex interactions with host DNA repair and recombination machinery (Stracker *et al.*, 2002; Carson *et al.*, 2003; Mohammadi *et al.*, 2004; Wu *et al.*, 2004). Furthermore, baculoviruses which also have linear genomes appear to utilize recombination-dependent replication (Martin and Weber, 1997; Kamita *et al.*, 2003) suggesting that DNA viruses with linear genomes may share the feature of using recombination-dependent DNA replication.

Role of ND10 in HSV DNA replication

Initially it was assumed that during the earliest stages of infection, the viral genomes traveled through the nucleus to existing ND10; however, it has recently been shown that ND10 are dynamic in nature, forming and reforming in response to environmental stimuli (Muratani *et al.*, 2002). Based on recent live cell imaging, a clearer picture of the early events in HSV infection is beginning to emerge (Everett *et al.*, 2003; Everett *et al.*, 2004; Everett and Murray, 2005). HSV genomes enter the nucleus where they undergo a round of immediate early gene expression. ICP4 and perhaps ICP27 are recruited to viral genomes forming a viral nucleoprotein complex. ND10 proteins may then be recruited perhaps because of the nucleoprotein complex which forms on the viral genome (Tang *et al.*, 2003). The recruitment of ND10 components may reflect a nuclear defense mechanism against foreign DNA nucleoprotein complexes which are seen as a threat. Under this scenario, the recruitment of ND10 proteins may act as a mechanism to repress expression of the viral genome. This is consistent with reports that many ND10 proteins are transcriptional repressors and are found at silenced regions of the chromatin (regions of heterochromatin) (Ishov *et al.*, 2002; Ayyanathan *et al.*, 2003; Ishov *et al.*, 2004). Alternatively, it is possible that ND10 components are attracted to viral genomes because the genome is seen as a damaged DNA molecule in need of repair. This is consistent with the observation that many proteins which associate with ND10 have functions in DNA repair and recombination (Dellaire and Bazett-Jones, 2004; Ching *et al.*, 2005). These two views may be unified by recent reports that DNA damage signals such as double strand breaks (DSBs) can recruit repair machinery and also cause silencing of nearby chromatin and formation of heterochromatin (Fernandez-Capetillo *et al.*, 2003a; Fernandez-Capetillo *et al.*, 2003b; Turner *et al.*, 2005). Thus, ND10 components may play dual roles in responding to damage and induction of silencing. It is possible that foreign nucleoprotein complexes are recognized and marked for silencing by host machinery (ND10 components) in a similar fashion to how cells recognize their own damaged chromosomes and silence them until they can be repaired (Rogakou *et al.*, 2000a; Rogakou *et al.*, 2000b; Fernandez-Capetillo *et al.*, 2003b). According to this scenario, ND10 could be thought of as "mobile repair centers" which could respond to sites of cellular damage or to invading viral genomes.

The recruitment of ND10 proteins to viral genomes is only transient; as soon as sufficient levels of the viral transactivator ICP0 are expressed, ND10 are disrupted by ICP0 which has been shown to degrade sumoylated PML and induce the disruption of ND10 and dispersal of associated proteins (Boutell *et al.*, 2002; Boutell and Everett, 2003). Disruption of ND10 is associated with the derepression of viral genomes and expression of early genes. Following disruption of ND10, it is possible that some ND10 proteins are recruited back to prerelicative sites and participate in HSV DNA replication (Burkham *et al.*, 2001). Components of this cellular pathway may be essential for efficient viral DNA replication, although the precise mechanism remains to be elucidated.

Regulation of viral DNA synthesis

Many cellular processes are regulated by various forms of post-translational modification. For example, cell cycle control, regulation of gene expression, differentiation, signal transduction, apoptosis, DNA repair, DNA replication, and the immune response are regulated by phosphorylation, ubiquitination and proteasomal degradation. Viruses have also evolved to use these mechanisms for their own benefit. By analogy with another nuclear DNA virus, SV40, it is likely that the initiation of viral DNA replication is regulated in some fashion. It turns out that phosphorylation of SV40 large T antigen plays a major role in regulation of SV40 DNA replication. We and others have recently demonstrated that UL9 may be phosphorylated and ubiquitnated and that its protein levels are modulated during infection at least in some cell types (Chattophadhahy and Weller, manuscript in preparation; Isler and Schaffer,

2001; Eom and Lehman, 2003; Eom *et al.*, 2004). As described above, we have proposed that HSV replication occurs in two stages, a UL9-dependent step involving the origins of replication and a later UL9-independent step involving either rolling circle replication or recombination (Fig. 5.6). We suggest that UL9 activity or protein levels may be regulated after the initial events at the origin.

Summary

In summary the herpesviruses have evolved an unusual strategy of DNA replication which utilizes viral and host recombination and repair machinery. HSV DNA replication is a complex process involving at least seven viral DNA replication proteins and several viral and cellular repair-recombination proteins. The complex interactions between viral and cellular DNA repair and recombination pathways are important for several reasons. First, viruses because of their relative simplicity and ease of study make excellent models for the study of these processes in mammalian cells. In fact, much of what we know about eukaryotic DNA replication has come from the study of viral systems. Even the identification of proteins required for cellular DNA replication did not occur until the development of the *in vitro* reaction system for SV40 DNA replication. By analogy with how SV40 facilitated the study of eukaryotic DNA replication, we expect that the study of HSV will provide important knowledge about replication, repair, and recombination in cells.

References

Arana, M.E., Haq, B., Tanguy Le Gac, N., and Boehmer, P.E. (2001). Modulation of the herpes simplex virus type-1 UL9 DNA helicase by its cognate single-strand DNA-binding protein, ICP8. J. Biol. Chem. *276*, 6840–6845.

Aslani, A., Olsson, M., and Elias, P. (2002). ATP-dependent unwinding of a minimal origin of DNA replication by the origin-binding protein and the single-strand DNA-binding protein ICP8 from herpes simplex virus type I. J. Biol. Chem. *277*, 41204–41212.

Ayyanathan, K., Lechner, M.S., Bell, P., Maul, G.G., Schultz, D.C., Yamada, Y., Tanaka, K., Torigoe, K., and Rauscher, F.J., 3rd (2003). Regulated recruitment of HP1 to a euchromatic gene induces mitotically heritable, epigenetic gene silencing: a mammalian cell culture model of gene variegation. Genes Dev. *17*, 1855–1869.

Baker, R.O., Murata, L.B., Dodson, M.S., and Hall, J.D. (2000). Purification and characterization of OF-1, a host factor implicated in herpes simplex replication. J. Biol. Chem. *275*, 30050–30057.

Barnard, E.C., Brown, G., and Stow, N.D. (1997). Deletion mutants of the herpes simplex virus type 1 UL8 protein: effect on DNA synthesis and ability to interact with and influence the intracellular localization of the UL5 and UL52 proteins. Virology *237*, 97–106.

Betz, U.A., Fischer, R., Kleymann, G., Hendrix, M., and Rubsamen-Waigmann, H. (2002). Potent *in vivo* antiviral activity of the herpes simplex virus primase-helicase inhibitor BAY 57–1293. Antimicrob. Agents Chemother. *46*, 1766–1772.

Biswas, N., and Weller, S.K. (1999). A mutation in the C-terminal putative Zn2+ finger motif of UL52 severely affects the biochemical activities of the HSV-1 helicase-primase subcomplex. J. Biol. Chem. *274*, 8068–8076.

Biswas, N., and Weller, S.K. (2001). The UL5 and UL52 subunits of the herpes simplex virus type 1 helicase-primase subcomplex exhibit a complex interdependence for DNA binding. J. Biol. Chem. *276*, 17610–17619.

Blumel, J., and Matz, B. (1995). Thermosensitive UL9 gene function is required for early stages of herpes simplex virus type 1 DNA synthesis. J. Gen. Virol. *76 (Pt 12)*, 3119–3124.

Blumel, J., Graper, S., and Matz, B. (2000). Structure of simian virus 40 DNA replicated by herpes simplex virus type 1. Virology *276*, 445–454.

Boehmer, P.E. (1998). The herpes simplex virus type-1 single-strand DNA-binding protein, ICP8, increases the processivity of the UL9 protein DNA helicase. J. Biol. Chem. *273*, 2676–2683.

Boehmer, P.E., and Lehman, I.R. (1993). Physical interaction between the herpes simplex virus 1 origin-binding protein and single-stranded DNA-binding protein ICP8. Proc. Natl. Acad. Sci. USA *90*, 8444–8448.

Boehmer, P.E., and Lehman, I.R. (1997). Herpes simplex virus DNA replication. Annu. Rev. Biochem. *66*, 347–384.

Boehmer, P.E., and Nimonkar, A.V. (2003). Herpes virus replication. IUBMB Life 55, 13–22.

Boehmer, P.E., and Villani, G. (2003). Herpes simplex virus type-1: a model for genome transactions. Prog. Nucleic Acid Res. Mol. Biol. *75*, 139–171.

Bortner, C., Hernandez, T.R., Lehman, I.R., and Griffith, J. (1993). Herpes simplex virus 1 single-strand DNA-binding protein (ICP8) will promote homologous pairing and strand transfer. J. Mol. Biol. *231*, 241–250.

Boutell, C., and Everett, R.D. (2003). The herpes simplex virus type 1 (HSV-1) regulatory protein ICP0 interacts with and Ubiquitinates p53. J. Biol. Chem. *278*, 36596–36602.

Boutell, C., Sadis, S., and Everett, R.D. (2002). Herpes simplex virus type 1 immediate-early protein ICP0 and is isolated RING finger domain act as ubiquitin E3 ligases *in vitro*. J. Virol. *76*, 841–850.

Bowden, R., Sakaoka, H., Donnelly, P., and Ward, R. (2004). High recombination rate in herpes simplex virus type 1 natural populations suggests significant co-infection. Infect. Genet. Evol. *4*, 115–123.

Burkham, J., Coen, D.M., and Weller, S.K. (1998). ND10 protein PML is recruited to herpes simplex virus type 1 prereplicative sites and replication compartments in the presence of viral DNA polymerase. J. Virol. *72*, 10100–10107.

Burkham, J., Coen, D.M., Hwang, C.B., and Weller, S.K. (2001). Interactions of herpes simplex virus type 1 with ND10 and recruitment of PML to replication compartments. J. Virol. *75*, 2353–2367.

Carrington-Lawrence, S.D., and Weller, S.K. (2003). Recruitment of polymerase to herpes simplex virus type 1 replication foci in cells expressing mutant primase (UL52) proteins. J. Virol. *77*, 4237–4247.

Carson, C.T., Schwartz, R.A., Stracker, T.H., Lilley, C.E., Lee, D.V., and Weitzman, M.D. (2003). The Mre11 complex is required for ATM activation and the G2/M checkpoint. EMBO J. *22*, 6610–6620.

Chabaud, S., Lambert, H., Sasseville, A.M., Lavoie, H., Guilbault, C., Massie, B., Landry, J., and Langelier, Y. (2003). The R1 subunit of herpes simplex virus ribonucleotide reductase has chaperone-like activity similar to Hsp27. FEBS Lett *545*, 213–218.

Chattopadyhay, S., and Weller, S.K. (2006). DNA binding activity of herpes simplex virus type I origin binding protein, UL9, can be modulated by sequences in the N terminus: correlation between transdominance and DNA binding. J. Virol. *80*, 4491–4500.

Chen, Y.M., and Knipe, D.M. (1996). A dominant mutant form of the herpes simplex virus ICP8 protein decreases viral late gene transcription. Virology *221*, 281–290.

Chen, Y., Carrington-Lawrence, S.D., Bai, P., and Weller, S.K. (2005). Mutations in the putative zinc binding motif of UL52 demonstrate a complex interdependence between the UL5 and UL52 subunits of the human herpes simplex virus type 1 helicase/primase complex. J of Virology *in press*.

Ching, R.W., Dellaire, G., Eskiw, C.H., and Bazett-Jones, D.P. (2005). PML bodies: a meeting place for genomic loci? J. Cell Sci. *118*, 847–854.

Coen, D.M. (1996). Viral DNA polymerases, In DNA replication in eukaryotic cells., M.L. D. (ed.), ed. (Cold Spring Harbor, N.Y.: Cold Spring Harbor Laboratory Press), pp. p. 495–523.

Coen, D.M., and Schaffer, P.A. (2003). Antiherpesvirus drugs: a promising spectrum of new drugs and drug targets. Nat. Rev. Drug Discov. *2*, 278–288.

Courcelle, C.T., Courcelle, J., Prichard, M.N., and Mocarski, E.S. (2001). Requirement for uracil-DNA glycosylase during the transition to late-phase cytomegalovirus DNA replication. J. Virol. *75*, 7592–7601.

Crute, J.J., Grygon, C.A., Hargrave, K.D., Simoneau, B., Faucher, A.M., Bolger, G., Kibler, P., Liuzzi, M., and Cordingley, M.G. (2002). Herpes simplex virus helicase-primase inhibitors are active in animal models of human disease. Nat. Med. *8*, 386–391.

Dellaire, G., and Bazett-Jones, D.P. (2004). PML nuclear bodies: dynamic sensors of DNA damage and cellular stress. Bioessays *26*, 963–977.

Dracheva, S., Koonin, E.V., and Crute, J.J. (1995). Identification of the primase active site of the herpes simplex virus type 1 helicase-primase. J. Biol. Chem. *270*, 14148–14153.

Eom, C.Y., Heo, W.D., Craske, M.L., Meyer, T., and Lehman, I.R. (2004). The neural F-box protein NFB42 mediates the nuclear export of the herpes simplex virus type 1 replication initiator protein (UL9 protein) after viral infection. Proc. Natl. Acad. Sci. USA *101*, 4036–4040.

Eom, C.Y., and Lehman, I.R. (2002). The human DnaJ protein, hTid-1, enhances binding of a multimer of the herpes simplex virus type 1 UL9 protein to oris, an origin of viral DNA replication. Proc. Natl. Acad. Sci. USA *99*, 1894–1898.

Eom, C.Y., and Lehman, I.R. (2003). Replication-initiator protein (UL9) of the herpes simplex virus 1 binds NFB42 and is degraded via the ubiquitin–proteasome pathway. Proc. Natl. Acad. Sci. USA *100*, 9803–9807.

Everett, R.D., and Murray, J. (2005). ND10 components relocate to sites associated with herpes simplex virus type 1 nucleoprotein complexes during virus infection. J. Virol. *79*, 5078–5089.

Everett, R.D., and Zafiropoulos, A. (2004). Visualization by live-cell microscopy of disruption of ND10 during herpes simplex virus type 1 infection. J. Virol. *78*, 11411–11415.

Everett, R.D., Freemont, P., Saitoh, H., Dasso, M., Orr, A., Kathoria, M., and Parkinson, J. (1998). The disruption of ND10 during herpes simplex virus infection correlates with the Vmw110- and proteasome-dependent loss of several PML isoforms. J. Virol. *72*, 6581–6591.

Everett, R.D., Sourvinos, G., and Orr, A. (2003). Recruitment of herpes simplex virus type 1 transcriptional regulatory protein ICP4 into foci juxtaposed to ND10 in live, infected cells. J. Virol. *77*, 3680–3689.

Everett, R.D., Sourvinos, G., Leiper, C., Clements, J.B., and Orr, A. (2004). Formation of nuclear foci of the herpes simplex virus type 1 regulatory protein ICP4 at early times of infection: localization, dynamics, recruitment of ICP27, and evidence for the de novo induction of ND10-like complexes. J. Virol. *78*, 1903–1917.

Fernandez-Capetillo, O., Celeste, A., and Nussenzweig, A. (2003a). Focusing on foci: H2AX and the recruitment of DNA-damage response factors. Cell Cycle *2*, 426–427.

Fernandez-Capetillo, O., Mahadevaiah, S.K., Celeste, A., Romanienko, P.J., Camerini-Otero, R.D., Bonner, W.M., Manova, K., Burgoyne, P., and Nussenzweig, A. (2003b). H2AX is required for chromatin remodeling and inactivation of sex chromosomes in male mouse meiosis. Dev. Cell *4*, 497–508.

Fu, X., Wang, H., and Zhang, X. (2002). High-frequency intermolecular homologous recombination during herpes simplex virus-mediated plasmid DNA replication. J. Virol. *76*, 5866–5874.

Haber, J.E. (2000). Partners and pathwaysrepairing a double-strand break. Trends Genet. *16*, 259–264.

Hamatake, R.K., Bifano, M., Hurlburt, W.W., and Tenney, D.J. (1997). A functional interaction of ICP8, the herpes simplex virus single-stranded DNA-binding protein, and the helicase-primase complex that is dependent on the presence of the UL8 subunit. J. Gen. Virol. *78 (Pt 4)*, 857–865.

Havas, K., Flaus, A., Phelan, M., Kingston, R., Wade, P.A., Lilley, D.M., and Owen-Hughes, T. (2000). Generation of superhelical torsion by ATP-dependent chromatin remodeling activities. Cell *103*, 1133–1142.

Hayward, G.S., Jacob, R.J., Wadsworth, S.C., and Roizman, B. (1975). Anatomy of herpes simplex virus DNA: evidence for four populations of molecules that differ in the relative orientations of their long and short components. Proc. Natl. Acad. Sci. USA *72*, 4243–4247.

Hoeijmakers, J.H. (2001). Genome maintenance mechanisms for preventing cancer. Nature *411*, 366–374.

Ishov, A.M., Vladimirova, O.V., and Maul, G.G. (2002). Daxx-mediated accumulation of human cytomegalovirus tegument protein pp71 at ND10 facilitates initiation of viral infection at these nuclear domains. J. Virol. *76*, 7705–7712.

Ishov, A.M., Vladimirova, O.V., and Maul, G.G. (2004). Heterochromatin and ND10 are cell-cycle regulated and phosphorylation-dependent alternate nuclear sites of the transcription repressor Daxx and SWI/SNF protein ATRX. J. Cell Sci. *117*, 3807–3820.

Isler, J.A., and Schaffer, P.A. (2001). Phosphorylation of the herpes simplex virus type 1 origin binding protein. J. Virol. *75*, 628–637.

Jackson, S.A., and DeLuca, N.A. (2003). Relationship of herpes simplex virus genome configuration to productive and persistent infections. Proc. Natl. Acad. Sci. USA *100*, 7871–7876.

Kamita, S.G., Maeda, S., and Hammock, B.D. (2003). High-frequency homologous recombination between baculoviruses involves DNA replication. J. Virol. *77*, 13053–13061.

Kato, M., Ito, T., Wagner, G., and Ellenberger, T. (2004). A molecular handoff between bacteriophage T7 DNA primase and T7 DNA polymerase initiates DNA synthesis. J. Biol. Chem. *279*, 30554–30562.

Kleymann, G., Fischer, R., Betz, U.A., Hendrix, M., Bender, W., Schneider, U., Handke, G., Eckenberg, P., Hewlett, G., Pevzner, V. *et al.* (2002). New helicase-primase inhibitors as drug candidates for the treatment of herpes simplex disease. Nat. Med. *8*, 392–398.

Klinedinst, D.K., and Challberg, M.D. (1994). Helicase-primase complex of herpes simplex virus type 1: a mutation in the UL52 subunit abolishes primase activity. J. Virol. *68*, 3693–3701.

Knipe, D.M. (1989). The role of viral and cellular nuclear proteins in herpes simplex virus replication. Adv. Virus Res. *37*, 85–123.

Kowalczykowski, S.C. (2000). Initiation of genetic recombination and recombination-dependent replication. Trends Biochem. Sci. *25*, 156–165.

Langelier, Y., Bergeron, S., Chabaud, S., Lippens, J., Guilbault, C., Sasseville, A.M., Denis, S., Mosser, D.D., and Massie, B. (2002). The R1 subunit of herpes simplex virus ribonucleotide reductase protects cells against apoptosis at, or upstream of, caspase-8 activation. J. Gen. Virol. *83*, 2779–2789.

Lee, S.S., Dong, Q., Wang, T.S., and Lehman, I.R. (1995). Interaction of herpes simplex virus 1 origin-binding protein with DNA polymerase alpha. Proc. Natl. Acad. Sci. USA *92*, 7882–7886.

Lees-Miller, S.P., Long, M.C., Kilvert, M.A., Lam, V., Rice, S.A., and Spencer, C.A. (1996). Attenuation of DNA-dependent protein kinase activity and its catalytic subunit by the herpes simplex virus type 1 transactivator ICP0. J. Virol. *70*, 7471–7477.

Lehman, I.R., and Boehmer, P.E. (1999). Replication of herpes simplex virus DNA. J. Biol. Chem. *274*, 28059–28062.

Liuzzi, M., Kibler, P., Bousquet, C., Harji, F., Bolger, G., Garneau, M., Lapeyre, N., McCollum, R.S., Faucher, A.M., Simoneau, B., and Cordingley, M.G. (2004). Isolation and characterization of herpes simplex virus type 1 resistant to aminothiazolylphenyl-based inhibitors of the viral helicase-primase. Antiviral Res. *64*, 161–170.

Macao, B., Olsson, M., and Elias, P. (2004). Functional properties of the herpes simplex virus type I origin-binding protein are controlled by precise interactions with the activated form of the origin of DNA replication. J. Biol. Chem. *279*, 29211–29217.

Makhov, A.M., Lee, S.S., Lehman, I.R., and Griffith, J.D. (2003). Origin-specific unwinding of herpes simplex virus 1 DNA by the viral UL9 and ICP8 proteins: visualization of a specific preunwinding complex. Proc. Natl. Acad. Sci. USA *100*, 898–903.

Malik, A.K., Martinez, R., Muncy, L., Carmichael, E.P., and Weller, S.K. (1992). Genetic analysis of the herpes simplex virus type 1 UL9 gene: isolation of a LacZ insertion mutant and expression in eukaryotic cells. Virology *190*, 702–715.

Malik, A.K., Shao, L., Shanley, J.D., and Weller, S.K. (1996). Intracellular localization of the herpes simplex virus type-1 origin binding protein, UL9. Virology *224*, 380–389.

Mapelli, M., Panjikar, S., and Tucker, P.A. (2004). The crystal structure of the HSV-1 ssDNA binding protein suggests the structural basis for flexible, cooperative single-stranded DNA binding. J. Biol. Chem.

Marintcheva, B., and Weller, S.K. (2001). A tale of two HSV-1 helicases: roles of phage and animal virus helicases in DNA replication and recombination. Prog. Nucleic Acid Res. Mol. Biol. *70*, 77–118.

Marintcheva, B., and Weller, S.K. (2003). Existence of transdominant and potentiating mutants of UL9, the herpes simplex virus type 1 origin-binding protein, suggests that levels of UL9 protein may be regulated during infection. J. Virol. *77*, 9639–9651.

Martin, D.W., and Weber, P.C. (1997). DNA replication promotes high-frequency homologous recombination during Autographa californica multiple nuclear polyhedrosis virus infection. Virology *232*, 300–309.

Maul, G.G. (1998). Nuclear domain 10, the site of DNA virus transcription and replication. Bioessays *20*, 660–667.

McNamee, E.E., Taylor, T.J., and Knipe, D.M. (2000). A dominant-negative herpesvirus protein inhibits intranuclear targeting of viral proteins: effects on DNA replication and late gene expression. J. Virol. *74*, 10122–10131.

Mohammadi, E.S., Ketner, E.A., Johns, D.C., and Ketner, G. (2004). Expression of the adenovirus E4 34k oncoprotein inhibits repair of double strand breaks in the cellular genome of a 293-based inducible cell line. Nucleic Acids Res. *32*, 2652–2659.

Murata, L.B., Dodson, M.S., and Hall, J.D. (2004). A human cellular protein activity (OF-1), which binds herpes simplex virus type 1 origin, contains the Ku70/Ku80 heterodimer. J. Virol. *78*, 7839–7842.

Muratani, M., Gerlich, D., Janicki, S.M., Gebhard, M., Eils, R., and Spector, D.L. (2002). Metabolic-energy-dependent movement of PML bodies within the mammalian cell nucleus. Nat. Cell Biol. *4*, 106–110.

Nguyen-Huynh, A.T., and Schaffer, P.A. (1998). Cellular transcription factors enhance herpes simplex virus type 1 oris-dependent DNA replication. J. Virol. *72*, 3635–3645.

Nimonkar, A.V., and Boehmer, P.E. (2002). *In vitro* strand exchange promoted by the herpes simplex virus type-1 single strand DNA-binding protein (ICP8) and DNA helicase-primase. J. Biol. Chem. *277*, 15182–15189.

Nimonkar, A.V., and Boehmer, P.E. (2003). On the mechanism of strand assimilation by the herpes simplex virus type-1 single-strand DNA-binding protein (ICP8). Nucleic Acids Res. *31*, 5275–5281.

Olesky, M., McNamee, E.E., Zhou, C., Taylor, T.J., and Knipe, D.M. (2005). Evidence for a direct interaction between HSV-1 ICP27 and ICP8 proteins. Virology *331*, 94–105.

Owen-Hughes, T., and Bruno, M. (2004). Molecular biology. Breaking the silence. Science *303*, 324–325.

Parkinson, J., Lees-Miller, S.P., and Everett, R.D. (1999). Herpes simplex virus type 1 immediate-early protein vmw110 induces the proteasome-dependent degradation of the catalytic subunit of DNA-dependent protein kinase. J. Virol. *73*, 650–657.

Perkins, D., Pereira, E.F., Gober, M., Yarowsky, P.J., and Aurelian, L. (2002). The herpes simplex virus type 2 R1 protein kinase (ICP10 PK) blocks apoptosis in hippocampal neurons, involving activation of the MEK/MAPK survival pathway. J. Virol. *76*, 1435–1449.

Perkins, D., Pereira, E.F., and Aurelian, L. (2003). The herpes simplex virus type 2 R1 protein kinase (ICP10 PK) functions as a dominant regulator of apoptosis in hippocampal neurons involving activation of the ERK survival pathway and upregulation of the antiapoptotic protein Bag-1. J. Virol. *77*, 1292–1305.

Pilger, B.D., Cui, C., and Coen, D.M. (2004). Identification of a small molecule that inhibits herpes simplex virus DNA Polymerase subunit interactions and viral replication. Chem Biol *11*, 647–654.

Ramirez-Aguilar, K.A., and Kuchta, R.D. (2004). Herpes simplex virus-1 primase: a polymerase with extraordinarily low fidelity. Biochemistry *43*, 9084–9091.

Ramirez-Aguilar, K.A., Low-Nam, N.A., and Kuchta, R.D. (2002). Key role of template sequence for primer synthesis by the herpes simplex virus 1 helicase-primase. Biochemistry *41*, 14569–14579.

Randell, J.C., and Coen, D.M. (2001). Linear diffusion on DNA despite high-affinity binding by a DNA polymerase processivity factor. Mol. Cell. *8*, 911–920.

Randell, J.C., and Coen, D.M. (2004). The herpes simplex virus processivity factor, UL42, binds DNA as a monomer. J. Mol. Biol. *335*, 409–413.

Reuven, N.B., Staire, A.E., Myers, R.S., and Weller, S.K. (2003). The herpes simplex virus type 1 alkaline nuclease and single-stranded DNA binding protein mediate strand exchange *in vitro*. J. Virol. *77*, 7425–7433.

Reuven, N.B., Antoku, S., and Weller, S.K. (2004a). The UL12.5 gene product of herpes simplex virus type 1 exhibits nuclease and strand exchange activities but does not localize to the nucleus. J. Virol. *78*, 4599–4608.

Reuven, N.B., Willcox, S., Griffith, J.D., and Weller, S.K. (2004b). Catalysis of strand exchange by the HSV-1 UL12 and ICP8 proteins: potent ICP8 recombinase activity is revealed upon resection of dsDNA substrate by nuclease. J. Mol. Biol. *342*, 57–71.

Rogakou, E.P., Nieves-Neira, W., Boon, C., Pommier, Y., and Bonner, W.M. (2000a). Initiation of DNA fragmentation during apoptosis induces phosphorylation of H2AX histone at serine 139. J. Biol. Chem. *275*, 9390–9395.

Rogakou, E.P., Redon, C., Boon, C., Johnson, K., and Bonner, W.M. (2000b). Rapid histone extraction for electrophoretic analysis. Biotechniques *28*, 38–40, 42, 46.

Stracker, T.H., Carson, C.T., and Weitzman, M.D. (2002). Adenovirus oncoproteins inactivate the Mre11-Rad50-NBS1 DNA repair complex. Nature *418*, 348–352.

Tang, Q., Li, L., Ishov, A.M., Revol, V., Epstein, A.L., and Maul, G.G. (2003). Determination of minimum herpes simplex virus type 1 components necessary to localize transcriptionally active DNA to ND10. J. Virol. *77*, 5821–5828.

Tanguy Le Gac, N., and Boehmer, P.E. (2002). Activation of the herpes simplex virus type-1 origin-binding protein (UL9) by heat shock proteins. J. Biol. Chem. *277*, 5660–5666.

Taylor, T.J., and Knipe, D.M. (2004). Proteomics of herpes simplex virus replication compartments: association of cellular DNA replication, repair, recombination, and chromatin remodeling proteins with ICP8. J. Virol. *78*, 5856–5866.

Trego, K.S., and Parris, D.S. (2003). Functional interaction between the herpes simplex virus type 1 polymerase processivity factor and origin-binding proteins: enhancement of UL9 helicase activity. J. Virol. *77*, 12646–12659.

Turner, J.M., Mahadevaiah, S.K., Fernandez-Capetillo, O., Nussenzweig, A., Xu, X., Deng, C.X., and Burgoyne, P.S. (2005). Silencing of unsynapsed meiotic chromosomes in the mouse. Nat. Genet. *37*, 41–47.

Walsh, D., and Mohr, I. (2006). Assembly of an active translation initiation factor complex by a viral protein. Genes Dev. *15*, 461–472.

Weller, S.K. (1995). Herpes simplex virus DNA replication and genome maturation. In Implications of the DNA provirus: Howard Temin's Scientific Legacy, G.M. C. B. Sugden, and R. Temin, ed. (Washington, D.C.: ASM Press), pp. pages 189–213.

Wilkinson, D.E., and Weller, S.K. (2003). The role of DNA recombination in herpes simplex virus DNA replication. IUBMB Life *55*, 451–458.

Wilkinson, D.E., and Weller, S.K. (2004). Recruitment of cellular recombination and repair proteins to sites of herpes simplex virus type 1 DNA replication is dependent on the composition of viral proteins within prereplicative sites and correlates with the induction of the DNA damage response. J. Virol. *78*, 4783–4796.

Wu, X., Avni, D., Chiba, T., Yan, F., Zhao, Q., Lin, Y., Heng, H., and Livingston, D. (2004). SV40 T antigen interacts with Nbs1 to disrupt DNA replication control. Genes Dev. *18*, 1305–1316.

Zhou, C., and Knipe, D.M. (2002). Association of herpes simplex virus type 1 ICP8 and ICP27 proteins with cellular RNA polymerase II holoenzyme. J. Virol. *76*, 5893–5904.

Zuccola, H.J., Filman, D.J., Coen, D.M., and Hogle, J.M. (2000). The crystal structure of an unusual processivity factor, herpes simplex virus UL42, bound to the C terminus of its cognate polymerase. Mol. Cell. *5*, 267–278.

6 Hailing the Arrival of the Messenger: Strategies for Translational Control in Herpes Simplex Virus-infected Cells

Ian Mohr

Abstract

Like all viruses, the translation of α-herpesvirus mRNAs relies completely upon components in their host cells. Thus, ensuring that viral mRNAs are able to effectively engage cellular translation factors and compete with cellular mRNAs for access to this equipment is of paramount importance. Notwithstanding the onslaught of a potent cellular antiviral response focused on incapacitating the translation machinery, the significance of viral protein production is further accentuated by viral functions devoted to neutralizing this host retort and preserving the activity of the host translational apparatus. Here, interactions between α-herpesviruses, exemplified by herpes simplex virus-1, with host translation initiation factors are discussed and recent developments concerning how viral functions control translation in infected cells are presented.

Introduction

To produce the polypeptides they encode which orchestrate their growth and development within eukaryotic cells, viruses depend upon the protein synthesis machinery resident in their host. Not only must they utilize this machinery, these obligate intracellular parasites need to have a strategy in place to ensure that their viral mRNAs are effectively translated in an environment filled with cellular mRNAs. While it represents a vital feature of their lifecycle, merely engaging the cellular machinery that produces polypeptides is not enough to sustain viral protein synthesis. To achieve this latter goal, viruses must surmount potent innate host defenses designed to inactivate the translational machinery and curtail viral replication. Failure to successfully exploit limiting amounts of translation initiation factors or fend off efforts of the host to incapacitate critical initiation factors would seriously compromise the production of viral polypeptides and adversely affect viral replication. Given the importance of translational control in the lifecycle of many viruses and the wealth of genetic tools available to investigators, it is no wonder that viruses provide attractive model systems to investigate the regulation of protein synthesis. In this chapter, we explore how herpes simplex virus 1, a representative member of the α-herpesvirus subfamily, interacts with the translational machinery in infected cells, and highlight recent developments concerning how the virus manipulates cellular translation initiation factors.

Regulation of translation and eIF4F

As eIF4F recognizes the 7-methyl guanine cap at the mRNA 5′ end and recruits the 40S ribosomal subunit bound to eIF3, it is a prime target for regulating translation initiation in response to a diverse array of stimuli

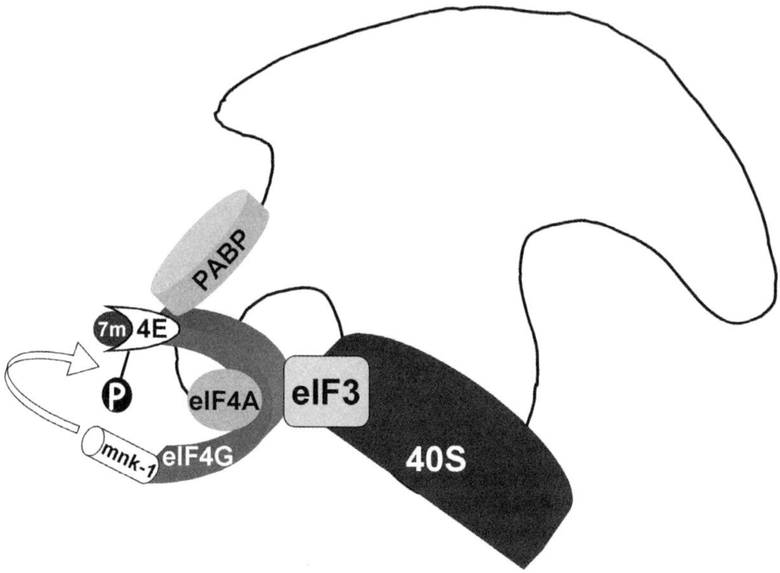

Figure 6.1 The eIF4F complex and associated factors bound to the mRNA. Upon binding the 7-methyl guanine cap (7m) at the 5′ end of the mRNA, eIF4E (4E) recruits eIF4G and eIF4A to assemble the eIF4F complex. The 40S ribosome is recruited through its association with eIF3. eIF4G also interacts with the polyA binding protein bound (PABP) to the 3′ end of the mRNA and the eIF4E kinase mnk-1, which phosphorylates eIF4E (depicted as a P).

including viral infection. Unlike viruses whose mRNAs do not contain a 5′ cap, initiate translation through alternate means, and promote the inactivation of eIF4F, the integrity of eIF4F must be preserved in HSV-1 infected cells since recruitment of this factor to the capped 5′ ends of HSV-1 mRNA is the first step in viral protein synthesis (reviewed in Schneider and Mohr, 2003). The translation initiation factor eIF4F (reviewed in Gingras *et al.*, 1999) is a complex of cellular polypeptides whose core is composed of the cap binding protein (eIF4E), a large scaffolding subunit (eIF4G), and an RNA helicase (eIF4A; Fig. 6.1). Two of three eIF4A isoforms (eIF4AI and II) encoded by discrete genes appear to function equivalently in translation initiation (Yoder Hill *et al.*, 1993; Li *et al.*, 1999), whereas eIF4AIII is required for nonsense mediated mRNA decay (Shibuya *et al.*, 2004). In addition, eIF4B or eIF4H stimulate the helicase activity of eIF4A, functioning as processivity factors for the helicase (reviewed in Rogers *et al.*, 2002). Depending upon their abundance, a family of small eIF4E binding proteins (4E-BPs) can sequester eIF4E in a phosphorylation dependent manner (reviewed in Gingras *et al.*, 2001). In their hypophosphorylated form, 4E-BPs inhibit cap dependent translation. Hyperphosphorylation of the 4E-BPs by the mTOR kinase in response to a variety of environmental cues results in the release of eIF4E, allowing the cap binding protein to associate with eIF4G and assemble an active eIF4F complex (Fig. 6.2; Haghihat *et al.*, 1995; Mader *et al.*, 1995; Marcotrigiano *et al.*, 1999). In addition, the cap binding protein eIF4E is phosphorylated by the eIF4G associated kinase mnk-1 (Fukunaga and Hunter, 1997; Waskiewicz *et al.*, 1997, 1999; Pyrronet *et al.*, 1999; Scheper *et al.*, 2001). While not essential for translation initiation, eIF4E phosphorylation appears to play a regulatory role, although the details of

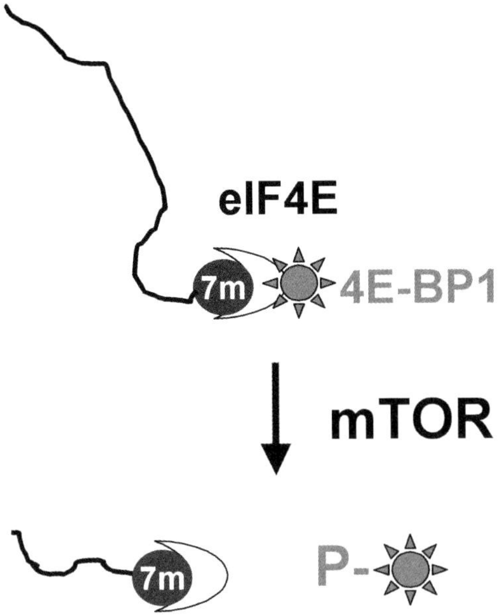

Figure 6.2 The translational repressor 4E-BP1 binds eIF4E and is regulated by phosphorylation. eIF4E (open crescent) bound to the 5′ mRNA cap (7m) associates with the translational repressor 4E-BP1. Upon phosphorylation by the cellular kinase mTOR, 4E-BP1 is released from eIF4E allowing the recruitment of eIF4G and eIF4F complex assembly.

how this is achieved and which mRNAs are most susceptible are poorly understood (reviewed in Scheper and Proud, 2002). Finally, the cellular polyA binding protein, while itself not an eIF4F core component physically interacts with eIF4G in the eIF4F initiation complex (Fig. 6.1; Imataja *et al.*, 1998). This association between the 5′ and 3′ ends of the mRNA generates a circular or "closed loop" topology that is thought to be important for translation (Wells *et al.*, 1998; Kahvejian *et al.*, 2001). It is possible that such interactions stimulate translation and act as a checkpoint, ensuring that the mRNA is both capped and polyadenylated prior to translation initiation. Finally, PABP can be found associated with two cellular polypeptides, Paip1, an eIF4G related protein that stimulates translation (Craig *et al.*, 1998), or Paip2, a translational repressor that prevents PABP from interacting with Paip1 and the mRNA polyA tail (Khaleghpour *et al.*, 2001).

All of these components are tethered in place at the 5′ end of the mRNA through an interaction with eIF4G (Fig. 6.1). In addition to providing a scaffold on which the eIF4F components eIF4E and eIF4A are assembled, eIF4G also binds to the eIF4E kinase mnk-1 and the polyA binding protein PABP (Imataja *et al.*, 1998; Pyronnet *et al.*, 1999). Aside from our perceived role for eIF4G as a molecular scaffold, little is known regarding its function. This is further highlighted by the existence of at least five eIF4GI isoforms, generated by translation initiation at alternate ATG codons, coupled with differential phosphorylation, all of which are of unknown functional significance (Raught et. al, 2000; Byrd *et al.*, 2002). Furthermore, mammalian cells contain a homolog of eIF4GI, eIF4GII, which appears to be selectively recruited to capped mRNA upon differentiation (Caron *et al.*, 2004). Finally, recent studies demonstrating that the N-terminus of eIF4G undergoes a conformational change upon engaging eIF4E suggest that the role of eIF4G may indeed go beyond that of an inert scaffold upon which other initiation factors assemble (Gross *et al.*, 2003).

CAP-turing eIF4F in HSV-1 infected cells

Given the importance of eIF4F in the initiation and regulation of cap-dependent translation, recent attention has focused on evaluating the impact of HSV-1 infection on this complex. Indeed, suppression of host polypeptide synthesis in cells infected with human adenovirus or VSV is mediated by dephosphorylation of the cap binding protein eIF4E (Cuesta *et al.*, 2000; Connor and Lyles, 2002). Since host protein synthesis is profoundly impaired in HSV-1 infected cells, it seemed plausible that a similar mechanism could be operative. In marked contrast to findings in most other viral systems, eIF4E is rapidly phosphorylated following infection of primary human cells or a variety of established cell lines with HSV-1. Phosphorylation occurred early in the viral lifecycle, was sustained throughout the productive growth cycle, and was sensitive to small molecule inhibitors of p38 and mnk, the latter being the cellular eIF4E kinase. More importantly, inhibiting the eIF4E kinase mnk impaired viral protein synthesis and reduced viral replication by 100-fold in primary human cells (Walsh and Mohr, 2004).

Not only does HSV-1 infection promote eIF4E phosphorylation, it also stimulates phosphorylation of the translational inhibitor 4E-BP1 early in its productive replication cycle. Phosphorylation of 4E-BP1 in cells infected with HSV-1 is sensitive to rapamycin, an inhibitor of the cellular kinase mTOR. Moreover, following its phosphorylation, steady state levels of 4E-BP1 were observed to precipitously decline in infected cells. Proteasome inhibitors or rapamycin effectively prevented the decrease in 4E-BP1 abundance. Thus, following its phosphorylation by mTOR, the translational repressor 4E-BP1 is degraded by the proteasome in HSV-1 infected cells (Walsh and Mohr, 2004). As eIF4E and 4E-BP1 phosphorylation have been reported to stimulate translation in other biological settings, this raised the possibility that the changes to eIF4F in HSV-1 infected cells might be required to stimulate the cellular translation machinery.

While the steady-state levels of eIF4F core components eIF4E, eIF4G, and eIF4A remain unchanged in HSV-1 infected cells, phosphorylation of 4E-BP1 relieves translational repression by liberating eIF4E, making it available to associate with eIF4G. Indeed, in growth arrested primary human cells infected with HSV-1, eIF4E is released from the 4E-BP1 repressor and the amount of eIF4G bound to eIF4E increases dramatically (Walsh and Mohr, 2004). This suggests that HSV-1 can promote the assembly of eIF4F complexes in quiescent resting cells. The ability to stimulate eIF4F complex assembly is likely important for productive viral growth in quiescent cells, as the capacity of resting cells to support high levels of protein synthesis are reduced relative to their exponentially growing counterparts (Stanners and Becker, 1971; Thomas and Gordon, 1979; Thomas *et al.*, 1981; Thomas and Thomas, 1986). Strikingly, although the amount of eIF4G bound to eIF4E increases substantially upon infection of quiescent cells with HSV-1, a concomitant increase in the amount of PABP bound to eIF4G was not observed (Walsh *et al.*, 2005). Furthermore, PABP has been found in association with the viral ICP27 protein, although this interaction is markedly diminished upon treatment with ribonuclease (Fontaine-Rodriguez *et al.*, 2004). While it is not a core component of the eIF4F complex, PABP bound to the polyA tail at the 3′ end of the mRNA is known to associate with eIF4G bound at the 5′ end of the mRNA; moreover, this association between eIF4G and PABP is capable of stimulating translation. Presently, the significance of the failure to recruit PABP into eIF4F complexes following HSV-1 infection of quiescent fibroblasts is not clear; however, upon infection of quiescent fibroblasts with HCMV, both eIF4G and PABP are recruited into eIF4E containing complexes (Walsh *et al.*, 2005). As host polypeptide synthesis is not impaired in HCMV infected cells, it is tempting to speculate that the differential association of PABP with eIF4F complexes might somehow play a role in host shut-off. Further work will hopefully illuminate the significance of this finding and either challenge or support the aforementioned hypothesis.

All of the changes to the eIF4F complex that occur in HSV-1 infected cells require viral gene expression (Walsh and Mohr, 2004). However, phosphorylation of eIF4E as well as 4E-BP1 was observed in infected cells treated with PAA, an inhibitor of viral DNA replication that blocks the expression of "true late" or γ_2 genes. Therefore, this class of viral genes is not required for phosphorylation of eIF4E together with 4E-BP1. Consistent with this finding, viruses with mutations in the immediate early ICP0 gene were unable to promote phosphorylation of eIF4E along with 4E-BP1 or stimulate the assembly of eIF4F complexes in quiescent cells (Walsh and Mohr, 2004). ICP0 is a master regulator of viral gene expression and one of the first viral polypeptides produced in productively infected cells (Elshiekh *et al.*, 1991; Lium and Silverstein, 1997). It localizes to nuclear ND-10 domains where it induces the redistribution and subsequent destruction of the cellular PML protein. In addition, ICP0 possesses an intrinsic ubiquitin protein ligase activity and functions as a promiscuous transactivator of viral gene expression (reviewed in Everett, 2000; Hagglund and Roizman. 2004). Although ICP0 is not essential for replication in cultured cells, ICP0 mutant viruses are deficient for growth at low MOIs in cultured cells (Stow and Stow, 1986; Sacks and Schaffer, 1987) and unable to reactivate from latency in a mouse model of infection (Cai *et al.*, 1993; Halford and Schaffer, 2001).

Although genetic analysis has established that the modifications to eIF4F observed in cells productively infected with HSV-1 depend upon the product of the multifunctional regulatory protein encoded by the ICP0 gene, exactly how ICP0 promotes these changes remains to be determined. Given that ICP0 is one of the first proteins produced in infected cells, it might positively regulate expression of one or more downstream genes whose products act directly on kinases upstream of mnk or mTOR to ultimately result in phosphorylation of eIF4E together with 4E-BP1. Likewise, a downstream effector might directly associate with one or more components of eIF4F to promote complex assembly. Alternatively, it is certainly possible that ICP0 acts directly and possesses an intrinsic ability to trigger the observed modifications of eIF4F. These may involve new, uncharacterized activities of this highly multifunctional protein or targeting known activities to previously unidentified substrates. Finally, it is important to keep in mind that models where ICP0 acts directly or indirectly are in no way mutually exclusive. For example, the product of an ICP0 responsive downstream gene might play a critical role to promote eIF4F complex assembly by associating with an eIF4F core component while ICP0 acts directly to promote inactivation of 4E-BP1. Once the players responsible for these changes have been identified, the exciting work to determine their mechanism of action can begin. Such mechanistic insights into how viral factors promote eIF4F complex assembly may not only have relevance for understanding the productive growth cycle of α-herpesviruses, but may contribute to our understanding of how translational is regulated during the lifecycle of related β and γ herpesviruses. Although the members of the herpesvirus family are certainly diverse, it is possible that the basic function of stimulating the cellular translation machinery has been conserved across subfamily lines. Indeed, eIF4F complex assembly along with eIF4E and 4E-BP1 phosphorylation are observed in HCMV infected cells; in addition, inhibiting the eIF4E kinase mnk reduces viral replication by 300-fold (Kudchodkar *et al.*, 2004; Walsh *et al.*, 2005). While promoting assembly of eIF4F complexes along with phosphorylation of eIF4E and 4E-BP1 is important for vegetative viral replication, herpesvirus biology raises additional considerations, as these viruses have adapted to colonize a variety of terminally differentiated cells (reviewed in Roizman and Knipe, 2001; Mocarski and Courcelle, 2001; Keiff and Rickinson, 2001). In many cases, cells latently infected with a herpesvirus produce either small or undetectable levels of viral polypeptides (reviewed in Roizman and Knipe, 2001; Mocarski and Courcelle, 2001). It is therefore conceivable that the switch to a vegetative mode of growth requiring abundant translation of viral gene products necessitates a mechanism to mobilize the translational machinery in quiescent cells.

Shutting down the host . . . albeit selectively

The preceding section reviewed evidence that HSV-1 can actually stimulate the host's translational machinery upon infection. How then is this reconciled with the observation that host polypeptide synthesis is seriously impaired in HSV-1 infected cells? Indeed, during its productive growth cycle, HSV-1 exerts a powerful effect on host macromolecular metabolism. Using multiple tactics requiring the action of several viral gene products, HSV-1 is able to suppress the translation of most host polypeptides. This process begins almost immediately after the viral tegument proteins are deposited into the host cell cytoplasm. Upon liberation from the tegument compartment in the virion, vhs, a FEN-1 related nuclease, enhances mRNA turnover (Everly *et al.*, 2002). *In vitro*, vhs possesses endonuclease activity and promotes degradation of mRNA substrates in the 5′ to 3′ direction (Elgadi *et al.* 1999; Karr and Read, 1999; Perez-Parada *et al.*, 2004). Vhs is specifically targeted to mRNA through an interaction with cellular translation initiation factors eIF4AII (G.S. Read, personal communication) together with eIF4H and eIF4B (Feng *et al.*, 2001; Doepker at. al., 2004; G.S. Read, personal communication). Moreover, vhs has recently been found tethered to at least a subset of eIF4F cap-binding complexes (G.S. Read, personal communication). Importantly, vhs is not selective for cellular mRNAs as opposed to viral mRNAs. Indeed, both cellular and viral mRNAs are substrates for vhs *in vitro* and *in vivo* (Kwong and Frenkel, 1987). Although not essential for growth in cultured cells (Read and Frenkel, 1983), host shut-off is delayed and the transition between different kinetic classes of viral mRNAs, such as IE (or α) to DE (or β), is not as discrete in cells infected with a vhs mutant virus (Kwong and Frenkel 1987; Oroskar and Read, 1987). The observed effects on viral and cellular mRNA stability support the contention that vhs accelerates the turnover of both viral and host mRNAs. Finally, vhs is also important for pathogenesis, as vhs mutants are attenuated in animal models (Strelow and Leib, 1995; Smith *et al.*, 2002). It has been suggested that enhanced turnover of host mRNAs important for innate responses, such as those encoding interferons, underlies the reduced virulence of these mutants (Leib *et al.*, 1999; Murphy *et al.*, 2003; Duerst *et al.*, 2004).

If vhs does not display any overt preference for cellular mRNAs, how then is host translation selectively silenced? While the mechanism underlying host shutoff is not completely understood, it clearly involves multiple viral components. Unlike vhs, which is carried in the virion, ICP27 is the product of an IE gene. In addition to its role promoting export of viral mRNA out of the nucleus late in the productive growth cycle (Soliman *et al.*, 1997), the essential, multifunctional ICP27 protein also inhibits mRNA splicing at early times (Hardy and Sandri-Golding, 1994). As most cellular mRNAs are spliced, while only 4 of the 84 known HSV-1 genes contain introns, ICP27 has a profound effect on the proper maturation of host mRNAs and in turn their subsequent translation. (Sciabica *et al.*, 2003; reviewed in Sandri-Goldin, 2004). In addition to inhibiting host mRNA processing, within the first 4 hours of infection transcription of host genes is reduced to levels less than 50% of those measured in uninfected cells (Preston and Newton, 1976). This may, perhaps in part, reflect the recruitment of the cellular transcription machinery to viral promoters along with modification of the cellular RNA pol II holoenzyme by viral IE gene products (Spencer *et al.*, 1997; Zhou and Knipe, 2002). Thus, while vhs enhances the turnover of all mRNAs indiscriminately, the inhibition of splicing together with the global reduction in cellular gene transcription ensures that the supply of host mRNA in the cytosol is not replenished. This stands in sharp contrast to the high level of transcripts being produced from viral genes. Finally, there may also be yet another uncharacterized factor that contributes to host shut-off whose function is to foster progressive movement of mRNAs from large polysomes to smaller ones (Laurent *et al.*, 1998).

Unlike Adenovirus, whose late mRNAs all contain a common sequence element termed the tri-partite leader found at their 5′ end that promotes their translation (Xi *et al.*, 2004), shared sequences at 5′ or 3′ end of HSV-1 mRNAs have not been identified in existing 5′ and 3′ untranslated regions. It is possible, however, given the GC rich nature of the HSV-1 genome, that the 5′ end of viral mRNAs are capable of adopting a related conformation and this structure might aid in recruiting ribosomes. Alternatively, the efficient translation of viral mRNAs in vhs mutants may simply be accounted for by the sheer abundance of viral transcripts against a pool of cellular mRNAs that are not being replenished.

Recently, it has become clear that while the inhibition of host mRNA translation is potent, it is not absolute as select cellular mRNAs are translated in HSV-1 infected cells. Further investigation will determine if these mRNAs are either important components of the host response or somehow contribute to viral replication. (Esclatine *et al.*, 2004).

eIF2α phosphorylation and translational control

While impairing host polypeptide synthesis and activating eIF4F in resting cells play an important role in enabling the production of viral proteins, these events in and of themselves are not sufficient to ensure sustained protein production throughout the course of the productive growth cycle. In particular, an innate host response designed to incapacitate the translation initiation factor eIF2 must be effectively thwarted.

The capacity of eukaryotic cells to respond to numerous forms of stress, ranging from thermal intolerance or nutrient deprivation to viral infection, involves, in part, their ability to regulate the activity of eIF2, a critical translation initiation factor. Upon phosphorylation on its α subunit by one of four known eIF2α kinases, eIF2 is inactivated and translation is inhibited. Composed of three subunits, eIF2 is a G-protein which functions to bring the initiator methionine tRNA to the 40S ribosomal subunit (Fig. 6.1A; reviewed in Dever, 2002). The association of eIF2 with the initiator tRNA and GTP forms what is known as the ternary complex, which subsequently engages eIF3 and eIF1A bound to a 40S ribosome giving rise to a 43S complex (Fig. 6.3A). After the 43S complex is loaded onto the 5′ end of the mRNA through an interaction between eIF3 and eIF4G, a component of the eIF4F cap-binding complex, the 40S ribosome along with eIF2, eIF3 and the RNA helicase eIF4A proceed to locate the AUG initiation codon through a unidirectional translocation process referred to as scanning. Recognition of the AUG codon is accompanied by the hydrolysis of GTP bound to eIF2 in a reaction that requires the GTPase activating factor eIF5 and is coupled to both 60S subunit joining and the release of inactive eIF2 bound to GDP (Fig. 6.3A). Translation elongation directed by the nascent 80S ribosome proceeds, while eIF2•GDP is recycled into its active GTP bound form through the actions of eIF2B, a guanine nucleotide exchange factor (Fig. 6.3A). Phosphorylation of eIF2 on its α subunit, however, stabilizes the interaction between eIF2B and eIF2 and prevents the exchange of GTP for GDP (Fig. 6.3B). Moreover, as eIF2B is present in limiting amounts, small changes in eIF2α phosphorylation can exert large effects on overall translation rates by sequestering available supplies of eIF2B.

Inactivation of translation initiation factors by host defenses is antagonized by viral functions

Not surprisingly, host innate defenses frequently target cellular translation initiation factors hoping to prevent viral protein synthesis, thus prematurely terminating the viral lifecycle and impeding the progress of the infection (reviewed in Schneider and Mohr, 2003). Copious quantities of dsRNA or RNA with extensive secondary or tertiary structure produced in virus-infected cells are thought to be responsible for alerting the cell to the viral intruder (reviewed in Pe'ery and Mathews, 2000). Some viral genomes are composed of dsRNA, while others such as single strand RNA viruses form dsRNA as replicative intermediates. However, dsDNA viruses, including HSV-1, in their efforts to accommodate large blocks of genetic information in limited spaces, contain overlapping open reading frames on opposite DNA strands that produce mRNA transcripts capable of forming dsRNA (Jacquemont and Roizman, 1975). Abundant viral dsRNA activates PKR, a cellular eIF2α kinase whose synthesis is upregulated by interferon α/β (reviewed by Kaufman, 2000). In addition to its kinase domain, PKR contains two canonical double-stranded RNA (dsRNA) binding domains. In the presence of dsRNA, two PKR molecules bind the same segment of dsRNA and each molecule phosphorylates the other, generating an activated kinase capable of phosphorylating eIF2α. Finally, the cellular protein PACT, which also binds to dsRNA, can activate PKR in a dsRNA independent fashion, resulting in apoptosis (Patel and Sen, 1998).

Unchecked, PKR activated by dsRNA in virus-infected cells would effectively deplete active eIF2, inhibit ternary complex formation, and prevent viral and cellular protein synthesis. Should cells initially infected succeed in inhibiting translation, the viral invader would effectively be stopped in its tracks, denied access to the cellular translational apparatus it needs to complete its life cycle. This arm of the innate host response then is designed to sacrifice the initially infected cells for the benefit of the larger population. However, numerous viruses, including HSV-1, have captured a variety of functions to counter this cellular response. This struggle for control of

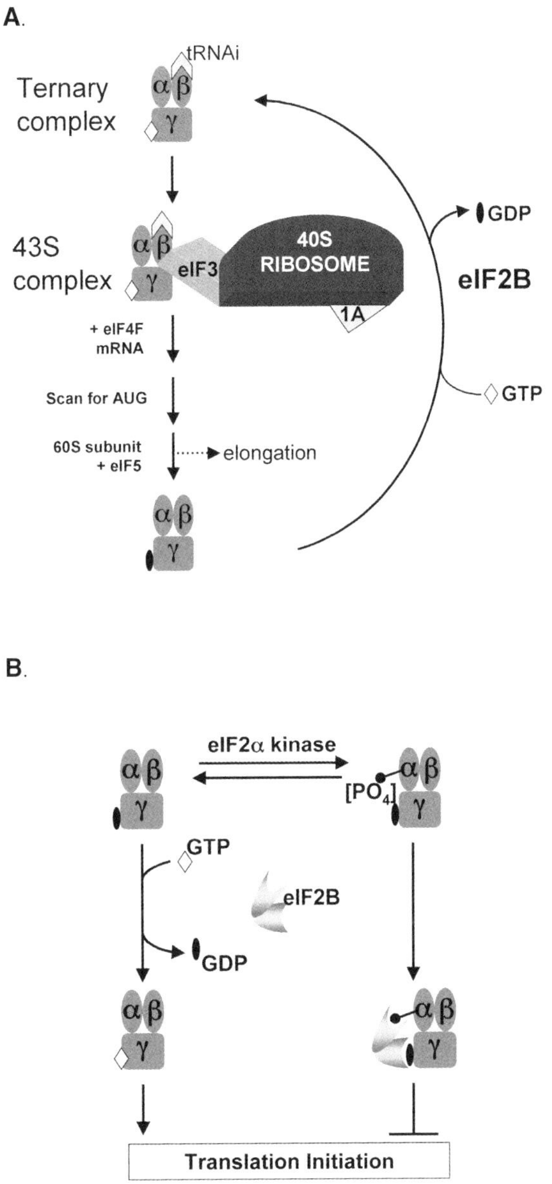

Figure 6.3 Regulation of translation by phosphorylation of eIF2, a critical translation initiation factor. A. Composed of α,β, and γ subunits, eukaryotic translation initiation factor 2 (eIF2) forms a ternary complex with GTP and the initiator tRNA (tRNAi). This complex associates with the 40S ribosomal subunit bound to eIF3, and recognizes the 5′ end of the mRNA through an association with eIF4F. Once the AUG codon in the mRNA has been identified by a unidirectional translocation process termed scanning, GTP hydroylsis stimulated by eIF5 and the subsequent release of the eIF2•GDP complex facilitate the joining of the 60S ribosome subunit and translation elongation commences. The guanine nucleotide exchange factor eIF2B is required to exchange the GDP bound to eIF2 and replace it with GTP, thus recycling the active form of eIF2. B. After phosphorylation of eIF2 on its alpha subunit by PKR, an eIF2α kinase, eIF2B remains tightly bound and cannot exchange the GDP bound to eIF2 for GTP. This failure to recycle eIF2 to its active, GTP bound form inhibits the initiation of translation.

the translational machinery is often an integral component of viral pathogenesis (Leib *et al*, 1999,2000; reviewed in Mohr, 2003).

Circumventing disaster: keeping eIF2 functioning in HSV-1-infected cells

To date, two viral gene products, acting at different stages of the HSV-1 lifecycle, that facilitate the abundant synthesis of viral proteins by preventing the cellular antiviral response from inactivating the translation initiation factor eIF2 have been described. The first of these was the product of the $\gamma_1$34.5 gene.

After initial reports established the $\gamma_1$34.5 gene was a major HSV-1 neurovirulence determinant and nonessential for growth in cultured monkey kidney cells (Chou *et al.*, 1990; Maclean *et al.*, 1991; Bolovan et. al, 1994), further investigation revealed that $\gamma_1$34.5 mutants actually behaved like classical viral host range mutants, exhibiting restricted growth in some lines of cultured cells but not others. Thus, while a standard line of monkey kidney cells were permissive or supported the replication $\gamma_1$34.5 mutants, many human tumor cells were in fact nonpermissive, or did not support the growth of $\gamma_1$34.5 mutant derivatives (Chou *et al.*, 1992). Upon infection of a nonpermissive human tumor cell with a $\gamma_1$34.5 mutant strain, all of the events in the viral lifecycle proceeded normally up to and including viral DNA replication and the accumulation of γ_2 late mRNA transcripts. These viral late mRNAs encoding key structural proteins required to complete the viral lifecycle and assemble the next generation of viral progeny, however, were never translated due to a block at the level of protein synthesis, effectively interrupting the viral lifecycle prior to the assembly and release of viral particles (Chou *et al.* 1992). Subsequent biochemical analysis demonstrated that the $\gamma_1$34.5 gene product prevents the accumulation of phosphorylated eIF2α by recruiting a cellular phosphatase, protein phosphatase 1α (PP1α), to remove phosphate from eIF2α (He *et al.*, 1997). Interestingly, the domain of the $\gamma_1$34.5 protein that contains this activity is homologous to a domain in the GADD34 protein, a cellular PP1α binding protein that promotes eIF2α dephosphorylation in response to other forms of cell stress (Chou and Roizman, 1994). Besides their defect in protein synthesis, $\gamma_1$34.5 mutant derivatives are hypersensitive to interferon α, and therefore more sensitive to this arm of innate host defenses that serve to limit viral replication in the host (Cheng *et al.*, 2001; Cerveny *et al.*, 2003). This hypothesis is supported by reports demonstrating that $\gamma_1$34.5 mutant viruses, while neuroattenuated in normal mice and mice with deficiencies in their acquired immune response, exhibit restored neurovirulence in mice with deficiencies in innate immunity (Leib *et al.*, 1999, 2000).

Ensuing genetic studies revealed that HSV-1 actually encodes multiple functions to control eIF2α phosphorylation, as the dsRNA binding protein specified by the true late Us11 gene prevents PKR activation (Mohr and Gluzman, 1996; Mulvey *et al.*, 1999; Poppers *et al.*, 2000; Khoo *et al.*, 2002). Analysis of a panel of $\gamma_1$34.5 and Us11 mutants established that both Us11 and $\gamma_1$34.5 gene products act at different times in the productive growth cycle to regulate eIF2α phosphorylation in infected cells (Fig. 6.4; Mulvey *et al.*, 2003). Importantly, Us11 expressed in its natural context as a late γ_2 protein is required to properly regulate PKR activation, eIF2α phosphorylation and viral translation (Mulvey *et al.*, 2003). Thus, since late proteins are not produced in nonpermissive tumor cells infected with a $\gamma_1$34.5 mutant virus, the $\gamma_1$34.5 mutant is actually doubly deficient in that it also fails to translate the γ_2 Us11 mRNA (Chou *et al.*, 1992). Moreover, the interferon sensitivity of $\gamma_1$34.5 mutants, previously attributed solely to the absence of

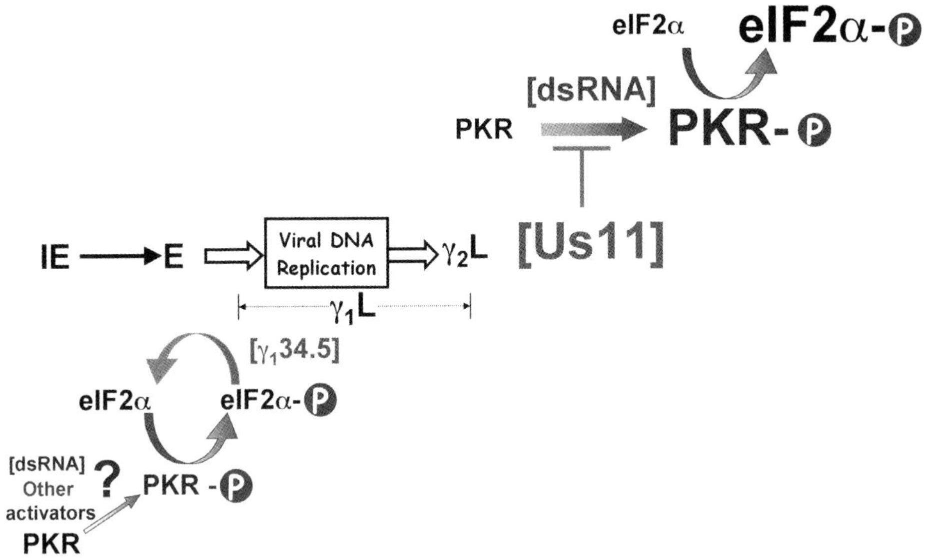

Figure 6.4 Regulation of eIF2α phosphorylation by different functions that act during discrete phases in the HSV-1 lifecycle. Early in the HSV-1 lifecycle, small quantities of dsRNA, or perhaps other effectors that remain to be identified activate the normally dormant cellular PKR kinase. After assembling a dimer of PKR on dsRNA, each subunit of the multimer phosphorylates the other (PKR-P). Since PKR is activated in cells infected with a $\gamma_1$34.5 mutant virus, preventing the translation of γ_2 mRNAs, we propose that the $\gamma_1$34.5 gene product, through its interaction with PP1α, is able to adequately dephosphorylate the quantities of phosphorylated eIF2α (eIF2α-P) produced from the period preceding the initiation of viral DNA synthesis and extending into the initial segment of the late phase (designated as γ_1L). However, late in the viral lifecycle, synthesis of mRNA from γ_2 genes (γ_2 L), many of which are transcribed from ORFs located on opposing DNA strands, results in the production of large quantities of viral dsRNA. In the absence of Us11, the increase in dsRNA concentration generates more activated PKR, which in turn phosphorylates eIF2α. The concentration of phosphorylated eIF2α quickly rises beyond the capacity of the $\gamma_1$34.5–PP1α complex to effectively reverse the reaction, accounting for the observed reduction in viral translation rates in cells infected with a Us11 mutant virus. We suggest that while the $\gamma_1$34.5 protein acts downstream of phosphorylated eIF2α and therefore has the potential to counter a variety of eIF2α kinases, Us11 acts late in infection to specifically antagonize PKR activation in response to the copious levels of dsRNA produced in virus infected cells. In this drawing, relative concentrations of PKR-P, dsRNA, and eIF2α-P at earlier, compared with later times in the viral lifecycle, are represented by character size. (reprinted from Mulvey *et al.*, 2003 with permission.)

$\gamma_1$34.5 function, likewise results from the absence of the $\gamma_1$34.5 protein and the failure to synthesize the Us11 polypeptide (Mulvey *et al.*, 2004).

As discussed in the preceding section, HSV-1 employs multiple tactics to prevent the inactivation of eIF2 by PKR, an important component of innate host defenses; however, at least three other kinases, distinct from PKR, with the capability to phosphorylate eIF2α have been identified in mammalian cells (reviewed in Dever, 2002). This family of kinases, by virtue of their intrinsic ability to monitor intracellular conditions, directly transduces stress signals to the translational machinery and thereby globally regulates rates of protein synthesis (Fig. 6.5). By virtue of the fact that the $\gamma_1$34.5 gene product acts downstream of eIF2α phosphorylation, it can reverse the consequences of an activated

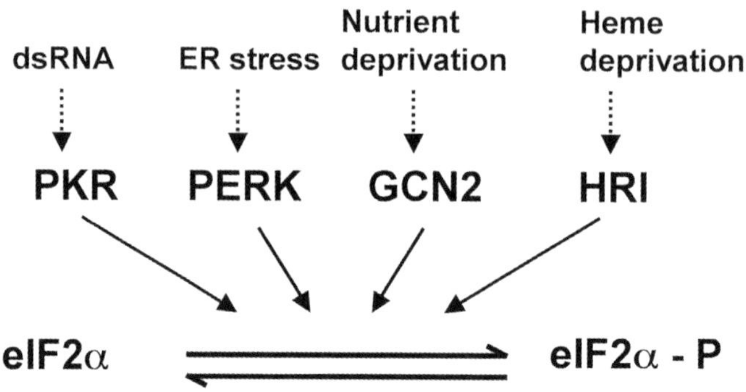

Figure 6.5 Activation of multiple eIF2α kinases by discrete stimuli. The four known eIF2α kinases share a common organization in that each has a kinase domain fused to a ligand binding domain and remains catalytically silent until encountering its respective activating ligand. While the accumulation of dsRNA activates PKR, reduced levels of heme induce HRI activity, nutrient deprivation or ultraviolet light exposure triggers GCN2, and the accrual of unfolded proteins in the ER lumen switches on PERK.

eIF2α kinase and should therefore be able to counter the effects of a broad range of eIF2α kinases. However, it is important to keep in mind that despite producing wild type levels of the $\gamma_1$34.5 protein, Us11 mutant viruses do not synthesize proteins at wild type rates and are sensitive to interferon α (Mulvey *et al.* 2003, 2004). Thus, even though they express the $\gamma_1$34.5 phosphatase component, an inhibitor of an upstream kinase is still required for normal rates of translation in HSV-1 infected cells, implying that recruiting PP1α may not be adequate to prevent the accumulation of phosphorylated eIF2α in HSV-1 infected cells. It is intriguing to speculate that just as the HSV-1 Us11 protein specifically antagonizes the dsRNA activated eIF2α kinase PKR and works in conjunction with the $\gamma_1$34.5 protein to preserve supplies of active eIF2, heretofore unidentified HSV-1 genes might act to target eIF2α kinases other than PKR to specifically prevent the accumulation of phosphorylated eIF2 in response to a diverse array of potential stresses.

Epilogue

Clearly, HSV-1 goes to great lengths to commandeer the translational machinery within its host cell and preserve its activity. To recruit 40S ribosomes to the 5′ end of its capped mRNAs (Barkowski and Roizman, 1976), HSV-1 must successfully engage the translation initiation factor eIF4F, while simultaneously preserving the activity of eIF2, a target of host innate defenses. Anything short of success in either of these endeavors will undoubtedly compromise the production of viral proteins and in turn, the subsequent assembly of viral progeny. Since the translation of viral proteins is important for viral replication and pathogenesis, an understanding of these pathways and how they function may expose a new Achilles' heel in herpesvirus biology that may one day be exploited to develop novel antiviral therapies capable of limiting vegetative growth and/or suppressing reactivation. Finally, continued exploration of how protein synthesis is regulated in HSV-1 infected cells will provide further insight into translational control strategies, a basic biological regulatory mechanism. While the tactics discussed herein may appear on the surface to be confined to cells infected with HSV-1, they may ultimately prove to be pivotal in our efforts to understand cellular translational control mechanisms in health and disease, and exactly how these mechanisms are manipulated so effectively by viral pathogens.

Acknowledgments

I would like to thank all the members of my laboratory (Derek Walsh, Matthew Mulvey, Cesar Perez, Joanna Notary, Carolina Arias) for many stimulating discussions, Derek Walsh for critically reading the manuscript, Sully Read for sharing unpublished data, and the NIH for their financial support of our research. Finally, I apologize to those investigators whose work could not be cited in this chapter due to space limitations.

References

Bartkowski, M. and Roizman, B. (1976). RNA synthesis in cells infected with herpes simplex virus XIII. Differences in the methylation patterns of viral RNA during the reproductive cycle. J. Virol. *20*, 583–588.

Bolovan, C.A., Sawtell, N.M., and Thompson, R.L. (1994). ICP34.5 mutants of herpes simplex virus type 1 strain 17syn+ are attenuated for neurovirulence in mice and for replication in confluent primary mouse embryo cell cultures. J. Virol., *68*, 48–55.

Byrd, M.P., Zamora, M., and Lloyd, R.E. (2002). Generation of multiple isoforms of eukaryotic translation initiation factor 4GI by use of alternate translation initiation codons. Mol. Cell. Biol. *22*, 4499–4511.

Cai, W., Astor, T.L., Liptak, L.M., Cho, C., Coen, D.M., and Schaffer, P.A. (1993). The herpes simplex virus type 1 regulatory protein ICP0 enhances virus replication during acute infection and reactivation from latency. J. Virol. *67*,7501–7512.

Caron, S., Charon, M., Cramer, E., Sonenberg, N., and Dusanter-Fourt, I. (2004), Selective modification of eukaryotic initiation factor 4F (eIF4F) at the onset of cell differentiation: recruitment of eIF4GII and long lasting phosphorylation of eIF4E. Mol. Cell. Biol. *24*, 4920–4928.

Cerveny, M., Hessefort, S., Yang, K., Cheng, G., Gross, M., and He, B. (2003). Amino acid substitutions in the effector domain of the $\gamma_1$34.5 protein of herpes simplex virus 1 have differential effects on viral response to interferon-α. Virology *307*, 290 -300.

Cheng, G., Brett, M.E., and He, B. (2001) Val193 and Phe195 of the $\gamma_1$34.5 protein of herpes simplex virus 1 are required for viral resistance to interferon α/β. Virology *290*, 115–120.

Chou, J. and Roizman, B. (1992). The γ34.5 gene of herpes simplex virus 1 precludes neuroblastoma cells from triggering total shutoff of protein synthesis characteristic of programmed cell death in neuronal cells. Proc. Natl. Acad. Sci USA *89*, 3266–3270.

Chou, J. and Roizman, B. (1994). Herpes simplex virus 1 gamma(1)34.5 gene function, which blocks the host response to infection, maps in the homologous domain of the genes expressed during growth arrest and DNA damage. Proc. Natl. Acad. Sci.USA *91*, 5247–55251.

Chou, J, Kern, E.R., Whitley, R.J., and Roizman, B. (1990). Mapping of herpes simplex virus-1 neurovirulence to gamma (1) 34.5, a gene nonessential for growth in culture. Science *250*,1262–1266.

Connor, J.H. and Lyles, D.S. (2002). Vesicular Stomatitis Virus infection alters the eIF4F translation initiation complex and causes dephosphorylation of the eIF4E binding protein 4E-BP1. J. Virol. *76*, 10177–19187.

Craig, A.W., Haghighat, A., Yu, A.T., and Sonenberg, N. (1998). Interaction of polyadenylate binding protein with the eIF4G homologue PAIP enhances translation. Nature *392*, 520–523.

Cuesta, R., Xi, Q., and Schneider, R.J. (2000). Adenovirus-specific translation by displacement of kinase Mnk1 from cap-initiation complex eIF4F. EMBO J. *19*,3465–3474.

Dever, T.E. (2002). Gene-specific regulation by general translation factors. Cell *108*, 545–556.

Doepker, R.C., Hsu, W-L, Saffran, H.A., and Smiley, J.R. (2004). Herpes simplex virus virion host shutoff protein is stimulated by translation initiation factors eIF4B and eIF4H. J. Virol. *78*, 4684–4699.

Duerst, R.J. and Morrison, L.A. (2004). Herpes simplex virus 2 virion host shutoff protein interferes with type I interferon production and responsiveness. Virology *322*,158–167.

Elgadi, M.M., Hayes, C.E., and Smiley, J.R. (1999). The herpes simplex virus vhs protein induces endoribonucleolytic cleavage of target RNAs in cell extracts. J. Virol. *73*, 7153–7164.

Elshiekh, N.A., Harris-Hamilton, E., and Bachenheimer, S.L. (1991). Differential dependence of herpes simplex virus immediate-early gene expression on de novo-infected cell protein synthesis. J. Virol. *65*, 6430–6437.

Esclatine, A., Taddeo, B., Evans, L., and Roizman, B. (2004). The herpes simplex virus 1 UL41 gene-dependent destabilization of cellular RNAs is selective and may be sequence specific. Proc. Natl. Acad. Sci. USA *101*, 3603–3608.

Everett, R.D. (2000). ICP0, a regulator of herpes simplex virus during lytic and latent infection. Bioessays 22, 761–770.

Everly, D.N. Jr., Feng, P., Mian, I.S., and Read, G.S. (2002). mRNA degradation by the virion host shutoff (vhs) protein of herpes simplex virus: genetic and biochemical evidence that Vhs is a nuclease. J. Virol. 76, 8560–8571.

Feng, P., Everly JR, D.N., and Read, G.S. (2001). mRNA decay during herpesvirus infections: interaction between a putative viral nuclease and a cellular translation factor. J. Virol. 75, 10272–10280.

Fontaine-Rodriguez, E.C., Taylor, T.J., Olesky, M., and Knipe, D.M. (2004). Proteomics of herpes simplex virus infected cell protein 27: association with translation initiation factors. Virology. 330, 487–492.

Fukunaga, R. and Hunter, T. (1997). MNK1, a new MAP kinase-activated protein kinase, isolated by a novel expression screening method for identifying protein kinase substrates. EMBO J. 16, 1921–1933.

Gingras, A.-C., Raught, B., and Sonenberg, N. (1999). eIF4 initiation factors: effectors of mRNA recruitment to ribosomes and regulators of translation. Annu. Rev. Biochem. 68, 913–963.

Gingras, A.-C., Raught, B., and Sonenberg, N. (2001). Regulation of translation initiation by FRAP/mTOR. Genes Dev. 15, 807–826.

Gross, JD, Moerke, NJ, von der Haar, T, Lugovskoy, AA, Sachs, AB, McCarthy, JE, and Wagner, G. (2003). Ribosome loading onto the mRNA cap is driven by conformational coupling between eIF4G and eIF4E. Cell. 115, 739–750.

Hagglund, R. and Roizman, B. (2004). Role of ICP0 in the strategy of conquest of the host cell by herpes simplex virus 1. J. Virol. 78, 2169–2178.

Haghighat A., Mader, S., Pause, A., and Sonenberg, N. (1995). Repression of cap-dependent translation by 4E-binding protein 1: competition with p220 for binding to eukaryotic initiation factor-4E. EMBO J. 14, 5701–5709.

Halford, W.P. and Schaffer, P.A. (2001). ICP0 is required for efficient reactivation of herpes simplex virus type 1 from neuronal latency. J. Virol. 75, 3240–3249.

Hardy, W.R., and Sandri-Goldin, R.M. (1994). Herpes simplex virus inhibits host cell splicing and regulatory protein ICP27 is required for this effect. J. Virol. 68, 7790–7799.

He, B., Gross, M., and Roizman, B. (1997). The gamma(1)34.5 protein of herpes simplex virus 1 complexes with protein phosphatase 1 alpha to dephosphorylate the alpha subunit of eukaryotic initiation factor 2 and preclude the shutoff of protein synthesis by double-stranded RNA-activated protein kinase. Proc. Natl. Acad. Sci. USA 94, 843–848.

Imataja, H., Gradi, A., and Sonenberg, N. (1998). A newly identified N-terminal amino acid sequence of human eIF4G binds poly(A)-binding protein and functions in poly(A)-dependent translation. EMBO J. 17, 7480–7489.

Jacquemont, B., and Roizman, B. (1975). RNA synthesis in cells infected with herpes simplex virus. X. Properties of viral symmetric transcripts and of double-stranded RNA prepared from them. J. Virol. 15, 707–713.

Kahvejian, A., Roy, G., and Sonenberg, N. (2001). The mRNA closed-loop model: the function of PABP and PABP-interacting proteins in mRNA translation. Cold Spring Harb. Symp. Quant. Biol. 66, 293–300.

Karr, B.M. and Read, G.S. (1999). The virion host shutoff function of herpes simplex virus degrades the 5′ end of a target mRNA before the 3′ end. Virology 264, 195–204.

Kaufman, R.J. (2000). Double-stranded RNA-activated protein kinase PKR. In: Translational control, N. Sonenberg, J.W. B. Hershey, and M. B. Mathews, eds. (Cold Spring Harbor, N.Y.; Cold Spring Harbor Laboratory Press), pp. 503–528.

Khaleghpour, K., Svitkin, Y.V., Craig, A.W., DeMaria, C.T., Deo, R.C., Burley, S.K., and Sonenberg, N. (2001). Translational repression by a novel partner of human poly(A) binding protein. Mol. Cell 7, 205–216.

Khoo, D., Perez, C., and Mohr, I. (2002). Characterization of RNA determinants recognized by the arginine- and proline-rich region of Us11, a herpes simplex virus type 1-encoded double-stranded RNA binding protein that prevents PKR activation. J. Virol. 76, 11971–11981.

Kieff, E., and Rickinson, A.B. (2001). Epstein-Barr virus and its replication. In: Fields virology, 4th edn, D.M. Knipe and P. M. Howley, eds. (Philadelphia, Pa; Lippincott/The Williams & Wilkins Co.), pp. 2511–2574

Kudchodkar, S.B., Yu, Y., Maguire, T.G., and Alwine, J.C. (2004).Human cytomegalovirus infection induces rapamycin-insensitive phosphorylation of downstream effectors of mTOR kinase. J. Virol. 78, 11030–11039.

Kwong, A.D. and Frenkel, N. (1987). Herpes simplex virus–infected cells contain a function(s) that destabilizes both host and viral mRNAs. Proc. Natl. Acad. Sci. USA 84, 1926–1930.

Laurent, A.M., Madjar, J.J., and Greco, A. (1998). Translational control of viral and host protein synthesis during the course of herpes simplex virus type 1 infection: evidence that initiation of translation is the limiting step. J. Gen. Virol. 79, 2765–2775.

Leib, D.A., Harrison, T.E., Laslo, K.M., Machalek, M.A., Moorman, N.J., and Virgin, H.W. (1999). Interferons regulate the phenotype of wild-type and mutant herpes simplex viruses *in vivo*. J. Exp. Med., *189*, 663–672.

Leib, D.A., Machalek, M.A., Williams, B.R., Silverman, R.H., and Virgin, H.W. (2000). Specific phenotypic restoration of an attenuated virus by knockout of a host resistance gene. Proc. Natl. Acad. Sci. USA *97*, 6097–6101.

Li, Q., Imataka, H., Morino, S., Rogers, G.W. Jr., Richter-Cook, N.J., Merrick, W.C., and Sonenberg, N. (1999). Eukaryotic translation initiation factor 4AIII (eIF4AIII) is functionally distinct from eIF4AI and eIF4AII. Mol. Cell. Biol. *19*, 7336–7346.

Lium, E.K., and Silverstein, S.J. (1997). Mutational analysis of the herpes simplex virus type 1 ICP0 C_3HC_4 zinc ring finger reveals a requirement for ICP0 in the expression of the essential α27 gene. J. Virol. *71*, 8602–8614.

Maclean, A.R., Ul-Fareed, M., Robertson, L., Harland, J., and Brown, S.M. (1991). Herpes simplex virus type 1 deletion variants 1714 and 1716 pinpoint neurovirulence-related sequences in Glasgow strain 17+ between immediate early gene 1 and the 'a' sequence. J. Gen. Virol. *72*, 631–639.

Mader, S., Lee, H., Pause, A., and Sonenberg, N. (1995). The translation initiation factor eIF-4E binds to a common motif shared by the translation factor eIF-4G and the translational repressors 4E-binding proteins. Mol. Cell. Biol. *15*, 4990–4997.

Marcotrigiano, J., Gingras, A.C., Sonenberg, N., and Burley, S.K. (1999). Cap-dependent translation initiation in eukaryotes is regulated by a molecular mimic of eIF4G. Mol. Cell 3, 707–716.

Mocarski, E.S. Jr., and Courcelle, C.T. 2001. Cytomegaloviruses and their replication, In Fields virology, 4th edn, D.M. Knipe and P. M. Howley, eds. (Philadelphia, Pa; Lippincott/The Williams & Wilkins Co.), pp. 2629–2673.

Mohr, I. (2003). Genetic metamorphosis of herpes simplex virus-1 into a biological therapeutic for human cancer. Exp. Opin. Biol. Ther. 3, 113–125.

Mohr, I., and Gluzman, Y. (1996). A herpesvirus genetic element which affects translation in the absence of the viral GADD34 function. EMBO J., *15*, 4759–4766.

Mulvey, M., Poppers, J., Ladd, A., and Mohr, I. (1999). A herpesvirus ribosome associated, RNA-binding protein confers a growth advantage upon mutants deficient in a GADD34-related function. J. Virol., *73*, 3375–3385.

Mulvey, M., Poppers, J., Sternberg, D., and Mohr, I. (2003). Regulation of eIF2 alpha phosphorylation by different functions that act during discrete phases in the herpes simplex virus type 1 life cycle. J. Virol. *77*,10917–10928.

Mulvey, M., Camarena, V. and Mohr, I. (2004). Full resistance of HSV-1 infected primary human cells to interferon α require both the Us11 and $\gamma_1$34.5 gene products. J. Virol. *78*, 10193–10196.

Murphy, J.A., Duerst, R.J., Smith, T.J., and Morrison, L.A. (2003). Herpes simplex virus type 2 virion host shutoff protein regulates alpha/beta interferon but not adaptive immune responses during primary infection *in vivo*. J. Virol. *77*, 9337–9345.

Oroskar, A.A. and Read, G.S. (1987). A mutant of herpes simplex virus type 1 exhibits increased stability of immediate-early (alpha) mRNAs. J. Virol. *61*, 604–606.

Patel, R.C., and Sen, G.C. (1998). PACT, a protein activator of the interferon-induced protein kinase, PKR. EMBO J. *17*, 4379–4390.

Pe'ery, T., and M. B. Mathews. (2000). Viral translational strategies and host defense mechanisms. *In* Translational Control. N. Sonenberg, J.W. B. Hershey, and M. B. Mathews, eds. (Cold Spring Harbor, N.Y.; Cold Spring Harbor Laboratory Press), pp. 371–424.

Perez-Parada, J., Saffran, H.A., and Smiley, J.R. (2004). RNA degradation induced by the herpes simplex virus vhs protein proceeds 5′ to 3′ *in vitro*. J. Virol. *78*, 13391–13394.

Peters G.A., Khoo, D., Mohr, I., and Sen, G.C. (2002). Inhibition of PACT-mediated activation of PKR by the herpes simplex virus type 1 Us11 protein. J. Virol., *76*, 11054–11064.

Poppers, J., Mulvey, M., Khoo, D., and Mohr, I. (2000). Inhibition of PKR activation by the proline rich RNA binding domain of the HSV-1 Us11 protein. J. Virol., *74*, 112315–11221.

Preston, C.M. and Newton, A.A. (1976). The effects of herpes simplex virus type 1 on cellular DNA-dependent RNA polymerase activities. J. Gen Virol. 33, 471–482.

Pyronnet, S., Imataka, H., Gingras, A.-C., Fukunaga, R., Hunter, T., and Sonenberg, N. (1999). Human eukaryotic translation initiation factor 4G (eIF4G) recruits mnk1 to phosphorylate eIF4E. EMBO J. *18*, 270–279.

Raught, B., Gingras, A.C., Gygy, S.P., Imataka, H., Morino, S., Gradi, A., Aebersold, R., and Sonenberg, N. (2000). Serum-stimulated, rapamycin-sensitive phosphorylation sites in the eukaryotic translation initiation factor 4GI. EMBO J. *19*, 434–444.

Read, G.S. and Frenkel, N. (1983). Herpes simplex virus mutants defective in the virion-associated shutoff of host polypeptide synthesis and exhibiting abnormal synthesis of a (immediate early) viral polypeptides. J. Virol. *46*, 498–512.

Rogers, G.W. Jr, Komar, A.A., and Merrick, W.C. (2002). eIF4A: the godfather of the DEAD box helicases. Prog. Nucleic Acid Res. Mol. Biol. 72, 307–331.

Roizman, B. and Knipe, D. (2001). Herpes simplex viruses and their replication. In: Fields' virology, 4th edn, D.M. Knipe and P. M. Howley, eds. (Philadelphia, Pa; Lippincott/The Williams & Wilkins Co.), pp. 2239–2459.

Sacks, W.R. and Schaffer, P.A. (1987). Deletion mutants in the gene encoding the herpes simplex virus type 1 immediate-early protein ICP0 exhibit impaired growth in cell culture. J. Virol. *61*, 829–839.

Sandri-Goldin, R.M. (2004). Viral regulation of mRNA export. J. Virol. *78*, 4389–4396.

Scheper, G.C. and Proud, C.G. (2002). Does phosphorylation of the cap-binding protein eIF4E play a role in translation initiation? Eur. J. Biochem. *269*, 5350–5359.

Scheper, G.C., Morrice, N.A., Kleijn, M., and Proud, C.G. (2001). The mitogen-activated protein kinase signal-integrating kinase Mnk2 is a eukaryotic initiation factor 4E kinase with high levels of basal activity in mammalian cells. Mol. Cell Biol. *21*, 743–754.

Schneider, R.J. and Mohr, I. (2003). Translation initiation and viral tricks. Trends Biochem. Sci. *28*, 130–136.

Sciabica, K.S., Dai, Q.J., and Sandri-Goldin, R.M. (2003). ICP27 interacts with SRPK1 to mediate HSV splicing inhibition by altering SR protein phosphorylation. EMBO J. *22*,1608–1619.

Shibuya, T., Tange, T.O., Sonenberg, N., and Moore, M.J. (2004). eIF4AIII binds spliced mRNA in the exon junction complex and is essential for nonsense-mediated decay. Nat. Struct. Mol. Biol. *11*, 346–351.

Smith, T.J., Morrison, L.A., and Leib, D.A. (2002). Pathogenesis of herpes simplex virus type 2 virion host shutoff (vhs) mutants. J. Virol. *76*, 2054–2061.

Soliman, T., Sandri-Goldin, R.M., and Silverstein, S.J. (1997). Shuttling of the herpes simplex virus type 1 regulatory protein ICP27 between the nucleus and cytoplasm mediates the expression of late proteins. J. Virol. *71*, 9188–9197.

Spencer, C.A., Dahmus, M.E., and Rice, S.A. (1997). Repression of host RNA polymerase II transcription by herpes simplex type 1. J. Virol. *71*, 2031–2040.

Stanners, C.P. and Becker, H. (1971). Control of macromolecular synthesis in proliferating and resting Syrian hamster cells in monolayer culture. I. Ribosome function. J. Cell. Physiol. *77*, 31–42.

Stow, N.D. and Stow, E.C. (1986). Isolation and characterization of a herpes simplex virus type 1 mutant containing a deletion within the gene encoding the immediate early polypeptide Vmw110. J. Gen. Virol. *67*, 2571–2585.

Strelow, L.I. and Leib, D.A. (1995). Role of the virion host shutoff (vhs) of herpes simplex virus type 1 in latency and pathogenesis. J. Virol. *69*, 6779–6786.

Thomas, G. and Gordon, J. (1979). Regulation of protein synthesis during the shift of quiescent animal cells into the proliferative state. Cell Biol. Int. Rep. *3*, 307–320.

Thomas, G. and Thomas, G. (1986). Translational control of mRNA expression during the early mitogenic response in Swiss mouse 3T3 cells: identification of specific proteins. J. Cell Biol. *103*, 2137–2144.

Thomas, G., Thomas, G., and Luther, H. (1981). Transcriptional and translational control of cytoplasmic proteins after serum stimulation of quiescent Swiss 3T3 cells. Proc. Natl. Acad. Sci. USA *78*, 5712–5716.

Walsh, D. and Mohr, I. (2004). Phosphorylation of eIF4E by mnk-1 enhances HSV-1 translation and replication in quiescent cells. Genes Dev. *18*, 660–672.

Walsh, D., Perez, C., Notary, J., and Mohr, I. (2005). Regulation of the translation initiation factor eIF4F by multiple mechanisms in human cytomegalovirus infected cells. J. Virol. *79*, 8057–8064.

Waskiewicz, A.J., Flynn, A., Proud, C.G. and Cooper, J.A. (1997). Mitogen-activated protein kinases activate the serine/threonine kinases mnk1 and mnk2. EMBO J. *16*,1909–1920.

Waskiewicz, A.J., Johnson, J.C., Penn, B., Mahalingam, M., Kimball, S.R., and Cooper, J.A. (1999). Phosphorylation of the cap-binding protein eukaryotic translation factor 4E by protein kinase mnk1 *in vivo*. Mol. Cell. Biol. *19*, 1871–1880.

Wells, S.E., Hillner, P.E., Vale, R.D., and Sachs, A.B. (1998). Circularization of mRNA by eukaryotic translation initiation factors. Mol. Cell *2*, 135–140.

Xi, Q., Cuesta, R., and Schneider, R.J. (2004). Tethering of eIF4G to adenoviral mRNAs by viral 100k protein drives ribosome shunting. Genes Dev. *18*, 1997–2009.

Yoder-Hill, J., Pause, A., Sonenberg, N., and Merrick, W.C. (1993). The p46 subunit of eukaryotic initiation factor (eIF)-4F exchanges with eIF-4A. J. Biol. Chem. *268*, 5566–5573.

Zhou, C., and Knipe, D.M. (2002). Association of herpes simplex virus type 1 ICP8 and ICP27 proteins with cellular RNA polymerase II holoenzyme. J. Virol. *76*, 5893–5904.

The Development and Use of Microarrays for Analyzing HSV Gene Expression

7

Edward K. Wagner

Abstract

The design and construction of long (75mer) oligonucleotide-based DNA microarray for quantitative analysis of herpes simplex virus transcripts expressed in infected cells and tissues is described. The HSV-1 probes used in the current array are able to specifically cross-hybridize with HSV-2 sequences, and thus, can be used to study both HSV-1 and HSV-2 transcripts. Data presentation and the sensitivity of several methods of detection hybrids on such microarrays are presented. The quantitative measure of HSV transcripts expressed in cells infected with mutations in specific regulatory genes is utilized to illustrate the application of the information gained towards the general understanding of the virus-host interaction.

Background

The application of robotic micro-arraying techniques and laser-based image analysis has lead to the development of DNA microarrays as a powerful tool for the global analysis of transcriptional responses of cells and microorganisms to perturbations in their environment such as stress. The global, quantitative measure of transcript levels of an organism or other biological entity defines the "transcriptome" for informatic analysis and evaluation. The relatively large HSV-1 transcriptome comprising in the order of 100 unique transcripts expressed at various levels at different stages of the infection cycle is clearly applicable to microarray-based analysis of viral mRNA populations in infected cells under a variety of conditions of infection. The availability of a large number of defined regulatory and kinetic mutations of the virus in which the timing of expression or the function of a single gene has been modified coupled with increasingly sensitive global transcript abundance analysis is allowing us to investigate the role of the kinetics of expression and transcript abundance upon the replication of the subject virus in cultured cells, and ultimately, in animal models. An added advantage to the study is the availability of a large body of research results in which a number of single and multigene transcript studies have been reported; thus, global data generated can be readily verified with complementary techniques.

Those studies represent a collaboration with P. Ghazal first at the Scripps Research Institute and now at the Genome Centre at the University of Edinburgh and follow the lead of his initial use of moderate length oligonucleotides in HCMV studies in designing and utilizing HSV-1 DNA microarrays and applying them to detailed global transcriptional analyses (Chambers *et al.*, 1999). Some of the properties and applications of HSV-1 microarrays are described in the following pages.

Properties of the chip and analysis of data

We have fully described the method of design of the 75-base oligomers we have used to construct our HSV-1 chip in a number of previous reports (Stingley *et al.*, 2000; Aguilar *et al.*, 2002; Wagner *et al.*, 2002). Briefly, each HSV transcription unit was surveyed near its 3′ end to find one to three 75-base regions of low internal structure and a G+C content as close to the average value for the genome as possible; this was to ensure efficient labeling of cDNA using oligo-dT primers and to preclude self-annealing. In the case of partially overlapping transcripts sharing the same polyadenylation site, oligonucleotides specific for the unique 5′ transcripts were chosen also—in this case, random hexamer-primed cDNA generated from purified poly(A) containing viral mRNA is used. The HSV-1 array contains 67 unique sets of HSV-1 probes of which 43 represent transcripts from single genes. Another 11 probes detect sets of viral transcripts within a single kinetic class that share polyadenylation signals. Thus, a total of 54 probes detect either individual or overlapping transcript sets whose kinetics can be unambiguously assigned, while an additional 13 probes detect overlapping transcripts of different kinetic classes.

Data presentation and representation of the HSV transcriptome

Since the analysis of each hybridized chip is subject to both systematic and nonsystematic errors, a self-consistent and defensible statistical analysis of data is required to provide robust numerical analysis of transcript abundance. The methods employed in analyzing HSV transcription have been described in some detail (Wagner *et al.*, 2002). Briefly, the median signal of three replicate probe values for each transcript on a single chip is the value for a single experiment. A minimum of three replicates is then carried out for statistical confidence, and the median of these three independent median values along with the standard deviation is calculated. To compare data from replicate experiments the 75th percentile rank for the total viral hybridization is calculated in which one experiment is arbitrarily chosen as the reference and the 75th percentile values of all other determinations adjusted to this value by appropriate factoring. Complete data can be presented in tabular form, and statistical analysis of variation can be carried out on such tabular values. Bar graphs provide readily interpretable graphical summaries. Most conveniently, transcripts can be clustered into the four general kinetic classes in which they are expressed, immediate-early, early, late, and latent associated and ordered by their location on the prototypical HSV-1 genome arrangement.

Specificity of HSV transcript probes

As described in (Wagner *et al.*, 2002) the absolute specificity of HSV signals detected on the chip (its receiver operating characteristics) is quite high. However, we have found that the probes used are able to efficiently and quantitatively detect the transcripts expressed during HSV-2 infections as well as HSV-1. The ability to efficiently detect HSV-2 sequences corresponding to those covered by the HSV-1 probes is graphically shown in Fig. 7.1 where hybridization of total HSV-1 DNA labeled with colloidal gold particles using nick translation as described in the following section to HSV-1 probes was compared with the hybridization of total HSV-2 DNA to the same probes. Although the hybridization efficiency is lower with the HSV-2 DNA, the patterns of hybridization are virtually indistinguishable. We have reported our preliminary analysis of HSV-2 transcription using a chip containing probes derived from both the HSV-1 and HSV-2 genomes, and have found only a very few significant differences in overall patterns of transcript abundance as a function of time after infection

(Aguilar *et al.*, 2004). The cross specificity of hybridization has allowed us to make reliable quantitative comparisons and a full analysis is nearing completion (Aguilar *et al.*, 2006).

Limits of detection of viral transcripts

The most widely used method of labeling material for oligonucleotide-based microarray transcription studies uses laser-induced fluorescence of tagged cDNA prepared from either total or poly(A)-containing RNA by oligo-dT or random hexamer/octomer-primed synthesis incorporating fluorescent nucleotides (Cy3dCTP and Cy5dCTP). As described in detail in (Wagner *et al.*, 2002), this method has the advantage of having a high reciprocity range of signal intensities as a function of laser power—thus both weak and strong signals can be reliably measured by doing multiple scans at increasing laser power. This approach is limited by the potential for laser bleaching of the chromophores, however, and chips are not stable to storage and rescanning following hybridization and analysis. Further, while extremely useful for the study of viral gene expression in cultured cells its overall sensitivity is not great enough to allow detection of low abundance signals from biological specimens without amplification of the RNA with the resulting potential for skewing distributions though differential amplification efficiencies of different RNA populations. Higher sensitivity detection of hybridized cDNA can be accomplished by using colloidal gold and silver labeling of cDNA and resonance light scattering (RLS) measurements—this method utilizes nano-sized gold and silver particles, which have the property of scattering polychromatic or white light to tag the hybridization probe (Yguerabide and Yguerabide, 1998a,b). The scattering is characterized by preferential radiation of specific resonance wavelength for each metal tag. Thus, in addition to increased sensitivity the method has the value of using significantly cheaper scanning equipment since white light is used for signal excitation; also, the hybridized chips are quite stable and can be rescanned repeatedly with no signal degradation. While the reciprocity between signal intensity and exposure time to the white light used to generate the scattering signal occurs over a much narrower range of conditions than with laser scanning, the increase in sensitivity inherent in this method has allowed us to detect viral signals in infected mouse brains with confidence.

With both methods, the general processes of scanning and data reduction are similar. Signals expressed in arbitrary units derived from each spotted oligonucleotide are adjusted by subtracting the background signal of an equivalent area within a concentric ring just outside the spotted sample. The data from individual spots is then expressed in a Microsoft Excel spreadsheet, and reduced net values can be determined by subtracting the average of values obtained from measuring a large number (ca. 100) of regions spotted with SSC alone. Typically, and depending on the exact intensity of the scan, this SSC background is less than 1—2% of the total signal, and can be ignored for all except the weakest signals. Finally, both provide an approximately linear response between RNA level and signal intensity between approximately 1,000 and 40,000 arbitrary signal units.

In comparative control experiments using RNA isolated from HSV-1 infected cells at various times after infection, we find that the RLS method provides a minimum of a 10-fold increase in sensitivity. A representative experiment in which poly(A) RNA from human fibroblasts was labeled with either method and hybridized to identical chips is shown in Fig. 7.2. The data in panel A was obtained with cDNA derived from a 150 ng sample of RNA labeled with colloidal gold and scanned with a 2 second exposure. That of panel B shows the signal resulting from the hybridization of Cy3-labeled cDNA generated from 2 μg of

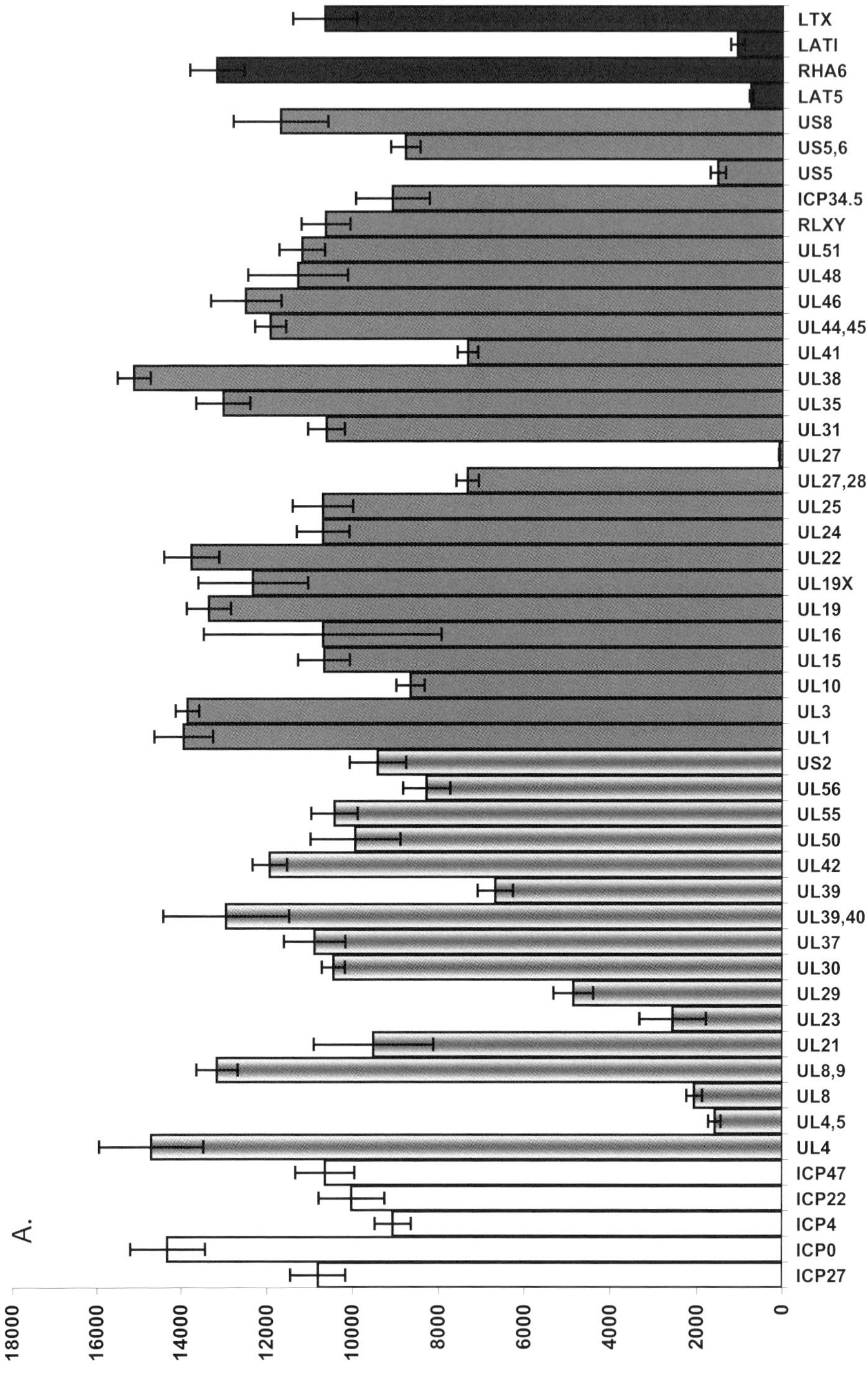

A.
LTX
LATI
RHA6
LAT5
US8
US5,6
US5
ICP34.5
RLXY
UL51
UL48
UL46
UL44,45
UL41
UL38
UL35
UL31
UL27
UL27,28
UL25
UL24
UL22
UL19X
UL19
UL16
UL15
UL10
UL3
UL1
US2
UL56
UL55
UL50
UL42
UL39
UL39,40
UL37
UL30
UL29
UL23
UL21
UL8,9
UL8
UL4,5
UL4
ICP47
ICP22
ICP4
ICP0
ICP27
0
2000
4000
6000
8000
10000
12000
14000
16000
18000

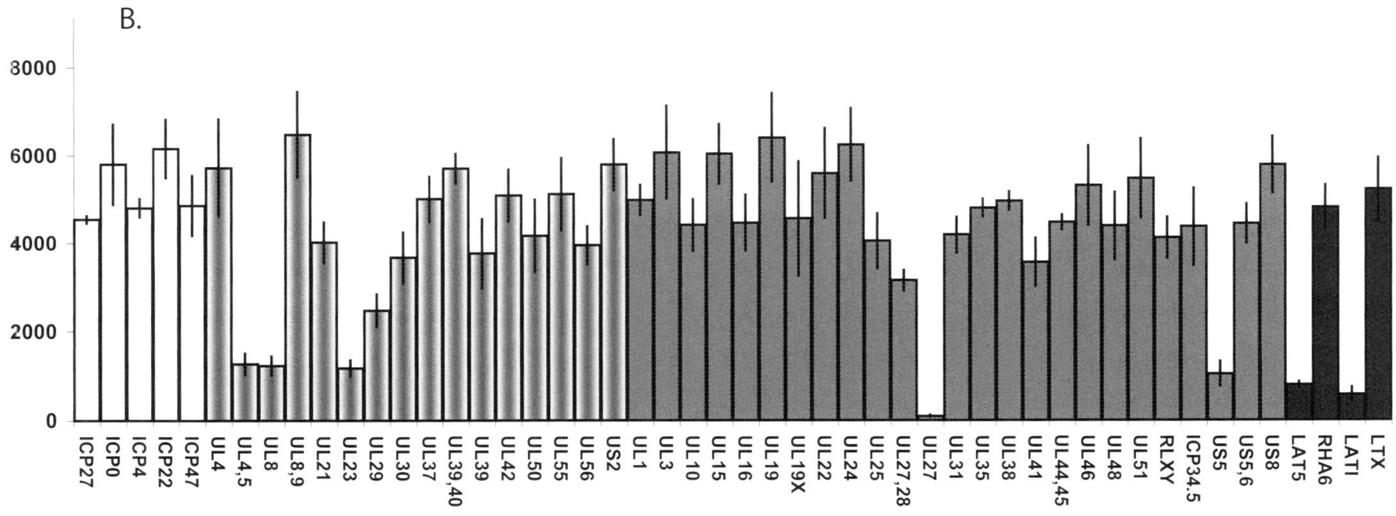

Figure 7.1 RLS detection of nick-translated HSV DNA hybridized to the HSV-1 DNA microarray. As described in the text, the HSV-1 DNA microarray contains 75-mer oligonucleotides complementary to each HSV transcription unit. A. In the experiment shown, an aliquot of 500 ng of HSV-1 DNA was nick translated and labeled with colloidal gold and hybridized to the chip, hybridization signals were detected by RLS. The general methods have been described (Wagner *et al*., 2002). The location of each probe is arranged on the ordinate according to its kinetics of expression (open bars, immediate early; partially filled bars, early; gray bars, late; black bars, latency associated transcript(s). In each kinetic class, the order is that of the transcript on the prototypical arrangement of the viral genome. The abscissa indicates the hybridization signal in arbitrary light scattering units. B. This panel shows the results of hybridization with an equivalent aliquot of HSV-2 DNA. Both chips were scanned for the same amount of time (0.5 second). Each panel shows a single representative experiment. Here, the error bars indicate the standard deviation of signals from each of the three replicate probe spots on the same chip.

the same RNA sample and scanned at 90% maximum laser power. While there is some variation, the patterns are generally equivalent, and sensitivity can be increased with the RLS method by scanning for longer periods of time (up to 15 seconds per chip).

An example of the value of this high sensitivity in transcriptional studies on tissues is seen in the following experiment carried out as part of a collaboration with D.C. Bloom of the U. of Florida (J.S. Aguilar, D.C. Bloom, and E.K. Wagner, in progress). Individual mice were injected stereotactically into the right hippocampus formation with 10^5 PFU of either replication competent HSV-1 or with a mutant (*in*8114), which expresses a nonfunctional ICP4. This mutant is transcriptionally highly restricted in tissue culture expressing low levels of some immediate-early transcripts and over expressing the transcript encoding the nonfunctional ICP4 protein due to its inability to auto regulate. Blocks of tissue (1 mm^3) were isolated at 3 days postinfection from the site of infection, and total RNA isolated. HSV transcription was assayed by generating oligo-dT primed cDNA labeled with colloidal gold from 100 μg of total RNA. As can be seen in Fig. 7.2D, the patterns of transcription in the tissue infected with replication competent virus is essentially that seen in productive infection in tissue culture cells with all kinetic classes represented. In contrast, only low levels of immediate-early transcripts (mainly ICP4 and ICP22) are seen in the tissue infected with nonreplicating virus (panel C). This ability to assay viral transcription in such an infection provides an important control point for the study of the effect of various aspects of virus infection upon tissue gene expression and host-response patterns.

Microarray analysis of HSV transcriptional "plasticity"

The patterns of HSV transcript abundance in productively infected cells (and, by inference, tissues) demonstrate a highly regulated process—one in which the importance of initial expression of five immediate early transcripts is apparent. The similarities between the detailed patterns following either HSV-1 or HSV-2 infection as well as other more genetically divergent herpesviruses supports the hypothesis that this regulated cascade is both diagnostic of herpesvirus infections and represents and important series of control points in the regulation of viral gene expression directed towards optimization of viral replication in the host. Workers in my laboratory as well as in those of many others have spent several decades determining the detailed nature of the cascade in terms of protein function, and, importantly for our continuing studies, the mechanistic nature of the control of the cascade. As reviewed elsewhere, indi-

Figure 7.2 (opposite) Sensitivity of various methods of labeling cDNA for detecting hybrids on the microarray. Panels A and B compare the hybridization of cDNA generated from HSV-1 infected cells isolated at 6 hours following infection. As described in the text, the same RNA was used to generate cDNA labeled either with colloidal gold for RLS detection or Cy3-CTP for laser-excited fluorescent detection. In panel A, 150 ng of poly(A) RNA was used to generate the cDNA using random hexamer primers, and exposure was for 2 seconds in a light scattering signal detector. In panel B, 2 μg of the same poly(A) RNA was used to generate the cDNA labeled with Cy3. Exposure was at 90% maximum laser power (the highest level normally used). Each panel shows a single experiment in which the error bars show the standard deviation of the signals for the three probes on the chip. Panels C and D show the hybridization of colloidal gold-labeled cDNA generated from 10 μg aliquots of total RNA isolated at 3 days following stereotactic injection of 105 PFU into the right hippocampus of a mouse brain. Panel C shows the pattern obtained following infection with a nonreplicating mutant of HSV-1 defective in ICP4 function (in8114). The data of panel D is the pattern seen following infection with a replicating virus isolated. The probe order is the same as that in Fig. 7.1; but the names of the probes are omitted for clarity.

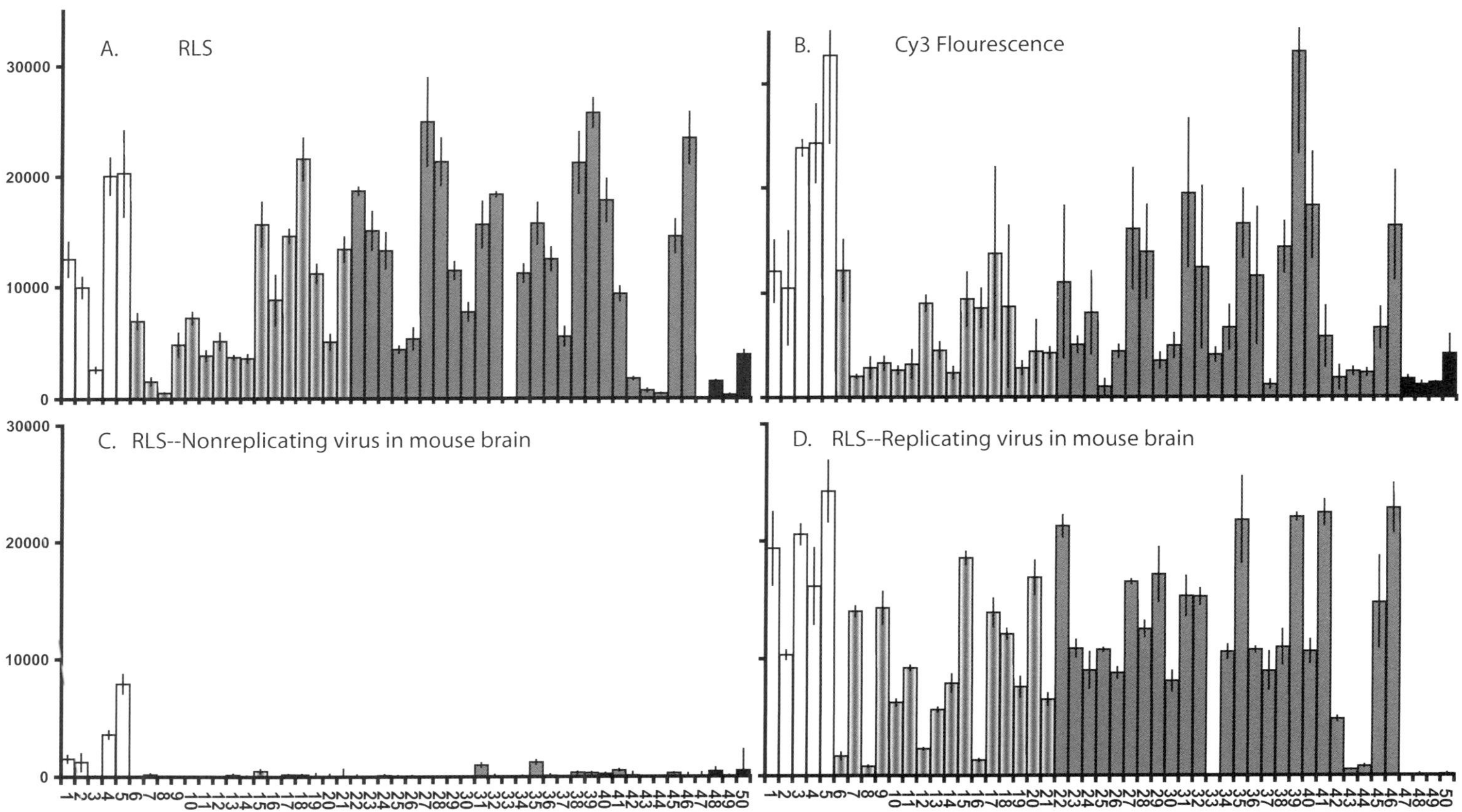
A. RLS
B. Cy3 Flourescence
C. RLS--Nonreplicating virus in mouse brain
D. RLS--Replicating virus in mouse brain
30000
20000
10000
0
1 2 3 4 5 6 7 8 9 10 11 12 13 14 15 16 17 18 19 20 21 22 23 24 25 26 27 28 29 30 31 32 33 34 35 36 37 38 39 40 41 42 43 44 45 46 47 48 49 50

vidual promoters of various kinetic signatures are a major factor in controlling expression of the various kinetic classes of viral transcripts, and suitable modification of these promoters can be used to generate viral mutants in which the kinetics of expression of any gene of interest can be modified (Lieu and Wagner, 2000; Sun *et al.*, 2004a,b; Tran *et al.*, 2004). Oligonucleotide based DNA microarrays provide an important tool for the quantitative transcriptional analysis of the results of modifying the timing of expression of critical viral regulatory genes. The analysis of several types of perturbations of the initial immediate-early stage of gene expression is reviewed in the following paragraphs.

The influence of the transcriptional state of the host cell on efficient viral replication

As reviewed in (Sun *et al.*, 2004a,b), the functions of four of the five immediate-early transcripts are fully consistent with the timing of their expression—ICP4 expression is absolutely required for subsequent genome transcription, and the loss of expression of functional ICP0 can also have profound effects on viral gene expression and replication. Two other genes, ICP22 and ICP47 appear to function to mediate both the range of differentiated cells in which the virus can replicate and in obviating antigen presentation upon reactivation. The RNA transport protein, ICP27 also has a profound role in productive infection, although, as shown below, the immediate-early timing of its expression presents a bit of a biological conundrum.

The importance of immediate-early expression is underlined by the complex interaction between host and viral proteins involved in the recognition of the immediate-early enhancers required to initiate the viral transcriptional cascade. We have studied the effect of altering this primary transcriptional "boost" in collaboration with S. Triezenberg of Michigan State University (Yang *et al.*, 2002). Here we studied the effect of blocking the activation of immediate early gene expression by use of a virus mutant, RP5, which lacks the activation domain of the virion-associated VP16 protein, and, thus, cannot active immediate early gene expression upon infection. An example of the effect of this blockage on viral transcription patterns is shown in Fig. 7.3A, where RNA isolated at 4 hours following moderate multiplicity of infection demonstrates low level expression of a limited number of viral transcripts. Further, careful quantitative analysis of transcript abundance levels at various times following infection

Figure 7.3 (opposite) The effect of host cell activation and multiplicity of infection on the transcription program of HSV-1 mutants deficient in immediate-early functions. Panels A and B compare the hybridization of cDNA generated from HSV-1 infected cells isolated at 4 hours following infection with the RP5 mutant of HSV-1 that lacks the ability to activate immediate-early transcription via the virion-associated transcription activator, VP16. As described in the text, RNA isolated from infected HeLa cells was used to generate cDNA labeled with Cy3 and Cy5 for laser-excited fluorescent detection of hybrids. Panel A shows the population of viral RNA present in cells infected with a multiplicity of 0.1 PFU/cell of the RP5 mutant (this is an equivalent number of virus particles to a nominal moi of 5 PFU/cell of *wt*-virus). Panel B shows the pattern of transcripts expressed when cells are treated with HMBA and infected at the same moi. These data are from (Yang *et al.*, 2002) and are the median values of three independent experiments. Panels C and D show the hybridization of Cy3 and Cy5-labeled cDNA generated from cells infected with an ICP0 null mutant. Panel C shows the patterns of infection at a nominal moi of 0.5 PFU/cell (as determined on complementing cells). Panel D shows the pattern of transcript abundance following infection at a 10-fold higher multiplicity. These data are from (Karaca *et al.*, 2004) and are the median values of three independent experiments. The probe order is the same as that in Fig. 7.1; but the names of the probes are omitted for clarity.

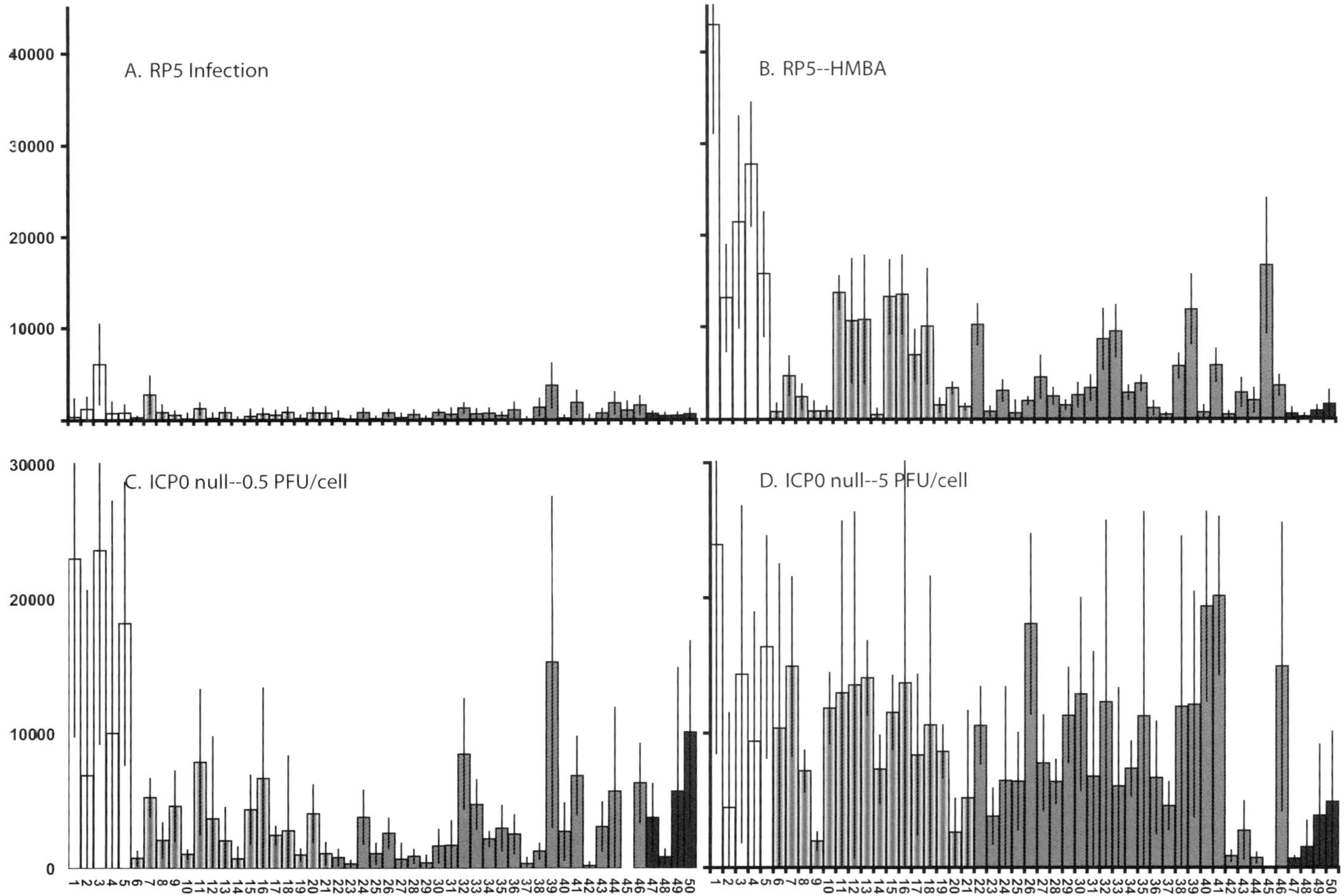
A. RP5 Infection
B. RP5--HMBA
C. ICP0 null--0.5 PFU/cell
D. ICP0 null--5 PFU/cell
40000
30000
20000
10000
0

failed to demonstrate any evidence of kinetic discrimination in those transcripts expressed at a detectable level—the kinetic cascade was fully abrogated.

Despite this profound (and fully expected) effect on the global transcription patterns of viral genes, stimulation of the host cell by the drug HMBA resulted in the re-establishment of a kinetically nominal cascade despite the lack of activation of the VP16–Oct 1 pathway (cf. Fig. 7.3B). A similar situation was seen by using high multiplicity infections, and treatment of the cells with protein synthesis inhibitors also resulted in re-establishment of ostensibly normal transcription patterns. We have interpreted such findings as evidence of a great deal of plasticity in the requirements for the initiation of the regulated viral gene expression cascade due to the ability of the host cell to provide functions provided by the virus during the earliest stages of productive infection. The importance of such a demonstration in the consideration of the mechanism of stress-induced reactivation is obvious. Similar analysis of the transcriptional effect of deletion of normal ICP0 function shows differences in detail (Karaca *et al.*, 2004); thus, as suggested in Fig. 7.3C, the abundance pattern of individual viral transcripts expressed at low multiplicity infection is different than seen with RP5 infections. Still, the reversal of these kinetic blocks by high multiplicity infection, and by inference, other cellular stress regimens are clear as shown in Fig. 7.3D.

The ability of HSV to utilize several pathways to ensure the initiation of its kinetic cascade of transcript abundance emphasizes its importance in initiating the productive replication cycle. Further, the fact that suitable cellular stress (i.e. cellular transcriptional environment) can initiate this cascade in the absence of the requirement for its activation mediated by the virion-associated transcriptional activator suggests that this latter pathway is an evolutionary adaptation showing unique features of the HSV interaction with its host. It seems clear that the route of infection, involving initial infection in differentiated and transcriptionally quiescent epidermal cells will be greatly enhanced by the presence of this initial transcriptional "boost." Just as clearly, the requirements for the initiation of the productive transcriptional cascade cannot be satisfied with a virion-associated transcription activator in neuronal cells in which latent genomes reside. Thus, the "primordial" ability of the virus to respond to stress-induced changes in the cellular transcriptional environment provides the key to the ability of the virus to maintain an efficient latency/reactivation cycle. Further, this ability of the virus to "sense" the transcriptional state of the host cell in regards to its own replication provides a reliable mechanism for insuring that viral genomes can reside in a state of transcriptional dormancy in the appropriate population of differentiated and transcriptionally quiescent cells—in the normal case neurons. The ability to detect latent transcription—as well as the ability of virus to reactivate from—other types of highly differentiated cells such as ocular keratinocytes (Devi-Rao *et al.*, 1997) is fully consistent with this scheme. Thus, the quantitative analysis of viral transcription patterns using micro-arrays affords a means of validating the potential minimum evolutionary and physiological requirements needed for the origination of herpesvirus neurotropism.

Analysis of the role of immediate-early expression of ICP27 function

While the above results demonstrate the biological flexibility of HSV in accommodating to the transcriptional environment of the cell, they also demonstrate the importance of the immediate-early transcriptional activation phase. This comes from the fact that the virus has evolved several paths for the entry into this phase of gene expression, where one could imagine alternate pathways might occur. Following this logic, it is of some interest to us to determine the replication and transcriptional consequences of interrupting the immediate-early phase of transcription at the level of individual gene function. In this light, the transcriptional role of the ICP27 protein at the earliest stages of infection had not been well characterized and we have generated several kinetic mutants of the gene to study it (Sun *et al.*, 2004a,b), and in light of the above discussion, the timing of expression of the immediate-early ICP27 protein stands as somewhat of a kinetic conundrum. Its well characterized activities as a mediator of splicing inhibition and transport of unspliced transcripts from the nucleus to the cytoplasm are required throughout the replication cycle; however, while viral mutants lacking this gene express at least the majority of early transcripts at normal or above normal levels, the levels of many late transcripts are significantly reduced. In order to investigate functions of ICP27 requiring expression immediately upon infection, we constructed two promoter mutations altering the immediate-early expression of the HSV-1 ICP27 transcript and their cognate *wt*-rescues in order to assess the role of the ICP27 protein in the earliest stages of viral infection using global transcriptional analysis with a DNA microarray. The first mutant, ICP27ΔSma, contained a deletion of the "TATGARAT" and surrounding sequences, but retained the rest of the ICP27 promoter. The second, ICP27/VP16, replaced the whole ICP27 promoter/enhancer with the VP16 promoter. Both mutants demonstrated loss of immediate-early expression of ICP27 using criteria of expression in the absence of *de novo* protein synthesis and earliest expression in the kinetic cascade—an example of the altered expression with the ΔSma mutant is shown in Fig. 7.4. While transcript abundance at 1 hour after infection at 1 PFU/cell in mouse embryo fibroblasts was significantly altered compared with infections with *wt*-rescues (compare panels A and B). By 2 hours after infection with1 PFU/cell these differences were clearly diminished, especially with the ICP27/VP16 mutant. Significant differences between mutant and *wt*-rescues were not observed with either at later times after infection—again see Fig. 7.4, panels C and D where the transcription profile of the ΔSma mutant is compared with its rescue. Consistent with this observation, single and multistep virus replication was equivalent with both mutants and rescues. Thus, altering the immediate early kinetics of ICP27 led to a suboptimal quantitative lag-phase in gene expression but without consequence to replication fitness *in vitro*. Infections *in vivo* also revealed the ability of mutant and rescue virus to invade the CNS of mice following footpad injections was equivalent.

This rather surprising set of results clearly demonstrate that abrogation of the immediate-early expression of ICP27 has an effect on the earliest patterns of transcript abundance in the viral replication cascade, but this does not result in any measurable biological effect. Thus, both kinetic mutants of ICP27 demonstrate altered patterns of transcript abundance at the earliest times studied, but, unlike the situation with a deletion mutant, delayed expression by as much as an hour has no measurable effect on virus replication *in vitro* or *in vivo*. This "robustness" in the interleaving of viral functions in the early stages of the replication cycle is different than that seen with the VP16 mutants because here, there is a loss of coordination of ICP27 expression with that of ICP4 and ICP0 at the earliest times after infection. To us, the complete lack of any discernable biological manifestation was unexpected.

Conclusions

The evolutionary stability of the immediate-early kinetic signature for the ICP27 transcript appears to be a strong argument in

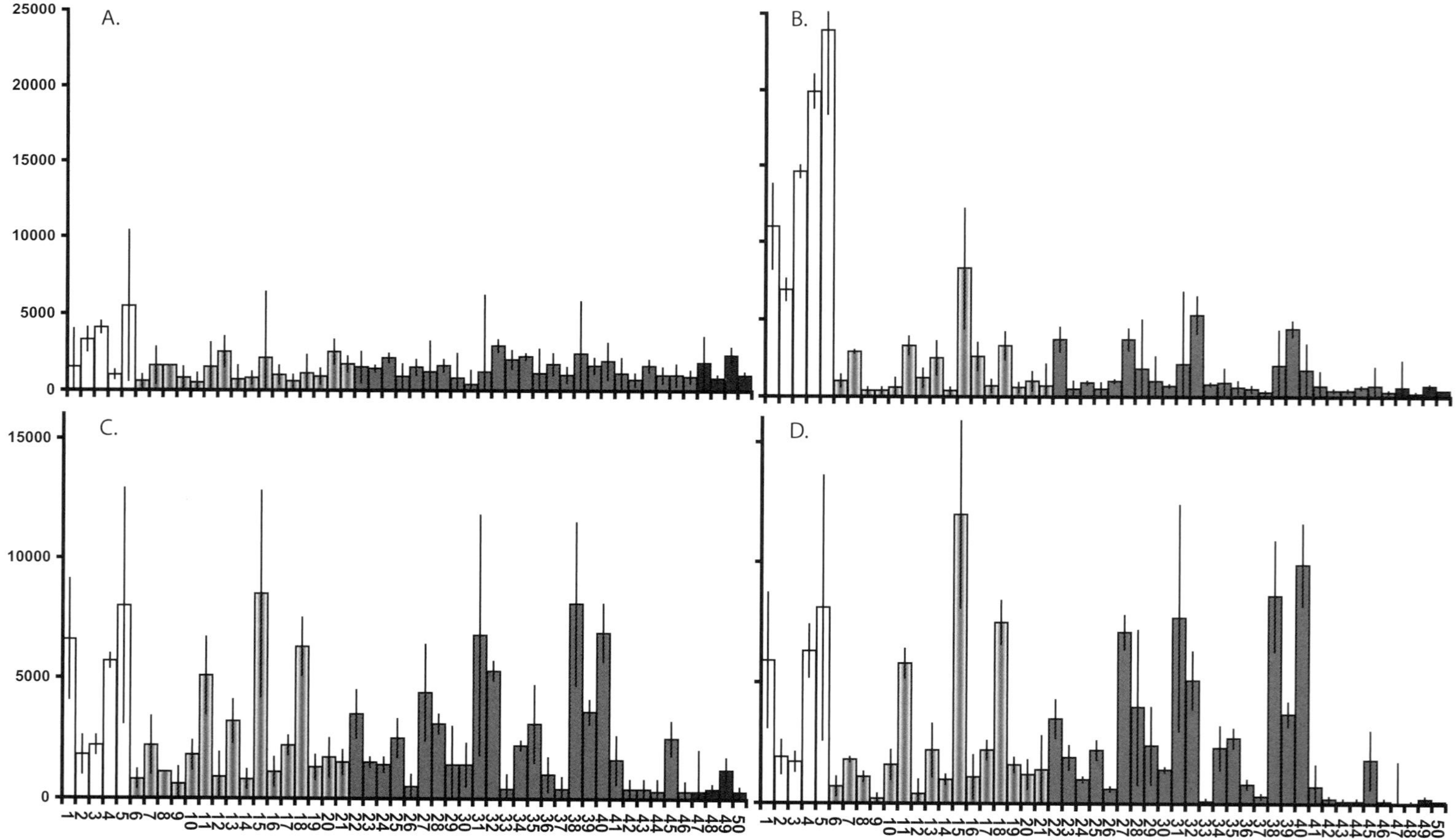
A.
B.
C.
D.
25000
20000
15000
10000
5000
0

favor of the timing of expression of this viral protein being critical and essential during the earliest window of viral gene expression in at least some tissue important in the natural history of HSV. Despite this however, any such restriction is subtler than measurable in several well-established pathogenicity models for HSV-1 infections including measures of the ability of the virus to invade and establish latent infections in mouse sensory nerve ganglia. The phenotype of time-based differences in quantitative transcript expression but neutral fitness strongly suggests a threshold model for activation of viral expression that incorporates a high level of regulatory-network robustness. Such robustness can be envisioned to play a major role in the continuing evolution and divergence of the various tissue specific adaptations of the "prototypical" herpesvirus that defines the proliferation of specific herpesvirus types.

Other applications of herpesvirus-specific DNA microarrays and concluding remarks

The approach that we have described in detail in earlier studies and outlined here provides rapid and relatively inexpensive approaches towards the design and utilization of micro chips for the characterization of any large virus "transcriptome." Clearly, microarray analysis allows the measure of mRNA levels from large numbers of genes in a population of cells—something that is not practicable using gel-based methods. It is important to understand such quantitative measure can be carried out in a statistically defensible quantitative way, and in concert with the analysis of cellular RNA populations in the same material. Further, continued increases in sensitivity of detection of hybridization signals will allow the analysis of smaller and smaller populations of cells—perhaps to reliable measures of single cell populations.

While there will always be practical limitations to the accuracy of these quantitative measures, the fact that analysis of infection of individual cells will be clearly subject to the stochastic variation in the population of susceptible cells being measured. The fact that the state of the host cell has such a major role on the transcription program of the virus brings this variation and the need to properly interpret the range of values measured into sharp focus. No method, no matter how precise and accurate can penetrate this limit of biological uncertainty, but the ability to define the limits of variation as well as the amounts of variation permissible provides an important parameter to understanding the biological environment in which HSV and any virus must operate. While gene detection and mutational analysis define complex interactive control loops, only quantitative understanding of the limits of permissible variation such as potentially afforded by microarray analysis can reveal how precisely such loops function. Since the reduplication and alteration of gene function is a primary mechanism of microevolution, the type of information gained with this ap-

Figure 7.4 (opposite) The effect of delaying the expression of the ICP27 on HSV-1 transcript abundance. Oligonucleotide-microarray hybridization analysis of RNA isolated from infected cells. These data are from (Sun *et al.*, 2004 a,b) and are the median values of three independent experiments. Panel A. shows the transcript abundance at 1 hour following infection in HFF cells infected with 1 PFU/cell of the ICP27ΔSma mutant in which the TATAGARAT box of the ICP27 promoter has been deleted. Panel B shows the abundance of transcripts in cells infected with a wt-rescue of the mutant and harvested at the same time following infection. Panels C and D compare the transcript abundance profiles at 4 hours following infection in cells infected with these two virus isolates. The probe order is the same as that in Fig. 7.1; but the names of the probes are omitted for clarity.

proach can provide a continuing picture of the evolution of the virus host interaction.

Acknowledgments

This work was supported by grants CA11861 and CA90287 from the NIH to my own laboratory as well as extramural grants to my collaborators. The full cooperation and participation of the Scottish Genome Center at the University of Edinburgh directed by Prof. Peter Ghazal, and the UCI Bioinformatics Center administered by Dr. Denis Hecht were invaluable in providing the technical resources to allow us to do the work.

References

Aguilar, J.S., Roy, D., Ghazal, P., and Wagner, E.K. (2002). Dimethyl sulfoxide blocks herpes simplex virus-1 productive infection *in vitro* acting at different stages with positive cooperativity. Application of micro-array analysis. BMC. Infect. Dis. *2*, 9.

Aguilar, J.S., Ghazal, P., and Wagner, E.K. (2004). Design of a Herpes Simplex Virus Type 2 Long Oligonucleotide-Based Microarray. In DNA Viruses: Methods and Protocols (P.M. Lieberman, Ed.), Vol. 292, Humana Press, Totowa.

Aguilar, J.S., Devi-Rao, G.V., Rice, M.K., Sunabe, J., Ghazal, P., and Wagner, E.K. (2006). Quantitative comparison of the HSV-1 and HSV-2 transcriptomes using DNA microarray analysis. Virology *348*, 233–241.

Chambers, J., Angulo, A., Amarantunga, D., Guo, H., Jian, Y., Wan, J.S., Bittner, A., Frueh, K., Jackson, M.R., Peterson, P.A., Erlander, M.G., and Ghazal, P. (1999). DNA microarrays of the complex human cytomegalovirus genome: profiling kinetic class with drug sensitivity of viral gene expression. J. Virol. *73*, 5757–5766.

Devi-Rao, G.B., Aguilar, J.S., Rice, M.K., Garza, H.H., Jr., Bloom, D.C., Hill, J.M., and Wagner, E.K. (1997). Herpes simplex virus genome replication and transcription during induced reactivation in the rabbit eye. J. Virol. *71*, 7039–7047.

Karaca, G., Hargett, D., McLean, T.I., Aguilar, J.S., Ghazal, P., Wagner, E.K., and Bachenheimer, S.L. (2004). Inhibition of the stress-activated kinase, p38, does not affect the virus transcriptional program of herpes simplex virus type 1. Virology *329*, 142–156.

Lieu, P.T. and Wagner, E.K. (2000). The kinetics of VP5 mRNA expression is not critical for viral replication in cultured cells. J. Virol. *74*, 2770–2776.

Stingley, S.W., Ramirez, J.J.G., Aguilar, S.A., Simmen, K., Sandri-Goldin, R.M., Ghazal, P., and Wagner, E.K. (2000). Global analysis of herpes simplex virus type 1 transcription using an oligonucleotide-based DNA microarray. J. Virol. *74*, 9916–9927.

Sun, A.-X., Devi-Rao, G.V., Rice, M.K., Gary, L.H., Bloom, D.C., Sandri-Goldin, R.M., Ghazal, P., and Wagner, E.K. (2004a). Immediate-early expression of the HSV-1 Icp27 transcript is not critical for efficient replication *in vitro* or *in vivo*. J. Virol. *78*, 10470–10478.

Sun, A-X, Devi-Rao, G.V., Rice, M.K., Gary, L.H., Bloom, D.C., Sandri-Goldin, R.M., Ghazal, P., and Wagner, E.K. (2004b) The TATATGARAT box of the HSV-1 ICP27 gene is essential for immediate early expression but not critical for efficient replication *in vitro* or *in vivo*. Virus Genes *29*, 334–343.

Tran, R.K., Lieu, P.T., Aguilar, J.S., Wagner, E.K., and Bloom, D.C. (2004). Altering the expression kinetics of VP5 results in altered virulence and pathogenesis of herpes simplex virus type 1 in mice. J. Virol. *76*, 2199–2205.

Wagner, E.K., Garcia, R.J.J., Stingley, S.W., Aguilar, J.S., Buehler, L., Devi-Rao, G.B., and Ghazal, P. (2002). Practical approaches to long oligonucleotide-based DNA microarrays: lessons from herpesviruses. Prog. Nucleic Acid Res. Mol. Biol. 71, 445–491.

Yang, W.C., Devi-Rao, G.V., Ghazal, P., Wagner, E.K., and Triezenberg, S.J. (2002). General and specific alterations in programming of global viral gene expression during infection by VP16 activation-deficient mutants of herpes simplex virus type 1. J. Virol. *76*, 12758–12774.

Yguerabide, J. and Yguerabide, E.E. (1998a). Light-scattering submicroscopic particles as highly fluorescent analogs and their use as tracer labels in clinical and biological applications. Anal. Biochem. *262*, 137–156.

Yguerabide, J. and Yguerabide, E.E. (1998b). Light-scattering submicroscopic particles as highly fluorescent analogs and their use as tracer labels in clinical and biological applications. Anal. Biochem. *262*, 157–176.

HSV Glycoproteins and their Roles in Virus Entry and Egress

8

Gabriella Campadelli-Fiume and Tatiana Gianni

Abstract

In the past several years there has been remarkable progress in our understanding of the events leading to herpes simplex virus (HSV) entry into the cells and its egress. Foremost was the realization that HSV entry requires a multigene system; this property differentiates HSV, and herpesviruses in general, from the majority of viruses. The glycoproteins absolutely required for the fusion that leads to HSV entry are gB gD, gH, gL. The receptor binding glycoprotein gD interacts with multiple alternative receptors that belong to unrelated protein families. They are nectin1 and 2, and the herpesvirus entry mediator. In addition, gD encodes a pro-fusion domain, deputed to trigger fusion. Fusion requires gB, gH, gL. gH is likely to be the real fusion glycoprotein, as its ectodomain carries a α-helix with attributes of an internal fusion peptide, and two coiled coil heptad repeats. As far as HSV egress is concerned, controversy still exists about the possible routes of egress (luminal *versus* de-envelopment re-envelopment). A growing consensus in favor of the latter rests on studies of tegument assembly and the possible role of nonessential glycoproteins in the cytoplasmic re-envelopment. is discussed. The lines of evidence in favor of, or against each route, as well as a novel possible route of egress are discussed.

HSV entry

Introduction

The study of herpes simplex virus (HSV) entry into the cell has been the study of the viral glycoproteins which perform this task. HSV glycoproteins were first recognized in the early 1970s, and some of them were soon identified as targets of neutralizing antibodies, suggesting a role in virus infection (Spear and Roizman, 1970; Spear, 1976; Spear *et al.*, 1978). Further milestones in the 1980s and 1990s were the development of monoclonal antibodies and the genetic engineering of HSV genome, the identification of the nucleotide sequence of the genes encoding the major glycoproteins, and, later on, of the entire HSV genome, along with the derivation of deletion viruses (Watson *et al.*, 1982; Noble *et al.*, 1983; Pellett *et al.*, 1985; Gompels and Minson, 1986; McGeoch *et al.*, 1988; Roizman, 1988; Hutchinson *et al.*, 1992a). It became thus apparent that HSV encodes a high number of glycoproteins and membrane proteins (more than 12), some of which can be deleted without hampering the ability of the virus to grow in cell cultures (nonessential proteins), while other cannot (essential proteins). The quartet of gB, gD, gH, gL, necessary for virus entry, belongs to the second group.

HSV glycoproteins and membrane proteins are named with letters, or by the ORFs encoding them. From a structural point of view, the majority of HSV glycoproteins are type I glycoproteins. Exceptions include gL, which is soluble (Hutchinson *et al.*, 1992a); gB, which may carry 2–3 α-helices in the transmembrane (TM) region (Pellett *et al.*, 1985). US9 and HSV-2 UL45 are type II glycoproteins with a C-terminal ectodomain. In addition, some glycoproteins and membrane proteins (gK, UL20, gM) are polytopic and traverse the membrane 3–4 or 8 times (McGeoch *et al.*, 1988). Visualized in cryo-electron tomograms of isolated virions, the envelope contains 600–750 glycoprotein spikes that vary in length, spacings, and in the angle at which they emerge from the membrane. Their distribution in the envelope suggests functional clustering (Grunewald *et al.*, 2003).

The requirement for a multigene system in order for HSV to enter the cells has been one of the most puzzling issues in the field, and one that differentiates HSV, and herpesviruses in general, from the majority of viruses. Thus, while most enveloped viruses require one or at most two glycoproteins in order to carry out receptor-recognition, triggering and execution of fusion, HSV needs a minimum of four, but perhaps more proteins. Assignment of a specific function to each of them has been a major task in the past 15 years. Still another major task has been the identification of the cellular proteins that serve as receptors for virus entry, a task made more complicated by the ability of the virus to use multiple alternative receptors.

The current model of HSV entry consists of the following steps; see (Campadelli-Fiume *et al.*, 2000; Spear *et al.*, 2000; Spear and Longnecker, 2003). First, the virus attaches to cell membranes by interaction of gC, and possibly gB, to glycosaminoglycans (GAGs) (Herold *et al.*, 1991). Thereafter, gD recognizes trough its receptor-binding domain one of the alternative receptors, nectin1, nectin2, HVEM (herpesvirus entry mediator) (Montgomery *et al.*, 1996; Cocchi *et al.*, 1998b; Geraghty *et al.*, 1998; Warner *et al.*, 1998; Lopez *et al.*, 2000). A distinct domain in gD, the pro-fusion domain (Cocchi *et al.*, 2004a), intervenes and possibly triggers fusion by recruiting (one of) the three additional glycoproteins, gB, gH, gL, responsible for execution of fusion (Cai *et al.*, 1988; Ligas and Johnson, 1988; Forrester *et al.*, 1992; Roop *et al.*, 1993). Fusion remains the least understood step of entry. The identification in gH ectodomain of a hydrophobic α-helix with attributes of an internal fusion peptide, and of two heptad repeats with predicted propensity to form coiled coils points to gH as the actual fusion glycoprotein (Gianni *et al.*, 2005a; Gianni *et al.*, 2005b). Fusion may take place either at the plasma membrane or in endocytic vesicles (Nicola *et al.*, 2003; Gianni *et al.*, 2004; Nicola and Straus, 2004). For a schematic representation of the entry process, see Fig. 8.1. Following fusion, the released tegumented nucleocapsid travels along microtubules to the nuclear pore, where the viral DNA is released into the nucleus (Sodeik *et al.*, 1997).

Attachment of HSV to cells

The clear differentiation between attachment and entry reflects a net differentiation of the viral glycoproteins which perform these tasks, as well as of their cognate cellular receptors. gC, the product of UL44 gene, is a highly N- and O-glycosylated protein, and the major determinant of HSV attachment to cells (Herold *et al.*, 1991). gB (encoded by UL27 gene) is also involved in attachment through a lysine-rich region (Laquerre *et al.*, 1998). The glycosaminoglycan (GAG) moieties of heparan sulfate (HS), or of chondroitin sulfate, serve as attachment receptor (Spear *et al.*, 1992). The gC, and gB interaction with

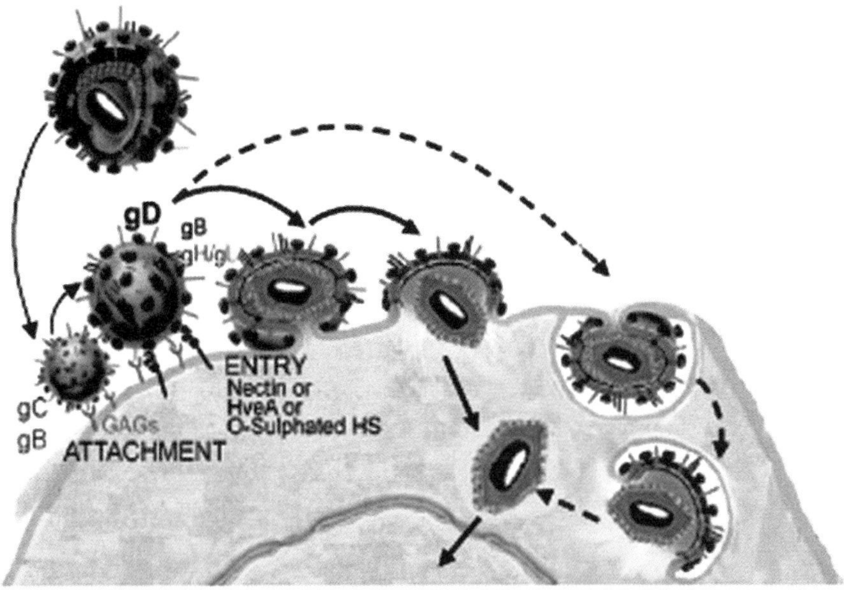

Figure 8.1 Schematic drawing showing HSV entry into the cell. Virus attaches to cell surface GAGs, and subsequently interacts with alternative gD receptors (thin solid arrows). Fusion of the viral envelope with cell membranes may take place at the plasma (solid arrows), or following endocytosis (dashed arrows).

GAGs likely creates multiple points of adhesion, is reversible, and the detached virus maintains its infectivity, indicating that fusion has yet to take place. Despite the fact that the gC-HS interaction is at high affinity (Rux *et al.*, 2002), gC is a nonessential glycoprotein, and cells defective in both HS and chondroitin sulfate are still infectable, although at low efficiency (Gruenheid *et al.*, 1993). Hence, attachment greatly enhances infectivity, but is not a strict requirement for it.

Receptor recognition

gD: structure and function

gD is the receptor-binding glycoprotein of HSV. It interacts with alternative receptors, HVEM, nectin1, nectin2, and also with modified sites on heparan sulfate (Montgomery *et al.*, 1996; Cocchi *et al.*, 1998b; Geraghty *et al.*, 1998; Warner *et al.*, 1998; Shukla *et al.*, 1999). The domain with receptor binding activity maps to the first 250–260 residues of the ectodomain (Krummenacher *et al.*, 1998). Downstream of it, lies the pro-fusion domain (Cocchi *et al.*, 2004a).

The receptor-binding activity of gD was recognized long before the receptors were actually identified. When constitutively expressed in cells, gD renders cells resistant to infection. This property (designated as resistance to infection, or gD-mediated interference) together with the ability of soluble gD to bind cells in a saturable manner and to block infectivity, provided the first indirect evidence that gD interacts with cellular receptor(s) (Campadelli-Fiume *et al.*, 1988; Johnson and Spear, 1989; Johnson *et al.*, 1990; Nicola *et al.*, 1997). HSV mutants able to overcome the gD-mediated resistance, named gD-unrestricted or *rid* mutants, all carry substitutions in gD at residue 25 or 27, and acquire the ability to interact with nectin 2 as an alternative receptor.

One of the key achievements of the past few years has been the resolution by the Cohen–Eisenberg and the Wiley–Carfi laboratories of the crystal structure of a soluble form of gD, up to amino acid residue 259 of the ectodomain (Carfi *et al.*, 2001). Parenthetically, gD is the first herpesvirus glycoprotein whose crystal structure has been solved, followed by Epstein Barr virus gp 42. The structure was not predicted by the primary sequence, and consists of a central core (residues 56–184) with immunoglobulin (Ig) folding, bracketed by N- and C-terminal extensions. The N-terminus (residues 1–37) was disordered in the crystal of gD alone, and formed a hairpin in the crystal of gD complexed with HEVM (Carfi *et al.*, 2001), documenting a conformation modification to gD upon receptor binding. The C-terminus folds back towards the N-terminus and forms an α-helix which supports the hairpin. The structure of the region downstream of residue 260 has been solved subsequently, and indicates that the membrane-proximal region contacts the N-terminus and the Ig core (Krummenacher *et al.*, 2005).

The crystal structure of gD in complex with nectin1 has not yet been solved. The nectin1-binding site on gD, determined by means of insertion-deletion or substitution mutants, appears to be more widespread than that of HVEM (Yoon *et al.*, 2003; Zhou *et al.*, 2003).

Recently, it has become apparent that the gD ectodomain encodes a second functional domain, the pro-fusion domain, located membrane-proximally, i.e. physically separate from the receptor-binding domain. The key observation that led to its discovery was that a soluble form of gD, deprived of the TM and C-tail regions, is sufficient to rescue the infectivity of a gDnull virus, provided that it carries the receptor-binding sites and the downstream membrane-proximal region (Cocchi *et al.*, 2004a). Its most likely role is to signal the encounter with the cell and the receptor-recognition, and to trigger fusion.

The gD-binding entry receptors

The interaction of gD with one of the entry receptor is an absolute requirement for HSV infectivity. The entry receptors known to date invariably interact with gD, in agreement with the notion that gD elicits potent neutralizing antibodies. The initial identification of HVEM as an HSV receptor, accomplished by the Spear laboratory through the screening of a c-DNA library transfected into receptor-negative cells, was a major breakthrough in the field, not only for its intrinsic relevance, but also because it paved the way to further search of entry receptors (Montgomery *et al.*, 1996). The receptors are listed in Table 8.1. In a few cases, gD-dependent pathways of entry, independent of above receptors, have been described, raising the possibility that the spectrum of entry receptors may be higher than known to date (Zago and Spear, 2003; Cocchi *et al.*, 2004b). The possibility that other glycoproteins, e.g. gB or gH, interact with cellular proteins can not discarded (Bender *et al.*, 2005). In particular, gB of other herpesviruses interacts with integrins (Akula *et al.*, 2002; Feire *et al.*, 2004).

HVEM belongs to the TNF (tissue necrosis factor) receptor family. Its ectodomain is made of 5 cysteine-rich domains (CRD) (Montgomery *et al.*, 1996). It serves as receptor for HSV-1 and HSV-2, as a weak receptor for the U10 HSV mutant (L25P), but not for HSV *rid*-1 and *rid*-2 unrestricted mutants (Q27P, Q27R) (Table 8.1) (Connolly *et al.*, 2003; Yoon *et al.*, 2003). It interacts physically with wild type gD. The affinity ranges from 10^{-6} to 10^{-8} Molar, is highest to forms of gD truncated at or after residue 250, and decreases by one hundred fold with gD_{306t} (Krummenacher *et al.*, 1998; Willis *et al.*, 1998; Whitbeck *et al.*, 1999). A systematic structure-based mutagenesis revealed 17 residues

Table 8.1 The human receptors which mediate HSV entry into cells

		Virus for which the molecule serves as entry receptor					Distribution	
Name	Alternative designation	Wt HSV-1	HSV-1 gD-Unr.	HSV-2	PrV	BHV-1	Cell lines	Tissues
HVEM	HvEA	+	–	+	–	N.D.	T-lymphocytic Monocytic	Lung, liver. kidney, heart, brain placenta
Nectin 1α and β	PRR1 HveC HIgR	+	+	+	+	+	Epithelial Endothelial Fibroblastic Keratinocytes Monocytic Myeloid Erythroid B-lymphocytic T-lymphocytic	Brain, spinal cord, trachea, prostate, placenta, skin, kidney, lung, pancreas, thyroid, liver, myelomonocytic cells
3-O-sulfated HS		+	N.D.	–	N.D.	N.D	*Human, rat and mouse endothelial Mouse mast cell	*Liver, placenta, heart, kidney, pancreas
Nectin 2α and β	PRR2α HveB PRR2β	–	+	+/–	+	–	Epithelial Endothelial Neuronal Monocytic Myeloid Erythroid B-lymphocytic	Placenta, prostate, kidney, lung, pancreas, thyroid, liver, myelomonocytic cells, endothelial cells

HSV-1 gD-Unr, gD-unrestricted or *rid* mutants.
N.D., not determined.
+, –, +/– indicate ability, lack of ability, or weak ability to mediate entry when cDNAs are transfected into receptor-negative cells.
*Distribution of heparan sulfate D-glucosaminyl 3-O-sulfotransferase (3-OST).

in CRD1 and 4 in CRD2 directly involved in the HVEM-gD interface. Residues 35–37 form an intermolecular antiparallel β-sheet with gD, thought to stabilize the complex (Carfi *et al.*, 2001; Connolly *et al.*, 2003). A number of mutations that abolish the gD binding to HVEM, but not to nectin1 have been described (Connolly *et al.*, 2002; Martinez and Spear, 2002; Yoon *et al.*, 2003)

While it was initially reported that HVEM expression is restricted mainly to cells of lymphoid origin, implying that it serves as the principal receptor for HSV in activated T lymphocytes, or in lymphoid organs (spleen and thymus), it was later recognized that HVEM is expressed in a number of additional cell types, some of which simultaneously express nectin1 (Montgomery *et al.*, 1996; Krummenacher *et al.*, 2004).

Nectins constitute a growing family of intercellular adhesion molecules with Ig structure. Together with cadherins, they represent key constituents of the adhrens junctions (Reymond *et al.*, 2000; Takai *et al.*, 2003). Three isoforms of nectin1 and 2 isoforms of nectin2 are known; the α and β isoforms are membrane-bound (Table 8.1) (Cocchi *et al.*, 1998a; Cocchi *et al.*, 1998b; Geraghty *et al.*, 1998; Warner *et al.*, 1998; Krummenacher *et al.*, 1999; Campadelli-Fiume *et al.*, 2000; Lopez *et al.*, 2000).

Nectin1 interacts physically with gD (Krummenacher *et al.*, 1999). The first 250 residues of gD and the CC′C″ ridge of the nectin1 V domain are required for the interaction (Krummenacher *et al.*, 2000; Cocchi *et al.*, 2001; Menotti *et al.*, 2002). The affinity of the binding is of the same order of magnitude as that of HVEM to gD (Whitbeck *et al.*, 1999). Residues critical for gD interaction with nectin1 include R134, P221, R222, F223, D215 (Milne *et al.*, 2003; Yoon *et al.*, 2003; Zhou *et al.*, 2003; Jogger *et al.*, 2004; Manoj *et al.*, 2004). Of note, residue Y38, located in the N-terminus of the molecule, and initially believed to represent the contact site to HVEM only, represents a critical residue for interaction with nectin1 (Connolly *et al.*, 2005). Remarkably, mutant forms of gD exhibiting low or undetectable binding to nectin1 maintain the ability to mediate HSV entry. This implies that gD functions in virus entry and cell fusion regardless of its receptor-binding affinity, and that as long as interaction with a functional receptor occurs, entry takes place (Milne *et al.*, 2003; Zhou *et al.*, 2003).

Human nectin1 also binds isoforms of gD from animal α-herpesviruses, with varying affinities (Table 8.1) (Connolly *et al.*, 2001).

With respect to nectin2, a single amino acid substitution in gD (L25P, Q27P, or Q27R) confers to HSV the ability to use this alternative receptor, without hampering its ability to use nectin1 (Table 8.1) (Warner *et al.*, 1998; Lopez *et al.*, 2000; Connolly *et al.*, 2003; Yoon *et al.*, 2003; Krummenacher *et al.*, 2004;). At the same time, this mutation abolishes the interaction with HVEM (Connolly *et al.*, 2003; Yoon *et al.*, 2003). The end result is that a single substitution in gD modifies the host range of the virus.

The availability of multiple alternative receptors raises the issue as to what receptor is preferred in cells that coexpress both nectin1 and HVEM, and which receptor serves as *bona fide* receptor in human tissues. Because the gD affinity to nectin1 and to HVEM is of the same order of magnitude, affinity can not constitute a basis for preferential usage (Krummenacher *et al.*, 1998). Density of receptors also does not seem to represent a critical parameter, as low density receptors suffice to mediate entry (Krummenacher *et al.*, 2004). The finding that clinical isolates can use both receptors suggests that such attribute facilitates successful infection and spread in the host (Krummenacher *et al.*, 2004).

The distribution of nectin 1, strikingly different from that of HVEM, is compatible with a major role of this receptor in the infection of sensory neurons and muco-epithelia in human. Thus, nectin1 is broadly expressed in human tissues, particularly tissues and organs targeted by HSV, like CNS, ganglia and muco-epithelia (Table 8.1) (Cocchi *et al.*, 1998b; Matsushima *et al.*, 2003; Richart *et al.*, 2003; Linehan *et al.*, 2004). It is also expressed in nearly every neuron of adult mice, rat sensory neurons, in some synapses, and serves as the primary receptor for HSV-1 infection in sensory neurons (Haarr *et al.*, 2001; Mata *et al.*, 2001; Richart *et al.*, 2003). Virtually all human cell lines, including epithelial, neuronal and fibroblastic cells express nectin1, see (Campadelli-Fiume *et al.*, 2000; Spear *et al.*, 2000; Krummenacher *et al.*, 2004). Nectin1-α is detected at the cell–cell adherens junctions in human skin (Matsushima *et al.*, 2003). Because of this restricted distribution, it has been proposed that nectin1 may be relevant for basolateral infection of epithelia (Schelhaas *et al.*, 2003). However, the β isoform is not restricted to basolateral domains of epithelial cells (Cocchi *et al.*, 1998b), and indeed human CaCo2 cells can be infected with HSV from the apical domain (Griffiths *et al.*, 1998). Nectin1 is also expressed in the epithelium of the human and murine vagina, where it mediates HSV-1 and -2 entry in the genital mucosa (Linehan *et al.*, 2004).

A third potential receptor for HSV gD is a modified form of heparan sulfate, 3-O-sulfated heparan sulfate (3-OS HS). This modification is catalyzed by the enzyme D-glucosaminyl 3-O-sulfotransferase, and results in the generation of gD-binding sites on HS (Shukla *et al.*, 1999). While the enzyme is present in a number of human cells, including neuronal and endothelial cells, it remains to be determined whether there are human cell types which become actually infected through this receptor.

The gD pro-fusion domain and the triggering of fusion

While the role of gD as the HSV receptor-binding glycoprotein has been known for more than a decade, the link between the receptor-binding activity of gD and the downstream events that culminate in fusion are still largely unknown. A step further has been the recognition that the gD ectodomain includes a distinct domain, the pro-fusion domain. At most, it encompasses residues 260–315. Evidence for this domain rests on the ability of soluble gD, truncated at or after residue 275, to substitute for membrane-bound gD, and to rescue the infectivity of a gDnull HSV. This property is in agreement with the observation that a glycosylphosphatidylinositol- (GPI-) anchored form of gD, or substitution of the TM and C-tail segments with heterologous sequences, leave gD function unaltered (Browne *et al.*, 2003; Cairns *et al.*, 2003). Further yet, substitution of the pro-fusion domain in full length gD with the membrane-proximal region of an unrelated molecule (CD8) abolishes HSV infectivity and cell fusion (Cocchi *et al.*, 2004a).

Biochemical and structural analyses concordantly indicate that the profusion domain is responsible for a major modification in gD conformation, which takes place at receptor binding. Specifically, prior to receptor binding, gD adopts a closed conformation where the C-terminus of the ectodomain, carrying the profusion domain, folds backs and wraps the gD N-terminus. At receptor binding, the C-terminus is displaced from its binding site on the N-terminus, and the latter interacts with the receptor. This modification is thought to be the key event in the triggering of fusion, and the opened gD is likely to create a structure for the recruitment of downstream glycoproteins (Fusco *et al.*, 2005; Krummenacher *et al.*, 2005).

Entry at the plasma membrane and through acidic endosomes

HSV entry is unusual in that fusion of virion with the target cell may take place either at the plasma membrane (or a neutral pH compartment) or in acidic endosomes (Nicola *et al.*, 2003; Nicola and Straus, 2004). Contrary to what happens with viruses like influenza viruses, that require the acidic compartment for activation of the fusion activity, HSV glycoproteins are capable of carrying out fusion both at neutral and acidic pH. The factors that determine the site of fusion are the cell lines and the structural features of the receptors. Specifically, the same receptor, nectin or HVEM, mediates a neutral pH- or a low pH-pathway of entry when expressed in the context of different cells (Nicola *et al.*, 2003; Gianni *et al.*, 2004). Furthermore, modified forms of nectin1, made competent for endosomal uptake, or sorted to lipid rafts, by means of GPI anchor, mediate a pathway of entry into J cells through acidic endosomes, and not through the plasma membrane or a neutral pH compartment (Gianni *et al.*, 2004).

The cell–cell fusion assay

A powerful tool in the studies of HSV entry and fusion has been the development by Minson, Browne, and coworkers of a virus-free fusion assay, called the cell–cell fusion assay (Turner *et al.*, 1998). In this assay, receptor-positive target cells fuse with effector cells transfected with the quartet of gD, gB, gH, gL. Because cell–cell fusion has the same primary requirements as virion-to-cell fusion (i.e. the quartet of glycoproteins plus a gD receptor), it is a relatively faithful mirror of the virion–cell fusion (Pertel, 2001). Its advantages are that (i) it is rapid, straightforward, and readily quantifiable, (ii) it is not influenced by other viral proteins that regulate the fusion activity of the quartet, (iii) it allows to study of the effect of substitutions or additional entities that affect fusion.

Fusion execution

Until recently, fusion execution was the least understood step in HSV entry. Part of the difficulties in elucidating this step stems from the fact that it requires three glycoproteins (gB, gH, gL). Therefore, it did not appear to fall into any known model of viral fusion. In particular, what is puzzling is why it requires three glycoproteins, and what are the respective roles of the members of the trio. Remarkably, the trio of gB, gH, gL is conserved throughout the Herpesviridae family; hence, the molecular mechanism of fusion appears to be conserved. What differs from one herpesvirus to the other is the receptor-binding glycoprotein, which confers specificity of the targeted cells and tissues.

Recent achievements suggest that gH may be the actual fusion glycoprotein in HSV, as it exhibits structural features typical of class I viral fusion glycoproteins, i.e. a hydrophobic α-helix with attributes of an internal fusion peptide, and two heptad repeats with propensity to form coiled coils, designated HR-1 and HR-2, respectively. Fusion peptides are hydrophobic segments located either at the N-terminus, or contained in the ectodomain of fusion glycoproteins. They penetrate the target membrane, such that the fusion glycoprotein is simultaneously anchored to the virion envelope and to the target membranes, an event critical to trigger fusion pore formation. Coiled-coil formation is a series of conformational changes that take place after insertion of the fusion peptide into the target cell, and lead to the hairpin structure of the fusion glycoprotein.

Structurally, gH is a type 1 glycoprotein, encoded in the UL22 gene (Gompels and Minson, 1986). It forms a heterodimer with gL, a soluble glycoprotein that enables gH maturation and trafficking through the exocytic pathway (Hutchinson *et al.*, 1992a; Roop *et al.*, 1993). Evidence that the hydrophobic α-helix in gH ectodomain has attributes of an internal fusion peptide is as follows. First, the α-helix has intrinsic ability to interact with membranes, as documented by the ability to convert a soluble into a membrane-bound glycoprotein. Second, the α-helix represents a critical domain: its partial or entire removal, or substitution of critical residues abolish fusion and infectivity. Third, and foremost, substitution of the native α-helix with the fusion peptide sequence from immunodeficiency virus gp41 or from vesicular stomatitis virus G rescues the fusion and infectivity activities of gH (Gianni *et al.*, 2005a). With respect to the heptad repeat sequences, mutations in HR-1, predicted to abolish the propensity to form coiled coils, reduce virus entry and fusion activity (Gianni *et al.*, 2005b). A peptide with the sequence of HR-1 specifically reduces HSV infectivity. Mutations in HR-2 reduce fusion and infectivity (Gianni *et al.*, 2006). HR-1 and -2 interact physically.

The respective roles of gB and gL in the execution of HSV fusion are not known.

Negative control on fusion

Fusion appears to be a very critical event in the life cycle of herpesviruses. A novel concept that has emerged from studies of varicella zoster virus and HSV glycoproteins (see Chapter 9) is that not only have herpesviruses evolved a number of gene products to trigger and execute fusion, but they also evolved a sophisticated mechanism for its negative control. The two basic mechanisms of control in HSV appear to be: (i) the steady state amount of viral fusion glycoproteins, in particular of gB, at the plasma membranes; (ii) the activity of additional entities, in particular of the polytopic membrane proteins gK, UL20 and gM. Such a sophisticated control system testifies how inimical fusion may be to the virus, if it occurs out of place and out of time.

gB

gB appears to play a dual role in HSV fusion. On one hand, it is required for execution of fusion, through the ectodomain. On the other hand, it negatively regulates fusion, via its long C-tail. The gB C-tail carries at least two functional endocytosis motifs, one di-leucine based (LL871) and one thyrosine-based (YTQV889–892) (Fan *et al.*, 2002; Avitabile *et al.*, 2004; Beitia Ortiz de Zarate *et al.*, 2004). They traffic gB from the plasma membrane to small endocytic vacuoles, which coalesce into large vacuoles in 1–2 hours (Avitabile *et al.*, 2004). The end result is a subtraction of the amount of gB that appears on the plasma membrane, and a concomitant decrease in the ability of the cells to fuse with adjacent cells (Avitabile *et al.*, 2003).

The gB C-tail is the site of *syn* mutations (*syn* 3). Cells infected with *syn* mutants fuse with adjacent cells; by contrast, cells infected with wild type viruses round up and aggregate, but do not form syncytia. *Syn* mutations are thought to affect genes whose products negatively regulate fusion in their wild type allelic form. The mechanism by which *syn* mutations disable the negative control on cell fusion is unknown.

gK, UL20, and gM

These three proteins share structural and functional properties. They all are multiple-spanning membrane proteins, i.e. they carry multiple TM segments, are extremely hydrophobic and embedded in the membranes. gK and UL20 are thought to have 4 TM segments, and gM is predicted to have 8 TM segments (Hutchinson *et al.*, 1992b; Foster *et al.*, 2001). Because of the high hydrophobicity, the characterization of these proteins has been a demanding task.

Together with gB, gK (the product of UL53 gene) is the most frequent locus of *syn* mutations (*syn* 1). Similarly, deletion of UL20 gene also generated a *syn* mutant (Baines *et al.*,

1991). When expressed singly, gK and UL20 localize at the endoplasmic reticulum (ER) (Hutchinson *et al.*, 1992b; Ward *et al.*, 1994; Avitabile *et al.*, 2003; Foster *et al.*, 2003a; Foster *et al.*, 2003b). When coexpressed, they enable each other to traffic and localize mainly at the Golgi apparatus and *trans* Golgi network (TGN) (Avitabile *et al.*, 2004). They are not detectable at the plasma membrane. At least for gK, this may be due to endocytosis from the plasma membrane (Foster *et al.*, 2004).

To investigate the role of gK and UL20 in fusion, the proteins were coexpressed with the quartet of gB, gD, gH, and gL in the cell–cell fusion assay. gK alone inhibits fusion in a cell line dependent fashion. The inhibition is augmented by coexpression of UL20 (Avitabile *et al.*, 2003; Avitabile *et al.*, 2004). Attempts to elucidate the mechanism by which gK and UL20 decrease fusion activity highlighted a modest decrease in cell surface expression of all fusion glycoproteins, in particular gD.

The phenotype of deletion mutant viruses in gK and UL20 is interesting. The first generation deletion viruses were reported to accumulate virions at the perinuclear space (Baines *et al.*, 1991; Hutchinson and Johnson, 1995). The second generation deletion mutants are still defective in virus egress, but the unenveloped nucleocapsids accumulate in the cytoplasm rather than in the perinuclear space (Foster and Kousoulas, 1999; Foster *et al.*, 2004). It has been postulated that a negative control of fusion is required not only at the plasma membranes, but also intracellularly, at the ER and Golgi apparatus. These membranes are decorated on their luminal sides with the viral fusion glycoproteins, yet do not fuse with each other. The gK and UL20 proteins are good candidates for this inhibition.

gM (the product of UL10 gene) exhibits interesting analogies to gK and UL20 (Baines and Roizman, 1993). In addition to being multiple spanning, gM is conserved throughout the Herpesviridae, despite of the fact that is defined as a nonessential glycoprotein in alpha-herpesviruses. It forms a complex with the UL49A (or UL49.5 gene product, also named gN in other herpesviruses). Recently, it has been reported that both HSV gM, as well pseudorabies virus (PrV) gM, which are themselves localized at the TGN, reduce fusion in the cell–cell fusion assay (Klupp *et al.*, 2000; Crump *et al.*, 2004). The coexpression of UL49A augments the inhibition. Remarkably, in gM-expressing cells gD and gH-gL were relocalized from the plasma membrane to the TGN compartment, raising the possibility that the fusion inhibition was the result of glycoprotein subtraction from the plasma membrane. Of note, the gM-UL49A-mediated decrease in cell surface expression is not restricted to the HSV fusion glycoproteins; this raises the possibility that trafficking out of the plasma membrane may go beyond preventing the infected cells from fusing with adjacent cells, and may include the maintenance of sufficient concentrations of envelope proteins at site of secondary envelopment, as well as mechanisms of immune evasions (Crump *et al.*, 2004).

HSV egress

Models of HSV egress

A distinctive property of the Herpesviridae family is capsid assembly in the nucleus. This is thought to be followed by envelopment and budding at the inner nuclear membranes (see chapter "HSV packaging and assembly"), and release of enveloped virus into the perinuclear space (the space between the inner and outer nuclear membrane). Because this is a closed compartment, virions need to exit it, in order to be ultimately released into the extracellular space and initiate a new round of infection.

Two models of egress have been extensively discussed in the literature, with controversy; for excellent reviews, see (Enquist *et al.*, 1998; Skepper *et al.*, 2001; Johnson and Huber, 2002; Mettenleiter, 2002). A novel model has been proposed recently for bovine herpesvirus 1 (BHV-1) and HSV-1; interestingly, it appears to overcome a major inconsistency of one of the two other models. For a recent debate on the strong and weak points of the pathways of HSV egress, see Mettenleiter and Minson (2006), Wild (2006), and Campadelli-Fiume and Roizman (2006). A schematic drawing is shown in Fig. 8.2.

In the luminal model (also referred to as the single envelopment model), credited to Johnson and Spear (Johnson and Spear, 1982), virions leave the perinuclear space as they become wrapped by a vesicle/vacuole derived from the outer nuclear membrane (Fig. 8.2, route 1, left hand). Thereafter, the vesicle-vacuole which surrounds the enveloped viruses interacts with membranes of the exocytic pathway, mainly the Golgi apparatus, which provide the machinery for glycoprotein maturation. Finally, the vesicle–vacuole fuses with the cytoplasmic face of the plasma membrane, thus releasing the mature virions into the extracellular space. According to this route of egress, virions maintain the envelope acquired at the inner nuclear membrane, always remain in luminal spaces, and their oligosaccharide moieties undergo maturation.

The de-envelopment re-envelopment model of egress (Stackpole, 1969) (Fig. 8.2, route 2, right hand) envisions that virions in the perinuclear space fuse their initial enve-

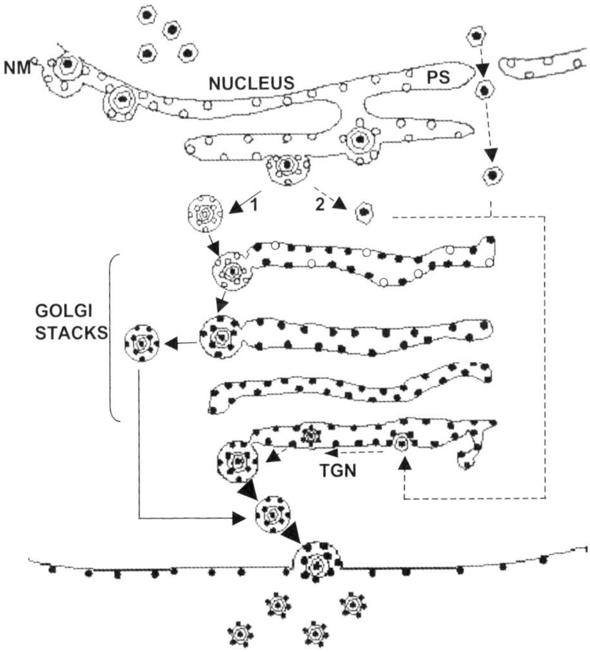

Figure 8.2 Three cited models of herpes simplex virus egress from infected cells. On the left, the luminal model is depicted (1, solid arrows). The de-envelopment re-envelopment initiates at the centre of the nuclear membrane (NM) and proceeds on the right (2, dashed arrows). Exit through the nuclear pores is depicted at the right of the nuclear membrane, and proceeds via the dotted arrows (3, right). Acquisition of the final envelope in the de-envelopment re-envelopment model and in the nuclear pore exit model is depicted at *trans*-Golgi network (TGN). Arrowheads refer to the putative final steps of egress, common to the three pathways. Viral glycoproteins are shown are small circles; hollow circles, immature forms of glycoproteins; full circles, mature forms of glycoproteins. PS, perinuclear space.

lope with the outer nuclear membrane (de-envelopment), such that unenveloped nucleocapsids are released into the cytoplasm. These travel to the late Golgi membranes, or TGN, where they undergo a secondary envelopment (re-envelopment), and acquire the final envelope with mature glycoproteins.

Electron microscopy studies of infected cells have added to the dispute. One source of disagreement centers on the significance of unenveloped nucleocapsids in the cytoplasm. In the luminal route of egress they are "dead-ends," formed by uncontrolled fusion of the virion envelope with the membrane of the surrounding vesicle-vacuole (Campadelli-Fiume *et al.*, 1991). In the de-envelopment re-envelopment route of egress the unenveloped nucleocapsids are the fundamental intermediates for secondary envelopment. Even though a growing preference is emerging in favor of the de-envelopment re-envelopment pathway, there remain unexplained features, as outlined below.

A third model of egress, recently described for BHV-1 and HSV-1, envisions that, in the late stages of infection, nucleocapsids exit through the modified nuclear pores and thus pass directly from the nucleoplasm to the cytoplasm (Wild *et al.*, 2004). Here they would undergo their first envelopment at late Golgi or TGN. Nucleocapsid budding at the inner nuclear membranes is a rare early event, not relevant for the final production of extracellular virus, because perinuclear virions never exit the perinuclear space.

This chapter is not intended to provide an extensive overview of all the pertinent literature. We emphasize here recent lines of evidence in favor and against each pathway.

The major lines of evidence in favor of the de-envelopment re-envelopment pathway of egress are as follows.

1 Electron microscopic studies of HSV assembly in neurons provide primary evidence for separate transport of capsids and viral glycoproteins. Capsids were observed migrating in anterograde direction within axons, whereas the viral glycoproteins (gD) were observed in vesicles which do not colocalize with nucleocapsids, suggesting that virion assembly in axons represents a final step that occurs at the axon termini (Penfold *et al.*, 1994). These observations were confirmed and extended in an immunocytochemical study of tegument assembly and envelope acquisition in rat dorsal neurons; it was observed that while the site of HSV tegument assembly is the cytoplasm of the neuronal cell body, the major sites of envelope acquisition are the vesicles of the Golgi and TGN. Evidence rested on the finding that the tegument proteins VP13/14, VP16, VP22, and US9 were readily detected in the nucleus, but almost absent from the budding virions at the nuclear membranes and from virions in the perinuclear space. By contrast, they were abundant on cytoplasmic unenveloped and enveloped nucleocapsids and in extracellular virions (Miranda-Saksena *et al.*, 2002). The same pattern of labeling was observed for gD.
2 The UL31 protein is a matrix-associated phosphoprotein that localizes to the nuclear membranes in the infected cells, and when coexpressed with the UL34 protein. The UL34 protein is a predicted type-2 membrane-bound phosphoprotein with the bulk of the protein constituting the endodomain exposed to the nucleoplasm. It requires the UL31 protein for localization at the nuclear envelope. The two proteins are required for virus envelopment at the inner nuclear membranes, and are constituents of the perinuclear virions. Remarkably, both proteins are absent (or undetectable) from extracellular virions, a finding inter-

preted as evidence of envelope substitution (Roller *et al.*, 2000; Ye and Roizman, 2000; Reynolds *et al.*, 2001; Reynolds *et al.*, 2002).

3 A form of gD carrying an ER-retrieval motif. The rationale of this study was to engineer a form of gD carrying an ER-retrieval motif. The ER-retrieved gD was expected to be present at the ER and the nuclear membranes, which are continuous with the ER, but not to travel along the exocytic pathway. Virions at the perinuclear space were expected to be decorated with gD; virions undergoing de-envelopment re-envelopment were expected to exchange this envelope, so that extracellular virions would be devoid of gD (Whiteley *et al.*, 1999; Skepper *et al.*, 2001). When analyzed by immunocytochemistry, the perinuclear virions and the nuclear membranes were indeed decorated with the ER-retargeted gD, whereas the extracellular virions and the plasma membrane were not (Whiteley *et al.*, 1999; Skepper *et al.*, 2001). These results are consistent with a double envelopment route of egress. However, in the same study quantification of the distribution of gD in wild type-virus infected cells showed that the plasma membranes contained 20-fold less, or even lower amount of gD than the nuclear and cytoplasmic membranes (Skepper *et al.*, 2001). This contrasts with the detection of gD in approximately the same amounts at the nuclear membranes and at the plasma membranes in a fracture-label study, with the abundant detection of gD at the plasma membrane by immunofluorescence, and with the detection of gD more abundantly at the plasma membrane than at the nuclear membranes in neurons and raises the possibility that the degree of detection may be affected by the specific reactivity of the antibodies employed (Torrisi *et al.*, 1992; Miranda-Saksena *et al.*, 2002).

4 Studies of tegument assembly performed with PrV indicate that intranuclear nucleocapsids do not contain a well-defined tegument, detected as an electrondense layer surrounding the capsid, whereas extracellular virions and cytoplasmic nucleocapsids contain an electrondense tegument, see (Mettenleiter, 2002). UL36 and UL37 proteins constitute the inner layer of the tegument. Cells infected with deletion mutant viruses in these tegument proteins show that capsids acquire the envelope at the inner nuclear membrane, are subsequently translocated into the cytoplasm, but do not mature into enveloped virus and do not exit the cell (Desai, 2000; Desai *et al.*, 2001).

By contrast with the effect of UL36 and UL37 deletions, virion morphogenesis is not hampered in a VP22 deletion mutant HSV, nor in a number of deletion mutant viruses in the genes encoding other tegument proteins, despite the fact that VP22 localizes predominantly at the cytoplasm see (Elliott and O'Hare, 1999; Pomeranz and Blaho, 2000; Mettenleiter, 2002). These findings have been interpreted as evidence that tegument proteins play redundant functions, and that the absence of a single tegument protein does not hamper tegument assembly (see Mettenleiter, 2002).

5 The phospholipid composition was determined in infected cells fractionated into two fractions – a nuclear fraction (which contains nuclei and virions at perinuclear space) and a cytoplasmic fraction (which contains the cytoplasmic membranes, the intracytoplasmic virions and the plasma membranes) – and compared with that of extracellular virions. The phospholipid composition of extracellular HS virions and of the cytoplasmic fraction differs

from that of nuclei; the former contains higher concentrations of lipids that are typically enriched in the Golgi apparatus and plasma membrane (van Genderen *et al.*, 1994). This difference implies that the lipid composition of the envelope acquired at the inner nuclear membrane differs from that of extracellular virions. It has been interpreted as evidence in favor of the de-envelopment re-envelopment pathway (Enquist *et al.*, 1998).

6 Multiple deletion of glycoprotein genes, especially nonessential glycoproteins. Practically all HSV glycoproteins—both essential and nonessential—have been deleted singly or in groups, and none of the deletion viruses is defective in virus transport out of the perinuclear space (Longnecker and Roizman, 1987; Longnecker *et al.*, 1987; Cai *et al.*, 1988; Ligas and Johnson, 1988; Forrester *et al.*, 1992; Baines and Roizman, 1993; Roop *et al.*, 1993; Dingwell *et al.*, 1994), suggesting that no single glycoprotein plays a critical role in virus exocytosis. The exceptions are the gK- and UL20-deletion mutant viruses, discussed above.

In contrast, viruses carrying multiple deletions in glycoprotein genes exhibited defects in virus morphogenesis, cumulatively suggesting that gE, gI and gD participate in secondary envelopment and that they act in a redundant manner, such that viruses carrying single or double deletions do not exhibit a marked phenotype, whereas viruses carrying a triple deletion do (Farnsworth *et al.*, 2003).

7 Studies of VZV egress, reviewed in the chapter on VZV glycoproteins, support the de-envelopment re-envelopment model. In particular gE, gB and gH all carry tyrosine based endocytosis motifs. They were shown to internalize from the cell surface, after biotin was attached. The envelopes of the virions made in the biotin-labeled cells were themselves labeled, suggesting that the glycoproteins transit from the cell surface to the site of envelopment (Maresova *et al.*, 2005).

The major weakness of the de-envelopment re-envelopment pathway is the inability to explain how virions leave the perinuclear space. This route of egress predicts that perinuclear virions fuse their envelopes with the outer nuclear membranes. However, deletion mutant viruses in the glycoproteins absolutely required for fusion (gB, gD, gH, gL) all produce extracellular virions, indicating that they have no defect in leaving the perinuclear space. To explain these findings it must be postulated that envelope proteins other than the quartet carry out fusion. This contrasts with observations that none of the HSV proteins, singly or in combination, are able to carry out fusion, at least when expressed at the plasma membranes (Turner *et al.*, 1998). Of note, the only viruses which exhibit defects in the release of extracellular virions are the mutants deleted in gK or UL20 genes. This phenotype would point to gK and UL20 as proteins able to carry out fusion at intracellular compartments. Because these proteins do not reach the plasma membrane, the hypothesis can not be verified in the cell–cell fusion assay. In addition, this prediction contrasts with the inhibition exerted by gK and UL20 on fusion mediated by the quartet (Avitabile *et al.*, 2004). The unanswered question is, therefore, how do the cytoplasmic nucleocapsids originate?

A second weakness of the double envelopment pathway is the failure to explain how the de-enveloped nucleocapsids travel from the cytoplasmic face of the outer nuclear membrane to the Golgi or TGN. Nucleocapsids that do not travel along microtubules move very slowly (Sodeik, 2000). However, transport along microtubules seems problematic because of the dramatic modifications in their architecture (Avitabile *et al.*, 1995). The infected cell microtubules form circular rings at the periphery of the cell, and seem rather unsuitable to export capsids from outside the outer nuclear membrane to the TGN. So far, this feature has not received any attention.

The major lines of evidence in favor of the luminal route of egress are as follows.

1 Glycoproteins (and glycolipids) present in the nuclear membranes and in ER carry immature type of oligosaccharides (glycomoieties), which are converted first to intermediate type, and then to fully mature oligosaccharides in the serial stacks of the Golgi apparatus. The single envelopment route of egress predicts that the perinuclear virions carry immature glycomoieties, the extracellular virions carry mature glycomoieties, and the cytoplasmic virions carry in part mature glycomoieties and in part intermediate types of glycomoieties. The latter cannot be present in virions formed in the de-envelopment re-envelopment pathway, unless the secondary envelopment takes place at the cis- or medium-Golgi. If this is the case, the question arises "how does the virion travel from cis- or medium-Golgi to farther compartments of the exocytic pathway?" When the HSV oligosaccharides were characterized in a cytochemical study, it was found that perinuclear virions carry immature oligosaccharides, the intracytoplasmic virions carry both intermediate and mature type of oligosaccharides, and extracellular virions carry exclusively mature oligosaccharides (Di Lazzaro *et al*., 1995).

2 A similar line of reasoning applies to the significance of the markers of the *cis*- and *medium*-Golgi in extracellular virions. It is a well-known phenomenon that viral envelopes carry cellular proteins that are constituents of the membranes where budding took place, or of the compartments which the virus transited. As a consequence, the cellular proteins which decorate the envelope of extracellular virions are indicators of the cellular membranes from which the envelope was derived, or through which the viruses transited. A detailed immunocytochemical study of virus assembly and maturation in neurons showed that extracellular virions carry giantin and mannosidase II, two markers typical of *cis*- and *medium*-Golgi, respectively (Miranda-Saksena *et al*., 2002).

3 A further issue centers on the glycoprotein composition of the perinuclear virions, and particularly whether the glycoproteins required for virion infectivity are acquired during envelopment at the inner nuclear membrane. The actual presence of the essential glycoproteins at nuclear membranes and at the perinuclear virions was established in *in situ* experiments for the two glycoproteins that were searched for, gD and gB (Torrisi *et al*., 1992; Skepper *et al*., 2001; Miranda-Saksena *et al*., 2002).

In essence, the evidence required to support the double envelopment model is the definition of the mechanism by which enveloped virions accumulating in the perinuclear space could become de-enveloped, resulting in the release of capsids into the cytoplasm. In turn, the key evidence required to support the single envelopment model and exit out of enlarged nuclear pores is identification of nuclear transport proteins that could interact with capsids and demonstration that the transport is abolished by mutagenesis of the capsid partner or genetic manipulation of the nuclear transport protein.

We note that most of the studies on HSV egress have dealt with the role of viral gene products; the contribution of cellular proteins, and of the exocytic compartment in general, has been neglected. Studies of the role of cellular functions is complicated by the fact that the cell is severely modified following HSV infection (Campadelli-Fiume *et al*.,

1993; Avitabile *et al.*, 1995; Scott and O'Hare, 2001; Reynolds *et al.*, 2004). These changes may well contribute to mislocalization of cellular markers. Further yet, the modifications are cell line-dependent, and may contribute, in part, to apparent discrepancies between different studies. Finally, most of the studies have been carried out in cultures of epithelial cells or fibroblasts. Exit from the neuronal cells (one of the key host cells in the natural infection) may well be different.

At present, we can only conclude that the biology of HSV egress is still an opened issue.

Concluding remarks

In the past years there has been remarkable progress in our understanding of the events leading to HSV entry into and egress out of the cell. Nevertheless, a great deal remains to be discovered. The key questions that remain to be answered are the structure of viral glycoproteins participating in the events leading to entry into cells both before and after their interaction with cellular counterparts or among each other. We still lack detailed information on the cascade of events that lead to the entry or egress and the signaling pathways that alter the physiology of the cell to activate host response to infection and initiate viral gene expression. It is a characteristic of science that every discovery that sheds light on a particular phenomenon raises more questions than it solves. It is in a way challenging that so many important questions still remain unresolved.

Acknowledgments

We thank Dr. C. Grose for critical reading of the manuscript. Work conducted in our laboratory was supported by FIRB autonomous and coordinated project, Cofin-MIUR 40%, CNR-Functional genomics, University of Bologna 60%, Fondo Pallotti.

References

Akula, S.M., Pramod, N.P., Wang, F.Z., and Chandran, B. (2002). Integrin alpha3beta1 (CD 49c/29) is a cellular receptor for Kaposi's sarcoma-associated herpesvirus (KSHV/HHV-8) entry into the target cells. Cell *108*, 407–419.

Avitabile, E., Di Gaeta, S., Torrisi, M.R., Ward, P.L., Roizman, B., and Campadelli-Fiume, G. (1995). Redistribution of microtubules and Golgi apparatus in herpes simplex virus-infected cells and their role in viral exocytosis. J. Virol. *69*, 7472–7482.

Avitabile, E., Lombardi, G., and Campadelli-Fiume, G. (2003). Herpes simplex virus glycoprotein K, but not its syncytial allele, inhibits cell–cell fusion mediated by the four fusogenic glycoproteins, gD, gB, gH and gL. J. Virol. *77*, 6836–6844.

Avitabile, E., Lombardi, G., Gianni, T., Capri, M., and Campadelli-Fiume, G. (2004). Coexpression of UL20p and gK inhibits cell–cell fusion mediated by herpes simplex virus glycoproteins gD, gH-gL, and wt- gB or an endocytosis-defective gB mutant, and downmodulates their cell surface expression. J. Virol. *78*, 8015–8025.

Baines, J.D., and Duffy, C. Nucleocapsid assembly and envelopment of herpes simplex virus. In, S. Goldin, ed.

Baines, J.D., and Roizman, B. (1993). The UL10 gene of herpes simplex virus 1 encodes a novel viral glycoprotein, gM, which is present in the virion and in the plasma membrane of infected cells. J. Virol. *67*, 1441–1452.

Baines, J.D., Ward, P.L., Campadelli-Fiume, G., and Roizman, B. (1991). The UL20 gene of herpes simplex virus 1 encodes a function necessary for viral egress. J. Virol. *65*, 6414–6424.

Beitia Ortiz de Zarate, I., Kaelin, K., and Rozenberg, F. (2004). Effects of mutations in the cytoplasmic domain of herpes simplex virus type 1 glycoprotein B on intracellular transport and infectivity. J. Virol. *78*, 1540–1551.

Bender, F.C., Whitbeck, J.C., Lou, H., Cohen, G.H. and Eisenberg, R.J. (2005). Herpes simplex virus glycoprotein B binds to cell surfaces independently of heparan sulfate and blocks virus entry. J. Virol. *79*, 11588–11597.

Brack, A.R., Dijkstra, J.M., Granzow, H., Klupp, B.G., and Mettenleiter, T.C. (1999). Inhibition of virion maturation by simultaneous deletion of glycoproteins E, I, and M of pseudorabies virus. J. Virol. *73*, 5364–5372.

Browne, H., Bruun, B., Whiteley, A., and Minson, T. (2003). Analysis of the role of the membrane-spanning and cytoplasmic tail domains of herpes simplex virus type 1 glycoprotein D in membrane fusion. J. Gen. Virol. *84*, 1085–1089.

Browne, H., Bell, S., and Minson, T. (2004). Analysis of the requirement for glycoprotein m in herpes simplex virus type 1 morphogenesis. J. Virol. *78*, 1039–1041.

Cai, W.H., Gu, B., and Person, S. (1988). Role of glycoprotein B of herpes simplex virus type 1 in viral entry and cell fusion [published erratum appears in J. Virol. 1988 Nov;62(11):4438]. J. Virol. *62*, 2596–2604.

Cairns, T.M., Milne, R.S., Ponce-de-Leon, M., Tobin, D.K., Cohen, G.H., and Eisenberg, R.J. (2003). Structure–function analysis of herpes simplex virus type 1 gD and gH-gL: clues from gDgH chimeras. J. Virol. *77*, 6731–6742.

Campadelli Fiume, G., and Roizman, B. (2006). The egress of herpesviruses from cells: the unanswered questions. J. Virol. *80*, in press.

Campadelli-Fiume, G., Arsenakis, M., Farabegoli, F., and Roizman, B. (1988). Entry of herpes simplex virus 1 in BJ cells that constitutively express viral glycoprotein D is by endocytosis and results in degradation of the virus. J. Virol. *62*, 159–167.

Campadelli-Fiume, G., Farabegoli, F., Di Gaeta, S., and Roizman, B. (1991). Origin of unenveloped capsids in the cytoplasm of cells infected with herpes simplex virus 1. J. Virol. *65*, 1589–1595.

Campadelli-Fiume, G., Brandimarti, R., Di Lazzaro, C., Ward, P.L., Roizman, B., and Torrisi, M.R. (1993). Fragmentation and dispersal of Golgi proteins and redistribution of glycoproteins and glycolipids processed through the Golgi apparatus after infection with herpes simplex virus 1. Proc. Natl. Acad. Sci. USA *90*, 2798–2802.

Campadelli-Fiume, G., Cocchi, F., Menotti, L., and Lopez, M. (2000). The novel receptors that mediate the entry of herpes simplex viruses and animal alphaherpesviruses into cells. Rev. Med. Virol. *10*, 305–319.

Carfi, A., Willis, S.H., Whitbeck, J.C., Krummenacher, C., Cohen, G.H., Eisenberg, R.J., and Wiley, D.C. (2001). Herpes simplex virus glycoprotein D bound to the human receptor HveA. Mol. Cell. *8*, 169–179.

Cocchi, F., Lopez, M., Menotti, L., Aoubala, M., Dubreuil, P., and Campadelli-Fiume, G. (1998a). The V domain of herpesvirus Ig-like receptor (HIgR) contains a major functional region in herpes simplex virus-1 entry into cells and interacts physically with the viral glycoprotein D. Proc. Natl. Acad. Sci. USA *95*, 15700–15705.

Cocchi, F., Menotti, L., Mirandola, P., Lopez, M., and Campadelli-Fiume, G. (1998b). The ectodomain of a novel member of the immunoglobulin superfamily related to the poliovirus receptor has the attibutes of a *bonafide* receptor for herpes simplex viruses 1 and 2 in human cells. J. Virol. *72*, 9992–10002.

Cocchi, F., Lopez, M., Dubreuil, P., Campadelli-Fiume, G., and Menotti, L. (2001). Chimeric nectin1-poliovirus receptor molecules identify a nectin1 region functional in herpes simplex virus entry. J. Virol. *75*, 7987–7994.

Cocchi, F., Fusco, D., Menotti, L., Gianni, T., Eisenberg, R.J., Cohen, G.H., and Campadelli-Fiume, G. (2004a). The soluble ectodomain of herpes simplex virus gD contains a membrane-proximal pro-fusion domain and suffices to mediate virus entry. Proc. Natl. Acad. Sci. USA *101*, 7445–7450.

Cocchi, F., Menotti, L., Di Ninni, V., Lopez, M., and Campadelli-Fiume, G. (2004b). The herpes simplex virus JMP mutant enters receptor-negative J cells through a novel pathway independent of the known receptors nectin1, HveA, and nectin2. J. Virol. *78*, 4720–4729.

Connolly, S.A., Whitbeck, J.J., Rux, A.H., Krummenacher, C., van Drunen Littel-van den Hurk, S., Cohen, G.H., and Eisenberg, R.J. (2001). Glycoprotein D homologs in herpes simplex virus type 1, pseudorabies virus, and bovine herpes virus type 1 bind directly to human HveC(nectin-1) with different affinities. Virology *280*, 7–18.

Connolly, S.A., Landsburg, D.J., Carfi, A., Wiley, D.C., Eisenberg, R.J., and Cohen, G.H. (2002). Structure-based analysis of the herpes simplex virus glycoprotein D binding site present on herpesvirus entry mediator HveA (HVEM). J. Virol. *76*, 10894–10904.

Connolly, S.A., Landsburg, D.J., Carfi, A., Wiley, D.C., Cohen, G.H., and Eisenberg, R.J. (2003). Structure-based mutagenesis of herpes simplex virus glycoprotein D defines three critical regions at the gD-HveA/HVEM binding interface. J. Virol. *77*, 8127–8140.

Connolly, S.A., Landsburg, D.J., Carfi, A., Whitbeck, J.C., Zuo, Y., Wiley, D.C., Cohen, G.H., and Eisenberg, R.J. (2005) Potential nectin-1 binding site on herpes simplex virus glycoprotein D. J. Virol. *79*, 1282–1295.

Crump, C.M., Bruun, B., Bell, S., Pomeranz, L.E., Minson, T., and Browne, H.M. (2004). Alphaherpesvirus glycoprotein M causes the relocalization of plasma membrane proteins. J. Gen. Virol. *85*, 3517–3527.

Desai, P.J. (2000). A null mutation in the UL36 gene of herpes simplex virus type 1 results in accumulation of unenveloped DNA-filled capsids in the cytoplasm of infected cells. J. Virol. *74*, 11608–11618.

Desai, P., Sexton, G.L., McCaffery, J.M., and Person, S. (2001). A null mutation in the gene encoding the herpes simplex virus type 1 UL37 polypeptide abrogates virus maturation. J. Virol. *75*, 10259–10271.

Di Lazzaro, C., Campadelli-Fiume, G., and Torrisi, M.R. (1995). Intermediate forms of glycoconjugates are present in the envelope of herpes simplex virions during their transport along the exocytic pathway. Virology *214*, 619–623.

Dingwell, K.S., Brunetti, C.R., Hendricks, R.L., Tang, Q., Tang, M., Rainbow, A.J., and Johnson, D.C. (1994). Herpes simplex virus glycoproteins E and I facilitate cell-to-cell spread *in vivo* and across junctions of cultured cells. J. Virol. *68*, 834–845.

Elliott, G., and O'Hare, P. (1999). Live-cell analysis of a green fluorescent protein-tagged herpes simplex virus infection. J. Virol. *73*, 4110–4119.

Enquist, L.W., Husak, P.J., Banfield, B.W., and Smith, G.A. (1998). Infection and spread of alphaherpesviruses in the nervous system. Adv. Virus Res. *51*, 237–347.

Fan, Z., Grantham, M.L., Smith, M.S., Anderson, E.S., Cardelli, J.A., and Muggeridge, M.I. (2002). Truncation of herpes simplex virus type 2 glycoprotein B increases its cell surface expression and activity in cell–cell fusion, but these properties are unrelated. J. Virol. *76*, 9271–9283.

Farnsworth, A., Goldsmith, K., and Johnson, D.C. (2003). Herpes simplex virus glycoproteins gD and gE/gI serve essential but redundant functions during acquisition of the virion envelope in the cytoplasm. J. Virol. *77*, 8481–8494.

Feire, A.L., Koss, H., and Compton, T. (2004). Cellular integrins function as entry receptors for human cytomegalovirus via a highly conserved disintegrin-like domain. Proc. Natl. Acad. Sci. USA *101*, 15470–15475.

Forrester, A., Farrell, H., Wilkinson, G., Kaye, J., Davis Poynter, N., and Minson, T. (1992). Construction and properties of a mutant of herpes simplex virus type 1 with glycoprotein H coding sequences deleted. J. Virol. *66*, 341–348.

Foster, T.P., and Kousoulas, K.G. (1999). Genetic analysis of the role of herpes simplex virus type 1 glycoprotein K in infectious virus production and egress. J. Virol. *73*, 8457–8468.

Foster, T.P., Rybachuk, G.V., and Kousoulas, K.G. (2001). Glycoprotein K specified by herpes simplex virus type 1 is expressed on virions as a Golgi complex-dependent glycosylated species and functions in virion entry. J. Virol. *75*, 12431–12438.

Foster, T.P., Alvarez, X., and Kousoulas, K.G. (2003a). Plasma membrane topology of syncytial domains of herpes simplex virus type 1 glycoprotein K (gK): the UL20 protein enables cell surface localization of gK but not gK-mediated cell-to-cell fusion. J. Virol. *77*, 499–510.

Foster, T.P., Rybachuk, G.V., Alvarez, X., Borkhsenious, O., and Kousoulas, K.G. (2003b). Overexpression of gK in gK-transformed cells collapses the Golgi apparatus into the endoplasmic reticulum inhibiting virion egress, glycoprotein transport, and virus-induced cell fusion. Virology *317*, 237–252.

Foster, T.P., Melancon, J.M., Baines, J.D., and Kousoulas, K.G. (2004). The herpes simplex virus type 1 UL20 protein modulates membrane fusion events during cytoplasmic virion morphogenesis and virus-induced cell fusion. J. Virol. *78*, 5347–5357.

Fusco, D., Forghieri, C., Campadelli-Fiume, G. (2005). The pro-fusion domain of herpes simplex virus glycoprotein D (gD) interacts with the gD N terminus and is displaced by soluble forms of viral receptors. Proc. Natl. Acad. Sci. USA. *102*, 9323–9328.

Geraghty, R.J., Krummenacher, C., Cohen, G.H., Eisenberg, R.J., and Spear, P.G. (1998). Entry of alphaherpesviruses mediated by poliovirus receptor-related protein 1 and poliovirus receptor. Science *280*, 1618–1620.

Gianni, T., Campadelli-Fiume, G., and Menotti, L. (2004). Entry of Herpes Simplex Virus Mediated by Chimeric Forms of Nectin1 Retargeted to Endosomes or to Lipid Rafts Occurs through Acidic Endosomes. J. Virol. *78*, 12268–12276.

Gianni, T., Martelli, P.L., Casadio, R., and Campadelli-Fiume, G. (2005a). The ectodomain of herpes simpex virus glycoprotein H contains a membrane alpha-helix with attributes of an internal fusion peptide, positionally conserved in the Herpesviridae family. J. Virol. *79*, 2931–2940.

Gianni, T., Menotti, L., and Campadelli-Fiume, G. (2005b). A heptad repeat in herpes simplex virus gH, located downstream of the alpha-helix with attributes of a fusion peptide, is critical for virus entry and fusion. J. Virol. *79*, 7042–7049.

Gianni, T., Piccoli, A., Bertucci, C. and Campadelli-Fiume, G. (2006) Heptad repeat 2 in herpes simplex virus 1 gH interacts with heptad repeat 1 and is critical for virus entry and fusion. J. Virol. *80*, 2216–2224.

Gompels, U., and Minson, A. (1986). The properties and sequence of glycoprotein H of herpes simplex virus type 1. Virology *153*, 230–247.

Griffiths, A., Renfrey, S., and Minson, T. (1998). Glycoprotein C-deficient mutants of two strains of herpes simplex virus type 1 exhibit unaltered adsorption characteristics on polarized or non-polarized cells. J. Gen. Virol. *79 (Pt 4)*, 807–812.

Gruenheid, S., Gatzke, L., Meadows, H., and Tufaro, F. (1993). Herpes simplex virus infection and propagation in a mouse L cell mutant lacking

heparan sulfate proteoglycans. J. Virol. *67*, 93–100.
Grunewald, K., Desai, P., Winkler, D.C., Heymann, J.B., Belnap, D.M., Baumeister, W., and Steven, A.C. (2003). Three-dimensional structure of herpes simplex virus from cryo-electron tomography. Science *302*, 1396–1398.
Haarr, L., Shukla, D., Rodahl, E., Dal Canto, M.C., and Spear, P.G. (2001). Transcription from the gene encoding the herpesvirus entry receptor nectin-1 (HveC) in nervous tissue of adult mouse. Virology *287*, 301–309.
Herold, B.C., WuDunn, D., Soltys, N., and Spear, P.G. (1991). Glycoprotein C of herpes simplex virus type 1 plays a principal role in the adsorption of virus to cells and in infectivity. J. Virol. *65*, 1090–1098.
Hutchinson, L., and Johnson, D.C. (1995). Herpes simplex virus glycoprotein K promotes egress of virus particles. J. Virol. *69*, 5401–5413.
Hutchinson, L., Browne, H., Wargent, V., Davis-Poynter, N., Primorac, S., Goldsmith, K., Minson, A.C., and Johnson, D.C. (1992a). A novel herpes simplex virus glycoprotein, gL, forms a complex with glycoprotein H (gH) and affects normal folding and surface expression of gH. J. Virol. *66*, 2240–2250.
Hutchinson, L., Goldsmith, K., Snoddy, D., Ghosh, H., Graham, F.L., and Johnson, D.C. (1992b). Identification and characterization of a novel herpes simplex virus glycoprotein, gK, involved in cell fusion. J. Virol. *66*, 5603–5609.
Jogger, C.R., Montgomery, R.I., and Spear, P.G. (2004). Effects of linker-insertion mutations in herpes simplex virus 1 gD on glycoprotein-induced fusion with cells expressing HVEM or nectin-1. Virology *318*, 318–326.
Johnson, D.C., and Huber, M.T. (2002). Directed egress of animal viruses promotes cell-to-cell spread. J. Virol. *76*, 1–8.
Johnson, D.C., and Spear, P.G. (1982). Monensin inhibits the processing of herpes simplex virus glycoproteins, their transport to the cell surface, and the egress of virions from infected cells. J. Virol. *43*, 1102–1112.
Johnson, R.M., and Spear, P.G. (1989). Herpes simplex virus glycoprotein D mediates interference with herpes simplex virus infection. J. Virol. *63*, 819–827.
Johnson, D.C., Burke, R.L., and Gregory, T. (1990). Soluble forms of herpes simplex virus glycoprotein D bind to a limited number of cell surface receptors and inhibit virus entry into cells. J. Virol. *64*, 2569–2576.
Jones, F., and Grose, C. (1988). Role of cytoplasmic vacuoles in varicella-zoster virus glycoprotein trafficking and virion envelopment. J. Virol. *62*, 2701–2711.
Klupp, B.G., Nixdorf, R., and Mettenleiter, T.C. (2000). Pseudorabies virus glycoprotein M inhibits membrane fusion. J. Virol. *74*, 6760–6768.
Krummenacher, C., Nicola, A.V., Whitbeck, J.C., Lou, H., Hou, W., Lambris, J.D., Geraghty, R.J., Spear, P.G., Cohen, G.H., and Eisenberg, R.J. (1998). Herpes simplex virus glycoprotein D can bind to poliovirus receptor-related protein 1 or herpesvirus entry mediator, two structurally unrelated mediators of virus entry. J. Virol. *72*, 7064–7074.
Krummenacher, C., Rux, A.H., Whitbeck, J.C., Ponce-de-Leon, M., Lou, H., Baribaud, I., Hou, W., Zou, C., Geraghty, R.J., Spear, P.G. *et al.* (1999). The first immunoglobulin-like domain of HveC is sufficient to bind herpes simplex virus gD with full affinity, while the third domain is involved in oligomerization of HveC. J. Virol. *73*, 8127–8137.
Krummenacher, C., Baribaud, I., Ponce De Leon, M., Whitbeck, J.C., Lou, H., Cohen, G.H., and Eisenberg, R.J. (2000). Localization of a binding site for herpes simplex virus glycoprotein D on herpesvirus entry mediator C by using antireceptor monoclonal antibodies. J. Virol. *74*, 10863–10872.
Krummenacher, C., Baribaud, F., Ponce De Leon, M., Baribaud, I., Whitbeck, J.C., Xu, R., Cohen, G.H., and Eisenberg, R.J. (2004). Comparative usage of herpesvirus entry mediator A and nectin-1 by laboratory strains and clinical isolates of herpes simplex virus. Virology *322*, 286–299.
Krummenacher, C., Supekar, V.M., Whitbeck, J.C., Lazear, E., Connolly, S.A., Eisenberg, R.J., Cohen, G.H., Wiley, D.C. and Carfi, A. (2005). Structure of unliganded HSV gD reveals a mechanism for receptor-mediated activation of virus entry. EMBO J. *24*, 4144–4153.
Laquerre, S., Argnani, R., Anderson, D.B., Zucchini, S., Manservigi, R., and Glorioso, J.C. (1998). Heparan sulfate proteoglycan binding by herpes simplex virus type 1 glycoproteins B and C, which differ in their contributions to virus attachment, penetration, and cell-to-cell spread. J. Virol. *72*, 6119–6130.
Ligas, M.W., and Johnson, D.C. (1988). A herpes simplex virus mutant in which glycoprotein D sequences are replaced by beta-galactosidase sequences binds to but is unable to penetrate into cells. J. Virol. *62*, 1486–1494.
Linehan, M.M., Richman, S., Krummenacher, C., Eisenberg, R.J., Cohen, G.H., and Iwasaki, A. (2004). *In vivo* role of nectin-1 in entry of herpes simplex virus type 1 (HSV-1) and HSV-2

through the vaginal mucosa. J. Virol. *78*, 2530–2536.

Longnecker, R., and Roizman, B. (1987). Clustering of genes dispensable for growth in culture in the S component of the HSV-1 genome. Science *236*, 573–576.

Longnecker, R., Chatterjee, S., Whitley, R.J., and Roizman, B. (1987). Identification of a herpes simplex virus 1 glycoprotein gene within a gene cluster dispensable for growth in cell culture. Proc. Natl. Acad. Sci. USA *84*, 4303–4307.

Lopez, M., Cocchi, F., Menotti, L., Avitabile, E., Dubreuil, P., and Campadelli-Fiume, G. (2000). Nectin2a (PRR2a or HveB) and nectin2d are low-efficiency mediators for entry of herpes simplex virus mutants carrying the Leu25Pro substitution in glycoprotein D. J. Virol. *74*, 1267–1274.

Manoj, S., Jogger, C.R., Myscofski, D., Yoon, M., and Spear, P.G. (2004). Mutations in herpes simplex virus glycoprotein D that prevent cell entry via nectins and alter cell tropism. Proc. Natl. Acad. Sci. USA *101*, 12414–12421.

Maresova, L., Pasieka, T.J., Homan, E., Gerday, E., and Grose, C. (2005). Incorporation of three endocytosed varicella-zoster virus glycoproteins, gE, gH, and gB, into the virion envelope. J. Virol. *79*, 997–1007.

Martinez, W.M., and Spear, P.G. (2002). Amino acid substitutions in the V domain of nectin-1 (HveC) that impair entry activity for herpes simplex virus types 1 and 2 but not for Pseudorabies virus or bovine herpesvirus 1. J. Virol. *76*, 7255–7262.

Mata, M., Zhang, M., Hu, X., and Fink, D.J. (2001). HveC (nectin-1) is expressed at high levels in sensory neurons, but not in motor neurons, of the rat peripheral nervous system. J. Neurovirol. *7*, 476–480.

Matsushima, H., Utani, A., Endo, H., Matsuura, H., Kakuta, M., Nakamura, Y., Matsuyoshi, N., Matsui, C., Nakanishi, H., Takai, Y., and Shinkai, H. (2003). The expression of nectin-1alpha in normal human skin and various skin tumours. Br. J. Dermatol. *148*, 755–762.

McGeoch, D.J., Dalrymple, M.A., Davison, A.J., Dolan, A., Frame, M.C., McNab, D., Perry, L.J., Scott, J.E., and Taylor, P. (1988). The complete DNA sequence of the long unique region in the genome of herpes simplex virus type 1. J. Gen. Virol. *69*, 1531–1574.

Menotti, L., Cocchi, F., and Campadelli-Fiume, G. (2002). Critical residues in the CC′ ridge of the human nectin1 receptor V domain enable herpes simplex virus entry into the cell and act synergistically with the downstream region. Virology *301*, 6–12.

Mettenleiter, T.C. (2002). Herpesvirus assembly and egress. J. Virol. *76*, 1537–1547.

Mettenleiter, T.C., and Minson, T. (2006). Egress of alphaherpesviruses. J. Virol. *80*, 1610–1611; author reply 1611–1612.

Milne, R.S., Hanna, S.L., Rux, A.H., Willis, S.H., Cohen, G.H., and Eisenberg, R.J. (2003). Function of herpes simplex virus type 1 gD mutants with different receptor-binding affinities in virus entry and fusion. J. Virol. *77*, 8962–8972.

Miranda-Saksena, M., Boadle, R.A., Armati, P., and Cunningham, A.L. (2002). In rat dorsal root ganglion neurons, herpes simplex virus type 1 tegument forms in the cytoplasm of the cell body. J. Virol. *76*, 9934–9951.

Montgomery, R.I., Warner, M.S., Lum, B.J., and Spear, P.G. (1996). Herpes simplex virus-1 entry into cells mediated by a novel member of the TNF/NGF receptor family. Cell *87*, 427–436.

Nicola, A.V., and Straus, S.E. (2004). Cellular and viral requirements for rapid endocytic entry of herpes simplex virus. J. Virol. *78*, 7508–7517.

Nicola, A.V., Peng, C., Lou, H., Cohen, G.H., and Eisenberg, R.J. (1997). Antigenic structure of soluble herpes simplex virus (HSV) glycoprotein D correlates with inhibition of HSV infection. J. Virol. *71*, 2940–2946.

Nicola, A.V., McEvoy, A.M., and Straus, S.E. (2003). Roles for endocytosis and low pH in herpes simplex virus entry into HeLa and Chinese hamster ovary cells. J. Virol. *77*, 5324–5332.

Noble, A.G., Lee, G.T., Sprague, R., Parish, M.L., and Spear, P.G. (1983). Anti-gD monoclonal antibodies inhibit cell fusion induced by herpes simplex virus type 1. Virology *129*, 218–224.

Pellett, P.E., Kousoulas, K.G., Pereira, L., and Roizman, B. (1985). Anatomy of the herpes simplex virus 1 strain F glycoprotein B gene: primary sequence and predicted protein structure of the wild type and of monoclonal antibody-resistant mutants. J. Virol. 53, 243–253.

Penfold, M.E., Armati, P., and Cunningham, A.L. (1994). Axonal transport of herpes simplex virions to epidermal cells: evidence for a specialized mode of virus transport and assembly. Proc. Natl. Acad. Sci. USA *91*, 6529–6533.

Pomeranz, L.E., and Blaho, J.A. (2000). Assembly of infectious Herpes simplex virus type 1 virions in the absence of full-length VP22. J. Virol. *74*, 10041–10054.

Reymond, N., Borg, J., Lecocq, E., Adelaide, J., Campadelli-Fiume, G., Dubreuil, P., and Lopez, M. (2000). Human nectin3/PRR3: a novel member of the PVR/PRR/nectin family that interacts with afadin. Gene 255, 347–355.

Reynolds, A.E., Ryckman, B.J., Baines, J.D., Zhou, Y., Liang, L., and Roller, R.J. (2001). U(L)31 and U(L)34 proteins of herpes simplex virus type 1 form a complex that accumulates at the nuclear

rim and is required for envelopment of nucleocapsids. J. Virol. *75*, 8803–8817.
Reynolds, A.E., Wills, E.G., Roller, R.J., Ryckman, B.J., and Baines, J.D. (2002). Ultrastructural localization of the herpes simplex virus type 1 UL31, UL34, and US3 proteins suggests specific roles in primary envelopment and egress of nucleocapsids. J. Virol. *76*, 8939–8952.
Reynolds, A.E., Liang, L., and Baines, J.D. (2004). Conformational changes in the nuclear lamina induced by herpes simplex virus type 1 require genes U(L)31 and U(L)34. J. Virol. *78*, 5564–5575.
Richart, S., Simpson, S., Krummenacher, C., Whitbeck, J., Pizer, L., Cohen, G., Eisenberg, R., and Wilcox, C. (2003). Entry of herpes simplex virus type 1 into primary sensory neurons *in vitro* is mediated by Nectin-1/HveC. J. Virol. *77*, 3307–3311.
Roizman, B. (1988). Molecular and genetic engineering: the principles, the power, and the promise. Science *239*, G110.
Roller, R.J., Zhou, Y., Schnetzer, R., Ferguson, J., and DeSalvo, D. (2000). Herpes simplex virus type 1 U(L)34 gene product is required for viral envelopment. J. Virol. *74*, 117–129.
Roop, C., Hutchinson, L., and Johnson, D.C. (1993). A mutant herpes simplex virus type 1 unable to express glycoprotein L cannot enter cells, and its particles lack glycoprotein H. J. Virol. *67*, 2285–2297.
Rux, A.H., Lou, H., Lambris, J.D., Friedman, H.M., Eisenberg, R.J., and Cohen, G.H. (2002). Kinetic analysis of glycoprotein C of herpes simplex virus types 1 and 2 binding to heparin, heparan sulfate, and complement component C3b. Virology *294*, 324–332.
Schelhaas, M., Jansen, M., Haase, I., and Knebel-Morsdorf, D. (2003). Herpes simplex virus type 1 exhibits a tropism for basal entry in polarized epithelial cells. J. Gen. Virol. *84*, 2473–2484.
Scott, E.S., and O'Hare, P. (2001). Fate of the inner nuclear membrane protein lamin B receptor and nuclear lamins in herpes simplex virus type 1 infection. J. Virol. *75*, 8818–8830.
Shukla, D., Liu, J., Blaiklock, P., Shworak, N.W., Bai, X., Esko, J.D., Cohen, G.H., Eisenberg, R.J., Rosenberg, R.D., and Spear, P.G. (1999). A novel role for 3-O-sulfated heparan sulfate in herpes simplex virus 1 entry. Cell *99*, 13–22.
Skepper, J.N., Whiteley, A., Browne, H., and Minson, A. (2001). Herpes simplex virus nucleocapsids mature to progeny virions by an envelopment –> deenvelopment –> reenvelopment pathway. J. Virol. *75*, 5697–5702.
Sodeik, B. (2000). Mechanisms of viral transport in the cytoplasm. Trends Microbiol. *8*, 465–472.
Sodeik, B., Ebersold, M.W., and Helenius, A. (1997). Microtubule-mediated transport of incoming herpes simplex virus 1 capsids to the nucleus. J. Cell Biol. *136*, 1007–1021.
Spear, P.G. (1976). Membrane proteins specified by herpes simplex viruses. I. Identification of four glycoprotein precursors and their products in type 1-infected cells. J. Virol. *17*, 991–1008.
Spear, P.G., and Longnecker, R. (2003). Herpesvirus entry: an update. J. Virol. *77*, 10179–10185.
Spear, P.G., and Roizman, B. (1970). Proteins specified by herpes simplex virus, IV. Site of glycosylation and accumulation of viral membrane proteins. Proc. Natl. Acad. Sci. USA *66*, 730–737.
Spear, P.G., Sarmiento, M., and Manservigi, R. (1978). The structural proteins and glycoproteins of herpesviruses: a review. IARC Sci. Publ. 157–167.
Spear, P.G., Shieh, M.T., Herold, B.C., WuDunn, D., and Koshy, T.I. (1992). Heparan sulfate glycosaminoglycans as primary cell surface receptors for herpes simplex virus. Adv. Exp. Med. Biol. *313*, 341–353.
Spear, P.G., Eisenberg, R.J., and Cohen, G.H. (2000). Three classes of cell surface receptors for alphaherpesvirus entry. Virology *275*, 1–8.
Takai, Y., Irie, K., Shimizu, K., Sakisaka, T., and Ikeda, W. (2003). Nectins and nectin-like molecules: roles in cell adhesion, migration, and polarization. Cancer Sci. *94*, 655–667.
Torrisi, M.R., Di Lazzaro, C., Pavan, A., Pereira, L., and Campadelli-Fiume, G. (1992). Herpes simplex virus envelopment and maturation studied by fracture label. J. Virol. *66*, 554–561.
Turner, A., Bruun, B., Minson, T., and Browne, H. (1998). Glycoproteins gB, gD, and gHgL of herpes simplex virus type 1 are necessary and sufficient to mediate membrane fusion in a Cos cell transfection system. J. Virol. *72*, 873–875.
van Genderen, I.L., Brandimarti, R., Torrisi, M.R., Campadelli, G., and van Meer, G. (1994). The phospholipid composition of extracellular herpes simplex virions differs from that of host cell nuclei. Virology *200*, 831–836.
Wang, Z.H., Gershon, M.D., Lungu, O., Zhu, Z., Mallory, S., Arvin, A.M., and Gershon, A.A. (2001). Essential role played by the C-terminal domain of glycoprotein I in envelopment of varicella-zoster virus in the trans-Golgi network: interactions of glycoproteins with tegument. J. Virol. *75*, 323–340.
Ward, P.L., Campadelli-Fiume, G., Avitabile, E., and Roizman, B. (1994). Localization and putative function of the UL20 membrane protein in cells infected with herpes simplex virus 1. J. Virol. *68*, 7406–7417.

Warner, M.S., Geraghty, R.J., Martinez, W.M., Montgomery, R.I., Whitbeck, J.C., Xu, R., Eisenberg, R.J., Cohen, G.H., and Spear, P.G. (1998). A cell surface protein with herpesvirus entry activity (HveB) confers susceptibility to infection by mutants of herpes simplex virus type 1, herpes simplex virus type 2, and pseudorabies virus. Virology *246*, 179–189.

Watson, R.J., Weis, J.H., Salstrom, J.S., and Enquist, L.W. (1982). Herpes simplex virus type-1 glycoprotein D gene: nucleotide sequence and expression in Escherichia coli. Science *218*, 381–384.

Whitbeck, J.C., Muggeridge, M.I., Rux, A.H., Hou, W., Krummenacher, C., Lou, H., van Geelen, A., Eisenberg, R.J., and Cohen, G.H. (1999). The major neutralizing antigenic site on herpes simplex virus glycoprotein D overlaps a receptor-binding domain. J. Virol. *73*, 9879–9890.

Whiteley, A., Bruun, B., Minson, T., and Browne, H. (1999). Effects of targeting herpes simplex virus type 1 gD to the endoplasmic reticulum and trans-Golgi network. J. Virol. *73*, 9515–9520.

Wild, P., Senn, C., Tobler, K., Schraner, E., Loepfe, E., Engels, M., Ackermann, M., Walther, P., Mueller, M., and Ziegler, U. (2004). Dissection of BOHV-1 envelopment by high resolution electron microscopy. Paper presented at: Reno, Nevada.

Willis, S.H., Rux, A.H., Peng, C., Whitbeck, J.C., Nicola, A.V., Lou, H., Hou, W., Salvador, L., Eisenberg, R.J., and Cohen, G.H. (1998). Examination of the kinetics of herpes simplex virus glycoprotein D binding to the herpesvirus entry mediator, using surface plasmon resonance. J. Virol. *72*, 5937–5947.

Ye, G.J., and Roizman, B. (2000). The essential protein encoded by the UL31 gene of herpes simplex virus 1 depends for its stability on the presence of UL34 protein. Proc. Natl. Acad. Sci. USA *97*, 11002–11007.

Yoon, M., Zago, A., Shukla, D., and Spear, P.G. (2003). Mutations in the N termini of herpes simplex virus type 1 and 2 gDs alter functional interactions with the entry/fusion receptors HVEM, nectin-2, and 3-O-sulfated heparan sulfate but not with nectin-1. J. Virol. *77*, 9221–9231.

Zago, A., and Spear, P.G. (2003). Differences in the N termini of herpes simplex virus type 1 and 2 gDs that influence functional interactions with the human entry receptor Nectin-2 and an entry receptor expressed in Chinese hamster ovary cells. J. Virol. *77*, 9695–9699.

Zhou, G., Avitabile, E., Campadelli-Fiume, G., and Roizman, B. (2003). The domains of glycoprotein D required to block apoptosis induced by herpes simplex virus 1 are largely distinct from those involved in cell- cell fusion and binding to nectin1. J. Virol. *77*, 3759–3767.

9

Endocytosis of Varicella Zoster Virus Glycoproteins: Virion Envelopment and Egress

Charles Grose, Lucie Maresova, Guruprasad Medigeshi, Gregory K. Scott, and Gary Thomas.

Abstract

Endocytosis from the infected cell surface is a property of the four major glycoproteins of varicella zoster virus (VZV), designated gE, gI, gB and gH. Each of these glycoproteins has a functional endocytosis motif in its cytoplasmic tail. Three have tyrosine based motifs, while gI has a dileucine motif. The YAGL motif of gE is located adjacent to an acidic cluster, which also contains serine and threonine residues variably phosphorylated by both a viral kinase (ORF 47 kinase) and casein kinase II. The acidic cluster interacts with a connector protein called prosphofurin acidic cluster sorting protein 1 (PACS-1) after clathrin mediated internalization. Although the function of endocytosis has been elusive, recent research suggests that the internalized VZV glycoproteins traffic to the site of virion assembly in the trans-Golgi network and are subsequently incorporated into the envelopes of nascent virions. In turn, the enveloped virions reside within vacuoles, which travel toward and fuse with the plasma membrane. A recombinant VZV genome containing a gE gene lacking only its endocytosis motif cannot replicate. Taken together, the above results indicate the importance of VZV glycoprotein endocytosis in the life cycle of this alpha-herpesvirus.

Introduction

Endocytosis is an important process by which proteins are continually internalized within the cell through specific interactions at the plasma membrane. Cell surface receptors typically undergo endocytosis from the cell membrane through interactions with clathrin-coated pits and are often recycled through the endosomes back to the cell membrane. Both clustering in clathrin coated pits and internalization of receptors have been shown to be dependent on internalization signals in the cytoplasmic tail, which have been defined as tyrosine motifs or dileucine motifs (Trowbridge, 1991). Endocytosis from the plasma membrane by a process similar to receptor-mediated endocytosis has now been demonstrated for numerous herpesvirus glycoproteins.

Varicella zoster virus (VZV) has four major glycoproteins gE, gI, gH and gB. Their endocytosis motifs are highly conserved among all sequenced VZV strains (Davison and Scott, 1986; Grose, 1990; Grose *et al.*, 2004). As discussed below, while the VZV glycoproteins share many characteristics with their alpha-herpesvirus homologs, such as HSV and PRV, there are also unanticipated differences. One of these remarkable differences relates to trafficking pathways. In the case of VZV, glycoproteins internalized from the cell membrane are delivered to sites of assembly in the cytoplasm, where they are incorporated into nascent virions. In contrast, not all homologous glycoproteins from the closely related alpha-herpesviruses appear to use the same pathways for transit to the site of assembly. For the above reasons, the subject of herpesviral glycoprotein endocytosis is reviewed. Also, the relationship of endocytosis to the deenvelopment–reenvelopment model of herpesvirus assembly and egress is discussed.

Glycoprotein gE

Endocytosis of VZV gE

Unlike other human herpesviruses, the predominant viral glycoprotein present on the envelope of the mature VZV virion is gE (Montalvo *et al.*, 1985). VZV gE is a type I transmembrane protein of 623 amino acids. Of these, 544 residues are found in the ectodomain, 17 in the transmembrane domain, and 62 in the C-terminal cytoplasmic tail (Fig. 9.1) (Grose, 1990). The gE protein traffics from the endoplasmic reticulum (ER) through the Golgi, where it is extensively processed, to the outer cell membrane. The extensive posttranslational modifications produce a mature gE glycoprotein of 98 kDa (Montalvo *et al.*, 1985). Both the ectodomain and endodomain of gE have important functions. Based on the fact that a single point mutation in the VZV-MSP gE ectodomain changes VZV MSP egress pattern from a typical "viral highways" phenotype to a diffuse pattern similar to that observed in HSV infected cells (Santos *et al.*, 1998; 2000), the ectodomain is critical for determining virus–cell interactions of the mature virion.

VZV gE was the first viral glycoprotein shown to undergo endocytosis in transfected cells (Olson and Grose, 1997). This process is mediated by a YAGL sequence (aa 582–585) located in the cytoplasmic tail of the protein (Fig. 9.1) (Olson and Grose, 1997). This sequence fits the consensus tyrosine based endocytosis motif known to mediate endocytosis of numerous cellular transmembrane proteins (Bonifacino and Dell'Angelica, 1999). The tetrapeptide of the tyrosine-based motif is recognized by the mu-2 subunit of AP-2, a clathrin-associated complex localized to the plasma membrane (Ohne *et al.*, 1995; Boll *et al.*, 1996). AP-2 facilitates the formation of clathrin-coated vesicles by acting as the adaptor between the membrane protein and clathrin. Endocytosis of gE was shown to be dependent on clathrin-coated vesicle formation within the cells. An example of the VZV endocytosis assay is shown in Fig. 9.2. In addition, colocalization studies showed that endocytosis of gE mimicked internalization of the transferrin receptor (Olson and Grose, 1997). Endocytic recycling of gE back to the plasma membrane postendocytosis has been demonstrated, but there is also a pathway for gE to recycle to the trans-Golgi network (TGN) postendocytosis (Alconada *et al.*, 1996; Olson and Grose, 1997). TGN recycling is mediated by a cluster of acidic amino acids in the gE cytoplasmic tail (Fig. 9.1) (Grose *et al.*, 1989; Alconada *et al.*, 1996; Wan *et al.*, 1998). The acidic cluster of gE interacts with a connector protein, phosphofurin acidic cluster sorting protein 1 (PACS-1), and directs VZV gE from endosomes to the TGN, a proposed site of tegument assembly and virion envelopment (Crump *et al.*, 2001; Wang *et al.*, 2001). Phosphorylation within this acidic cluster motif mediates PACS-1 interaction (Fish *et al.*, 1998; Scott *et al.*, 2003). VZV gE is phosphorylated on serine and threonine residues of its cytoplasmic tail (Montalvo and Grose, 1986; Grose *et al.*, 1989; Yao *et al.*, 1993; Olson *et al.*, 1997).

In addition, differential phosphorylation of the serine and threonine residues adjacent to the gE acidic cluster by either the VZV ORF 47 protein serine kinase or the cellular casein kinase II targets endocytosed gE to

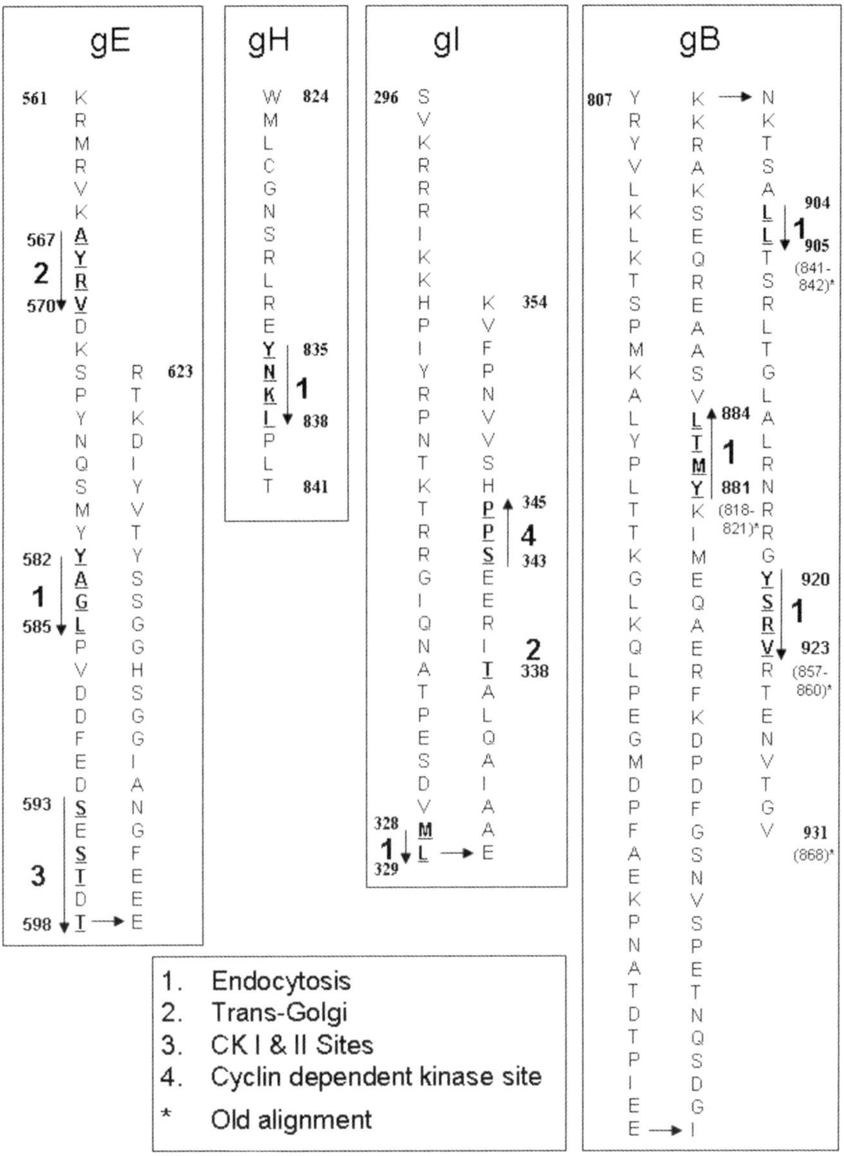

Figure 9.1 Amino acid sequences of the cytoplasmic tails of VZV glycoproteins gE, gH, gI and gB. The sequences are based on published VZV sequences (Davison and Scott, 1986; Grose *et al.*, 2004). The amino terminal of gB has been recently realigned, so the enumeration of amino acids in the cytoplasmic tail differs from the earlier descriptions (Maresova *et al.*, 2003). Both sets of numbers are given. The cytoplasmic tail of gH has been recently realigned also (Pasieka *et al.*, 2003). Domains discussed in the text are annotated.

either the TGN or cell surface, thereby increasing virus assembly or cell-to-cell spread, respectively (Montalvo and Grose, 1986; Kenyon *et al.*, 2002; Kenyon *et al.*, 2003). For example, in cells infected with a VZV ORF47-null mutant, internalized VZV gE recycles to the plasma membrane and causes an increased syncytial formation. Thus, en-

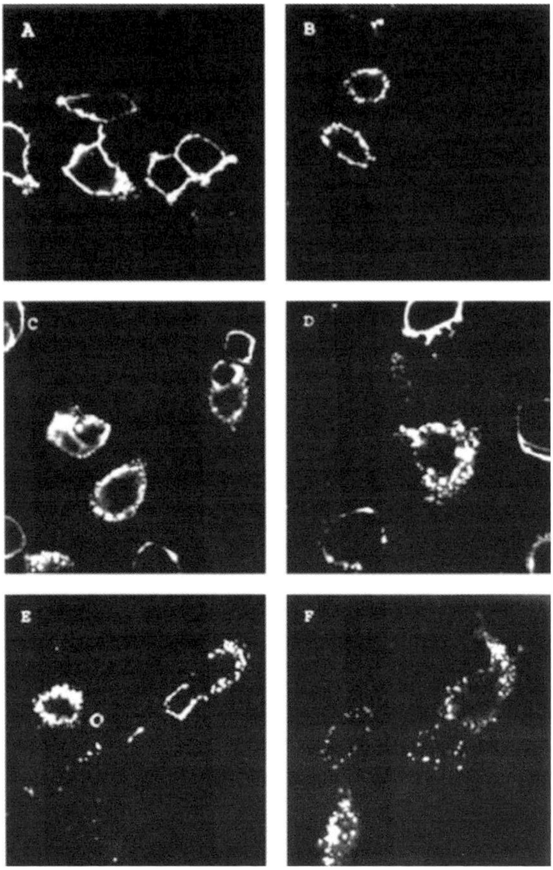

Figure 9.2 Endocytosis assay. Endocytosis of the transfected VZV gE was gauged by a confocal microscopy assay. The assay was photographed at the following timepoints: 0 min (A), 5 min (B), 10 min (C), 15 min (D), 30 min (E) and 60 min (F). The characteristic profile of endocytosis includes multiple vesicles in the cytoplasm; see panels D, E and F. Endocytosis is more pronounced in assays of transfected cells than in VZV infected cells.

docytosis is also linked to the mechanisms of fusion in infected cells.

Endocytosis of alpha-herpesvirus gE homologs

Similar to gE of VZV, pseudorabies virus (PrV) gE and herpes simplex type 1 (HSV-1) gE undergo endocytosis (Tirabassi and Enquist, 1998; Alconada *et al.*, 1999; Tirabassi and Enquist, 1999). It has been demonstrated that PrV gE cytoplasmic domain is required for gE internalization from the plasma membrane (Tirabassi and Enquist, 1998). HSV-1 gE cycles between TGN and the cell surface. This TGN localization is determined by a tyrosine-based motif and an acidic stretch of amino acids in the gE cytoplasmic tail that are homologous to the VZV gE sites (Alconada *et al.*, 1999). Given the high degree of conservation of the different C-terminal sorting signals in all of the alpha-herpesvirus gE homologs, it would seem likely that these molecules traffic inside the cell following identical signal-mediated transport processes. Yet, studies described below will show that this assumption appears not to be true.

VZV Glycoprotein gI

Endocytosis of VZV gI

VZV gI is a type I transmembrane glycoprotein of 354 amino acids, 59 of which are cytoplasmic (Grose, 2002). Similar to gE, VZV gI also contains trafficking motifs in its cytoplasmic tail that mediate similar functions; however, the motifs are different (Fig. 9.1). While a YAGL tyrosine motif mediates gE endocytosis, a di-leucine motif ML at amino acids 328–329 mediates gI endocytosis (Olson and Grose, 1998). And instead of an acidic cluster for TGN targeting, gI has a threonine at position 338 that when mutated disrupts gI TGN targeting (Wang *et al.*, 2000). VZV gE and gI form a complex in infected cells or when cotransfected. This noncovalent complex forms a biologically active human Fc receptor (Litwin *et al.*, 1990; 1992). Although gI can internalize in the absence of gE, complex formation between gE:gI increases the efficiency of gE:gI endocytosis over that of gE or gI alone (Olson and Grose, 1998). Complex formation between gE and gI is also able to direct internalization of a gE endocytosis deficient mutant, which suggests that the gI motif may be more efficient than that of gE (Olson and Grose, 1998). Further, this result suggests that VZV gI behaves as an accessory component by facilitating the endocytosis of the major constituent gE and thereby modulating the trafficking of the entire cell surface gE:gI complex. The fact that a VZV gI null mutant spreads very poorly in cell culture reinforces the importance of the VZV gE:gI complex (Mallory *et al.*, 1997).

Endocytosis of alpha-herpesvirus gI homologs

Complex formation and subsequent endocytosis between gE and gI and endocytosis as a complex has been shown not only in VZV, but also PrV and HSV (Alconada *et al.*, 1998; Olson and Grose, 1998; Tirabassi and Enquist, 1998; Wang *et al.*, 1998). In contrast to VZV, however, the PRV gI protein cannot be internalized from the plasma membrane on its own. Thus, the PrV and VZV gI proteins seem to have functionally different roles in endocytosis, as the VZV gI protein increases the rate of internalization of the VZV gE protein, while the PrV gE protein directs endocytosis of the PrV gI protein.

VZV glycoprotein B

Endocytosis of VZV gB

VZV gB is the second most abundant VZV glycoprotein and also the most conserved among the herpesvirus glycoproteins. VZV gB shares a higher degree of homology (49%) with gB of HSV-1 than that found between any other two VZV- HSV homologous glycoproteins (which range from 24 to 27%) (Edson *et al.*, 1985; Davison and Scott, 1986). The alpha-herpesvirus gB homologs are required for virus replication, attachment to and penetration of the host membrane, and fusion (Cai *et al.*, 1988; Rauh and Mettenleiter, 1991; Peeters *et al.*, 1992; Baghian *et al.*, 1993; Spear, 1993; Pereira, 1994; Maresova *et al.*, 2001). VZV gB is a type 1 membrane protein consisting of a large ectodomain, a hydrophobic transmembrane region, and a cytoplasmic domain (Davison and Scott, 1986). The mature 140 kDa form of gB is proteolytically cleaved into a disulfide linked heterodimer consisting of 66 and 68 kDa components (Keller *et al.*, 1986; Montalvo *et al.*, 1987). The length of the signal sequence has been recently corrected to include 71 amino acids, thereby giving gB a length of 931 amino acids (Maresova *et al.*, 2003). Since the gB protein cytoplasmic domain is predicted to contain 125 amino acids, it has the longest cytoplasmic tail of any VZV encoded membrane protein (Fig. 9.1) (Elliott and O'Hare, 1999). VZV gB accumulates in the Golgi of infected cells (Wang *et al.*, 1998; Heineman *et al.*, 2000), and the cytoplasmic domain of gB contains specific sequences that are required for both its ER to Golgi transport and post-Golgi transport (Heineman and Hall, 2002).

The VZV gB protein contains three consensus internalization motifs within its cytoplasmic domain: two tyrosine (YMTL at aa 818–821 and YSRV at aa 857–860) and one di-leucine (LL at aa 841–842). The gB protein containing the mutation Y857 failed to be internalized, while gB mutated in Y818 was endocytosed but did not subsequently accumulate in the Golgi (Heineman and Hall, 2001). The LL motif within the cytoplasmic domain is not required for either VZV gB internalization or Golgi localization. Therefore, the membrane-distal YSRV motif is most responsible for gB endocytosis (Heineman and Hall, 2001). However, a VZV recombinant virus (VZV gB-36) that expresses a truncated form of gB lacking the C-terminal 36 amino acids of its cytoplasmic domain, despite lacking the YSRV endocytosis motif, is internalized from the plasma membrane but fails to localize to the Golgi (Heineman and Hall, 2002). Thus, when one of the internalization motifs is removed, the others may subsume some of the lost functions.

In addition to the post-Golgi transport signals, the cytoplasmic domain of VZV gB has been shown to contain specific ER-to-Golgi transport signals. The transmembrane domain proximal tyrosine motif (YMTL) is within a nine-amino acid region (YMTLVSAAE) required by a sequence-dependent mechanism for efficient ER-to-Golgi targeting of the gB protein (Heineman *et al.*, 2000).

Endocytosis of alpha-herpesvirus gB homologs

A similar triad of endocytosis motifs have been identified in HSV-1 and PRV gB. As for VZV gB, the homologous gB proteins contain two tyrosine motifs and one di-leucine based motif, arranged in a similar order in C-terminus. Endocytosis of HSV-1 gB depends principally on the membrane-distal YTQV motif, since either removal or disruption of this signal totally abolishes gB internalization from the cell surface to endosomes (Beitia Ortiz de Zarate *et al.*, 2004). Analysis of the LL motif showed that its replacement did not hinder the process of gB endocytosis. However, the subsequent retrograde gB trafficking from early or recycling endosomes to the TGN was impaired while recycling from the endosomes to the plasma membrane was enhanced.

The PRV gB homolog is internalized from the cell surface and colocalizes precisely with the endocytosed PRV gE:gI complex (Tirabassi and Enquist, 1998). A mutated PrV gB form lacking the membrane-distal tyrosine and the di-leucine motifs remains primarily membrane associated because endocytosis is abolished (Tirabassi and Enquist, 1998). Furthermore, mutation of the one tyrosine Y902 residue in the tyrosine motif (YQRL) inhibits functioning of gB during antibody induced gB internalization in PRV infected monocytes (Favoreel *et al.*, 2002). This same tyrosine mutation also diminishes efficient PrV cell-to-cell spread. However, mutating the dileucine motif in the gB tail has no effect on antibody triggered endocytosis but does promote an increased cell-to-cell spread phenotype with formation of larger syncytia, presumably through increased recycling of gB back to the cell surface (Favoreel *et al.*, 2002).

VZV glycoprotein H

Endocytosis of VZV gH

VZV gH is the third most abundant glycoprotein and the second most conserved. The gH glycoprotein of VZV is a major fusogen. Complete maturation of VZV gH in Golgi and its expression at the cell surface requires coexpression of gL (Forghani *et al.*, 1994). When cotransfected with gE or gI, in the absence of gL, VZV gH is capable of exiting the ER in its immature form (Duus and Grose, 1996). However, unlike coexpression with gL, gH + gE or gH + gI does not result in complete gH maturation or subsequent syncytium formation. Within a transfection system, coexpression of VZV gH and gL is sufficient to mediate polykaryocytosis (Duus *et al.*, 1995). While the requirement for gH and gL association is conserved among herpesvirus gH homologs, HSV cell-to-cell fusion requires the quartet of gH, gL, gB and gD.

Like the three previously described VZV glycoproteins, VZV gH is internalized in infected cells. Further, when expressed alone with its chaperone gL in a transfected system, gH is capable of endocytosis independent of gE, gI, or gB. This observation was unexpected because all herpesvirus gH homologs have very short endodomains. Also, in the original sequencing, the cytoplasmic tail was assigned only 12 residues. Upon further inspection of the gH cytoplasmic tails of all alpha-herpesviruses, however, a putative tyrosine-based YNKI endocytosis motif (aa 835–838) in proper context was discovered if the cytoplasmic tail was realigned to include 18 amino acids (Table 9.1). When the tyrosine was replaced with an alanine, endocytosis of gH was blocked. Thus, endocytosis of VZV glycoprotein gH is mediated by a tyrosine based YNKI motif in its short cytoplasmic tail (Fig. 9.1) (Pasieka *et al.*, 2003).

Endocytosis of alpha-herpesvirus gH homologs

Alignment analysis of the VZV gH cytoplasmic tail to other alpha-herpesvirus gH homologs revealed two important findings (Table 9.1). HSV-1 and 2 gH proteins lacked an endocytosis motif, while all other alpha-herpesvirus gH homologs contained a potential motif. The VZV gH and simian varicella virus gH cytoplasmic tails were likely longer in length (18 amino acids) than predicted in the original sequence analyses (12 and 16 amino acids, respectively). The longer tails provide the proper context for a functional endocytosis motif with at least 5 intervening amino acids between the transmembrane domain and the tyrosine residue.

Table 9.1 Endocytosis motifs in herpesvirus gH homologs

Virus	Tail length	Motif in cytoplasmic tail
HSV-1	14	KVLRTSVPFFWRRE (None predicted)
HSV-2	14	RVVRTCVPFLWRRE (None predicted)
VZV	18	WMLCGNSRLREYNKIPLT(Realigned)
SVV	18	WMLCGSPRNIEYTAVPLV (Realigned)
PRV	19	KMLCSFSSEGYSRLINARS
FHV-1	17	KMLCSFTPDVRYTLLNN
BHV-1	19	KMLCSSVPLARGYSSVPVF
BHV-5	19	KMLCSSVPIARGYSAVPAF
EHV-1	19	KMLCGGVTNDGYKLLLSYE
EHV-4	19	KMLCGGVINNDYSLLLNSE

Underlined sequences indicate potential endocytosis motifs in the alpha-herpesvirus gH homologs. Only the VZV gH endocytosis motif has been proven to be functional. See reference by Pasieka *et al.* (2003) for greater detail and accession numbers. S=simian; F=feline; B=bovine; and E=equine herpesvirus.

Incorporation of endocytosed VZV glycoproteins into virion envelopes

Since endocytosis can target viral glycoproteins to the TGN, the proposed site of viral assembly, one attractive hypothesis is that endocytosis is a mechanism for delivering viral glycoproteins to the site of final envelopment (Brideau *et al.*, 2000b; Marsh and Pelchen-Matthews, 2000). The deenvelopment-reenvelopment model for acquisition of the final viral envelope has been proposed for VZV (Jones and Grose, 1988; Gershon *et al.*, 1994; Gershon *et al.*, 1994; Zhu *et al.*, 1996). An analogous mechanism has been postulated to be involved in the envelopment of other herpesviruses (Card *et al.*, 1993; Tooze *et al.*, 1993; Fuchs *et al.*, 1996; Radsak *et al.*, 1996; Granzow *et al.*, 1997; Sanchez *et al.*, 2000). The model predicts that capsids obtain a primary envelope as they pass through the inner nuclear membrane. As they exit the outer nuclear membrane, they undergo de-envelopment. Subsequently capsids released into the cytoplasm are enclosed by membranes of a post-Golgi compartment, such as the TGN. As previously reported, VZV apparently obtains its final envelope from these trans-Golgi network vesicles (Gershon *et al.*, 1994; Zhu *et al.*, 1995; Wang *et al.*, 2001). The glycoproteins of VZV are targeted to the TGN and, late in infection, all appear to become concentrated within this organelle (Olson and Grose, 1997; 1998). Thus, for the assembly of VZV a high degree of organization is required to localize all the viral envelope components to the final compartment. TGN sorting implies that TGN targeting signals must be present in the sequences of at least some VZV glycoproteins.

In a recently reported study, the role of viral glycoprotein endocytosis within the VZV life cycle was further assessed (Maresova *et al.*, 2005). VZV differs from other alphaherpesviruses in that gE is not only the major glycoprotein (as measured by percentage of total viral glycoprotein production in an infected cell) but also an essential protein for viral growth in cultured cells. Furthermore, the essential nature of the YAGL motif within gE itself was firmly documented by production of a recombinant VZV genome containing gE lacking its endocytosis motif but otherwise intact (Moffat *et al.*, 2004). This mutated virus did not replicate. When other mutations were introduced into the gE cytoplasmic tail, the virus was able to replicate. Because of this remarkable demonstration of the apparent importance of endocytosis in the VZV life cycle, we subsequently analyzed the postendocytosis trafficking pathways of glycoproteins in VZV infected cells. The methodology included biotinylation of cell surface glycoproteins followed by a glutathione cleavage assay. This assay measures the extent of endocytosis of a surface protein based on the proportion of a disulfide-linked biotinylated protein that becomes inaccessible to extracellular glutathione cleavage buffer added at the end of the endocytosis timepoint. Under the above conditions, viral glycoproteins are internalized efficiently during the first 24 hours after infection, but endocytosis is markedly decreased as infection advances over the next 24 hour time period.

To determine the extent to which VZV glycoproteins retrieved from the cell surface were incorporated into the virion particles, an uninfected monolayer was inoculated with infected cells (1:2 ratio) and incubated for 24 hours at 37°C. After surface biotinylation with cleavable sulfo-NHS-SS-biotin, the infected monolayer was incubated for an additional 5 hours to allow for endocytosis of the surface glycoproteins and their subsequent incorporation into virions (Fig. 9.3). After several washes, the infected monolayer was harvested and sonically disrupted in order to break apart the infected cells. The released cell

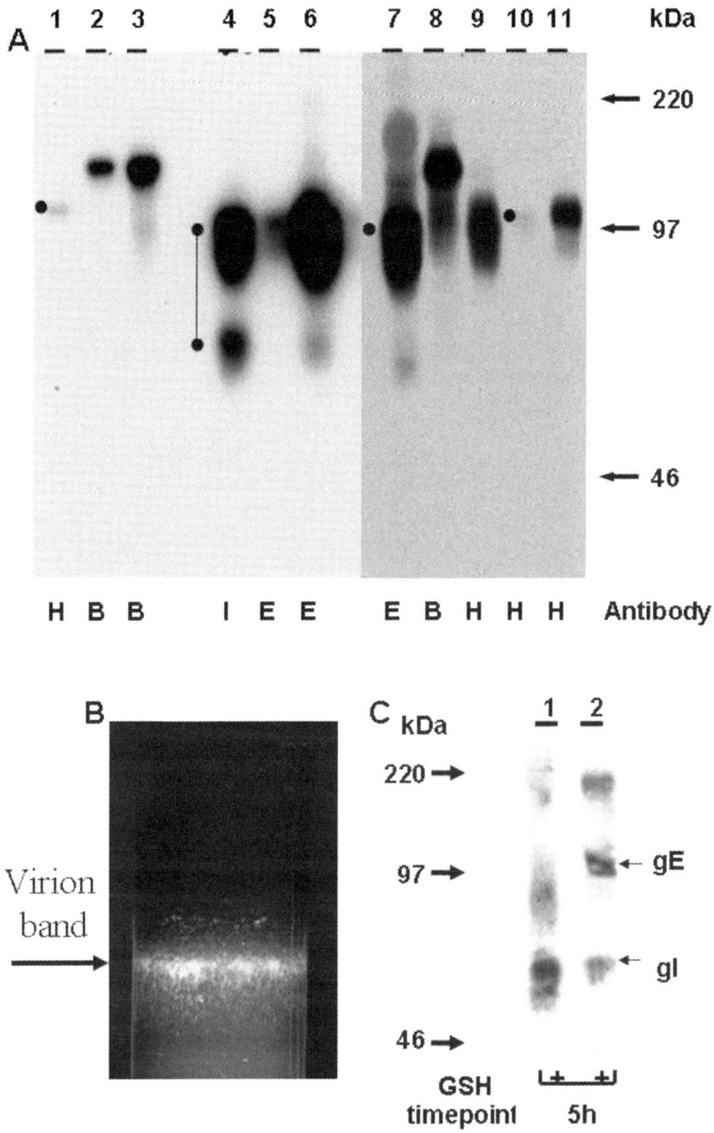

Figure 9.3 Identification of biotinylated VZV glycoproteins on infected cells and purified virions. (A) Cultured cells were infected and processed for the surface biotinylation assay as described in the text. The VZV glycoproteins were precipitated with monoclonal antibodies as designated at the bottom of each lane: gH (lanes 1, 9, 10 and 11); gB (lanes 2, 3 and 8); gE (lanes 5, 6 and 7) and gI (lane 4). Precipitated samples were resolved in an 8% gel under nonreducing conditions prior to Western blotting with Streptavidin-horse radish peroxidase. Closed circles indicate locations of some glycoproteins; in lane 4, closed circles also indicate the location of gE coprecipitated by antibody to gI. Lanes 1–6 and 7–11 represent two separate gels. The corresponding molecular mass markers (kilodaltons) are indicated on the right. (B) Virion band after ultracentrifugation in the second density gradient. The band was subsequently solubilized for immunoprecipitation (see panel C). (C). Biotinylated VZV gI in the virion band. As described in the text, biotinylated gI was immunoprecipitated from the virion band and stained by Western blotting with Streptavidin-horse radish peroxidase. Similar experiments for VZV gE, gH and gB are illustrated in the reference by Maresova *et al*. (2005).

free virus was purified by ultracentrifugation in two sequential potassium-tartrate density gradients. The particulate virion band was removed and solubilized in a buffer for immunoprecipitation. Precipitations were performed with monoclonal antibodies specific for each of the four VZV glycoproteins. After the samples were separated by polyacrylamide gel electrophoresis, individual biotinylated glycoproteins were detected by Western blotting with streptavidin-horse radish peroxidase staining (Fig. 9.3). The results demonstrated that the four glycoproteins retrieved from the cell surface were incorporated into virus particles isolated from infected cells by density gradient sedimentation (Fig. 9.3). Furthermore, additional glycoprotein complexes were present on the virion envelope, for example, the gE:gI complex and a higher molecular mass gE. Interestingly, the recently described gE:gH complex was also found in the virion envelope (Pasieka *et al.*, 2004). This result supported a mechanism of gE:gH complex formation as a means by which gH was shuttled to the TGN after internalization.

To confirm and expand the above observations, the virions were also examined directly by transmission electron microscopy (TEM). To this end, an aliquot of the gradient-purified virion fraction was placed on a copper formvar grid and incubated at 24°C with ultra-small gold beads (0.8 nm) conjugated to antibiotin antibody. Silver enhancement was performed in order to allow visualization of the ultra-small beads. In the next step, the sample was negatively stained with 1% ammonium molibdate and viewed by TEM. The biotin label was easily detected in the virion fraction (Maresova *et al.*, 2005). When the particles with biotinylated glycoproteins on their envelope were counted, the positively labeled virions represented approximately 27% of the total number of counted viral particles. This value corresponded to the percentage of intact enveloped virions expected after two successive gradient sedimentations, during which many particles lose their envelopes (Grose *et al.*, 1979). When considered in this regard, the TEM visualization experiment indicated that a majority of enveloped virions were biotinylated. Thus, in the process of VZV assembly, endocytosis is one of the pathways by which viral glycoproteins are transported to the site of envelopment in the cytoplasm.

Summary of trafficking pathways

As is apparent from the previous sections, the replication and egress of viruses require that viral proteins be correctly modified and targeted to specific intracellular compartments. Many of these modification and targeting steps rely on the communication between the cell surface and a dynamic and highly regulated network of late secretory pathway organelles that comprise the TGN/endosomal system. Strategically located at the boundary of the biosynthetic and endocytic pathways, the TGN acts as a sorting station for membrane proteins and lipids moving through the biosynthetic pathway by segregating them into nascent vesicles bound for specific target organelles including the endosomes, lysosomes and plasma membrane (Gu *et al.*, 2001; Bonifacino and Glick, 2004; Robinson, 2004). In addition, the TGN receives molecules internalized from the cell surface via a series of complex and highly dynamic early/sorting and late endosomal compartments (Fig. 9.4).

Membrane proteins endocytosed from the cell surface may traverse various endosomal paths depending on the protein and its function. For example, transferrin receptors constitutively recycle between the plasma membrane and recycling endosomes via early endosomes (van Dam and Stoorvogel, 2002), whereas growth factor receptors, such as the epidermal growth factor receptor (EGFR),

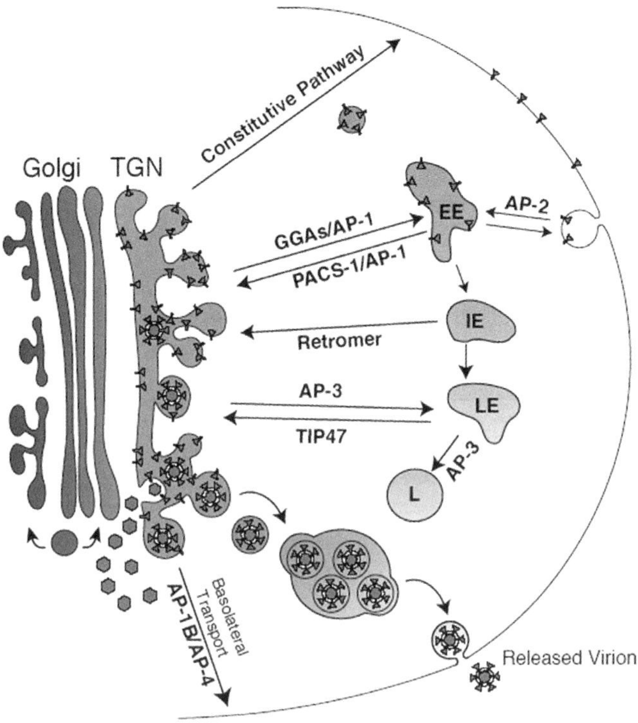

Figure 9.4 Endosomal trafficking of the VZV glycoproteins. VZV glycoproteins (triangles) may follow several late secretory trafficking pathways to the TGN, where viral envelopment occurs. As reported for HCMV gB (Crump *et al.*, 2003), retrograde transport from the early endosomes (EE) to the TGN may occur from early endosomes (EE) via PACS-1 and AP-1. In addition, transport through sorting pathways from intermediate endosomes (IE) via the retromer complex or from late endosomes (LE) via TIP47 may also occur. Anterograde transport from the TGN to early endosomes is mediated by GGA proteins and AP-1 or from the TGN to late endosomes by AP-3. VZV capsids (hexagons) bud into the TGN, forming a viral envelope of membranes enriched with viral glycoproteins. In turn, individual TGN-derived vesicles appear to form multivesicular organelles (vacuoles) containing several virions; these vacuoles fuse with the plasma membrane, exposing the virions on the surface of the cell. See Fig. 9.5 also.

are segregated into the sorting compartment of early endosomes and are delivered to lysosomes via the late endosome/multivesicular body (MVB) pathway for degradation (Raiborg *et al.*, 2003; Gruenberg and Stenmark, 2004). In addition, the two mannose 6-phosphate receptors (cation-dependent and cation-independent MPRs), recycle between the TGN, plasma membrane and endosomal compartments as they sort newly synthesized lysosomal hydrolases to lysosomes (Ghosh *et al.*, 2003). These complex protein trafficking events require the orchestrated interaction of lipid and protein components of the vesicular trafficking machinery, the cytoskeleton and multiple regulatory factors (Bonifacino and Traub, 2003; Gruenberg, 2003; Roth, 2004).

The localization of many membrane proteins within the TGN/endosomal system relies upon canonical sorting motifs within their cytoplasmic domains, the best characterized of which are the tyrosine, dileucine and acidic cluster-based motifs. These sorting signals act as "address tags," which are recognized by components of the vesicular trafficking machinery that direct sorting to specific subcellular compartments (Bonifacino and Traub, 2003). The four major VZV envelope glycoproteins have retained through evolution one or more types of these sorting signals, thus mimicking cellular membrane proteins to reach endosomes and the TGN. Together with the findings that gE and gI may mediate the endocytosis of other VZV glycoproteins (Wang *et al.*, 1998; Pasieka *et al.*, 2003), these results strongly demonstrate the importance of sorting motifs for the trafficking of VZV glycoproteins to sites of virus assembly.

The function of tyrosine-based signals arises from the interaction of these motifs with the heterotetrameric adaptor complexes AP-1, AP-2, AP-3 and AP-4 (Robinson, 2004). Adaptor complexes connect the vesicular coat protein clathrin to sorting signals in the cytoplasmic domains of membrane proteins thereby promoting their concentration into clathrin-coated pits (Fig. 9.4). AP-1 (subunits: γ, β1, μ1 and σ1), which is localized mainly at the TGN, mediates sorting of membrane proteins in both anterograde and retrograde pathways between the TGN and endosomes. AP-1B (subunits: γ, β1, μ1B and σ1), an epithelial cell specific isoform of AP-1, directs transport to the basolateral surface from the TGN and endosomes. AP-2 (subunits: α, β2, μ2 and σ2), which is localized to the plasma membrane, plays a prominent role in clathrin-mediated endocytosis of cell surface receptors. AP-3 (subunits: δ, β3, μ3 and σ3) is localized to the TGN and endosomes and mediates sorting of membrane proteins to the lysosomes and melanosomes. AP-4 (subunits: ε, β4, μ4 and σ4), which is localized to the TGN and endosomes, mediates the sorting of membrane proteins to basolateral membrane in polarized cells. The γ/α/δ subunits bind clathrin whereas the β1–4 and μ1–4 subunits bind to [D/E]XXXL[L/I] and YXXL motifs in the cytoplasmic domain of the cargo proteins, respectively. The σ subunit is predicted to play a structural role by stabilizing the tetrameric complex. Adaptor complex induced formation of vesicles from clathrin coated pits requires membrane fission, which is mediated by the GTPase dynamin at the plasma membrane and a dynamin homolog, dynamin 2, at the TGN (Cao *et al.*, 2000). Together these adaptor complexes and coat proteins orchestrate intracellular transport of membrane proteins in the TGN/endosomal system (Collins *et al.*, 2002; Robinson, 2004).

A number of cellular and viral membrane proteins containing acidic cluster sorting motifs have been identified including furin, a proprotein convertase that cleaves proteins in the secretory pathway (Thomas, 2002), the MPRs (Ghosh *et al.*, 2003), as well as the herpesvirus glycoproteins VZV gE (Grose *et al.*, 1989), HCMV gB (Norais *et al.*, 1996), HSV gB, gE (Alconada *et al.*, 1999; Beitia Ortiz de Zarate *et al.*, 2004) and the PrV glycoprotein Us9 (Brideau *et al.*, 2000). Acidic cluster motifs often contain a serine or threonine residue that can be phosphorylated by casein kinase II, a ubiquitous, heterotetrameric serine/threonine kinase involved in a plethora of phosphorylation events including the regulation of protein sorting (Grose, 1990; Kenyon *et al.*, 2003; Meggio and Pinna, 2003). Phosphofurin acidic cluster sorting protein-1 (PACS-1) connects phosphorylated acidic cluster containing proteins to AP-1 and directs the delivery of membrane proteins from endosomes to the TGN (Fig. 9.4). This function of PACS-1 is required for the sorting of several integral membrane proteins such as furin, the MPRs and HCMV gB (Wan *et al.*, 1998; Crump *et al.*, 2003; Scott *et al.*, 2003). In addition, PACS-1 binds to an acidic cluster (that contains a phosphorylation site for casein kinase II) within the cytoplasmic tail of VZV gE (Wan *et al.*, 1998) and may mediate delivery of VZV gE to the TGN, though this function has not yet been proven. Considering the importance for the TGN localization of VZV glycoproteins in the final envelopment of VZV virions, PACS-1 may play a critical role for the formation of VZV virions that bud from the TGN, as PACS-1 does for HCMV (Crump *et al.*, 2003).

A newer class of sorting motifs, the acidic-dileucine (DxxLL) motifs, are recognized by a family of monomeric adaptors consisting of GGA-1, -2 and -3 (Golgi-localized, γ-adaptin ear homology domain, ARF-binding protein 1–3, GGAs) (Bonifacino, 2004) (Fig. 9.4). The GGAs are localized to the TGN where they cooperate with AP-1 to link membrane proteins with acidic-dileucine motifs directly to clathrin. This function concentrates several endogenous membrane proteins, including the MPRs, into clathrin-coated buds at the TGN and activates anterograde sorting of the MPRs to the early endosomes (Puertollano *et al.*, 2001; Bonifacino, 2004). Like acidic cluster motifs, many acidic-dileucine motifs contain a serine/threonine residue that can be phosphorylated by casein kinase II (Doray *et al.*, 2002). Interestingly, the cytoplasmic domain of VZV gI contains a potential GGA binding site (ESDVML; not canonical, but highly similar to the GGA binding motif of the neuronal SorLA receptor), implicating GGAs in the sorting of VZV containing vesicles from the TGN to endosomes.

Two additional sorting molecules, TIP47 and the retromer complex, may also direct transport of membrane proteins such as MPRs, from endosomes to the TGN (Fig. 9.4). TIP47 binds to a diaromatic amino acid motif in the cytoplasmic tail of cargo proteins like the CD-MPR (FW) and HIV-gp160 (YW), mediating the late endosome to TGN transport of these proteins (Diaz and Pfeffer, 1998). This trafficking step has been shown to be essential for the incorporation of gp160 into HIV virions, however no VZV glycoprotein contains a TIP47 binding motif. The mammalian retromer complex, which consists of five subunits (vps35, vps29, vps26, SNX1 and SNX2), was recently shown to localize to the tubular-vesicular profiles that emanate from either early endosomes or from intermediates in the maturation of early endosomes to late endosomes. From these endosomal compartments, retromer mediates the recycling of CI-MPR to the TGN (Seaman, 2004). The cargo recognition motif for retromer is unknown and proposed to be conformational in nature.

The deenvelopment–reenvelopment model for acquisition of the final viral envelope proposed for herpesviruses suggests that mature viral glycoproteins must traffic to the TGN or to endosomal structures before being incorporated into the viral envelope (Jones and Grose, 1989; Tooze *et al.*, 1993; Gershon *et al.*, 1994). This scenario requires, at least in case of VZV, endocytosis of mature viral glycoproteins from the plasma membrane via a clathrin/AP-2/dynamin dependent process and delivery of these proteins to early endosomes. Endocytosed viral glycoproteins can reach the site of assembly, the TGN, either directly from early endosomes or indirectly from early endosomes via late endosomes. These retrograde transport pathways have been shown to be utilized not only by cellular and viral proteins like the MPRs, furin and HCMV gB, but also by bacterial toxins like Shiga toxin, which is internalized in clathrin coated vesicles and traverses the early endosome-to-TGN pathway before reaching the ER (Mallard *et al.*, 1998; Mallard *et al.*, 2002; Sandvig and van Deurs, 2002). The final event in VZV assembly and egress is exocytosis at the cell surface (Fig. 9.5).

The failure in viral assembly in cells expressing either mutant VZV glycoproteins with trafficking defects or mutant VZV serine protein kinases with phosphorylation defects demonstrates the importance of

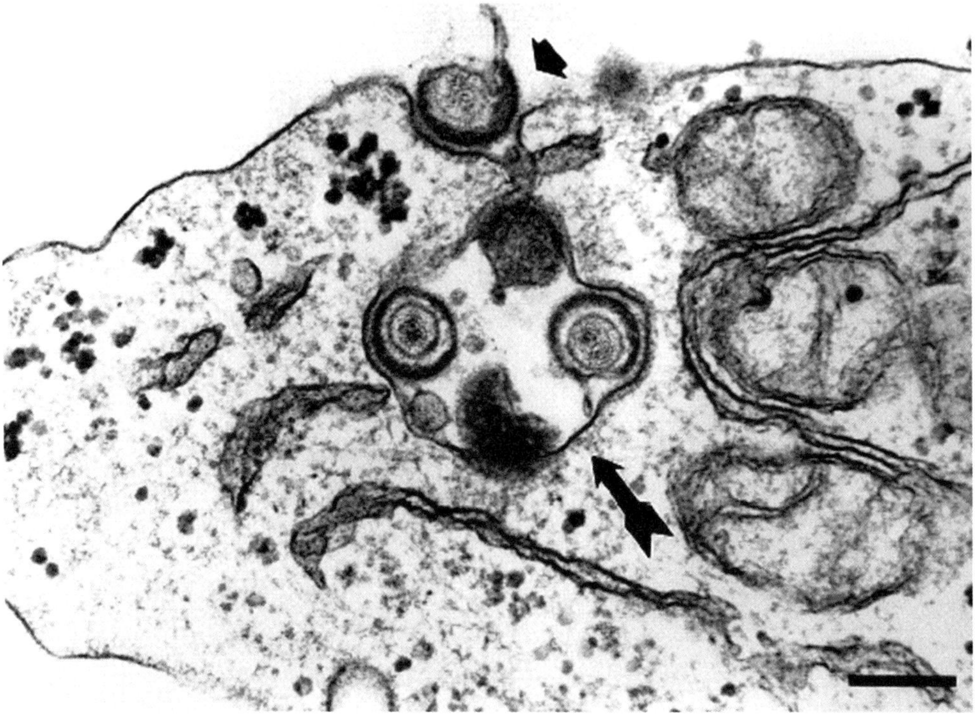

Figure 9.5 Egress and exocytosis of virions from a cytoplasmic vacuole. This electron micrograph shows a cytoplasmic vacuole with four enveloped virions (arrow). The micrograph also shows one virion in the final stage of egress at the cell surface (arrowhead). Magnification bar=200nm.

these events for virus maturation (Wang *et al.*, 2001; Grose, 2002; Heineman and Hall, 2002; Kenyon *et al.*, 2002). In particular, the fact that a recombinant VZV genome containing only a single mutation, namely, the endocytosis motif within the gE gene, cannot replicate emphasizes the pivotal role of gE endocytosis in the life cycle of this alpha-herpesvirus (Moffat *et al.*, 2004; Maresova *et al.*, 2005).

Acknowledgments

The authors thank all colleagues whose names appear on our cited articles from previous years. Research described in this chapter was supported by NIH grants AI 22795, DK 37274, AI 48585 and AI 49793.

References

Alconada, A., Bauer, U., and Hoflack, B. (1996). A tyrosine-based motif and a casein kinase II phosphorylation site regulate the intracellular trafficking of the varicella-zoster virus glycoprotein I, a protein localized in the trans-Golgi network. EMBO J. *15*, 6096–6110.

Alconada, A., Bauer, U., Baudoux, L., Piette, J., and Hoflack, B. (1998). Intracellular transport of the glycoproteins gE and gI of the varicella-zoster virus. gE accelerates the maturation of gI and determines its accumulation in the trans-Golgi network. J. Biol. Chem. *273*, 13430–13436.

Alconada, A., Bauer, U., Sodeik, B., and Hoflack, B. (1999). Intracellular traffic of herpes simplex virus glycoprotein gE: characterization of the sorting signals required for its trans-Golgi network localization. J. Virol. *73*, 377–387.

Baghian, A., Huang, L., Newman, S., Jayachandra, S., and Kousoulas, K.G. (1993). Truncation of the carboxy-terminal 28 amino acids of glycoprotein B specified by herpes simplex virus type 1 mutant amb1511–7 causes extensive cell fusion. J. Virol. *67*, 2396–2401.

Beitia Ortiz de Zarate, I., Kaelin, K., and Rozenberg, F. (2004). Effects of mutations in the cytoplasmic domain of herpes simplex virus type 1 glycoprotein B on intracellular transport and infectivity. J. Virol. *78*, 1540–1551.

Boll, W., Ohno, H., Songyang, Z., Rapoport, I., Cantley, L.C., Bonifacino, J.S., and Kirchhausen, T. (1996). Sequence requirements for the recognition of tyrosine-based endocytic signals by clathrin AP-2 complexes. EMBO J. *15*, 5789–5795.

Bonifacino, J.S. (2004). The GGA proteins: adaptors on the move. Nat. Rev. Mol. Cell. Biol. 5, 23–32.

Bonifacino, J.S., and Dell'Angelica, E.C. (1999). Molecular bases for the recognition of tyrosine-based sorting signals. J. Cell Biol. *145*, 923–926.

Bonifacino, J.S., and Glick, B.S. (2004). The mechanisms of vesicle budding and fusion. Cell *116*, 153–166.

Bonifacino, J.S., and Traub, L.M. (2003). Signals for sorting of transmembrane proteins to endosomes and lysosomes. Annu. Rev. Biochem. *72*, 395–447.

Brideau, A.D., Eldridge, M.G., and Enquist, L.W. (2000a). Directional transneuronal infection by pseudorabies virus is dependent on an acidic internalization motif in the Us9 cytoplasmic tail. J. Virol. *74*, 4549–4561.

Brideau, A.D., Enquist, L.W., and Tirabassi, R.S. (2000b). The role of virion membrane protein endocytosis in the herpesvirus life cycle. J. Clin. Virol. *17*, 69–82.

Cai, W.H., Gu, B., and Person, S. (1988). Role of glycoprotein B of herpes simplex virus type 1 in viral entry and cell fusion. J. Virol. *62*, 2596–2604.

Cao, H., Thompson, H.M., Krueger, E.W., and McNiven, M.A. (2000). Disruption of Golgi structure and function in mammalian cells expressing a mutant dynamin. J. Cell Sci. *113 (Pt 11)*, 1993–2002.

Card, J.P., Rinaman, L., Lynn, R.B., Lee, B.H., Meade, R.P., Miselis, R.R., and Enquist, L.W. (1993). Pseudorabies virus infection of the rat central nervous system: ultrastructural characterization of viral replication, transport, and pathogenesis. J. Neurosci. *13*, 2515–2539.

Collins, B.M., McCoy, A.J., Kent, H.M., Evans, P.R., and Owen, D.J. (2002). Molecular architecture and functional model of the endocytic AP2 complex. Cell *109*, 523–535.

Crump, C.M., Xiang, Y., Thomas, L., Gu, F., Austin, C., Tooze, S.A., and Thomas, G. (2001). PACS-1 binding to adaptors is required for acidic cluster motif-mediated protein traffic. EMBO J. *20*, 2191–2201.

Crump, C.M., Hung, C.H., Thomas, L., Wan, L., and Thomas, G. (2003). Role of PACS-1 in trafficking of human cytomegalovirus glycoprotein B and virus production. J. Virol. *77*, 11105–11113.

Davison, A.J., and Scott, J.E. (1986). The complete DNA sequence of varicella-zoster virus. J. Gen. Virol. *67*, 1759–1816.

Diaz, E., and Pfeffer, S.R. (1998). TIP47: a cargo selection device for mannose 6-phosphate receptor trafficking. Cell 93, 433–443.

Doray, B., Bruns, K., Ghosh, P., and Kornfeld, S. (2002). Interaction of the cation-dependent mannose 6-phosphate receptor with GGA proteins. J. Biol. Chem. *277*, 18477–18482.

Duus, K.M., and Grose, C. (1996). Multiple regulatory effects of varicella-zoster virus (VZV) gL on trafficking patterns and fusogenic properties of VZV gH. J. Virol. *70*, 8961–8971.

Duus, K.M., Hatfield, C., and Grose, C. (1995). Cell surface expression and fusion by the varicella-zoster virus gH:gL glycoprotein complex: analysis by laser scanning confocal microscopy. Virology *210*, 429–440.

Edson, C.M., Hosler, B.A., Respess, R.A., Waters, D.J., and Thorley-Lawson, D.A. (1985). Cross-reactivity between herpes simplex virus glycoprotein B and a 63,000-dalton varicella-zoster virus envelope glycoprotein. J. Virol. 56, 333–336.

Elliott, G., and O'Hare, P. (1999). Live-cell analysis of a green fluorescent protein-tagged herpes simplex virus infection. J. Virol. *73*, 4110–4119.

Favoreel, H.W., Van Minnebruggen, G., Nauwynck, H.J., Enquist, L.W., and Pensaert, M.B. (2002). A tyrosine-based motif in the cytoplasmic tail of pseudorabies virus glycoprotein B is important for both antibody-induced internalization of viral glycoproteins and efficient cell-to-cell spread. J. Virol. *76*, 6845–6851.

Fish, K.N., Soderberg-Naucler, C., and Nelson, J.A. (1998). Steady-state plasma membrane expression of human cytomegalovirus gB is determined by the phosphorylation state of Ser900. J. Virol. *72*, 6657–6664.

Forghani, B., Ni, L., and Grose, C. (1994). Neutralization epitope of the varicella-zoster virus gH:gL glycoprotein complex. Virology *199*, 458–462.

Fuchs, W., Klupp, B.G., Granzow, H., Rziha, H.J., and Mettenleiter, T.C. (1996). Identification and characterization of the pseudorabies virus UL3.5 protein, which is involved in virus egress. J. Virol. *70*, 3517–3527.

Gershon, A.A., Sherman, D.L., Zhu, Z., Gabel, C.A., Ambron, R.T., and Gershon, M.D. (1994). Intracellular transport of newly synthesized varicella-zoster virus: final envelopment in the trans-Golgi network. J. Virol. *68*, 6372–6390.

Ghosh, P., Dahms, N.M., and Kornfeld, S. (2003). Mannose 6-phosphate receptors: new twists in the tale. Nat. Rev. Mol. Cell. Biol. *4*, 202–212.

Granzow, H., Weiland, F., Jons, A., Klupp, B.G., Karger, A., and Mettenleiter, T.C. (1997). Ultrastructural analysis of the replication cycle of pseudorabies virus in cell culture: a reassessment. J. Virol. *71*, 2072–2082.

Grose, C. (1990). Glycoproteins encoded by varicella-zoster virus: biosynthesis, phosphorylation, and intracellular trafficking. Annu. Rev. Microbiol. *44*, 59–80.

Grose, C. (2002). The predominant varicella-zoster virus gE and gI glycoprotein complex. In Structure–Function Relationships of Human Pathogenic Viruses. Holzenburg, E., and Bogner, E., ed. (New York, N.Y., Kluwer Academic Press), pp. 195–223.

Grose, C., Perrotta, D.M., Brunell, P.A., and Smith, G.C. (1979). Cell-free varicella-zoster virus in cultured human melanoma cells. J. Gen. Virol. *43*, 15–27.

Grose, C., Jackson, W., and Traugh, J.A. (1989). Phosphorylation of varicella-zoster virus glycoprotein gpI (gE) by mammalian casein kinase II and casein kinase I. J. Virol. *63*, 3912–3918.

Grose, C., Tyler, S., Hiebert, J., Stephens, G., Ruyechan, W., Jackson, W., Storlie, J., and Tipples, G.A. (2004) Complete DNA sequence analyses of the first two varicella-zoster virus glycoprotein E (D150N) mutant viruses found in North America: evolution of genotypes with an accelerated cell spread phenotype. J. Virol. 78, 6799–807.

Gruenberg, J. (2003). Lipids in endocytic membrane transport and sorting. Curr. Opin. Cell Biol. *15*, 382–388.

Gruenberg, J., and Stenmark, H. (2004). The biogenesis of multivesicular endosomes. Nat. Rev. Mol. Cell. Biol. 5, 317–323.

Gu, F., Crump, C.M., and Thomas, G. (2001). Trans-Golgi network sorting. Cell. Mol. Life Sci. 58, 1067–1084.

Heineman, T.C., and Hall, S.L. (2001). VZV gB endocytosis and Golgi localization are mediated by YXXphi motifs in its cytoplasmic domain. Virology *285*, 42–49.

Heineman, T.C., and Hall, S.L. (2002). Role of the varicella-zoster virus gB cytoplasmic domain in gB transport and viral egress. J. Virol. *76*, 591–599.

Heineman, T.C., Krudwig, N., and Hall, S.L. (2000). Cytoplasmic domain signal sequences that mediate transport of varicella-zoster virus gB from the endoplasmic reticulum to the Golgi. J. Virol. *74*, 9421–9430.

Jones, F., and Grose, C. (1988). Role of cytoplasmic vacuoles in varicella-zoster virus glycoprotein trafficking and virion envelopment. J. Virol. *62*, 2701–2711.

Keller, P.M., Davison, A.J., Lowe, R.S., Bennett, C.D., and Ellis, R.W. (1986). Identification and structure of the gene encoding gpII, a major glycoprotein of varicella-zoster virus. Virology *152*, 181–191.

Kenyon, T.K., Cohen, J.I., and Grose, C. (2002). Phosphorylation by the varicella-zoster virus ORF47 protein serine kinase determines whether endocytosed viral gE traffics to the trans-Golgi network or recycles to the cell membrane. J. Virol. *76*, 10980–10993.

Kenyon, T.K., Homan, E., Storlie, J., Ikoma, M., and Grose, C. (2003) Comparison of varicella-zoster virus ORF47 protein kinase and casein kinase II and their substrates. J. Med. Virol. S95-S102.

Litwin, V., Sandor, M., and Grose, C. (1990). Cell surface expression of the varicella-zoster virus glycoproteins and Fc receptor. Virology *178*, 263–272.

Litwin, V., Jackson, W., and Grose, C. (1992). Receptor properties of two varicella-zoster virus glycoproteins, gpI and gpIV, homologous to herpes simplex virus gE and gI. J. Virol. *66*, 3643–3651.

Mallard, F., Antony, C., Tenza, D., Salamero, J., Goud, B., and Johannes, L. (1998). Direct pathway from early/recycling endosomes to the Golgi apparatus revealed through the study of shiga toxin B-fragment transport. J. Cell Biol. *143*, 973–990.

Mallard, F., Tang, B.L., Galli, T., Tenza, D., Saint-Pol, A., Yue, X., Antony, C., Hong, W., Goud, B., and Johannes, L. (2002). Early/recycling endosomes-to-TGN transport involves two SNARE complexes and a Rab6 isoform. J. Cell Biol. *156*, 653–664.

Mallory, S., Sommer, M., and Arvin, A.M. (1997). Mutational analysis of the role of glycoprotein I in varicella-zoster virus replication and its effects on glycoprotein E conformation and trafficking. J. Virol. *71*, 8279–8288.

Maresova, L., Pasieka, T.J., and Grose, C. (2001). Varicella-zoster Virus gB and gE coexpression, but not gB or gE alone, leads to abundant fusion and syncytium formation equivalent to those from gH and gL coexpression. J. Virol. *75*, 9483–9492.

Maresova, L., Pasieka, T., Wagenaar, T., Jackson, W., and Grose, C. (2003). Identification of the authentic varicella-zoster virus gB (gene 31) initiating methionine overlapping the 3′ end of gene 30. J. Med. Virol. *70 Suppl 1*, S64–70.

Maresova, L., Pasieka, T., Homan, E., Gerday, E., and Grose, C. (2005) Incorporation of three endocytosed varicella-zoster virus glycoproteins gE,

gH and gB into the virion envelope. J. Virol. *79*, 997–1007.
Marsh, M., and Pelchen-Matthews, A. (2000). Endocytosis in viral replication. Traffic *1*, 525–532.
Meggio, F., and Pinna, L.A. (2003). One-thousand-and-one substrates of protein kinase CK2? FASEB J. *17*, 349–368.
Moffat, J., Mo, C., Cheng, J. Sommer, M. Zerboni, L., Stamatis, S., and Arvin, A.M. (2004) Functions of the C-terminal domain of varicella-zoster virus glycoprotein E in replication *in vitro* and skin and T-cell tropism *in vivo*. J. Virol. *78*, 12406–12415.
Montalvo, E.A., and Grose, C. (1986). Varicella zoster virus glycoprotein gpI (gE) is selectively phosphorylated by a virus-induced protein kinase. Proc. Natl. Acad. Sci. USA *83*, 8967–8971.
Montalvo, E.A., and Grose, C. (1987). Assembly and processing of the disulfide-linked varicella-zoster virus glycoprotein gpII(gB). J. Virol. *61*, 2877–2884.
Montalvo, E.A., Parmley, R.T., and Grose, C. (1985). Structural analysis of the varicella-zoster virus gp98-gp62 complex: posttranslational addition of N-linked and O-linked oligosaccharide moieties. J. Virol. 53, 761–770.
Nixdorf, R., Klupp, B.G., Karger, A., and Mettenleiter, T.C. (2000). Effects of truncation of the carboxy terminus of pseudorabies virus glycoprotein B on infectivity. J. Virol. *74*, 7137–7145.
Norais, N., Hall, J.A., Gross, L., Tang, D., Kaur, S., Chamberlain, S.H., Burke, R.L., and Marcus, F. (1996). Evidence for a phosphorylation site in cytomegalovirus glycoprotein gB. J. Virol. *70*, 5716–5719.
Ohno, H., Stewart, J., Fournier, M.C., Bosshart, H., Rhee, I., Miyatake, S., Saito, T., Gallusser, A., Kirchhausen, T., and Bonifacino, J.S. (1995). Interaction of tyrosine-based sorting signals with clathrin-associated proteins. Science *269*, 1872–1875.
Olson, J.K., and Grose, C. (1997). Endocytosis and recycling of varicella-zoster virus Fc receptor glycoprotein gE: internalization mediated by a YXXL motif in the cytoplasmic tail. J. Virol. *71*, 4042–4054.
Olson, J.K., and Grose, C. (1998). Complex formation facilitates endocytosis of the varicella-zoster virus gE:gI Fc receptor. J. Virol. *72*, 1542–1551.
Olson, J.K., Bishop, G.A., and Grose, C. (1997). Varicella-zoster virus Fc receptor gE glycoprotein: serine/threonine and tyrosine phosphorylation of monomeric and dimeric forms. J. Virol. *71*, 110–119.
Olson, J.K., Santos, R.A., and Grose, C. (1998). Varicella-zoster virus glycoprotein gE: endocytosis and trafficking of the Fc receptor. J. Infect. Dis. *178 Suppl 1*, S2–6.
Pasieka, T.J., Maresova, L., and Grose, C. (2003). A functional YNKI motif in the short cytoplasmic tail of varicella-zoster virus glycoprotein gH mediates clathrin-dependent and antibody-independent endocytosis. J. Virol. *77*, 4191–4204.
Pasieka, T.J., Maresova, L., Shiraki, K., and Grose, C. (2004). Regulation of varicella-zoster virus-induced cell-to-cell fusion by the endocytosis-competent glycoproteins gH and gE. J. Virol. *78*, 2884–2896.
Peeters, B., de Wind, N., Hooisma, M., Wagenaar, F., Gielkens, A., and Moormann, R. (1992). Pseudorabies virus envelope glycoproteins gp50 and gII are essential for virus penetration, but only gII is involved in membrane fusion. J. Virol. *66*, 894–905.
Pereira, L. (1994). Function of glycoprotein B homologues of the family herpesviridae. Infect. Agents Dis. *3*, 9–28.
Puertollano, R., Aguilar, R.C., Gorshkova, I., Crouch, R.J., and Bonifacino, J.S. (2001). Sorting of mannose 6-phosphate receptors mediated by the GGAs. Science *292*, 1712–1716.
Radsak, K., Eickmann, M., Mockenhaupt, T., Bogner, E., Kern, H., Eis-Hubinger, A., and Reschke, M. (1996). Retrieval of human cytomegalovirus glycoprotein B from the infected cell surface for virus envelopment. Arch. Virol. *141*, 557–572.
Raiborg, C., Rusten, T.E., and Stenmark, H. (2003). Protein sorting into multivesicular endosomes. Curr. Opin. Cell Biol. *15*, 446–455.
Rauh, I., and Mettenleiter, T.C. (1991). Pseudorabies virus glycoproteins gII and gp50 are essential for virus penetration. J. Virol. *65*, 5348–5356.
Robinson, M.S. (2004). Adaptable adaptors for coated vesicles. Trends Cell Biol. *14*, 167–174.
Roth, M.G. (2004). Phosphoinositides in constitutive membrane traffic. Physiol. Rev. *84*, 699–730.
Sanchez, V., Greis, K.D., Sztul, E., and Britt, W.J. (2000). Accumulation of virion tegument and envelope proteins in a stable cytoplasmic compartment during human cytomegalovirus replication: characterization of a potential site of virus assembly. J. Virol. *74*, 975–986.
Sandvig, K., and van Deurs, B. (2002). Transport of protein toxins into cells: pathways used by ricin, cholera toxin and Shiga toxin. FEBS Lett. *529*, 49–53.
Santos, R.A., Padilla, J.A., Hatfield, C., and Grose, C. (1998). Antigenic variation of varicella zoster virus Fc receptor gE: loss of a major B cell epitope in the ectodomain. Virology *249*, 21–31.
Santos, R.A., Hatfield, C.C., Cole, N.L., Padilla, J.A., Moffat, J.F., Arvin, A.M., Ruyechan, W.T., Hay, J., and Grose, C. (2000). Varicella-zoster virus gE

escape mutant VZV-MSP exhibits an accelerated cell-to-cell spread phenotype in both infected cell cultures and SCID-hu mice. Virology *275*, 306–317.

Scott, G.K., Gu, F., Crump, C.M., Thomas, L., Wan, L., Xiang, Y., and Thomas, G. (2003). The phosphorylation state of an autoregulatory domain controls PACS-1-directed protein traffic. EMBO J. 22, 6234–6244.

Seaman, M.N. (2004). Cargo-selective endosomal sorting for retrieval to the Golgi requires retromer. J. Cell Biol. *165*, 111–122.

Spear, P.G. (1993). Entry of alphaherpesviruses into cells. Semin. Virol. *4*, 167–180.

Thomas, G. (2002). Furin at the cutting edge: from protein traffic to embryogenesis and disease. Nat. Rev. Mol. Cell. Biol. 3, 753–766.

Tirabassi, R.S., and Enquist, L.W. (1998). Role of envelope protein gE endocytosis in the pseudorabies virus life cycle. J. Virol. *72*, 4571–4579.

Tirabassi, R.S., and Enquist, L.W. (1999). Mutation of the YXXL endocytosis motif in the cytoplasmic tail of pseudorabies virus gE. J. Virol. *73*, 2717–2728.

Tooze, J., Hollinshead, M., Reis, B., Radsak, K., and Kern, H. (1993). Progeny vaccinia and human cytomegalovirus particles utilize early endosomal cisternae for their envelopes. Eur J. Cell Biol. *60*, 163–178.

Trowbridge, I.S. (1991). Endocytosis and signals for internalization. Curr. Opin. Cell Biol. 3, 634–641.

van Dam, E.M., and Stoorvogel, W. (2002). Dynamin-dependent transferrin receptor recycling by endosome-derived clathrin-coated vesicles. Mol. Biol. Cell *13*, 169–182.

Wan, L., Molloy, S.S., Thomas, L., Liu, G., Xiang, Y., Rybak, S.L., and Thomas, G. (1998). PACS-1 defines a novel gene family of cytosolic sorting proteins required for trans-Golgi network localization. Cell *94*, 205–216.

Wang, Z., Gershon, M.D., Lungu, O., Panagiotidis, C.A., Zhu, Z., Hao, Y., and Gershon, A.A. (1998). Intracellular transport of varicella-zoster glycoproteins. J. Infect. Dis. *178 Suppl 1*, S7–12.

Wang, Z.H., Gershon, M.D., Lungu, O., Zhu, Z., and Gershon, A.A. (2000). Trafficking of varicella-zoster virus glycoprotein gI: T(338)-dependent retention in the trans-Golgi network, secretion, and mannose 6-phosphate-inhibitable uptake of the ectodomain. J. Virol. *74*, 6600–6613.

Wang, Z.H., Gershon, M.D., Lungu, O., Zhu, Z., Mallory, S., Arvin, A.M., and Gershon, A.A. (2001). Essential role played by the C-terminal domain of glycoprotein I in envelopment of varicella-zoster virus in the trans-Golgi network: interactions of glycoproteins with tegument. J. Virol. *75*, 323–340.

Yao, Z., Jackson, W., and Grose, C. (1993). Identification of the phosphorylation sequence in the cytoplasmic tail of the varicella-zoster virus Fc receptor glycoprotein gpI. J. Virol. *67*, 4464–4473.

Zhu, Z., Gershon, M.D., Hao, Y., Ambron, R.T., Gabel, C.A., and Gershon, A.A. (1995). Envelopment of varicella-zoster virus: targeting of viral glycoproteins to the trans-Golgi network. J. Virol. *69*, 7951–7959.

Zhu, Z., Hao, Y., Gershon, M.D., Ambron, R.T., and Gershon, A.A. (1996). Targeting of glycoprotein I (gE) of varicella-zoster virus to the trans-Golgi network by an AYRV sequence and an acidic amino acid-rich patch in the cytosolic domain of the molecule. J. Virol. *70*, 6563–6575.

Nucleocapsid Assembly and Envelopment of Herpes Simplex Virus

10

Joel D. Baines and Carol Duffy

Abstract

Herpes simplex virus capsids assemble in the nuclei of infected cells in a number of ordered steps. The spherical procapsid is the first capsid formed and consists of an inner and outer shell. Proteolytic cleavage of the inner capsid shell, followed by subsequent expulsion or degradation of the inner shell and insertion of genomic DNA, results in a dramatic conformational change in which the outer shell of the procapsid angularizes and matures into a stable icosahedron. The completed, DNA-filled nucleocapsid is then actively transported to the nuclear membrane where it attaches and buds into the perinuclear space to become a virion. These events are detailed in this chapter.

Introduction

Herpesvirus assembly is important to understand because the events are targets of novel antiviral drugs and the information will likely shed light on cellular processes that are modified by the virus to augment virion assembly.

At late times in herpesvirus infection, progeny virions are assembled through a multistep process. The first stages of virus assembly occur in the nucleus of the infected cell. Capsid proteins are expressed and assembled into empty capsid precursors, called procapsids. The viral genome undergoes both rolling circle replication and recombination to yield a highly branched DNA concatemer. Viral packaging proteins cleave individual genomes from this concatemeric substrate and translocate them into procapsids during the process of DNA packaging. DNA-filled capsids then egress through the nucleoplasm to the nuclear membrane, first acquiring a primary envelope at the inner nuclear membrane (INM). Much has been learned about these stages of herpesvirus assembly. The goal of this chapter is to describe what is known about these processes, and present models based on that knowledge.

Capsid assembly

Capsid types

Four capsid types, called procapsids, A-, B-, and C-capsids, are found in herpesvirus-infected cells (Gibson and Roizman, 1972; Newcomb *et al.*, 1996). These four capsid types are morphologically distinguishable by electron microscopy and differ in their protein and/or genome composition. Much of the current knowledge of herpesvirus capsid maturation and DNA packaging has been gained through the study of these capsid types.

Procapsids are precursors of mature capsids. These unstable, porous capsids contain an uncleaved inner shell, or scaffold, not found in the mature virion, and an outer capsid shell that is a component of all capsid types and infectious virions (Trus *et al.*, 1996; Newcomb *et al.*, 1996). Procapsids do not contain DNA and are less angular, morphologically, than the other capsid types. The presence of procapsids is required for the initial DNA cleavage event to occur during DNA packaging and packaging proteins are found associated with purified procapsids (Desai *et al.*, 1993; McNab *et al.*, 1998; Newcomb *et al.*, 2001; Ogasawara *et al.*, 2001; Sheaffer *et al.*, 2001). Based on this information and comparisons to similar bacteriophage systems, it is widely believed that herpesvirus DNA packaging is initiated in association with procapsids that lose their scaffolds and angularize as packaging progresses.

C-capsids contain packaged viral genomes and eventually mature into infectious virions. During DNA packaging the scaffold is cleaved by the viral protease and exits the capsid or is degraded (Rixon *et al.*, 1988; Person *et al.*, 1993; Weinheimer *et al.*, 1993). Thus, C-capsids lack this inner capsid shell and the external shell is morphologically more angular than that of procapsids.

Angular-shaped B-capsids do not contain DNA, but do contain cleaved scaffold (Rixon *et al.*, 1988; Newcomb and Brown, 1991). These capsids are thought to form through either scaffold cleavage in the absence of DNA packaging, or from early-aborted packaging events in which the scaffold is cleaved but does not exit the capsid.

A-capsids are also angular in shape. These capsids have lost the internal scaffold, but do not contain DNA. A-capsids are thought to form through packaging events that are aborted after part or all of the genome is initially packaged and the cleaved scaffold is lost. In these events, the packaged DNA is not retained (Sherman and Bachenheimer, 1988).

While procapsids and C-capsids are the capsid types most relevant to the formation of infectious virions, much of our understanding of capsid structure and maturation comes from protein composition studies of all the capsid types. Thus, we will consider the various capsid forms, as applicable, in our discussion of capsid structure, protein composition, assembly, and maturation.

Capsid structure

The majority of work on *alpha-herpesvirinae* capsid structure and assembly has been performed with herpes simplex virus. HSV-1 capsids are T = 16 icosahedral protein shells composed of 162 capsomers (12 pentons and 150 hexons), and 320 connecting structures called triplexes. The pentons possess five-fold rotational-symmetry and one is present at each of the icosahedral vertices except the portal vertex. The hexons possess six-fold rotational-symmetry and make up the edges and faces of the icosahedron with 12 hexons per icosahedral facet. Each of the triplexes connects three capsomers.

The three-dimensional structure of mature HSV-1 capsids has been determined through computer reconstructions of cryo-electron micrographs of A-, B- and C-capsids (Fig. 10.1) (Schrag *et al.*, 1989; Newcomb *et al.*, 1993; Zhou *et al.*, 1994). These analyses show mature capsids to be ~125 nm in diameter and the outer shell to be ~16 nm thick. The pentons and hexons contain open axial channels that extend from the capsid exterior, through the capsomer, to the capsid interior. The triplexes lie on the floor of the capsid while the capsomers have tower-like domains that protrude outward. The inner face of the capsid appears relatively smooth.

In contrast, electron micrographs of cryo-preserved procapsids show these capsids are spherical in shape, ~124 nm in diameter, and contain a scaffold (Fig. 10.1) (Newcomb *et al.*, 1999; Newcomb *et al.*, 2000; Heymann *et al.*, 2003). While reconstructed density maps based on icosahedral averaging show very little contrast inside the outer capsid shell, radial distribution density profiles identify a scaffold region that extends ~30 nm inward from the inner surface of the outer shell (Newcomb

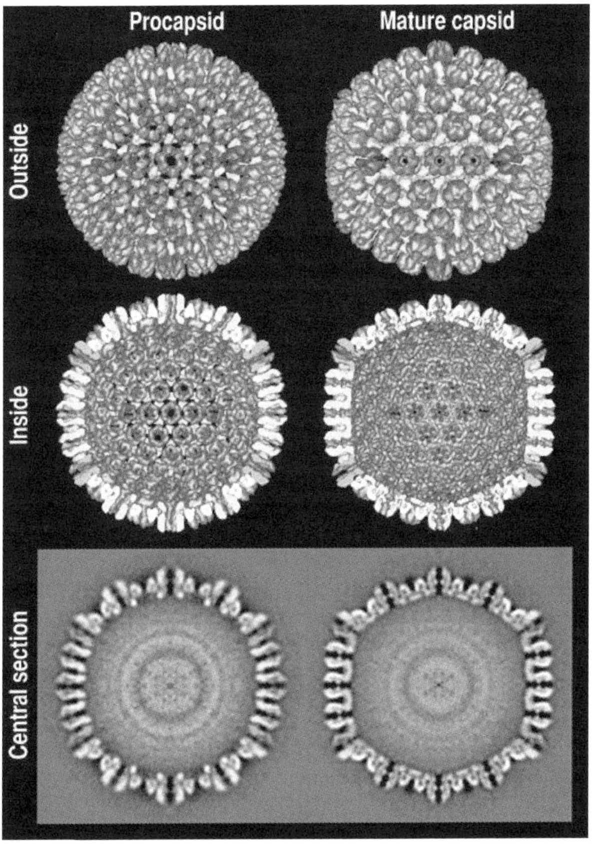

Figure 10.1 Structures of the HSV-1 procapsid (left) and mature capsid (right), color-coded to distinguish the various components. The particles are viewed along a 2-fold axis of symmetry. The resolution is ~ 22 Å. VP5 hexons are light blue; VP5 pentons are dark blue; triplexes, heterotrimers of VP23 and VP19c, are green. Adapted from Heymann *et al.* (2003). Top: outer surfaces; middle: inner surfaces; bottom: central sections. These reconstructions were calculated from isolated cryo-micrographs of procapsids that were allowed to mature *in vitro* at moderate ionic strength at room temperature (Newcomb *et al.*, 1996; Trus *et al.*, 1996). Accordingly, they did not acquire VP26 which does not bind to hexon VP5 in the procapsid and was not available when the particles matured *in vitro*. The procapsid is rounder and more porous than other capsid types; procapsid capsomers do not make direct contact with their neighbors; interactions between them are mediated by the triplexes. After maturation, the role of the triplexes switches to that of reinforcing clamps that stabilize the interactions between adjacent capsomers. The mechanism underlying maturation is the rotation of domains in the hexon VP5 subunits together with altered interactions of the triplexes. The procapsid has an inner shell or scaffold consisting (major component) of the precursor form of the scaffolding protein, pre-VP22a, and (minor component) of a fusion of pre-VP22a with the viral protease, VP24 and a connecting polypeptide. Layers of density in the scaffold are discernible in the section (bottom left) but individual scaffolding molecules are not resolved in the plane of the shell: it is not yet clear whether the scaffold conforms to icosahedral symmetry like the outer shell or whether it is less ordered, forming a micelle with a preferred curvature. After assembly, the protease is activated and processed at the C-termini of the scaffolding proteins. The processing relaxes the interaction between the scaffold and the outer shell. In the normal pathway, the scaffolding proteins are expelled concomitant with DNA packaging. However, they are retained in B-capsids (devoid of DNA) *in vivo* or in this experiment and form a smaller and less ordered proteinaceous core. Courtesy of B. Heymann, N. Cheng W, Newcomb, B. Trus, J. Brown, A. Steven.

et al., 2000). These reconstructions also show a large peak of density near the center of the procapsid that may correspond to the unique N-terminus of the viral protease (Newcomb *et al.*, 2000). Although the composition of procapsid and mature capsid outer shells are the same in terms of capsomer and triplex number and position, cryoelectron reconstructions show there are structural differences between the capsid types. For example, while the hexons of A-capsids are hexagonal in shape, the hexons in procapsids are asymmetric: hexons along the edges of icosahedral facets are oval, and those in the facet faces are triangular (Trus *et al.*, 1995; Trus *et al.*, 1996; Newcomb *et al.*, 1999; Newcomb *et al.*, 2000; Heymann *et al.*, 2003). Also, each of the pentons and hexons in a procapsid is surrounded by a set of five or six, respectively, holes which are not present in mature capsids.

Both procapsids and mature capsids contain an important structural feature not identified by cryoelectron reconstructions where computer averaging is used. This structure, called the portal, is a dodecamer of the HSV-1 U_L6 gene product (pU_L6) located at one five-fold vertex. The portal serves as the conduit through which the genome is packaged into the capsid (Newcomb *et al.*, 2001; Tvus *et al.*, 2004).

The knowledge gained from the above studies on the structure of HSV-1 capsids and procapsids has aided greatly in our understanding of capsid assembly, maturation, and DNA packaging.

Protein composition of herpesvirus capsids

Table 10.1 shows the HSV-1 proteins present in various capsid types. There are four proteins common to the outer capsid shells of both procapsids and C-capsids: VP5, VP19C, VP23, and pU_L6 (Gibson and Roizman, 1972; Newcomb *et al.*, 1993; Patel *et al.*, 1996; Sheaffer *et al.*, 2001). VP5 is the major capsid protein. It is the structural subunit of the pentons and hexons, each of which contain five and six copies, respectively, of VP5. The triplexes are composed of one copy of VP19C and two copies of VP23. Each outer shell also contains twelve copies of pU_L6 arranged in a ring to form the portal. A fifth protein, pU_L25, is found in greater amounts in C-capsids than in procapsids and, thus, may function in retaining packaged DNA inside the capsid (Sheaffer *et al.*, 2001).

In addition to the above proteins, procapsids contain the uncleaved form of the scaffolding protein, called preVP22a (Fig. 10.2), and a lesser amount of full-length viral protease (Newcomb *et al.*, 2000). In contrast, C-capsids do not contain either preVP22a or full-length protease, but do contain VP24, the N-terminal cleavage product of the viral protease. C-capsids also contain VP26, a small protein found on the hexon tips of angularized capsids at a ratio of six VP26 molecules per hexon (Post *et al.*, 1980; Booy *et al.*, 1994; Zhou *et al.*, 1995).

B-capsids are considered intermediates in the capsid maturation process in which the scaffold and viral protease have been cleaved, but are retained within the capsid. In addition to the four basic structural proteins of the outer shell, B-capsids contain VP24 and VP21, resulting from cleavage of the viral protease, and VP22a, resulting from cleavage of preVP22a (Gibson and Roizman, 1972; Newcomb *et al.*, 1993). Like C-capsids, B-capsids also contain VP26, located on the hexon tips, and pU_L25.

Procapsid assembly

Along with genetic studies, *in vitro* capsid assembly systems, in which capsids are formed from purified components, have been instrumental to our understanding of herpesvirus capsid assembly. Wild type procapsids are very unstable both *in vivo* and *in vitro* and quickly mature into B-capsids (Newcomb *et*

Table 10.1. Protein components of HSV-1 capsids

Protein	Gene	Component of	Location in Capsid	Approximate copy number per capsid
VP5	U_L19	All capsid types	Capsomers	955 (assuming a penton is replaced by portal)
VP19C	U_L38	All capsid types	Triplexes	320
VP23	U_L18	All capsid types	Triplexes	640
pU_L6	U_L6	All capsid types	Portal vertex	12
pU_L25	U_L25	All capsid types	Outside on multiple facets	27–42
pU_L17	U_L17	A, B, and C capsids	Outside on multiple facets	19.2 +/– 5.0 in B capsids
pre-VP22a	$U_L26.5$	Procapsids	Scaffold	> 1000
Protease	U_L26	Procapsids	Inside near scaffold	87–147
VP22a	$U_L26.5$	B-capsids	Inside (=cleaved scaffold)	1153
VP21	U_L26	B-capsids	Inside (=cleaved protease)	87–147
VP24	U_L26	A(?)*, B, and C	Inside (=cleaved protease)	87–147
VP26	U_L35	A, B, and C	Hexon tips	952
pU_L15	U_L15	All capsids types**	Interacts with pU_L6	1 in B capsids, 12.0 (+/–5.63) in A capsids
pU_L28	U_L28	Procapsids, B, and C**	Outside with portal	2.4 (+/– 1) in B capsids. $\leq$ 1 in A capsids
pU_L33	U_L33	Procapsids(?)*, A, B, and C(?)*	Outside with pU_L15 and pU_L28	?

*The presence of VP24 in A capsids and of pU_L33 in procapsids and C capsids has not been studied. **pU_L15 and pU_L28 are present in C capsids, but in lower amounts compared to the other capsids types.

al., 1999; Newcomb *et al.*, 2000). The use of HSV-1 viruses and recombinant baculoviruses containing lesions in the U_L26 gene that render the protease inactive have allowed the study of procapsid structure and assembly. As with procapsid assembly in dsDNA bacteriophages such as T4, P22, and λ, the scaffold has been found to play a critical role in the assembly of herpesvirus capsids. In the absence of scaffold, aberrant capsids are formed (Thomsen *et al.*, 1994; Tatman *et al.*, 1994; Desai *et al.*, 1994; Homa and Brown, 1997). The HSV-1 scaffold protein, preVP22a, associates with the portal protein (pU_L6), the major capsid protein (VP5), the viral protease (U_L26), and itself to promote proper capsid formation (Thomsen *et al.*, 1995; Hong *et al.*, 1996; Desai and Person, 1996; Pelletier *et al.*, 1997; Newcomb *et al.*, 1999; Preston and McDougall, 2002; Newcomb *et al.*, 2003).

Capsids morphologically similar to B-capsids form both *in vitro* and *in vivo* in the absence of portal protein. While these capsids lacking a portal are not viable for DNA packaging, they pose the interesting question of how and when the portal is added in the assembly pathway. With dsDNA bacteriophages, procapsid assembly begins with the copolymerization of scaffold-major capsid protein complexes onto a portal ring initiator complex. Using purified components, Newcomb *et al.* showed HSV-1 scaffold-portal ring complexes form *in vitro*, and these complexes, when combined with major capsid protein, triplexes, and viral protease, form capsids with stably incorporated portal proteins (Newcomb *et al.*, 2003). Thus, as with the dsDNA bacteriophages, herpesvirus portal rings may be added at an early stage in procapsid assembly.

Procapsid assembly progresses through the polymerization of the various procapsid proteins. This process depends on the ability of preVP22a to interact with itself, the viral protease, and the major capsid protein. The procapsid scaffold is composed of preVP22a and viral protease at a ratio of ~10:1. The overlapping, in-frame genes U_L26 and $U_L26.5$ encode the viral protease and preVP22a, respectively (Fig. 10.2). The preVP22a protein is identical to the C-terminal half of the viral protease and both proteins possess the scaffold oligomerization and VP5-binding domains. Mutational studies have identified several scaffold oligomerization domains throughout the VP22a protein that appear to function independently of one another (Preston and McDougall, 2002). The C-terminal 25 residues of preVP22a and the viral protease contain the VP5-interaction domain. Specifically, residues F-V/A and M-M within the "AXXFV/AXQMMXXR" domain, conserved amongst the *alpha-herpesvirinae*, are essential for interaction with VP5 (Hong *et al.*, 1996; Oien *et al.*, 1997). Mutational studies have shown that the N-terminus of VP5, specifically the regions encompassing residues 27–45 and 63–84, interact with the C-terminus of the scaffold protein (Warner *et al.*, 2001).

Two different models can be visualized for the process of procapsid assembly. In one, scaffold and protease proteins form a spherical inner core onto which outer shell proteins assemble to create a two-layered procapsid of correct size and shape. In the other model, individual scaffold-outer shell complexes assemble first and then polymerize onto a growing capsid until it is closed. To determine the process by which soluble capsid proteins assemble into capsids, Thomsen *et al.* studied the *in vitro* assembly of capsids over time (Thomsen *et al.*, 1994). Capsid proteins expressed in insect cells were mixed and capsid-related structures were followed over time by electron microscopy. Partial capsids, procapsids, and angular B-capsids were observed after 1 min, 90 min, and 8 hours, respectively. The partial capsids observed very early in the assembly process contained both an outer shell of capsomers and an inner scaffold. Thus, herpesvirus capsid assembly proceeds through the polymerization of individual scaffold-outer shell complexes and individual capsomers can form before capsid assembly is complete. The scaffold plays a crucial role in assuring capsids formed through this process are the correct size and shape. In the absence of preVP22a and the viral protease, incomplete shells containing capsomeres become organized into hexagonal networks (Tatman *et al.*, 1994; Thomsen *et al.*, 1994; Desai *et al.*, 1994). Thus, while the scaffold is not required for VP5, VP19C, and VP23 interactions, it is vital in promoting correct capsid curvature and closure. In contrast to the incomplete shells formed from VP5, VP19C, and VP23, shells formed from only VP5 and VP19C are small, closed spheres (Thomsen *et al.*, 1994; Rixon *et al.*, 1996). Thus, VP19C is sufficient to link capsomers together, but VP23 is apparently required to modulate the interactions between VP5 and VP19C and/or between VP5 and adjacent capsomers.

An additional role of the scaffold that ensures maximal capsid production during infection has been observed using a $U_L26.5/U_L26$ double null virus. In $U_L26.5^-/U_L26^-$ HSV-1 infected cells, only aberrant particles are produced, while in wild type HSV-1 in-

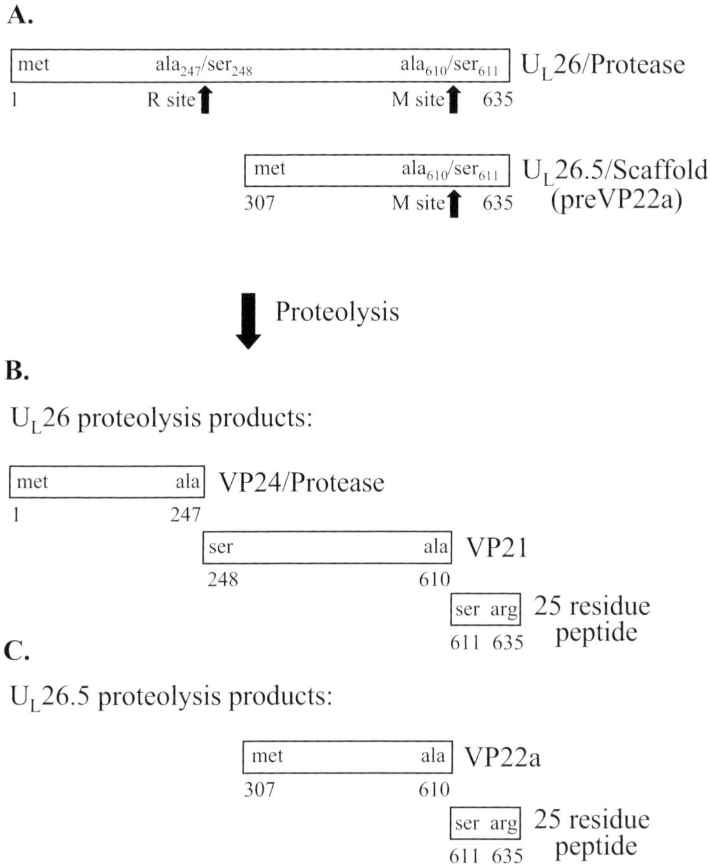

Figure 10.2 HSV-1 protease and scaffold proteins. A. The overlapping, in-frame genes U_L26 and $U_L26.5$ encode the viral protease and preVP22a, respectively, such that preVP22a is identical to residues 307–635 of the viral protease. B. Autoproteolysis cleaves the viral protease into the active protease called VP24, VP21, and the C-terminal 25-residue peptide. C. Proteolysis of preVP22a by the viral protease produces VP22a and the C-terminal 25-residue peptide.

fected cells very few, if any, aberrant particles are produced (Desai *et al.*, 1994). Thus, the scaffold not only enables correct assembly, it prevents incorrect assembly. preVP22a's chaperone/nuclear import function may aid in this role. While preVP22a can enter the nucleus on its own, VP5 requires a chaperone for nuclear import. Both preVP22a and VP19C can import VP5 to the nucleus (Nicholson *et al.*, 1994; Rixon *et al.*, 1996). Cytoplasmic interaction of preVP22a with VP5 and subsequent nuclear import may prevent scaffold-free VP5 interactions, and thus preclude incorporation of capsid proteins into aberrant particles.

Capsid maturation

During the process of herpesvirus capsid maturation the spherical, porous, unstable procapsid is transformed into an angular, stable capsid through an irreversible conformational change initiated by cleavage of the procapsid scaffold.

The scaffold must be cleaved by the viral protease to release molecular constraints that preclude capsid angularization. preVP22a contains one cleavage site, M, that divides the C-terminal 25 residues necessary for VP5-interaction from the rest of the protein (Fig. 10.2). The viral protease, whose C-terminal half is identical to preVP22a, contains the M site and an additional site, R, which divides the N-terminal 247 residues, termed VP24, from the rest of the protein (DiIanni *et al.*, 1993a). VP24 mediates the proteolytic activity and upon activation cleaves the U_L26 gene product at both the R and M sites releasing VP24, VP21, and the C-terminal 25 residues (Weinheimer *et al.*, 1993; Liu and Roizman, 1993; DiIanni *et al.*, 1993b). The protease also cleaves preVP22a within the scaffold, releasing VP22a and the C-terminal 25 residues (DiIanni *et al.*, 1993b). While VP22a, VP21, and the C-terminal 25 residue peptide are lost during maturation, VP24 is retained in DNA-filled C capsids.

Cryoelectron reconstructions have identified the major conformational changes that take place during angularization of the capsid shell (Fig. 10.1) (Newcomb *et al.*, 1999; Newcomb *et al.*, 2000; Heymann *et al.*, 2003). In the procapsid, the triplexes form the only linkage between adjacent capsomers. The floor (innermost) domains of the capsomer rings are surrounded by gaps ~25 Å in size that separate individual capsomers. During capsid maturation, the capsomer floor domains rotate ~40° relative to the outer domains. As a result of this rotation, the gaps separating capsomers disappear and the floor domains of adjacent capsomers contact one another. Meanwhile, the axial channels within each capsomer constrict. Although the triplexes remain in place and change very little in structure, their interactions with VP5 alter to reinforce capsomer–capsomer interactions at points of the capsid floor where three capsomers come together. Conformational changes are also seen in the outer domains of the capsomers. The hexon protrusion domains, which appear as ovals along facet edges and triangles on facet faces of procapsids, rotate such that all the hexons acquire six-fold symmetry. The end result is an angular, stable capsid designed to optimally protect the herpesvirus genome.

DNA cleavage and packaging

The transformation of empty procapsids into DNA-filled C-capsids is the result of an elegant molecular machine that cleaves individual genomes from a concatemeric substrate and translocates them into the capsid precursor. Proper functioning of this machine requires specific cleavage sequences on the DNA substrate, a DNA packaging enzyme called the terminase, and capsid structural elements such as the portal protein.

The vast majority of *alpha-herpervirinae* work on DNA cleavage and packaging has been performed with HSV-1. However, the packaging proteins and the processes of DNA cleavage and packaging are highly conserved amongst all Herpesviridae. Significant contributions to the field have been made through the study of various cytomegaloviruses of the *beta-herpesvirinae* and those findings will be discussed as appropriate.

Herpesvirus packaging sequences

Individual herpesvirus genomes are cleaved from the concatemeric substrate in a sequence-specific manner. Minimal sequences required for correct cleavage have been identified and are contained within the *a* sequences located in the inverted repeats flanking the U_L and U_S genomic segments (Fig. 10.3) (Spaete and Frenkel, 1982; Vlazny *et al.*, 1982; Mocarski and Roizman, 1982; Stow and McMonagle, 1983; Stow *et al.*, 1983). The *a* sequences of various HSV strains contain the subsequences *DR1-Uc-DR4*(n)-*DR2*(m)-*Ub-DR1* where n and m represent variable numbers of repeats. The sizes of these subsequences vary from strain to strain, but appear

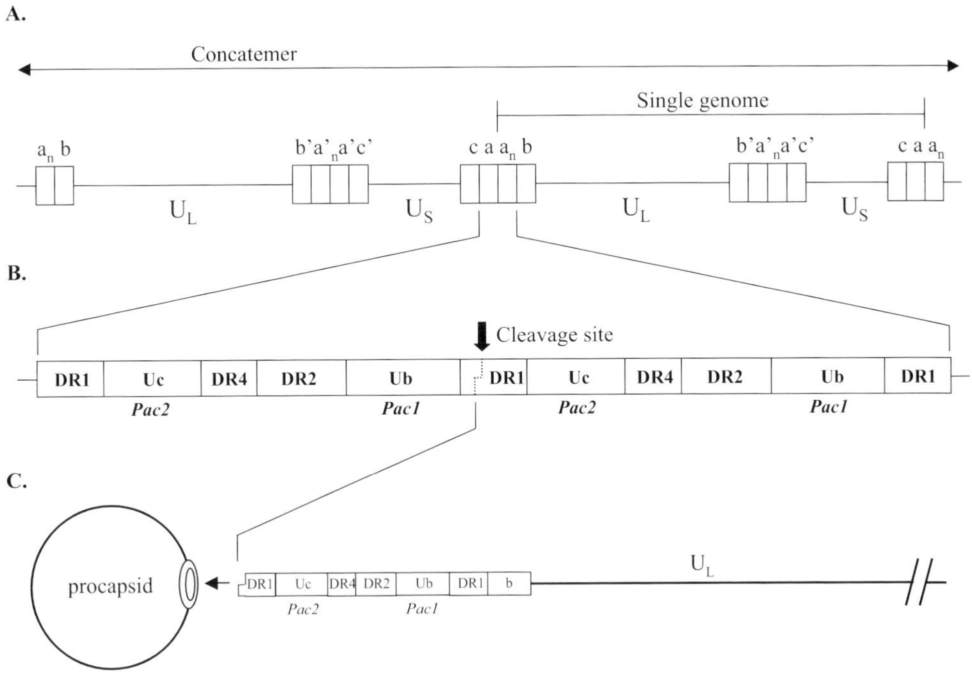

Figure 10.3 Organization of the HSV-1 genome and required cleavage/packaging sites. A. Genome concatemers comprise the DNA cleavage/packaging substrate and are composed of head-to-tail multimers of HSV-1 chromosomes. Packaged genomes may contain more than one *a* sequence on the U_L end, but only one *a* sequence on the U_S end. B. The *a* sequences present in the inverted repeats flanking the U_L and U_S genome segments contain the subsequences DR1-Uc-DR4-DR2-Ub-DR1. Two adjacent *a* sequences are shown. Single *a* sites are flanked by DR1 sites, while adjacent *a* site share an intervening DR1 sequence. DR1 subsequences are 20 bp long. The Uc sequence is also called *Pac2* and is composed of a 5–10 bp A-rich region near a CGCGGCG motif. DR4 sites contain 2–3 copies of a 37 bp sequence while DR2 sites contain 19–22 copies of a 12 bp sequence. The Ub sequence is also termed *Pac1* and is composed of a 3–7 bp A- or T-rich sequence flanked by a poly C sequence. The *Pac1* and *Pac2* sequences, along with the intervening DR1 sequence are required for production of genomic termini. During cleavage/packaging, the site-specific, staggered cleavage event takes place in the DR1 sequence shared by adjacent *a* sequences leaving 17 bp plus a 3′ overhang on the U_L side of the cleavage. C. The newly cleaved concatemeric end adjacent to the U_L genome segment is utilized for packaging which proceeds in a U_L to U_S direction.

to be consistent within strains (Davison and Wilkie, 1981; Mocarski and Roizman, 1981). Adjacent *a* sequences at concatemeric genome junctions share intervening *DR1* subsequences (Davison and Wilkie, 1981; Mocarski and Roizman, 1981; Mocarski and Roizman, 1982). The *Ub* and *Uc* subsequences have also been termed *Pac1* and *Pac2*, respectively (Deiss *et al.*, 1986; Nasseri and Mocarski, 1988). *Pac1*, *Pac2*, *DR1*, and at least some repeated sequences comprise a cleavage signal sequence sufficient for generation of genomic termini during cleavage and packaging (Mocarski *et al.*, 1985; Varmuza and Smiley, 1985; Deiss *et al.*, 1986; Deiss and Frenkel, 1986). *Pac1* sequences contain a 3–7 bp A- or T-rich sequence flanked by a poly C sequence. *Pac2* sequences contain a 5–10 bp A-rich region near a CGCGGCG motif. Studies with murine cytomegalovirus (MCMV) have shown that the poly C of *Pac1* and the A-rich and CGCGGCG motifs of *Pac2* are required for production of genomic termini (McVoy *et al.*, 1998).

Concatemeric cleavage occurs within the *DR1* sequence shared by adjacent *a* sequences. Cleavage events are staggered, producing single bp 3′ overhangs, and always occur at the same site within *DR1* such that upon circularization of the genome early in infection, a single and complete *a* sequence is generated (Davison and Wilkie, 1981; Mocarski and Roizman, 1982; Poffenberger and Roizman, 1985). The inverted repeats flanking U_L are designated *ab* and *b′a′*, and the repeats flanking U_S are designated *a′c′* and *ca* (Wadsworth *et al.*, 1975; Wagner and Summers, 1978). The fact that *a* sequences at genomic termini are in the opposite orientation of *a* sequences at L/S junctions is key to proper recognition and cleavage of complete genomes from the concatemeric substrate during packaging. During cleavage and packaging two separate double-stranded cleavage events take place (Varmuza and Smiley, 1985). The first cleavage event is believed to initiate packaging with subsequent translocation of the U_L genomic segment into the capsid followed by the U_L/U_S junction, and then U_S. The second cleavage occurs at the end of the U_S segment releasing the packaged genome from the rest of the concatemer. While some packaged HSV genomes contain multiple *a* sequences at their L termini, only a single *a* sequence is ever present at S termini (Wagner and Summers, 1978). This evidence suggests a model in which the initial cleavage event can take place in the *DR1* site of any *a* sequence on the concatemer, but the second cleavage event always takes place at the first *DR1* site of proper orientation encountered following genome packaging.

Herpesvirus packaging machinery

Genetic studies have shown that seven viral proteins are essential for the cleavage and/or packaging of herpesvirus DNA. The HSV-1 genes encoding these proteins are U_L6, U_L15, U_L17, U_L25, U_L28, U_L32, and U_L33 (Weller *et al.*, 1983; Addison *et al.*, 1984; Weller *et al.*, 1987; Sherman and Bachenheimer, 1988; Addison *et al.*, 1990; Al-Kobaisi *et al.*, 1991; Tengelsen *et al.*, 1993; Poon and Roizman, 1993; Patel *et al.*, 1996; Baines *et al.*, 1997; Yu *et al.*, 1997; Salmon *et al.*, 1998; McNab *et al.*, 1998). Though not required for the assembly of empty B-capsids, DNA-filled C-capsids are not produced in the absence of any of these proteins. Further studies and/or analogies to phage systems have enabled the functional grouping of these proteins. While U_L17 and U_L33 appear to play indirect roles in DNA cleavage/packaging (discussed below), most of the remaining proteins are thought to be directly involved in the cleavage/packaging process as components of the packaging machinery. These proteins are thought to function as the terminase (U_L15, U_L28, and possibly U_L33), the portal vertex (U_L6), and in packaging termination/DNA retention (U_L25). Though not essential for viral growth, an alkaline exonuclease (encoded by U_L12) plays an important role in aiding the packaging process.

The terminase

Viral packaging enzymes, called terminases, are usually heteromultimers composed of two different proteins that together possess the different activities required for separating individual genomes from the concatemeric substrate and translocating them into the procapsid. The process of DNA cleavage requires both sequence-specific DNA-binding and endonuclease activities. With analogous phage systems, one terminase subunit typically functions in site-specific binding and anchors the holoenzyme to the DNA substrate at the cleavage site, while the other subunit performs the actual cleavage event and functions in procapsid binding. The process of translocating the DNA substrate into the procapsid is very energy-intensive. Studies with analogous phage systems estimate ~1 molecule of ATP is hydrolyzed per 2 bps of DNA packaged (Morita *et al.*, 1993). As with those analogous systems, the packaging of HSV-1 DNA requires energy obtained through ATP hydrolysis (Dasgupta and Wilson, 1999). In all packaging systems studied to date, viral terminases are (1) the only packaging proteins that possess ATPase activity and (2) involved in ATP hydrolysis-dependent DNA packaging activities. Thus, the definition of a terminase as a cleavage enzyme (producing genomic *termini*) has been extended to include the ATPase activity that powers the packaging motor.

The HSV-1 terminase is thought to be composed of proteins encoded by U_L15 and U_L28, and possibly U_L33. HSV-1 genomes are neither cleaved nor packaged in the absence of these proteins. As with other terminases, pU_L15, pU_L28, and U_L33 have been shown to interact both *in vitro* and *in vivo*. Evidence supporting this interaction include (1) coimmunoprecipitation experiments performed with lysates of both HSV-infected cells and insect cells expressing pU_L15, pU_L28 and pU_L33 (Koslowski *et al.*, 1999; Abbotts *et al.*, 2000; Beard *et al.*, 2002) (2) copurification of the pU_L15/pU_L28 complex from lysates of infected cells as a 1:1 heterodimer by ion exchange and DNA affinity chromatography (Koslowski *et al.*, 1999), and (3) the observation that under some circumstances, pU_L28 requires pU_L15 for nuclear import (Koslowski *et al.*, 1999; Abbotts *et al.*, 2000). In addition, indirect evidence supporting the pU_L15–pU_L28 interaction comes from the study of antiviral drugs directed against the human cytomegalovirus (HCMV) homologs of these proteins in which mutations in both proteins were found to promote increased resistance to BDCRB (Underwood *et al.*, 1998; Krosky *et al.*, 1998).

The U_L28 protein has been shown to bind novel DNA structures adopted by the *Pac1* sequence of the herpesvirus cleavage site *in vitro* and is thus thought to comprise the DNA-binding component of the viral terminase (Adelman *et al.*, 2001). The aberrant electophoretic mobility of the bound substrates suggests that pU_L28 recognizes single-stranded regions present in various DNA oligomers bearing the *Pac1* sequence. This finding along with evidence that (1) the *Pac* sequences of human herpesvirus 6 contain S1 nuclease-hypersensitivity areas indicating a single-stranded DNA structure (Deng and Dewhurst, 1998), and (2) the *DR2* sequences adjacent to *Pac1* can adopt novel non-B DNA structures with deformed, single-stranded regions (Wohlrab *et al.*, 1987), suggest that these adjacent sequences exert torsional stresses that expose single strands within *Pac1* that are recognized by pU_L28. The U_L28 protein can also bind structured DNA containing the *Pac1* element of HCMV with only slightly reduced affinity (Adelman *et al.*, 2001). Point mutations within the C-rich element of *Pac1* that eliminate cleavage/packaging of HCMV DNA *in vivo* preclude both formation of the structures recognized by pU_L28 and formation of the pU_L28-DNA complex itself. These findings provide indirect evidence that recognition of *Pac1* structured DNA is relevant *in vivo*. In addition, the HCMV homolog of pU_L28 was reported to bind the HCMV *Pac* motif and possess nonspecific nuclease activity (Bogner *et al.*, 1998; Scheffczik *et al.*, 2002).

The ATPase component of the HSV-1 terminase has yet to be clearly defined. The U_L15 protein was originally suggested to comprise part of the HSV-1 terminase based on sequence similarity to the ATPase subunit of the bacteriophage T4 terminase (McGeoch *et al.*, 1988; Davison, 1992). U_L15 encodes a partial match to the conserved Walker A and Walker B motifs present in the majority of studied terminase and helicase ATPase domains (Mitchell *et al.*, 2002). The Walker A motif comprises the phosphate-binding loop, or P-loop, of the ATPase (Walker *et al.*, 1982). Substitution of an alanine for the conserved glycine of pU_L15's putative P-loop failed to complement the growth of U_L15 insertion mutant viruses (Yu and Weller, 1998a). Thus, this conserved residue within the putative ATP binding site of pU_L15 is essential to the protein's structure and/or function. However, direct biochemical evidence that pU_L15 can hydrolyze ATP has not been obtained to date. Other evidence suggests that pU_L28 may comprise the ATPase component of HSV-1's terminase. Specifically, partially purified preparations of the HCMV pU_L28 homolog have been shown to possess ATPase activity (Hwang and Bogner, 2002).

While both pU_L15 and pU_L28 are consistently associated with procapsids and B-capsids, lesser amounts are associated with C-capsids inferring that these proteins dissociate from the capsid once packaging is complete (Salmon and Baines, 1998; Sheaffer *et al.*, 2001). Thus, like other terminases, the putative pU_L15–pU_L28 terminase is thought to be only transiently associated with capsids. The observation that up to 15 copies of pU_L15 are associated with A capsids, whereas pU_L28 is absent from A capsids (Beard *et al.*, 2004), suggests that pU_L15 is more intimately associated with the capsid or portal whereas pU_L28 is more tightly associated with DNA. Thus, pU_L28 may be lost when packaged DNA escapes the capsid during A capsid formation. Forms of pU_L15 that are truncated at the amino terminus by approximately 35 amino acids are associated with capsids derived from cells in which DNA packaging is allowed to at least initiate (Salmon and Baines, 1998; Salmon *et al.*, 1999). Such forms are absent from capsids derived from cells in which DNA packaging does not occur (e.g. in cells infected with most DNA packaging mutants). The exact function of the pU_L15 truncation is not known but it is tempting to speculate that it serves to modulate terminase activity.

Like pU_L28 and pU_L15, pU_L33 is associated with capsids (Salmon and Baines, 1998; Taus and Baines, 1998; Yu and Weller, 1998b; Beard and Baines, 2004). It has also been shown that pU_L33 coimmunoprecipitates with pU_L15 and pU_L28 from lysates of HSV-1 infected cells as well as from lysates of insect cells overexpressing these proteins (Beard *et al.*, 2002). In HSV-1 infected cells, U_L33 protein forms a complex with pU_L28, is rapidly degraded by the proteosome in the absence of pU_L28, and requires pU_L28 for coimmunoprecipitation with pU_L15 (Yang and Baines, 2006). Thus, pU_L28 and pU_L33 interact directly, whereas pU_L15 and pU_L33 interact indirectly, presumably by their mutual interaction with pU_L28. On the other hand, the presence of pU_L33 enhances the coimmunoprecipitation of pU_L15 with pU_L28 (Yang and Baines, 2006). These data suggest several possible functions for pU_L33, including that of a third terminase subunit. Whatever its role, studies have shown that neither DNA cleavage nor packaging take place in the absence of pU_L33 (Al-Kobaisi *et al.*, 1991).

The portal protein

The capsids of all analogous packaging systems contain a ring-shaped structure, called the portal protein, at one five-fold vertex. The portal provides a docking site for terminase on the capsid during packaging and serves as the conduit through which the DNA substrate is translocated into the capsid. Much evidence gained over recent years points to pU_L6 forming the HSV-1 portal protein. First, while B-capsids form in U_L6 null virus-infected cells, neither cleavage nor packaging take place (Patel *et al.*, 1996). Also, pU_L6 has been shown to interact with the putative terminase subunits pU_L15 and pU_L28 by both coimmunoprecipitation and intracellular colocalization experiments (White *et al.*, 2003). Studies using immunoelectron microscopy with intact capsids and an antibody directed against epitopes in the C-terminus of pU_L6 showed that, in general, this protein is located at one pentameric vertex on each capsid (Newcomb *et al.*, 2001; Taus *et al.* 1998). Results from immunoblot experiments estimate that 12–17 pU_L6 molecules are present in each capsid, consistent with the dodecameric portal protein found in analogous systems (Newcomb *et al.*, 2001). The masses of individual pU_L6 rings, as measured by scanning transmission electron microscopy, correspond to an oligomeric state of 12, again consistent with both the above immunoblot experiments and the oligomeric state of analogous portal proteins. To examine the structure of the pU_L6 complex, purified soluble pU_L6 expressed in insect cells was examined by cryoelectron microscopy and resolved to a resolution of 16Å. The reconstruction revealed 13nm long rings with 12-fold symmetry and maximal external diameter of 15.5nm containing an internal channel with diameter of 6.5nm (Trus *et al.*, 2004). The morphology of these structures are consistent with those of bacteriophage portal rings which have internal diameters in the 2.5–4.5nm range and outside diameters in the 14.5–17.5nm range (Bazinet and King, 1985; Valpuesta *et al.*, 1999).

Packaging termination/DNA retention

Genetic studies have shown that pU_L25 performs functions likely to be critical for late stages of DNA cleavage and packaging. Although this protein is required for viral growth, DNA cleavage has been shown to take place in cells infected with a U_L25 null mutant (McNab *et al.*, 1998; Stow, 2001). However, almost all the DNA remains DNase sensitive, indicating that very little is successfully packaged. Studies performed at the nonpermissive temperature with two different *ts* lesions in U_L25 have shown that this protein is required for the production of DNA-filled C-capsids (Addison *et al.*, 1984). Additional studies have shown that in the absence of pU_L25, genomic L-termini are correctly generated, but considerably fewer S-termini are detectable and A-capsids accumulate to abnormally high levels (Stow, 2001). These findings indicate that in the absence of pU_L25, the packaging process is aborted at some point between the first and second cleavage events. The presence of pU_L25 is less critical for the second cleavage event if the genome size is reduced (Stow, 2001). Thus, pU_L25 must perform some function critical for packaging full-length genomes.

There are several possibilities for what this function might be. For example, pU_L25 may be involved in DNA retention within the capsid. In C-capsids containing full-length genomes, the DNA is packed within the capsid at a near-crystalline density (Booy *et al.*, 1991). Thus, as packaging nears completion, the repulsive forces within the capsid become very large, making DNA translocation more difficult. The U_L25 protein may serve to retain packaged DNA within the capsid by altering the conformation of DNA-containing capsids to stabilize them, or by physically plugging holes in the capsid through which the DNA might escape. Consistent with these hypotheses are (1) the observation that pU_L25 is located on the external surface of the viral capsid (Thurlow *et al.*, 2006) and (2) the existence of an inverse correlation between the amount of pU_L25 and the levels of scaffold processing within capsids (Sheaffer *et al.*, 2001). Very little pU_L25 is present in procapsids containing unprocessed scaffold, more is present in B-capsids, and the largest amounts are present in C- and A-capsids that lack scaffold

(Sheaffer *et al.*, 2001). Thus, as the scaffold is processed and lost and the genome is packaged, more pU_L25 associates with capsids, possibly to retain packaged DNA within the capsid. Analogous bacteriophage packaging systems contain proteins required for packaging termination/DNA retention (Catalano and Tomka, 1995). The U_L25 protein may comprise HSV-1's functional homolog to those proteins.

Alkaline nuclease

Genetic studies have shown that unlike the other packaging proteins listed above, the U_L12 protein (pU_L12) is not essential for virus assembly (Martinez *et al.*, 1996). However, in the absence of pU_L12, viral yields are reduced 100- to 1000-fold. As with the U_L25 mutants, U_L12 mutant viruses produce cleaved concatemeric DNA and abnormally large numbers of A-capsids. The substrate for DNA packaging is not only concatemeric, but is also highly branched. The U_L12 protein is an alkaline nuclease that may act to resolve these DNA branches during DNA packaging allowing the packaging motor to efficiently translocate the genome into the capsid.

Capsid trafficking in the nucleus

It is controversial whether DNA packaging and procapsid assembly occur in the same or different places in the infected cell nucleus. Late in infection, the nuclear architecture of infected cells becomes considerably altered as the cellular chromatin marginates to abut the INM. Within the voided center of the nucleus, a virus-specific compartment forms that has been termed the DNA replication compartment (Quinlan *et al.*, 1984). Viral DNA and enzymes necessary for its synthesis, such as the DNA polymerase, helicase, primase, and single stranded DNA binding protein (ICP8), accumulate almost exclusively within the DNA replication compartment (Quinlan *et al.*, 1984; Goodrich *et al.*, 1990; Malik *et al.*, 1996; Lukonis and Weller, 1996; Liptak *et al.*, 1997). In its mature form, the replication compartment appears as a large globular region devoid of host chromatin. Static electron microscopic and fluorescent images have shown that angularized viral capsids and some viral DNA cleavage/packaging proteins (e.g. those encoded by U_L15 and U_L33) also accumulate in the viral DNA replication compartment, suggesting the possibility that DNA packaging also occurs there (Trus *et al.*, 1992; Baines *et al.*, 1994; Taus *et al.*, 1998; Lamberti and Weller, 1998; de Bruyn *et al.*, 1998; Reynolds *et al.*, 2000). Moreover, deletion of U_L17 or incubation of cells infected with a temperature sensitive protease mutant at the nonpermissive temperature cause capsids to accumulate within punctate aggregates on the margin of, rather than within, the DNA replication compartment (Preston *et al.*, 1983; Wilson and Church, 1997; Taus *et al.*, 1998). Upon shift to the permissive temperature, the protease mutant capsids then move to the DNA replication compartment and eventually to the nuclear membrane (Wilson and Church, 1997).

In Vero cells late in infection, viral capsid proteins and some tegument proteins have also been shown to accumulate in punctate areas along the peripheral margins of DNA replication compartments. These areas, termed assemblons, have been hypothesized to represent areas of capsid assembly (Ward *et al.*, 1996).

In considering these separate observations, it is presumed, but unproven, that the punctate accumulations of procapsids seen in cells infected with the protease mutant, or B-like capsids in cells infected with the U_L17 deletion mutant, represent a failure to disperse from areas of capsid assembly. A hypothesis consistent with these observations is that capsids assemble in assemblons, and are transported in some way that requires U_L17

to the DNA replication compartment where DNA is packaged. Another possibility is that assemblons are not a productive part of the life cycle, but represent an overexpression phenomenon where capsids and tegument proteins stick together creating aggregates. In this case, the aberrant localization of capsids and other assembly proteins observed in cells infected with a U_L17 deletion mutant may reflect its role in assuring proper folding of other proteins (Taus *et al.*, 1998). That chaperone like functions are relevant to HSV DNA packaging is supported by studies of the U_L32 gene (Kosinski and Weller, unpublished data). These workers have observed that in the absence of U_L32 a population of the capsid triplex protein VP19C is localized in cytoplasmic aggregates. VP19C in these foci colocalize with the cellular chaperone Hsc70, and the formation of these foci in U_L32-mutant infected cells is inhibited by the microtubule-depolymerizing drug nocodazole, supporting the possibility that the aggregates are similar to aggresomes. These results suggest that in the absence of U_L32 some VP19C proteins are misfolded and aggregate in the cytoplasm. Although capsids are assembled in the absence of U_L32, these capsids may not be competent for encapsidation owing to the incorporation of VP19C proteins that are misfolded or in the wrong conformation.

Examination of living cells infected with a virus encoding VP26 fused to GFP has indicated that at least some capsids exit intranuclear punctate regions containing VP26-GFP (Forest and Baines, 2005). These punctate regions resemble the previously described assemblons, and are slowly dynamic inasmuch as fluorescence recovery after photobleaching of VP26-GFP occurs within 90 minutes (Forest and Baines, unpublished data). These assemblon-like regions also stain with an antibody specific for actin-related protein ARP 3, a protein that normally localizes in the cytoplasm as an essential component of the ARP 2/3 complex that serves to nucleate and set branch points for actin polymerization (Millard *et al.*, 2004).

Interestingly, mean square displacement analysis of the capsid trajectories indicates that intranuclear capsids move at rates of 18–62 μm per minute (average rate of 32 μm per minute), in a fashion most consistent with directional transport (Forest and Baines, 2005). Moreover, the movement is ATP and temperature dependent, and is inhibited by latrunculin A (which binds G actin) but is resistant to cytochasin D (which binds the barbed ends of F actin). Thus, it seems likely that capsids are actively transported in some way that requires both energy and nuclear actin.

The observation that ARP 3 colocalizes with assemblon-like regions suggests mechanisms of transport that resemble those employed by *Listeria* and *Shigella* and Vaccinia virus in which actin polymerization transports these organisms through the cytoplasm (Goldberg, 2001). In the case of herpes simplex virus, however, the nature of the transport mechanism may be different because there is no evidence that filamentous (F) actin (as observed by binding to phalloidin) accumulates in uninfected or infected cell nuclei. Thus, if actin filaments are employed for movement of herpesvirus capsids, they are presumably at very low concentrations, or shorter than the 13–14 monomers required for phalloidin binding. The fact that movement occurs normally in the presence of cytochalasin D further suggests that the barbed end of F actin in the nucleus is either absent (as would be the case with monomeric actin) or somehow sequestered from the drug. The latter might occur if capping protein competed for cytochalasin D binding at the barbed ends (Wear and Cooper, 2004). Alternatively, it is possible that the HSV packaging motor could drive capsid movement as capsids move progressively along concatameric DNA, presumably in search of their cognate cleavage signals.

Envelopment

Envelopment at the nuclear membrane

It has been known for some time that herpesvirus nucleocapsids engage and bud through electron dense regions of the inner nuclear membrane to become virions (Morgan *et al.*, 1959). Through a poorly understood mechanism, capsids containing DNA are preferentially enveloped whereas capsids lacking DNA are not enveloped (Roizman and Furlong, 1974). The suggestion that U_L15 and U_L28 have something to do with this discrimination follows from the observation that enveloped B-like capsids are readily observed in the cytoplasm of cells infected with U_L15 and U_L28 deletion mutants (Tengelsen *et al.*, 1993; Baines *et al.*, 1997). This apparently indiscriminant envelopment is not observed as frequently upon deletion of other genes required for viral DNA packaging. Because U_L15 and U_L28 are less abundant in C capsids (containing DNA), as opposed to B capsids (Salmon and Baines, 1998; Yu and Weller, 1998b), it is logical to hypothesize that the relative absence of the U_L15 and U_L28 proteins from C capsids might facilitate their selection for budding. In this model, the association of U_L15 and U_L28 proteins would act to preclude envelopment of B capsids, perhaps by masking some region of the capsid that engages the envelopment apparatus at the nuclear membrane.

Because it would reveal important information about the nucleocapsid envelopment reaction, the biochemical nature of perinuclear virions is of considerable interest. Unfortunately, efforts to characterize perinuclear virions have been hampered by a lack of means to purify them from other virions within cellular lysates. Moreover, use of mutants that cause virions to aberrantly accumulate in the perinuclear space is potentially unsatisfying because the characterized particles may or may not reflect the biochemical nature of wild type perinuclear virions. Although an exhaustive study has yet to be done, *in situ* immunogold electron microscopy has identified a subset of proteins that associate with perinuclear virions. Thus, at least glycoproteins B, D, and M, the myristylated tegument protein encoded by U_L11, the integral membrane protein encoded by U_L34, and the U_L34-interacting protein encoded by U_L31 (see below) associate with the nuclear membrane and perinuclear virions (Torrisi *et al.*, 1992; Baines *et al.*, 1995; Jensen and Norrild, 1998; Reynolds *et al.*, 2002; Baines and Roizman, unpublished data). Because integral membrane proteins with extraluminal domains less than 60–70 kDa normally migrate past the nuclear pore membrane and into the INM (Soullam and Worman, 1995), all predicted HSV integral membrane proteins should at least transiently localize at the INM (McGeoch *et al.*, 1988). It is assumed that some tegument proteins interact with nucleocapsids and are incorporated into virions during envelopment at the INM, but the identity of these proteins is unknown. On the other hand, the sparse appearance of the tegument in perinuclear virions compared with the more electron dense tegument of extracellular virions suggests that the bulk of the tegument is applied after the nucleocapsid exits the perinuclear space (Granzow *et al.*, 2001).

There is intense interest in whether perinuclear virions are infectious but the evidence on this point is not conclusive. For example, cells infected with a virus encoding a U_L20 gene fused to a portion of $U_L20.5$ produces large amounts of virions that accumulate in what appears to be the perinuclear space of Vero cells, and these virions are infectious (Baines *et al.*, 1991). Similarly, deletion of the U_S3 protein causes virions to accumulate in the perinuclear space (Reynolds *et al.*, 2002). The correlation between the slight delay in infectious virus production and the presence of perinuclear virions in cells infected with the U_S3 deletion mutant suggests that the perinuclear virions are not infectious (Reynolds *et al.*, 2002), but cannot rule out the possibility that U_S3 plays other roles required for efficient production of infectious virus.

Genetic studies have shown that the absence of either U_L11, gK, U_L31 or U_L34 impede nucleocapsid envelopment at the INM under different circumstances and to different degrees. U_L11 is ultimately dispensable for envelopment, but nucleocapsids are readily observed abutting the nuclear membrane in its absence, whereas capsids are only rarely observed in this stage in cells infected with wild type viruses (Baines and Roizman, 1992). This observation suggests that while the U_L11 protein minimally affects engagement with the INM, the protein somehow facilitates the completion of the envelopment reaction. gK also appears to play a role in envelopment. This appears most obvious in non-dividing cells infected with a gK null mutant in which capsids fail to become enveloped as viewed by electron microscopy (Jayachandra *et al.*, 1997).

The roles of U_L31 and U_L34 proteins are arguably the best understood of the proteins implicated in nucleocapsid envelopment at the INM. Amino acids 137–181 of U_L34 protein are necessary and sufficient to mediate an interaction with U_L31 protein that has been documented *in vitro* and in infected cells (Ye *et al.*, 2000; Reynolds *et al.*, 2001; Liang and Baines, 2004). A series of charged cluster mutations in U_L34 protein (pU_L34) showed that mislocalization of pU_L34 invariably led to mislocalization of pU_L31 to the same regions (Bjerke *et al.*, 2003).

The U_L31 and U_L34 proteins colocalize exclusively at the nuclear rim as viewed by indirect immunofluorescence, and ultrastructural analysis has indicated that the U_L31/U_L34 protein complex localizes in both lamellae of the nuclear membrane and within virions located in the perinuclear space (Reynolds *et al.*, 2001; Reynolds *et al.*, 2002). Similar analyses have indicated that the pseudorabies virus homologs of the U_L31 and U_L34 proteins localize similarly (Fuchs *et al.*, 2002). In many cell lines, proper localization of the U_L31 and U_L34 proteins at the nuclear rim depends on the presence of the interacting counterpart. Thus, in Hep2 and Vero cells in the absence of U_L34, the U_L31 protein localizes within the nucleoplasm, and in the absence of U_L31, the U_L34 protein localizes not only at the nuclear rim, but also in the nucleoplasm and cytoplasm (Reynolds *et al.*, 2001). The latter distribution resembles that of the endoplasmic reticulum. In rabbit skin cells and human 143 cells however, U_L31 is dispensable for localization of pU_L34 and largely dispensable for production of infectious virus (Liang *et al.*, 2004). Therefore, some cellular proteins can provide functions normally required of U_L31. The U_L34 protein, however, is virtually essential for envelopment in all cell lines tested. For example, in even the permissive rabbit skin cell line, peak titers are reduced at least 100 fold in the absence of U_L34 compared with wild type viruses (Liang and Baines, 2004).

Envelopment and the nuclear lamina

The nuclear lamina is a fibrous meshwork of filaments that lines the inner surface of the nuclear membrane and might therefore represent a barrier precluding nucleocapsids from access to the INM (Gruenbaum *et al.*, 2000). The lamina is comprised of lamins, which are type V intermediate filaments. There are three types of lamins in differentiated cells: type A and C lamins are derived from alternative splicing of the *Lmna* transcript whereas B lamins are derived from different genes (Fisher *et al.*, 1986; Hoger *et al.*, 1988; Lin and Worman, 1993). Like all intermediate filaments, lamins are comprised of a globular head domain, a rod domain and a gobular tail domain

(Fisher *et al.*, 1986). Filaments are generated partly by intertwining of the rod domains and are linked in a head to tail fashion. A remarkable feature of the lamina is its dissolution prior to mitosis and reassembly afterwards. The dissolution of the lamina prior to mitosis correlates with phosphorylation of lamins by Cdc2. In turn, dephosphorylation of lamins correlates with lamina reassembly (Ottaviano and Gerace, 1985; Peter *et al.*, 1990; Collas, 1999). Apoptosis also requires lamina disassembly, but in this case it is irreversible as lamins are proteolytically cleaved following their phosphorylation by protein kinase C.

Several lines of evidence support the possibility that the U_L31 and U_L34 protein complex interacts with the nuclear lamina. First, both U_L31 and U_L34 have been shown to interact with lamin A/C using GST pull down reactions (Reynolds *et al.*, 2004). Second, the solubility of U_L31 closely parallels that of lamin A/C. In particular, extraction conditions that solubilize lamin A/C, also solubilize U_L31 protein (Chang and Roizman, 1993; Reynolds *et al.*, 2004). Third, lamins and the U_L31 and U_L34 proteins colocalize in a rim-like pattern surrounding the nucleus (Reynolds *et al.*, 2001). Fourth, epitopes within the lamin A/C tail are precluded from recognition by a monoclonal antibody in cells infected with wild type viruses, whereas these epitopes become available for immunoreaction in cells infected with either the U_L31 or U_L34 deletion viruses (Reynolds *et al.*, 2004). Thus, U_L31 and U_L34 are both required for the masking of the lamin A/C epitope that occurs in infected cells. Lamin A/C epitopes in the rod domain are not masked or bound by U_L31 protein.

Remarkably, overexpression of the U_L31 protein can completely relocalize lamin A/C to the nucleoplasm. Mapping studies revealed that lamin A/C amino acids 339–519 and 460–660 (i.e. the lamin A/C tail domain) were sufficient to interact with U_L31 protein *in vitro* (Reynolds *et al.*, 2004). The immunoglobulin-like fold within this region of the lamin A/C tail has been shown to bind chromatin and to mediate lamin/lamin interactions (Hoger *et al.*, 1991; Taniura *et al.*, 1995; Dhe-Paganon *et al.*, 2002; Krimm *et al.*, 2002; Stierle *et al.*, 2003). Thus, it seems plausible that the U_L31 protein might mediate lamina depolymerization by competing for lamin/lamin and lamin/chromatin interactions, and thereby cause lamin A/C to accumulate aberrantly in the nucleoplasm. That U_L31 might disrupt lamin/chromatin interactions is consistent with the observation that the U_L31 and U_L34 proteins are necessary for partitioning of chromatin away from regions of the nuclear membrane (Simpson-Holley *et al.*, 2004). Thus, it seems likely that the dramatic displacement of lamin A/C to the nucleoplasm induced by U_L31 overexpression is an artifact that nevertheless reflects a normal function of U_L31: specifically, to locally disrupt the lamina by competing for lamin/lamin and lamin/chromatin interactions.

Intercalation into the lamina by U_L31 protein would also be expected to serve as an anchoring point for the U_L34 protein in the INM of infected cells, similar to the way that lamin receptors are anchored to the INM by interaction with lamins. Thus, in the absence of U_L31 protein, pU_L34 drifts between the INM, and the continuous ONM and ER membranes, giving rise to the ER-like distribution seen in Vero and Hep2 cells infected with U_L31 null viruses (Reynolds *et al.*, 2001).

Why does the massive lamin A/C displacement mediated by U_L31 overexpression not occur in infected cells? Although a definitive answer is not known, one might expect that the level of expression in HSV infected cells is lower than when U_L31 protein expression is driven transiently by the immediate early cytomegalovirus promoter. Less competition for lamin/lamin interactions would lead to less dramatic changes in the lamina. It should be also be noted that whereas U_L31 might be free to diffuse in the nucleus when solely expressed, localization of U_L31 protein is restricted to regions containing U_L34 protein in an HSV infected cell. Thus, the lamina-disrupting effects of U_L31 protein would be essentially constrained to localized regions and U_L31 protein with bound lamins would still be expected to localize at the nuclear membrane. Maintaining the nuclear lamina might play to the advantage of the virus inasmuch as the lamina is not only required for structural integrity of the nucleus, but also for DNA replication and transcription (Spann *et al.*, 1997; Spann *et al.*, 2002). Maintaining the lamina in some semblance of normalcy might thereby ensure the biochemical integrity of the infected cell, and thereby promote optimal virus production.

It is also likely that the conformation of lamin B is altered in HSV infected cells. This is consistent with the observations that (1) the lamin B receptor has been shown to become less tightly tethered to the nuclear membrane in infected cells as opposed to uninfected cells and (2) lamin B is displaced to a limited extent into the cytoplasm or into punctate regions of the nuclear rim in HSV-infected cells (Scott and O'Hare, 2001; Park and Baines, 2006; Bjerke and Roller, 2006). Both U_L31 and U_L34 are required for this limited lamin B disruption (Park and Baines, 2006).

Other methods may also be employed by the U_L31/U_L34 protein complex to partition the lamina. It has been shown, for example, that lamin B is more highly phosphorylated in HSV-infected cells than in uninfected cells (Park and Baines, 2006). This phosphorylation correlates with recruitment of some activated protein kinase C alpha and delta to the nuclear membrane. This recruitment is dependent on HSV-1 U_L31/U_L34. Similarly, cytomegalovirus (CMV) proteins homologous to U_L31 and U_L34 are required for recruitment of protein kinase C to the nuclear membrane and for phosphorylation of lamin A (Muranyi *et al.*, 2002). Examinations of lamin A/C in HSV infected cells indicate that, unlike in CMV infected cells, the electrophoretic mobility of lamin A/C does not change upon infection with HSV (Radsak *et al.*, 1989; Radsak *et al.*, 1991; Reynolds *et al.*, 2004). However, phosphorylation of lamin A/C in HSV infected cells remains a possibility if lamin A/C is phosphorylated to an extent that does not dramatically alter its electrophoretic mobility. Given the results with lamin B, more extensive analysis of lamin A/C phosphorylation in HSV infected cells seems warranted.

It should also be kept in mind that in addition to lamina and chromatin displacement, the U_L31/U_L34 protein complex may function as a receptor that interacts with nucleocapsids upon envelopment. This is supported by the observations that (i) at least U_L34 protein has been shown to interact with ICP5 in immunoprecipitation reactions (Ye *et al.*, 2000) and (ii) both the U_L31 and U_L34 proteins become incorporated into the tegument of virions located in the perinuclear space (Reynolds *et al.*, 2002; Fuchs *et al.*, 2002).

Viral kinases involved in nuclear egress of HSV-1 virions

While U_S3 is not essential for virion budding or egress, in cells infected with a U_S3 null mutant, virions accumulate aberrantly in punctate extensions of the nuclear membrane (Klupp *et al.*, 2001; Reynolds *et al.*, 2002; Ryckman and Roller, 2004). The HSV-1 U_L31 and U_L34 proteins localize within these punctate regions and in the virions located within them (Reynolds *et al.*, 2002). The U_S3-encoded kinase phosphorylates a serine or theronine residue in a hydrophilic region of the U_L34 protein and an unknown residue(s) in pU_L31 (Purves *et al.*, 1991; Kato *et al.*, 2005; Poon and Roizman, 2005). The increased number of virions within the perinuclear space in the absence of U_S3 has been interpreted to mean that the U_S3 kinase activity somehow accelerates egress from the perinuclear space. Although U_S3 kinase activity is also required for virion egress, a viral mutant that encodes an unphosphorylatable U_L34 protein grows normally (i.e. produces no punctate extensions of the nuclear membrane) (Ryckman and Roller, 2004). These observations indicate that the function of U_S3 that facilitates virion egress is something other than its phosphorylation of pU_L34, leaving pU_L31 as an obvious candidate for such an effector. On the other hand, it is possible that other substrates may be involved.

To complicate matters further, the U_L13 kinase of HSV can phosphorylate the U_S3 kinase. Some cells infected with a viral deletion mutant lacking U_L13 display nuclear extensions similar in appearance to those caused by the lack of U_S3 (Kato *et al.*, 2006). Clearly, key questions remain concerning the role of the U_S3 and U_L13 encoded kinases and their respective substrates in regulating nuclear egress of HSV virions. It seems likely that these studies will represent an area of intense future research.

Model of herpesvirus DNA packaging, capsid transport and envelopment

The following model (Fig. 10.4) is based on the current herpesvirus literature as well as comparisons drawn to analogous bacteriophage packaging systems. Procapsids are assembled in specific sites of the nucleus and are the capsid type utilized for DNA packaging. To initiate packaging, the putative terminase, composed of pU_L15, pU_L28, and pU_L33, recognizes and binds the DNA substrate at an *a* sequence joining adjacent chromosomes of the concatemer via pU_L28's interaction with *Pac1*. Next, terminase introduces a staggered double-stranded cleavage at a precise location in *DR1* generating a U_L concatemeric end containing 17 bps of *DR1* and a 3′ overhang. Terminase remains bound to the newly generated U_L end, engages a procapsid through interactions with the portal protein, and the DNA is inserted through the portal ring into the procapsid. Then, in a reaction requiring ATP hydrolysis, the DNA substrate is translocated into the procapsid through the portal protein. At some point after the initial cleavage event, the scaffold of the procapsid undergoes proteolysis and exits the capsid, providing internal space for the viral genome. Loss of the scaffold allows the procapsid to mature into a strong, protective shell. DNA translocation continues until an entire chromosome has been packaged. During this process pU_L12 and pU_L25 provide assistance via their nuclease and DNA retention/packaging termination activities, respectively. Once an entire genome has been packaged, the terminase performs a second cleavage in the first *DR1* sequence encountered downstream of the packaged U_S genomic segment, separating the packaged chromosome from the rest of the concatemer. Terminase dissociates from the DNA-filled C-capsid upon packaging termination and may remain bound to the newly generated U_L concatemeric end to initiate packaging the next chromosome on the concatemer in a processive manner. VP26 is applied after capsid angularization and capsids engage in an actin-dependent transport system that eventually delivers them to the nuclear membrane (Fig. 10.5).

Envelopment sites in the INM are generated by (i) nuclear import of U_L31 protein via its nuclear localization signal and intercalation into the nuclear lamina by interaction with lamin A/C followed by (ii) migration of U_L34 protein past the nuclear pore membrane such that its nucleoplasmic domain engages U_L31 protein within the lamina (Fig. 10.6). Other viral membrane proteins are also attracted to sites containing the U_L31/

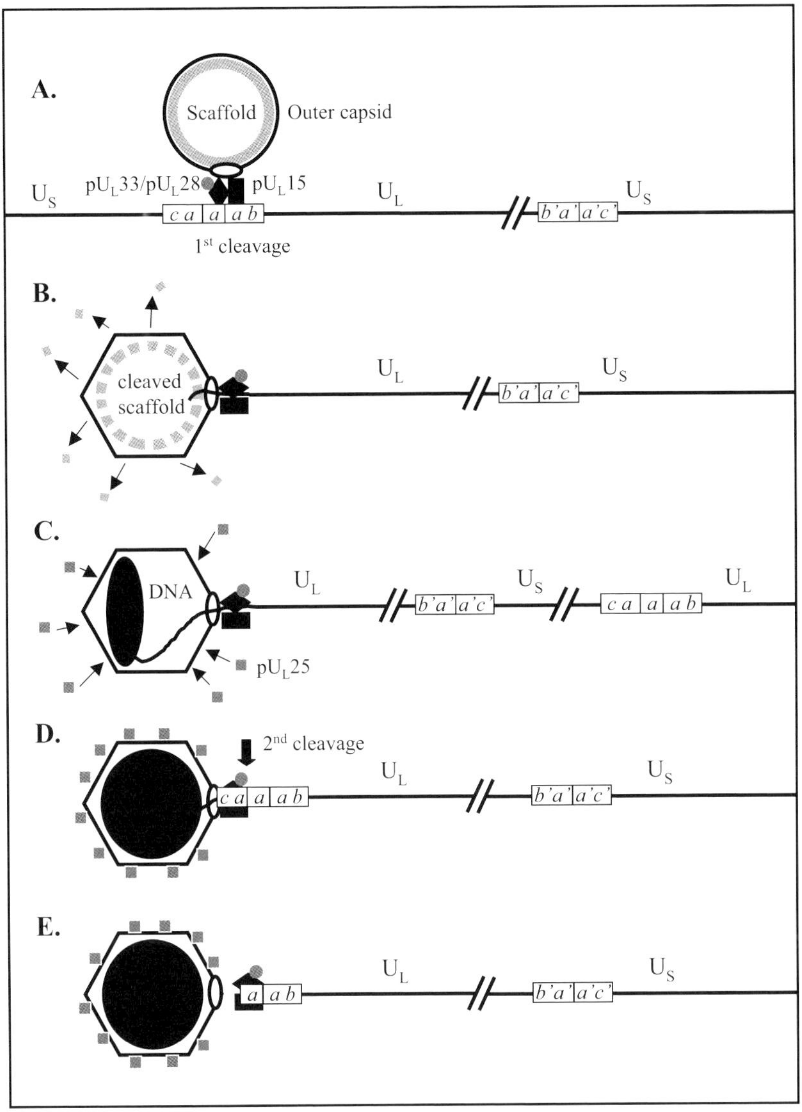

Figure 10.4 Model of capsid maturation and DNA packaging. A. DNA packaging is initiated with a procapsid that contains an outer capsid shell and inner scaffold. pU$_L$28 (black diamond), pU$_L$15 (black rectangle) and pU$_L$33 (grey circle) are associated with procapsids, pU$_L$28 and pU$_L$15 are known to interact with the pU$_L$6 portal ring and with pU$_L$33. To initiate cleavage and packaging, the terminase complex composed of pU$_L$28, pU$_L$15, and pU$_L$33 cleave the concatemeric DNA substrate in the DR1 sequence that joins *a* sequences between the U$_S$ and U$_L$ segments. B. After the initial cleavage event, the newly formed end adjacent to U$_L$ is inserted through the portal into the procapsid. The scaffold is cleaved by the viral protease and exits the procapsid allowing the capsid to angularize. C. The concatemeric substrate is translocated into the capsid in a reaction powered by ATP hydrolysis. The U$_L$25 protein (grey squares) becomes associated with the maturing capsid in greater amounts and possibly stabilizes the capsid. D. DNA translocation continues such that the U$_L$, U$_L$-U$_S$, and U$_S$ genomic segements are packaged. A second cleavage event takes place when the packaging machinery encounters the first *a* sequence following the packaged U$_S$ genomic segment. E. Upon packaging termination, the terminase complex dissociates from the filled capsid.

U_L34 protein complex by interactions that are somewhat obscure. The U_L11 protein associates with these sites partly through its N-terminal myristylation. The UL31 protein also locally modifies the nuclear lamina by competing with lamin/lamin and lamin/chromatin interactions. Both U_L31 and U_L34 also recruit protein kinase C to phosphorylate lamins, thereby mediating their localized disassembly. Nucleocapsids can then engage the nucleoplasmic domain of U_L34 protein at the INM and this engagement initates the budding reaction. Completion of the budding reaction occurs through a function of the U_L11 protein. The U_L31 and U_L34 proteins become incorporated into the nascent virion after budding is complete. Fusion of the perinuclear virion envelope with the ONM occurs in a process that is facilitated by the U_S3 and U_L13 encoded kinases. The consequence of the fusion reaction is that the U_L34 protein is essentially delivered to the ONM, and the U_L31 protein remains attached to pU_L34. It

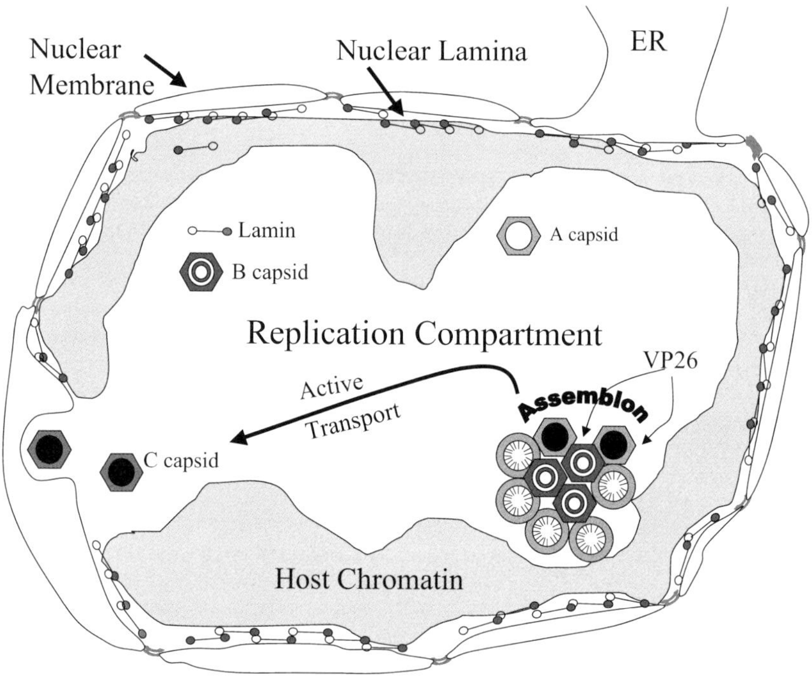

Figure 10.5 Diagram of capsid maturation in HSV infected nuclei. The infected cell nucleus consists of a central DNA replication compartment that is largely the consequence of virus induced host chromatin margination to abut the nuclear membrane. Assemblons are accumulations of spherical procapsids, and angularized types A, B and C capsids that accumulate at the periphery of the replication compartment. Assemblons appear after 6 hours post infection and increase in number with time. It is unclear where DNA packaging occurs, but capsid angularization and the addition of VP26 to capsid hexons can occur in or near assemblons. Capsids bearing VP26/GFP can also be observed in living cells exiting assemblon-like clusters of capsids. The somewhat erratic capsid transport is active, mostly directional, and actin dependent. As a result of transport, single A, B and C capsids can be seen in the DNA replication compartment at any given time. The DNA replication compartment communicates with small regions of the nuclear membrane that are devoid of chromatin and nuclear lamina. This parting of the lamina and chromatin barrier requires the U_L31 and U_L34 proteins. C Capsids eventually find their way to these denuded regions of the nuclear membrane to begin the process of envelopment.

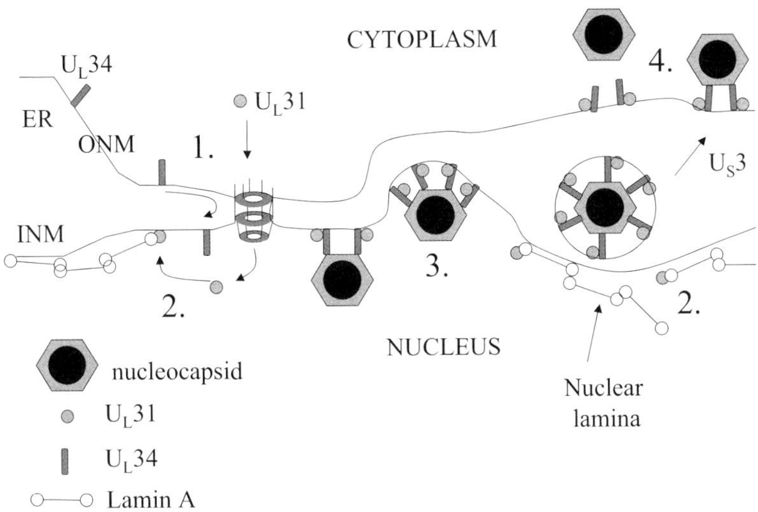

Figure 10.6 Diagram of nucleocapsid envelopment at the inner nuclear membrane (INM). The U_L31 protein contains a nuclear localization signal and enters the nucleus through the nuclear pore (1) and interacts with lamin A/C to intercalate into the lamina (2). This intercalation competes for lamin interactions with lamins, causing regions of the nuclear membrane to become devoid of lamina. U_L34 protein is a type II membrane protein that becomes incorporated into the endoplasmic reticulum (ER) membrane, diffuses into the continuous outer nuclear membrane (ONM), past the nuclear pore membrane, and into the INM where it interacts with U_L31 protein in the lamina. The U_L31 and U_L34 protein complex is proposed to interact with nucleocapsids (3) during envelopment. Once the virion envelope fuses with the outer nuclear membrane (4), the U_L34 and U_L31 proteins remain associated with the latter. U_S3 is important to regulate this fusion event which releases nucleocapsids without envelopes into the cytoplasm. The nucleocapsids eventually receive a new envelope derived from the golgi apparatus or trans-golgi network (not shown).

is possible that the U_L31/U_L34 complex is then recycled back to the INM.

Several aspects of this model remain speculative. In particular, the nature of the actin-based transport system and the mechanism of virion budding are particularly devoid of molecular detail. It is hoped that further studies will contribute information to clarify the picture.

References

Abbotts, A.P., Preston, V.G., Hughes, M., Patel, A.H., and Stow, N.D. (2000). Interaction of the herpes simplex virus type 1 packaging protein UL15 with full-length and deleted forms of the UL28 protein. J. Gen. Virol. *81 Pt 12*, 2999–3009.

Addison, C., Rixon, F.J., Palfreyman, J.W., O'Hara, M., and Preston, V.G. (1984). Characterisation of a herpes simplex virus type 1 mutant which has a temperature-sensitive defect in penetration of cells and assembly of capsids. Virology *138*, 246–259.

Addison, C., Rixon, F.J., and Preston, V.G. (1990). Herpes simplex virus type 1 UL28 gene product is important for the formation of mature capsids. J. Gen. Virol. *71*, 2377–2384.

Adelman, K., Salmon, B., and Baines, J.D. (2001). Herpes simplex DNA packaging sequences adopt novel structures that are specifically recognized by a component of the cleavage and packaging machinery. Proc. Natl. Acad. Sci. USA *98*, 3086–3091.

Al-Kobaisi, M.F., Rixon, F.J., McDougall, I., and Preston, V.G. (1991). The herpes simplex virus UL33 gene product is required for the assembly of full capsids. Virology *180*, 380–388.

Baines, J.D. and Roizman, B. (1992). The U_L11 gene of herpes simplex virus 1 encodes a function that facilitates nucleocapsid envelopment and egress from cells. J. Virol. *66*, 5168–5174.

Baines, J.D., Ward, P.L., Campadelli-Fiume, G., and Roizman, B. (1991). The U_L20 gene of herpes simplex virus 1 encodes a function necessary for viral egress. J. Virol. *65*, 6414–6424.

Baines, J.D., Poon, A.P.W., Rovnak, J., and Roizman, B. (1994). The U_L15 gene of herpes simplex virus encodes two proteins and is required for cleavage of viral DNA. J. Virol. *68*, 8118–8124.

Baines, J.D., Jacob, R.J., Simmerman, L., and Roizman, B. (1995). The U_L11 gene products of herpes simplex virus 1 are present in the nuclear and cytoplasmic membranes, and intranuclear dense bodies of infected cells. J. Virol. *69*, 825–833.

Baines, J.D., Cunningham, C., Nalwanga, D., and Davison, A.J. (1997). The U_L15 gene of herpes simplex virus type 1 contains within its second exon a novel open reading frame that is translated in frame with the U_L15 gene product. J. Virol. *71*, 2666–2673.

Bazinet, C. and King, J. (1985). The DNA translocating vertex of dsDNA bacteriophage. Annu. Rev. Microbiol. *39*, 109–129.

Beard, P.M. and Baines, J.D. (2004). The DNA cleavage and packaging protein encoded by the UL33 gene of herpes simplex virus 1 associates with capsids. Virology *324*, 475–482.

Beard, P.M., Taus, N.S., and Baines, J.D. (2002). The DNA cleavage and packaging proteins encoded by genes U_L28, U_L15, and U_L33 of herpes simplex virus 1 form a complex in infected cells. J. Virol. *76*, 4785–4791.

Beard, P.M., Duffy, C., and Baines, J.D. (2004). Quantification of the DNA cleavage and packaging proteins U(L)15 and U(L)28 in A and B capsids of herpes simplex virus type 1. J. Virol. *78*, 1367–1374.

Bjerke, S.L. and Roller, R.J. (2006). Roles for herpes simplex virus type 1 U(L)34 and U(S)3 proteins in disrupting the nuclear lamina during herpes simplex virus type 1 egress. Virology *347*, 261–276.

Bjerke, S.L., Cowan, J.M., Kerr, J.K., Reynolds, A.E., Baines, J.D., and Roller, R.J. (2003). Effects of charged cluster mutations on the function of herpes simplex virus type 1 UL34 protein. J. Virol. *77*, 7601–7610.

Bogner, E., Radsak, K., and Stinski, M.F. (1998). The gene product of human cytomegalovirus open reading frame UL56 binds the pac motif and has specific nuclease activity. J. Virol. *72*, 2259–2264.

Booy, F.P., Newcomb, W.W., Trus, B.L., Brown, J.C., Baker, T.S., and Steven, A.C. (1991). Liquid crystalline, phage-like packaging of encapsidated DNA in herpes simplex virus. Cell *64*, 1007–1015.

Booy, F.P., Trus, B.L., Newcomb, W.W., Brown, J.C., Conway, J.F., and Steven, A.C. (1994). Finding a needle in a haystack: detection of a small protein (the 12-kDa VP26) in a large complex (the 200-MDa capsid of herpes simplex virus). Proc. Natl. Acad. Sci. USA *91*, 5652–5656.

Catalano, C.E. and Tomka, M.A. (1995). Role of gpFI protein in DNA packaging by bacteriophage lambda. Biochemistry *34*, 10036–10042.

Chang, Y.E. and Roizman, B. (1993). The product of the U_L31 gene of herpes simplex virus 1 is a nuclear phosphoprotein which partitions with the nuclear matrix. J. Virol. *67*, 6348–6356.

Collas, P. (1999). Sequential PKC- and Cdc2-mediated phosphorylation events elicit zebrafish nuclear envelope disassembly. J. Cell Sci. *112 (Pt 6)*, 977–987.

Dasgupta, A. and Wilson, D.W. (1999). ATP depletion blocks herpes simplex virus DNA packaging and capsid maturation. J. Virol. 73, 2006–2015.

Davison, A.J. (1992). Channel catfish virus: a new type of herpesvirus. Virology *186*, 9–14.

Davison, A.J. and Wilkie, N.M. (1981). Nucleotide sequence of the joint between the L and S segments of herpes simplex virus types 1 and 2. J. Gen. Virol. 55, 315–331.

de Bruyn, K.A., Uprichard, S.L., Chen, M., and Knipe, D.M. (1998). Comparison of the intranuclear distributions of herpes simplex virus proteins involved in various viral functions. Virology *252*, 162–178.

Deiss, L.P. and Frenkel, N. (1986). Herpes simplex virus amplicon: cleavage of concatameric DNA is linked to packaging and involves amplification of the terminally reiterated *a* sequence. J. Virol. *57*, 933–941.

Deiss, L.P., Chou, J., and Frenkel, N. (1986). Functional domains within the a sequence involved in the cleavage-packaging of herpes simplex virus DNA. J. Virol. 59, 605–618.

Deng, H. and Dewhurst, S. (1998). Functional identification and analysis of cis-acting sequences which mediate genome cleavage and packaging in human herpesvirus 6. J. Virol. 72, 320–329.

Desai, P. and Person, S. (1996). Molecular interactions between the HSV-1 capsid proteins as measured by the yeast two-hybrid system. Virology *220*, 516–521.

Desai, P., DeLuca, N.A., Glorioso, J.C., and Person, S. (1993). Mutations in herpes simplex virus type 1 genes encoding VP5 and VP23 abrogate capsid formation and cleavage of replicated DNA. J. Virol. *67*, 1357–1364.

Desai, P., Watkins, S.C., and Person, S. (1994). The size and symmetry of B capsids of herpes simplex virus type 1 are determined by the gene products of the UL26 open reading frame. J. Virol. *68*, 5365–5374.

Dhe-Paganon, S., Werner, E.D., Chi, Y.I., and Shoelson, S.E. (2002). Structure of the globular tail of nuclear lamin. J. Biol. Chem. *277*, 17381–17384.

DiIanni, C.L., Drier, D.A., Deckman, I.C., McCann, P.J., Liu, F., Roizman, B., Colonno, R.J., and Cordingley, M.G. (1993a). Identification of the herpes simplex virus-1 protease cleavage sites by direct sequence analysis of autoproteolytic cleavage products. J. Biol. Chem. *268*, 2048–2051.

DiIanni, C.L., Mapelli, C., Drier, D.A., Tsao, J., Natarajan, S., Riexinger, D., Festin, S.M., Bolgar, M., Yamanaka, G., and Weinheimer, S.P. (1993b). *In vitro* activity of the herpes simplex virus type 1 protease with peptide substrates. J. Biol. Chem. *268*, 25449–25454.

Fisher, D.Z., Chaudhary, N., and Blobel, G. (1986). cDNA sequencing of nuclear lamins A and C reveals primary and secondary structural homology to intermediate filament proteins. Proc. Natl. Acad. Sci. USA *83*, 6450–6454.

Forest, T., Barnard, S., and Baines, J.D. (2005). Active intranuclear movement of herpesvirus capsids. Nat. Cell Biol. *7*, 429–431.

Fuchs, W., Klupp, B.G., Granzow, H., Osterrieder, N., and Mettenleiter, T.C. (2002). The interacting UL31 and UL34 gene products of pseudorabies virus are involved in egress from the host-cell nucleus and represent components of primary enveloped but not mature virions. J. Virol. *76*, 364–378.

Gibson, W. and Roizman, B. (1972). Proteins specified by herpes simplex virus VIII. Characterization and composition of multiple capsid forms of subtypes 1 and 2. J. Virol. *10*, 1044–1052.

Goldberg, M.B. (2001). Actin-based motility of intracellular microbial pathogens. Microbiol. Mol. Biol. Rev. *65*, 595–626, table.

Goodrich, L.D., Schaffer, P.A., Dorsky, D.I., Crumpacker, C.S., and Parris, D.S. (1990). Localization of the herpes simplex virus type 1 65-kilodalton DNA-binding protein and DNA polymerase in the presence and absence of viral DNA synthesis. J. Virol. *64*, 5738–5749.

Granzow, H., Klupp, B.G., Fuchs, W., Veits, J., Osterrieder, N., and Mettenleiter, T.C. (2001). Egress of alphaherpesviruses: comparative ultrastructural study. J. Virol. *75*, 3675–3684.

Gruenbaum, Y., Wilson, K.L., Harel, A., Goldberg, M., and Cohen, M. (2000). Review: nuclear lamins--structural proteins with fundamental functions. J. Struct. Biol. *129*, 313–323.

Heymann, J.B., Cheng, N., Newcomb, W.W., Trus, B.L., Brown, J.C., and Steven, A.C. (2003). Dynamics of herpes simplex virus capsid maturation visualized by time-lapse cryo-electron microscopy. Nat. Struct. Biol. *10*, 334–341.

Hoger, T.H., Krohne, G., and Franke, W.W. (1988). Amino acid sequence and molecular characterization of murine lamin B as deduced from cDNA clones. Eur. J. Cell Biol. *47*, 283–290.

Hoger, T.H., Krohne, G., and Kleinschmidt, J.A. (1991). Interaction of Xenopus lamins A and LII with chromatin *in vitro* mediated by a sequence element in the carboxyterminal domain. Exp. Cell Res. *197*, 280–289.

Homa, F.L. and Brown, J.C. (1997). Capsid assembly and DNA packaging in herpes simplex virus. Rev. Med. Virol. *7*, 107–122.

Hong, Z., Beaudet-Miller, M., Durkin, J., Zhang, R., and Kwong, A.D. (1996). Identification of a minimal hydrophobic domain in the herpes simplex virus type 1 scaffolding protein which is required for interaction with the major capsid protein. J. Virol. *70*, 533–540.

Hwang, J.S. and Bogner, E. (2002). ATPase activity of the terminase subunit pUL56 of human cytomegalovirus. J. Biol. Chem. *277*, 6943–6948.

Jayachandra, S., Baghian, A., and Kousoulas, K.G. (1997). Herpes simplex virus type 1 glycoprotein K is not essential for infectious virus production in actively replicating cells but is required for efficient envelopment and translocation of infectious virions from the cytoplasm to the extracellular space. J. Virol. *71*, 5012–5024.

Jensen, H.L. and Norrild, B. (1998). Herpes simplex virus type 1-infected human embryonic lung cells studied by optimized immunogold cryosection electron microscopy. J. Histochem. Cytochem. *46*, 487–496.

Kato, A., Yamamoto, M., Ohno, T., Kodaira, H., Nishiyama, Y., and Kawaguchi, Y. (2005). Identification of proteins phosphorylated directly by the Us3 protein kinase encoded by herpes simplex virus 1. J. Virol. *79*, 9325–9331.

Kato, A., Yamamoto, M., Ohno, T., Tanaka, M., Sata, T., Nishiyama, Y., and Kawaguchi, Y. (2006). Herpes simplex virus 1-encoded protein kinase UL13 phosphorylates viral Us3 protein kinase and regulates nuclear localization of viral envelopment factors UL34 and UL31. J. Virol. *80*, 1476–1486.

Klupp, B.G., Granzow, H., and Mettenleiter, T.C. (2001). Effect of the pseudorabies virus US3 protein on nuclear membrane localization of the UL34 protein and virus egress from the nucleus. J. Gen. Virol. *82*, 2363–2371.

Koslowski, K.M., Shaver, P.R., Casey, J.T.I., Wilson, T., Yamanaka, G., Sheaffer, A.K., Tenny, D.J., and Pedersen, N.E. (1999). Physical and functional interactions between the herpes simplex virus

UL15 and UL28 DA cleavage and packaging proteins. J. Virol. *73*, 1704–1707.

Krimm, I., Ostlund, C., Gilquin, B., Couprie, J., Hossenlopp, P., Mornon, J.P., Bonne, G., Courvalin, J.C., Worman, H.J., and Zinn-Justin, S. (2002). The Ig-like structure of the C-terminal domain of lamin A/C, mutated in muscular dystrophies, cardiomyopathy, and partial lipodystrophy. Structure (Camb.) *10*, 811–823.

Krosky, P.M., Underwood, M.R., Turk, S.R., Feng, K.W., Jain, R.K., Ptak, R.G., Westerman, A.C., Biron, K.K., Townsend, L.B., and Drach, J.C. (1998). Resistance of human cytomegalovirus to benzimidazole ribonucleosides maps to two open reading frames: UL89 and UL56. J. Virol. *72*, 4721–4728.

Lamberti, C. and Weller, S.K. (1998). The herpes simplex virus type 1 cleavage/packaging protein, UL32, is involved in efficient localization of capsids to replication compartments. J. Virol. *72*, 2463–2473.

Liang, L. and Baines, J. D. (2005). Identification of an essential domain in the Herpes Simplex Virus 1 UL34 protein that is necessary and sufficient to interact with UL31 protein. J. Virol. *79*, 3797–3806.

Liang, L., Tanaka, M., Kawaguchi, Y., and Baines, J.D. (2004). Cell lines that support replication of a novel herpes simplex 1 UL31 deletion mutant can properly target UL34 protein to the nuclear rim in the absence of UL31. Virology *329*, 68–76.

Lin, F. and Worman, H.J. (1993). Structural organization of the human gene encoding nuclear lamin A and nuclear lamin C. J. Biol. Chem. *268*, 16321–16326.

Liptak, L.M., Uprichard, S.L., and Knipe, D.M. (1997). Functional order of assembly of herpes simplex virus DNA replication proteins into prereplicative site structures. J. Virol. *70*, 1759–1767.

Liu, F. and Roizman, B. (1993). Characterization of the protease and other products of amino-terminus-proximal cleavage of the herpes simplex virus 1 UL26 protein. J. Virol. *67*, 1300–1309.

Lukonis, C.J. and Weller, S.K. (1996). Characterization of nuclear structures in cells infected with herpes simplex virus type 1 in the absence of viral DNA replication. J. Virol. *70*, 1751–1758.

Malik, A.K., Shao, L., Shanley, J.D., and Weller, S.K. (1996). Intracellular localization of the herpes simplex virus type-1 origin binding protein, UL9. Virology *224*, 380–389.

Martinez, R., Sarisky, R.T., Weber, P.C., and Weller, S.K. (1996). Herpes simplex virus type 1 alkaline nuclease is required for efficient processing of viral DNA replication intermediates. J. Virol. *70*, 2075–2085.

McGeoch, D.J., Dalrymple, M.A., Davison, A.J., Dolan, A., Frame, M.C., McNab, D., Perry, L.J., Scott, J.E., and Taylor, P. (1988). The complete DNA sequence of the long unique region in the genome of herpes simplex virus type 1. J. Gen. Virol. *69*, 1531–1574.

McNab, A.R., Desai, P., Person, S., Roof, L.L., Thomsen, D.R., Newcomb, W.W., Brown, J.C., and Homa, F.L. (1998). The product of the herpes simplex virus type 1 UL25 gene is required for encapsidation but not for cleavage of replicated DNA. J. Virol. *72*, 1060–1070.

McVoy, M.A., Nixon, D.A., Adler, S.P., and Mocarski, E.S. (1998). Sequences within the herpesvirus-conserved *pac*1 and *pac*2 motifs are required for cleavage and packaging of the murine cytomegalovirus genome. J. Virol. *72*, 48–56.

Millard, T.H., Sharp, S.J., and Machesky, L.M. (2004). Signalling to actin assembly via the WASP (Wiskott-Aldrich syndrome protein)-family proteins and the Arp2/3 complex. Biochem. J. *380*, 1–17.

Mitchell, M.S., Matsuzaki, S., Imai, S., and Rao, V.B. (2002). Sequence analysis of bacteriophage T4 DNA packaging/terminase genes 16 and 17 reveals a common ATPase center in the large subunit of viral terminases. Nucleic Acids Res. *30*, 4009–4021.

Mocarski, E.S. and Roizman, B. (1981). Site-specific inversion sequence of the herpes simplex virus genome: domain and structural features. Proc. Natl. Acad. Sci. USA *78*, 7047–7051.

Mocarski, E.S. and Roizman, B. (1982). Structure and role of the herpes simplex virus DNA termini in inversion, circularization and generation of virion DNA. Cell *31*, 89–97.

Mocarski, E.S., Deiss, L.P., and Frenkel, N. (1985). The nucleotide sequence and structural features of a novel U_S-*a* junction present in a defective herpes simplex virus genome. J. Virol. 55, 140–146.

Morgan, C., Rose, H.M., Holden, M., and Jones, E.P. (1959). Electron microscopic observations on the development of herpes simplex virus. J. Exp. Med. *110*, 643–656.

Morita, M., Tasaka, M., and Fujisawa, H. (1993). DNA packaging ATPase of bacteriophage T3. Virology *193*, 748–752.

Muranyi, W., Haas, J., Wagner, M., Krohne, G., and Koszinowski, U.H. (2002). Cytomegalovirus recruitment of cellular kinases to dissolve the nuclear lamina. Science *297*, 854–857.

Nasseri, M. and Mocarski, E.S. (1988). The cleavage recognition signal is contained within sequences surrounding an *a*-*a* junction in herpes simplex virus DNA. Virology *167*, 25–30.

Newcomb, W.W. and Brown, J.C. (1991). Structure of the herpes simplex virus capsid: effects of extraction with guanidine hydrochloride and partial reconstitution of extracted capsids. J. Virol. *65*, 613–620.

Newcomb, W.W., Trus, B.L., Booy, F.P., Steven, A.C., Wall, J.S., and Brown, S.C. (1993). Structure of the herpes simplex virus capsid molecular composition of the pentons and the triplexes. J. Mol. Biol. *232*, 499–511.

Newcomb, W.W., Homa, F.L., Thomsen, D.R., Booy, F.P., Trus, B.L., Steven, A.C., Spencer, J.V., and Brown, J.C. (1996). Assembly of the herpes simplex virus capsid: characterization of intermediates observed during cell-free capsid formation. J. Mol. Biol. *263*, 432–446.

Newcomb, W.W., Homa, F.L., Thomsen, D.R., Trus, B.L., Cheng, N., Steven, A., Booy, F., and Brown, J.C. (1999). Assembly of the herpes simplex virus procapsid from purified components and identification of small complexes containing the major capsid and scaffolding proteins. J. Virol. *73*, 4239–4250.

Newcomb, W.W., Trus, B.L., Cheng, N., Steven, A.C., Sheaffer, A.K., Tenney, D.J., Weller, S.K., and Brown, J.C. (2000). Isolation of herpes simplex virus procapsids from cells infected with a protease-deficient mutant virus. J. Virol. *74*, 1663–1673.

Newcomb, W.W., Juhas, R.M., Thomsen, D.R., Homa, F.L., Burch, A.D., Weller, S.K., and Brown, J.C. (2001). The UL6 gene product forms the portal for entry of DNA into the herpes simplex virus capsid. J. Virol. *75*, 10923–10932.

Newcomb, W.W., Thomsen, D.R., Homa, F.L., and Brown, J.C. (2003). Assembly of the herpes simplex virus capsid: identification of soluble scaffold-portal complexes and their role in formation of portal-containing capsids. J. Virol. *77*, 9862–9871.

Nicholson, P., Addison, C., Cross, A.m., Kennard, J., Preston, V.G., and Rixon, F.J. (1994). Localization of the herpes simplex virus type 1 major capsid protein VP5 to the cell nucleus requires the abundant scaffolding protein VP22a. J. Gen. Virol. *75*, 1091–1099.

Ogasawara, M., Suzutani, T., Yoshida, I., and Azuma, M. (2001). Role of the UL25 gene product in packaging DNA into the herpes simplex virus capsid: location of UL25 product in the capsid and demonstration that it binds DNA. J. Virol. *75*, 1427–1436.

Oien, N.L., Thomsen, D.R., Wathen, M.W., Newcomb, W.W., Brown, J.C., and Homa, F.L. (1997). Assembly of herpes simplex virus capsids using the human cytomegalovirus scaffold protein: critical role of the C terminus. J. Virol. *71*, 1281–1291.

Ottaviano, Y. and Gerace, L. (1985). Phosphorylation of the nuclear lamins during interphase and mitosis. J. Biol. Chem. *260*, 624–632.

Park, R. and Baines, J.D. (2006). Herpes simplex virus type 1 infection induces activation and recruitment of protein kinase C to the nuclear membrane and increased phosphorylation of lamin B. J. Virol. *80*, 494–504.

Patel, A.H., Rixon, F.J., Cunningham, C., and Davison, A.J. (1996). Isolation and characterization of herpes simplex virus type 1 mutants defective in the UL6 gene. Virology *217*, 111–123.

Pelletier, A., Do, F., Brisebois, J.J., Lagace, L., and Cordingley, M.G. (1997). Self-association of herpes simplex virus type 1 ICP35 is via coiled-coil interactions and promotes stable interaction with the major capsid protein. J. Virol. *71*, 5197–5208.

Person, S., Laquerre, S., Desai, P., and Hempel, J. (1993). Herpes simplex virus type 1 capsid protein, VP21, originates within the UL26 open reading frame. J. Gen. Virol. *74*, 2269–2273.

Peter, M., Nakagawa, J., Doree, M., Labbe, J.C., and Nigg, E.A. (1990). *In vitro* disassembly of the nuclear lamina and M phase-specific phosphorylation of lamins by cdc2 kinase. Cell *61*, 591–602.

Poffenberger, K.L. and Roizman, B. (1985). A noninverting genome of a viable herpes simplex virus 1: presence of head-to-tail linkages in packaged genomes and requirements for circularization after infection. J. Virol. 53, 587–595.

Poon, A.P.W. and Roizman, B. (1993). Characterization of a temperature-sensitive mutant of the U_L15 open reading frame of herpes simplex virus 1. J. Virol. *67*, 4497–4503.

Poon, A.P. and Roizman, B. (2005). Herpes simplex virus 1 ICP22 regulates the accumulation of a shorter mRNA and of a truncated US3 protein kinase that exhibits altered functions. J. Virol. *79*, 8470–8479.

Post, L.E., Conley, A.J., Mocarski, E.S., and Roizman, B.R. (1980). Cloning of reiterated and non reiterated herpes simplex virus 1 sequences as BamHI fragments. Proc. Natl. Acad. Sci. USA *77*, 4201–4205.

Preston, V.G. and McDougall, I.M. (2002). Regions of the herpes simplex virus scaffolding protein that are important for intermolecular self-interaction. J. Virol. *76*, 673–687.

Preston, V.G., Coates, A.M., and Rixon, F.J. (1983). Identification and characterization of a herpes simplex virus gene product required for encapsidation of viral DNA. J. Virol. *45*, 1056–1064.

Purves, F.C., Spector, D., and Roizman, B. (1991). The herpes simplex virus protein kinase encoded by the U_S3 gene mediates posttranslational modi-

fication of the phosphoprotein encoded by the U_L34 gene. J. Virol. 65, 5757–5764.
Quinlan, M.P., Chen, L.B., and Knipe, D.M. (1984). The intranuclear location of a herpes simplex virus DNA-binding protein is determined by the status of viral DNA replication. Cell 36, 857–868.
Radsak, K., Schneider, D., Jost, E., and Brucher, K.H. (1989). Alteration of nuclear lamina protein in human fibroblasts infected with cytomegalovirus (HCMV). Arch. Virol. 105, 103–112.
Radsak, K.D., Brucher, K.H., and Georgatos, S.D. (1991). Focal nuclear envelope lesions and specific nuclear lamin A/C dephosphoylation during infection with human cytomegalovirus. Eur. J. Cell Biol. 54, 299–304.
Reynolds, A.E., Fan, Y., and Baines, J.D. (2000). Characterization of the U_L33 gene product of herpes simplex virus 1. Virology 266, 310–318.
Reynolds, A.E., Ryckman, B., Baines, J.D., Zhou, Y., Liang, L., and Roller, R.J. (2001). U_L31 and U_L34 proteins of herpes simplex virus type 1 form a complex that accumulates at the nuclear rim and is required for envelopment of nucleocapsids. J. Virol. 75, 8803–8817.
Reynolds, A.E., Wills, E.G. Roller, R.J., Ryckman, B.J., and Baines, J.D. (2002). Ultrastructural localization of the HSV-1 U_L31, U_L34, and U_S3 proteins suggests specific roles in primary envelopment and egress of nucleocapsids. J. Virol. 76, 8939–8952.
Reynolds, A.E., Liang, L., and Baines, J.D. (2004). Confomational changes in the nuclear lamina of cells infected with herpes simplex virus 1 requires genes UL31 and UL34. J. Virol. 78, 5564–5575.
Rixon, F.J., Cross, A.m., Addison, C., and Preston, V.G. (1988). The products of herpes simplex virus type 1 gene UL26 which are involved in DNA packaging are strongly associated with empty but not with full capsids. J. Gen. Virol. 69, 2879–2891.
Rixon, F.J., Addison, C., McGregor, A., Macnab, S.J., Nicholson, P., Preston, V.G., and Tatman, J.D. (1996). Mutiple interactions control the intracellular localization of the herpes simplex virus type 1 capsid proteins. J. Gen. Virol. 77, 2251–2260.
Roizman, B. and Furlong, D. (1974). The replication of herpesviruses. In Comprehensive Virology, H.Fraenkel-Conrat and R.R.Wagner, eds. (New York: Plenum Press), pp. 229–403.
Ryckman, B.J. and Roller, R.J. (2004). Herpes simplex virus type 1 primary envelopment: UL34 protein modification and the US3-UL34 catalytic relationship. J. Virol. 78, 399–412.
Salmon, B. and Baines, J.D. (1998). Herpes simplex virus DNA cleavage and packaging: association of multiple forms of U_L15-encoded proteins with B capsids requires at least the U_L6, U_L17 and U_L28 genes. J. Virol. 72, 3045–3050.
Salmon, B., Cunningham, C., Davison, A.J., Harris, W.J., and Baines, J.D. (1998). The Herpes Simplex Virus 1 U_L17 Gene Encodes Virion Tegument Proteins that are Required for Cleavage and Packaging of Viral DNA. J. Virol. 72, 3779–3788.
Salmon, B., Nalwanga, D., Fan, Y., and Baines, J.D. (1999). Proteolytic cleavage of the amino-terminus of the U_L15 gene product of herpes simplex virus 1 U_L15 protein is coupled with maturation of viral DNA into unit length genomes. J. Virol. 73, 8338–8348.
Scheffczik, H., Savva, C.G., Holzenburg, A., Kolesnikova, L., and Bogner, E. (2002). The terminase subunits pU_L56 and pU_L89 of human cytomegalovirus are DNA- metabolizing proteins with toroidal structure. Nucleic Acids Res. 30, 1695–1703.
Schrag, J.D., Prasad, B.V., Rixon, F.J., and Chiu, W. (1989). Three-dimensional structure of the HSV1 nucleocapsid. Cell 56, 651–660.
Scott, E.S. and O'Hare, P. (2001). Fate of the inner nuclear membrane protein lamin b receptor and nuclear lamins in herpes simplex virus type 1 infection. J. Virol. 75, 8818–8830.
Sheaffer, A.K., Newcomb, W.W., Gao, M., Yu, D., Weller, S.K., Brown, J.C., and Tenney, D.J. (2001). Herpes simplex virus DNA cleavage and packaging proteins associate with the procapsid prior to its maturation. J. Virol. 75, 687–698.
Sherman, G. and Bachenheimer, S.L. (1988). Characterization of intranuclear capsids made by ts morphogenetic mutants of HSV-1. Virology 163, 471–480.
Simpson-Holley, M., Baines, J., Roller, R., and Knipe, D.M. (2004). Herpes simplex virus 1 U(L)31 and U(L)34 gene products promote the late maturation of viral replication compartments to the nuclear periphery. J. Virol. 78, 5591–5600.
Soullam, B. and Worman, H.J. (1995). Signals and structural features involved in integral membrane protein targeting to the inner nuclear membrane. J. Cell Biol. 130, 15–27.
Spaete, R.R. and Frenkel, N. (1982). The herpes simplex virus amplicon: a new eucaryotic defective-virus cloning-amplifying vector. Cell 30, 295–304.
Spann, T.P., Moir, R.D., Goldman, A.E., Stick, R., and Goldman, R.D. (1997). Disruption of nuclear lamin organization alters the distribution of replication factors and inhibits DNA synthesis. J. Cell Biol. 136, 1201–1212.
Spann, T.P., Goldman, A.E., Wang, C., Huang, S., and Goldman, R.D. (2002). Alteration of nuclear lamin organization inhibits RNA polymer-

ase II- dependent transcription. J. Cell Biol. *156*, 603–608.
Stierle, V., Couprie, J., Ostlund, C., Krimm, I., Zinn-Justin, S., Hossenlopp, P., Worman, H.J., Courvalin, J.C., and Duband-Goulet, I. (2003). The carboxyl-terminal region common to lamins A and C contains a DNA binding domain. Biochemistry *42*, 4819–4828.
Stow, N.D. (2001). Packaging of genomic and amplicon dna by the herpes simplex virus type 1 ul25-null mutant kul25ns. J. Virol. *75*, 10755–10765.
Stow, N.D. and McMonagle, E.C. (1983). Characterization of the TRs/IRs origin of DNA replication of herpes simplex virus type 1. Virology *130*, 427–438.
Stow, N.D., McMonagle, E.C., and Davison, A.J. (1983). Fragments from both termini of the herpes simplex virus type 1 genome contain signals required for the encapsidation of viral DNA. Nucleic Acids Res. *11*, 8205–8220.
Taniura, H., Glass, C., and Gerace, L. (1995). A chromatin binding site in the tail domain of nuclear lamins that interacts with core histones. J. Cell Biol. *131*, 33–44.
Tatman, J.D., Preston, V.G., Nicholson, P., Elliott, R.M., and Rixon, F.J. (1994). Assembly of herpes simplex virus type 1 capsids using a panel of recombinant baculoviruses. J. Gen. Virol. *75*, 1101–1113.
Taus, N.S. and Baines, J.D. (1998). Herpes simplex virus DNA cleavage and packaging: the UL28 gene product is a minor component of B capsids. Virology *252*, 443–449.
Taus, N.S., Salmon, B., and Baines, J.D. (1998). The herpes simplex virus 1 U_L17 gene is required for localization of capsids and major and minor capsid proteins to intranuclear sites where viral DNA is cleaved and packaged. Virology *252*, 115–125.
Tengelsen, L.A., Pedersen, N.E., Shaver, P.R., Wathen, M.W., and Homa, F.L. (1993). Herpes simplex virus type 1 DNA cleavage and capsidation require the product of the UL28 gene: isolation and characterization of two UL28 deletion mutants. J. Virol. *67*, 3470–3480.
Thomsen, D.R., Roof, L.L., and Homa, F.L. (1994). Assembly of herpes simplex virus (HSV) intermediate capsids in insect cells infected with recombinant baculoviruses expressing HSV capsid proteins. J. Virol. *68*, 2442–2457.
Thomsen, D.R., Newcomb, W.W., Brown, J.C., and Homa, F.L. (1995). Assembly of the herpes simplex virus capsid: requirement for the carboxy-terminal twenty-five amino acids of the proteins encoded by the UL26 and UL26.5 genes. J. Virol. *69*, 3690–3703.
Torrisi, M.R., Di Lazzaro, C., Pavan, A., Pereira, L., and Campadelli-Fiume, G. (1992). Herpes simplex virus envelopment and maturation studies by fracture label. J. Virol. *66*, 554–561.
Trus, B.L., Newcomb, W.W., Booy, F.P., Brown, J.C., and Steven, A.C. (1992). Distinct monoclonal antibodies separately label the hexons or the pentons of herpes simplex virus capsid. Proc. Natl. Acad. Sci. USA *89*, 11508–11512.
Trus, B.L., Homa, F.L., Booy, F.P., Newcomb, W.W., Thomsen, D.R., Cheng, N., Brown, J.C., and Steven, A.C. (1995). Herpes simplex virus capsids assembled in insect cells infected with recombinant baculoviruses: structural authenticity and localization of VP26. J. Virol. *69*, 7362–7366.
Trus, B.L., Booy, F.P., Newcomb, W.W., Brown, J.C., Homa, F.L., Thomsen, D.R., and Steven, A.C. (1996). The herpes simplex virus procapsid: structure, conformational changes upon maturation, and roles of the triplex proteins VP19C and VP23 in assembly. J. Mol. Biol. *263*, 447–462.
Trus, B.L., Cheng, N., Newcomb, W.W., Homa, F.L., Brown, J.C., and Steven, A.C. (2004). Structure and polymorphism of the UL6 portal protein of herpes simplex virus type 1. J. Virol. *78*, 12668–12671.
Underwood, M.R., Harvey, R.J., Stanat, S.C., Hemphill, M.L., Miller, T., Drach, J.C., Townsend, L.B., and Biron, K.K. (1998). Inhibition of a human cytomegalovirus DNA maturation by a benzimidizole ribonucleoside is mediated through the UL89 gene product. J. Virol. *72*, 717–725.
Valpuesta, J.M., Fernandez, J.J., Carazo, J.M., and Carrascosa, J.L. (1999). The three-dimensional structure of a DNA translocating machine at 10 A resolution. Structure Fold. Des. *7*, 289–296.
Varmuza, S.L. and Smiley, J.R. (1985). Signals for site-specific cleavage of HSV DNA: maturation involves two separate cleavage events at sites distal to the recognition sequences. Cell *41*, 793–802.
Vlazny, D.A., Kwong, A., and Frenkel, N. (1982). Site-specific cleavage/packaging of herpes simplex virus DNA and the selective maturation of nucleocapsids containing full length viral DNA. Proc. Natl. Acad. Sci. USA *79*, 1423–1427.
Wadsworth, S., Jacob, R.J., and Roizman, B. (1975). Anatomy of herpes simplex virus DNA. II. Size, composition and arrangement of inverted terminal repetitions. J. Virol. *15*, 1487–1497.
Wagner, M.J. and Summers, W.C. (1978). Structures of the joint region and the termini of the DNA of herpes simplex virus type 1. J. Virol. *27*, 374–387.
Walker, J.E., Saraste, M., Runswick, M.J., and Gay, N.J. (1982). Distantly related sequences in the alpha- and beta-subunits of ATP synthase, myosin, kinases and other ATP-requiring enzymes and a common nucleotide binding fold. EMBO J. *1*, 945–951.

Ward, P.L., Ogle, W.O., and Roizman, B. (1996). Assemblons: nuclear structures defined by aggregation of immature capsids and some tegument proteins of herpes simplex virus 1. J. Virol. *70*, 4623–4631.

Warner, S.C., Chytrova, G., Desai, P., and Person, S. (2001). Mutations in the N-terminus of VP5 alter its interaction with the scaffold proteins of herpes simplex virus type 1. Virology *284*, 308–316.

Wear, M.A. and Cooper, J.A. (2004). Capping protein: new insights into mechanism and regulation. Trends Biochem. Sci. *29*, 418–428.

Weinheimer, S.P., McCann III, P.J., O'Boyle II, D.R., Stevens, J.T., Boyd, B.A., Drier, D.A., Yamanaka, G.A., Dilanni, C.L., Deckman, I.C., and Cordingly, M.G. (1993). Autoproteolysis of herpes simplex virus type 1 protease releases an active catalytic domain found in intermediate capsid particles. J. Virol. *67*, 5813–5822.

Weller, S.K., Aschman, D.P., Sacks, W.R., Coen, D.M., and Schaffer, P.A. (1983). Genetic analysis of temperature-sensitive mutants of HSV-1: the combined use of complementation and physical mapping for cistron assignment. Virology *130*, 290–305.

Weller, S.K., Carmichael, E.P., Aschman, D.P., Goldstein, D.J., and Schaffer, P.A. (1987). Genetic and phenotypic characterization of mutants in four essential genes that map to the left half of HSV-1 U_L DNA. Virology *161*, 198–210.

White, C.A, Stow, N.D., Patel, A.H., Hughes, M., and Preston, V.G. (2003). Herpes simplex virus type 1 portal protein UL6 interacts with the putative terminase subunits UL15 and UL28. J. Virol *77*, 6351–6358.

Wilson, D.W. and Church, G.A. (1997). Study of herpes simplex virus maturation during a synchronous wave of assembly. J. Virol. *71*, 3603–3612.

Wohlrab, F., McLean, M.J., and Wells, R.D. (1987). The segment inversion site of herpes simplex virus type 1 adopts a novel DNA structure. J. Biol. Chem. *262*, 6407–6416.

Yang, K. and Baines, J. D. (2006). The putative terminase subunit of herpes simplex virus 1 encoded by UL28 is necessary and sufficient to mediate interaction between pU_L15 and pU_L33. J. Virol. (in press).

Ye, G.J., Vaughan, K.T., Vallee, R.B., and Roizman, B. (2000). The herpes simplex virus 1 UL34 protein interacts with a cytoplasmic dynein intermediate chain and targets nuclear membrane. J. Virol. *74*, 1355–1363.

Yu, D. and Weller, S.K. (1998a). Genetic analysis of the UL 15 gene locus for the putative terminase of herpes simplex virus type 1. Virology *243*, 32–44.

Yu, D. and Weller, S.K. (1998b). Herpes simplex virus type 1 cleavage and packaging proteins UL15 and UL28 are associated with B but not C capsids during packaging. J. Virol. *72*, 7428–7439.

Yu, D., Shaeffer, A.K., Tenny, D.J., and Weller, S.K. (1997). Characterization of ICP6::*Lac Z* insertion mutants of the UL15 gene of herpes simplex virus type 1 reveals the translation of two proteins. J. Virol. *71*, 2656–2665.

Zhou, Z.H., Prasad, B.V., Jakana, J., Rixon, F.J., and Chiu, W. (1994). Protein subunit structures in herpes simplex virus A-capsid determined from 400 kV spot-scan electron cryomicroscopy. J. Mol. Biol. *242*, 456–469.

Zhou, Z.H., He, J., Jakana, J., Tatman, J.D., Rixon, F.J., and Chiu, W. (1995). Assembly of VP26 in herpes simplex virus-1 inferred from structures of wild-type and recombinant capsids. Nat. Struct. Biol. 2, 1026–1030.

The Development and Use of Alpha-herpesviruses-expressing Fluorescent Proteins

11

Gregory A. Smith and Bruce W. Banfield

Abstract

The development of fluorescent proteins as tools to monitor gene expression and protein trafficking within living cells and intact organisms has had a tremendous impact on virtually every discipline of the biological sciences including virology. In this chapter we review the development and application of recombinant alpha-herpesviruses that express a variety of fluorescent proteins. First, we cover some general considerations, concerns and caveats associated with the use of these viruses. Next, we describe the use of viruses that express fluorescent protein fusions to further understand the trafficking and assembly of subvirion components within living cells. Finally, we discuss the use of recombinant alpha-herpesviruses expressing fluorescent proteins to monitor virus infections in experimental animal model systems.

Fluorescent proteins, mostly derived from hydrozoa and anthozoa species, have been recognized for many decades. However, it was not until the seminal papers of Prasher, Chalfie, Inouye and colleagues in the mid-1990s, using the green fluorescent protein from the jellyfish *Aequorea victoria* as a marker for monitoring gene expression in live cells that the "green revolution" began (Prasher *et al.*, 1992; Chalfie *et al.*, 1994; Inouye and Tsuji, 1994). Improvements in the rate at which GFP folds, the quantum yield of fluorescence, and the optimization of codon preferences towards mammalian cells followed quickly resulting in the enhanced GFP (EGFP) molecule most commonly used today. Color shifted variants of EGFP are now widely used to distinguish multiple proteins within individual living cells, and new fluorescent proteins continue to be discovered and enhanced. There now exists a remarkable palate of colors to choose from.

Researchers in the virology community did not take long to catch on. The vast majority of viruses are too small to image directly by light microscopy. Fusion of fluorescent proteins to viral particles made investigating the dynamics of the entire viral infectious cycle possible for the first time. EGFP, its derivatives, and fluorescent proteins from other organisms have now been used in a wide variety of paradigms for monitoring virus-infected cells in culture and in animals. This chapter will focus on the use of fluorescent proteins in the analysis of alpha-herpesvirus infections.

General considerations

It should first be stated that insertion of a fluorescent protein coding sequence into a viral genome results in a mutant virus. This is a fundamental principle that cannot be ignored. Fluorescent viruses are exceptionally valuable reagents that require careful evaluation. Insertion of foreign sequences into the herpesvirus genome can effect expression of the gene of interest, as well as neighboring genes. The consequences of introducing an AT rich fluorescent protein coding sequence into a often GC rich herpesvirus genome is not immediately clear. Of course, the principle concern is whether the resulting tagged viral protein maintains proper function.

In the face of these hazards, an effort to confirm that the viral property under investigation is not adversely affected by the presence of the fluorescent protein is warranted. Viral replication can often be assayed by single-step growth curve analysis, and viral transmission between cells can be approximated by measuring plaque diameters. In some cases, alternative fluorescent protein fusions may allow for confirmation of an observed phenotype. However, a virus that appears phenotypically wild type in cultured cells may not be appropriate for studies of virulence or pathogenesis. Fusion of EGFP to the pp38 phosphoprotein of Marek's disease virus (MDV) results in a fluorescent virus that grows with wild type single-step growth kinetics in cultured primary chicken cells, but is severely attenuated in chickens (Prigge *et al.*, 2004). Fusion of EGFP to the VP26 capsid protein of pseudorabies virus (PRV) has no impact on PRV single-step growth kinetics in cultured cells and the recombinant virus retains wild type neuroinvasive properties in mice. In addition, fusion of EGFP two capsid associated protein, VP1/2 and UL37, provides supporting evidence that capsid dynamics in cells are maintained in the presence of fluorescent protein fusions (Luxton *et al.*, 2005; Antinone, Shubeita, Smith and Gross, unpublished). However, the appearance of symptoms in animals are delayed (Brittle, Feierbach, Reynolds and Enquist, personal communication; Pickard and Banfield, unpublished). In fact most, if not all, fluorescent herpesviruses are attenuated to some degree in animal models of infection. The reason for attenuation is not always obvious.

Choice of fluorescent protein

There are several considerations in choosing a fluorescent protein for studies of herpesvirus infections. Protein fluorescence is available essentially across the visible spectrum. Proteins with long excitation wavelengths are typically preferable, as the lower energy excitation light is less damaging to cells and the fluorescence emissions are relatively free of background cellular autofluorescence. Imaging of herpesvirus infections can be particular susceptible to background fluorescence, as cells develop increasing levels of fluorescence as infection progresses (Smith *et al.*, 2001; del Rio *et al.*, 2005). Therefore, applications requiring detection of a dim fluorescent signal, or imaging for a long duration, are benefited from the use of red fluorescent proteins. In contrast, short wavelength fluorescence proteins provide increased spatial resolution, and the cyan and yellow fluorescence proteins are most commonly used for fluorescence resonance energy transfer (FRET).

Fluorescent proteins can be monomeric or multimeric. Whereas multimers (i.e. ZsGreen and tdTomato) can provide remarkably bright fluorescence emissions that are advantageous for tracking virus infection in animals, monomeric proteins (i.e. EGFP, mRFP1 and mCherry) are desirable for studies involving protein fusions (Jain *et al.*, 2001; Campbell *et al.*, 2002; Shaner *et al.*, 2004; Stepanenko *et al.*, 2004). Additional considerations are required for design of translational fusions. Fusing a fluorescent protein to the N- or C-terminus of a herpesvirus protein can have differing results, and the best fusion can only be determined empirically. Fusion of EGFP to the MDV UL34 nuclear protein results in wild type viral growth only when EGFP is placed at the C-terminus (i.e.

UL34-EGFP) (Neubauer *et al.*, 2002). Of three tegument proteins examined in PRV, two tolerate EGFP only at the C-terminus (UL37 and VP16) and one only at the N-terminus (VP22) (del Rio *et al.*, 2005; Luxton *et al.*, 2005). In some instances, inclusion of a linker at the fusion junction may also benefit the preservation of protein function. We have noticed this to be true when fusing EGFP to the C-terminus of the US3 protein kinase of PRV (Randall, Coller, Banfield and Smith, unpublished). The fluorescent protein can also be compromised when fused to a viral protein. While proteins can be fused to either the N- or C-terminus of EGFP, only fusions to the C-terminus of mRFP1 are routinely compatible with fluorescence whereas N-terminal fusions to mRFP1 are often associated with considerable loss of fluorescence. An improved derivative of mRFP1, mCherry, has recently been described that in addition to having brighter fluorescence, replaces the N-terminal seven residues of mRFP1 with the corresponding residues of EGFP and a four amino acid spacer sequence. These changes allow mCherry to maintain substantial fluorescence in the context of N-terminal fusions (Shaner *et al.*, 2004).

Perhaps more surprising is the observation that different fluorescent proteins can impact viral protein function to differing degrees when expressed as a fusion. The herpes simplex virus (HSV) ICP4 protein tolerates CFP fused to the N-terminus (CFP-ICP4), but placing YFP at the ICP4 N-terminus results in aggregation in transfected cells (Everett *et al.*, 2004). Although aggregation of YFP-ICP4 is not seen in cells infected with a recombinant virus, the YFP-ICP4 HSV-1 nevertheless grows to lower titer than the equivalent CFP-ICP4 virus (Everett *et al.*, 2003).

To date, insertion of EGFP into genomes of herpesviruses remains the *de facto* standard. However, the beneficial spectral properties of the new generations of monomeric red fluorescent proteins will undoubtedly result in their supplanting EGFP for many applications. The dynamics of envelope, tegument, capsid and nonstructural proteins have now been examined in living cells (Fig. 11.1 and Table 11.1). Although not all initial reports of fluorescently tagged herpesviruses took advantage of time-lapse imaging, the trend of studying infection in living cells is clearly gaining momentum. Following initial characterization of viral protein dynamics in living cells, research groups are now examining the interactions of viral proteins with cellular structures (Everett *et al.*, 2003). Mutational analysis is a natural complement to live-cell imaging, and studies combining these two approaches are also being reported (Desai *et al.*, 2001; Everett *et al.*, 2004).

Working with two fluorescent proteins

If expression of more than one fluorescent protein from a single virus is required, then the fluorescent reporters must have distinct spectral properties, and preferably should be distinct genetically as well. CFP and YFP were the first used fluorescent protein pair for imaging of two herpesvirus proteins in living infected cells (Hutchinson *et al.*, 2002; Everett *et al.*, 2003). Because the coding sequences of CFP and YFP are 98% identical, recombination between the two would result in interchanges of the CFP and YFP fusions such that a heterogeneous viral population with differently tagged viruses could arise. The advent of monomeric red fluorescent proteins has improved both the spectral separation and sequence homology issues. The combination of EGFP and mRFP1 was recently used to monitor the composition of intracellular capsid complexes as they move within sensory neurons. Changes in composition of tegument proteins associated with moving capsids was determined to be coupled to retrograde

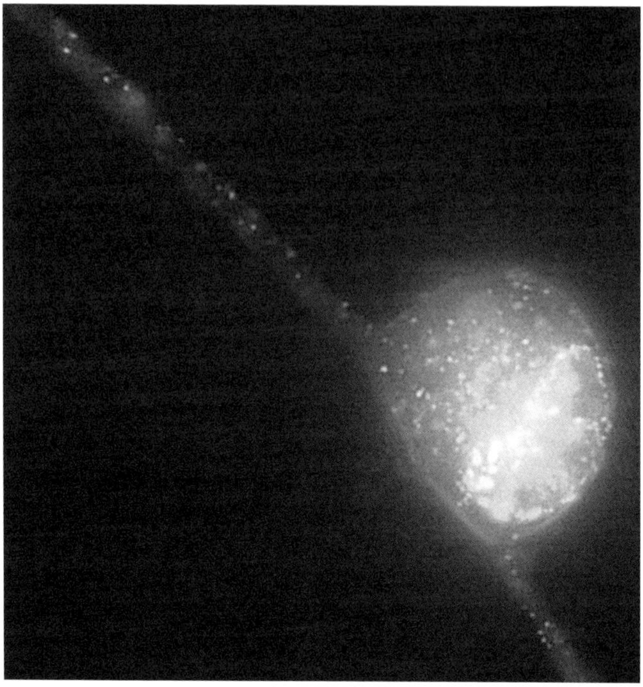

Figure 11.1 Chicken dorsal root sensory ganglion neuron infected with PRV expressing EGFP-capsids (vGS443) in culture. Two axons extending from the neural soma are filled with capsids emitting GFP fluorescence. The brightest fluorescence originates from large capsid inclusions that form in the nucleus. Time lapse imaging has revealed that capsids transport to distal regions of the axon at average velocities of ~2 μm/s (Smith *et al.*, 2001).

and anterograde capsid trafficking in axons, and the composition of individual viral particles was found to be highly variable (del Rio *et al.*, 2005; Luxton *et al.*, 2005).

Co-infection of two recombinant viruses, each expressing a different fluorescent protein, has also proven useful for studies of infection in cell culture and neuroinvasion in animals. However, once a cell has been infected with two similar viruses, inter-strain recombination will occur resulting in a mixed population of viruses and potentially false positive results. Because of this, mixed infections must be examined early after infection before recombination takes place, or the two viruses must be inoculated at distinct sites in animal models of infection. Using PRV encoding a human cytomegalovirus major immediate early promoter (CMV IEP) driving expression of either EGFP or mRFP1, a dynamic study of the herpesvirus induced block to superinfection in cultured sensory neurons was achieved (Banfield *et al.*, 2003). Exposure of infected neurons to a second virus was found to result in superinfection only when the neurons had been infected for less than five hours. PRV expressing either EGFP or mRFP1 fused to the VP26 capsid protein were also used to co-infect cultured sensory neurons. In this experiment, trafficking of mRFP1-capsids newly deposited in the axon were determined to proceeded unimpeded to the nucleus in the presence of *de novo* assembled capsids that were actively trafficking in the opposing direction (Smith *et al.*, 2004). This demonstrated that capsid targeting in

Table 11.1 Recombinant alpha-herpesviurses expressing fluorescent protein fusions

Protein	Virus	Frame rates	Summary of findings	Reference
Envelope proteins				
GFP-gB	HSV-1	–	Proof of concept: virus is viable but grows poorly	(Potel *et al*., 2002)
gB-GFP	HSV-1	–	Proof of concept: virus is viable but grows poorly	(Potel *et al*., 2003)
gB(GFP)	BHV1	–	Proof of concept: gB/furin as a recombinant secretion system	(Keil *et al*., 2005)
GFP-gH	HSV-1	–	Proof of concept: virus has wild type growth and gH function	(Lorentzen *et al*., 2001)
gK-GFP	EHV1	–	Used for static imaging	(Neubauer and Osterrieder, 2004)
US9-GFP	PRV	–	Used for circuit tracing in rodents	(Billig *et al*., 2000)
Tegument proteins				
GFP-ICP0	HSV-1	–	Used for fluorescence-activated cell sorting	(Lomonte and Everett, 1999)
YFP-ICP0	HSV-1	–	Dynamics of ICP0 and ICP4 association studied	(Everett *et al*., 2003)
CFP-ICP4	HSV-1	1 f/30min	ICP4 forms nuclear foci that develop into replication compartments	(Everett *et al*., 2003)
YFP-ICP4	HSV-1	1 f/10s	ICP4 nuclear foci appear rapidly	(Everett *et al*., 2004)
UL37-GFP	PRV	5 f/s	Capsids always transport in axons in association with UL37	(Luxton *et al*., 2005)
GFP-VP1/2	PRV	5 f/s	Capsids always transport in axons in association with VP1/2	(Luxton *et al*., 2005)
VP11/12-GFP	HSV-1	2 f/s	Large structures move in cytoplasm	(Willard, 2002)
YFP-VP13/14	HSV-1	1 f/5min	Large structures move in cytoplasm	(Donnelly and Elliott, 2001a)
GFP-VP13/14	HSV-1	–	Used for static imaging	(Donnelly and Elliott, 2001b)
	PRV	5 f/s	VP13/14 transports with capsids in axons during egress but not entry	(Luxton *et al*., 2005)

Table 11.1 Continued

Protein	Virus	Frame rates	Summary of findings	Reference
VP16-GFP	HSV-1	1 f/3–6s	Small punctae move retrograde in axons after microinjection	(Bearer *et al.*, 2000)
	HSV-1	1 f/5min	Large structures move into induced cellular protrusions	(La Boissiere *et al.*, 2004)
	PRV	5 f/s	VP16 transports with capsids in axons during egress but not entry	(Luxton *et al.*, 2005)
GFP-VP22	HSV-1	1 f/5min	Large structures move in cytoplasm	(Elliott and O'Hare, 1999)
	HSV-1	–	Used for static imaging	(Martin *et al.*, 2002)
	PRV	1 f/3s	VP22 transports in axons during egress	(del Rio *et al.*, 2005)
	PRV	5 f/s	VP22 transports with capsids in axons during egress but not entry	(Luxton *et al.*, 2005)
Capsid proteins				
GFP-VP26	HSV-1	–	Used for static imaging; capsids can be imaged by light microscopy	(Desai and Person, 1998)
	HSV-1	1.75 f/s	Capsids move in nuclei by an energy-dependent mechanism	(Forest *et al.*, 2005)
	PRV	1–10 f/s	Capsid entry transport in axons described	(Smith *et al.*, 2001)
	PRV	1–10 f/s	Capsid egress transport in axons described	(Smith *et al.*, 2004)
	PRV	1 f/10s	Capsids move into induced cellular protrusions	(Favoreel *et al.*, 2005)
mRFP1-VP26	PRV	1 f/s	Capsid targeting in axons occurs intrinsically	(Smith *et al.*, 2004)
	PRV	5 f/s	Capsid trafficking is coupled to differential tegument associations	(Luxton *et al.*, 2005)
Nonstructural proteins				
ICP8-GFP	HSV-1	1 f/3min	Replication compartments move and aggregate in nucleus	(Taylor *et al.*, 2003)
YFP-ICP27	HSV-1	–	ICP27 partly colocalizes with ICP4 in nuclei	(Everett *et al.*, 2004)
pp38-GFP	MDV	–	Proof of concept: pp38-GFP localization same as unmodified pp38	(Prigge *et al.*, 2004)

cells is achieved by an intrinsic component of the transporting capsid, and not due to changes in the cellular environment.

Site of fluorescent protein gene insertion in the virus genome

Fortunately, the genomes of many alpha-herpesviruses are fairly easy to manipulate. Ideally, the site of fluorescent protein gene insertion should not influence the ability of viruses to cause disease—either indirectly by attenuating virus replication or by directly affecting virus transport in the nervous system. Early studies using PRV as a self-amplifying neural tracer adopted the US4 gene (encodes the nonessential gG envelope protein), as a site to insert fluorescent protein expression cassettes. The US4 gene was selected as viruses lacking gG expression had no apparent defects in cell culture or in animals. We now know that a number of gG homologs act as broad-spectrum chemokine binding proteins (Bryant *et al.*, 2003). Moreover, expression of heterologous proteins from the PRV US4 locus results in polar effects on the expression of the upstream US3 gene (Demmin *et al.*, 2001). Insertions in US4 are now recognized to result in ~60-fold increases in LD50 and ~15% increases in survival times in animals (Kim *et al.*, 1999; Demmin *et al.*, 2001). While this attenuation may provide some benefit in neuroanatomical tract tracing studies because animal survival times are increased, virulence studies are compromised.

Utilization of alpha-herpesviruses-expressing fluorescent proteins to identify infected cells *in vivo*

Recombinant alpha-herpesviruses that express fluorescent proteins are powerful tools for studying virus infection in a variety of experimental paradigms and animal models. In this section we will discuss four such systems where the use of strains expressing fluorescent proteins have proven invaluable.

Using fluorescent PRV for neuronal circuit tracing

Neurobiologists have used alpha-herpesviruses as neuroanatomical tract tracers for over twenty years. The utility of alpha-herpesviruses for this purpose stems from the propensity of these viruses to spread between synaptically linked chains of neurons with exquisite specificity. Many excellent reviews on this subject have been published and we direct readers interested in learning more about this topic to them (Enquist, 1999; Aston-Jones and Card, 2000; Enquist and Card, 2003; Mettenleiter, 2003; Boldogkoi *et al.*, 2004).

The Bartha strain of PRV is the most widely used neuroanatomical tract tracer. PRV-Bartha is an attenuated vaccine strain that is transsynaptically transported in only the retrograde direction (i.e. from postsynaptic to presynaptic neuron) (Pickard *et al.*, 2002). Additional observations indicate that three membrane proteins gI, gE and US9, which are absent in the Bartha strain, are required for anterograde spread (Brideau *et al.*, 2000; Husak *et al.*, 2000; Tomishima and Enquist, 2001; Enquist, 2002). Historically, PRV-Bartha infected neurons have been identified using anti-PRV antibodies and standard immunocytochemical procedures. Although this technique yields excellent labeling, PRV infected neurons can only be identified after tissue fixation and tissue sectioning. Recombinant PRV strains that express EGFP allow for the direct examination of infected tissues and cells, thereby eliminating the need for labor-intensive immunocytochemical tissue processing.

The first EGFP expressing PRV-Bartha strain was described in 1997 (Jons and Mettenleiter, 1997). As with previous PRV reporter strains that were engineered to express beta-galactosidase (Mettenleiter and Rauh, 1990), the EGFP gene is fused to the first seven amino acids of glycoprotein gG and is expressed under the control of the early gG promoter (see Site of fluorescent protein gene insertion in the virus genome, above). This virus, originally called PrV-B80eGFP but often referred to as GFP-Bartha, can be detected in cultured cells by 12 hours post infection and by 24 hours plaques containing 10–20 fluorescent cells are present. PrV-B80eGFP has been used extensively in neuroanatomical tracing studies, however, in many cases antibodies against EGFP are used for detection of virally infected cells (Jansen *et al.*, 1995; Ueyama *et al.*, 1999; Geerling *et al.*, 2003; Krout *et al.*, 2003). This suggests that the intrinsic fluorescence of EGFP provided by PrV-B80eGFP is not robust. Subsequently, fluorescent PRV-Bartha derivatives were improved by driving EGFP expression with the HCMV IE promoter. One such virus, PRV152, harbors the EGFP expression cassette as an insertion in the gG locus (Smith *et al.*, 2000). In neurons infected with PRV152, EGFP diffuses throughout the cell and fills the dendritic arbor and the majority of the axon. The EGFP signal is very strong when viewed by standard epifluorescence light microscopy and no additional tissue processing is required to visualize infected neurons (Pickard *et al.*, 2002). In addition to being efficient and inexpensive to use, PRV152 labeled neurons can be identified in brain slices *in vitro* for physiological studies (Smith *et al.*, 2000). Remarkably, PRV152 infected neurons in the retina and suprachiasmatic nucleus, and PrV-B80eGFP infected cardiac vagal neurons, retain healthy physiological properties (Smith *et al.*, 2000; Irnaten *et al.*, 2001; Wang *et al.*, 2001). Collectively, these studies demonstrate that PRV-Bartha derivatives expressing EGFP allow for the identification of synaptically linked neurons, and for subsequent electrophysiological recording from these infected cells.

Double-virus transneuronal labeling

The goal of double-virus tracing studies is to identify a common set of neurons that are synaptically linked to two distinct populations of neurons (Kim *et al.*, 1999; Ter Horst, 2000; Enquist and Card, 2003). By using two, antigenically distinguishable virus strains and injecting them at two different sites one can identify neurons that become infected by both viruses on the basis of reactivity with both antigens. The first demonstration of this method was described by Jansen and colleagues in 1995 (Jansen *et al.*, 1995). In this study, and the majority of double-virus labeling studies that have followed, PRV-Bartha derivatives that expressed either EGFP or beta-galactosidase were used and each antigen detected by immunohistochemical procedures (Ueyama *et al.*, 1999; Billig *et al.*, 2000; Travers and Rinaman, 2002; Geerling *et al.*, 2003; Kerman *et al.*, 2003; Cano *et al.*, 2004). Clearly, the availability of PRV-Bartha derivatives that express a red fluorescent protein that would complement EGFP expressing Bartha strains would be useful for double-virus transneuronal labeling applications.

A red fluorescent protein (RFP) was highly sought after for use in multicolor labeling or fluorescence resonance energy transfer experiments. The RFP cloned from *Discosoma* coral, drFP583, commercially known as DsRed, appeared to be a good candidate for a spectrally distinct companion for EGFP (Matz *et al.*, 1999). However, the slow and incomplete maturation and obligate tetramerization have hampered the evolution of DsRed as a useful tool (Baird *et al.*, 2000; Yarbrough *et al.*, 2001). Attempts to address the rate and/or extent of maturation of DsRed provided only marginal improvements (Clontech, 2001). In 2002, Campbell and colleagues (Campbell *et al.*, 2002) described a monomeric RFP (mRFP1) derived from a rapidly maturing DsRed derivative, DsRed T1, isolated by Bevis and Glick (Bevis and Glick, 2002). The mRFP1 overcomes three critical problems associated with the wild type tetramer of DsRed: (1) it is a monomer; (2) it matures roughly 10 times faster than DsRed; and (3) the excitation and emission peaks are about 25 nm red-shifted from DsRed, conferring greater spectral separation from autofluorescence and having minimal emission when excited at wavelengths optimal for EGFP (Campbell *et al.*, 2002). A PRV-Bartha strain, PRV614, that expresses mRFP1 from the US4 locus has been reported and appears to be an ideal companion for PRV152 in double-virus transneuronal labeling experiments (Banfield *et al.*, 2003). An example of neuronal infection by PRV614 is shown in Fig. 11.2.

A number of potential problems with double-virus labeling experiments exist and have been reviewed in some detail (Kim *et al.*, 1999; Ter Horst, 2000; Enquist and Card, 2003). Several factors can influence the dual labeling of a single neuron by two different reporter viruses. Depending on the reporter, the promoter used to drive reporter expression, and the site at which the reporter is inserted into the viral genome, the viruses constructed may not be equally fit. If one reporter virus has reduced transport or replication capabilities compared with the second reporter virus these deficiencies will be amplified as the number of neurons in the multisynaptic circuit increases. This may result in the arrival of the more fit virus to the population of neurons under study before the less fit virus arrives. To complicate things further, the rate at which the reporter "matures" or is detectable is also an important consideration. The delayed arrival or detection of the second reporter virus has important ramifications for dual virus tract-tracing experiments.

Prior infection of a cell with one virus has the potential to inhibit subsequent infection by another, a process referred to as superinfection inhibition. Indeed, investigators have shown that BHK cells constitutively expressing the alpha-herpesvirus envelope glycoprotein, gD, are resistant to infection by virus (Campadelli-Fiume *et al.*, 1988; Brandimarti *et al.*, 1994). This inhibition of infection may occur through gD sequestration of critical herpesvirus receptors. It is likely that other mechanisms for superinfection inhibition downstream of the virus entry pathway also exist. Although the "window" for double infections is probably small (i.e. on the order

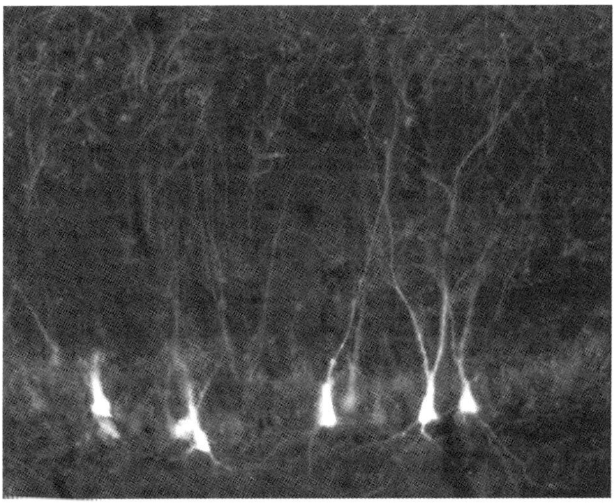

Figure 11.2 Hamster hippocampal pyramidal cells retrogradely labeled by PRV expressing unfused mRFP1 (PRV614). mRFP1 is imaged from a fixed coronal brain slice. Note the robust labeling of dendrites and axons. Image courtesy of Dr. G.E. Pickard, Colorado State University.

of a few hours), there are no data in the literature addressing this issue in an established experimental transsynaptic-tracing paradigm *in vivo*. *In vitro* studies using cultured sensory neurons affirm the small temporal window available for dual infections (Banfield *et al.*, 2003).

Conditional transneuronal labeling studies

In 2001, DeFalco and colleagues described a powerful new method for identifying CNS neurons that are in synaptic contact with a restricted subset of neurons (DeFalco *et al.*, 2001). This technique is based on Cre-loxP site-specific recombination and requires expression of the Cre recombinase for the production of EGFP-expressing virus capable of trans-synaptic spread. The recombinant PRV-Bartha virus used in these studies is unable to replicate in neurons and other non-dividing cells due to lack of thymidine kinase (TK) expression. However, upon infection of Cre-expressing neurons, site-specific recombination between two virally encoded loxP sites removes a DNA sequence that blocks expression of both TK and EGFP. The resulting virus is now capable of replication and transport between synaptically connected neurons. By infecting transgenic mice that express the Cre recombinase under the control of a specific neuronal promoter, labeling of neurons that express either the leptin receptor or neuropeptide Y, and all synaptically linked neurons in retrograde circuits, was achieved.

Use of fluorescent proteins to monitor herpesvirus-based therapeutics

HSV derivatives are currently being used as vectors for gene therapy applications and as therapeutic agents for the treatment of a variety of cancers (Advani *et al.*, 2002; Glorioso and Fink, 2004). One such strain, NV1066, has a ~15 kbp deletion between the UL56 and US1 genes and expresses EGFP at the deletion site. This virus has proven to be effective against a variety of tumor types in rodent model systems (Wong *et al.*, 2002; Stiles *et al.*, 2003; Stanziale *et al.*, 2004). Stanziale and colleagues have demonstrated the remarkable specificity of NV1066 infection for human gastric cancer cells in live animals by using laparoscopy to detect EGFP fluorescence *in situ* (Stanziale *et al.*, 2004). Thus, the inclusion of fluorescent protein genes in therapeutic HSV strains has the potential to allow physicians to identify tissues that have received the recombinant virus and follow the course of therapy using minimally invasive procedures.

Closing remarks

Where electron microscopy was once used to subjectively infer the nature of a dynamic processes in an infected cell, fluorescent proteins have now opened the door to studying these processes in real time. Technology to efficiently image fluorescent emission continues to improve, and the advent of diverse spectrums of fluorescent proteins is making new applications, such as triple and quadruple labeling paradigms, possible (Zhang *et al.*, 2002; Shaner *et al.*, 2004; Wang *et al.*, 2004). The coming years will undoubtedly see the illumination of many previously unrecognized aspects of herpesvirus replication and transmission. The only remaining limitation seems to be one's imagination.

References

Advani, S.J., Weichselbaum, R.R., Whitley, R.J., and Roizman, B. (2002). Friendly fire: redirecting herpes simplex virus-1 for therapeutic applications. Clin. Microbiol. Infect. *8*, 551–563.

Aston-Jones, G., and Card, J.P. (2000). Use of pseudorabies virus to delineate multisynaptic circuits in brain: opportunities and limitations. J. Neurosci. Methods *103*, 51–61.

Baird, G.S., Zacharias, D.A., and Tsien, R.Y. (2000). Biochemistry, mutagenesis, and oligomerization of DsRed, a red fluorescent protein from coral. Proc. Natl. Acad. Sci. USA 97, 11984–11989.

Banfield, B.W., Kaufman, J.D., Randall, J.A., and Pickard, G.E. (2003). Development of pseudorabies virus strains expressing red fluorescent proteins: new tools for multisynaptic labeling applications. J. Virol. *77*, 10106–10112.

Bearer, E.L., Breakefield, X.O., Schuback, D., Reese, T.S., and LaVail, J.H. (2000). Retrograde axonal transport of herpes simplex virus: Evidence for a single mechanism and a role for tegument. Proc. Natl. Acad. Sci. USA *97*, 8146–8150.

Bevis, B.J., and Glick, B.S. (2002). Rapidly maturing variants of the *Discosoma* red fluorescent protein (DsRed). Nat. Biotechnol. *20*, 83–87.

Billig, I., Foris, J.M., Enquist, L.W., Card, J.P., and Yates, B.J. (2000). Definition of neuronal circuitry controlling the activity of phrenic and abdominal motoneurons in the ferret using recombinant strains of pseudorabies virus. J. Neurosci. *20*, 7446–7454.

Boldogkoi, Z., Sik, A., Denes, A., Reichart, A., Toldi, J., Gerendai, I., Kovacs, K.J., and Palkovits, M. (2004). Novel tracing paradigms – genetically engineered herpesviruses as tools for mapping functional circuits within the CNS: present status and future prospects. Prog. Neurobiol. *72*, 417–445.

Brandimarti, R., Huang, T., Roizman, B., and Campadelli-Fiume, G. (1994). Mapping of herpes simplex virus 1 genes with mutations which overcome host restrictions to infection. Proc. Natl. Acad. Sci. USA *91*, 5406–5410.

Brideau, A.D., Card, J.P., and Enquist, L.W. (2000). Role of pseudorabies virus Us9, a type II membrane protein, in infection of tissue culture cells and the rat nervous system. J. Virol. *74*, 834–845.

Bryant, N.A., Davis-Poynter, N., Vanderplasschen, A., and Alcami, A. (2003). Glycoprotein G isoforms from some alphaherpesviruses function as broad-spectrum chemokine binding proteins. EMBO J. *22*, 833–846.

Campadelli-Fiume, G., Arsenakis, M., Farabegoli, F., and Roizman, B. (1988). Entry of herpes simplex virus 1 in BJ cells that constitutively express viral glycoprotein D is by endocytosis and results in degradation of the virus. J. Virol. *62*, 159–167.

Campbell, R.E., Tour, O., Palmer, A.E., Steinbach, P.A., Baird, G.S., Zacharias, D.A., and Tsien, R.Y. (2002). A monomeric red fluorescent protein. Proc. Natl. Acad. Sci. USA *99*, 7877–7882.

Cano, G., Card, J.P., and Sved, A.F. (2004). Dual viral transneuronal tracing of central autonomic circuits involved in the innervation of the two kidneys in rat. J. Comp. Neurol. *471*, 462–481.

Chalfie, M., Tu, Y., Euskirchen, G., Ward, W.W., and Prasher, D.C. (1994). Green fluorescent protein as a marker for gene expression. Science *263*, 802–805.

Clontech (2001). Living Colors ™. CLONTECHniques *XVI*, 2–3.

DeFalco, J., Tomishima, M., Liu, H., Zhao, C., Cai, X., Marth, J.D., Enquist, L., and Friedman, J.M. (2001). Virus-assisted mapping of neural inputs to a feeding center in the hypothalamus. Science *291*, 2608–2613.

del Rio, T., Ch'ng, T.H., Flood, E.A., Gross, S.P., and Enquist, L.W. (2005). Heterogeneity of a fluorescent tegument component in single pseudorabies virus virions and enveloped axonal assemblies. J. Virol. *79*, 3903–3919.

Demmin, G.L., Clase, A.C., Randall, J.A., Enquist, L.W., and Banfield, B.W. (2001). Insertions in the gG gene of pseudorabies virus reduce expression of the upstream Us3 protein and inhibit cell-to-cell spread of virus infection. J. Virol. *75*, 10856–10869.

Desai, P., and Person, S. (1998). Incorporation of the green fluorescent protein into the herpes simplex virus type 1 capsid. J. Virol. *72*, 7563–7568.

Desai, P., Sexton, G.L., McCaffery, J.M., and Person, S. (2001). A null mutation in the gene encoding the herpes simplex virus type 1 UL37 polypeptide abrogates virus maturation. J. Virol. *75*, 10259–10271.

Donnelly, M., and Elliott, G. (2001a). Fluorescent tagging of herpes simplex virus tegument protein VP13/14 in virus infection. J. Virol. *75*, 2575–2583.

Donnelly, M., and Elliott, G. (2001b). Nuclear localization and shuttling of herpes simplex virus tegument protein VP13/14. J. Virol. *75*, 2566–2574.

Elliott, G., and O'Hare, P. (1999). Live-cell analysis of a green fluorescent protein-tagged herpes simplex virus infection. J. Virol. *73*, 4110–4119.

Enquist, L.W. (2002). Exploiting circuit-specific spread of pseudorabies virus in the central nervous system: insights to pathogenesis and circuit tracers. J. Infect. Dis. *186 Suppl 2*, S209–214.

Enquist, L.W., and Card, J.P. (2003). Recent advances in the use of neurotropic viruses for circuit analysis. Curr. Opin. Neurobiol. *13*, 603–606.

Enquist, L.W., P.J. Husak, B.W. Banfield, and G.A. Smith (1999). Infection and spread of alphaherpesviruses in the nervous system. Adv. Virus Res. *51*, 237–347.

Everett, R.D., Sourvinos, G., and Orr, A. (2003). Recruitment of herpes simplex virus type 1 transcriptional regulatory protein ICP4 into foci juxtaposed to ND10 in live, infected cells. J. Virol. *77*, 3680–3689.

Everett, R.D., Sourvinos, G., Leiper, C., Clements, J.B., and Orr, A. (2004). Formation of nuclear foci of the herpes simplex virus type 1 regulatory protein ICP4 at early times of infection: localization, dynamics, recruitment of ICP27, and evidence for the de novo induction of ND10-like complexes. J. Virol. *78*, 1903–1917.

Favoreel, H.W., Minnebruggen, G., Adriaensen, D., and Nauwynck, H.J. (2005). Cytoskeletal rearrangements and cell extensions induced by the US3 kinase of an alphaherpesvirus are associated with enhanced spread. Proc. Natl. Acad. Sci. USA *102*, 8990–8995.

Forest, T., Barnard, S., and Baines, J.D. (2005). Active intranuclear movement of herpesvirus capsids. Nat. Cell Biol. *7*, 429–431.

Geerling, J.C., Mettenleiter, T.C., and Loewy, A.D. (2003). Orexin neurons project to diverse sympathetic outflow systems. Neuroscience *122*, 541–550.

Glorioso, J.C., and Fink, D.J. (2004). Herpes vector-mediated gene transfer in treatment of diseases of the nervous system. Annu. Rev. Microbiol. *58*, 253–271.

Husak, P.J., Kuo, T., and Enquist, L.W. (2000). Pseudorabies virus membrane proteins gI and gE facilitate anterograde spread of infection in projection-specific neurons in the rat. J. Virol. *74*, 10975–10983.

Hutchinson, I., Whiteley, A., Browne, H., and Elliott, G. (2002). Sequential localization of two herpes simplex virus tegument proteins to punctate nuclear dots adjacent to ICP0 domains. J. Virol. *76*, 10365–10373.

Inouye, S., and Tsuji, F.I. (1994). Aequorea green fluorescent protein. Expression of the gene and fluorescence characteristics of the recombinant protein. FEBS Lett. *341*, 277–280.

Irnaten, M., Neff, R.A., Wang, J., Loewy, A.D., Mettenleiter, T.C., and Mendelowitz, D. (2001). Activity of cardiorespiratory networks revealed by transsynaptic virus expressing GFP. J. Neurophysiol. *85*, 435–438.

Jain, R.K., Joyce, P.B., Molinete, M., Halban, P.A., and Gorr, S.U. (2001). Oligomerization of green fluorescent protein in the secretory pathway of endocrine cells. Biochem. J. *360*, 645–649.

Jansen, A.S., Nguyen, X.V., Karpitskiy, V., Mettenleiter, T.C., and Loewy, A.D. (1995). Central command neurons of the sympathetic nervous system: basis of the fight-or-flight response. Science *270*, 644–646.

Jons, A., and Mettenleiter, T.C. (1997). Green fluorescent protein expressed by recombinant pseudorabies virus as an *in vivo* marker for viral replication. J. Virol. Methods *66*, 283–292.

Keil, G.M., Hohle, C., Giesow, K., and Konig, P. (2005). Engineering glycoprotein B of bovine herpesvirus 1 to function as transporter for secreted proteins: a new protein expression approach. J. Virol. *79*, 791–799.

Kerman, I.A., Enquist, L.W., Watson, S.J., and Yates, B.J. (2003). Brainstem substrates of sympathomotor circuitry identified using trans-synaptic tracing with pseudorabies virus recombinants. J. Neurosci. *23*, 4657–4666.

Kim, J.S., Enquist, L.W., and Card, J.P. (1999). Circuit-specific coinfection of neurons in the rat central nervous system with two pseudorabies virus recombinants. J. Virol. *73*, 9521–9531.

Krout, K.E., Mettenleiter, T.C., and Loewy, A.D. (2003). Single cns neurons link both central motor and cardiosympathetic systems: a double-virus tracing study. Neuroscience *118*, 853–866.

La Boissiere, S., Izeta, A., Malcomber, S., and O'Hare, P. (2004). Compartmentalization of VP16 in cells infected with recombinant herpes simplex virus expressing VP16-green fluorescent protein fusion proteins. J. Virol. *78*, 8002–8014.

Lomonte, P., and Everett, R.D. (1999). Herpes simplex virus type 1 immediate-early protein Vmw110 inhibits progression of cells through mitosis and from G(1) into S phase of the cell cycle. J. Virol. *73*, 9456–9467.

Lorentzen, E.U., Eing, B.R., Hafezi, W., Manservigi, R., and Kuhn, J.E. (2001). Replication-competent herpes simplex virus type 1 mutant expressing an autofluorescent glycoprotein H fusion protein. Intervirology *44*, 232–242.

Luxton, G.W., Haverlock, S., Coller, K.E., Antinone, S.E., Pincetic, A., and Smith, G.A. (2005). Targeting of herpesvirus capsid transport in axons is coupled to association with specific sets of tegument proteins. Proc. Natl. Acad. Sci. USA *102*, 5832–5837.

Martin, A., O'Hare, P., McLauchlan, J., and Elliott, G. (2002). Herpes simplex virus tegument protein VP22 contains overlapping domains for cytoplasmic localization, microtubule interaction, and chromatin binding. J. Virol. *76*, 4961–4970.

Matz, M.V., Fradkov, A.F., Labas, Y.A., Savitsky, A.P., Zaraisky, A.G., Markelov, M.L., and Lukyanov, S.A. (1999). Fluorescent proteins from nonbioluminescent anthozoa species. Nat. Biotechnol. *17*, 969–973.

Mettenleiter, T.C. (2003). Pathogenesis of neurotropic herpesviruses: role of viral glycoproteins in neuroinvasion and transneuronal spread. Virus Res. *92*, 197–206.

Mettenleiter, T.C., and Rauh, I. (1990). A glycoprotein gX-beta-galactosidase fusion gene as insertional marker for rapid identification of pseudorabies virus mutants. J. Virol. Methods *30*, 55–65.

Neubauer, A., and Osterrieder, N. (2004). Equine herpesvirus type 1 (EHV-1) glycoprotein K is required for efficient cell-to-cell spread and virus egress. Virology *329*, 18–32.

Neubauer, A., Rudolph, J., Brandmuller, C., Just, F.T., and Osterrieder, N. (2002). The equine herpesvirus 1 UL34 gene product is involved in an early step in virus egress and can be efficiently replaced by a UL34-GFP fusion protein. Virology *300*, 189–204.

Pickard, G.E., Smeraski, C.A., Tomlinson, C.C., Banfield, B.W., Kaufman, J., Wilcox, C.L., Enquist, L.W., and Sollars, P.J. (2002). Intravitreal injection of the attenuated pseudorabies virus PRV Bartha results in infection of the hamster suprachiasmatic nucleus only by retrograde transsynaptic transport via autonomic circuits. J. Neurosci. *22*, 2701–2710.

Potel, C., Kaelin, K., Gautier, I., Lebon, P., Coppey, J., and Rozenberg, F. (2002). Incorporation of green fluorescent protein into the essential envelope glycoprotein B of herpes simplex virus type 1. J. Virol. Methods *105*, 13–23.

Potel, C., Kaelin, K., Danglot, L., Triller, A., Vannier, C., and Rozenberg, F. (2003). Herpes simplex virus type 1 glycoprotein B sorting in hippocampal neurons. J. Gen. Virol. *84*, 2613–2624.

Prasher, D.C., Eckenrode, V.K., Ward, W.W., Prendergast, F.G., and Cormier, M.J. (1992). Primary structure of the Aequorea victoria green-fluorescent protein. Gene *111*, 229–233.

Prigge, J.T., Majerciak, V., Hunt, H.D., Dienglewicz, R.L., and Parcells, M.S. (2004). Construction and characterization of Marek's disease viruses having green fluorescent protein expression tied directly or indirectly to phosphoprotein 38 expression. Avian Dis. *48*, 471–487.

Shaner, N.C., Campbell, R.E., Steinbach, P.A., Giepmans, B.N., Palmer, A.E., and Tsien, R.Y. (2004). Improved monomeric red, orange and yellow fluorescent proteins derived from Discosoma sp. red fluorescent protein. Nat. Biotechnol. *22*, 1567–1572.

Smith, B.N., Banfield, B.W., Smeraski, C.A., Wilcox, C.L., Dudek, F.E., Enquist, L.W., and Pickard, G.E. (2000). Pseudorabies virus expressing enhanced green fluorescent protein: A tool for *in vitro* electrophysiological analysis of transsynaptically labeled neurons in identified central nervous system circuits. Proc. Natl. Acad. Sci. USA *97*, 9264–9269.

Smith, G.A., Gross, S.P., and Enquist, L.W. (2001). Herpesviruses use bidirectional fast-axonal transport to spread in sensory neurons. Proc. Natl. Acad. Sci. USA *98*, 3466–3470.

Smith, G.A., Pomeranz, L., Gross, S.P., and Enquist, L.W. (2004). Local modulation of plus-end transport targets herpesvirus entry and egress in sensory axons. Proc. Natl. Acad. Sci. USA *101*, 16034–16039.

Stanziale, S.F., Stiles, B.M., Bhargava, A., Kerns, S.A., Kalakonda, N., and Fong, Y. (2004). Oncolytic herpes simplex virus-1 mutant expressing green fluorescent protein can detect and treat peritoneal cancer. Hum. Gene Ther. *15*, 609–618.

Stepanenko, O.V., Verkhusha, V.V., Kazakov, V.I., Shavlovsky, M.M., Kuznetsova, I.M., Uversky, V.N., and Turoverov, K.K. (2004). Comparative studies on the structure and stability of fluorescent proteins EGFP, zFP506, mRFP1, "dimer2", and DsRed1. Biochemistry *43*, 14913–14923.

Stiles, B.M., Bhargava, A., Adusumilli, P.S., Stanziale, S.F., Kim, T.H., Rusch, V.W., and Fong, Y. (2003). The replication-competent oncolytic herpes simplex mutant virus NV1066 is effective in the treatment of esophageal cancer. Surgery *134*, 357–364.

Taylor, T.J., McNamee, E.E., Day, C., and Knipe, D.M. (2003). Herpes simplex virus replication compartments can form by coalescence of smaller compartments. Virology *309*, 232–247.

Ter Horst, G.J. (2000). Transneuronal retrograde dual viral labelling of central autonomic circuitry: possibilities and pitfalls. Auton. Neurosci. *83*, 134–139.

Tomishima, M.J., and Enquist, L.W. (2001). A conserved alpha-herpesvirus protein necessary for axonal localization of viral membrane proteins. J. Cell Biol. *154*, 741–752.

Travers, J.B., and Rinaman, L. (2002). Identification of lingual motor control circuits using two strains of pseudorabies virus. Neuroscience *115*, 1139–1151.

Ueyama, T., Krout, K.E., Nguyen, X.V., Karpitskiy, V., Kollert, A., Mettenleiter, T.C., and Loewy, A.D. (1999). Suprachiasmatic nucleus: a central autonomic clock. Nat. Neurosci. *2*, 1051–1053.

Wang, J., Irnaten, M., Neff, R.A., Venkatesan, P., Evans, C., Loewy, A.D., Mettenleiter, T.C., and Mendelowitz, D. (2001). Synaptic and neurotransmitter activation of cardiac vagal neurons in the nucleus ambiguus. Ann. N.Y. Acad. Sci. *940*, 237–246.

Wang, L., Jackson, W.C., Steinbach, P.A., and Tsien, R.Y. (2004). Evolution of new nonantibody proteins via iterative somatic hypermutation. Proc. Natl. Acad. Sci. USA *101*, 16745–16749.

Willard, M. (2002). Rapid directional translocations in virus replication. J. Virol. *76*, 5220–5232.

Wong, R.J., Joe, J.K., Kim, S.H., Shah, J.P., Horsburgh, B., and Fong, Y. (2002). Oncolytic herpesvirus effectively treats murine squamous cell carcinoma and spreads by natural lymphatics to treat sites of lymphatic metastases. Hum. Gene Ther. *13*, 1213–1223.

Yarbrough, D., Wachter, R.M., Kallio, K., Matz, M.V., and Remington, S.J. (2001). Refined crystal structure of DsRed, a red fluorescent protein from coral, at 2.0-A resolution. Proc. Natl. Acad. Sci. USA 98, 462–467.

Zhang, J., Campbell, R.E., Ting, A.Y., and Tsien, R.Y. (2002). Creating new fluorescent probes for cell biology. Nat. Rev. Mol. Cell. Biol. 3, 906–918.

The Induction of Apoptosis by HSV-1

12

Christine M. Sanfilippo and John A. Blaho

Abstract

Herpes simplex virus type 1 (HSV-1) is a cytolytic alpha-herpesvirus which profoundly impacts its host cells. It is now widely recognized that consequences of HSV-1 infection include the induction of programmed cell death, also known as apoptosis, and the concomitant synthesis of proteins which act to block this process. While our current understanding of the precise mechanism by which apoptosis induction occurs remains elusive, several breakthrough studies have revealed much about how HSV-1 modulates the apoptotic process. In this chapter, we will review recent data focusing on the triggering of apoptosis by HSV-1, as well as evidence showing how this important human pathogen interferes with the fundamental cell death process.

Introduction

As a consequence of the cell's intrinsic ability to undergo apoptosis, several distinct viruses have developed mechanisms to block the premature death of infected cells (Kieff, 1998; Tschopp *et al.*, 1998; Koyama *et al.*, 2000). The most likely reason for this viral prevention of apoptosis is to prolong cell survival so that the production of the viral progeny can be maximized. Human herpes simplex virus type 1 (HSV-1) had been shown to interfere with the process of apoptosis in infected cells (Leopardi and Roizman, 1996; Koyama and Adachi, 1997; Koyama and Miwa, 1997; Leopardi *et al.*, 1997; Galvan and Roizman, 1998; Koyama *et al.*, 1998; Aubert and Blaho, 1999; Aubert *et al.*, 1999; Galvan *et al.*, 1999; Jerome *et al.*, 1999; Aubert *et al.*, 2001) and this phenomenon will be the basis of this chapter. The findings presented in this chapter emphasize the fact that modulation of apoptosis by HSV-1 during infection is a multicomponent phenomenon. The current data accumulated thus far in this field suggest that just as programmed cell death is an innate cell physiological process, it is likely that active prevention of apoptosis is a fundamental process common to almost all viruses (Koyama *et al.*, 2000).

Apoptotic signaling

The phenomenon of cell death has been studied extensively and it has now become possible to classify cell death into two discrete categories, apoptosis or necrosis, based on the morphological changes that occur (Wyllie *et al.*, 1980). Apoptosis is the purposeful process of cell suicide, characterized by condensation of the cell, maintenance of organelle integrity, and plasma membrane blebbing (Kerr *et al.*, 1972; Wyllie *et al.*, 1980). The dying cell then detaches into globular membrane-bound "apoptotic bodies" that become engulfed by neighboring cells, thus eliminating any inflammation from surrounding tissues. On the other hand, necrosis, the term currently used for nonapoptotic, accidental cell death, involves the swelling and subsequent rupture of cell and organelle membranes, resulting in the destruction of all organized structure and cellular contents. While both processes are involved in cell deletion, apoptosis is tightly controlled and involves signaling through a variety of death effector pathways.

Several morphological and biochemical changes are associated with, and have become the hallmarks of, apoptosis. The process of apoptosis involves nuclear alterations, including condensation of chromatin and convolution/fragmentation of the nuclear membrane, concurrent with cytosolic shrinkage and plasma membrane protuberances or "blebbing" (Kerr *et al.*, 1972; Wyllie *et al.*, 1980; Kerr *et al.*, 1994). Another such alteration is the apparent fragmentation of cellular DNA by endogenous endonucleases (Wyllie, 1980). Chromatin segments appear to be randomly excised from the genome, causing a "laddering" phenomenon when electrophoresed following staining with DNA intercalating dyes. This fragmentation can also be detected by the terminal deoxynucleotidyltransferase-mediated incorporation of labeled nucleotides at DNA breakage sites (TUNEL assay) (Gavrieli *et al.*, 1992). Other measurable changes involve the mitochondria; specifically (1) the swelling of the outer membrane (Vander Heiden *et al.*, 1997), (2) the loss of the inner membrane potential (Petit *et al.*, 1996), and (3) the subsequent release of apoptotic factors from the intermembrane space (Green and Reed, 1998; Green and Evan, 2002). The composition of the plasma membrane bilayer is also altered during the process of apoptosis. In particular, the lipid component phosphatidyl serine (PS), normally present on the inner leaflet, "flip-flops" to the outer leaflet (Fadok and Henson, 1998). Apoptotic cells containing this surface-exposed PS can then be detected by staining with Annexin V (Koopman *et al.*, 1994; Vanags *et al.*, 1996). Additionally, several cytoskeletal constituents are redistributed in dying cells, contributing to the morphological changes that occur during the formation of apoptotic bodies (Kayalar *et al.*, 1996; Orth *et al.*, 1996; Takahashi *et al.*, 1996; Brown *et al.*, 1997; Kothakota *et al.*, 1997; Mashima *et al.*, 1997). These morphological and biochemical changes represent uniform criteria which distinguish apoptotic cells from those undergoing other forms of cell death (Ellis *et al.*, 1991; Hengartner and Horvitz, 1994; Vaux and Strasser, 1996).

Apoptotic signaling pathways involve a common machinery comprised of caspases (cysteinyl aspartases), which are synthesized as zymogens and are normally present in the cell in their inactive proenzyme form (Thornberry and Lazebnik, 1998). Full enzymatic activity is achieved upon cleavage at internal aspartate residues by either autocatalysis (as with initiator caspases) or caspase-mediated processing of downstream, executioner (effector) caspases (Shi, 2002). Processing of upstream, initiator caspases (e.g. caspase-8), culminates in the activation of effector caspases, such as caspase-3. The proteolytic processing of these executioner caspases then leads to the cleavage and subsequent inactivation of various vital cellular proteins. Known substrates of effector caspases include (1) structural proteins (e.g. actin, gelsolin, and nuclear lamin), (2) protein kinases (such as the DNA-dependent protein kinase and the receptor-interacting protein RIP), (3) inhibitory proteins [e.g. the p53-inhibitor Mdm-2 and the DNAse-inhibitor DNA fragmentation factor 45 (DFF45/ICAD)], (4) enzymes including caspases and the DNA repair enzyme poly(ADP-ribose) polymerase (PARP), which is cleaved from its 116 kDa form to an 85 kDa product, as well as other (5) proapoptotic proteins (Nicholson and Thornberry, 1997; Thornberry and Lazebnik,

1998; Saraste and Pulkki, 2000; Shi, 2002). Cleavage of these downstream molecules results in the morphological and biochemical changes observed during apoptosis.

Two major pathways of caspase activation induced by proapoptotic stimuli have been characterized in detail. One is initiated by the extracellular binding of a death ligand to its receptor on the cell surface, while the other pathway integrates intracellular apoptotic signals through a mitochondrial route. These signaling pathways are depicted in Fig. 12.1.

Receptor-mediated (extrinsic) apoptotic pathway

Members of the tumor necrosis factor (TNF) receptor family are involved in the transduction of extracellular death signals from the plasma membrane to downstream effectors within the cell (Ashkenazi and Dixit, 1998). TNF receptor family members can trigger a variety of processes, such as proliferation, survival, differentiation, as well as cell death, depending on the cell type and the reception of simultaneous signals (Nagata, 1997; Ashkenazi and Dixit, 1998; Wallach *et al.*, 1998). The death receptors, which include Fas/CD95 and TNF-R1, contain an extracellular ligand-binding region characterized by variable numbers of cysteine-rich repeats, and an intracellular region essential for apoptosis signaling (Smith *et al.*, 1994; Gruss and Dower, 1995; Nagata, 1997). The ligands for these receptors, FasL and TNFα respectively, induce trimerization and extensive cross-link-

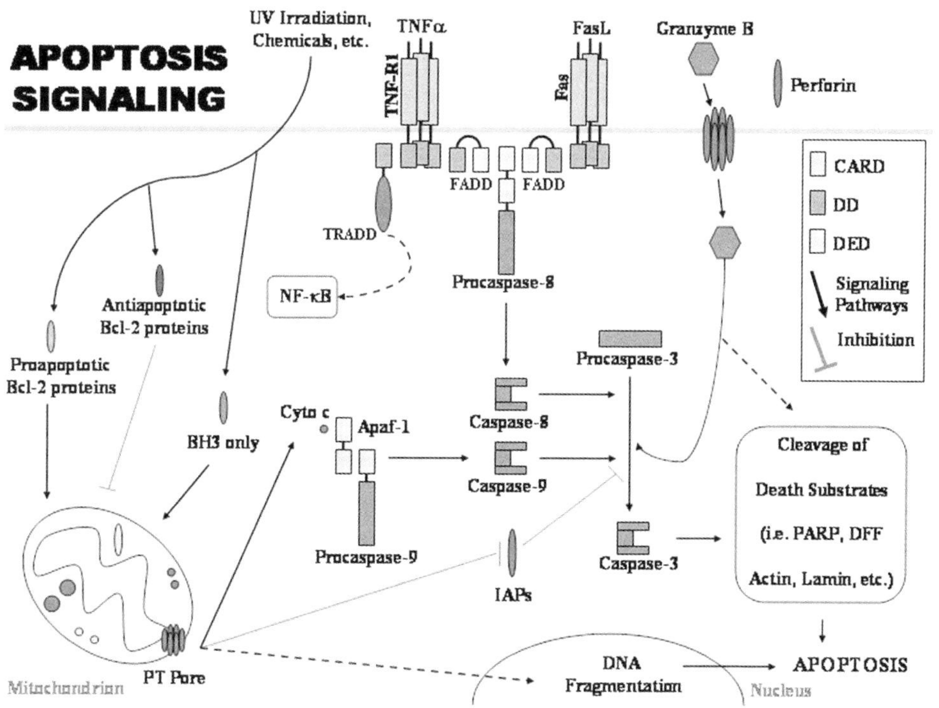

Figure 12.1 Major molecular pathways of apoptosis and the relationships between participating components. Ligand-binding to death receptors (TNF-R1 and Fas) and release of caspase-activating factors (cyto c) from the mitochondria commence death cascades via initiator caspase-8 and caspase-9, respectively, which activate downstream effector caspase-3. Caspase activation leads to the processing of various nuclear and cytoplasmic death substrates, culminating in apoptosis. Adapted from (Sanfilippo and Blaho, 2003).

ing of receptors upon binding (Smith *et al.*, 1994; Gruss and Dower, 1995).

The activated receptor complex then serves as a docking station for various adaptor proteins, which link initiator caspases with the upstream regulators of apoptosis (Ashkenazi and Dixit, 1998; Strasser *et al.*, 2000). These associations occur through protein–protein interactions between domains (Tartaglia *et al.*, 1993; Hofmann *et al.*, 1997), such as the death domain (DD), the death effector domain (DED), and the caspase recruitment domain (CARD). The cytoplasmic regions of Fas and TNF-R1, as well as the adaptor proteins Fas-associating death domain protein (FADD) and TNF-R1-associated death domain protein (TRADD), contain and interact via their DDs (Boldin *et al.*, 1995; Chinnaiyan *et al.*, 1995; Hsu *et al.*, 1995). Adaptor proteins having DEDs then recruit the initiator caspase containing tandem repeats of DEDs in their prodomains, and the cascade culminating in the activation of effector caspases (i.e. caspase-3) ensues.

TNF receptor family members can also stimulate the transcriptional regulator NF-κB (Bharti and Aggarwal, 2002). NF-κB is induced by a variety of stimuli, including TNFα, and plays a critical role in cell survival, cell adhesion, cell growth, inflammation, and differentiation (Collins *et al.*, 1995; Chen *et al.*, 2001). The active NF-κB complex is usually a heterodimer, comprised of members of the Rel family of proteins, such as p50, p52, Rel A(p65), Rel B, c-Rel, and v-Rel (Thanos and Maniatis, 1995; Baeuerle and Baltimore, 1996). These proteins contain an N-terminal amino acid sequence responsible for its dimerization, nuclear translocation, DNA binding properties, and association with other transcription factors. In most resting cells, inactive NF-κB is found in a complex with the inhibitory IκB proteins, which block its nuclear localization sequence (NLS) and sequester NF-κB in the cytoplasm (Baldwin, 1996). In response to various stimuli, NF-κB is released from its inhibitory restraints by IκB kinase (IKK) complex-mediated phosphorylation of IκB, which leads to IκB's subsequent ubiquitination and proteasome degradation (Karin, 1999). With its NLS exposed, NF-κB translocates to the nucleus and stimulates the transcription of target genes (Pahl, 1999). Since TNFα ligand binding thus induces FADD-mediated caspase-8 activation as well as cell survival promotion through NF-κB activation, apoptosis is enhanced under conditions in which transcribed NF-κB-stimulated genes are not synthesized.

Mitochondrial (intrinsic) apoptotic pathway

In addition to the death receptor-mediated pathway of cell death, apoptotic signals can also be transmitted through the mitochondria. Upon the reception of an apoptotic signal, such as growth factor withdrawal, ultraviolet (UV) irradiation, or exposure to toxic chemicals, a large nonselective conductance channel composed of both inner and outer membrane proteins is opened (Green and Reed, 1998). This mitochondrial permeability transition (PT) pore allows ions of the matrix and intermembrane space to equilibrate and causes the mitochondrial inner transmembrane potential ($\Delta\Psi m$) to collapse by dissipating the H+ gradient (Petit *et al.*, 1996; Qian *et al.*, 1997). Opening of the PT pore also induces volume expansion of the matrix, eventually resulting in the rupture of the outer membrane and the release of caspase activators, such as cytochrome c (cyto c) and other caspase-independent factors (Green and Reed, 1998; Green and Evan, 2002). Cyto c, once in the cytosol, binds to and activates the apoptotic protease activating factor-1 (Apaf-1), inducing its oligomerization (Zou *et al.*, 1997). Apaf-1 then associates with the initiator caspase 9 in an interaction mediated by their shared CARD domains, resulting in caspase-9, followed by caspase-3, activation. The active cyto c/Apaf-1/caspase-9 complex has been nicknamed the "apoptosome" (Li *et al.*, 1997). Downstream effector caspase activation can be thwarted by a family of endogenous caspase inhibitors called the inhibitor of apoptosis proteins (IAPs), which bind to and inhibit active caspase-9, as well as target active caspases for degradation (Suzuki *et al.*, 2001).

Members of the Bcl-2 family of proteins have proven instrumental in regulating the mitochondrial apoptotic pathway (Adams and Cory, 1998; Chao and Korsmeyer, 1998; Green and Evan, 2002). All members possess at least one of four conserved motifs known as Bcl-2 homology domains (BH1 to BH4). They are classified into one of three categories: antiapoptotic, proapoptotic "BH-123," and proapoptotic "BH3 only" proteins. Antiapoptotic members include Bcl-2 and Bcl-x_L, while the proapoptotic BH-123 proteins, which share three of the BH domains with their antiapoptotic counterparts, include Bax and Bcl-x_S (a splice variant of the *bcl-x* gene). The most potent proapoptotic proteins contain only a BH3 domain, and include Bad and Bik. The mechanisms by which the Bcl-2 family members exert their activities are quite varied. Some, such as Bax, have pore-forming abilities and can insert into the outer mitochondrial membrane to allow the passage of caspase-activating factors into the cytosol (Martinou and Green, 2001). Others, like Bcl-2 and Bcl-x_L, can bind to Apaf-1 and inhibit its association with procaspase-9, while proapoptotic relatives, such as Bik, may serve to free Apaf-1 from these inhibitory restraints (Chinnaiyan *et al.*, 1997; Chaudhary *et al.*, 1998; Hu *et al.*, 1998; Pan *et al.*, 1998). Pro- and antiapoptotic family members can heterodimerize via interaction between their various BH domains and seemingly balance one another's function (Oltvai *et al.*, 1993). Whatever their mechanism of action, this diverse family somehow integrates competing signals and, based on the final outcome, determines the fate of the cell.

Accomplices of apoptosis

Several defined subsidiary pathways, including signaling through granzymes and p53, can act as "accomplices" of apoptosis, representing other mechanisms by which cells can execute the apoptotic machinery. Granule enzymes, or granzymes, are a family of serine proteases expressed exclusively by specialized cells of the immune system, such as cytotoxic T lymphocytes (CTLs) and natural killer (NK) cells (Jenne and Tschopp, 1988; Trapani, 2001). Granzymes are synthesized as zymogens, yet are processed and stored in cytolytic granules as active enzymes in preparation for use on a potentially harmful or virally infected cell. CTLs release the contents of these granules, which also contain perforin, upon contacting a target cell. Perforin is capable of forming transmembrane pores in a process requiring calcium and allows the passage of granzymes into the cytoplasm of the cell, where they proceed to induce cell death. Granzyme B is the most powerful proapoptotic member of the granzyme family, due to its ability to cleave substrates at Asp residues. It can activate several procaspases (e.g. procaspase-3) directly, as well as process downstream death substrates, such as ICAD. Additionally, it has been shown that granzyme B can truncate the proapoptotic BH3-only protein Bid, resulting in mitochondrial disruption in a caspase-independent manner (Sutton *et al.*, 2000). Thus, granzyme B serves as a potent inducer of apoptosis, utilizing the pre-existing cell death machinery to cause destruction of the cell.

DNA-damaging agents, which can be environmental or endogenous in nature, cause chromosomal aberrations that require correction before cell cycle progression ensues. Checkpoint controls are thus present at key stages of the cell cycle in order to assess the severity of the damage and to determine the appropriate cellular response, whether it be repair, growth arrest, or apoptosis. The tetrameric transcription factor and tumor suppressor p53 plays a crucial role in reacting to DNA damage in higher eukaryotes (Evan and Littlewood, 1998; Norbury and Hickson, 2001). The p53 protein is normally maintained at low levels in resting cells through association with its negative modulator Mdm-2, which acts to destabilize p53 by targeting it for degradation via the ubiquitin-mediated proteasome pathway (Haupt *et al.*, 1997). DNA-damage-induced serine phosphorylation of either p53 or Mdm-2 prevents their interaction, and contributes to the stability and activation of p53. These complex post-translational modifications (Caspari, 2000) regulate the sequence-specific DNA-binding activity p53, and allow for transcriptional stimulation of proapoptotic p53 target genes, which include *bax* and *Fas* (Miyashita *et al.*, 1994; Zhan *et al.*, 1994; Muller *et al.*, 1998).

Productive HSV-1 infection

The cell's innate ability to self destruct by the process of apoptosis serves as an important mechanism of host cell defense against viral infections, including HSV-1. HSV-1, the prototypical alpha-herpesvirus, lytically infects epithelial cells, establishes latency in neuronal cells, and reactivates from these sites as a result of physical and emotional stresses (Stevens and Cook, 1971). During productive lytic infection in cultured cells, regulation of HSV-1 replication occurs mainly at the transcriptional level and involves the coordination of three phases of gene expression which occur in cascade fashion (Honess and Roizman, 1974; Honess and Roizman, 1975). The first viral genes expressed during infection, termed the immediate early (IE) or alpha (α) genes, are highly stimulated by the α-trans-inducing function of the incoming tegument protein VP16 (Post *et al.*, 1981; Campbell *et al.*, 1984; Pellett *et al.*, 1985), and are transcribed in the absence of de novo viral protein synthesis (Batterson and Roizman, 1983). The resultant α gene products, which include infected cell proteins (ICPs) 4, 22, and 27, are responsible for regulating viral protein expression during subsequent phases of the infection cycle (Roizman, 2001). The genes transcribed next are the early (E) or beta (β) genes and encode viral proteins principally involved in DNA synthesis, such as thymidine kinase (tk). The last phase of transcription involves expression of the late (L) or gamma (γ) genes, which mainly encode structural virion components, such as VP16 and gC. During productive replication, LAT (or, lytic LAT) is transcribed as a late gene (Aubert *et al.*, 2001). HSV-1 generally completes its replication cycle within 12–18 hours of infection.

HSV-1-infected cells undergo major structural and biochemical alterations (Roizman, 1974) collectively termed cytopathic effect (CPE), evident almost from the beginning of the reproductive cycle and becoming more severe as infection progresses. Manifestations of HSV-1-induced CPE include (1) nucleolar enlargement and disaggregation, (2) chromatin margination and/or damage, (3) cytoskeletal destabilizations and the loss of matrix binding proteins causing the rounding up of cells, (4) duplication/folding of organelle membranes, (5) membrane modifications due to the insertion of viral glycoproteins, (6) formation of fused cells (polykaryocytosis), and (7) a decrease in cellular macromolecular synthesis (Hampar, 1961; Roizman, 1962; Roizman, 1964; Shipkey *et al.*, 1967; Schwartz and Roizman, 1969; Heeg *et al.*, 1986; Avitabile *et al.*, 1995; Roizman, 2001). Such alterations ultimately lead to the destruction of cells in culture, presumably through a necrotic mechanism involving cell lysis. However, recent data have demonstrated that one aspect of cell death due to HSV-1 infection is apoptosis.

A breakthrough study performed in the Koyama laboratory in 1997 cracked the field of HSV-1-induced apoptosis wide open. Data from this report (Koyama and Adachi, 1997) demonstrated that wild type HSV-1 could, in fact, induce the characteristic apoptotic morphological changes (i.e. cell shrinkage and membrane blebbing) and endonucleosomal DNA cleavages when infections were performed under conditions in which *de novo* viral protein synthesis is prevented, as by addition of the translation inhibitor cycloheximide (CHX). This finding was significant because it implied that HSV-1 induces apoptosis early in infection and infected cell polypeptides are produced which specifically block apoptosis. Additional studies (Leopardi and Roizman, 1996; Leopardi *et al.*, 1997; Aubert and Blaho, 1999; Aubert *et al.*, 1999; Aubert *et al.*, 2001) demonstrated that infection by replication-defective HSV-1 viruses that do not make late proteins (i.e. the ICP27-deletion virus Δ27, and the ICP4/U_S3-deletion virus *d*120) results in apoptosis in infected human cells. Infected cell proteins synthesized between 3 and 6 hpi ("the apoptosis prevention window") were found to be responsible for blocking apoptosis (Aubert and Blaho, 1999; Aubert *et al.*, 2001). More comprehensive discussions detailing the specific experimental history of HSV-1-induced apoptosis can be found elsewhere (Aubert and Blaho, 2001; Goodkin *et al.*, 2003a).

Induction of apoptosis during HSV-1 infection

While the process of apoptosis during HSV-1 infection had been clearly demonstrated (reviewed in Aubert and Blaho, 2001; Goodkin *et al.*, 2003a), the specific mechanism(s) of its induction remained elusive. What was known was that the triggering event occurs early and *de novo* protein synthesis is not required to induce the process (Koyama and Adachi, 1997; Aubert and Blaho, 1999; Aubert *et al.*, 1999). This narrowed the possible induction events to those depicted in Fig. 12.2. These potential triggering mechanisms include virus binding, membrane fusion/de-envelopment, capsid translocation, viral DNA injection/interactions with the nuclear pore complex, and α/IE gene expression/RNA metabolic events associated with viral transcription. A key preliminary report (Aubert *et al.*, 1999) showed that no apoptosis occurs upon infection of human epithelial HEp-2 cells with UV-inactivated HSV-1, suggesting that virus binding and entry are not required for the induction process. In that study, the virion protein VP16 but not the viral IE protein ICP22 was detected in the nuclei of infected cells at 2 hpi. The absence of ICP22 confirmed that following UV treatment of HSV-1, no viral IE gene translation occurred. However, during later use of this technique, the UV treatment time originally used was found to be extremely excessive and possibly generated unwanted damage. Additionally, while VP16 derived from UV-irradiated HSV-1 retained its ability to translocate to the nucleus, it was unknown whether it remained functional for the transactivation of viral IE genes. Thus, it was necessary to rigorously reevaluate the data generated on the apoptotic induction event.

In a further attempt to discern the viral component(s) responsible for the induction of apoptosis by HSV-1, an important subsequent study revealed much of our current understanding of the induction process (Sanfilippo *et al.*, 2004). In this report, two key experiments were performed which identified the viral process responsible for the induction of apoptosis and the respective data is presented in Fig. 12.3. The first experiment involved the development of an optimized UV treatment system; exposing wild type HSV-1 (KOS) to UV light for 4 min. Under these

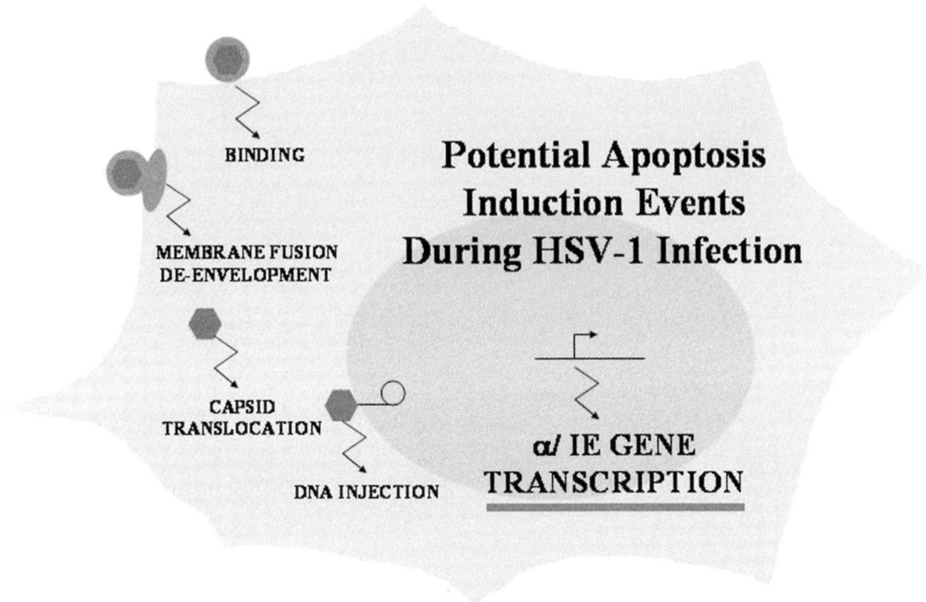

Figure 12.2 Potential apoptotic induction events during HSV-1 infection. An early viral event which precedes the translation of HSV-1 IE proteins triggers the induction of apoptosis (Koyama and Adachi, 1997; Aubert and Blaho, 1999). Possible candidates for this event include virus binding, membrane fusion/de-envelopment, capsid translocation, viral DNA injection into nuclei, and the transcription of α/IE genes. Current data indicates that an event associated with the expression of HSV-1 α/IE genes is responsible for the induction of apoptosis during infection (Sanfilippo *et al.*, 2004).

conditions, incoming virion VP16 remained functional for transcriptional transactivation and virions were deemed replication defective as evidenced by the absence of viral protein synthesis and a reduction [< 10 plaque forming units (PFU) per ml] in viral titers. In contrast to wild type virus infections which, when carried out under CHX treatment conditions, resulted in the cleavage of the death factors caspase-3 and PARP, indicating apoptosis, UV-inactivated virus generated by the optimized system had clearly lost its ability to trigger the process (Fig. 12.3A, compare lane 4 with 6). These data demonstrated that the viral processes of virus binding and de-envelopment, capsid translocation, and viral DNA injection are not responsible for inducing apoptosis.

At this point, it was suggested that viral IE gene expression was required to trigger apoptosis. The second important experiment in this study (Sanfilippo *et al.*, 2004) utilized the HSV-1 recombinant virus *d*109, a GFP-expressing virus which contains deletions in all five α genes (Samaniego *et al.*, 1998), to address this issue. While cells infected with the ICP27-deletion virus Δ27 displayed classic features of apoptosis, including death factor processing and apoptotic morphologies (Aubert and Blaho, 1999), *d*109-infected cells exhibited no PARP or caspase-3 cleavage and little to no cell shrinkage, membrane blebbing, or condensed chromatin when visualized by phase-contrast microscopy and fluorescence microscopy (Fig. 12.3B, compare lane 3 with 4, and Fig. 12.3C). These data in-

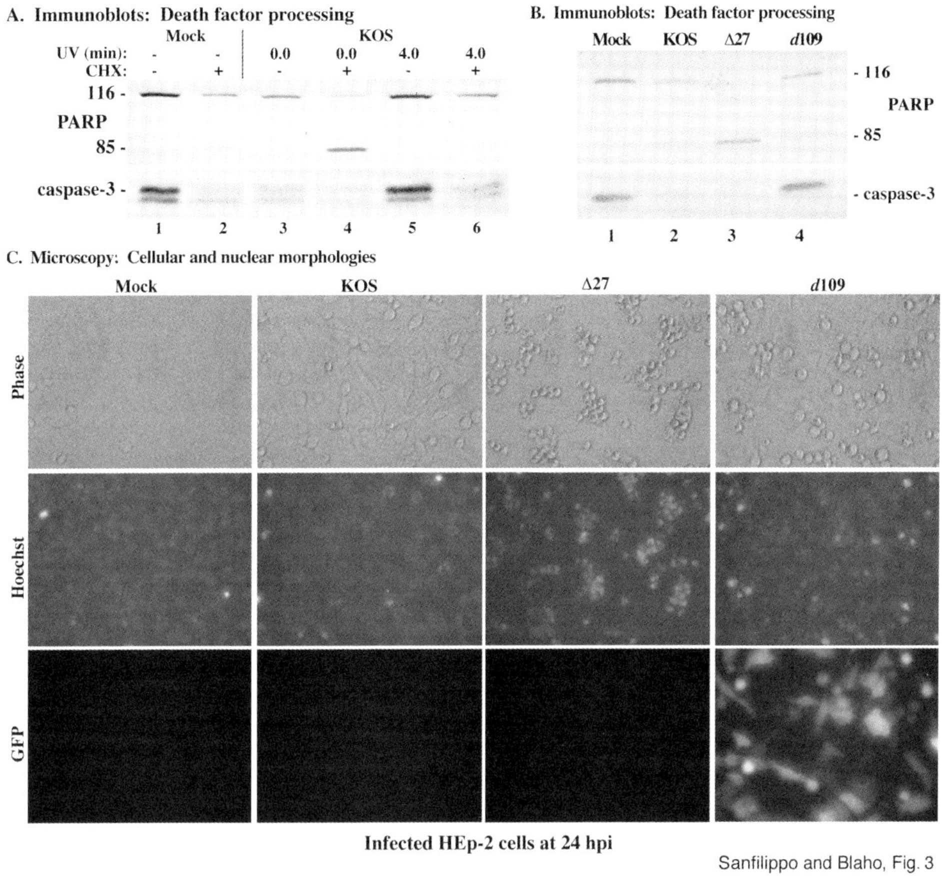

Figure 12.3 HSV-1 α/IE gene expression is required for the induction of apoptosis in human cells. (A) Immune reactivities of the cellular death factors PARP and caspase-3 at 24 hpi in cells infected with untreated (0.0 min UV exposure) and UV-treated (4.0 min UV exposure) HSV-1(KOS) in the absence (–) or presence (+) of CHX (Sanfilippo *et al.*, 2004). (B and C) Immune reactivities of the cellular death factors PARP and caspase-3 (B) and corresponding phase-contrast (Phase) and fluorescence (Hoechst and GFP) microscopic images (C) of HSV-1(KOS), Δ27, and *d*109-infected cells at 24 hpi (Sanfilippo *et al.*, 2004). Infected cell nuclei were visualized by staining with Hoechst 33258 (Aubert and Blaho, 1999).

dicate that *d*109 does not induce apoptosis in HEp-2 cells due to the absence of α gene expression. Together, these findings form the basis of the proposal that viral α gene expression is required for apoptosis induction by HSV-1. An interesting supportive piece of evidence for this theory is the finding that virions devoid of glycoprotein D can be taken up by endocytosis when added at extremely high levels (Zhou *et al.*, 2000). This unorthodox method of placing virions in cells actually leads to the induction of apoptosis (Zhou *et al.*, 2000; Zhou and Roizman, 2001; Zhou *et al.*, 2003; Zhou and Roizman, 2002b). This finding again shows that in the absence of virus binding to receptor and normal entry, viral gene expression can still trigger apoptosis.

One interesting finding from this experiment (Fig. 12.3B and C) was the manifestation of an altered cellular morphology in *d*109-infected HEp-2 cells. One possible explanation for this observation may be due to the establishment of a host antiviral state upon HSV-1 infection. Previous studies have shown that HSV-1 induces a host antiviral response, and that inhibition of this response is dependent on the production of ICP0 (Mossman *et al.*, 2000; Nicholl *et al.*, 2000; Mossman *et al.*, 2001; Eidson *et al.*, 2002). Details of this phenomenon are being reviewed in the chapters of this series dealing with ICP0 and interferon-induced barriers to viral replication. Because ICP0 is not synthesized in *d*109-infected cells, the activation of interferon-stimulated genes may bring about the morphology changes observed in these cells. The finding that cells neighboring the GFP-expressing, *d*109-infected cells, rather than the *d*109-infected cells themselves (Fig. 12.3C, compare *d*109 phase and GFP), exhibit this altered morphology may suggest a type of paracrine effect in which infected cells "alert" noninfected cells to the presence of virus. However, the reason why this effect is not seen in cells infected with UV-irradiated HSV-1 (Sanfilippo *et al.*, 2004), which has been shown to be capable of triggering this Antiviral response (Mossman *et al.*, 2001), remains elusive at this time. More importantly, the altered morphology observed in *d*109-infected cells, which may well be due to the establishment of an antiviral state, is clearly not a result of HSV-1-induced apoptosis. Further experimentation on the cellular biochemical and morphological alterations caused by the HSV-1-induced antiviral activity may offer some clarification on this issue.

We have now systematically evaluated the contribution of each α/IE gene to stimulate apoptosis (Sanfilippo and Blaho, 2006). Using a series of viruses individually deleted for α27, α4, and α22, we determined that these genes are not required for apoptosis induction but rather their products play roles in its prevention, likely through regulatory effects. Sole expression of α0 acted as an apoptoxic inducer that was necessary and sufficient to trigger the cell death cascade. Importantly, results using a recominant virus which contains a stop codon in α0 showed that it is not the ICP0 protein which was acting as the apoptoxic inducer. Based on these findings, we propose that ICP0 gene expression acts as an initial inducer of apoptosis during HSV-1 infection. Of course, later proteins likely also act to facilitate optimal apoptosis progression during infection. Expression of apoptoxic viral genes represents a unique mechanism through which human pathogens may modulate interactions with their host cells. Thus, the repetoire of tools that pathogenic human viruses utilize to modulate the apoptoxic cell death process to their advantage now includes defined viral ribonucleic acid molecules.

It remains unclear whether α0 gene mRNA itself is sufficient for the induction event. Other potential viral stimuli beyond the actual mRNA molecules themselves might involve any one of the steps associated with the expression of the α0 genes. These possible induction mechanisms include splicing of the α0 gene transcripts, export of the α0 mRNAs from the nucleus, initial interactions of these RNAs with the cellular translational machinery, and the general stability of transcripts in infected cells. One must even consider the global effects of viral gene expression on cell nuclei, such as the impact of chromatin margination (Aubert and Blaho, 1999). Development of appropriate biochemical and molecular genetic systems are necessary to elucidate the exact molecular mechanism(s) responsible for the triggering of apoptosis in HSV-1-infected HEp-2 cells.

Prevention of apoptosis during HSV-1 infection

The upset of normal cell physiology can trigger apoptosis and it follows that metabolic inhibitors of vital cell processes can act as potent apoptotic inducers. In fact, several studies have investigated HSV-1 infection following treatment of cells with such metabolic inhibitors and have demonstrated that HSV-1 is able to block chemically induced apoptosis by CHX, the protein kinase inhibitor staurosporine, the lipid-derived second messenger ceramide, the osmotic shock inducer sorbitol, and the topoisomerase inhibitor etoposide, albeit in a cell type-specific fashion (Galvan and Roizman, 1998; Aubert *et al.*, 1999; Perng, 2000). It has also been shown that HSV-1 can prevent apoptosis induced by granzyme B, UV irradiation, and hyperthermia (Leopardi and Roizman, 1996; Galvan and Roizman, 1998; Jerome *et al.*, 2001). Similarly, the triggering of apoptosis by TNFα or Fas antibody treatment can also be inhibited by HSV-1 infection (Galvan and Roizman, 1998; Jerome *et al.*, 2001; Goodkin *et al.*, 2003b).

These breakthrough discoveries prompted further studies attempting to discern the viral proteins which participate in HSV-1 apoptosis prevention. Up to this point, eight HSV-1 gene products have been implicated as candidates for inhibitors of cell death. These factors can be separated into two groups. The first are those that are required for viral replication in cell culture. The second group contains viral proteins that are not essential for productive viral replication in cultured cells and can thus be defined as accessory factors. The essential regulatory proteins ICP4 and ICP27 have been shown to possess anti-apoptotic functions (Leopardi and Roizman, 1996; Aubert and Blaho, 1999). Thus, they are examples of viral proteins absolutely required for apoptosis prevention. However, it is still unknown whether it is these proteins themselves or the proteins whose synthesis they regulate that are actually responsible for the inhibition of apoptosis. The latter explanation seems most likely, since apoptosis prevention has been shown to correlate with the accumulation of E and leaky-L viral proteins (Aubert *et al.*, 2001).

Other suspected viral gene products with proposed roles in inhibition of cell death are dispensable for growth in cell culture. It follows that while they may be necessary, they are not sufficient to optimally block apoptosis. The nonessential IE protein ICP22 and the latency-associated transcript (LAT) are examples of such accessories which have been reported to contribute to apoptosis prevention, but their respective roles are not well understood (Aubert *et al.*, 1999; Perng, 2000; Inman *et al.*, 2001; Thompson and Sawtell, 2001; Ahmed *et al.*, 2002; Henderson *et al.*, 2002; Jin *et al.*, 2003; Peng *et al.*, 2003). For a more detailed perspective on LAT and its role in virus-host interactions, please see the HSV-1 latency chapter in this series.

The viral glycoproteins gJ (U_S5) and gD (U_S6) have also been reported to be capable of inhibiting apoptosis (Jerome *et al.*, 1999; Zhou *et al.*, 2000; Jerome *et al.*, 2001; Zhou and Roizman, 2001; Zhou and Roizman, 2002b; Zhou *et al.*, 2003). While little is known about the mechanisms of action of these glycoproteins, gD has been proposed to block the influx of lysosomal enzymes through interaction with the cation-independent mannose 6-phosphate receptor (Zhou and Roizman, 2002a). A recent report has also suggested that the U_L14 gene product, a minor component of the virion tegument which is thought to play a role in viral egress (Cunningham *et al.*, 2000), may also have antiapoptotic activity (Yamauchi *et al.*, 2003).

The involvement of the E protein kinase U_S3 in apoptosis prevention has been more thoroughly studied (Leopardi *et al.*, 1997; Galvan and Roizman, 1998; Aubert *et al.*, 1999; Jerome *et al.*, 1999; Munger *et al.*, 2001; Munger and Roizman, 2001; Hagglund *et al.*, 2002; Benetti *et al.*, 2003; Cartier *et al.*, 2003). While the precise mechanism of its action remains unclear, several reports suggest that U_S3 may interfere with the function of proapoptotic Bcl-2 family members, such as Bad, Bid, and Bax (Munger and Roizman, 2001; Benetti *et al.*, 2003; Cartier *et al.*, 2003; Ogg *et al.*, 2004). Also a topic of debate is whether U_S3 exerts its antiapoptotic effects at a premitochondrial stage (Munger *et al.*, 2001) or downstream of the mitochondria (Ogg *et al.*, 2004). Additional studies have suggested that U_S3 and gJ cooperatively act to prevent apoptotic cell death (Jerome *et al.*, 1999). Because the U_S3 gene is dispensable for growth in cell culture (Purves *et al.*, 1987), it is unlikely that U_S3 alone is sufficient for optimal apoptosis inhibition (Aubert *et al.*, 1999). Of course, each viral protein plays a necessary and sufficient role in very specific situations *in vivo* and thus no gene product is truly nonessential. Together, these findings indicate that more than one protein may likely be required for optimal prevention of apoptosis.

Role of the cell during HSV-1-induced apoptosis

The suggestion that HSV-1 is able to functionally interact with the cellular apoptotic machinery, as with U_S3 and proapoptotic Bcl-2 family members (Munger and Roizman, 2001; Benetti *et al.*, 2003; Cartier *et al.*, 2003; Ogg *et al.*, 2004), is an indication of the types of mechanisms by which HSV-1 likely modulates infection-induced apoptosis. It also seems logical that HSV-1 would recruit cellular antiapoptotic factors, as well, to aid in the prevention of apoptosis.

NF-κB

NF-κB is one of the most important antiapoptotic factors found in cells (reviewed in (Barkett and Gilmore, 1999). It is activated by various stimuli, such as the cytokine TNFα and is known to be involved in antiviral responses (Kohase *et al.*, 1986; Mestan *et al.*, 1986; Wong and Goeddel, 1986). One of the receptors for HSV-1 is the herpesvirus entry mediator (HVEM) HveA, which shares homology with TNF receptor family members (Montgomery *et al.*, 1996; Terry-Allison *et al.*, 1998). This brings up the question of whether signaling of the extrinsic apoptotic pathway is induced during HSV-1 infection. While evidence of HSV-1 induced signaling through the receptor is unclear (Marsters *et al.*, 1997; Qie *et al.*, 1999), other herpesviruses, such as EBV and CMV, are known to stimulate such signaling (Mosialos *et al.*, 1995; Yurochko *et al.*, 1997; Cahir-McFarland *et al.*, 2000). However, it has been shown that during HSV-1 infection, NF-κB becomes activated and translocates to the nucleus, contributing to increases in viral yield and suggesting that NF-κB activation is necessary for efficient HSV-1 replication (Patel *et al.*, 1998; Amici *et al.*, 2001). The subsequent finding that NF-κB activation and nuclear translocation during the apoptosis prevention window correlates with the ability of HSV-1 to inhibit both apoptosis induced by TNFα plus CHX treatment, as well as infection-induced apoptosis indicates that NF-κB is necessary, at least in part, for apoptosis prevention in HSV-1-infected human cells (Goodkin *et al.*, 2003b). While a key role for NF-κB in apoptosis prevention during HSV-1 infection has recently been confirmed (Medici *et al.*, 2003), an additional study has suggested that the NF-κB activation observed during HSV-1 infection is dependent on PKR activation and is not necessarily involved in the inhibition of apoptosis (Taddeo, 2003; Taddeo *et al.*, 2004). Further studies investigating NF-κB's potential antiapoptotic activity during HSV-1 infection may help clarify this issue.

Oncoapoptosis

As stated previously, the transcription factor p53 becomes modified and activated in response to various types of stress, such as DNA damage, resulting in a proapoptotic response (Caspari, 2000; Norbury and Hickson, 2001). Because p53 is a major determinant of whether a cell will undergo apoptosis (Levine, 1997; Prives and Hall, 1999), it is thus not surprising that mutations in the p53 gene are found in greater than half of all human tumors (Hainaut *et al.*, 1998; Hainaut and Hollstein, 2000). Additionally, since cells whose p53 has been inactivated are associated with resistance to anticancer drugs (Lowe *et al.*, 1994; Weinstein *et al.*, 1997), new strategies aimed at killing human tumor cells are a hot topic of investigation. Recombinant HSV-1 strains that initiate but cannot complete their replication cycle due to a reduction in their expression of early and late genes have been defined as apoptotic viruses (Aubert *et al.*, 2001). A recent report (Aubert and Blaho, 2003) has evaluated the use of such replication-defective viruses as potential antioncogenic agents based on their ability to trigger apoptosis in transformed human cells and this ability was found to be dependent on the abundance and distribution of modified p53 protein. In this particular study, tumor cells, which possess high levels of markedly undermodified p53, were shown to be susceptible to HSV-1-induced apoptosis and this process was termed oncoapoptosis. In contrast, primary human and mouse fibroblasts that produce modified p53 were found to be resistant to killing by HSV-1 apoptotic viruses, but not to apoptosis induced by nonviral environmental factors (Aubert and Blaho, 2003). Thus, it appears that primary and tumor cell types react differently to HSV-1-induced apoptosis. Recent results also indicate that while transformed cells are susceptible to oncoapoptosis by HSV-1, immortalized cells require synthesis of one or more proteins which facilitate apoptosis induction by HSV-1 (Nguyen *et al.*, 2005). Importantly, the finding that an HSV-1 virus, which is unable to complete its replication cycle, initiates an abortive infection that specifically triggers tumor cells to undergo apoptosis has significant implications in the future development of novel virus-based anticancer therapies.

Summary and perspective

Our current understanding of HSV-1 apoptosis modulation is summarized in Fig. 12.4. The field of apoptosis during HSV-1 infection is now growing out of its infancy. Attention is turning towards the development of specific experimental models to define exactly how the virus modulates the cell death program. Determining the feature of viral infection that triggers apoptotic cell death will provide fundamental insights into human cell biology which might be translated into developing strategies for limiting uncontrolled cell proliferation. Identification of viral induced antiapoptotic factors may provide information that could be used in developing novel antiviral therapies. Finally, it would be beneficial to determine whether viral-induced antiapoptotic factors are capable of blocking cell death triggered by other (nonviral) environmental stimuli. It would then be possible to use them to limit certain proapoptotic "toxicities" associated with currently available therapeutics. The next several years should continue to be very fruitful in this rapidly maturing research area.

Acknowledgments

We thank Natalie F.W. Chirimuuta (University of the West of England, Bristol, UK) and Renzo C. Lombardozzi (Saint Francis Preparatory School, Fresh Meadows, NY, USA). for their assistance. We thank Marie L. Nguyen for providing Fig. 12.4. Studies reviewed above were supported in part by grants from the United States Public Health Service (AI 38873 and AI 48582 to J.A.B). C.M.S. is a predoctoral trainee and was supported in part by a United States Public Health Service Institutional Research Training Award (AI 07647).

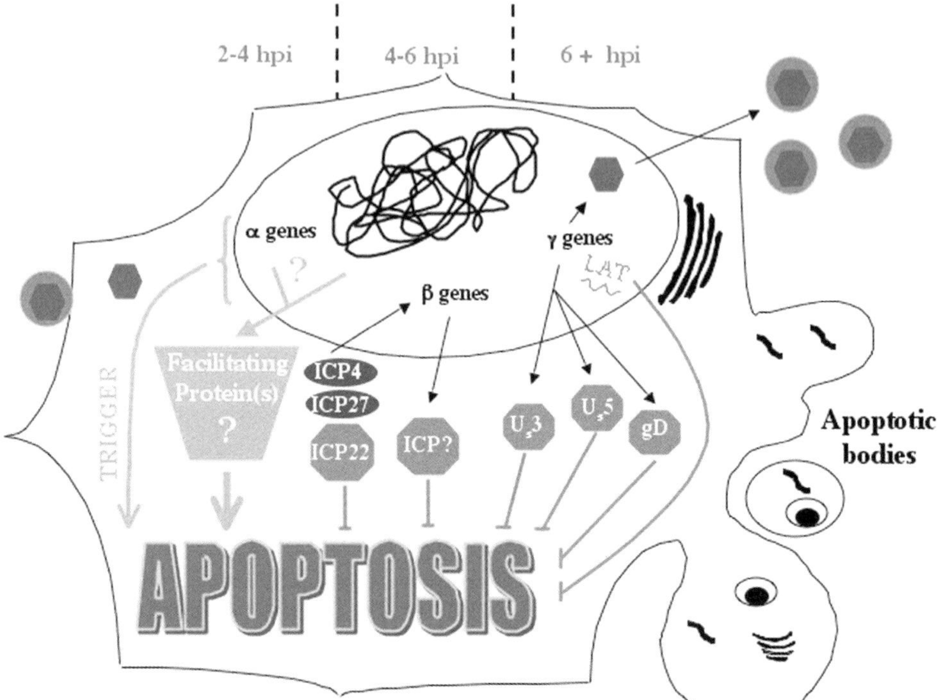

Figure 12.4 Model for apoptosis modulation during HSV-1 infection. An event associated with the expression of HSV-1 α/IE genes triggers the induction of apoptosis, and this induction is facilitated by the production of one or more proteins in a cell type-specific manner. Infected cell proteins are produced which prevent the process from killing the cells. The α regulatory ICP4 and ICP27 proteins are essential for precluding apoptosis, and the accumulation of viral β and γ gene products correlates with apoptosis prevention.

References

Adams, J.M., and Cory, S. (1998). The Bcl-2 protein family: arbiters of cell survival. Science *281*, 1322–1326.

Ahmed, M., Lock, M., Miller, C.G., and Fraser, N.W. (2002). Regions of the herpes simplex virus type 1 latency-associated transcript that protect cells from apoptosis *in vitro* and protect neuronal cells *in vivo*. J. Virol. *76*, 717–729.

Amici, C., Belardo, G., Rossi, A., and Santoro, M.G. (2001). Activation of I kappa b kinase by herpes simplex virus type 1. A novel target for anti-herpetic therapy. J. Biol. Chem. *276*, 28759–28766.

Ashkenazi, A., and Dixit, V.M. (1998). Death receptors: signaling and modulation. Science *281*, 1305–1308.

Aubert, M., and Blaho, J.A. (1999). The herpes simplex virus type 1 regulatory protein ICP27 is required for the prevention of apoptosis in infected human cells. J. Virol. *73*, 2803–2813.

Aubert, M., and Blaho, J.A. (2001). Modulation of apoptosis during herpes simplex virus infection in human cells. Microbes Infect. *3*, 859–866.

Aubert, M., and Blaho, J.A. (2003). Viral oncoapoptosis of human tumor cells. Gene Ther. *10*, 1437–1445.

Aubert, M., O'Toole, J., and Blaho, J.A. (1999). Induction and prevention of apoptosis in human HEp-2 cells by herpes simplex virus type 1. J. Virol. *73*, 10359–10370.

Aubert, M., Rice, S.A., and Blaho, J.A. (2001). Accumulation of herpes simplex virus type 1 early and leaky-late proteins correlates with apoptosis prevention in infected human HEp-2 cells. J. Virol. *75*, 1013–1030.

Avitabile, E., Di Gaeta, S., Torrisi, M.R., Ward, P.L., Roizman, B., and Campadelli-Fiume, G. (1995). Redistribution of microtubules and Golgi apparatus in herpes simplex virus-infected cells and their role in viral exocytosis. J. Virol. *69*, 7472–7482.

Baeuerle, P.A., and Baltimore, D. (1996). NF-kappa B: ten years after. Cell *87*, 13–20.

Baldwin, A.S., Jr. (1996). The NF-kappa B and I kappa B proteins: new discoveries and insights. Annu. Rev. Immunol. *14*, 649–683.

Barkett, M., and Gilmore, T.D. (1999). Control of apoptosis by Rel/NF-kappaB transcription factors. Oncogene *18*, 6910–6924.

Batterson, W., and Roizman, B. (1983). Characterization of the herpes simplex virion-associated factor responsible for the induction of alpha genes. J. Virol. *46*, 371–377.

Benetti, L., Munger, J., and Roizman, B. (2003). The herpes simplex virus 1 US3 protein kinase blocks caspase-dependent double cleavage and activation of the proapoptotic protein BAD. J. Virol. *77*, 6567–6573.

Bharti, A.C., and Aggarwal, B.B. (2002). Nuclear factor-kappa B and cancer: its role in prevention and therapy. Biochem. Pharmacol. *64*, 883–888.

Boldin, M.P., Mett, I.L., Varfolomeev, E.E., Chumakov, I., Shemer-Avni, Y., Camonis, J.H., and Wallach, D. (1995). Self-association of the "death domains" of the p55 tumor necrosis factor (TNF) receptor and Fas/APO1 prompts signaling for TNF and Fas/APO1 effects. J. Biol. Chem. *270*, 387–391.

Brown, S.B., Bailey, K., and Savill, J. (1997). Actin is cleaved during constitutive apoptosis. Biochem. J. *323*, 233–237.

Cahir-McFarland, E.D., Davidson, D.M., Schauer, S.L., Duong, J., and Kieff, E. (2000). NF-kappa B inhibition causes spontaneous apoptosis in Epstein-Barr virus-transformed lymphoblastoid cells. Proc. Natl. Acad. Sci. USA *97*, 6055–6060.

Campbell, M.E., Palfreyman, J.W., and Preston, C.M. (1984). Identification of herpes simplex virus DNA sequences which encode a trans-acting polypeptide responsible for stimulation of immediate early transcription. J. Mol. Biol. *180*, 1–19.

Cartier, A., Komai, T., and Masucci, M.G. (2003). The Us3 protein kinase of herpes simplex virus 1 blocks apoptosis and induces phosporylation of the Bcl-2 family member Bad. Exp Cell Res *291*, 242–250.

Caspari, T. (2000). How to activate p53. Curr. Biol. *10*, R315–317.

Chao, D.T., and Korsmeyer, S.J. (1998). BCL-2 family: regulators of cell death. Annu. Rev. Immunol. *16*, 395–419.

Chaudhary, D., O'Rourke, K., Chinnaiyan, A.M., and Dixit, V.M. (1998). The death inhibitory molecules CED-9 and CED-4L use a common mechanism to inhibit the CED-3 death protease. J. Biol. Chem. *273*, 17708–17712.

Chen, F., Castranova, V., and Shi, X. (2001). New insights into the role of nuclear factor-kappaB in cell growth regulation. Am. J. Pathol. *159*, 387–397.

Chinnaiyan, A.M., O'Rourke, K., Tewari, M., and Dixit, V.M. (1995). FADD, a novel death domain-containing protein, interacts with the death domain of Fas and initiates apoptosis. Cell *81*, 505–512.

Chinnaiyan, A.M., O'Rourke, K., Lane, B.R., and Dixit, V.M. (1997). Interaction of CED-4 with CED-3 and CED-9: a molecular framework for cell death. Science *275*, 1122–1126.

Collins, T., Read, M.A., Neish, A.S., Whitley, M.Z., Thanos, D., and Maniatis, T. (1995). Transcriptional regulation of endothelial cell adhesion molecules: NF- kappa B and cytokine-inducible enhancers. FASEB J. *9*, 899–909.

Cunningham, C., Davison, A.J., MacLean, A.R., Taus, N.S., and Baines, J.D. (2000). Herpes simplex virus type 1 gene UL14: phenotype of a null mutant and identification of the encoded protein. J. Virol. *74*, 33–41.

Eidson, K.M., Hobbs, W.E., Manning, B.J., Carlson, P., and DeLuca, N.A. (2002). Expression of herpes simplex virus ICP0 inhibits the induction of interferon-stimulated genes by viral infection. J. Virol. *76*, 2180–2191.

Ellis, R.E., Yuan, J.Y., and Horvitz, H.R. (1991). Mechanisms and functions of cell death. Annu. Rev. Cell Biol. *7*, 663–698.

Evan, G., and Littlewood, T. (1998). A matter of life and cell death. Science *281*, 1317–1322.

Fadok, V.A., and Henson, P.M. (1998). Apoptosis: getting rid of the bodies. Curr. Biol. *8*, R693–695.

Galvan, V., and Roizman, B. (1998). Herpes simplex virus 1 induces and blocks apoptosis at multiple steps during infection and protects cells from exogenous inducers in a cell- type-dependent manner. Proc. Natl. Acad. Sci. USA *95*, 3931–3936.

Galvan, V., Brandimarti, R., and Roizman, B. (1999). Herpes simplex virus 1 blocks caspase-3-independent and caspase-dependent pathways to cell death. J. Virol. *73*, 3219–3226.

Gavrieli, Y., Sherman, Y., and Ben-Sasson, S.A. (1992). Identification of programmed cell death *in situ* via specific labeling of nuclear DNA fragmentation. J. Cell Biol. *119*, 493–501.

Goodkin, M.L., Morton, E.R., and Blaho, J.A. (2003a). Herpes simplex virus infection and apoptosis. Int. Rev. Immunol. *23*, 141–172.

Goodkin, M.L., Ting, A.T., and Blaho, J.A. (2003b). NF-kappaB is required for apoptosis prevention during herpes simplex virus type 1 infection. J. Virol. *77*, 7261–7280.

Green, D.R., and Evan, G.I. (2002). A matter of life and death. Cancer Cell *1*, 19–30.

Green, D.R., and Reed, J.C. (1998). Mitochondria and apoptosis. Science *281*, 1309–1312.

Gruss, H.J., and Dower, S.K. (1995). Tumor necrosis factor ligand superfamily: involvement in the

pathology of malignant lymphomas. Blood *85*, 3378–3404.
Hagglund, R., Munger, J., Poon, A.P., and Roizman, B. (2002). U(S)3 protein kinase of herpes simplex virus 1 blocks caspase 3 activation induced by the products of U(S)1.5 and U(L)13 genes and modulates expression of transduced U(S)1.5 open reading frame in a cell type-specific manner. J. Virol. *76*, 743–754.
Hainaut, P., and Hollstein, M. (2000). p53 and human cancer: the first ten thousand mutations. Adv. Cancer Res. *77*, 81–137.
Hainaut, P., Hernandez, T., Robinson, A., Rodriguez-Tome, P., Flores, T., Hollstein, M., Harris, C.C., and Montesano, R. (1998). IARC Database of p53 gene mutations in human tumors and cell lines: updated compilation, revised formats and new visualisation tools. Nucleic Acids Res. *26*, 205–213.
Hampar, B., and S. A. Elison (1961). Chromosomal abberations induced by an animal virus. Nature *192*, 145–147.
Haupt, Y., Maya, R., Kazaz, A., and Oren, M. (1997). Mdm2 promotes the rapid degradation of p53. Nature *387*, 296–299.
Heeg, U., Dienes, H.P., Muller, S., and Falke, D. (1986). Involvement of actin-containing microfilaments in HSV-induced cytopathology and the influence of inhibitors of glycosylation. Arch. Virol. *91*, 257–270.
Henderson, G., Peng, W., Jin, L., Perng, G.C., Nesburn, A.B., Wechsler, S.L., and Jones, C. (2002). Regulation of caspase 8- and caspase 9-induced apoptosis by the herpes simplex virus type 1 latency-associated transcript. J. Neurovirol. *8 Suppl 2*, 103–111.
Hengartner, M.O., and Horvitz, H.R. (1994). Programmed cell death in Caenorhabditis elegans. Curr. Opin. Genet. Dev. *4*, 581–586.
Hofmann, K., Bucher, P., and Tschopp, J. (1997). The CARD domain: a new apoptotic signalling motif. Trends Biochem. Sci. *22*, 155–156.
Honess, R.W., and Roizman, B. (1974). Regulation of herpesvirus macromolecular synthesis. I. Cascade regulation of the synthesis of three groups of viral proteins. J. Virol. *14*, 8–19.
Honess, R.W., and Roizman, B. (1975). Regulation of herpesvirus macromolecular synthesis: sequential transition of polypeptide synthesis requires functional viral polypeptides. Proc. Natl. Acad. Sci. USA *72*, 1276–1280.
Hsu, H., Xiong, J., and Goeddel, D.V. (1995). The TNF receptor 1-associated protein TRADD signals cell death and NF-kappa B activation. Cell *81*, 495–504.
Hu, Y., Benedict, M.A., Wu, D., Inohara, N., and Nunez, G. (1998). Bcl-XL interacts with Apaf-1 and inhibits Apaf-1-dependent caspase-9 activation. Proc. Natl. Acad. Sci. USA *95*, 4386–4391.
Inman, M., Perng, G.C., Henderson, G., Ghiasi, H., Nesburn, A.B., Wechsler, S.L., and Jones, C. (2001). Region of herpes simplex virus type 1 latency-associated transcript sufficient for wild-type spontaneous reactivation promotes cell survival in tissue culture. J. Virol. *75*, 3636–3646.
Jenne, D.E., and Tschopp, J. (1988). Granzymes, a family of serine proteases released from granules of cytolytic T lymphocytes upon T cell receptor stimulation. Immunol. Rev. *103*, 53–71.
Jerome, K.R., Fox, R., Chen, Z., Sears, A.E., Lee, H., and Corey, L. (1999). Herpes simplex virus inhibits apoptosis through the action of two genes, Us5 and Us3. J. Virol. *73*, 8950–8957.
Jerome, K.R., Chen, Z., Lang, R., Torres, M.R., Hofmeister, J., Smith, S., Fox, R., Froelich, C.J., and Corey, L. (2001). HSV and glycoprotein J inhibit caspase activation and apoptosis induced by granzyme B or Fas. J. Immunol. *167*, 3928–3935.
Jin, L., Peng, W., Perng, G.C., Brick, D.J., Nesburn, A.B., Jones, C., and Wechsler, S.L. (2003). Identification of herpes simplex virus type 1 latency-associated transcript sequences that both inhibit apoptosis and enhance the spontaneous reactivation phenotype. J. Virol. *77*, 6556–6561.
Karin, M. (1999). The beginning of the end: IkappaB kinase (IKK) and NF-kappaB activation. J. Biol. Chem. *274*, 27339–27342.
Kayalar, C., Ord, T., Testa, M.P., Zhong, L.T., and Bredesen, D.E. (1996). Cleavage of actin by interleukin 1 beta-converting enzyme to reverse DNase I inhibition. Proc. Natl. Acad. Sci. USA *93*, 2234–2238.
Kerr, J.F., Wyllie, A.H., and Currie, A.R. (1972). Apoptosis: a basic biological phenomenon with wide-ranging implications in tissue kinetics. Br. J. Cancer *26*, 239–257.
Kerr, J.F., Winterford, C.M., and Harmon, B.V. (1994). Apoptosis. Its significance in cancer and cancer therapy. Cancer 73, 2013–2026.
Kieff, E. (1998). Current perspectives on the molecular pathogenesis of virus-induced cancers in human immunodeficiency virus infection and acquired immunodeficiency syndrome. J. Natl. Cancer Inst. Monogr. 7–14.
Kohase, M., Henriksen-DeStefano, D., May, L.T., Vilcek, J., and Sehgal, P.B. (1986). Induction of beta 2-interferon by tumor necrosis factor: a homeostatic mechanism in the control of cell proliferation. Cell *45*, 659–666.
Koopman, G., Reutelingsperger, C.P., Kuijten, G.A., Keehnen, R.M., Pals, S.T., and van Oers, M.H. (1994). Annexin V for flow cytometric detection of phosphatidylserine expression on B cells undergoing apoptosis. Blood *84*, 1415–1420.

Kothakota, S., Azuma, T., Reinhard, C., Klippel, A., Tang, J., Chu, K., McGarry, T.J., Kirschner, M.W., Koths, K., Kwiatkowski, D.J., and Williams, L.T. (1997). Caspase-3-generated fragment of gelsolin: effector of morphological change in apoptosis. Science *278*, 294–298.

Koyama, A.H., and Adachi, A. (1997). Induction of apoptosis by herpes simplex virus type 1. J. Gen. Virol. *78*, 2909–2912.

Koyama, A.H., and Miwa, Y. (1997). Suppression of apoptotic DNA fragmentation in herpes simplex virus type 1-infected cells. J. Virol. *71*, 2567–2571.

Koyama, A.H., Akari, H., Adachi, A., Goshima, F., and Nishiyama, Y. (1998). Induction of apoptosis in HEp-2 cells by infection with herpes simplex virus type 2. Arch. Virol. *143*, 2435–2441.

Koyama, A.H., Fukumori, T., Fujita, M., Irie, H., and Adachi, A. (2000). Physiological significance of apoptosis in animal virus infection. Microbes Infect. *2*, 1111–1117.

Leopardi, R., and Roizman, B. (1996). The herpes simplex virus major regulatory protein ICP4 blocks apoptosis induced by the virus or by hyperthermia. Proc. Natl. Acad. Sci. USA *93*, 9583–9587.

Leopardi, R., Van Sant, C., and Roizman, B. (1997). The herpes simplex virus 1 protein kinase US3 is required for protection from apoptosis induced by the virus. Proc. Natl. Acad. Sci. USA *94*, 7891–7896.

Levine, A.J. (1997). p53, the cellular gatekeeper for growth and division. Cell *88*, 323–331.

Li, P., Nijhawan, D., Budihardjo, I., Srinivasula, S.M., Ahmad, M., Alnemri, E.S., and Wang, X. (1997). Cytochrome c and dATP-dependent formation of Apaf-1/caspase-9 complex initiates an apoptotic protease cascade. Cell *91*, 479–489.

Lowe, S.W., Bodis, S., McClatchey, A., Remington, L., Ruley, H.E., Fisher, D.E., Housman, D.E., and Jacks, T. (1994). p53 status and the efficacy of cancer therapy *in vivo*. Science *266*, 807–810.

Marsters, S.A., Ayres, T.M., Skubatch, M., Gray, C.L., Rothe, M., and Ashkenazi, A. (1997). Herpesvirus entry mediator, a member of the tumor necrosis factor receptor (TNFR) family, interacts with members of the TNFR-associated factor family and activates the transcription factors NF-kappaB and AP-1. J. Biol. Chem. *272*, 14029–14032.

Martinou, J.C., and Green, D.R. (2001). Breaking the mitochondrial barrier. Nat. Rev. Mol. Cell. Biol. *2*, 63–67.

Mashima, T., Naito, M., Noguchi, K., Miller, D.K., Nicholson, D.W., and Tsuruo, T. (1997). Actin cleavage by CPP-32/apopain during the development of apoptosis. Oncogene *14*, 1007–1012.

Medici, M.A., Sciortino, M.T., Perri, D., Amici, C., Avitabile, E., Ciotti, M., Balestrieri, E., De Smaele, E., Franzoso, G., and Mastino, A. (2003). Protection by herpes simplex virus glycoprotein D against Fas-mediated apoptosis: role of nuclear factor kappaB. J. Biol. Chem. *278*, 36059–36067.

Mestan, J., Digel, W., Mittnacht, S., Hillen, H., Blohm, D., Moller, A., Jacobsen, H., and Kirchner, H. (1986). Antiviral effects of recombinant tumour necrosis factor *in vitro*. Nature *323*, 816–819.

Miyashita, T., Krajewski, S., Krajewska, M., Wang, H.G., Lin, H.K., Liebermann, D.A., Hoffman, B., and Reed, J.C. (1994). Tumor suppressor p53 is a regulator of bcl-2 and bax gene expression *in vitro* and *in vivo*. Oncogene *9*, 1799–1805.

Montgomery, R.I., Warner, M.S., Lum, B.J., and Spear, P.G. (1996). Herpes simplex virus-1 entry into cells mediated by a novel member of the TNF/NGF receptor family. Cell *87*, 427–436.

Mosialos, G., Birkenbach, M., Yalamanchili, R., VanArsdale, T., Ware, C., and Kieff, E. (1995). The Epstein-Barr virus transforming protein LMP1 engages signaling proteins for the tumor necrosis factor receptor family. Cell *80*, 389–399.

Mossman, K.L., Saffran, H.A., and Smiley, J.R. (2000). Herpes simplex virus ICP0 mutants are hypersensitive to interferon. J. Virol. *74*, 2052–2056.

Mossman, K.L., Macgregor, P.F., Rozmus, J.J., Goryachev, A.B., Edwards, A.M., and Smiley, J.R. (2001). Herpes simplex virus triggers and then disarms a host Antiviral response. J. Virol. *75*, 750–758.

Muller, M., Wilder, S., Bannasch, D., Israeli, D., Lehlbach, K., Li-Weber, M., Friedman, S.L., Galle, P.R., Stremmel, W., Oren, M., and Krammer, P.H. (1998). p53 activates the CD95 (APO-1/Fas) gene in response to DNA damage by anticancer drugs. J. Exp. Med. *188*, 2033–2045.

Munger, J., and Roizman, B. (2001). The US3 protein kinase of herpes simplex virus 1 mediates the posttranslational modification of BAD and prevents BAD-induced programmed cell death in the absence of other viral proteins. Proc. Natl. Acad. Sci. USA *98*, 10410–10415.

Munger, J., Chee, A.V., and Roizman, B. (2001). The U(S)3 protein kinase blocks apoptosis induced by the d120 mutant of herpes simplex virus 1 at a premitochondrial stage. J. Virol. *75*, 5491–5497.

Nagata, S. (1997). Apoptosis by death factor. Cell *88*, 355–365.

Nguyen, M.L., Kraft, R.M., and Blaho, J.A. (2005). African green monkey kidney Vero cells require *de novo* protein synthesis for efficient herpes simplex virus 1-dependent apoptosis. Virology, *336*, 274–290.

Nicholl, M.J., Robinson, L.H., and Preston, C.M. (2000). Activation of cellular interferon-responsive genes after infection of human cells with herpes simplex virus type 1. J. Gen. Virol. *81*, 2215–2218.

Nicholson, D.W., and Thornberry, N.A. (1997). Caspases: killer proteases. Trends Biochem. Sci. *22*, 299–306.

Norbury, C.J., and Hickson, I.D. (2001). Cellular responses to DNA damage. Annu Rev Pharmacol Toxicol *41*, 367–401.

Ogg, P.D., McDonell, P.J., Ryckman, B.J., Knudson, C.M., and Roller, R.J. (2004). The HSV-1 Us3 protein kinase is sufficient to block apoptosis induced by overexpression of a variety of Bcl-2 family members. Virology *319*, 212–224.

Oltvai, Z.N., Milliman, C.L., and Korsmeyer, S.J. (1993). Bcl-2 heterodimerizes *in vivo* with a conserved homolog, Bax, that accelerates programmed cell death. Cell *74*, 609–619.

Orth, K., Chinnaiyan, A.M., Garg, M., Froelich, C.J., and Dixit, V.M. (1996). The CED-3/ICE-like protease Mch2 is activated during apoptosis and cleaves the death substrate lamin A. J. Biol. Chem. *271*, 16443–16446.

Pahl, H.L. (1999). Activators and target genes of Rel/NF-kappaB transcription factors. Oncogene *18*, 6853–6866.

Pan, G., O'Rourke, K., and Dixit, V.M. (1998). Caspase-9, Bcl-XL, and Apaf-1 form a ternary complex. J. Biol. Chem. *273*, 5841–5845.

Patel, A., Hanson, J., McLean, T.I., Olgiate, J., Hilton, M., Miller, W.E., and Bachenheimer, S.L. (1998). Herpes simplex type 1 induction of persistent NF-kappa B nuclear translocation increases the efficiency of virus replication. Virology *247*, 212–222.

Pellett, P.E., McKnight, J.L., Jenkins, F.J., and Roizman, B. (1985). Nucleotide sequence and predicted amino acid sequence of a protein encoded in a small herpes simplex virus DNA fragment capable of trans- inducing alpha genes. Proc. Natl. Acad. Sci. USA *82*, 5870–5874.

Peng, W., Henderson, G., Perng, G.C., Nesburn, A.B., Wechsler, S.L., and Jones, C. (2003). The gene that encodes the herpes simplex virus type 1 latency-associated transcript influences the accumulation of transcripts (Bcl-x(L) and Bcl-x(S)) that encode apoptotic regulatory proteins. J. Virol. *77*, 10714–10718.

Perng, G.C., Jones, C., Ciacci-Zanella, J., Stone, M., Henderson, G., Yukht, A., Slanina, S.M., Hofman, F.M., Ghiasi, H., Nesburn, A.B., Wechsler, S.L. (2000). Virus-induced neuronal apoptosis blocked by the herpes simplex virus latency-associated transcript. Science *287*, 1500–1503.

Petit, P.X., Susin, S.A., Zamzami, N., Mignotte, B., and Kroemer, G. (1996). Mitochondria and programmed cell death: back to the future. FEBS Lett. *396*, 7–13.

Post, L.E., Mackem, S., and Roizman, B. (1981). Regulation of alpha genes of herpes simplex virus: expression of chimeric genes produced by fusion of thymidine kinase with alpha gene promoters. Cell *24*, 555–565.

Prives, C., and Hall, P.A. (1999). The p53 pathway. J. Pathol. *187*, 112–126.

Purves, F.C., Longnecker, R.M., Leader, D.P., and Roizman, B. (1987). Herpes simplex virus 1 protein kinase is encoded by open reading frame US3 which is not essential for virus growth in cell culture. J. Virol. *61*, 2896–2901.

Qian, T., Nieminen, A.L., Herman, B., and Lemasters, J.J. (1997). Mitochondrial permeability transition in pH-dependent reperfusion injury to rat hepatocytes. Am. J. Physiol. *273*, C1783–1792.

Qie, L., Marcellino, D., and Herold, B.C. (1999). Herpes simplex virus entry is associated with tyrosine phosphorylation of cellular proteins. Virology *256*, 220–227.

Roizman, B. (1962). Polykaryocytosis induced by viruses. Proc. Natl. Acad. Sci. USA *48*, 228–234.

Roizman, B., and Furlong, D. (1974). The replication of herpesviruses. In Comprehensive virology, H.F.-C. a. R. R. Wagner, ed. (New York, N.Y., Plenum), pp. 229–403.

Roizman, B., and Knipe, D. M. (2001). Herpes simplex viruses and their replication. In Virology, a.P. M. H. D. M. Knipe, ed. (Philadelphia, Pa., Lippincott-Raven), pp. 2399–2459.

Roizman, B., and Roanne, P. R. (1964). Multiplication of herpes simplex virus. II. The relationship between protein synthesis and the duplication of viral DNA in infected HEp-2 cells. Virology *22*, 262–269.

Samaniego, L.A., Neiderhiser, L., and DeLuca, N.A. (1998). Persistence and expression of the herpes simplex virus genome in the absence of immediate-early proteins. J. Virol. *72*, 3307–3320.

Sanfilippo, C.M., and Blaho, J.A. (2003). The facts of death. Int. Rev. Immunol. *22*, 327–340.

Sanfilippo, C.M., and Blaho, J.A. (2006). ICP0 gene expression is herpes simplex virus type 1 apoptoxic trigger. J. Virol. (*In press*).

Sanfilippo, C.M., Chirimuuta, F.N., and Blaho, J.A. (2004). Herpes simplex virus type 1 immediate-early gene expression is required for the induction of apoptosis in human epithelial HEp-2 cells. J. Virol. *78*, 224–239.

Saraste, A., and Pulkki, K. (2000). Morphologic and biochemical hallmarks of apoptosis. Cardiovasc. Res. *45*, 528–537.

Schwartz, J., and Roizman, B. (1969). Concerning the egress of herpes simplex virus from infected cells: electron and light microscope observations. Virology *38*, 42–49.

Shi, Y. (2002). Mechanisms of caspase activation and inhibition during apoptosis. Mol. Cell. *9*, 459–470.

Shipkey, F.H., Erlandson, R.A., Bailey, R.B., Babcock, V.I., and Southam, C.M. (1967). Virus biographies. II. Growth of herpes simplex virus in tissue culture. Exp. Mol. Pathol. *6*, 39–67.

Smith, C.A., Farrah, T., and Goodwin, R.G. (1994). The TNF receptor superfamily of cellular and viral proteins: activation, costimulation, and death. Cell *76*, 959–962.

Stevens, J.G., and Cook, M.L. (1971). Latent herpes simplex virus in spinal ganglia of mice. Science *173*, 843–845.

Strasser, A., O'Connor, L., and Dixit, V.M. (2000). Apoptosis signaling. Annu. Rev. Biochem. *69*, 217–245.

Sutton, V.R., Davis, J.E., Cancilla, M., Johnstone, R.W., Ruefli, A.A., Sedelies, K., Browne, K.A., and Trapani, J.A. (2000). Initiation of apoptosis by granzyme B requires direct cleavage of bid, but not direct granzyme B-mediated caspase activation. J. Exp. Med. *192*, 1403–1414.

Suzuki, Y., Nakabayashi, Y., and Takahashi, R. (2001). Ubiquitin-protein ligase activity of X-linked inhibitor of apoptosis protein promotes proteasomal degradation of caspase-3 and enhances its anti-apoptotic effect in Fas-induced cell death. Proc. Natl. Acad. Sci. USA *98*, 8662–8667.

Taddeo, B., T.R. Luo, W. Zhang, and B. Roizman (2003). Activation of NF-kappaB in cells productively infected with HSV-1 depends on activated protein kinase R and plays no apparent role in blocking apoptosis. Proc. Natl. Acad. Sci. USA *100*, 12408–12413.

Takahashi, A., Alnemri, E.S., Lazebnik, Y.A., Fernandes-Alnemri, T., Litwack, G., Moir, R.D., Goldman, R.D., Poirier, G.G., Kaufmann, S.H., and Earnshaw, W.C. (1996). Cleavage of lamin A by Mch2 alpha but not CPP32: multiple interleukin 1 beta-converting enzyme-related proteases with distinct substrate recognition properties are active in apoptosis. Proc. Natl. Acad. Sci. USA *93*, 8395–8400.

Tartaglia, L.A., Ayres, T.M., Wong, G.H., and Goeddel, D.V. (1993). A novel domain within the 55 kD TNF receptor signals cell death. Cell *74*, 845–853.

Terry-Allison, T., Montgomery, R.I., Whitbeck, J.C., Xu, R., Cohen, G.H., Eisenberg, R.J., and Spear, P.G. (1998). HveA (herpesvirus entry mediator A), a coreceptor for herpes simplex virus entry, also participates in virus-induced cell fusion. J. Virol. *72*, 5802–5810.

Thanos, D., and Maniatis, T. (1995). NF-kappa B: a lesson in family values. Cell *80*, 529–532.

Thompson, R.L., and Sawtell, N.M. (2001). Herpes simplex virus type 1 latency-associated transcript gene promotes neuronal survival. J. Virol. *75*, 6660–6675.

Thornberry, N.A., and Lazebnik, Y. (1998). Caspases: enemies within. Science *281*, 1312–1316.

Trapani, J.A. (2001). Granzymes: a family of lymphocyte granule serine proteases. Genome Biol. 2.

Tschopp, J., Thome, M., Hofmann, K., and Meinl, E. (1998). The fight of viruses against apoptosis. Curr. Opin. Genet. Dev. *8*, 82–87.

Vanags, D.M., Porn-Ares, M.I., Coppola, S., Burgess, D.H., and Orrenius, S. (1996). Protease involvement in fodrin cleavage and phosphatidylserine exposure in apoptosis. J. Biol. Chem. *271*, 31075–31085.

Vander Heiden, M.G., Chandel, N.S., Williamson, E.K., Schumacker, P.T., and Thompson, C.B. (1997). Bcl-xL regulates the membrane potential and volume homeostasis of mitochondria. Cell *91*, 627–637.

Vaux, D.L., and Strasser, A. (1996). The molecular biology of apoptosis. Proc. Natl. Acad. Sci. USA *93*, 2239–2244.

Wallach, D., Kovalenko, A.V., Varfolomeev, E.E., and Boldin, M.P. (1998). Death-inducing functions of ligands of the tumor necrosis factor family: a Sanhedrin verdict. Curr. Opin. Immunol. *10*, 279–288.

Weinstein, J.N., Myers, T.G., O'Connor, P.M., Friend, S.H., Fornace, A.J., Jr., Kohn, K.W., Fojo, T., Bates, S.E., Rubinstein, L.V., Anderson, N.L. *et al.* (1997). An information-intensive approach to the molecular pharmacology of cancer. Science *275*, 343–349.

Wong, G.H., and Goeddel, D.V. (1986). Tumour necrosis factors alpha and beta inhibit virus replication and synergize with interferons. Nature *323*, 819–822.

Wyllie, A.H. (1980). Glucocorticoid-induced thymocyte apoptosis is associated with endogenous endonuclease activation. Nature *284*, 555–556.

Wyllie, A.H., Kerr, J.F., and Currie, A.R. (1980). Cell death: the significance of apoptosis. Int. Rev. Cytol. *68*, 251–306.

Yamauchi, Y., Daikoku, T., Goshima, F., and Nishiyama, Y. (2003). Herpes simplex virus UL14 protein blocks apoptosis. Microbiol. Immunol. *47*, 685–689.

Yurochko, A.D., Hwang, E.S., Rasmussen, L., Keay, S., Pereira, L., and Huang, E.S. (1997). The human cytomegalovirus UL55 (gB) and UL75 (gH) glycoprotein ligands initiate the rapid activation of Sp1 and NF-kappaB during infection. J. Virol. *71*, 5051–5059.

Zhan, Q., Fan, S., Bae, I., Guillouf, C., Liebermann, D.A., O'Connor, P.M., and Fornace, A.J., Jr. (1994). Induction of bax by genotoxic stress in human cells correlates with normal p53 status and apoptosis. Oncogene *9*, 3743–3751.

Zhou, G., and Roizman, B. (2001). The domains of glycoprotein D required to block apoptosis depend on whether glycoprotein D is present in the virions carrying herpes simplex virus 1 genome lacking the gene encoding the glycoprotein. J. Virol. *75*, 6166–6172.

Zhou, G., and Roizman, B. (2002a). Cation-independent mannose 6-phosphate receptor blocks apoptosis induced by herpes simplex virus 1 mutants lacking glycoprotein D and is likely the target of antiapoptotic activity of the glycoprotein. J. Virol. *76*, 6197–6204.

Zhou, G., and Roizman, B. (2002b). Truncated forms of glycoprotein D of herpes simplex virus 1 capable of blocking apoptosis and of low-efficiency entry into cells form a heterodimer dependent on the presence of a cysteine located in the shared transmembrane domains. J. Virol. *76*, 11469–11475.

Zhou, G., Galvan, V., Campadelli-Fiume, G., and Roizman, B. (2000). Glycoprotein D or J delivered in trans blocks apoptosis in SK-N-SH cells induced by a herpes simplex virus 1 mutant lacking intact genes expressing both glycoproteins. J. Virol. *74*, 11782–11791.

Zhou, G., Avitabile, E., Campadelli-Fiume, G., and Roizman, B. (2003). The domains of glycoprotein D required to block apoptosis induced by herpes simplex virus 1 are largely distinct from those involved in cell–cell fusion and binding to nectin1. J. Virol. *77*, 3759–3767.

Zou, H., Henzel, W.J., Liu, X., Lutschg, A., and Wang, X. (1997). Apaf-1, a human protein homologous to C. elegans CED-4, participates in cytochrome c-dependent activation of caspase-3. Cell *90*, 405–413.

Mechanisms of Subversion of Type I Interferon Responses by Alpha-herpesviruses

13

Karen L. Mossman

Abstract

Interferons are pleiotropic cytokines that were first discovered based on their ability to interfere with virus replication. *In vivo*, interferons have been shown to play a central role in limiting the spread of alpha-herpesviruses in that strains of mice that are resistant to virus infection rapidly produce high levels of this cytokine whereas strains that are susceptible to infection do not. Given the evolutionary pressure for alpha-herpesviruses to delay or subvert the antiviral activities of interferons, it is not surprising that multiple viral proteins have been identified that collectively function to limit interferon production, signaling and effector function. This chapter will summarize the current knowledge surrounding mechanisms utilized by alpha-herpesviruses to ensure efficient replication and spread in the face of an active interferon-mediated host antiviral response.

Introduction

Herpesviruses and interferon

One of the primary events of the innate immune response to virus infection is the production of type I interferon (IFN), a family of cytokines that includes IFN-β and multiple IFN-α species, along with IFN-ε, -δ, -κ, -λ, -τ and -ω (Kontsek *et al.*, 2003). In general, type I IFNs are produced in a wide variety of cells during the early phases of the immune response whereas type II IFN (IFN-γ) is produced by activated NK and T lymphocytes following the recognition of virally infected cells during the adaptive phase of the immune response. Early *in vivo* mouse experiments identified strains of mice that were both resistant (e.g. C57BL/6) and susceptible (e.g. BALB/c) to HSV-1 infection, with subsequent experiments linking resistance to the rapid production of type I IFN (refer to Mossman, 2002; Halford *et al.*, 2004 and references therein). These studies were confirmed using anti-IFN-α/β serum, which eliminated both natural and biologic response modifier-induced resistance to HSV-1 (Lopez, 1985; Su *et al.*, 1990; Kunder *et al.*, 1993; Halford *et al.*, 1997). A recent, thorough re-evaluation of the natural resistance of mice to HSV-1 found that HSV-1 can enter, replicate, spread and establish latency with similar efficiencies in both C57BL/6 and BALB/c strains of mice; this study also confirmed that the resistant phenotype can be explained by a type I IFN-mediated delay in the spread of HSV-1 from the site of inoculation to the peripheral nervous system (Halford *et al.*, 2004).

While both types I and II IFNs affect the overall pathogenesis of HSV *in vivo*, several lines of evidence suggest that type I IFN plays the predominant role in the control of early acute virus replication whereas IFN-γ appears to control persistent infections and reactivation events within the central nervous system (Leib *et al.*, 1999; Mossman, 2002; Halford *et al.*, 2004; Vollstedt *et al.*, 2004 and references therein). However, IFN-γ mRNA is detected in HSV-1 infected tissues at early times of infection (Halford and Schaffer, 2000) and a number of professional antigen-presenting cells have been found to elicit IFN-γ in significant quantities early in the host immune response (Suzue *et al.*, 2003). A recent study using *in vivo* bioluminescence imaging to ascertain the effects of the loss of type I and/or type II IFN receptor(s) on HSV-1 replication and tropism found that although the replication and spread of HSV-1 was similar in wild type and type II IFN receptor null mice, deletion of the type II IFN receptor in combination with the type I IFN receptor significantly increased the virulence of HSV-1 *in vivo* (Luker *et al.*, 2003). Collectively, these studies indicate that while IFN-γ is classically defined as a mediator of the adaptive immune response, it most likely also contributes to host defense early on.

Not surprisingly, the alpha-herpesviruses have evolved multiple strategies to evade the effects of IFN, from blocking IFN production to circumventing the activities of IFN-stimulated genes (ISGs). This chapter will focus on the induction of type I IFN in response to alpha-herpesvirus infection and the mechanisms used by these viruses to counteract IFN-mediated antiviral pathways. Although the majority of research in this field has focused on herpes simplex viruses, there is some information, albeit limited, on the relationship between type I IFN and the human alpha-herpesvirus varicella zoster virus (VZV). While the mode of replication of these two human alpha-herpesviruses differs, it is appreciated that IFN is important to limiting VZV replication (Ito *et al.*, 1992; Tovey and Maury, 1999). Indeed, administration of IFN-α to immunocompromised children with VZV aids in lessening the severity of the infection (Arvin, 1996).

Virus-mediated interferon production

Most mammalian cells do not constitutively produce IFN, but instead rapidly induce IFN-α/β following infection. IFN induction was originally described following infection with RNA viruses and it is now clear that the main inducing element is double-stranded RNA (dsRNA). During the life cycle of RNA viruses both positive and negative strands of RNA are produced, resulting in dsRNA intermediates. DNA viruses such as the alpha-herpesviruses are also capable of producing dsRNA, through symmetrical transcription (Jacquemont and Roizman, 1975), although the levels are certainly not as high as those produced by RNA viruses. As opposed to fibroblasts and epithelial cells that predominantly produce IFN-β upon recognition of dsRNA, in response to enveloped viruses and in particular viral glycoproteins, specialized leukocytes called "natural IFN producing cells" are the main IFN-α producers *in vivo* (Fitzgerald-Bocarsly, 1993; Bjorck, 2001; Vollstedt *et al.*, 2003). These cells have been identified as $CD4^{+}CD11c^{-}$ type 2 dendritic cell precursors (Siegal *et al.*, 1999) and subsequent research found that viral infection not only stimulates IFN production but also induces differentiation into mature dendritic cells capable of triggering adaptive T cell-mediated immunity (Kadowaki *et al.*, 2000). While lymphoid cells appear to produce type I IFN in response to viral glycoproteins (Ito, 1994), the precise mechanism of induction in response to alpha-herpesviruses is unclear. Viral glycoprotein D, cellular mannose and/or chemokine receptor(s), protein kinases A and C and tyrosine kinases have all been implicated (Li *et al.*, 1996; Ankel *et al.*, 1998; Milone and Fitzgerald-Bocarsly, 1998), but the precise mechanisms remain to be elucidated.

At the molecular level, induction of type I IFN in virus-infected cells has been extensively studied and the current model suggests a three-step mode of induction (Fig. 13.1) (Juang *et al.*, 1998; Yoneyama *et al.*, 1998; Sato *et al.*, 2000). In the immediate early or sensitization phase, virus infection leads to the activation of a number of latently expressed transcription factors, including interferon regulatory factor 3 (IRF3), IRF1, ATF-2/c-Jun and NF-κB, which collectively form an "enhanceasome" leading to production of low levels of IFN-β (Juang *et al.*, 1998; Yoneyama *et al.*, 1998; LePage *et al.*, 2000; Sato *et al.*, 2000). During the second or inductive phase, secreted IFN-β binds to its cellular receptor and initiates a cascade of phosphorylation events of specific Janus (Jak) kinases and signal transducer of activated transcription (Stat) molecules, culminating in translocation

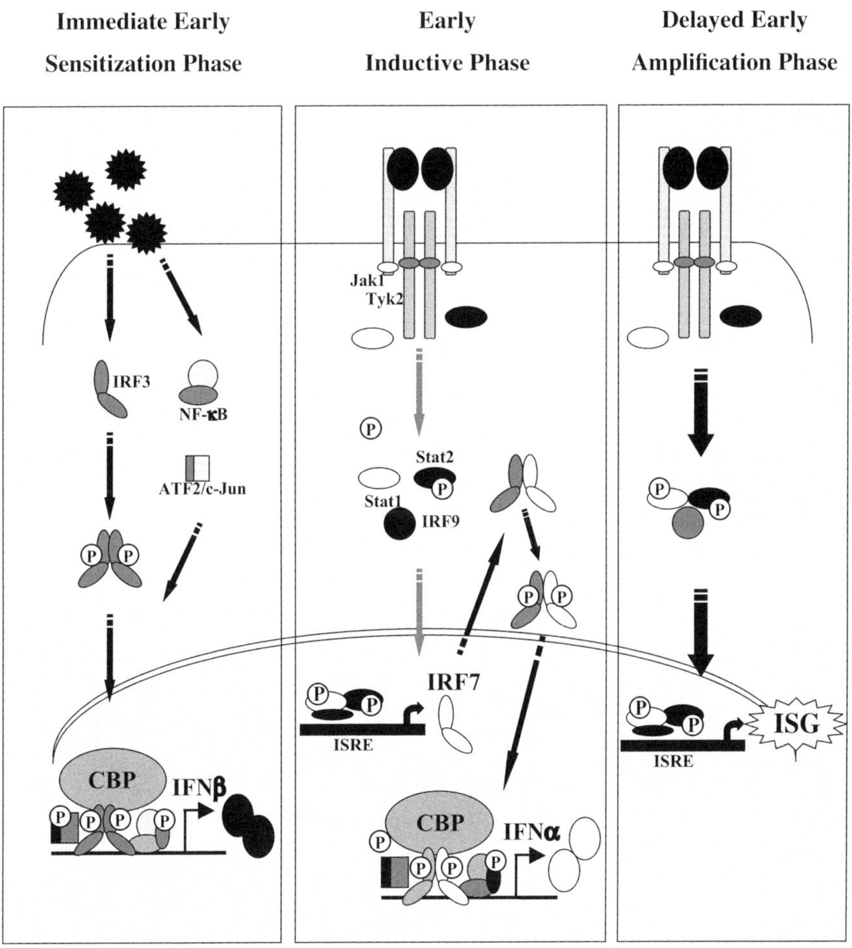

Figure 13.1 The tri-phasic model of virus-mediated interferon production. In the sensitization phase, virus infection activates latent transcription factors, causing low level induction of IFN-β. In the inductive phase, secreted IFN-β binds to its cognate receptor on surrounding cells, activating the Jak/Stat pathway and eliciting the induction of a subset of ISGs including IRF7. Virus-activated IRF7 and IRF3 subsequently dimerize and induce the full complement of IFN-α species. In the amplification phase, IFN-α/β function in concert to induce the full range of ISGs.

of a phosphorylated Stat-containing complex into the nucleus (Levy, 1995; Williams and Haque, 1997; Leonard, 2001). This complex binds the IFN stimulated response element (ISRE) within the promoter region of interferon-stimulated genes (ISGs) and initiates gene transcription. One of the genes induced in the context of low-level stimulation with IFN-β is IRF7. Like IRF3, cytoplasmic IRF7 becomes activated following virus infection (Lin *et al.*, 2000) and, in the delayed early or amplification phase, IRF3 and IRF7 co-operate to induce the full array of IFN subtypes, leading to the induction of a large subset of ISGs (Marie *et al.*, 1998; Levy *et al.*, 2002).

As mentioned above, a number of transcription factors are constitutively expressed and become activated upon viral infection to form an enhanceasome on the IFN-β promoter. Studies with mice harboring deletions in IRF3 demonstrated the requirement for this transcription factor in eliciting IFN-β production (Sato *et al.*, 1998; Sato *et al.*, 2000; Sato *et al.*, 2001). The current model of virus-induced IRF3 activation consists of 5 sequential steps: (1) phosphorylation by a cellular "virus-activated" kinase(s), (2) a conformational change resulting in protein dimerization, (3) translocation to the nucleus, (4) association with the co-activators CBP and/or p300, and (5) ISRE binding (Lin *et al.*, 1998; Kumar *et al.*, 2000; Servant *et al.*, 2001). To date, three cellular kinases with the capacity to phosphorylate IRF3 have been identified. The first, DNA-PK, a cellular DNA-depended protein kinase involved in DNA repair (Lees-Miller and Meek, 2003; Meek *et al.*, 2004), was found to phosphorylate IRF3 on Thr135 (Karpova *et al.*, 2002). The significance of N-terminal phosphorylation of IRF3, however, remains elusive as both the Fujita and Hiscott laboratories have published evidence that C-terminal phosphorylation is essential for its activity (Lin *et al.*, 1998; Yoneyama *et al.*, 1998). Recent determination of the crystal structure of IRF3 by two independent groups provided further evidence of a requirement for C-terminal phosphorylation (Qin *et al.*, 2003; Takahasi *et al.*, 2003). While the cellular kinases IKK-ε and TBK-1 have been shown to phosphorylate multiple C-terminal serine residues of IRF3 (Fitzgerald *et al.*, 2003; Sharma *et al.*, 2003), identification of the critical phosphorylation event(s) remains controversial (Servant *et al.*, 2001; Servant *et al.*, 2003; Mori *et al.*, 2004).

IRF family members can bind several DNA motifs, including the ISRE, indicating that IRFs also play a direct role in induction of ISGs. It is now clear that ISGs can be induced by virus or dsRNA in the absence of IFN (Tiwari *et al.*, 1987; Wathelet *et al.*, 1988; Wathelet *et al.*, 1992; Bandyopadhyay *et al.*, 1995; Guo *et al.*, 2000b). Using IFN-resistant HEC-1B cells, Daly and Reich identified two dsRNA activated ISRE-binding complexes, dsRNA activating factor 1 and 2 (DRAF1 and DRAF2) (Daly and Reich, 1993). DRAF1 was found to contain IRF3, CBP and p300 (Weaver *et al.*, 1998); the components of DRAF2 have not been reported. With respect to alpha-herpesviruses, HSV-1 was found to induce a small subset of ISGs in the absence of both IFN production and virus replication (Nicholl *et al.*, 2000; Mossman *et al.*, 2001) that is sufficient to induce a complete antiviral state in infected cells (Mossman *et al.*, 2001). These experiments utilized either genetically or UV-inactivated viruses, or wild type virus in the presence of actinomycin D or cycloheximide, inhibitors of transcription and translation, respectively. IRF3 was not only activated under these conditions (Preston *et al.*, 2001), but was required for IFN-independent ISG and antiviral state induction (Collins *et al.*, 2004). The limited set of ISGs that were induced by incoming HSV-1 particles (Table 13.1) overlapped

Table 13.1 ISGs induced by incoming herpesvirus particles in human fibroblasts in the absence of IFN production and viral gene expression

Gene	Function
Mx 1 and 2 homologs	Inhibition of RNA virus replication
ISG15	Ubiquitin-like immunomodulator
ISG56	Protein synthesis inhibition
IP10	Chemokine
STAT1	Transcription factor in IFN pathway
9–27	Antiproliferative/antiviral activity
Viperin	ER-associated antiviral protein
OAS 2 and 3	2–5A synthesis; RNase L pathway activation
SP100	ND10 component
ABC B2	Antigen peptide transporter
β_2 microglobulin	Antigen presentation
ADAR	RNA editing

with those found to be induced by a constitutively activated form of IRF3 (Mossman *et al.*, 2001; Grandvaux *et al.*, 2002).

Inhibition of virus replication by interferon-stimulated genes

The precise mechanisms by which IFN influences herpesvirus replication are poorly understood, due to the complexity of the virus, the number of IFN subtypes, and the *in vivo* role of multiple host genes that collectively influence the host antiviral response. In general terms, IFN activation of immune cells, including macrophages, NK cells and B and T lymphocytes, and immune events, such as apoptosis, collectively influence herpesvirus replication and reduce viral load (Borysiewicz and Sissons, 1994; Tanaka *et al.*, 1998; Biron *et al.*, 1999; Deshpande *et al.*, 2000; Goodbourne *et al.*, 2000). Inhibition of viral immediate early and early gene expression and alteration of viral glycoprotein processing and trafficking are activities ascribed to IFN (Panet and Falk, 1983; Mittnacht *et al.*, 1988; Oberman and Panet, 1988; Chatterjee and Burns, 1990; DeStasio and Taylor, 1990; Maheshwari *et al.*, 1994; Holberg-Petersen *et al.*, 1996; Wintergerst *et al.*, 1996; O'Brien *et al.*, 1998; Taylor *et al.*, 1998). However, at the level of an infected cell, the precise antiviral cellular program remains to be elucidated.

IFN has been estimated to stimulate the induction of upwards of 2000 genes (de Veer *et al.*, 2001). Several of the well-characterized ISGs are key in establishing intracellular antiviral pathways that inhibit viral transcription and translation, including the PKR, 2–5A synthetase (OAS) and Mx genes (Fig. 13.2) (reviewed in Sen, 2001; Sen and Ransohoff, 1993; Stark *et al.*, 1998). PKR is a dsRNA dependent protein kinase that resides within cells in a latent form and its levels are elevated following treatment of cells with IFN (Stark *et al.*, 1998). Production of dsRNA during virus infection causes the autophosphorylation and subsequent activation of PKR, leading to the intermolecular phosphorylation of a number of cellular protein structures, including the critical translation initiation factor eIF2α, the transcription factor inhibitor IκB, the M-phase specific phosphopro-

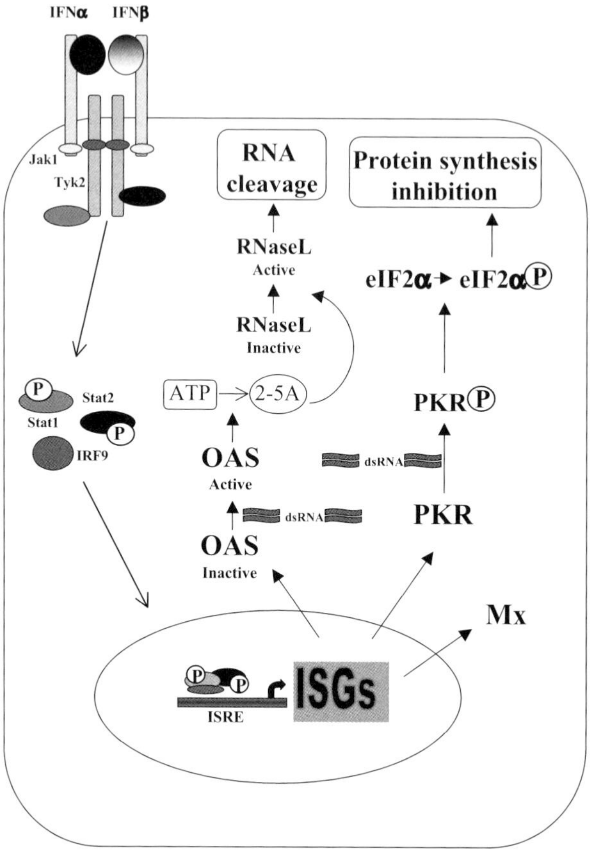

Figure 13.2 Induction of intracellular antiviral pathways that are amplified by interferon. OAS and PKR are latent cytoplasmic proteins whose levels increase significantly upon exposure of cells to IFN. Both enzymes become activated upon binding to dsRNA, a by-product of virus infection. Although the Mx GTPases also function within infected cells to inhibit virus replication, to date their activity has been restricted to RNA viruses.

tein MPP4 and the 90kDa NFAT protein (Samuel, 2001). Phosphorylation of eIF2α prevents the initiation of protein synthesis by impairing the eIF-2B-mediated guanine nucleotide exchange reaction; in addition to specifically blocking virus protein synthesis, a general block in cellular protein synthesis ensues, leading to apoptosis or programmed cell death. Apoptosis has been shown to be an antiviral defense mechanism to limit virus replication and propagation of the infection (reviewed in Barber, 2001).

The OAS pathway has many similarities to that of the PKR pathway. Like PKR, OAS is found latent within cells, it is highly induced following treatment with IFN, it becomes activated upon binding to dsRNA and it functions as an important intracellular mediator of IFN antiviral resistance (reviewed in Castelli *et al.*, 1997; Player and Torrence, 1998). Activated OAS catalyzes the synthesis of 2′-5′-linked oligoadenylates referred to as 2–5A. Upon binding to 2–5A, the latent monomeric enzyme RNase L dimerizes and acquires endoribonuclease activity. Activated RNase L catalyzes the degradation of both viral and cellular RNA species, including cellular ribosomal RNA, which leads to apoptosis

(Castelli *et al.*, 1998). Despite the similarities in the two IFN-mediated intracellular antiviral pathways, only PKR has a clearly defined role in mediating the inhibition of HSV-1 in resistant strains of mice. The role played by the OAS pathway is more controversial, as there is evidence for and against its ability to block the replication and spread of HSV (Fujihara *et al.*, 1989; Taylor *et al.*, 1995; Khabar *et al.*, 2000; Zheng *et al.*, 2001; Al-Khatib *et al.*, 2002; Harle *et al.*, 2002a; Al-khatib *et al.*, 2003; Al-Khatib *et al.*, 2004). Furthermore, a recent investigation into the role of the RNase L pathway as a cellular defense mechanism against Varicella zoster virus infection revealed that this pathway fails to significantly affect the replication of this alpha-herpesvirus (Desloges *et al.*, 2004). The Mx family of IFN-inducible antiviral factors belongs to the superfamily of dynamin-like GTPases and functions to block the replication of a number of RNA viruses (Haller and Kochs, 2002). Although the human counterpart to Mx1 was shown by microarray analysis to be highly induced in fibroblasts in response to incoming HSV-1 particles (Mossman *et al.*, 2001), there is no evidence in the literature demonstrating it ability to limit HSV-1 replication. In fact, an older study demonstrated that over expression of Mx1 had no effect on the replication of HSV-1 (Pavlovic *et al.*, 1990).

For many years, IFN-mediated cellular antiviral responses were studied almost exclusively in the context of these three pathways. Although these classical ISGs play fundamental roles in limiting virus replication at a single cell level, ISGs also aid in the clearance of virally infected cells and establishment of immunity to subsequent infection by modulating the immune system through inhibition of cell growth, control of apoptosis and development of cellular and humoral immunity (Stark *et al.*, 1998). Not surprisingly, mice triply deficient for PKR, RNase L and Mx1 retain partial responsiveness to IFN-induced antiviral effects (Zhou *et al.*, 1999), confirming the presence of multiple ISGs that collectively conspire to limit virus replication and spread. Recent studies have begun to characterize additional antiviral ISGs such as ISG15, ISG56 and IP-10. ISG15 amplifies the immunomodulatory effects of the IFNs by conjugating to cellular targets in a ubiquitin-like manner (D'Cunha *et al.*, 1996; Taylor *et al.*, 1996). ISG56 binds to a subunit of eukaryotic initiation factor 3 and inhibits protein synthesis (Guo *et al.*, 2000a) and IP-10 functions as a secreted chemokine with pleiotropic functions (Neville *et al.*, 1997). Of particular interest, ISG15, ISG56 and IP-10, along with OAS and Mx, are highly induced following entry of HSV-1 particles into cells in the absence of IFN production (Table 13.1) (Nicholl *et al.*, 2000; Mossman *et al.*, 2001).

Subversion of interferon responses by alpha-herpesviruses

Given the importance of IFN in controlling virus replication and spread, most, if not all, viruses have developed strategies to counteract this antiviral cytokine (reviewed in Alcami and Koszinowski, 2000; Cebulla *et al.*, 1999; Goodbourne *et al.*, 2000; Haig, 2001; Levy and Garcia-Sastre, 2001; Sen, 2001; Katze *et al.*, 2002). Subject to inhibition are all aspects of IFN signaling, from IFN production to secretion to signaling to ISG function. Typically, DNA viruses with large coding capacities develop multiple measures to counteract IFN function. For example, poxviruses, which encode upwards of 200 proteins, block IFN production through inhibition of NF-κB and IRF3 (DiPerna *et al.*, 2004), IFN signaling through secretion of soluble IFN receptor homologs (Upton *et al.*, 1992; Mossman *et al.*, 1995a; Mossman *et al.*, 1995b; Mossman *et al.*, 1996) and intracellular antiviral pathways by producing viral dsRNA binding proteins and eIF2α decoys (Davies *et al.*, 1992; Davies *et al.*, 1993; Xiang *et al.*, 2002). Owing to the large coding capacity of herpesviruses, it was hypothesized that these viruses would also elicit multiple proteins with anti-IFN activity and, indeed, *in vitro* infection with HSV-1 failed to elicit the accumulation of ISGs, confirming the notion that HSV-1 encodes one or more proteins capable of inhibiting the antiviral IFN response (Nicholl *et al.*, 2000; Mossman *et al.*, 2001). Within the last several years, a number of viral proteins have been characterized (Leib, 2002; Mossman, 2002), and will be discussed in more detail in the proceeding sections (Table 13.2). A recent report using in1814, an HSV-1 recombinant lacking the tegument-derived transactivator VP16, demonstrated that impairment of the replication competence of HSV-1 leads to enhanced production of not only IFN-α/β, but a number of proinflammatory cytokines as well (Mogensen *et al.*, 2004).

Inhibition of interferon production

In the triphasic model of virus-mediated IFN production, the cellular transcription factor IRF3 plays a central role in the immediate early and amplification phases. As with other viruses, infection of cells with HSV-1 leads to IRF3 activation (Preston *et al.*, 2001). However, activation of this transcription factor was only observed when *de novo* protein synthesis was blocked with cycloheximide. Experimental evidence had also determined that the IRF3-dependent, IFN-independent induction of ISGs and an antiviral state following HSV-1 particle entry occurred in the absence of HSV-1 replication (Nicholl

Table 13.2 Inhibition of IFN-mediated antiviral immune responses by alpha-herpesviruses

Inhibition of	Mechanism	Virus	Gene Product(s)
IFN production	Inhibition of IRF3 nuclear translocation and function	HSV-1	ICP0
	Unknown; higher levels of IFN-α/β detected upon infection of mice with mutant	HSV-1	UL13§
	Unknown; enhanced IFN and cytokine production observed with mutant virus	HSV-1	VP16§
	Unknown; viral gene expression required	VZV	Unknown
IFN signaling	Inhibition of Jak/Stat phosphorylation	HSV-1	Vhs§ and UL13§
	Rapid induction of SOCS-3	HSV-1	Vhs§ and UL13§
	Degradation of Jak1 and Stat2	HSV-1	Vhs§
	Degradation of type I IFN receptor	HSV-1	Vhs§
	Inhibition of nuclear localization of Stat1	VZV	Unknown
ISG production and function	Degradation of ISG mRNA; increased virulence in IFN receptor null mice	HSV-1	Vhs§
	Enhance proteasome-dependent degradation of ISGs (e.g. PML and Sp100)	HSV-1	ICP0
	Inhibition of PKR pathway through:		
	dephosphorylation of eIF2α	HSV-1	ICP34.5
	inhibition of PKR autophosphorylation	HSV-1	Us11
	Inhibition of RNase L through production of 2–5A derivatives	HSV-1	Unknown
	Overcome IFN-induced block to viral transcription	HSV-1	ICP0§
	Unknown; increased virulence of mutants in type I IFN receptor knock-out mice	HSV-1 HSV-2	RR§ and TK§ Vhs§

§Gene products that are implicated based on phenotype of gene deletion mutants.

et al., 2000; Mossman et al., 2001; Collins et al., 2004). Collectively, these observations suggested that activation of IRF3 is a target for viral interference. The HSV-1 immediate early protein, ICP0, had previously been implicated in protecting cells from the antiviral effects of IFN, in that ICP0 mutants were hypersensitive to pretreatment of cells with this cytokine (Mossman et al., 2000; Harle et al., 2002b). Thus, it was not surprising when ICP0 was found to block induction of ISGs following infection of human fibroblast cells with a replication-defective HSV-1 recombinant (Eidson et al., 2002). ICP0 is a fascinating, multifunctional protein that is produced rapidly upon infection of cells with HSV-1 (Everett, 2000), and is the subject of an entire chapter within this book (R. Everett, Chapter 3). Although ICP0 can associate with a number of cellular proteins and affect multiple aspects of the HSV-1 lytic and latent replication cycles (Everett, 2000), the N-terminal RING finger domain appears to be important for most, if not all, biological activities of ICP0. This domain has been shown to possess E3 ubiquitin ligase activity (Boutell et al., 2002; Hagglund et al., 2002); accordingly, many of the functions of ICP0, including the inhibition of ISG accumulation (Eidson et al., 2002), have been found to require a functional proteasome (Everett, 2000).

In determining that ICP0 can block accumulation of ISGs, DeLuca and colleagues pretreated human fibroblasts with ICP0 using an adenovirus delivery system and demonstrated that under these conditions, infection with d109, a replication-defective HSV-1 isolate, failed to accumulate ISGs (Eidson et al., 2002). ICP0 was subsequently shown to block IFN-β and RANTES promoter activation mediated by infection with Sendai virus, a small RNA virus that potently induces IFN production, or transfection with constitutively activated forms of IRF3 and IRF7 (Lin et al., 2004). Within the context of these transient luciferase assays, ICP0 was more efficient at blocking the activity of IRF3 than IRF7. Given that IRF3 plays an essential role in launching the IFN signaling cascade, while IRF7 induction and activation is observed at a later phase, it is likely that ICP0 does indeed possess greater inhibitory effects on IRF3 than on IRF7 *in vivo*. Not surprisingly, the RING finger motif of ICP0 was shown to be essential for inhibition of promoter activation and ISG induction, both in the context of transient transfection assays and following infection of human cells (Lin et al., 2004). DeLuca and colleagues had demonstrated the requirement of a functional proteasome for ICP0 activity, suggesting that ICP0 elicits the degradation of a component of the IRF3 pathway. However, the only known protein associated with this pathway that is degraded following infection with HSV-1 or transfection with ICP0 is DNA-PK (Collins et al., 2004). Degradation of this kinase was demonstrated several years previously (Lees-Miller et al., 1996; Parkinson et al., 1999), although the significance is still not completely understood. While DNA-PK has been shown to phosphorylate IRF3 (Karpova et al., 2002), the absence of this enzyme in either immortalized (Melroe et al., 2004) or primary (Noyce and Mossman, 2006) cells fails to preclude accumulation of ISGs following HSV-1 infection. Although the mechanism by which the RING finger of ICP0 mediates the inhibition of IRF3 is unknown, a recent study demonstrated that ICP0 blocks the nuclear accumulation of phosphorylated IRF3 following infection of cells with Sendai virus (Melroe et al., 2004). Studies are currently under way in a number of laboratories to determine the precise mechanism of ICP0-mediated inhibition of IRF3.

In addition to ICP0, HSV-1 encodes at least one additional protein that facilitates inhibition of IRF3-mediated ISG and/or IFN-β production. Northern blot and RT-

PCR analyses of mRNA harvested from HSV-1 infected cells demonstrated that in the absence of ICP0, ISG mRNA species were detected at early, but not late, times postinfection, whereas in the complete absence of virus replication, ISG mRNA levels were maintained (Mossman *et al.*, 2001; Lin *et al.*, 2004). HSV-1 particles contain within their tegument the virion-host shut-off protein (vhs) that either is, or is part of, a ribonuclease that causes the degradation of both viral and cellular mRNAs (Read *et al.*, 1993; Strelow and Leib, 1995; Elgadi *et al.*, 1999). Infection of human fibroblasts with a mutant HSV-1 harboring deletions in both ICP0 and vhs led to sustained accumulation of ISG mRNA species, demonstrating that at late times postinfection, vhs functions to degrade ISG messages, presumably in a nonspecific fashion (Lin *et al.*, 2004). A similar result was observed in a series of co-infection studies using Sendai virus as an activator of the IRF3 pathway (Melroe *et al.*, 2004). Co-infection of fibroblasts with Sendai virus and either *d*109 or *d*106, HSV-1 recombinants engineered to express either no viral proteins or only the ICP0 protein, respectively, led to a marginal (*d*109) or significant (*d*106) decrease, but not inhibition, of Sendai virus-mediated IFN production. The slight inhibition of IFN production following co-infection with *d*109 was suggested to occur due to the presence of vhs within the incoming viral tegument. The inability of *d*106 to completely suppress IFN production led the authors to suggest that an additional component of HSV-1 may also function in an inhibitory fashion.

With respect to the related alpha-herpesvirus VZV, a recent investigation using severe combined immunodeficiency mice with human skin grafts (SCIDhu mice) to study transfer of infectious virus from T cells to the skin demonstrated that the pathology of VZV is modulated by IFN-α (Ku *et al.*, 2004). Naïve epidermal cells were found to spontaneously express IFN-α, whereas epidermal cells infected with VZV blocked IFN-α production. Immunostaining of VZV-infected skin grafts revealed that a VZV gene product functions to block the nuclear accumulation of phosphorylated Stat 1. Collectively, these observations suggest that downregulation of IFN-α production by VZV allows focal replication to occur to levels sufficient to allow propagation of the virus from one individual to the next, whereas surrounding, un-infected cells that produce IFN-α form an innate barrier against uncontrolled spread. Identification of the VZV gene product will be of extreme interest and importance.

Inhibition of interferon signaling

Once produced and secreted, IFN-α/β elicits an antiviral program in surrounding cells by binding to the type I IFN receptor and activating the IFN signal transduction pathway (Darnell *et al.*, 1994; Darnell, 1998; Stark *et al.*, 1998; Goodbourne *et al.*, 2000). Receptor binding stimulates phosphorylation of the associated Janus kinases Jak1 and Tyk2, which in turn phosphorylate Stat1 and Stat2. The phosphorylated Stat molecules subsequently associate with IRF9 (also known as p48 or ISGF3γ) to form the transcription factor complex ISGF3, which translocates to the nucleus and binds to ISRE elements in promoter regions of ISGs. Many viruses suppress the IFN signal transduction pathway (Alcami and Koszinowski, 2000; Goodbourne *et al.*, 2000), and alpha-herpesviruses are no exception. During the early stages of infection, HSV-1 was found to limit IFN-mediated induction of a reporter plasmid due to prevention of Jak1, Tyk2, Stat1 and Stat2 phosphorylation (Yokota *et al.*, 2001). At later times of infection, a significant decrease in the basal levels of Jak1 and Stat2 protein was also observed. HSV-1-mediated suppression of Jak/Stat phosphorylation was suggested to occur through the activation of suppressor of cytokine signaling 3 (SOCS-3), a negative regulator of IFN signaling, since HSV-1 infection led to a significant accumulation of SOCS-3 between 1 and 2 hours postinfection, when Jak/Stat phosphorylation was shown to decline (Yokota *et al.*, 2004). Viral gene expression was found to be required for these activities (Yokota *et al.*, 2001), and studies with gene deletion HSV-1 recombinants implicated the tegument proteins vhs and UL13 (Yokota *et al.*, 2004). UL13 encodes a protein kinase shown to regulate the function of vhs (Overton *et al.*, 1994). In the absence of either vhs or UL13, the level of Stat1 phosphorylation and the accumulation of SOCS-3 protein were intermediate in nature compared with treatment with IFN alone or in combination with wild type HSV-1 infection. These data suggest that at early times of infection, tegument derived vhs and UL13 play a role, albeit limited, in suppressing IFN-mediated signaling. Since SOCS-3 production and inhibition of Jak/Stat phosphorylation are detected within 1–2 hours of infection, an additional viral gene product, presumably an immediate early or early protein, likely functions to aid in this inhibition. The HSV-1 mediated decline in Jak1 and Stat2 levels and the involvement of vhs in this decline were subsequently confirmed by an independent group (Chee and Roizman, 2004). At late times postinfection the level of the type I IFN receptor was also found to decrease, although it is not clear if this decrease is significant at these late times of infection when cellular protein synthesis in general is known to be shut-down by HSV-1.

Inhibition of interferon-stimulated gene function

Aside from encoding proteins that block IRF3 activity and IFN signal transduction, alpha-herpesviruses also code for proteins that function to overcome IFN-mediated blocks to virus transcription and translation. ICP0-deficient mutants are hypersensitive to pretreatment of cells with IFN, leading to decreases in viral mRNA accumulation and subsequent virus propagation (Mossman *et al.*, 2000; Nicholl *et al.*, 2000; Eidson *et al.*, 2002; Mossman and Smiley, 2002). Using transcriptional run-on assays, it was shown that ICP0 overcomes an IFN-mediated block to virus transcription (Mossman and Smiley, 2002). The precise nature of the transcriptional block is currently not understood, although the cellular protein PML has been implicated. PML is an essential component of nuclear bodies known as PML bodies, Kremer bodies, ND10 or PODs (for PML oncogenic domains) (Ascoli and Maul, 1991; Hodges *et al.*, 1998; Maul *et al.*, 2000). Of particular interest, PML has been identified as a tumor suppressor (Salomoni and Pandolfi, 2002) and as a mediator of IFN function and immune surveillance, in that IFN dramatically increases the expression of PML and other ND10 components, including Sp100 and ISG20, and overexpression of PML induces resistance to infection of a number of viruses (Zhong *et al.*, 2000). Incoming viral genomes of several DNA virus families localize to the periphery of ND10 prior to viral gene expression and the early steps of transcription and replication are thought to occur at these sites (Ishov and Maul, 1996). Subsequent viral protein production results in the dramatic disturbance of ND10 and, in the case of HSV-1, dispersal of ND10 components early in infection is mediated by ICP0 (Everett and Maul, 1994; Maul and Everett, 1994). In particular, ICP0 induces the proteasome-mediated degradation of SUMO-modified forms of PML and Sp100, via its RING finger domain (Everett *et al.*, 1998; Muller and Dejean, 1999; Parkinson and Everett, 2000). Experiments with mouse fibroblasts derived from embryos of wild type or PML deficient mice demonstrated that in the absence of exogenous IFN, wild type and ICP0 null HSV-1 viruses replicated with equal efficiency regardless of PML status whereas in the presence of exogenous IFN, the antiviral effects of this cytokine were exaggerated following infection of PML-containing cells with ICP0 null HSV-1 (Chee *et al.*, 2003). While these findings suggest that ICP0 blocks IFN-induced antiviral effects through degradation of PML, the mechanism is unclear as PML has not been shown to have inhibitory effects on HSV-1 transcription.

Aside from blocking HSV-1 transcription, several ISGs also function in blocking HSV-1 translation (Altinkilic and Brandner, 1988; Harle *et al.*, 2001). Although the role of the OAS pathway in controlling herpesvirus infection is unclear, HSV-1 induces the synthesis of 2–5A derivatives that bind and inhibit RNase L (Cayley *et al.*, 1984). The mechanism by which HSV-1 induces these derivatives is unknown. Aside from degrading viral mRNA species, another strategy used by host cells is to incapacitate the cellular translational machinery on which the synthesis of herpesvirus proteins completely depends. As outlined above, this strategy is mediated in part through the phosphorylation of eIF2α by specific cellular protein kinases. Genetic and biochemical studies have shown that HSV-1 encodes at different stages in the virus replication cycle at least two proteins, ICP34.5 and Us11, that function to circumvent phosphorylation of eIF2α (reviewed in Mohr, 2004). The HSV-1 ICP34.5 protein, which shows homology to the cellular DNA damage protein GADD34, was originally shown to be a critical determinant of virus infection and HSV-induced virulence *in vivo* (Chou *et al.*, 1990). Deletion of ICP34.5 renders the virus non-neurovirulent and imposes a block to viral replication in neuronal cells. Subsequent work demonstrated that this protein overcomes IFN-induced blocks to virus replication *in vivo* (Leib *et al.*, 1999), specifically by circumventing IFN-mediated inhibition of viral protein translation (Cheng *et al.*, 2001; Mossman and Smiley, 2002). Following infection of cells with both wild type and ICP34.5 null strains of HSV-1, PKR activation is detected, although phosphorylation of eIF2α is only observed in the absence of ICP34.5 (Chou *et al.*, 1995). Investigations into this observation led to the discovery that ICP34.5 recruits cellular protein phosphatase 1 and forms a high molecular weight complex that dephosphorylates eIF2α, enabling the maintenance of protein synthesis (He *et al.*, 1997b; He *et al.*, 1998). Consistent with these observations, HSV-1 mutants lacking ICP34.5 are virulent in PKR-null mice but not wild type mice (Leib *et al.*, 2000).

Serial passage of ICP34.5 null viruses led to the selection of suppressor mutants capable of surmounting the translational block imposed by activated PKR (Mohr and Gluzman, 1996). The suppressor mutants contained spontaneous deletions resulting in the juxtaposition of the ICP47 immediate early promoter to the coding domain of the Us11 gene (He *et al.*, 1997a). This juxtaposition shifted expression of Us11 from the late to the immediate early stage of infection and led to restoration of protein synthesis. Although immediate early expression of Us11 blocked the downstream function of activated PKR, the suppressor mutants remained avirulent *in vivo* (Mohr and Gluzman, 1996; Cassady *et al.*, 1998a; Cassady *et al.*, 1998b; Mohr *et al.*, 2001). Us11 is a multifunctional protein that is involved in posttranscriptional regulation of gene expression and in biological processes that relate to cell survival following environmental stress. Us11 is abundantly expressed as a late protein under normal conditions and is packaged into the tegument of virus particles (Roller and Roizman, 1992). While Us11 has been found to interact with a number of cellular proteins and bind to DNA, in the context of the IFN response its ability to bind dsRNA and associate with ribosomes, PKR and a protein activator of

PKR (PACT) (Peters *et al.*, 2001a), is of fundamental importance (Roller and Roizman, 1990; Roller and Roizman, 1992; Peters *et al.*, 2001b; Cassady and Gross, 2002; Khoo *et al.*, 2002; Giraud *et al.*, 2004). *In vitro* biochemical assays demonstrated that Us11, if expressed prior to PKR activation, could bind PKR, preclude its autophosphorylation and hence block PKR-mediated phosphorylation of eIF2α (Cassady *et al.*, 1998a; Cassady and Gross, 2002). Recent evidence suggests that both ICP34.5 and Us11 are required for complete regulation of eIF2α phosphorylation and full resistance to IFN-mediated translational arrest (Mulvey *et al.*, 2003; Mulvey *et al.*, 2004). Of interest, a second series of suppressor mutants were selected following serial propagation of ICP34.5 null viruses that fail to map to the Us11 region (Cassady *et al.*, 2002). These mutants display sustained late protein synthesis due to reduced levels of PKR phosphorylation, which appears to be phosphatase dependent. While the location of the second-site, compensatory mutation remains to be mapped, the overall conclusion from this discovery is that blocking the IFN pathway is greatly advantageous to the virus.

Confirmation of importance of herpesvirus proteins in blocking IFN pathways *in vivo*

Identification of ICP0, ICP34.5, vhs, UL13 and Us11 as viral modulators of type I IFN production and activity primarily resulted from genetic and biochemical analyses *in vitro*. In many cases, however, the importance of these viral proteins in counteracting the host innate immune response has been confirmed *in vivo* in a variety of mouse models. An elegant set of experiments to test the hypothesis that null mutations in host genes can override attenuated phenotypes of null mutations in viral genes confirmed the importance of ICP0, ICP34.5 and vhs in blocking host IFN responses (Leib *et al.*, 1999). Wild type and null mutants of HSV-1 were tested for their ability to replicate in the corneas and trigeminal ganglia of mice with null mutations in type I IFN receptors, type II IFN receptors or both type I and II IFN receptors. Replication of a vhs or an ICP0 null mutant within mouse corneas was enhanced 2 to 3 logs or 4 to 5 logs, respectively, in IFN receptor null mice versus wild type mice, with virus titres in the IFN receptor null mice approaching that of wild type virus. Within trigeminal ganglia, ICP0 and vhs null viruses replicated to readily detectable levels in IFN receptor null mice, but were undetectable in wild type mice. The ICP34.5 mutant demonstrated a slightly different phenotype with little enhancement of virus replication in the corneas of IFN receptor null mice versus the parental controls. However, replication of the ICP34.5 mutant in the trigeminal ganglia of IFN receptor null mice was robust and led to uniform death of the mice due to viral encephalitis. This investigation also revealed a role for the viral proteins ribonucleotide reductase and thymidine kinase in counteracting IFN responses. A similar investigation by the same group utilizing wild type and ICP34.5 mutant viruses in mice expressing full-length or truncated PKR clearly demonstrated that the *in vivo* target of ICP34.5 is the IFN-inducible PKR protein (Leib *et al.*, 2000).

The importance of vhs in mediating resistance to host IFN responses was corroborated in at least two separate studies. In the first study, BALB/c mice were found to be fairly resistant to encephalitis following intracranial inoculation with an HSV-1 vhs null mutant, due to a 20-fold increase in sensitivity of the mutant virus to IFN-α (Suzutani *et al.*, 2000). In the second study, BALB/c or congenic SCID mice were infected intravaginally with an HSV-2 vhs null mutant; even in the SCID mice, which retain innate antiviral effectors, the vhs null virus remained severely attenuated (Murphy *et al.*, 2003). When the same vhs mutant was used to infect type I IFN receptor null mice, it replicated to nearly wild type levels and demonstrated all of the hallmarks of a virulent virus, including severe genital disease and hind leg paralysis. Infection of BALB/c mice with an HSV-1 UL13 null virus resulted in similar findings to the first study outlined above, in that HSV-1 viruses lacking UL13 demonstrated a higher sensitivity to IFN-α and elicited elevated IFN production in mice, resulting in rapid clearance of the mutant virus (Shibaki *et al.*, 2001). The similarities within these two studies are not surprising given the apparent role of UL13 in modulating the activity of vhs (Overton *et al.*, 1994).

Conclusions

Evolutionary pressures demand that both viruses and their hosts develop strategies to ensure survival and maintenance of their species. The host immune response is complex and requires the co-ordinate activation of multiple immune modulators and cells types. The type I IFNs play an important role within the early phase of the immune response to suppress virus replication and spread. In response, viruses such as the herpesviruses have evolved mechanisms to delay suppression by their hosts. These mechanisms target all aspects of the IFN pathway and function in concert to ensure the maintenance and propagation of the virus within its host. As more viral gene products are studied and the function of more ISGs are identified, it is likely that the list of herpesvirus proteins that function to suppress the host IFN response will increase.

Acknowledgments

I would like to thank William Halford and Jeffrey Hummel for critical review of this book chapter. The Canadian Institutes of Health Research funded work from my laboratory.

References

Alcami, A., and Koszinowski, U.H. (2000). Viral mechanisms of immune evasion. Immunol. Today *21*, 447–455.

Al-Khatib, K., Williams, B.R., Silverman, R.H., Halford, W.P., and Carr, D.J. (2002). Absence of PKR attenuates the anti-HSV-1 activity of an adenoviral vector expressing murine IFN-beta. J. Interferon Cytokine Res. *22*, 861–871.

Al-Khatib, K., Williams, B.R., Silverman, R.H., Halford, W., and Carr, D.J. (2003). The murine double-stranded RNA-dependent protein kinase PKR and the murine 2′,5′-oligoadenylate synthetase-dependent RNase L are required for IFN-beta-mediated resistance against herpes simplex virus type 1 in primary trigeminal ganglion culture. Virology *313*, 126–135.

Al-Khatib, K., Williams, B.R., Silverman, R.H., Halford, W., and Carr, D.J. (2004). Distinctive roles for 2′,5′-oligoadenylate synthetases and double-stranded RNA-dependent protein kinase R in the *in vivo* antiviral effect of an adenoviral vector expressing murine IFN-beta. J. Immunol. *172*, 5638–5647.

Altinkilic, B., and Brandner, G. (1988). Interferon inhibits herpes simplex virus-specific translation: a reinvestigation. J. Gen. Virol. *69*, 3107–3112.

Ankel, H., Westra, D.F., Welling-Wester, S., and Lebon, P. (1998). Induction of interferon-alpha by glycoprotein D of herpes simplex virus: a possible role of chemokine receptors. Virology *251*, 317–326.

Arvin, A.M. (1996). Immune responses to varicella-zoster virus. Infect. Dis. Clin. North Am. *10*, 529–570.

Ascoli, C.A., and Maul, G.G. (1991). Identification of a novel nuclear domain. J. Cell Biol. *112*, 785–795.

Bandyopadhyay, S.K., Leonaard, G.T., Bandyopadhyay, T., Stark, G.R., and Sen, G.C. (1995). Transcriptional induction by double-stranded RNA is mediated by interferon-stimulated response elements without activation of interferon-stimulated gene factor 3. J. Biol. Chem. *270*, 19624–19629.

Barber, G.N. (2001). Host defense, viruses and apoptosis. Cell Death Differ. *8*, 113–126.

Biron, C.A., Nguyen, K.B., Pien, G.C., Cousens, L.P., and Salazar-Mather, T.P. (1999). Natural killer cells in antiviral defense: function and regulation by innate cytokines. Annu. Rev. Immunol. *17*, 189–220.

Bjorck, P. (2001). Isolation and characterization of plasmacytoid dendritic cells from Flt3 ligand and granulocyte-macrophage colony-stimulating factor-treated mice. Blood *98*, 3520–3526.

Borysiewicz, L.K., and Sissons, J.G. (1994). Cytotoxic T cells and human herpes virus infections. Curr. Topics Microbiol. Immunol. *189*, 123–150.

Boutell, C., Sadis, S., and Everett, R.D. (2002). Herpes simplex virus type 1 immediate-early protein ICP0 and its isolated RING finger domain act as ubiquitin E3 ligases *in vitro*. J. Virol. *76*, 841–850.

Cassady, K.A., and Gross, M. (2002). The herpes simplex virus type 1 U(S)11 protein interacts with protein kinase R in infected cells and requires a 30-amino-acid sequence adjacent to a kinase substrate domain. J. Virol. *76*, 2029–2035.

Cassady, K.A., Gross, M., and Roizman, B. (1998a). The herpes simplex virus US11 protein effectively compensates for the gamma1(34.5) gene if present before activation of protein kinase R by precluding its phosphorylation and that of the alpha subunit of eukaryotic translation initiation factor 2. J. Virol. *72*, 8620–8626.

Cassady, K.A., Gross, M., and Roizman, B. (1998b). The second-site mutation in the herpes simplex virus recombinants lacking the $\gamma_1$34.5 genes precludes shutoff of protein synthesis by blocking the phosphorylation of eIF-2α. J. Virol. *72*, 7005–7011.

Cassady, K.A., Gross, M., Gillespie, G.Y., and Roizman, B. (2002). Second-site mutation outside of the U(S)10–12 domain of Deltagamma(1)34.5 herpes simplex virus 1 recombinant blocks the shutoff of protein synthesis induced by activated protein kinase R and partially restores neurovirulence. J. Virol. *76*, 942–949.

Castelli, J.C., Hassel, B.A., Wood, K.A., Li, X.L., Amemiya, K., Dalakas, M.C., Torrence, P.F., and Youle, R.J. (1997). A study of the interferon antiviral mechanism: apoptosis activation by the 2–5A system. J. Exp. Med. *186*, 967–972.

Castelli, J.C., Hassel, B.A., Maran, A., Paranjape, J., Hewitt, J.A., Li, X.L., Hsu, Y.T., Silverman, R.H., and Youle, R.J. (1998). The role of 2′-5′ oligoadenylate-activated ribonuclease L in apoptosis. Cell Death Differ. *5*, 313–320.

Cayley, P.J., Davies, J.A., McCullagh, K.G., and Kerr, I.M. (1984). Activation of the ppp(A2′p)nA system in interferon-treated, herpes simplex virus-infected cells and evidence for novel inhibitors of the ppp(A2′p)nA-dependent RNase. Eur. J. Biochem. *143*, 165–174.

Cebulla, C.M., Miller, D.M., and Sedmak, D.D. (1999). Viral inhibition of interferon signal transduction. Intervirology *42*, 325–330.

Chatterjee, S., and Burns, P. (1990). Expression of herpes simplex virus type 1 glycoproteins in interferon-treated human neuroblastoma cells. J. Virol. *64*, 5209–5213.

Chee, A.V., and Roizman, B. (2004). Herpes simplex virus 1 gene products occlude the interferon signaling pathway at multiple sites. J. Virol. *78*, 4185–4196.

Chee, A.V., Lopez, P., Pandolfi, P.P., and Roizman, B. (2003). Promyelocytic leukemia protein mediates interferon-based anti-herpes simplex virus 1 effects. J. Virol. *77*, 7101–7105.

Cheng, G., Brett, M., and He, B. (2001). Val^{193} and Phe^{195} of the $\gamma_1$34.5 protein of herpes simplex virus 1 are required for viral resistance to interferon-α/β. Virology *290*, 115–120.

Chou, J., Kern, E.R., Whitley, R.J., and Roizman, B. (1990). Mapping of herpes simplex virus-1 neurovirulence to gamma 134.5, a gene nonessential for growth in culture. Science *250*, 1262–1266.

Chou, J., Chen, J.J., Gross, M., and Roizman, B. (1995). Association of a M(r) 90,000 phosphoprotein with protein kinase PKR in cells exhibiting enhanced phosphorylation of translation initiation factor eIF-2 alpha and premature shutoff of protein synthesis after infection with gamma 1 34.5 mutants of herpes simplex virus 1. Proc. Natl. Acad. Sci. USA *92*, 10516–10520.

Collins, S.E., Noyce, R.S., and Mossman, K.L. (2004). The innate cellular response to virus particle entry requires IRF3 but not virus replication. J. Virol. *78*, 1706–1717.

Daly, C., and Reich, N.C. (1993). Double-stranded RNA activates novel factors that bind to the interferon-stimulated response element. Mol. Cell. Biol *13*, 3756–3764.

Darnell, J.E., Jr. (1998). Studies of IFN-induced transcriptional activation uncover the Jak-Stat pathway. J. Interferon Cytokine Res. *18*, 549–554.

Darnell, J.E., Jr., Kerr, I.M., and Stark, G.R. (1994). Jak-STAT pathways and transcriptional activation in response to IFNs and other extracellular signaling proteins. Science *264*, 1415–1421.

Davies, M.V., Elroy-Stein, O., Jagus, R., Moss, B., and Kaufman, R.J. (1992). The vaccinia virus K3L gene product potentiates translation by inhibiting double-stranded-RNA-activated protein kinase and phosphorylation of the alpha subunit of eukaryotic initiation factor 2. J. Virol. *66*, 1943–1950.

Davies, M.V., Chang, H.W., Jacobs, B.L., and Kaufman, R.J. (1993). The E3L and K3L vaccinia virus gene products stimulate translation through inhibition of the double-stranded RNA-dependent protein kinase by different mechanisms. J. Virol. *67*, 1688–1692.

D'Cunha, J., Knight, E., Jr., Haas, A.L., Truitt, R.L., and Borden, E.C. (1996). Immunoregulatory properties of ISG15, an interferon-induced cytokine. Proc. Natl. Acad. Sci. USA *93*, 211–215.

de Veer, M.J., Holko, M., Frevel, M., Walker, E., Der, S., Paranjape, J.M., Silverman, R.H., and Williams, B.R. (2001). Functional classification of interferon-stimulated genes identified using microarrays. J. Leukoc. Biol. *69*, 912–920.

Deshpande, S.P., Kumaraguru, U., and Rouse, B.T. (2000). Dual role of B cells in mediating innate and acquired immunity to herpes virus infections. Cell. Immunol. *202*, 79–87.

Desloges, N., Rahaus, M., and Wolff, M.H. (2004). Varicella-zoster virus does not significantly induce the cell defence mechanism mediated by the 2–5A/RNase L pathway during its replication cycle. Med. Microbiol. Immunol. (Berl).

DeStasio, R.P., and Taylor, M.W. (1990). Specific effect of interferon on the herpes simplex virus type 1 transactivation event. J. Virol. *64*, 2588–2593.

DiPerna, G., Stack, J., Bowie, A.G., Boyd, A., Kotwal, G., Zhang, Z., Arvikar, S., Latz, E., Fitzgerald, K.A., and Marshall, W.L. (2004). Poxvirus protein N1L targets the I-kappa B kinase complex, inhibits signaling to NF-kappa B by the tumor necrosis factor superfamily of receptors, and inhibits NF-kappa B and IRF3 signaling by toll-like receptors. J. Biol. Chem., M400567200.

Eidson, K.M., Hobbs, W.E., Manning, B.J., Carlson, P., and DeLuca, N.A. (2002). Expression of herpes simplex virus ICP0 inhibits the induction of interferon-stimulated genes by viral infection. J. Virol. *76*, 2180–2191.

Elgadi, M.M., Hayes, C.E., and Smiley, J.R. (1999). The herpes simplex virus vhs protein induces endoribonucleoltyic cleavage of target RNAs in cell extracts. J. Virol. *73*, 7153–7164.

Everett, R.D. (2000). ICP0, a regulator of herpes simplex virus during lytic and latent infection. Bioessays *22*, 761–770.

Everett, R.D., and Maul, G.G. (1994). HSV-1 IE protein Vmw110 causes redistribution of PML. EMBO J. *13*, 5062–5069.

Everett, R.D., Freemont, P., Saitoh, H., Dasso, M., Orr, A., Kathoria, M., and Parkinson, J. (1998). The disruption of ND10 during herpes simplex virus infection correlates with the Vmw110- and proteasome-dependent loss of several PML isoforms. J. Virol. *72*, 6581–6591.

Fitzgerald, K.A., McWhirter, S.M., Faia, K.L., Rowe, D.C., Latz, E., Golenbock, D.T., Coyle, A.J., Liao, S.M., and Maniatis, T. (2003). IKKepsilon and TBK1 are essential components of the IRF3 signaling pathway. Nat. Immunol. *4*, 491–496.

Fitzgerald-Bocarsly, P. (1993). Human natural interferon-α producing cells. Pharmacol. Ther. *60*, 39–62.

Fujihara, M., Milligan, J.R., and Kaji, A. (1989). Effect of 2′,5′-oligoadenylate on herpes simplex virus-infected cells and preventive action of 2′,5′-oligoadenylate on the lethal effect of HSV-2. J. Interferon Res. *9*, 691–707.

Giraud, S., Diaz-Latoud, C., Hacot, S., Textoris, J., Bourette, R.P., and Diaz, J.-J. (2004). US11 of herpes simplex virus type 1 interacts with HIPK2 and antagonizes HIPK2-induced cell growth arrest. J. Virol. *78*, 2984–2993.

Goodbourne, S., Didcock, L., and Randall, R.E. (2000). Interferons: cell signalling, immune modulation, Antiviral responses and virus countermeasures. J. Gen. Virol. *81*, 2341–2364.

Grandvaux, N., Servant, M.J., tenOever, B., Sen, G.C., Balachandran, S., Barber, G.N., Lin, R., and Hiscott, J. (2002). Transcriptional profiling of interferon regulatory factor 3 target genes: direct involvement in the regulation of interferon-stimulated genes. J. Virol. *76*, 5532–5539.

Guo, J., Hui, D.J., Merrick, W.C., and Sen, G.C. (2000a). A new pathway of translational regulation mediated by eukaryotic initiation factor 3. EMBO J. *19*, 6891–6899.

Guo, J., Peters, K.L., and Sen, G.C. (2000b). Induction of the human protein P56 by interferon, double-stranded RNA, or virus infection. Virology *267*, 209–219.

Hagglund, R., Van Sant, C., Lopez, P., and Roizman, B. (2002). Herpes simplex virus 1-infected cell protein 0 contains two E3 ubiquitin ligase sites specific for different E2 ubiquitin-conjugating enzymes. Proc. Natl. Acad. Sci. USA *99*, 631–636.

Haig, D.M. (2001). Subversion and piracy: DNA viruses and immune evasion. Res. Vet. Sci. *70*, 205–219.

Halford, W.P., and Schaffer, P.A. (2000). Optimized viral dose and transient immunosuppression enable herpes simplex virus ICP0-null mutants To establish wild-type levels of latency *in vivo*. J. Virol. *74*, 5957–5967.

Halford, W.P., Veress, L.A., Gebhardt, B.M., and Carr, D.J. J. (1997). Innate and acquired immunity to herpes simplex virus type 1. Virology *236*, 328–337.

Halford, W.P., Balliet, J.W., and Gebhardt, B.M. (2004). Re-evaluating natural resistance to herpes simplex virus type 1. J. Virol. *78*, 10086–10095.

Haller, O., and Kochs, G. (2002). Interferon-induced Mx proteins: dynamin-like GTPases with antiviral activity. Traffic *3*, 710–717.

Harle, P., Lauret, E., Pitha, P.M., De Maeyer, E., and Carr, D.J. (2001). Expression of human and macaque type I IFN transgenes interferes with HSV-1 replication at the transcriptional and translational levels: IFN-beta is more potent than IFN-alpha 2. Virology *290*, 237–248.

Harle, P., Cull, V., Agbaga, M.P., Silverman, R., Williams, B.R., James, C., and Carr, D.J. (2002a). Differential effect of murine alpha/beta interferon transgenes on antagonization of herpes simplex virus type 1 replication. J. Virol. *76*, 6558–6567.

Harle, P., Jr., B.S., Carr, D.J., and Halford, W.P. (2002b). The immediate-early protein, ICP0, is essential for the resistance of herpes simplex virus to interferon-alpha/beta. Virology *293*, 295–304.

He, B., Chou, J., Brandimarti, R., Mohr, I., Gluzman, Y., and Roizman, B. (1997a). Suppression of the phenotype of $\gamma_1 34.5^-$ herpes simplex virus 1: failure of activated RNA-dependent protein kinase to shut off protein synthesis is associated with a deletion in the domain of the α47 gene. J. Virol. *71*, 6049–6054.

He, B., Gross, M., and Roizman, B. (1997b). The $\gamma_1$34.5 protein of herpes simplex virus 1 complexes with protein phosphatase 1 α to dephosphorylate the α subunit of the eukaryotic translation initiation factor 2 and preclude the shutoff of protein synthesis by double-stranded RNA-activated protein kinase. Proc. Natl. Acad. Sci. USA *94*, 843–848.

He, B., Gross, M., and Roizman, B. (1998). The gamma134.5 protein of herpes simplex virus 1 has the structural and functional attributes of a protein phosphatase 1 regulatory subunit and is present in a high molecular weight complex with the enzyme in infected cells. J. Biol. Chem. *273*, 20737–20743.

Hodges, M., Tissot, C., Howe, K., Grimwade, D., and Freemont, P.S. (1998). Structure, organization, and dynamics of promyelocytic leukemia protein nuclear bodies. Am. J. Hum. Genet. *63*, 297–304.

Holberg-Petersen, M., Bukholm, G., Haarr, L., Langeland, N., and Degre, M. (1996). Human interferon reduces surface expression but not total production of herpes simplex virus type 1 glycoproteins gC and gE in heterologous hamster cells. J. Interferon Cytokine Res. *16*, 717–724.

Ishov, A.M., and Maul, G.G. (1996). The periphery of nuclear domain 10 (ND10) as site of DNA virus deposition. J. Cell Biol. *134*, 815–826.

Ito, Y. (1994). Induction of interferon by virus glycoprotein(s) in lymphoid cells through interactions with the cellular receptors via lectin-like action: an alternative interferon induction mechanism. Arch. Virol. *138*, 187–198.

Ito, M., Mizutani, K., Nakano, T., Yasuda, N., Kamiya, T., Kitamura, K., Torigoe, S., Ihara, T., Kamiya, H., and Sakurai, M. (1992). Flow cytometric analysis of effects of cytokines on the expression of varicella-zoster virus glycoproteins. Antiviral Res. *18*, 317–326.

Jacquemont, B., and Roizman, B. (1975). RNA synthesis in cells infected with herpes simplex virus. X. Properties of viral symmetric transcripts and of double-stranded RNA prepared from them. J. Virol. *15*, 707–713.

Juang, Y., Lowther, W., Kellum, M., Au, W.C., Lin, R., Hiscott, J., and Pitha, P.M. (1998). Primary activation of interferon A and interferon B gene transcription by interferon regulatory factor 3. Proc. Natl. Acad. Sci. USA *95*, 9837–9842.

Kadowaki, N., Antonenko, S., Lau, J.Y., and Liu, Y.J. (2000). Natural interferon alpha/beta-producing cells link innate and adaptive immunity. J. Exp. Med. *192*, 219–226.

Karpova, A.Y., Trost, M., Murray, J.M., Cantley, L.C., and Howley, P.M. (2002). Interferon regulatory factor-3 is an *in vivo* target of DNA-PK. Proc. Natl. Acad. Sci. USA *99*, 2818–2823.

Katze, M.G., He, Y., and Gale, M., Jr. (2002). Viruses and interferon: a fight for supremacy. Nat. Rev. Immunol. *2*, 675–687.

Khabar, K.S. A., Dhalla, M., Siddiqui, Y., Zhou, A., Al-Ahdal, M.N., Der, S.D., Silverman, R.H., and Williams, B.R. G. (2000). Effect of deficiency of the double-stranded RNA dependent protein kinase, PKR, on antiviral resistance in the presence or absence of ribonuclease L: HSV-1 replication is particularly sensitive to deficiency of the major IFN-mediated enzymes. J. Interferon Cytokine Res. *20*, 653–659.

Khoo, D., Perez, C., and Mohr, I. (2002). Characterization of RNA determinants recognized by the arginine- and proline-rich region of Us11, a herpes simplex virus type 1-encoded double-stranded RNA binding protein that prevents PKR activation. J. Virol. *76*, 11971–11981.

Kontsek, P., Karayianni-Vasconcelos, G., and Kontsekova, E. (2003). The human interferon system: characterization and classification after discovery of novel members. Acta Virol. *47*, 201–215.

Ku, C.-C., Zerboni, L., Ito, H., Graham, B.S., Wallace, M., and Arvin, A.M. (2004). Varicella-zoster virus transfer to skin by T cells and modulation of viral replication by epidermal cell interferon-α. J. Exp. Med., jem.20040634.

Kumar, K.P., McBride, K.M., Weaver, B.K., Dingwall, C., and Reich, N.C. (2000). Regulated nuclear-cytoplasmic localization of interferon regulatory factor 3, a subunit of double-stranded RNA-activated factor 1. Mol. Cell. Biol *20*, 4159–4168.

Kunder, S.C., Kelly, K.M., and Morahan, P.S. (1993). Biological response modifier-mediated resistance to herpesvirus infections requires induction of alpha/beta interferon. Antiviral Res. *21*, 129–139.

Lees-Miller, S.P., and Meek, K. (2003). Repair of DNA double strand breaks by non-homologous end joining. Biochimie *85*, 1161–1173.

Lees-Miller, S.P., Long, M.C., Kilvert, M.A., Lam, V., Rice, S.A., and Spencer, C.A. (1996). Attenuation of DNA-dependent protein kinase activity and its catalytic subunit by the herpes simplex virus type I transactivator ICP0. J. Virol. *70*, 7471–7477.

Leib, D.A. (2002). Counteraction of interferon-induced Antiviral responses by herpes simplex viruses. Curr. Topics Microbiol. Immunol. *269*, 171–185.

Leib, D.A., Harrison, T.E., Laslo, K.M., Machalek, M.A., Moorman, N.J., and Virgin, H.W. (1999). Interferons regulate the phenotype of wild-type and mutant herpes simplex viruses *in vivo*. J. Exp. Med. *189*, 663–672.

Leib, D.A., Machalek, M.A., Williams, B.R., Silverman, R.H., and Virgin, H.W. (2000). Specific phenotypic restoration of an attenuated virus by knockout of a host resistance gene. Proc. Natl. Acad. Sci. USA *97*, 6097–6101.

Leonard, W.J. (2001). Role of Jak kinases and STATs in cytokine signal transduction. Int. J. Hematol. *73*, 271–277.

LePage, C., Genin, P., Baines, M.G., and Hiscott, J. (2000). Interferon activation and innate immunity. Rev. Immunogenet. *2*, 374–386.

Levy, D.E. (1995). Interferon induction of gene expression through the Jak-Stat pathway. Semin. Virol. *6*, 181–189.

Levy, D.E., and Garcia-Sastre, A. (2001). The virus battles: IFN induction of the antiviral state and mechanisms of viral evasion. Cytokine Growth Factor Rev. *12*, 143–156.

Levy, D.E., Marie, I., Smith, E., and Prakash, A. (2002). Enhancement and diversification of IFN induction by IRF-7-mediated positive feedback. J. Interferon Cytokine Res. *22*, 87–93.

Li, Q., Feldman, M., Harmon, C., and Fitzgerald-Bocarsly, P. (1996). Role of tyrosine kinases, protein kinase C, and protein kinase A in the regulation of interferon-alpha production induced by herpes simplex virus type 1. J. Interferon Cytokine Res. *16*, 109–118.

Lin, R., Heylbroeck, C., Pitha, P.M., and Hiscott, J. (1998). Virus-dependent phosphorylation of the IRF-3 transcription factor regulates nuclear translocation, transactivation potential, and proteasome-mediated degradation. Mol. Cell. Biol *18*, 2986–2996.

Lin, R., Mamane, Y., and Hiscott, J. (2000). Multiple Regulatory Domains Control IRF-7 Activity in Response to Virus Infection. J. Biol. Chem. *275*, 34320–34327.

Lin, R., Noyce, R.S., Collins, S.E., Everett, R.D., and Mossman, K.L. (2004). The herpes simplex virus ICP0 RING finger domain inhibits IRF3- and IRF7-mediated activation of interferon-stimulated genes. J. Virol. *78*, 1675–1684.

Lopez, C. (1985). Natural resistance mechanisms in herpes simplex virus infections. In The Herpesviruses, B. Roizman, and C. Lopez, eds. (New York, Plenum Press), pp. 37–68.

Luker, G.D., Prior, J.L., Song, J., Pica, C.M., and Leib, D.A. (2003). Bioluminescence imaging reveals systemic dissemination of herpes simplex virus type 1 in the absence of interferon receptors. J. Virol. *77*, 11082–11093.

Maheshwari, R.K., Sidhu, G., Singh, A.K., Sivaram, S.S., Kinchington, P.R., Hay, J., and Friedman, R.M. (1994). Defective transport of herpes simplex virus glycoprotein in interferon-treated cells: role of intracellular pH. J. Interferon Res. *14*, 319–324.

Marie, I., Durbin, J.E., and Levy, D.E. (1998). Differential viral induction of distinct interferon-alpha genes by positive feedback through interferon regulatory factor-7. EMBO J. *17*, 6660–6669.

Maul, G.G., and Everett, R.D. (1994). The nuclear location of PML, a cellular member of the C_3HC_4 zinc-binding domain protein family, is rearranged during herpes simplex virus infection by he C_3HC_4 viral protein ICP0. J. Gen. Virol. *75*, 1223–1233.

Maul, G.G., Negorev, D., Bell, P., and Ishov, A.M. (2000). Review: properties and assembly mechanisms of ND10, PML bodies, or PODs. J. Struct. Biol. *129*, 278–287.

Meek, K., Gupta, S., Ramsden, D.A., and Lees-Miller, S.P. (2004). The DNA-dependent protein kinase: the director at the end. Immunol. Rev. *200*, 132–141.

Melroe, G.T., DeLuca, N.A., and Knipe, D.M. (2004). Herpes simplex virus 1 has multiple mechanisms for blocking virus-induced interferon production. J. Virol. *78*, 8411–8420.

Milone, M.C., and Fitzgerald-Bocarsly, P. (1998). The mannose receptor mediates induction of IFN-alpha in peripheral blood dendritic cells by enveloped RNA and DNA viruses. J. Immunol. *161*, 2391–2399.

Mittnacht, S., Straub, P., Kirchner, H., and Jacobsen, H. (1988). Interferon treatment inhibits onset of herpes simplex virus immediate-early transcription. Virology *164*, 201–210.

Mogensen, T.H., Melchjorsen, J., Malmgaard, L., Casola, A., and Paludan, S.R. (2004). Suppression of proinflammatory cytokine expression by herpes simplex virus type 1. J. Virol. *78*, 5883–5890.

Mohr, I. (2004). Neutralizing innate host defenses to control viral translation in HSV-1 infected cells. Int. Rev. Immunol. *23*, 199–220.

Mohr, I., and Gluzman, Y. (1996). A herpes virus genetic element which affects translation in the absence of the viral GADD34 function. EMBO J. *15*, 4759–4766.

Mohr, I., Sternberg, D., Ward, S., Leib, D., Mulvey, M., and Gluzman, Y. (2001). A herpes simplex virus type 1 gamma34.5 second-site suppressor mutant that exhibits enhanced growth in cultured glioblastoma cells is severely attenuated in animals. J. Virol. *75*, 5189–5196.

Mori, M., Yoneyama, M., Ito, T., Takahashi, K., Inagaki, F., and Fujita, T. (2004). Identification of Ser-386 of interferon regulatory factor 3 as critical target for inducible phosphorylation that determines activation. J. Biol. Chem. *279*, 9698–9702.

Mossman, K.L. (2002). Activation and inhibition of virus and interferon: the herpesvirus story. Viral Immun. *In press*.

Mossman, K.L., and Smiley, J.R. (2002). Herpes simplex virus ICP0 and ICP34.5 counteract distinct interferon-induced barriers to virus replication. J. Virol. *76*, 1995–1998.

Mossman, K., Upton, C., Buller, R.M., and McFadden, G. (1995a). Species specificity of ectromelia virus and vaccinia virus interferon-gamma binding proteins. Virology *208*, 762–769.

Mossman, K., Upton, C., and McFadden, G. (1995b). The myxoma virus-soluble interferon-gamma receptor homolog, M-T7, inhibits interferon-gamma in a species-specific manner. J. Biol. Chem. *270*, 3031–3038.

Mossman, K., Nation, P., Macen, J., Garbutt, M., Lucas, A., and McFadden, G. (1996). Myxoma virus M-T7, a secreted homolog of the interferon-gamma receptor, is a critical virulence factor for the development of myxomatosis in European rabbits. Virology *215*, 17–30.

Mossman, K.L., Saffran, H.A., and Smiley, J.R. (2000). Herpes simplex virus ICP0 mutants are hypersensitive to interferon. J. Virol. *74*, 2052–2056.

Mossman, K.L., Macgregor, P.F., Rozmus, J.J., Goryachev, A.B., Edwards, A.M., and Smiley, J.R. (2001). Herpes simplex virus triggers and then disarms a host Antiviral response. J. Virol. *75*, 750–758.

Muller, S., and Dejean, A. (1999). Viral immediate-early proteins abrogate the modification by SUMO-1 of PML and Sp100 proteins, correlating with nuclear body disruption. J. Virol. *73*, 5137–5143.

Mulvey, M., Poppers, J., Sternberg, D., and Mohr, I. (2003). Regulation of eIF2alpha phosphorylation by different functions that act during discrete phases in the herpes simplex virus type 1 life cycle. J. Virol. *77*, 10917–10928.

Mulvey, M., Camarena, V., and Mohr, I. (2004). Full resistance of herpes simplex virus type 1-infected primary human cells to alpha interferon requires both the Us11 and gamma(1)34.5 gene products. J. Virol. *78*, 10193–10196.

Murphy, J.A., Duerst, R.J., Smith, T.J., and Morrison, L.A. (2003). Herpes simplex virus type 2 virion host shutoff protein regulates alpha/beta interferon but not adaptive immune responses during primary infection *in vivo*. J. Virol. *77*, 9337–9345.

Neville, L.F., Mathiak, G., and Bagasra, O. (1997). The immunobiology of interferon-gamma inducible protein 10 kD (IP-10): a novel, pleiotropic member of the C-X-C chemokine superfamily. Cytokine Growth Factor Rev. *8*, 207–219.

Nicholl, M.J., Robinson, L.H., and Preston, C.M. (2000). Activation of cellular interferon-response genes after infection of human cells with herpes simplex virus type 1. J. Gen. Virol. *81*, 2215–2218.

Noyce, R., Collins, S, and Mossman, K.L. (2006). Signal transduction pathways involved in the interferon-independent antiviral response. J. Virol. *80*, 226–235.

O'Brien, W.J., Narasimhan, J., Guy, J., Tom, P., and Taylor, J.L. (1998). The effects of interferon-α and acyclovir on herpes simplex virus type-1 ribonucleotide reductase. Antiviral Res. *38*, 107–116.

Oberman, F., and Panet, A. (1988). Inhibition of transcription of herpes simplex virus immediate early genes in interferon-treated human cells. J. Gen. Virol. *69*, 1167–1177.

Overton, H., McMillan, D., Hope, L., and Wong-Kai-In, P. (1994). Production of host shutoff-defective mutants of herpes simplex virus type 1 by inactivation of the UL13 gene. Virology *202*, 97–106.

Panet, A., and Falk, H. (1983). Inhibition by interferon of herpes simplex virus thymidine kinase and DNA polymerase in infected and biochemically transformed cells. J. Gen. Virol. *64*, 1999–2006.

Parkinson, J., and Everett, R.D. (2000). Alphaherpesvirus proteins related to herpes simplex virus type 1 ICP0 affect cellular structures and proteins. J. Virol. *74*, 10006–10017.

Parkinson, J., Lees-Miller, S.P., and Everett, R.D. (1999). Herpes simplex virus type 1 immediate-early protein Vmw110 induces the proteasome-dependent degradation of the catalytic subunit of DNA-dependent protein kinase. J. Virol. *73*, 650–657.

Pavlovic, J., Zurcher, T., Haller, O., and Staeheli, P. (1990). Resistance to influenza virus and vesicular stomatitis virus conferred by expression of human MxA protein. J. Virol. *64*, 3370–3375.

Peters, G.A., Hartmann, R., Qin, J., and Sen, G.C. (2001a). Modular structure of PACT: distinct domains for binding and activating PKR. Mol. Cell. Biol *21*, 1908–1920.

Peters, G.A., Poppers, J., and Mohr, I. (2001b). PACT-mediated PKR activation is blocked by HSV-1 US11 protein. Paper presented at:

Annual Meeting of the International Society for Interferon and Cytokine Research (Cleveland, Ohio).

Player, M.R., and Torrence, P.F. (1998). The 2–5A system: modulation of viral and cellular processes through acceleration of RNA degradation. Pharmacol. Ther. *78*, 55–113.

Preston, C.M., Harman, A.N., and Nicholl, M.J. (2001). Activation of interferon response factor-3 in human cells infected with herpes simplex virus type 1 or human cytomegalovirus. J. Virol. *75*, 8909–8916.

Qin, B.Y., Liu, C., Lam, S.S., Srinath, H., Delston, R., Correia, J.J., Derynck, R., and Lin, K. (2003). Crystal structure of IRF-3 reveals mechanism of autoinhibition and virus-induced phosphoactivation. Nat. Struct. Biol. *10*, 913–921.

Read, G.S., Bradley, B.M., and Knight, K. (1993). Isolation of a herpes simplex virus type 1 mutant with a deletion in the virion host shutoff gene and identification of multiple forms of the vhs (UL41) polypeptides. J. Virol. *67*, 7149–7160.

Roller, R.J., and Roizman, B. (1990). The herpes simplex virus US11 open reading frame encodes a sequence-specific RNA binding protein. J. Virol. *64*, 3463–3470.

Roller, R.J., and Roizman, B. (1992). The herpes simplex virus 1 RNA binding protein US11 is a virion component and associates with ribosomal 60S subunits. J. Virol. *66*, 3624–3632.

Salomoni, P., and Pandolfi, P.P. (2002). The role of PML in tumor suppression. Cell *108*, 165–170.

Samuel, C.E. (2001). Antiviral actions of interferons. Clin. Microbiol. Rev. *14*, 778–809.

Sato, M., Tanaka, N., Hata, N., Oda, E., and Taniguchi, T. (1998). Involvement of the IRF family transcription factor IRF-3 in virus-induced activation of the IFN-beta gene. FEBS Lett *425*, 112–116.

Sato, M., Suemori, H., Hata, N., Asagiri, M., Ogasawara, K., Nakao, K., Nakay, T., Katsuki, M., Noguchi, S., Tanaka, N., and Taniguchi, T. (2000). Distinct and essential roles of transcription factors IRF-3 and IRF-7 in response to viruses for IFN-α/β gene induction. Immunity *13*, 539–548.

Sato, M., Taniguchi, T., and Tanaka, N. (2001). The interferon system and interferon regulatory factor transcription factors-studies from gene knockout mice. Cytokine Growth Factor Rev. *12*, 133–142.

Sen, G.C. (2001). Viruses and interferon. Annu. Rev. Microbiol. 55, 255–281.

Sen, G.C., and Ransohoff, R.M. (1993). Interferon-induced antiviral actions and their regulation. Adv. Virus Res. 42, 57–102.

Servant, M.J., ten Oever, B., LePage, C., Conti, L., Gessani, S., Julkunen, I., Lin, R., and Hiscott, J. (2001). Identification of distinct signaling pathways leading to the phosphorylation of interferon regulatory factor 3. J. Biol. Chem. *276*, 355–363.

Servant, M.J., Grandvaux, N., tenOever, B.R., Duguay, D., Lin, R., and Hiscott, J. (2003). Identification of the minimal phosphoacceptor site required for *in vivo* activation of interferon regulatory factor 3 in response to virus and double-stranded RNA. J. Biol. Chem. *278*, 9441–9447.

Sharma, S., tenOever, B.R., Grandvaux, N., Zhou, G.P., Lin, R., and Hiscott, J. (2003). Triggering the interferon Antiviral response through an IKK-related pathway. Science *300*, 1148–1151.

Shibaki, T., Suzutani, T., Yoshida, I., Ogasawara, M., and Azuma, M. (2001). Participation of type I interferon in the decreased virulence of the UL13 gene-deleted mutant of herpes simplex virus type 1. J. Interferon Cytokine Res. *21*, 279–285.

Siegal, F.P., Kadowaki, N., Shodell, M., Fitzgerald-Bocarsly, P.A., Shah, K., Ho, S., Antonenko, S., and Liu, Y. (1999). The nature of the principal type 1 interferon-producing cells in human blood. Science *284*, 1835–1837.

Stark, G.R., Kerr, I.M., Williams, B.R. G., Silverman, R.H., and Schreiber, R.D. (1998). How cells respond to interferons. Annu. Rev. Biochem. *67*, 227–264.

Strelow, L.I., and Leib, D.A. (1995). Role of the virion host shutoff (vhs) of herpes simplex virus type 1 in latency and pathogenesis. J. Virol. *69*, 6779–6786.

Su, Y.H., Oakes, J.E., and Lausch, R.N. (1990). Ocular avirulence of a herpes simplex virus type 1 strain is associated with heightened sensitivity to alpha/beta interferon. J. Virol. *64*, 2187–2192.

Suzue, K., Asai, T., Takeuchi, T., and Koyasu, S. (2003). *In vivo* role of IFN-gamma produced by antigen-presenting cells in early host defense against intracellular pathogens. Eur. J. Immunol. *33*, 2666–2675.

Suzutani, T., Nagamine, M., Shibaki, T., Ogasawara, M., Yoshida, I., Daikoku, T., Nishiyama, Y., and Azuma, M. (2000). The role of the UL41 gene of herpes simplex virus type 1 in evasion of non-specific host defense mechanisms during primary infection. J. Gen. Virol. *81 Pt 7*, 1763–1771.

Takahasi, K., Suzuki, N.N., Horiuchi, M., Mori, M., Suhara, W., Okabe, Y., Fukuhara, Y., Terasawa, H., Akira, S., Fujita, T., and Inagaki, F. (2003). X-ray crystal structure of IRF-3 and its functional implications. Nat. Struct. Biol. *10*, 922–927.

Tanaka, N., Sato, M., Lamphier, M.S., Nozawa, H., Oda, E., Noguchi, S., Schreiber, R.D., Tsujimoto, Y., and Taniguchi, T. (1998). Type I interferons are essential mediators of apoptotic death in virally infected cells. Genes Cells *3*, 29–37.

Taylor, J.L., Witt, P.L., Irizarry, A., Tom, P., and O'Brien, W.J. (1995). 2′,5′-oligoadenylate synthetase in interferon-alpha- and acyclovir-treated herpes simplex virus-infected cells. J. Interferon Cytokine Res. *15*, 27–30.

Taylor, J.L., D'Cunha, J., Tom, P., O'Brien, W.J., and Borden, E.C. (1996). Production of ISG-15, an interferon-inducible protein, in human corneal cells. J. Interferon Cytokine Res. *16*, 937–940.

Taylor, J.L., Tom, P., and O'Brien, W.J. (1998). Combined effects of interferon-α and acyclovir on herpes simplex virus type 1 DNA polymerase and alkaline DNase. Antiviral Res. *38*, 95–106.

Tiwari, R.K., Kusari, J., and Sen, G.C. (1987). Functional equivalents of interferon-mediated signals needed for induction of an mRNA can be generated by double-stranded RNA and growth factors. EMBO J. *6*, 3373–3378.

Tovey, M.G., and Maury, C. (1999). Oromucosal interferon therapy: marked antiviral and antitumor activity. J. Interferon Cytokine Res. *19*, 145–155.

Upton, C., Mossman, K., and McFadden, G. (1992). Encoding of a homolog of the IFN-gamma receptor by myxoma virus. Science *258*, 1369–1372.

Vollstedt, S., Franchini, M., Hefti, H.P., Odermatt, B., O'Keeffe, M., Alber, G., Glanzmann, B., Riesen, M., Ackermann, M., and Suter, M. (2003). Flt3 ligand-treated neonatal mice have increased innate immunity against intracellular pathogens and efficiently control virus infections. J. Exp. Med. *197*, 575–584.

Vollstedt, S., Arnold, S., Schwerdel, C., Franchini, M., Alber, G., Di Santo, J.P., Ackermann, M., and Suter, M. (2004). Interplay between alpha/beta and gamma interferons with B, T, and natural killer cells in the defense against herpes simplex virus type 1. J. Virol. *78*, 3846–3850.

Wathelet, M.G., Clauss, I.M., Content, J., and Huez, G.A. (1988). Regulation of two interferon-inducible human genes by interferon, poly(rI).poly(rC) and viruses. Eur. J. Biochem. *174*, 323–329.

Wathelet, M.G., Berr, P.M., and Huez, G.A. (1992). Regulation of gene expression by cytokines and virus in human cells lacking the type-I interferon locus. Eur. J. Biochem. *206*, 901–910.

Weaver, B.K., Kumar, K.P., and Reich, N.C. (1998). Interferon regulatory factor 3 and CREB-binding protein/p300 are subunits of double-stranded RNA-activated transcription factor DRAF1. Mol. Cell. Biol *18*, 1359–1368.

Williams, B.R. G., and Haque, S.J. (1997). Interacting pathways of interferon signaling. Semin. Oncol. *24*, S9-70-S79–77.

Wintergerst, U., Gangemi, J.D., Kern, E.R., Whitley, R.J., and Chatterjee, S. (1996). Differential effect of recombinant human and mouse interferons on replication of herpes simplex virus type 1 in mouse cells. Antiviral Res. 32, 27–34.

Xiang, Y., Condit, R.C., Vijaysri, S., Jacobs, B., Williams, B.R., and Silverman, R.H. (2002). Blockade of interferon induction and action by the E3L double-stranded RNA binding proteins of vaccinia virus. J. Virol. *76*, 5251–5259.

Yokota, S., Yokosawa, N., Kubota, T., Suzutani, T., Yoshida, I., Miura, S., Jimbow, K., and Fujii, N. (2001). Herpes simplex virus type 1 suppresses the interferon signaling pathway by inhibiting phosphorylation of STATs and janus kinases during an early infection stage. Virology *286*, 119–124.

Yokota, S.-i., Yokosawa, N., Okabayashi, T., Suzutani, T., Miura, S., Jimbow, K., and Fujii, N. (2004). Induction of suppressor of cytokine signaling-3 by herpes simplex virus type 1 contributes to inhibition of the interferon signaling pathway. J. Virol. *78*, 6282–6286.

Yoneyama, M., Suhara, W., Fukuhara, Y., Fukuda, M., Nishida, E., and Fujita, T. (1998). Direct triggering of the type I interferon system by virus infection: activation of a transcription factor complex containing IRF-3 and CBP/p300. EMBO J. *17*, 1087–1095.

Zheng, X., Silverman, R.H., Zhou, A., Goto, T., Kwon, B.S., Kaufman, H.E., and Hill, J.M. (2001). Increased severity of HSV-1 keratitis and mortality in mice lacking the 2–5A-dependent RNase L gene. Invest. Ophthalmol. Vis. Sci. *42*, 120–126.

Zhong, S., Salomoni, P., and Pandolfi, P.P. (2000). The transcriptional role of PML and the nuclear body. Nat. Cell Biol. 2, E85-E90.

Zhou, A., Paranjape, J.M., Der, S.D., Williams, B.R. G., and Silverman, R.H. (1999). Interferon action in triply deficient mice reveals the existence of alternative antiviral pathways. Virology *258*, 435–440.

HSV-1 and the Host Cell: A Story of Global Conspiracies, Plots and Counterplots

14

Bernard Roizman and Brunella Taddeo

Les faits ne parlent pas.
Jules Henri Poincaré

Man is the measure of all things.
Attributed by Socrates to Protagoras

Abstract

Studies on viral gene function revealed that infected cells mount an unsuccessful multilayered defense that includes attempts to silence the viral genome at early stages of infection, activation of interferon pathways, turnover of cellular proteins required by the virus, and finally attempts to commit suicide through programmed cell death to salvage the infected host. The surprising finding is that multiple viral functions expressed by different viral proteins or domains thereof target the same cellular pathways to render them dysfunctional. The grand design of viral conquest of the host cell is to sequester and redirect cellular proteins to (1) preclude the infected cells from signaling it status to the environment or be the recipient of signals from the environment, (2) disrupt the interferon defense mechanisms, and (3) perform novel functions required by the virus. Considering that current knowledge is based on analyses of the functions of a few viral proteins, a wealth of data awaits to be discovered through analyses of the many multifunctional proteins expressed by virus.

Introduction

Viruses have evolved two fundamental strategies. The first one is fairly obvious. To infect and replicate in cells the virus has to attach, penetrate, and express its proteins, replicate its nucleic acid, assemble, egress, and reinitiate this process at a nearby or distant site. Virologists have known for many decades that viruses have diverged in the mechanisms and the basic structural components by which they accomplish these steps. Biologically, it is a remarkable process: whereas many genes in the human genome perform functions similar to those found not just in the mouse but in organisms as distant as yeast, there is little conservation of genetic material even between families of DNA or RNA viruses infecting humans. The systematic studies of the diversity of genes encoded by different viruses and their function in viral replication have fueled the growth of virology for the more than five decades, ever since availability of penicillin and streptomycin enabled the maintenance of sterile eukaryotic cell cultures.

The discovery of the second strategy is of a more recent vintage. Its existence became apparent as site specific insertion and deletions of viral genes enabled studies of gene functions. In the past two decades it became clear that the host cell is not a helpless victim of invasion by a relatively naive aggressor whose genes are concerned with synthesis of viral components, assembly and egress. The discovery of viral functions designed to block specific cellular functions led to the realization that the cellular functions blocked by viral gene products constitute an armamentarium of defense. In essence, viruses armed to their gills enter cells fully prepared to do battle.

It is useful to put the magnitude of armamentaria in prospective: HSV-1 encodes 84 different proteins and several RNAs (reviewed in Roizman, 1996; and Roizman and Knipe, 2001). The numbers is misleading since virtually every protein analyzed in detail has been found to perform several functions and hence the actual number of functions may exceed several hundred. Of the 84 proteins, at least 47 are not essential for viral replication at least in some cells in culture. A few of the 47 are known to play a role increasing efficiency of virus assembly or spread from cell to cell. Many of the remainder appear to target host responses to infection.

At this time we are very far from understanding the full range of viral functions designed to thwart the host and conversely both the range and mechanisms by which the host attempts to thwart the invader. Nevertheless, the basic strategy evolved by HSV has become apparent. It appears to consist of two parts. Foremost, HSV-1, unlike members of the Beta- or Gamma-herpeviruses, does not encode orthologs of cellular genes designed to mislead or block a host response to infection. Instead, HSV encodes a large number of adaptor proteins that bind and redirect cellular proteins to perform functions that they do not normally carry out. From an anthropomorphic point of view, this design enables HSV-1 to encode fewer proteins than the beta- and gamma-herpesviruses and suffer few host-range restrictions. The second aspect of the basic strategy is more insidious in that HSV has evolved a functional redundancy. Thus, functions targeting specific host responses to infection are encoded in more than one viral protein. The end result is that the absence of a specific host protein from differentiated cells does not impede viral replication since another viral protein-host protein recruit steps into the breach.

In this chapter it is impossible to address all of the viral functions and the host responses with which they interface. The primary object of this chapter is to examine a few viral proteins that illustrate the strategies evolved by HSV-1 to thwart host responses to infection. Since viral proteins seem to work in concert very much like the pieces in a game of chess, it is appropriate at this point to introduce the major players discussed in this chapter.

HSV-1 like all other herpesviruses brings into cells a large number of proteins packaged in layer designated tegument and which is located between the capsid and the envelope. Three of the tegument proteins encoded by U_L47, U_L49 and U_S11 bind and transport mRNA from the producer cells to the newly infected cells (Roller and Roizman, 1990; Sciortino *et al.*, 2002). Another, encoded by U_L41, also designated the virion host shutoff protein or *vhs*, degrades mRNA in newly infected cells (Strom and Frenkel, 1987; Kwong *et al.*, 1988; and reviewed in Smiley, 2004). This process continues for several hours until another viral tegument protein encoded by U_L48 and designated VP16 or αTIF (Pellett *et al.*, 1985; Kristie and Roizman, 1987) binds and neutralizes its activity (Smibert *et al.*, 1994). Two of the products of the first round of transcription of viral genome are of particular interest. Thus, infected cell protein no. 0 (ICP0) is a multifunctional protein which is not essential for viral growth in cell culture, but plays a prominent role in viral replication *in vivo* (reviewed in Hagglund and Roizman, 2004). The second protein, ICP27, also performs multiple functions (Soliman *et al.*, 1997; Mears and Rice, 1998; Sandri-Goldin, 1998), but its major role in this story is to block splicing of mRNA early in infection

(Hardy and Sandri-Goldin, 1994; Bryant *et al.*, 2001; Linberg and Kreivi, 2002). The other viral proteins play an eminent role in abrogating host defenses. These are the $\gamma_1$34.5, a protein whose synthesis is augmented by the onset of viral DNA synthesis and U_S11, a protein totally dependent on viral DNA synthesis for its accumulation. It is in fact convenient to thread the viral efforts to degrade host responses with the function of these two proteins.

Initiation of viral gene expression: the first skirmish

At least two competing sets of events take place in the few minutes after the encounter of HSV with a susceptible cell. The first, not yet fully understood, is the competition between cellular responses activated by the attachment and penetration of virus into the cell and functions designed to abrogate the host from mounting a response to infection (Nicholl *et al.*, 2000; Mossman *et al.*, 2001; Preston *et al.*, 2001). The second competing event occurs at the time of entry of the viral DNA into the nucleus, between cellular proteins, with a propensity to bind and silence the naked DNA, and viral proteins, whose functions it is to ensure viral gene expression. The counteracting events are orchestrated initially by a few viral proteins. Thus, the primary function of αTIF (VP16) is to actively engage cellular proteins and cognate response elements in α gene promoters to enhance the transcription of these genes and in the process to denude these promoters from nucleosomal proteins (Kristie and Roizman, 1987; McKnight *et al.*, 1987; Kristie and Roizman, 1988; Herrera and Triezenberg, 2004). VP16 has been reported to bind TAF-1 proteins and inhibit nucleosome assembly of the promoter domains of the genes (Gamble *et al.*, 2005). ICP4, the product of the α4 gene acts both as a repressor and activator of transcription (Gu *et al.*, 1993; Gu *et al.*, 1995; Leopardi *et al.*, 1995; Xiao *et al.*, 1997; Kim *et al.*, 2002) and is essential for the expression of β and γ genes (Knipe *et al.*, 1978; Dixon and Schaffer, 1980; reviewed in Roizman and Knipe, 2001). However, ICP4 is not sufficient to block silencing inasmuch as in cells infected at low multiplicity with ICP0 null mutant viruses (ΔICP0) a large fraction of cells express the remaining α genes but not β or γ genes. What is then the function of ICP0 particularly in light of the evidence that at high multiplicities of infection ΔICP0 mutants replicate as well as wild type viruses?

Few viral proteins have evolved as much mystique and controversy as ICP0 (reviewed in Hagglund and Roizman, 2004). It was first described in 1974–1975 then promptly renamed vmw110 as was the case for all proteins discovered outside Glasgow. The fundamental properties of this protein are straightforward: it is made immediately after infection, it is one of only four protein translated from spliced mRNA and for the purposes of this chapter, it is convenient to refer to domains of ICP0 in terms of their derivation. Thus, the first 19 amino acids are derived from exon 1, the next 221 from exon 2 and the last 533 from exon 3. Intron 1 is translocated into the cytoplasm and is stable. The topology of ICP0 is particularly interesting: it contain a RING finger in the middle of exon 2 and several SH3 binding sites and a nuclear translocation signal in exon 3. The gene encoding ICP0 is present in two copies in the repeats flanking the long component of viral DNA. ICP0 is extensively post-translationally modified by U_L13 and cellular (cdc2) protein kinases and nucletidylylated by casein kinase II (reviewed in Hagglund and Roizman, 2004). Evidence abounds that the functions of ICP0 are related to its post-translational modifications.

ICP0 interacts with or forms complexes with a large number of proteins. In the order in which they were discovered, these include the ubiquitin specific protease 7 (USP7) (Everett *et al.*, 1997), the translation elongation factor EF-1δ (Kawaguchi *et al.*, 1997a), cyclin D3 (Kawaguchi *et al.*, 1997b), a protein designated p60 (Bruni *et al.*, 1999), BMAL1 (Kawaguchi *et al.*, 2001), cdc34 (Van Sant *et al.*, 2001b) and more recently CoREST (Gu *et al.*, 2005; Roizman *et al.*, 2005). Several other interactions have not yet been fully characterized. The major function of ICP0 is that of a promiscuous transactivator. In transduced cells, ICP0 transactivates genes introduced by infection or transfection. ICP0 has also been reported to transactivate resident cellular genes (reviewed in Hagglund and Roizman, 2004). This property of ICP0 mirrors the observation that in most cell lines exposed to low ratios of ΔICP0 mutant virus per cell, replication is frequently arrested prior to β gene expression. Several lines of investigation indicate that ICP0 may play an important role in preventing the silencing of viral genes. Specifically:

1. In cells infected at low multiplicities of infection with a mutant in which the native genes encoding ICP0 were replaced with cDNA copies, the expression of viral genes was delayed by several hours although ultimately the virus yields were the same as those of wild type virus (Poon *et al.*, 2002). In these cells viral gene expression could be accelerated by inhibitors of histone deacetylases such as sodium butyrate, *Helminthosporium carbonum* toxin, or trichostatin A, whereas overexpression of HDAC1 extended the delay in viral gene expression. It was also noted that HDAC1 and HDAC2 but not HDAC3 were post-translationally modified between 3 and 6 hours after infection with wild type virus (Poon *et al.*, 2003).
2. ICP0 was reported to interact physically and functionally with HDACs 4, 5 and 7. These enzymes are members of class II HDACs reported to play a role in controlling cell differentiation (Lomonte *et al.*, 2004).
3. A more recent study initiated on the basis of an observation that residues 537 to 613 of ICP0 are homologous without gaps to those of amino acids 3–79 of CoREST (Gu *et al.*, 2005; Roizman *et al.*, 2005). CoREST is a co-repressor for REST (also called NRSF), a transcriptional repressor that binds, in non-neuronal cells, to a consensus sequence present in a large number of neuronal genes (Chong *et al.*, 1995; Shi *et al.*, 2003). The CoREST complex in non-neuronal cells contains, in addition to REST, HDACs 1 and 2 (Humphrey *et al.*, 2001; You *et al.*, 2001; Lunyak *et al.*, 2002; Shi *et al.*, 2003; Roopra *et al.*, 2004). In neurons, where REST is absent, CoREST is expressed to high levels and exists in complexes with HDACs 1 and 2 (Ballas *et al.*, 2001). CoREST thus constitutes a predominant repressor mechanism in different cellular contexts. The fundamental finding of these studies is that in uninfected cells CoREST/REST-HDAC1 or HDAC2 form a complex that is predominantly in the nucleus (Gu and Roizman, unpublished studies). In wild type virus infected cells, a fraction of the complex is translocated to the cytoplasm and furthermore. In these cells, a fraction of the HDAC1 or HDAC2 is dissociated from the complex and CoREST, just like HDAC1 and HDAC2 reported earlier, is postranslationally modified by phosphorylation. The role of ICP0 in this process may be deduced from two observations. First, anti ICP0 antibody pulls down CoREST. Second, in the cells infected with ΔICP0 mutant virus, HDAC1 or HDAC2 are not dissociated from the CoREST/REST complex, although a fraction of the complex is trans-

located to the cytoplasm. It is noteworthy that the translocation of CoREST/REST complexes to the cytoplasm, while occurring *pari-passu* with the translocation of ICP0, are delayed in cells exposed to inhibitor of viral DNA synthesis and does not occur in cells infected with a mutant lacking ICP4.

The model that currently fits the available data is that on entry into cells, viral DNA aggregates with components of the ND10 nuclear structures. ICP4 and ICP0 aggregate also to these sites (reviewed in Hagglund and Roizman, 2004). The model proposes that in wild type virus infected cells ICP0 interacts with CoREST in the CoREST/REST HDAC complex and causes the dissociation of HDAC1 or HDAC2. This event would block repression of the viral genome enabling expression of post α genes. Further post-translational modification and export of the components of the complex would be mediated by post α protein, in which viral protein kinases may play a significant role. The synthesis of ICP8 would then enhance transcription and enable viral DNA synthesis. In cells infected at low multiplicity of infection with ΔICP0 mutant viruses β and γ genes are largely silenced. In cells infected with multiple genomes, escape of one genome from repression would result in the synthesis of ICP8 and initiation of viral replication.

This is the first rational model of the role of ICP0 based on its interactions with cellular repressors of gene expression.

mRNA synthesis and translation: the second front

The $\gamma_1$34.5 protein and protein kinase R (PKR)

A study of viral RNA conducted many years ago indicated that after the onset of viral DNA synthesis a large fraction of the viral genome was represented in abundant amounts of complementary mRNA that could anneal and form stable structures with a T_m characteristic of dsRNA (Jacquemont and Roizman, 1975; Kozak and Roizman, 1975). It is not surprising that in the course of HSV-1 infection PKR is activated. Among other consequences, the activated autophosphorylated PKR dimmers phosphorylated the α subunit of the translation initiation factor 2 (eIF-2α), which promptly and effectively shuts off proteins synthesis (reviewed in Williams, 2001). The transactivation of PKR is common to many viruses, and most viruses encode functions designed to block the shutoff of protein synthesis. The HSV solution is a multifunctional protein named $\gamma_1$34.5. This protein encodes the motif of the phosphatase 1α accessory protein in its carboxyl terminal domain. $\gamma_1$34.5 protein binds phosphatase 1α, redirecting it to dephosphorylate eIF-2α and thereby precluding the shutoff of protein synthesis by the activated PKR (He *et al.*, 1997a; He *et al.*, 1997b; He *et al.*, 1998). It is of interest to note that the carboxyl-terminal 60+ amino acids of $\gamma_1$34.5 are virtually identical to the corresponding carboxyl-terminal amino acids of GADD34 (Chou and Roizman, 1994). The GADD34 sequence functionally substitutes for the corresponding $\gamma_1$34.5 sequence.

The discovery of $\gamma_1$34.5 protein raised two questions. First, $\gamma_1$34.5 protein is encoded by HSV-1, HSV-2 and the simian B virus, but not by other viruses. In an important experiment, Mohr and colleagues selected a mutant $\Delta\gamma_1$34.5 virus that did not shutoff protein synthesis (Mohr and Gluzman, 1996). In this mutant U_S11, a late gene, was converted to an early gene by the spontaneous juxtaposition of the α47 gene promoter upstream of the U_S11 open reading frame. Subsequent studies showed that U_S11 binds PKR and that expression of U_S11 early in infection blocks the activation of PKR (Cassady *et al.*, 1998). This is probably not the end of the story since U_S11 is not a highly conserved gene. Another mutation that led to resistance to PKR in the background of $\Delta U_S11/\Delta\gamma_134.5$ mutants could not be mapped to a single locus.

The second question is why the virus has selected to evolve the $\gamma_134.5$ gene, which blocks activated PKR from shutting of protein synthesis, but has left the active PKR floating unencumbered in the milieu of the infected cell. A logical solution would have been to change the promoter of U_S11 protein. Does the virus need activated PKR?

PKR, NF-κB, and the transactivation of cellular genes

The puzzling stability of activated PKR for at least several hours after infection raised the obvious question whether HSV required a function expressed by this protein. The notion that this must be the case was supported by the observation that murine $PKR^{-/-}$ cells yielded 50 fold more wild type virus than sibling, $PKR^{+/+}$ cells (Taddeo *et al.*, 2004a). One explanation emerged from studies on DNA microarray based analyses of cellular genes transactivated after infection (Taddeo *et al.*, 2002). A surprising finding was that a substantial fraction of the upregulated mRNAs was NF-κB dependent. The activation of NF-κB in the course of HSV-1 infection was described by several laboratories, but the mechanism of activation was unclear (Patel *et al.*, 1998; Amici *et al.*, 2001; Gregory *et al.*, 2004). Two lines of investigation indicated that a major stimulus of NF-κB activation is activated PKR (Taddeo *et al.*, 2003b) and that the virus requires NF-κB proteins (Taddeo *et al.*, 2004a).

As noted earlier in the text, U_S11 protein expressed early in infection blocks the activation of PKR (Cassady *et al.*, 1998). U_S11 is not an essential protein and viruses lacking U_S11 replicate as well as wild type virus in cultured cells. In these cells, NF-κB is not activated (Taddeo *et al.*, 2003b). NF-κB is also not activated in $PKR^{-/-}$ cells infected with wild type virus but is activated in sibling, $PKR^{+/+}$ cells (Taddeo *et al.*, 2003b). Finally, murine cell lines lacking p50, p65 or both NF-κB genes yielded 10 to 30 fold less wild type virus than $p65^{+/+}/p50^{+/+}$ sibling cells (Taddeo *et al.*, 2004a). For HSV-1, activation of PKR was a way of insuring that NF-κB is activated. These results are concordant with earlier observations that cells expressing a dominant negative form of IκBα protein yield less virus than the wild type cells (Patel *et al.*, 1998; Amici *et al.*, 2001).

A central question is the function of NF-κB in the context of HSV-1 replication. It has been argued that NF-κB activates antiapoptotic functions (Goodkin *et al.*, 2003; Gregory *et al.*, 2004), but this is not the likely explanation, at least in the systems tested to date.

The life-expectancy of cellular mRNAs and the U_L41 protein

The general consensus held for over two decades is that the U_L41 protein mediates the degradation of mRNA transcribed by both host and virus in a nonspecific manner (Strom and Frenkel, 1987; Kwong *et al.*, 1988; and reviewed in Smiley, 2004). The translation of residual host mRNA is further disabled by the appearance of ICP27 which blocks splicing but not the transport of unspliced mRNA into the cytoplasm (Hardwicke and Sandri-Goldin, 1994; Sandri-Goldin, 1998). The prevailing notion was that HSV mRNA escape total destruction because of the sheer output of viral mRNA. This notion was finally tested with the DNA microarrays and the results validated by Northern analyses, real-time PCR and analyses of cellular protein accumulations (Taddeo *et al.*, 2002; Taddeo *et* al., 2003a; Esclatine *et al.*, 2004a). U_L41 does indeed mediate the degradation of mRNAs, but the process is more complex. Specifically:

1 The DNA microarrays indicated that a substantial fraction of total mRNAs was up or down regulated. A surprising finding that a large number of the up-regulated mRNAs was NF-κB dependent and contained AU-rich elements (AREs) in their 3′ untranslated domains (Esclatine *et al.*, 2004a). Although the up regulation of several of these mRNAs (e.g. IEX-1, IκBα, etc) was verified by Northern analyses and real-time PCR, the proteins encoded by these mRNAs did not accumulate (Taddeo *et al.*, 2003a; Esclatine *et al.*, 2004a). Two aspects of the results of these studies are of particular interest. First, in uninfected cells, the mRNAs containing AREs (ARE-mRNAs) are sequestered by members of the tristetraprolin (TTP) family in exosomes—structures containing numerous nucleases—and degraded very rapidly in the 3′ to 5′ direction (Chen *et al.*, 2001; Wilusz *et al.*, 2001; Blackshear, 2002; Butler, 2002). The process of degradation is extremely fast, in fact so fast that intermediate, partially degraded forms do not accumulate. The rapid degradation also takes place in ΔU_L41-infected cells. In wild type virus-infected cells, the degradation is orders of magnitude slower (Esclatine *et al.*, 2004c). The cytoplasm of infected cells contains deadenylated spliced and unspliced forms of the mRNAs along with truncated RNA fragments containing the 5′ termini and arising from endonucleolytic cleavage and 3′ to 5′ degradation (Taddeo *et al.*, 2003a; Esclatine *et al.*, 2004a). It would seem at first glance that U_L41 impairs the degradation of ARE-mRNAs even though the net effect is still failure of cellular translation products to accumulate in infected cells.
2 Curiously, some induced mRNAs that do not contain AREs and at least one ARE-mRNA escape this fate and are in fact translated (Esclatine *et al.*, 2004a; Esclatine *et al.*, 2004b; Taddeo *et al.*, 2004b). This is the case for GADD45β mRNA which lacks AREs and TTP mRNA, which contain AREs (Brooks *et al.*, 2004). No evidence has emerged that these mRNA are degraded in the 3′ to 5′ direction or deadenylated in wild type virus-infected cells. In contrast, TTP could not be demonstrated in ΔU_L41 mutant virus-infected cells (Esclatine *et al.*, 2004a).
3 Finally, the fate of two constitutively expressed mRNA is still different. Thus, both GAPDH and β-actin mRNAs are rapidly degraded in wild type virus-infected cells. In ΔU_L41 infected cells both mRNAs show the same stability as in mock-infected cells (Esclatine *et al.*, 2004c).

In essence, HSV-1 discriminates mRNAs into three distinct classes. Some induced mRNAs, both NF-κB dependent and some independent, with and without AREs, are expressed, and some, particularly ARE-mRNAs are degraded, but in a process much slower than that observed in mock-infected cells. Finally, constitutively expressed mRNAs tested to date are rapidly degraded. It has been suggested from *in vitro* studies that U_L41 forms complexes with cellular proteins and degrades mRNA in slow motion 5′ to 3′ (Karr and Read., 1999; Everly *et al.*, 2002; Doepeker *et al.*, 2004; Perez-Parada *et al.*, 2004). In more recent studies, U_L41 was found in association with TTP (Esclatine *et al.*, 2004b). Conceivably, the complex TTP and U_L41 would target ARE-mRNAs to the exosomes. Furthermore, it is conceivable that the slow degradation of ARE-mRNAs may reflect displacement of one or more exosomal nucleases by U_L41 protein. Since the studies described above centered on degradation of ARE-mRNAs that are degraded 3′ to 5′, it cannot be excluded that mRNAs lacking AREs are degraded 5′ to 3′.

The studies presented above raise many questions but two are paramount. There is no satisfactory answer to the question as to how HSV-1 discriminates between the various classes of mRNAs. There is, however, a better understanding why so many genes upregulated after infection are NF-κB dependent.

NF-κB-dependent gene expression constitutes a large fraction of stress-induced activities of eukaryotic cells (reviewed in Pahl, 1999). Among NF-κB-induced genes are those that yield ARE-mRNAs encoding growth factors, cytokines, pro- and antiapoptotic proteins but also proteins that degrade the ARE-mRNAs. The activated cell reacts to the stress internally and also signals its status to its immediate environment—a feature likely to be inimical to viral replication (reviewed in Santoro *et al.*, 2003). An additional problem is that NF-κB can be induced by factors other than activated PKR. Indeed, transient activation of NF-κB within the first few minutes of infection has been reported (Patel *et al.*, 1998; Mogensen and Paludan, 2001). By activating NF-κB, the infected cells insure that ARE-mRNAs, induced by whatever source, are degraded. A model of NF-κB-dependent and -independent RNA synthesis and U_L41-dependent selective degradation in cells infected with HSV-1 is represented in Fig. 14.1.

The function of U_L41 has been redefined, but its role fully appreciated early in the discovery of this protein, cannot be understated. One example is the role of U_L41 in deterring the activation of the interferon pathway (Chee and Roizman, 2004). A key protein in the interferon cascade is Jak1, a kinase activated by binding of interferon to its receptor. Jak1 is responsible for the phosphorylation of Stat1. The half life of Jak1 is relatively short. In uninfected cells, and cells infected with the ΔU_L41 mutant Jak1 levels are maintained, whereas in wild type virus-infected cells the pool of Jak1 is rapidly depleted (Chee and Roizman, 2004).

Degradation of cellular proteins

The role of ICP0 as an ubiquitin ligase

ICP0 interacts with and requires cyclin D3, and the amino acid substitution in residues 199 (D199A) abolishes this interaction (Kawaguchi *et al.*, 1997b; Van Sant *et al.*, 1999; Van Sant *et al.*, 2001a). Cyclin D3 colocalizes with ICP0 at or near the ND10 nuclear structures discussed in detail below. At later times, between 3 and 9 hours after infection, depending on the cell line, the wild type ICP0 is translocated to the cytoplasm or rather it shuttles between nucleus and cytoplasm and is dynamically associated with protesomes (Van Sant *et al*, 2001a, 2001b). In cells infected with the mutant carrying the D199A, ICP0 is restricted to the cytoplasm. The D199A substitution mutant and the repaired virus replicate as well as wild type virus in dividing cells but the yields decrease 10 fold in resting cells—possibly a reflection of the absence of cyclin D3.

Cyclin D3 has a short half life in mock-infected cells. In wild type virus infected cells there is an initial decrease in the cyclin D3 level, but then it stabilizes and is detectable for several hours longer than in cells infected with the D199A mutant or ΔICP0 mutant (Kawaguchi *et al.*, 1997b; Van Sant *et al.*, 1999; Van Sant *et al.*, 2001a; Hagglund and Roizman, 2003). The observation that cyclin D1 follows the same kinetics as cyclin D3, even though it does not interact with ICP0, led to an examination of the mechanism by

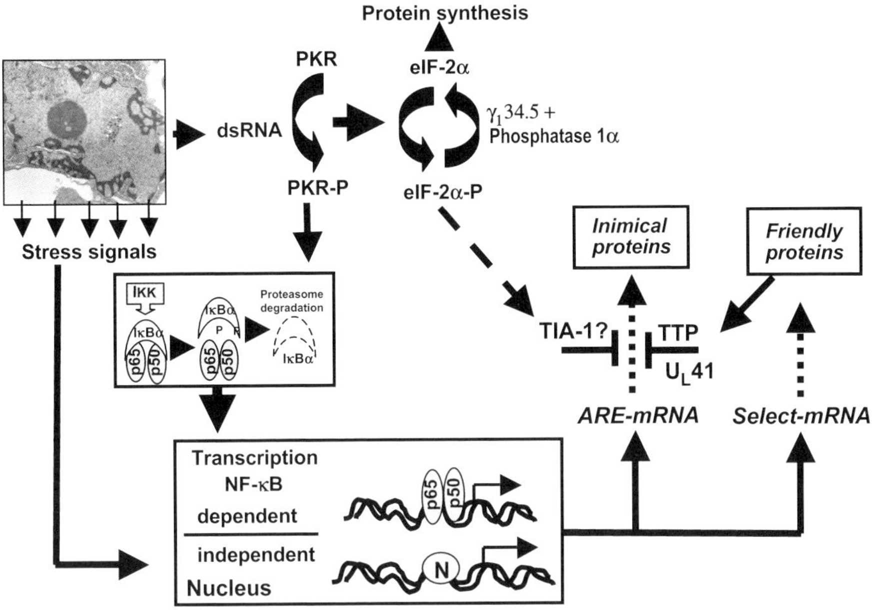

Figure 14.1 Schematic representation of NF-κB-dependent and –independent RNA synthesis and *vhs*-dependent selective degradation in HSV-1-infected cells. Double stranded RNA synthesis during HSV-1 infection leads to the activation of Protein kinase R (PKR). On activation, PKR dimerizes and phosphorylates key proteins, among which is the subunit of eukaryotic translation initiation factor 2α (eIF-2α). The net effect is a total shutoff of protein synthesis. To circumvent the effects of activated PKR, HSV-1 encodes a protein, $\gamma_1$34.5, which binds the protein phosphatase 1α and redirects it to dephosphorylate eIF-2α, enabling uninterrupted viral protein synthesis. Furthermore, activation of PKR after HSV-1 infection leads to the activation of the IκB kinase (IKK) complex, which prophosphorylates IκBα targeting it for proteasome-dependent degradation. Degradation of IκBα releases NF-κB, which can then migrate to the nucleus and act as a transcriptional activator of numerous genes. Among them, several ARE-containing mRNAs (ARE-mRNA). ARE-mRNAs generally encode growth factors, cytokines, pro- and antiapoptotic which could likely be inimical to viral replication. The translation of these mRNAs is indeed blocked in HSV-1 infected cells in a U_L41dependent manner. At the same time, selected mRNAs, resistant to U_L41 action, are translated during HSV-1 infection. Among these proteins, TTP, which is not activated by NF-κB, is largely expressed after HSV-1 infection and considerably less induced with ΔU_L41 deletion mutant. TTP, as TIA-1 and TIAR, is an ARE-binding protein involved in the normal degradation pathway of ARE-mRNAs.

which cyclins D1 and D3 turn over. In brief, these cyclins are degraded by SFC E3 ubiquitin ligase in conjunction with the cdc34 (UbcH3) ubiquitin ligase. The indication that ICP0 acts as an E3 ligase in degrading cdc34 (Fig. 14.2A) rests on several lines of evidence. Specifically:

1 Antibody to proteasomal proteins pulls down cdc34 from lysates of cells infected with wild type virus and treated with MG132, but not from lysates of untreated cells or cells infected with the mutant virus containing the D199A substitution in ICP0. Consistent with that observation, cdc34 binds wild type ICP0 and to a much lesser extent the D199A mutant. Moreover, cdc34 is degraded in wild type virus infected cells.
2 Sequences encoded by exon 3 polyubiquitylate cdc34 *in vitro*. The reaction has been mapped to a short segment of

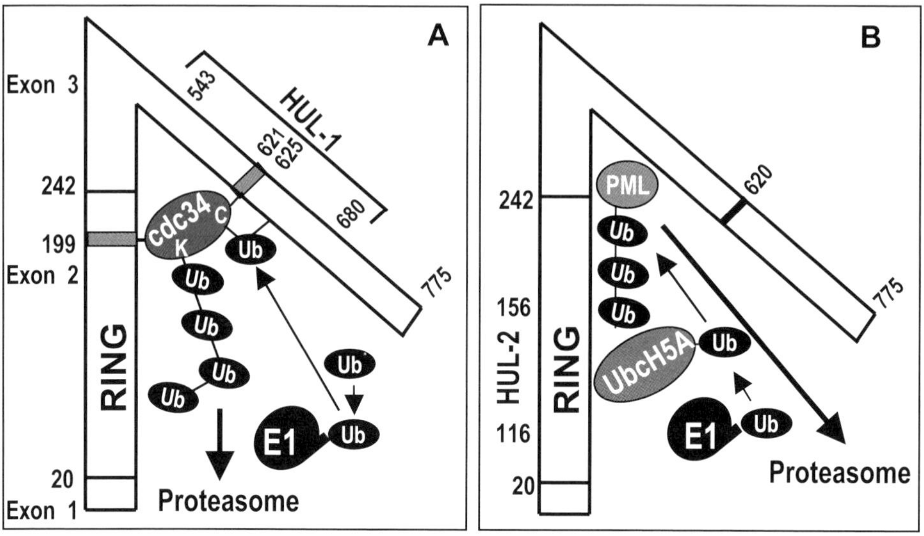

Figure 14.2 Ubiquitin ligase functions of ICPO. (A) Ubiquitylation of cdc34. The ubiquitylation of cdc34 requires the participation of the exon 3 domain of ICPO. Residues 543–680 are sufficient to polyubiqutylate cdc34 *in vitro*. Deletion of amino acids 521–625 abolishes this reaction. Cdc34 binds to exon 2 and exon 3 domains *in vitro*. Substitution of residue 199, or deletion of residues 621–625, precludes binding to respective domains. Cdc34 was degraded in cells infected with a virus mutant in which the ring finger was disrupted, but not in cells infected with a mutant lacking residues 621–625. (B). Ubiquitylation of PML. Ubiquitylation of and degradation of PML involves the ring finger domain of exon 2 and the UbcH5A ubiquitin conjugating enzyme. PML does not physically interact with ICP0. The degradation of SP100 and DNA-dependent protein kinase is also dependent on the ring-finger domain of ICPO and also involves the UbcH5A E2 enzyme.

exon 3. Exon 3 polypetides lacking amino acids 621 through 625 no longer polyubiquitylate cdc34. Consistent with this finding, cdc34 binds intact exon 3 but not polypeptides lacking amino acids 621 to 625.

3 An HSV-1 mutant, in which residues 621 through 625 of ICP0 were deleted, acts in a predictable manner in that cdc34 is readily detectable, whereas cyclin D3 disappears. In contrast, cdc34 was degraded in cells infected with a virus in which the ring-finger domain had mutagenized. Everett has repeatedly advanced the argument that intact ICP0 *in vitro* does not polyubiquitylate cdc34 without a detailed analysis of post transcriptional modifications required for this reaction (Everett, 2004). In infected cells, degradation of cdc34 requires the presence of functional U_L13 protein kinase—one of many enzymes that posttranslationally modify ICP0 in infected cells (Hagglund and Roizman, unpublished studies).

ICP0, as noted above, has also been linked to the dispersal of ND10 nuclear structures known by many alternative names (PODs etc). ND10 contain numerous proteins including the promyelocitic leukemia (PML) protein, Daxx, SP100, etc. PML appears to be the organizing protein of ND10, since in its absence ND10 becomes dispersed. A review of the literature attributes to PML, and by extension to the ND10 structure, a large number of functions that fundamentally boil down to transcriptional regulation (reviewed in Zhong *et al.*, 2000; Negorev and Maul,

2001; Regad and Chelbi-Alix, 2001). In addition to the dispersal of ND10, ICP0 mediates the degradation of PML and SP100. ICP0 has also been shown to degrade the centromeric proteins A and C (CNP-A and CNP-C) and the DNA-dependent protein kinase (DNA-PK) (Gu and Roizman, 2003). Dispersal of ND10 and the degradation of these proteins have been mapped to the RING finger domain, a structure compatible with known ubiquitin ligases. The evidence that RING finger domain acts as an ubiquitin ligase (Fig. 14.2B) emerged from studies showing that it polyubiquitylates UbcH5A and UbcH6, but not an assortment of other E2 enzymes, in *in vitro* substrate-independent ubiquitylation assays. The evidence that UbcH5A actually targets PML and SP100 proteins for degradation emerged from studies carried out in cells transduced with either wild type or dominant negative UbcH5A, UbcH6 or UbcH7. In these studies PML and SP100 were spared in cells transduced with UbcH5A but not in cells transduced with the other constructs (Gu and Roizman, 2003). It is noteworthy that DNA-PK was also spared in assays with dominant negative UbcH5A. However, none of the different dominant negative constructs of UbcH6 tested spared CNP-C and CNP-A from degradation (Suzuki and Roizman, unpublished data).

Why is HSV-1 targeting ND10 and PML for degradation? Ectopically overexpressed PML led to creation of engorged ND10 structures many times their normal size. In these cells the peregrinations of ICP0 did not differ from those taking place in wild type parental cells. PML was still present in large amounts late in infection and the overexpression of PML had no effect on viral replication (Lopez *et al.*, 2002). Studies on $PML^{-/-}$ cells yielded similar data in that the yields of virus from $PML^{-/-}$ cells were virtually identical to those of the wild type sibling cells (Chee *et al.*, 2003). The difference, however, emerged in the $PML^{-/-}$ or $PML^{+/+}$ exposed to α- or γ-interferon prior to infection with wild type virus. In $PML^{+/+}$ cells the virus yield decreased as expected approximately 1000 fold. In $PML^{-/-}$ cells the decrease was approximately 10 fold. It appears that PML, and by extension ND10, mediates the interferon response. By degrading ND10 structures, HSV-1 preempts the effects of exogenous interferon on infected cells. The preemptive anti-interferon strike may not be the only reason why HSV and other viruses target ND10 structures. As noted earlier in the text, HSV-1 targets the interferon pathway in other ways, through U_L41 and $\gamma_134.5$ proteins.

Targeting cell surface receptors

The cell surface is studded with receptors to growth factors, cytokines, etc. Recent studies showed that HSV-1 may negatively regulate at least the receptor tyrosine kinases (Liang *et al.*, 2005). The evidence supporting this conclusion stemmed from the observation that uncharted domains of exon 3 of ICP0 contain SH3 domain binding sites, and particularly, sites with a motif similar to that of CIN85, the M_R 85,000 protein that interacts with the E3 ubiqutin ligase c-Cbl (Kurakin *et al.*, 2003; Thien and Langdon, 2001). CIN85 and c-Cbl are involved in endocytosis and negative regulation of numerous receptor tyrosine kinases (Petrelli *et al.*, 2002; Soubeyran *et al.*, 2002; Dikic and Giordano, 2003). ICP0 physically interacts with CIN85 and it also pulls down c-Cbl (Liang *et al.*, 2005). The role of ICP0 emerged from studies of epidermal growth factor receptor (EGFR) as a prototype of receptor tyrosine kinases (Fig. 14.3). In transfection assays, ICP0, in absence of other viral gene products, down regulated the EGF-dependent expression of a reporter gene (luciferase). In addition, ICP0 down regulated both total and cell surface levels of EGFR, in EGF-independent manner. In wild type virus

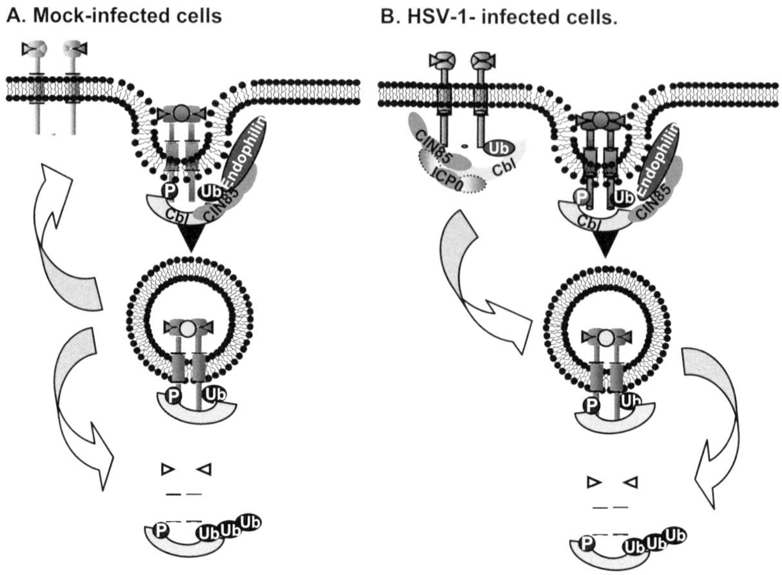

Figure 14.3 (A) In mock-infected cells, stimulation of EGFR receptor tyrosine kinases with EGF leads to phosphorylation of the receptor, binding and phosphorylation of Cbl, and endocytosis and monounbiquitylation of the receptor mediated by the Cbl/CIN85/endophylin complex. The receptor may be recycled or polyubiquitylated and degraded. (B) In cells transfected with ICPO, the receptor is degraded in both ligand-dependent and –independent manner. This could be due to inhibition of recycling of the receptor or degradation of receptor by a mechanism that is independent of stimulation by the ligand.

infected-cells, the surface levels of EGFR also decreased in the absence of EGF stimulation, but the decrease was more pronounced in cells stimulated by EGF.

The significance of these studies is based on evidence that c-Cbl interacts with over 40 partners. Among a large number of receptor tyrosine kinases, the list includes both growth factor and cytokine receptors (Thien and Langdon, 2001). Preliminary studies showed that the γ-interferon receptor is also down regulated in infected cells (Liang and Roizman, unpublished data). The mechanism whereby ICP0 and infected cells down regulate receptor tyrosine kinases is at this moment unknown. One of the functions of c-Cbl is to either degrade or recycle the receptors. It is conceivable that ICP0 may simply divert the receptors from recycling to degradation. An alternative mechanism is that, by altering the conformation of the proteins in the complex, c-Cbl targets nonactivated receptors for internalization and ultimately for degradation.

The obvious ultimate objective in targeting cell surface receptor tyrosine kinases is to isolate the infected cells from signaling pathway that may trigger undesirable antiviral responses.

HSV and apoptosis—the supreme self-sacrifice

Viruses have been known to block apoptosis for some time. The earliest and most striking studies on HSV-1-infected cells emerged from reports that HSV-1 blocked apoptosis induced by sorbitol (Koyama and Miwa, 1997) and Fas ligand (Sieg *et al.*, 1996; Galvan and Roizman, 1998; Jerome *et al.*, 1999; Jerome *et al.*, 2001). The evidence that HSV blocks the consequences of osmotic shock induced by sorbitol were particularly striking with respect to the appearance of exposed cells and metabolic consequence. It is worthwhile pointing out that HSV blocks apoptosis induced by a large number of nonphysiologic proapoptotic drugs and also by heat shock (Leopardi and Roizman, 1996; Galvan and Roizman, 1998). These studies served an important function: they pointed out the fact that HSV blocked apoptosis at least until very late in infection and that the virus was able to block apoptosis induced by both natural and purely synthetic compounds capable of damaging cells.

Subsequent studies diverged in three directions. The first and mildly significant were studies designed to enlarge the scope of proapoptotic functions that are blocked by HSV. The second focused on the biology of several mutants which were defective with respects to viral replication but capable to induce apoptosis in infected cells. Lastly, an interesting line of investigation centered on the mechanisms by which HSV blocked apoptosis. This chapter will focus on the latter two studies. It should be preceded, however, by a brief definition or disclaimer, as the case may be.

Even the most cursory review of the literature cannot fail to impress the reader with number of metabolic pathways whose failure or activation leads to cell death. At the heart of the problem is the definition of apoptosis since it can manifest itself differently in different cells. Perhaps the most contentions but also simplistic definition suitable for the purposes of this chapter is that apoptosis is cell death caused by an activated cellular function and that either the activation of this function or the effect of this function is abrogated by a viral gene product.

Defective viruses and apoptosis

The list of mutants that induce apoptosis in infected cells is relatively short. It includes ΔICP4 (Leopardi and Roizman, 1996; Galvan and Roizman, 1998; Munger *et al.*, 2001), ΔICP27 (Aubert and Blaho, 1999), ΔICP6 (ribonucleotide reductase) (Langelier *et al.*, 2002), $\Delta\gamma_1 34.5$ (Chou and Roizman, 1992), and mutants lacking glycoprotein D (ΔgD) (Zhou *et al.*, 2000; Zhou and Roizman, 2001). Very little is known regarding the events transpiring in ΔICP6 mutants. $\Delta\gamma_1 34.5$ mutants were described in detail elsewhere in this chapter and both the mechanism of death and the activation of the cellular function (PKR) that leads to cell death differ from those of other viral mutants.

ΔICP4 and ΔICP27 are defective in early functions. ΔICP4 mutants express predominantly α genes although with time they tend to express small amounts of β and γ proteins. ΔICP27 mutants express subsets of post α genes in cell-dependent fashion. ΔICP4 mutants are proapoptotic in a large variety of different types of cells although the proapoptotic manifestations tend to vary depending on the cell line. The proapoptotic activities induced by the ΔICP27 mutants is cell-type dependent: whereas ΔICP4 mutant virus induces apoptosis in Vero cells, ΔICP27 mutant does not (Leopardi and Roizman, 1996; Aubert and Blaho, 1999). Very early on it became apparent that apoptosis induced by ΔICP4 mutant could be blocked by transfection of cells with DNA fragments encoding post α genes (Leopardi *et al.*, 1997). It seems likely, on the basis of current knowledge, that both ICP4 and ICP27 do not themselves block apoptosis, but merely enable the expression of viral genes whose function it is to block apoptosis. The vexing puzzle is the mechanism by which these mutants induce apoptosis rather than the identity of viral genes that block apoptosis. Attempts to induce apoptosis by transfection of individual α genes have not yielded useful results. Studies of the published literature suggest that the culprit may be ICP0, in the context of infection by otherwise defective mutants.

The most interesting mutants capable of inducing apoptosis are those lacking gD. ΔgD mutants come in two flavors. To produce a stock of mutant viruses, ΔgD mutants are grown in cells expressing complementing gD. In these cells, the progeny virus lacks gD in its genome but picks up the protein in its envelope and hence it is $gD^{-/+}$ virus stock. In complementing cells, this virus will enter cells in the conventional manner, replicate, acquire gD in its envelope again, and egress cells. In noncomplementing cells, the virus enters in conventional manner, replicates, but does not acquire gD in its envelope ($gD^{-/-}$ virus stock) and does not egress form cells. The $gD^{-/-}$ virus cannot enter cells by fusion of the envelope with the plasma membrane—it enters cells only by endocytosis and is rapidly degraded. Both $gD^{-/+}$ and $gD^{-/-}$ viruses induce apoptosis but apparently by different mechanisms. This is evident from two different sets of studies. $gD^{-/-}$ virus induce apoptosis only in cells exposed to high ratios of particles per cell. Apoptosis induced by $gD^{-/-}$ virus is blocked by transduction of cells with intact gD or a heterodimer consisting of the ecto- and transmembrane domains, in one molecule, coupled, by S-S bonds, to the transmembrane domain and cytoplasmic tail, in another molecule (Zhou *et al.*, 2000; Zhou and Roizman, 2002b). In addition, apoptosis is blocked by chloroquine (Zhou and Roizman, 2002a). In contrast, apoptosis is induced by low ratios of $gD^{-/+}$ particles per cell. Moreover, apoptosis induced by $gD^{-/+}$ stocks is blocked by intact gD or the ectodomain of gD (Zhou *et al.*, 2000; Zhou and Roizman, 2002b). Neither transmembrane nor cytoplasmic domain is required and apoptosis is not affected by chloroquine. Finally, apoptosis induced by both $gD^{-/-}$ and $gD^{-/+}$ viruses is blocked by transduction of cells with the cation-independent mannose-6-phosphate receptor (Zhou and Roizman, 2002a), reported to colocalize with gD and one time thought to be a receptor for gD (Brunetti *et al.*, 1994; Brunetti *et al.*, 1998).

The model that best fits the available data is that $gD^{-/-}$ virions enter cells by endocytosis. Fusion of endocytic vesicle with lysosomes leads to the degradation of the particles. At high multiplicities of infection the huge influx of lysosomal enzymes results in cell death with typical proapoptotic symptoms. gD present in the virion and in the endocytic vesicle blocks apoptosis by interacting with the cation-independent mannose 6 phosphate—a regulator of lysosomal enzymes. Intact gD, the product of transfected gD or the heterodimer described above, may block lysosomal discharge by incorporation in the membrane of the endocytic vesicle. It is likely that $gD^{-/+}$ stocks induce apoptosis at egress rather than entry into the cells. Thus, the transport of virions in the exocytic pathway may lead to lysosomal discharge that is block this time by secreted soluble gD or by intact gD contained in the membranes of the exocytic vesicles. The studies led to two conclusions. Foremost, it showed that gD has an additional, novel function: it blocks apoptosis possibly at different stages of infection. Second, gJ, a small glycoprotein mapping next to gD and whose function was not know, has been shown to block apoptosis (Jerome *et al.*, 1999; Jerome *et al.*, 2001). The mechanism by which gJ blocks apoptosis may be different from that of gD.

Inhibition of apoptosis by viral gene products

Viral gene products reported to block apoptosis fall into three categories. The first involves products that complement the defective virus. These include gD, ICP6 and $\gamma_1 34.5$. None of the published studies indicate that they block apoptosis induced by cells of the immune system, exogenous agents, or defects in gene other than those that they complement. gJ has been reported to block cytotoxic activity of cells of the immune system by a mechanism that is not understood (Jerome *et al.*, 1998).

The role of the latency associated transcript (LAT) is unclear (for recent review on LAT see Peng *et al.*, 2004). There is no question that in the absence of LAT there is a decrease in the number of neurons in dorsal root ganglia that harbor latent virus as well as a decrease in the overall number of neurons. The data also indicate that LAT blocks cell death. What is less clear is the function that causes the neuronal cells to perish or the mechanism by which LAT blocks cell death. If, for example, LAT blocks expression of ICP0—often implied but never proven, the absence of LAT could cause the neurons to undergo apoptosis. But this model does not exclude a plethora of others.

U_S3 is the closest bonafide antiapoptotic gene that still remains somewhat of a puzzle. U_S3 protein kinase has multiple functions. The 481-residues U_S3 protein phosphorylates a large number of viral and cellular proteins and, in the infected cell, it plays a key role in viral egress from nuclei (Reynolds *et al.*, 2002). U_S3 blocks apoptosis induced by exogenous agents (Jerome *et al.*, 1999), by overexpression of proapoptotic proteins (e.g. BAD) (Munger and Roizman, 2001; Benetti *et al.*, 2003; Ogg *et al.*, 2004), and by defective viruses such as ΔICP4 (Leopardi *et al.*, 1997; Munger *et al.*, 2001).

Studies designed to identify the target of the antiapoptotic activity of U_S3 led to the discovery that the motif phosphorylated by U_S3 is similar to that of cyclic AMP-dependent protein kinase A (PKA). U_S3 phophorylated *in vitro* PKA substrates including the PKA regulatory subunit RII. Lastly, activation of PKA by forskolin blocked apoptosis induced by exogenous agents or by ΔICP4 virus infection, and this effect was reversed by a peptide that blocks the phosphorylation of the PKA RII subunit (Poon *et al.*, 2006).

The primary evidence that U_S3 activates PKA, but is independently active in blocking apoptosis, rests on the observation that is localizes with mitochondria (Van Minnebruggen *et al.*, 2003; Calton *et al.*, 2004; Poon *et al.*, 2006). The target of U_S3 is unknown.

It is of interest to note that the U_S3 open reading frames yields two mRNAs (McGeoch *et al.*, 1985; McGeoch *et al.*, 1988). Until lately, it has been thought that both mRNAs encode the same protein. Recent studies indicate that the abundance of the two mRNA is determined by ICP22 and that in the absence of ICP22 larger amounts of the smaller ($U_S3.5$) mRNAs and larger quantities of a truncated protein accumulate in the infected cells (Poon and Roizman, 2005). The data indicate that the truncated $U_S3.5$ protein initiates at met77. Like the larger protein, it retains kinase activity and mediates phosphorylation of several viral and cellular proteins, but does not block apoptosis. The question why HSV has evolved both a full length and a truncated versions of the same kinase remains for the moment unanswered. It is conceivable that these kinases target different proteins or accumulate in different compartments of the infected cells.

Conclusions

The studies on antiapoptotic functions of HSV revealed three significant aspects of the interaction of HSV with its host. The first is that HSV blocks apoptosis induced by nonphysiologic drugs (e.g. sorbitol). The implications of this observation is that HSV targets a myriad of pathways activated by osmotic shock or at the very least a key target common to all pathways.

The second aspect of particular interest is that expression of some viral proteins is proapoptotic. While the expression of many proteins in uninfected cells appears to be toxic (e.g. U_L41), the identity of the proapototic protein in either the $gD^{-/-}$ virions or in the defective ΔICP4 or ΔICP27 viruses remain unclear.

Lastly, cells do succumb to infection by HSV. The evidence that proapoptotic functions contribute to cell death emerged from studies showing that a HEp-2 cell line transduced and constitutively expressing Bcl-2 exhibited delayed cytopathic effects, even though it produced the same amount of wild type virus as the parent cell line (Galvan *et al.*, 2000).

The studies described in this chapter revealed the following:

1 Viral proteins are multifunctional. They are also exquisite examples of protein engineering. It is difficult to envision the construction of proteins with as many diverse functions, as for example ICP0, ICP22 or ICP27. The implications of these findings is that many proteins, for which only one function is known, may surprise us with many additional functions. To reiterate, introductory remarks, the sum total of viral functions may well reach many hundreds.

2 Viral functions described in this chapter are designed to block or evade cellular functions. It seems appropriate to evaluate the importance of specific cellular function by the number of diverse viral functions directed against it. By this criterion, the interferon pathway wins by a wide margin. For example, the interferon pathway is attacked by ICP0 (dispersal of ND10, degradation of interferon γ receptor), $\gamma_1$34.5, U_L41, etc. The next in line would be interference with signaling by the infected cell to its environment and conversely, that of the environment to the virally infected cell. Again this is the role of ICP0—in its role as a mediator of degradation of receptor tyrosine kinases (Liang and Roizman, 2006), ICP27, the virion host shutoff protein U_L41, but also ICP47—the viral protein that blocks presentation of antigenic peptides by MHC class I proteins.

3 Lastly, it seems appropriate to point out that the studies described in this report have opened up a world of cellular functions designed to block viral replication. Surprisingly, the functions described in this chapter and designed to subjugate the host cell are in fact expressed by a relatively small number of proteins. These include ICP0, ICP27, U_L41, $\gamma_1$34.5, U_S11, gD, gJ, U_S3 and to a lesser extent by αTIF, ICP4. It is very likely that many more viral proteins, especially proteins which do not play a direct role in viral replication, also express functions designed to speed the conquest of the infected cells. Studies of these proteins are likely to enrich our knowledge of host defenses to viral infections.

Acknowledgments

When not writing chapters, our research is supported by grants from the National Cancer Institute (CA87661, CA83939, CA71933, CA78766, and CA88860,), the United States Public Heath Service.

References

Amici, C., Belardo, G., Rossi, A., and Santoro, M.G. (2001). Activation of IκB kinase by herpes simplex virus type 1. A novel target for anti-herpetic therapy. J. Biol. Chem. *276*, 28759–28766.

Aubert, M., and Blaho, J.A. (1999). The Herpes simplex virus type 1 regulatory protein ICP27 is required for the prevention of apoptosis in infected human cells. J. Virol. *73*, 2803–2813.

Ballas, N., Battaglioli, E., Atouf, F., Andres, M.E., Chenoweth, J., Anderson, M.E., Burger, C., Moniwa, M., Davie, J.R., Bowers, W.J., Federoff, H.J., Rose, D.W., Rosenfeld, M.G., Brehm, P., and Mandel, G. (2001). Regulation of neuronal traits by a novel transcriptional complex. Neuron *31*, 353–365.

Benetti, L., and Roizman, B. (2003). Herpes simplex virus protein kinase U_S3 activates and functionally overlaps protein kinase A to block apoptosis. Proc. Natl. Acad. Sci. USA *101*, 9411–9416.

Benetti, L., Munger, J., and Roizman, B. (2003). The herpes simplex virus 1 U_S3 protein kinase blocks caspase dependent double cleavage and activation of the pro-apoptotic protein BAD. J. Virol. *77*, 6567–6573.

Blackshear, P.J. (2002). Tristetraprolin and other CCCH tandem zinc-finger proteins in the regulation of mRNA turnover. Biochem. Soc. Trans. *30*, 945–952.

Brooks, S.A., Connolly, J.E., and Rigby, W.F. (2004). The role of mRNA turnover in the regulation of tristetraprolin expression: evidence for an extracellular signal-regulated kinase-specific, AU-rich element-dependent, autoregulatory pathway. J. Immunol. *172*, 7263–7271.

Brunetti, C.R., Brurke, R.L., Kornfeld, S., Gregory, W., Masiarz, F.R., Dingwell, K.S., and Johnson, D.C. (1994). Herpes simplex virus glycoprotein D acquires mannose 6-phosphate residues and binds to mannose 6-phosphate receptors. J. Biol. Chem. *269*, 17067–17074.

Brunetti, C.R., Dingwell, K.S., Wale, C., Graham, F.L., and Johnson, D.C. (1998). Herpes simplex virus gD and virions accumulate in endosomes by mannose 6-phosphate-dependent and -independent mechanisms. J. Virol. *72*, 3330–3339.

Bruni, R., Fineschi, B., Ogle, W.O., and Roizman, B. (1999). A novel cellular protein, p60, interacting with both herpes simplex virus 1 regulatory proteins ICP22 and ICP0 is modified in a cell-type-specific manner and is recruited to the nucleus after infection. J. Virol. *73*, 3810–3817.

Bryant, H.E., Wadd, S., Lamond, A.I., Silverstein, S.J., and Clements, J.B. (2001). Herpes simplex virus IE63 (ICP27) protein interacts with spliceosome-associated protein 145 and inhibits splicing prior to the first catalytic step. J. Virol. *75*, 4376–4385.

Butler, J.S. (2002). The yin and yang of the exosome. Trends Cell. Biol. *12*, 90–96.

Calton, C.M., Randall, J.A., Adkins, M.W., and Banfield, B.W. (2004). The pseudorabies virus serine/threonine kinase U_S3 contains mitochondrial, nuclear and membrane localization signals. Virus Genes *29*, 131–145.

Cassady, K.A., Gross, M., and Roizman, B. (1998). The herpes simplex virus U_S11 protein effectively compensates for the $\gamma_134.5$ gene if present before activation of protein kinase R by precluding its phosphorylation and that of the alpha subunit of eukaryotic translation initiation factor 2. J. Virol. *72*, 8620–8626.

Chee, A.V., and Roizman, B. (2004). Herpes simplex virus 1 gene products occlude the interferon signaling pathway at multiple sites. J. Virol. *78*, 4185–4196.

Chee, A.V., Lopez, P., Pandolfi, P.P., and Roizman, B. (2003). Promyleocytic leukemia protein mediates interferon-based anti-herpes simplex virus 1 effects. J. Virol. *77*, 7101–7105.

Chen, C.Y., Gherzi, R., Ong, S.E., Chan, E.L., Raijmakers, R., Pruijn, G.J., Stoecklin, G., Moroni, C., Mann, M., and Karin, M. (2001). AU binding proteins recruit the exosome to degrade ARE-containing mRNAs. Cell *107*, 451–464.

Chong, J.A., Tapia-Ramirez, J., Kim, S., Toledo-Aral, J.J., Zheng, Y., Boutros, M.C., Altshuller, Y.M., Frohman, M.A., Kraner, S.D., and Mandel, G. (1995). REST: a mammalian silencer protein that restricts sodium channel gene expression to neurons. Cell *80*, 949–957.

Chou, J., and Roizman, B. (1992). The $\gamma_134.5$ gene of herpes simplex virus 1 precludes neuroblastoma cells from triggering total shutoff of protein synthesis characteristic of programed cell death in neuronal cells. Proc. Natl. Acad. Sci. USA *89*, 3266–3270.

Chou, J., and Roizman, B. (1994). Herpes simplex virus 1 $\gamma_134.5$ gene function, which blocks the host response to infection, maps in the homologous domain of the genes expressed during growth arrest and DNA damage. Proc. Natl. Acad. Sci. USA *91*, 5247–5251.

Dikic, I., and Giordano, S. (2003). Negative receptor signaling. Curr. Opin. Cell. Biol. *15*, 128–135.

Dixon, R.A., and Schaffer, P.A. (1980). Fine-structure mapping and functional analysis of temperature-sensitive mutants in the gene encoding the herpes simplex virus type 1 immediate early protein VP175. J. Virol. *36*, 189–203.

Doepker, R.C., Hsu, W.L., Saffran, H.A., and Smiley, J.R. (2004). Herpes simplex virus virion host shutoff protein is stimulated by translation initiation factors eIF4B and eIF4H. J. Virol. *78*, 4684–4699.

Esclatine, A., Taddeo, B., Evans, L., and Roizman, B. (2004a). The herpes simplex virus 1 U_L41 gene-dependent destabilization of cellular RNAs is selective and may be sequence-specific. Proc. Natl. Acad. Sci. USA *101*, 3603–3608.

Esclatine, A., Taddeo, B., and Roizman, B. (2004b). Herpes simplex virus 1 induces cytoplasmic accumulation of TIA-1/TIAR and both synthesis and cytoplasmic accumulation of tristetraprolin, two cellular proteins that bind and destabilize AU-rich RNAs. J. Virol. *78*, 8582–8592.

Esclatine, A., Taddeo, B., and Roizman, B. (2004c). The U_L41 protein of herpes simplex virus mediates selective stabilization or degradation of cellular mRNAs. Proc. Natl. Acad. Sci. USA *101*, 18165–18170.

Everett, R.D. (2004). Herpes simplex virus type 1 regulatory protein ICP0 does not protect cyclins D1 and D3 from degradation during infection. J. Virol. *78*, 9599–9604.

Everett, R.D., Meredith, M., Orr, A., Cross, A., Kathoria, M., and Parkinson J. (1997). A novel ubiquitin-specific protease is dynamically associated with the PML nuclear domain and binds to a herpesvirus regulatory protein. EMBO J. *16*, 1519–1530.

Everly, D.N. Jr., Feng, P., Mian, I.S., and Read, G.S. (2002). mRNA degradation by the virion host shutoff (Vhs) protein of herpes simplex virus: genetic and biochemical evidence that Vhs is a nuclease. J. Virol. *76*, 8560–8571.

Galvan, V. and Roizman, B. (1998). Herpes simplex virus 1 induces and blocks apoptosis at multiple steps during infection and protects cells from exogenous inducers in a cell-type-dependent manner. Proc. Natl. Acad. Sci. USA 95, 3931–3936.

Galvan, V., Brandimarti, R., Munger, J., and Roizman, B. (2000). Bcl-2 blocks a caspase-dependent pathway of apoptosis activated by herpes simplex virus 1 infection in HEp-2 cells. J. Virol. *74*, 1931–1938.

Gamble, M.J., Erdjument-Bromage, H., Tempst, P., Freedman, L.P., and Fisher, R.P. (2005). The histone chaperone TAF-I/SET/INHAT is required for transcription *in vitro* of chromatin templates. Mol. Cell. Biol. *25*, 797–807.

Goodkin, M.L., Ting, A.T., and Blaho, J.A. (2003). NF-κB is required for apoptosis prevention during herpes simplex virus type 1 infection. J. Virol. *77*, 7261–7280.

Gregory, D., Hargett, D., Holmes, D., Money, E., and Bachenheimer, S.L. (2004). Efficient replication by herpes simplex virus type 1 involves activation of the IκB kinase-IκB-p65 pathway. J. Virol. *78*, 13582–13590.

Gu, H., and Roizman, B. (2003). The degradation of promyelocytic leukemia and Sp100 proteins by herpes simplex virus 1 is mediated by the ubiquitin-conjugating enzyme UbcH5a. Proc. Natl. Acad. Sci. USA *100*, 8963–8968.

Gu, B., Rivera-Gonzalez, R., Smith, C.A., and DeLuca, N.A. (1993). Herpes simplex virus infected cell polypeptide 4 preferentially represses Sp1-activated over basal transcription from its own promoter. Proc. Natl. Acad. Sci. USA *90*, 9528–9532.

Gu, B., Kuddus, R., and DeLuca, N.A. (1995). Repression of activator-mediated transcription by herpes simplex virus ICP4 via a mechanism involving interactions with the basal transcription factors TATA-binding protein and TFIIB. Mol. Cell. Biol. *15*, 3618–3626.

Gu, H., Liang, Y., Mandel, G., and Roizman, B. (2005). Components of the REST/CoREST/histone deacetylase repressor complex are disrupted, modified, and translocated in HSV-1-infected cells. Proc. Natl. Acad. Sci. USA *102*, 7571–7576.

Hagglund, R., and Roizman, B. (2003). Herpes simplex virus 1 mutant in which ICP0 HUL-1 E3 ubiquitin ligase site is disrupted stabilizes cdc34 but degrades D-type cyclins and exhibits diminished neurotoxicity. J. Virol. *77*, 13194–13202.

Hagglund, R., and Roizman, B. (2004). Role of ICP0 in the strategy of conquest of the host cell by herpes simplex virus 1. J. Virol. *78*, 2169–2178.

Hardy, W.R., and Sandri-Goldin, R.M. (1994). Herpes simplex virus inhibits host cell splicing, and regulatory protein ICP27 is required for this effect. J. Virol. *68*, 7790–7799.

He, B., Chou, J., Brandimarti, R., Mohr, I., Gluzman, Y., and Roizman. B. (1997a). Suppression of the phenotype of $\gamma_1 34.5^-$ herpes simplex virus 1: failure of activated RNA-dependent protein kinase to shut off protein synthesis is associated with a deletion in the domain of the α47 gene. J. Virol. *71*, 6049–6054.

He, B., Gross, M., and Roizman, B. (1997b). The $\gamma_1 34.5$ protein of herpes simplex virus 1 complexes with protein phosphatase 1 α to dephosphorylate the α subunit of the eukaryotic translation initiation factor 2 and preclude the shutoff of protein synthesis by double-stranded RNA-activated protein kinase. Proc. Natl. Acad. Sci. USA *94*, 843–848.

He, B., Gross, M., and Roizman, B. (1998). The $\gamma_1 34.5$ protein of herpes simplex virus 1 has the structural and functional attributes of a protein phosphatase 1 regulatory subunit and is present in a high molecular weight complex with the enzyme in infected cells. J. Biol. Chem. *273*, 20737–20743.

Herrera, F.J., and Triezenberg, S.J. (2004). VP16-dependent association of chromatin-modifying coactivators and underrepresentation of histones at immediate-early gene promoters during herpes simplex virus infection. J. Virol. *78*, 9689–9696.

Humphrey, G.W., Wang, Y., Russanova, V.R., Hirai, T., Qin, J., Nakatani, Y., and Howard, B.H. (2001). Stable histone deacetylase complexes distinguished by the presence of SANT domain proteins CoREST/kiaa0071 and Mta-L1. J. Biol. Chem. *276*, 6817–6824.

Jacquemont, B., and Roizman, B. (1975). RNA synthesis in cells infected with herpes simplex virus. X. Properties of viral symmetric transcripts and of double-stranded RNA prepared from them. J. Virol. *15*, 707–713.

Jerome, K.R., Tait, J.F., Koelle, D.M., and Corey, L. (1998). Herpes simplex virus type 1 renders infected cells resistant to cytotoxic T-lymphocyte-induced apoptosis. J. Virol. *72*, 436–441.

Jerome, K.R., Fox, R., Chen, Z., Sears, A.E., Lee, H.-Y., and Corey, L. (1999). Herpes simplex virus inhibits apoptosis through the action of two genes, U_S5 and U_S3. J. Virol. *73*, 8950–8957.

Jerome, K.R., Chen, Z., Lang, R., Torres, M.R., Hofmeister, J., Smith, S., Fox, R., Froelich, C.J., and Corey, L. (2001). HSV and glycoprotein J inhibit caspase activation and apoptosis induced by granzyme B or Fas. J. Immunol. *167*, 3928–3935.

Karr, B.M., and Read, G.S. (1999). The virion host shutoff function of herpes simplex virus degrades the 5′ end of a target mRNA before the 3′ end. Virology *264*, 195–204.

Kawaguchi, Y., Bruni, R., and Roizman, B. (1997a). Interaction of herpes simplex virus 1 α regulatory protein ICP0 with elongation factor 1δ: ICP0 affects translational machinery. J. Virol. *71*, 1019–1024.

Kawaguchi, Y., Van Sant, C., and Roizman, B. (1997b). Herpes simplex virus 1 regulatory protein ICP0 interacts with and stabilizes the cell cycle regulator cyclin D3. J. Virol. *71*, 7328–7336.

Kawaguchi, Y., Tanaka, M., Yokoymama, A., Matsuda, G., Kato, K., Kagawa, H., Hirai, K., and Roizman, B. (2001). Herpes simplex virus 1 regulatory protein ICP0 functionally interacts with cellular transcription factor BMAL1. Proc. Natl. Acad. Sci. USA *98*, 1877–1882.

Kim, D.B., Zabierowski, S., and DeLuca, N.A. (2002). The initiator element in a herpes simplex virus type 1 late-gene promoter enhances activation by ICP4, resulting in abundant late-gene expression. J. Virol. *76*, 1548–1558.

Knipe, D.M., Ruyechan, W.T., Roizman, B., and Halliburton, I.W. (1978). Molecular genetics of herpes simplex virus: demonstration of regions of obligatory and nonobligatory identity within diploid regions of the genome by sequence replacement and insertion. Proc. Natl. Acad. Sci. USA *75*, 3896–3900.

Koyama, A.H., and Miwa, Y. (1997). Suppression of apoptotic DNA fragmentation in herpes simplex virus type 1-infected cells. J. Virol. *71*, 2567–2571.

Kozak, M., and Roizman, B. (1975). RNA synthesis in cells infected with herpes simplex virus. IX. Evidence for accumulation of abundant symmetric transcripts in nuclei. J. Virol. *15*, 36–40.

Kristie, T.M., and Roizman, B. (1987). Host cell proteins bind to the *cis*-acting site required for virion-mediated induction of herpes simplex virus 1 α genes. Proc. Natl. Acad. Sci. USA *84*, 71–75.

Kristie, T.M., and Roizman, B. (1988). Differentiation and DNA contact points of host proteins binding at the *cis* site for virion-mediated induction of alpha genes of herpes simplex virus 1. J. Virol. *62*, 1145–1157.

Kurakin, A.V., Wu, S., and Bredesen, D.E. (2003). Atypical recognition consensus of CIN85/SETA/Ruk SH3 domains revealed by target-assisted iterative screening. J. Biol. Chem. *278*, 34102–34109.

Kwong, A.D., Kruper, J.A., and Frenkel, N. (1988). Herpes simplex virus virion host shutoff function. J. Virol. *62*, 912–921.

Langelier, Y., Bergeron, S., Chabaud, S., Lippens, J., Guilbault, C., Sasseville, A.M., Denis, S., Mosser, D.D., and Massie, B. (2002). The R1 subunit of herpes simplex virus ribonucleotide reductase protects cells against apoptosis at, or upstream of, caspase-8 activation. J. Gen. Virol. *83*, 2779–2789.

Leopardi, R., and Roizman, B. (1996). The herpes simplex virus major regulatory protein ICP4 blocks apoptosis induced by the virus or by hyperthermia. Proc. Natl. Acad. Sci. USA *93*, 9583–9587.

Leopardi, R., Michael, N., and Roizman, B. (1995). Repression of the herpes simplex virus 1 alpha 4 gene by its gene product (ICP4) within the context of the viral genome is conditioned by the distance and stereoaxial alignment of the ICP4 DNA binding site relative to the TATA box. J. Virol. *69*, 3042–3048.

Leopardi, R., Van Sant, C., and Roizman, B. (1997). The herpes simplex virus 1 protein kinase US3 is required for protection from apoptosis induced by the virus. Proc. Natl. Acad. Sci. USA *94*, 7891–7896.

Liang, Y, and Roizman, B. (2006). State and role of SRC family kinases in replication of herpes simplex virus 1. J. Virol. *80*, 3349–3359.

Liang, Y., Kourakin, A., and Roizman, B. (2005). Herpes simplex virus 1 infected cell protein 0 forms a complex with CIN85 and Cbl and mediates the degradation of EGF receptor from cell surfaces. Proc. Natl. Acad. Sci. USA *102*, 58738–5843.

Linberg, A., and Kreivi, J.P. (2002). Splicing inhibition at the level of spliceosome assembly in the presence of herpes simplex virus protein ICP27. Virology *294*, 189–198.

Lomonte, P., Thomas, J., Texier, P., Caron, C., Khochbin, S., and Epstein, A.L. (2004). Functional interaction between class II histone deacetylases and ICP0 of herpes simplex virus type 1. J. Virol. *78*, 6744–6757.

Lopez, P., Jacob, R.J., and Roizman, B. (2002). Overexpression of promyelocytic leukemia protein precludes the dispersal of ND10 structures and has no effect on accumulation of infectious herpes simplex virus 1 or its proteins. J. Virol. *76*, 9355–9367.

Lunyak, V.V., Burgess, R., Prefontaine, G.G., Nelson, C., Sze, S.H., Chenoweth, J., Schwartz, P., Pevzner, P.A., Glass, C., Mandel, G., and Rosenfeld, M.G. (2002). Corepressor-dependent silencing of chromosomal regions encoding neuronal genes. Science *298*, 1747–1752.

McGeoch, D.J., Dolan, A., Donald, S., and Rixon, F.J. (1985). Sequence determination and genetic content of the short unique region in the genome of herpes simplex virus type 1. J. Mol. Biol. *181*, 1–13.

McGeoch, D.J., Dalrymple, M.A., Davison, A.J., Dolan, A., Frame, M.C., McNab, D., Perry, L.J., Scott, J.E., and Taylor, P. (1988). The complete DNA sequence of the long unique region in the genome of herpes simplex virus type I. J. Gen. Virol. *69*, 1531–1574.

McKnight, J.L., Kristie, T.M., and Roizman, B. (1987). Binding of the virion protein mediating alpha gene induction in herpes simplex virus 1-infected cells to its *cis* site requires cellular proteins. Proc. Natl. Acad. Sci. USA *84*, 7061–7065.

Mears, W.E., and Rice, S.A. (1998). The herpes simplex virus immediate-early protein ICP27 shuttles between the nucleus and cytoplasm. Virology *242*, 128–137.

Mogensen, T.H., and Paludan, S.R. (2001). Molecular pathways in virus-induced cytokine production. Microbiol. Mol. Biol. Rev. *65*, 131–150.

Mohr, I., and Gluzman, Y. (1996). A herpesvirus genetic element which affects translation in the absence of the viral GADD34 function. EMBO J. *15*, 4759–4766.

Mossman, K.L., Macgregor, P.F., Rozmus, J.J., Goryachev, A.B., Edwards, A.M., and Smiley, J.R. (2001). Herpes simplex virus triggers and then disarms a host Antiviral response. J. Virol. *75*, 750–758.

Munger, J., and Roizman, B. (2001). The U_S3 protein kinase of herpes simplex virus 1 mediates the posttranslational modification of BAD and prevents BAD-induced programmed cell death in the absence of other viral proteins. Proc. Natl. Acad. Sci. USA *98*, 10410–10415.

Munger, J., Chee, A.V., and Roizman, B. (2001). The U_S3 protein kinase blocks apoptosis induced by the d120 mutant of herpes simplex virus 1 at a premitochondrial stage. J. Virol. 75, 5491–5497.

Negorev, D., and Maul, G.G. (2001). Cellular proteins localized at and interacting within ND10/PML nuclear bodies/PODs suggest function of nuclear depot. Oncogene *20*, 7234–7242.

Nicholl, M.J., Robinson, L.H., and Preston, C.M. (2000). Activation of cellular interferon-responsive genes after infection with herpes simplex virus type 1. J. Gen. Virol. *81*, 2215–2218.

Ogg, P.D., McDonell, P.J., Ryckman, B.J., Knudson, C.M., and Roller, R.J. (2004). The HSV-1 U_S3 protein kinase is sufficient to block apoptosis induced by overexpression of a variety of Bcl-2 family members. Virology *319*, 212–224.

Pahl, H.L. (1999). Activators and target genes of Rel/NF-κB transcription factors. Oncogene *18*, 6853–6866.

Patel, A., Hanson, J., McLean, T.I., Olgiate, J., Hilton, M., Miller, W.E., and Bachenheimer, S.L. (1998). Herpes simplex type 1 induction of persistent NF-κB nuclear translocation increases the efficiency of virus replication. Virology *247*, 212–222.

Pellett, P.E., McKnight, J.L., Jenkins, F.J., and Roizman, B. (1985). Nucleotide sequence and predicted amino acid sequence of a protein encoded in a small herpes simplex virus DNA fragment capable of trans-inducing alpha genes. Proc. Natl. Acad. Sci. USA *82*, 5870–5874.

Peng, W., Jin, L., Henderson, G., Perng, G.C., Brick, D.J., Nesburn, A.B., Wechsler, S.L., and Jones, C. (2004). Mapping herpes simplex virus type 1 latency-associated transcript sequences that protect from apoptosis mediated by a plasmid expressing caspase-8. J. Neurovirol. *10*, 260–265.

Perez-Parada, J., Saffran, H.A., and Smiley, J.R. (2004). RNA degradation induced by the herpes simplex virus vhs protein proceeds 5′ to 3′ *in vitro*. J. Virol. *78*, 13391–13394.

Petrelli, A., Gilestro, G.F., Lanzardo, S., Comoglio, P.M., Migone, N., and Giordano, S. (2002). The endophilin-CIN85-Cbl complex mediates ligand-dependent downregulation of c-Met. Nature *416*, 187–190.

Poon, A.P., and Roizman, B. (2005). Herpes simplex virus 1 ICP22 regulates the accumulation of a shorter mRNA and of a truncated U_S3 protein kinase that exhibits altered functions. J. Virol. *79*, 8470–8479.

Poon, A.P., Silverstein, S.J., and Roizman, B. (2002). An early regulatory function required in a cell type-dependent manner is expressed by the genomic but not the cDNA copy of the herpes simplex virus 1 gene encoding infected cell protein 0. J. Virol. *76*, 9744–9755.

Poon, A.P., Lyang, Y., and Roizman, B. (2003). Herpes simplex virus 1 gene expression is accelerated by inhibitors of histone deacetylases in rabbit skin cells infected with a mutant carrying a cDNA copy of the infected-cell protein no. 0. J. Virol. *77*, 12671–12678.

Poon, A.P., Benetti, L., and Roizman, B. (2006). U_S3 and $U_S3.5$ protein kinases of herpes simplex virus

1 differ with respect to their functions in blocking apoptosis and in virion maturation and egress. J. Virol. *80*, 3752–3764.

Preston, C.M., Harman, A.N., and Nicholl, M.J. (2001). Activation of interferon response factor-3 in human cells infected with herpes simplex virus type 1 or human cytomegalovirus. J. Virol. *75*, 8909–8916.

Regad, T., and Chelbi-Alix, M.K. (2001). Role and fate of PML nuclear bodies in response to interferon and viral infections. Oncogene *20*, 7274–7286.

Reynolds, A.E., Wills, E.G. Roller, R.J., Ryckman, B.J., and Baines, J.D. (2002). Ultrastructural localization of the herpes simplex virus type 1 U_L31, U_L34, and U_S3 proteins suggests specific roles in primary envelopment and egress of nucleocapsids. J. Virol. *76*, 8939–8952.

Roizman, B. (1996). The function of herpes simplex virus genes: a primer for genetic engineering of novel vectors. Proc. Natl. Acad. Sci. USA *93*, 11307–11312.

Roizman, B., and Knipe, D.M. (2001). Herpes simplex viruses and their replication. In Virology, Knipe, D.M. and Howley, P.M., ed. (Philadelphia, PA, USA: Lippincott, Williams & Wilkins), pp 2399–2459.

Roller, R.J., and Roizman, B. (1990). The herpes simplex virus U_S11 open reading frame encodes a sequence specific RNA-binding protein. J. Virol. *64*, 3463–3470.

Roopra, A., Qazi, R., Schoenike, B., Daley, T.J., and Morrison, J.F. (2004). Localized domains of g9a-mediated histone methylation are required for silencing of neuronal genes. Mol. Cell *14*, 727–738.

Sandri-Goldin, R.M. (1998). ICP27 mediates HSV RNA export by shuttling through a leucine-rich nuclear export signal and binding viral intronless RNAs through an RGG motif. Genes Dev. *12*, 868–879.

Santoro, M.G., Rossi, A., and Amici, C. (2003). NF-κB and virus infection: who controls whom. EMBO J. *22*, 2552–2260.

Sciortino, M.T., Taddeo, B., Poon, A.P., Mastino, A., and Roizman, B. (2002). Of the three tegument proteins that package mRNA in herpes simplex virions, one (VP22) transports the mRNA to uninfected cells for expression prior to viral infection. Proc. Natl. Acad. Sci. USA *99*, 8318–8323.

Shi, Y., Sawada, J., Sui, G., Affar, E.B., Whetstine, J.R., Lan, F., Ogawa, H., Luke, M.P., Nakatani, Y., and Shi, Y. (2003). Coordinated histone modifications mediated by a CtBP co-repressor complex. Nature *422*, 735–738.

Sieg, S., Yildirim, Z., Smith, D., Kayagaki, N., Yagita, H., Huang, Y., and Kaplan, D. (1996). Herpes simplex virus type 2 inhibition of Fas ligand expression. J. Virol. *70*, 8747–8751.

Smibert, C.A., Popova, B., Xiao, P., Capone, J.P., and Smiley, J.R. (1994). Herpes simplex virus VP16 forms a complex with the virion host shutoff protein vhs. J. Virol. *68*, 2339–2346.

Smiley, J.R. (2004). Herpes simplex virus virion host shutoff protein: immune evasion mediated by a viral RNase? J. Virol. *78*, 1063–1068.

Soliman, T.M., Sandri-Goldin, R.M., and Silverstein, S.J. (1997). Shuttling of the herpes simplex virus type 1 regulatory protein ICP27 between the nucleus and cytoplasm mediates the expression of late proteins. J. Virol. *71*, 9188–9197.

Soubeyran, P., Kowanetz, K., Szymkiewicz, I., Langdon, W.Y., and Dikic, I. (2002). Cbl-CIN85-endophilin complex mediates ligand-induced downregulation of EGF receptors. Nature *416*, 183–187.

Strom, T., and Frenkel, N. (1987). Effects of herpes simplex virus on mRNA stability. J. Virol. *61*, 2198–2207.

Taddeo, B., Esclatine, A., and Roizman, B. (2002). The patterns of accumulation of cellular RNAs in cells infected with a wild-type and a mutant herpes simplex virus 1 lacking the virion host shutoff gene. Proc. Natl. Acad. Sci. USA *99*, 17031–17036.

Taddeo, B., Esclatine, A., Zhang, W., and Roizman, B. (2003a). The stress-inducible immediate-early responsive gene IEX-1 is activated in cells infected with herpes simplex virus 1, but several viral mechanisms, including 3′ degradation of its RNA, preclude expression of the gene. J. Virol. *77*, 6178–6187.

Taddeo, B., Luo, T.R., Zhang, W., and Roizman, B. (2003b). Activation of NF-κB in cells productively infected with HSV-1 depends on activated protein kinase R and plays no apparent role in blocking apoptosis. Proc. Natl. Acad. Sci. USA *100*, 12408–12413.

Taddeo, B., Zhang, W., Lakeman, F., and Roizman, B. (2004a). Cells lacking NF-κB or in which NF-κB is not activated vary with respect to ability to sustain herpes simplex virus 1 replication and are not susceptible to apoptosis induced by a replication-incompetent mutant virus. J. Virol. *78*, 11615–11621.

Taddeo, B., Esclatine, A., and Roizman, B. (2004b). Post-transcriptional processing of cellular RNAs in herpes simplex virus-infected cells. Biochem. Soc .Trans. 32, 697–701.

Thien, C.B., and Langdon, W.Y. (2001). Cbl: many adaptations to regulate protein tyrosine kinases. Nat. Rev. Mol. Cell. Biol. 2, 294–307.

Van Minnebruggen, G., Favoreel, H.W., Jacobs, L., and Nauwynck, H.J. (2003). Pseudorabies virus U_S3 protein kinase mediates actin stress fiber breakdown. J. Virol. *77*, 9074–9080.

Van Sant, C., Kawaguchi, Y., and Roizman, B. (1999). A single amino acid substitution in the cyclin D binding domain of the infected cell protein no. 0 abrogates the neuroinvasiveness of herpes simplex virus without affecting its ability to replicate. Proc. Natl. Acad. Sci. USA 96, 8184–8189.

Van Sant, C., Lopez, P., Advani, S.J., and Roizman B. (2001a). Role of cyclin D3 in the biology of herpes simplex virus 1 ICP0. J. Virol. 75, 1888–1898.

Van Sant, C., Hagglund, R., Lopez, P., and Roizman, B. (2001b). The infected cell protein 0 of herpes simplex virus 1 dynamically interacts with proteasomes, binds and activates the cdc34 E3 ubiquitin-conjugating enzyme, and possesses *in vitro* E3 ubiquitin ligase activity. Proc. Natl. Acad. Sci. USA 98, 8815–8820.

Williams, B.R. (2001). Signal integration via PKR. Sci. STKE 89, RE2.

Wilusz, C.J., Wormington, M., and Peltz, S.W. (2001). The cap-to-tail guide to mRNA turnover. Nat. Rev. Mol. Cell. Biol. 2, 237–246.

Xiao, W., Pizer, L.I., and Wilcox, K.W. (1997). Identification of a promoter-specific transactivation domain in the herpes simplex virus regulatory protein ICP4. J. Virol. 71, 1757–1765.

You, A., Tong, J.K., Grozinger, C.M., and Schreiber, S.L. (2001). CoREST is an integral component of the CoREST- human histone deacetylase complex. Proc. Natl. Acad. Sci. USA 98, 1454–1458.

Zhong, S., Salomoni, P. and Pandolfi, P.P. (2000). The transcriptional role of PML and the nuclear body. Nat. Cell. Biol. 2, E85–90.

Zhou, G., and Roizman, B. (2001). The domains of glycoprotein D required to block apoptosis depend on whether glycoprotein D is present in the virions carrying herpes simplex virus 1 genome lacking the gene encoding the glycoprotein. J. Virol. 75, 6166–6172.

Zhou, G., and Roizman, B. (2002a). Cation-independent mannose 6-phosphate receptor blocks apoptosis induced by herpes simplex virus 1 mutants lacking glycoprotein D and is likely the target of antiapoptotic activity of the glycoprotein. J. Virol. 76, 6197–6204.

Zhou, G., and Roizman, B. (2002b). Truncated forms of glycoprotein D of herpes simplex virus 1 capable of blocking apoptosis and of low-efficiency entry into cells form a heterodimer dependent on the presence of a cysteine located in the shared transmembrane domains. J. Virol. 76, 11469–11475.

Zhou, G., Galvan, V., Campadelli-Fiume, G., and Roizman, B. (2000). Glycoprotein D or J delivered in trans blocks apoptosis in SK-N-SH cells induced by a herpes simplex virus 1 mutant lacking intact genes expressing both glycoproteins. J. Virol. 74, 11782–11791.

Investigations of the Molecular Mechanisms of Varicella Zoster Virus Pathogenesis

15

Ann M. Arvin, Anne C. Schaap, Chia Chi Ku, Jeremy O. Jones, Marvin Sommer, and Leigh Zerboni

Abstract

Two major advances have provided new opportunities for understanding the molecular mechanisms of VZV pathogenesis. First, VZV cosmids permit construction of VZV recombinant viruses with targeted genetic mutations. Second, the development of the SCIDhu mouse model, in which skin, T cell and dorsal root ganglia xenografts are infected *in vivo* makes it possible to investigate VZV replication in differentiated human cells within their unique tissue microenvironments. These two approaches have yielded new insights about the molecular mechanisms of VZV pathogenesis that are involved in viral tropism for skin, T cells and neurons *in vivo* and about the host cell response to VZV infection.

Introduction

Varicella zoster virus (VZV) is a medically important human alpha-herpesvirus that causes varicella and herpes zoster in healthy and immunocompromised patients (Arvin, 2001; Gilden *et al.*, 2000). Primary VZV infection, or varicella, is almost always symptomatic and is characterized by a cell-associated viremia and skin lesions. VZV establishes latency in sensory ganglia, and causes zoster upon reactivation. The VZV genome is ~125,000 base pairs, encoding at least 70 unique open reading frames (ORFs) (Cohen and Straus, 2001). VZV resembles the other two human alpha-herpesviruses, herpes simplex virus (HSV) 1 and 2, in the linear arrangement of open reading frames (ORFs) and in similarities among "core" genes that are characteristic of the herpesviruses in general, as well as having substantial sequence homologies between some ORFs. However, VZV pathogenesis is associated with unique clinical manifestations that must result from its genetic differences from HSV-1 and HSV-2. Importantly, VZV infects T cells, which we propose to be the mechanism for its transport from respiratory epithelial sites of inoculation to sites of replication in dermal and epidermal cells during primary infection (Ku *et al.*, 2002; Ku *et al.*, 2004). VZV and HSV-1 and HSV-2 also differ in that the VZV genome has no latency-associated transcripts (LATs).

Two major advances have provided new opportunities for understanding the molecular mechanisms of VZV pathogenesis. First, VZV cosmids permit construction of VZV recombinant viruses with targeted genetic mutations (Cohen *et al.*, 1993; Mallory *et al.*, 1997; Kemble *et al.*, 2000). Whereas generating mutant VZV strains by homologous recombination was hampered by the extreme difficulty of plaque-purifying VZV, cosmid mutagenesis resolves this problem. Our VZV cosmids were made from VZV DNA recovered from cells infected with the vaccine Oka virus (vOka) or with the original wild type parent Oka virus (pOka), and prepared as four or five overlapping fragments that correspond to the complete VZV genome (Kemble *et al.*, 2000; Niizuma *et al.*, 2003; Besser *et al.*, 2003; Zerboni *et al.*, 2005a) (Fig. 15.1). To make VZV mutants, the gene or promoter region of interest is targeted for PCR mutagenesis, using a series of subcloning steps from the cosmid in which it is located. After the mutation has been inserted into the cosmid and confirmed by sequencing, melanoma cells are transfected with the mutated cosmid along with the other three or four intact cosmids. When the mutation is compatible with VZV replication, recombinant virus plaques usually appear within 7–10 days. If infectious virus is not recovered, the intact gene is inserted into a non-native AvrII cloning site between ORF65 and ORF66 and transfections are done using the mutated and "rescue" cosmids or with complementing cell lines. The results of cosmid mutagenesis experiments, with the necessary controls, establish whether the mutation is lethal or compatible with viral replication *in vitro*.

The second major advance is the development of the SCIDhu mouse model, in which skin and T cell xenografts are infected *in vivo*. This approach makes it possible to define effects of VZV mutations on virulence in differentiated human cells that are targets of VZV infection within their unique tissue microenvironments (review Ku *et al.*, 2005). The SCIDhu model overcomes obstacles to studies of VZV pathogenesis that result from the very restricted infectivity of the virus in non-human species. In recent work, we have

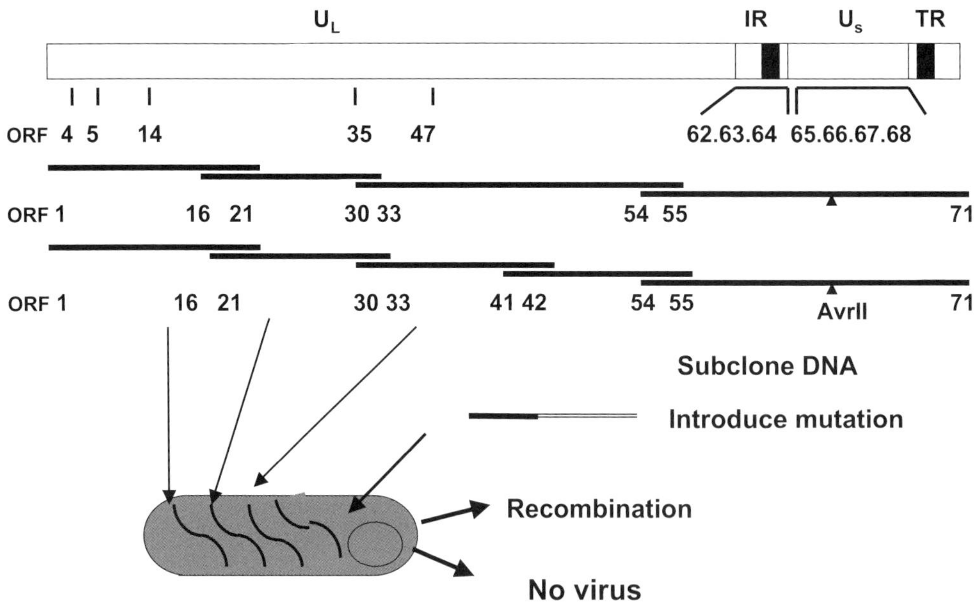

Figure 15.1 Cosmid Mutagenesis. Overlapping fragments corresponding to the complete VZV genome yield infectious VZV when transfected into melanoma cells. PCR mutagenesis is used to create deletions or targeted mutations in the genome through a series of subcloning steps from the relevant cosmid and inserted back into the cosmid for transfection with the other three or four intact cosmids. Recombinant virus is generated if the mutation is compatible with VZV replication. When infectious virus is not recovered, the intact gene is inserted into a non-native AvrII cloning site between ORF65 and ORF66 to demonstrate rescue.

also established xenografts of human dorsal root ganglia (DRG) in SCIDhu mice and demonstrated VZV tropism for neurons within these tissues (Zerboni *et al.*, 2005b). SCIDhu experiments showed that VZV infection of skin xenografts produced clusters of multinucleated, fused cells in the epidermis and dermis in the pattern that is typical of polykaryocyte formation seen in biopsies of varicella and zoster lesions (Fig. 15.2). Furthermore, VZV inoculation of thymus/liver xenografts resulted in productive infection of differentiated human T cells, with yields of infectious virus increasing over 7–21 days (Moffat *et al.*, 1995). VZV proteins were expressed by CD4, CD8 and dual CD4/CD8 T cells. Single VZV-infected cells were distributed throughout T cell xenografts, with no evidence of cell fusion and T cells that were infected *in vivo* released infectious VZV. T cells infected with VZV *in vivo* showed extensive virion formation in nuclei and many fully enveloped virions in the cytoplasm (Fig. 15.2) (Schaap *et al.*, 2005). These experiments in the SCIDhu model provided direct evidence that VZV has a robust tropism for T cells, suggesting that these migratory immune cells have the capacity to support viral transfer from respiratory sites of inoculation to skin during primary VZV infection.

In clinical practice, live attenuated vOka vaccines are effective for preventing varicella healthy children and are being evaluated for zoster prevention in the elderly (Gershon 2001). However, the attenuation of vOka, which was achieved by passing the clinical isolate, pOka, in guinea pig and human fibroblasts, has not eliminated its capacity to cause varicella in immunocompromised children. The vaccine virus can also establish latency and reactivate in healthy and immunocompromised patients. Combining the mutational analysis of VZV genes and promoters using cosmids with the evaluation of mutant VZV recombinant viruses in SCIDhu skin, T cell and DRG xenografts has the potential to identify options for the design of a recombinant "second generation" varicella vaccine. The criteria for an improved VZV vaccine virus would be that it retains infectivity in skin but has restricted tropism for T cells and sensory ganglia.

This chapter summarizes new observations about VZV pathogenesis that have emerged from the use of VZV cosmids for targeted mutagenesis of the VZV genome

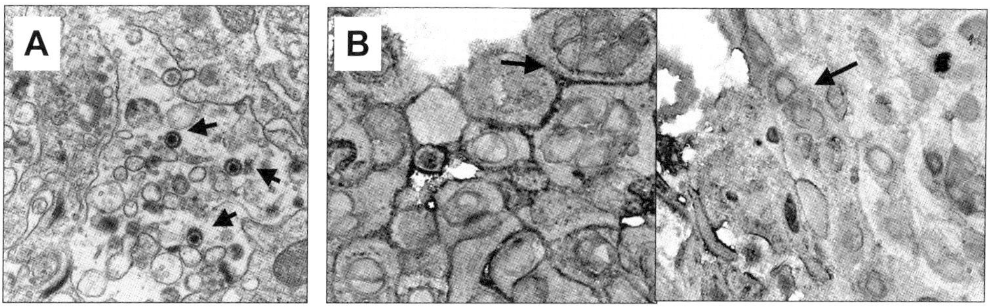

Figure 15.2 VZV infection of T cells and skin. A. Complete enveloped virions can be seen within vesicles in the cytoplasm of a T cell infected in a SCIDhu T cell xenograft. B. These sections were prepared from skin implants harvested at Day 21 after infection with rOka (left) or a ORF47 kinase deficient mutant (right) and stained with anti-gE mAb and counterstained with hematoxylin. Arrows indicate polykaryocytes with dark staining of gE in plasma membranes of rOka-infected cells and persisent formation of polykaryocytes in skin infected with the ORF47 kinase mutant despite minimal gE localization to plasma membranes. (Adapted from Besser *et al.*, 2004; Schaap *et al.*, 2005).

and investigations in the SICDhu model to establish the functions of VZV genes in tropism for skin, T cells and neurons *in vivo*.

Correlations between observations about the clinical characteristics of VZV strains in the human host and assessments of virulence in the SCIDhu skin model

Evidence that VZV infection in skin xenografts reproduces the pathogenicity of VZV isolates as defined from clinical experience, has been obtained by analyses of the vOka vaccine (Oka-Merck), chimeric vOka/pOka recombinants and VZV-MSP, which is a naturally occurring VZV escape mutant (Santos *et al.*, 2000). To assess the attenuation of vOka, the pathogenic potential of vOka in skin was compared with pOka, a second low-passage clinical isolate, VZV-S, and VZV-Ellen, a standard, high-passage laboratory strain (Moffat *et al.*, 1998a). All of these VZV strains replicated well in tissue culture, but only wild type pOka and VZV-S were fully virulent in skin, based on infectious virus yields and the extent of lesion formation. The impaired growth of vOka in skin xenografts is consistent with its failure to produce cutaneous lesions when given to healthy children by subcutaneous inoculation. Since adaptive immunity is not elicited in SCID mice, these experiments provided the first evidence that the clinical attenuation of vOka is intrinsic, reflecting changes in the viral genome, rather than being a consequence of the artificial route of inoculation or the dose of infectious virus (Arvin, 2001a). Infection of skin xenografts with pOka and VZV-S was associated with release of cell-free virus, which occurs in VZV skin lesions in the human host, whereas vOka replication remained cell-associated. Notably, VZV-Ellen, which has been passed > 100 times in cultured human cells, was avirulent in skin, indicating that adapting VZV for growth *in vitro* can completely eliminate its pathogenicity *in vivo*.

In order to further investigate the genetic basis of vOka attenuation, we used our five fragment pOka and vOka cosmids to generate chimeric pOka and vOka recombinant viruses and evaluated these chimeric viruses in SCIDhu skin xenografts (Zerboni *et al.*, 2005a). The growth kinetics and plaque morphologies of the six chimeric pOka/vOka viruses, pOka and vOka did not differ *in vitro* but the pOka/vOka chimeric viruses exhibited variable virulence in skin xenografts *in vivo*. Wild type growth was preserved when the chimeric viral genome contained ORFs 30–55 from the pOka genome. However, chimeric viruses that included a number of combinations of vOka components had reduced growth in skin, suggesting that vOka attenuation is mediated by changes in genes from several regions of the vOka genome.

The VZV genome has been considered to be highly stable, based on predictable antigenic cross-reactivity between geographically unrelated isolates. However, Grose and colleagues identified the VZV-MSP variant, isolated from an immunocompromised child with varicella. VZV-MSP encodes a mutant gE protein (D150N) that lacks the epitope recognized by the monoclonal antibody, 3B3, and has an unusual pattern of viral egress and substantially enhanced cell-to-cell spread *in vitro* (Santos *et al.*, 2000). This altered phenotype was associated with accelerated growth and enhanced lesion formation in skin xenografts *in vivo*.

Experiments in the SCIDhu model have also revealed differences between VZV and HSV-1 infection that correlate with clinical observations about pathogenesis in the human host (Moffat *et al.*, 1998a). The clinical isolates, pOka and VZV-S, caused skin lesions involving the epidermis and dermis, with extensive destruction of the basement membrane, while HSV-1 produced small foci of infected cells in the epidermis only. Whereas VZV infects CD4 and CD8 T cells in SCIDhu T cell xenografts, HSV-1 was detected only in stromal cells, with no evidence of T cell tropism. This difference is consistent with the clinical pattern of localized mucocutaneous HSV-1 lesions, whereas the hallmark of primary VZV infection is dissemination to multiple cutaneous sites by a cell-associated viremia.

Mutagenesis of VZV glycoprotein genes and effects on VZV replication *in vitro* and on pathogenesis in the SCIDhu model

As is characteristic of herpesviruses, viral glycoproteins are predominant constituents of the final VZV envelope. VZV encodes glycoproteins designated gB, gC, gE, gH, gI, gK, gL, gM and gN. However, information about the functions of most of these proteins is limited. Our investigations of VZV glycoproteins have focused on gE, encoded by ORF68, and gI, encoded by ORF67, located in the short unique (U_S) segment of the VZV genome. gE and gI are of particular interest for understanding VZV pathogenesis because they are the only glycoproteins encoded by the U_S region, whereas HSV has gD, gE, gG, gI and gJ genes. One of the most obvious differences between the VZV and HSV genomes is that VZV lacks gD, a glycoprotein which is essential for HSV entry into host cells. Since VZV gE is one third larger than HSV gD, it may incorporate some gD-like functions. VZV gE and gI form noncovalently linked heterodimers in infected cells and appear to function together at critical stages of VZV replication, including cell–cell spread and envelopment (review Cole and Grose 2003). The trafficking and post-translational modification of gE are abnormal without gI and gI facilitates gE endocytosis and accumulation in the trans-Golgi network (TGN), where it may bind to tegument proteins for secondary envelopment.

Glycoprotein E

VZV gE is the most abundant viral glycoprotein in infected cells and is a major envelope protein. Deleting ORF68 using the VZV cosmids was not compatible with viral replication, whereas insertion of ORF68 at the non-native AvrII site yielded VZV recombinants (Mo *et al.*, 2002). These experiments indicated that gE was an essential glycoprotein in VZV replication, whereas eliminating gE is not lethal for other alpha-herpesviruses. When gE was expressed as a *tet*-inducible protein in melanoma cells (Met-gE cells), gE colocalized with a TGN marker in perinuclear sites, but did not reach plasma membranes. We used these cells to show that the ORF68 deletion was rescued by complementation in Met-gE cells and to investigate how VZV replication affected intracellular trafficking of gE. Since gE from VZV-MSP does not bind the 3B3 monoclonal antibody, we made a recombinant virus, vOka-MSPgE, in which MSP-gE was substituted for native gE and followed intracellular localization of Met-gE when Met-gE cells were infected with this recombinant. VZV infection resulted in re-location of Met-gE from the perinuclear spaces to other cytoplasmic sites and to plasma membranes. This approach was used to assess the role of gI in gE trafficking by deriving a second recombinant, vOka-MSPgEΔgI, that lacked gI. Met-gE trafficking to plasma membranes was severely inhibited in the absence of gI and gE maturation to the 94K late form was also delayed. These observations demonstrated that gE:gI interactions were critical for correct intracellular localization and maturation of gE during VZV replication and were consistent with results about these effects that have been obtained using gE and gI expression methods.

Having established that gE was essential for VZV replication, we mapped the domains of the gE C-terminus by mutagenesis of pOka cosmids in order to determine which of the functional elements that have been identified by transient expression experiments were critical for infection *in vitro* and to mediate the skin and T cell tropisms of VZV (Moffat *et al.*, 2004). The complete 62 amino acid (aa) C-terminal domain was deleted and targeted mutations were made changing the YAGL motif (aa 582–585), which is known to mediate gE endocytosis (Olson and Grose, 1998), to GAGL. The AYRV domain (aa 568–571), which directs gE to the trans-Golgi network (TGN) (Zhu *et al.*, 1996) was altered to AARV. Phosphorylation of the gE C-terminus has been mapped to an acid cluster, SSTT, at aa 588–601 (Yao *et al.*, 1993), which were mutated by alanine substitution. The deletion of the gE C-terminus was lethal, which was explained by the finding that disruption of the YAGL endocytosis motif was incompatible with VZV replication *in vitro*. These experiments established that gE endocytosis is necessary for VZV replication, in contrast to other alpha-herpesviruses. Both the C-terminal deletion and YAGL mutations were rescued by inserting intact ORF68 into the non-native site. Infectious virus was recovered despite mutagenesis of the AYRV and SSTT motifs. Interestingly, the AYRV mutation in rOka-gE-AYRV was associated with accelerated cell–cell spread and yields of infectious virus were higher with rOka-gE-SSTT compared with pOka in cultured cells. Nevertheless, these effects did not predict the consequences of gE C-terminal mutations on VZV virulence in skin and T cells *in vivo*. Blocking TGN trafficking resulted in significant attenuation, particularly in skin xenografts. Disrupting phosphorylation of the gE C-terminus in rOka-gE-SSTT did not affect VZV tropism for skin or T cells in SCIDhu xenografts. This comparative analysis by mutagenesis in the context of the viral genome demonstrated a hierarchy in functions of the gE C-terminus, with endocytosis being essential, TGN targeting contributing to optimal replication and phosphorylation being dispensible for VZV virulence in skin and T cells in differentiated human cells *in vivo*.

Since VZV pathogenesis involves inoculation of respiratory epithelial cells and cell–cell fusion, we have also examined how gE, with and without gI, contributed to tight junction (TJ) formation in polarized epithelial cells using Madin-Darby canine kidney (MDCK) cells that expressed gE, gI, or gE/gI (Mo *et al.*, 2000). VZV gE induced partial TJ formation and co-localized with ZO-1, a TJ protein, under conditions of low calcium and low cell density; these effects were observed with or without gI. By sequence analysis, the gE ectodomain had two regions, amino acids (aa) 278 to 355 and aa 467 to 498, with similarity to regions in the ectodomains of the cell adhesion molecules, E-cadherin and desmocollin. Functionally, gE expression accelerated the establishment of maximum transepithelial electrical resistance (TER) in MDCK-gE cells. While gE expression altered F-actin organization and lipid distribution in these cells, the presence of gI modulated the changes induced by gE. Thus, gE and gE/gI may contribute to cell–cell spread of VZV by facilitating epithelial cell–cell contacts at mucosal sites of inoculation.

Glycoprotein I

Our initial mutational analysis of VZV gI showed that deleting ORF67 was compatible with VZV replication but yielded mutants with a small plaque phenotype and reduced growth in cultured cells (Mallory *et al.*, 1997). Recombinant viruses with partial deletions of ORF67 that eliminated 60% of the N-terminus or 40% of the C-terminus, also had decreased plaque size and produced less infectious virus. Syncytia formation is the hallmark of VZV infection in cultured cells and in skin lesions, but polykaryocytes produced by rOkaΔgI mutants lacked the typical symmetrical central organization of nuclei. VZV gE reached plasma membranes in the absence of gI but it became localized in an aberrant punctate distribution. These effects on polykaryocyte morphology and gE trafficking were also observed with the gI N-terminal and C-terminal truncation mutants. Cell-to-cell spread and growth kinetics were restored when ORF67 was inserted into the cosmid AvrII site. These experiments indicated that although VZV encodes other conserved fusogenic proteins, including gB and the gH:gL complex, these proteins do not compensate for the effects of deleting gI on cell–cell spread. Subsequent EM analyses of rOkaΔgI and the N terminal and C terminal deletion mutants demonstrated a requirement for gI to achieve normal VZV envelopment in the TGN (Wang *et al.*, 2001). Without gI, VZV tegument proteins were mislocalized, the TGN cisternae, where VZV envelopment occurs, became disorganized and virions did not egress from the Golgi.

Although gI was not essential in cultured cells, no replication was detected when rOkaΔgI mutants were evaluated in SCIDhu skin and T cell xenografts *in vivo* (Moffat *et al.*, 2002). The repaired rOkaΔgI recombinants were indistinguishable from intact rOka in their virulence. That gI is essential for VZV infection of skin suggests that gI and gE must function together to achieve VZV cell–cell spread *in vivo*, mediating "fusion from within" functions. Since VZV infection of T cells *in vivo* does not induce cell fusion, the failure of gI deletion mutants to replicate in T cell xenografts implies that VZV entry may also require gE:gI heterodimer functions to form intact enveloped virions that can mediate "fusion from without."

Glycoprotein I promoter

Since gI was not essential *in vitro*, we used cosmids to make recombinants in which binding sites for cellular tranactivators within the gI promoter were disrupted (Ito *et al.*, 2003). These VZV mutants produced intact gI but in reduced amounts, allowing a further evaluation of the requirement for gI in skin and T cell xenografts, as well as an assessment of the contributions of particular cellular transactivating factors on VZV infection of differentiated human cells *in vivo*. The gI promoter contains an activating upstream sequence that binds the cellular transactivators specificity factor 1 (Sp1) and upstream stimulatory factor (USF) and an open reading frame 29 (ORF29)-responsive element (29RE), which mediates enhancement of IE62-induced transcription by the single strand ORF29 DNA binding protein. Recombinants, rOKAgI-Sp1 and rOKAgI-USF, with two base pair substitutions in Sp1 or USF sites, replicated like rOKA *in vitro*, but infectivity of rOKAgI-Sp1 was significantly impaired in skin and T cells *in vivo*. A double mutant, rOKAgI-Sp1/USF, did not replicate in skin but yielded some infectious virus in T cells. These observations suggest that gI concentrations necessary to support cell–cell spread in skin and those required for virion formation and release in T cells differ. The repaired virus, rOKAgI:rep-Sp1/USF, was as infectious as rOKA. Thus, disrupting gI promoter sites for cellular transactivators altered VZV virulence *in vivo*, with variable consequences related to the cellular factor and the host cell type. Mutations in the 29RE of the gI promoter were made by substituting each of four 10-bp blocks in this region with a 10-bp sequence, GATAACTACA, which was predicted to interfere with enhancer effects of the ORF29 protein (He *et al.*, 2001). One of these mutants, designated rOKAgI-29RE-3, had diminished replication in skin and T cells, indicating that the ORF29 protein-mediated enhancement of gI expression induced by IE62, contributes to VZV virulence. These experiments suggest that mutations in promoters of viral genes that are not essential *in vitro* may allow construction of recombinant herpesviruses that have altered virulence in specific host cells *in vivo* and may be useful for designing attenuated viral vaccines.

Glycoprotein K

VZV ORF5 encodes glycoprotein K (gK), which is conserved among alpha-herpesviruses. Using cosmid mutagenesis, we demonstrated that gK is essential for VZV replication, as it is in HSV-1 and pseudorabiesvirus (Mo *et al.*, 1999); partial gK deletions also prevented VZV replication. VZV gK was detected as a 40K protein in cell-free virus preparations and was present in enveloped virions within infected cells by immunogold EM. Substituting the HSV-1 (KOS) gK gene for ORF5 did not compensate for the deletion of VZV gK but inserting ORF5 at the non-native site in the VZV genome restored replication as did VZV gK-complementing cells.

Mutagenesis of VZV immediate early (IE) genes and effects on VZV replication *in vitro* and on pathogenesis in the SCIDhu model

The unique long (U_L) and unique short (U_S) segments of the VZV genome are flanked by internal (IR) and terminal (TR) repeat sequences that contain the duplicated genes, ORF62/71, ORF63/70 and ORF64/69. The gene products of ORFs 62 and 63 are phosphoproteins that have immediate early (IE) regulatory functions and are components of the virion tegument. VZV ORF4, which is conserved in the alpha-herpesviruses, also encodes a protein, IE4, that has transcriptional effects. IE62, a large 1310 aa protein, is a predominant tegument protein and has a nuclear localization signal, an N-terminal activation domain and a DNA-binding domain; its functional HSV homolog is ICP4, although sequence similarities are limited. As the major VZV transactivator, IE62 induces transcription of all VZV genes that have been tested and elicits viral protein synthesis shortly after cell entry (Cohen and Straus, 2001). The functions of IE62 and IE63 proteins are influenced by their phosphorylation state, which can be altered by one or both of the VZV kinases, the ORF47 and ORF66 proteins, as well as by cellular kinases. IE62 can be phosphorylated by the ORF47 viral kinase. Negative regulation of IE62 is achieved by the VZV ORF66 viral kinase (Kinchington *et al.*, 2000). IE63, a much smaller 278 aa protein, localizes primarily to the nucleus via a nuclear localization signal (Kinchington *et al.*, 1995). IE63 has been described as the homolog of HSV ICP22 (U_S1), but we have noted that it is related most closely to U_S1.5 protein, which is expressed co-linearly with ICP22 (Baiker *et al.*, 2004). IE62 and IE63 form heterodimers and, as kinase substrates for ORF47 or ORF66 kinases, also bind to these proteins. Thus, interactions between IE62, IE63, ORF47 and ORF66 proteins are likely to be pivotal for regulating viral replication as well as for the formation of virions during VZV pathogenesis.

IE62 protein

As predicted from its role as the major VZV transactivator, cosmid mutagenesis experiments in which ORF62, ORF71 or the ORF62/71 gene pairs were deleted showed that one copy of the gene was essential for VZV replication. Infectious virus was recovered when pOka cosmids that lacked ORF62/71 were repaired by inserting ORF62 from pOka at the non-native U_S site, indicating that expression from the repeat regions was not necessary (Sato *et al.*, 2003a). Although vOka ORF62/71 has some sequence differences from the pOka ORF62/71 gene pair, inserting vOka ORF62 also permitted replication. Nevertheless, plaque size and infectious virus yields of this repaired rOkaΔORF62/71 were diminished and IE62 as well as gE protein expression was decreased. Interestingly, transfections of cosmids with single deletions of ORF62 or ORF71 produced a subpopulation of progeny viruses in which the deleted copy had been restored to the native site during viral replication *in vitro*. Mutagenesis of residues in IE62 protein that have been reported to affect interactions with the IE4 protein and a putative ORF9 protein binding site (Spengler *et al.*, 2001) were not lethal. In skin xenografts, as in cultured cells, mixtures of single copy rOkaΔ62 or rOkaΔ71 and repaired rOka were generated by recombination events and emerged after replication in some skin xenografts that had been inoculated with rOkaΔ62 or rOkaΔ71. Whereas IE62 expression from one copy of the gene in the non-native U_S site was compatible with limited replication in cultured cells, expression from the native site was required for VZV growth in differentiated human skin tissues *in vivo*. The small plaque phenotype of repaired rOkaΔORF62/71 was associated with decreased virulence in epidermal cells in which cell–cell spread is necessary for pathogenesis.

IE63 protein

As was observed with the dual deletion of ORF62/71, removing both ORF63 and ORF70 from our cosmids blocked VZV replication and inserting ORF63 into the non-native U_S site restored growth (Baiker *et al.*, 2004). In other work by Cohen *et al.*, VZV has been recovered when an ORF62 plasmid was transfected along with ORF63/70 deletion cosmids, suggesting that providing the IE62 transactivtor can compensate for the lack of IE63, or when mutations were introduced into ORF63/70 at both native sites (Cohen *et al.*, 2005). In our experiments, transfection of cosmids that had a single copy of the gene at either native site was compatible with replication. The single copy ORF63 mutants retained full infectivity compared with rOka in skin xenografts and replicated efficiently in T cells that were infected *in vitro*.

In order to further analyze the regions of IE63 necessary for replication *in vitro* and for T cell and skin tropism, twenty-two ORF63 domains were targeted for mutagenesis. Only three of 22 mutations in ORF63 at the non-native site were compatible with recovery of infectious VZV from cosmids. Infectivity of the other mutated cosmids was restored by inserting intact ORF63. The IE63 mutations were characterized for effects on IE63 phosphorylation, nuclear localization and binding to IE62 using plasmids or recombinant viruses. Alanine substitutions within the IE63 center region showed that S165, S173 and S185 were phosphorylated by cellular kinases. Four mutations that changed two putative nuclear localization signal (NLS) sequences altered the distribution of IE63 to a cytoplasmic/nuclear pattern. The N-terminal domain required for IE63 binding to IE62 was aa55–67, with R59/L60 being critical residues. Alanine substitutions of T171, S181 or S185 were compatible with replication of VZV with ORF63 at the non-native site, yielding rOKA/ORF63rev[T171], rOKA/ORF63rev[S181] and rOKA/ORF63rev[S185]. All of these IE63 mutants had a small plaque phenotype, produced less infectious virus and had reduced ORF47 protein and gE synthesis *in vitro*, indicating that IE63 facilitates IE62-induced transcription of early and late VZV gene products. The IE63 mutants exhibited significantly impaired virulence in skin xenografts whereas T cell tropism was not diminished, suggesting that the requirement for IE63 functional domains depends on the cell type targeted for infection during pathogenesis *in vivo*.

IE4 protein

Deleting ORF4 from the VZV genome was lethal but IE4 expression from the U_S site, which is quite distant from its native location, permitted VZV replication (Sato *et al.*, 2003b). IE4 forms homodimers through a C-terminal domain, KYFKC (Baudoux *et al.*, 2000). Disrupting this functional domain with an ORF4 substitution designed to yield K443S, did not permit replication, indicating that IE4 protein–protein interactions are essential. Whereas the domain for IE62 binding to IE4 was not required, based on the IE62 mutagenesis experiments, expression of intact IE4 protein was necessary for VZV replication.

ORF64 protein

In addition to the ORF62/71 and ORF63/70 gene pairs, VZV has a third set of duplicated genes, ORF64/69 located within the inverted repeat region (Cohen and Straus, 2001). Although the HSV-1 Us10 gene is not duplicated, ORF64/69 has some sequence similarity to this HSV gene, which is located adjacent to the HSV repeats flanking the U_S region. When ORF64 and ORF69 were deleted individually using the VZV cosmid system, single deletions of ORF64 or ORF69 yielded viral plaques that appeared with the same kinetics and had the same morphology as VZV recombinants generated with intact cosmids (Sommer *et al.*, 2001). The dual deletion of ORF64 and ORF69 was compatible with VZV replication but produced mutants that had an abnormal plaque phenotype characterized by very large multinucleated syncytia. All of the ORF64/69 mutants retained infectivity for human T cells *in vitro* and replicated efficiently in human skin in the SCIDhu mouse model of VZV pathogenesis. Thus, like HSV U_S10, the role of the ORF64 protein in VZV pathogenesis remains undefined.

Mutagenesis of VZV kinase genes and effects on VZV replication *in vitro* and on pathogenesis in the SCIDhu model

VZV has two genes, ORF47 and ORF66, that encode serine/threonine kinases. ORF47 protein is the homolog of HSV U_L13, and ORF66 protein is homologous to HSV U_S3. Genes related to ORF47 are present in all herpesviruses, whereas ORF66 and its homologs are unique to the alpha-herpesviruses. ORF47 and ORF66 proteins, like IE62 and IE63, are components of the virion tegument.

ORF47 protein

Our first investigations of the contributions of ORF47 protein to VZV pathogenesis were done with ROKA47S, generated by Heineman *et al.*, in which ORF47 expression is blocked by a stop codon insertion (Heineman and Cohen, 1995). The ORF47 protein was essential for T cell tropism and for VZV infection of skin xenografts (Moffat *et al.*, 1998). Replication in T cell and skin xenografts was restored when ORF47 was replaced at the AvrII site in the ROKA47S genome. These experiments indicated that the presence of cellular kinases from the CKII family that are related to ORF47 and the second VZV kinase, ORF66 protein, did not compensate for functions provided by ORF47 protein *in vivo*.

Given this evidence for the importance of the ORF47 protein, we designed VZV recombinants with an ORF47 C-terminal truncation and with targeted mutations in two putative kinase motifs of ORF47 protein, DYS and PPE, located in the ORF47 C-terminus (Besser *et al.*, 2003). Like ROKA47S, these recombinants, designated rOka47ΔC, rOka47D-N and rOkaP-S, showed no difference in virus yields from cultured cells when compared with rOka in an infectious focus assay *in vitro*. ORF47 protein was expressed but the C-terminal truncation eliminated its kinase activity. Targeted mutagenesis of the DYS motif demonstrated that this C-terminal motif was required for ORF47 kinase function. Whereas ORF47 autophosphorylation and IE62 phosphorylation were blocked, ORF47 homodimer formation and the binding of ORF47 protein to IE62 protein were preserved in rOkaΔC and rOkaD-N-infected cells. These interactions were mapped to the N-terminus of ORF47 protein. ORF47 and IE62 proteins showed extensive nuclear expression at early and late times in cultured cells infected with rOka47ΔC and rOka47D-N, in contrast to the normal late cytoplasmic accumulation. Mutation of the second putative kinase domain in rOka47P-N did not affect ORF47 kinase activity or the intracellular localization of ORF47 or IE62 proteins. When these ORF47 mutants were evaluated for effects on VZV pathogenesis, eliminating ORF47 kinase activity was associated with a severe impairment of VZV growth and lesion formation in skin xenografts. Nevertheless, some infectious virus was made, indicating that VZV replication could occur in skin in the absence of ORF47 kinase function, as long as ORF47 protein binding to IE62 protein was preserved. ROKA47S did not replicate in skin while rOkaDC and rOkaD-N retained some pathogenic capacity. Thus, the formation of ORF47 and IE62 protein complexes, both of which are VZV tegument components, appeared to be the essential contribution of ORF47 protein to VZV replication in skin *in vivo*.

Further investigation of the characteristics of rOka47ΔC and rOka47D-N infection in skin demonstrated that epidermal cell fusion was preserved and typical VZV polykaryocytes were generated by these ORF47 kinase-defective mutants. ORF47 kinase is known to be important for gE phosphorylation and for its correct intracellular trafficking to the plasma membranes and accumulation in the TGN for virion envelopment (Kenyon *et al.*, 2002). Cell fusion was evident in rOka47ΔC and rOka47D-N infected skin, even though gE localization to intracellular junctions was very restricted compared with rOka (Fig. 15.2). Thus, other fusogenic proteins appeared to compensate for the interference with gE trafficking that resulted from blocking the phosphorylating functions of ORF47 protein. However, examination of skin tissues by EM revealed a dramatic reduction in virion production and transport to cell surfaces in xenografts infected with infected with rOka47ΔC or rOka47D-N. This aberrant virion formation and trafficking was also evident in cultured cells even though plaque numbers and plaque morphology were indistinguishable from rOka-infected cells in the infectious focus assay. The evidence of disrupted intracellular trafficking of gE as well as the abnormal localization of ORF47 and IE62 proteins in epidermal cells infected with the ORF47 kinase-defective mutants may explain the diminished production of complete virions. Nevertheless, the preservation of some cell fusion and polykaryocyte formation appeared to be sufficient for VZV replication in skin.

In contrast to their limited growth in skin, rOka47ΔC and rOka47D-N failed to replicate at all in human T cell xenografts *in vivo* (Besser *et al.*, 2004). The pathogenesis of VZV infection of T cells *in vivo* is associated with production and release of infectious virus particles but does not result in cell fusion (Moffat *et al.*, 1995). The fact that the ORF47 kinase-defective mutants did not replicate in T cell xenografts suggests that the T cell tropism of VZV requires assembly and release of complete VZ virions whereas cell–cell spread in skin can occur despite very restricted virion assembly.

ORF66 protein

ORF66 encodes a second VZV serine-threonine protein kinase that has similarities to HSV-1 U_S3. Initial evaluations of ROKA66S, a stop codon mutant made by Heineman *et al.* using vOka cosmids, showed that this recombinant replicated as well as intact ROka in tissue culture (Heineman *et al.*, 1996). However, experiments in the SCIDhu model demonstrated that eliminating ORF66 expression diminished VZV infection of T cell xenografts while infection of skin xenografts was equivalent when compared with recombinant vOka (Moffat *et al.*, 1998b). ORF66 stop codon mutants made using our pOka cosmids exhibited a slight decrease in replication of pOka66S in cultured cells and in skin xenografts which could not be detected when ORF66 expression was blocked in the attenuated vOka genome (Schaap *et al.*, 2005). However, growth of pOka66S was impaired much more significantly in T cell xenografts. Very few intact VZ virions were made in rOka66S-infected T cells whereas pOka-infected T cells exhibited robust virion production by EM analysis of infected xenografts. pOka infection of T cells *in vitro* revealed that wild type VZV modulated the induction of both apoptosis and interferon (IFN) pathways within these differentiated target cells. In contrast, the absence of ORF66 increased the susceptibility of infected T cells to apoptosis and reduced the capacity of the virus to interfere with upregulation of IFN signaling following exposure to IFN-γ. These observations suggest that the ORF66 kinase plays a unique role during infection of T cells during VZV pathogenesis and supports VZV T cell tropism by contributing to immune evasion and enhancing the survival of infected T cells.

Summary: modeling the requirements for VZV infection of T cells and skin *in vivo*

These evaluations of VZV recombinants with mutations that block expression, alter the levels of expression or disrupt functional domains of gE, gI, IE62, IE63, and ORF47 and ORF66 proteins have provided new insights about the tropisms of VZV for T cells and skin. Preserving adequate cell fusion appears to be the most important factor for skin infection whereas sufficient levels of virion production and release seem to be critical in T cells (Fig. 15.3). The evidence suggests that gE and gI are necessary for achieving optimal VZV cell–cell spread, associated with cell fusion and polykaryocyte formation, in skin. VZV does not grow in skin when gI is deleted, although small plaques with aberrant gE localization appear in cultured cells. The requirement for gI exhibits a dose effect, in that mutation of the SpI motif in the gI promoter caused only a minor decrease in replication in skin while the Sp1/USF dual promoter mutation was lethal; further, these experiments indicated that VZV depends upon cellular transactivating factors to enhance replication in the epidermis. Downregulating gI expression by blocking the enhancing effects of the ORF29 protein on IE62 transactivaiton of gI also inhibited VZV growth in skin. In addition, reducing expression of gE, and while it was not assessed, presumably also of gI, by mutations that affected IE62 function prevented skin replication. VZV gE traffics to plasma membranes, where it contributes to cell fusion, and is retrieved to TGN sites where final envelopment of virions appears to occur (Zhu *et al.*, 1996; Cole and Grose, 2001). Mutagenesis of the gE TGN motif, resulted in enlarged VZV plaques whereas cell–cell spread was reduced in skin, showing that effects on plaque morphology *in vitro* do not predict the role of a glycoprotein functional domain in VZV pathogenesis *in vivo*. VZV gE trafficking and polykaryocyte formation were severely disrupted in skin when ORF47 kinase functions were blocked, but some cell fusion and production of infectious virus was preserved. Taken together, these experiments indicated that a small plaque phenotype in cultured cells correlated with impaired or no growth in skin, but the converse, i.e. preservation of cell fusion *in vitro*, did not consistently predict normal VZV cell–cell spread *in vivo*. In contrast to skin, our obser-

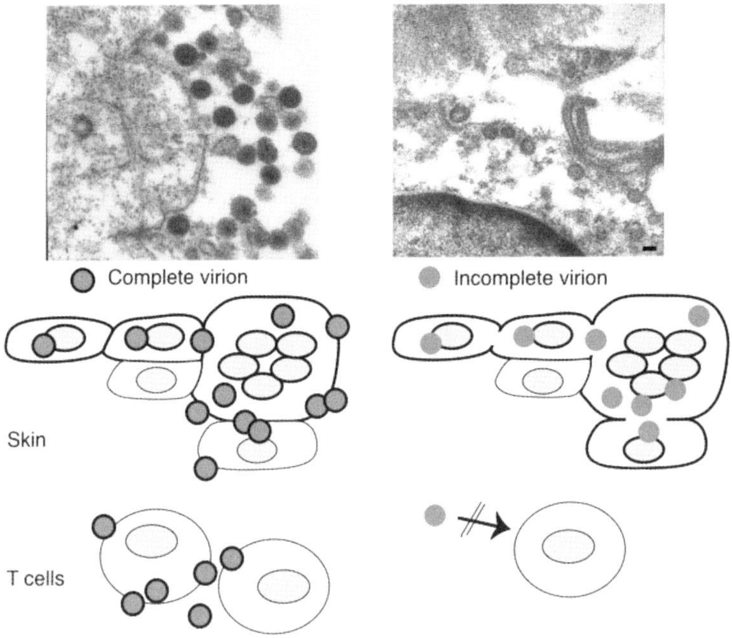

Figure 15.3 Differential requirement for cell fusion and virion formation in VZV infection of skin and T cells. The upper panels show virion formation in skin xenografts infected with VZV and kinase-defective ORF47 mutants. Transmission microscopy demonstrated many intact viral particles in rOka-infected skin cells. Viral particles were incomplete and fewer in rOka47D-N infected skin cells. As illustrated, virus-induced cell fusion, with the characteristic formation of polykaryocytes, supports VZV infection in skin. Cell–cell spread in skin progressed despite limited virion formation, as shown with kinase-defective ORF47 mutants. VZV infection of T cells is not associated with cell fusion. ORF47 kinase mutants do not infect T cells, suggesting that VZV T cell tropism depends upon efficient assembly and release of complete virions (adapted from Besser *et al*., 2004; Ku *et al*, 2005).

vations suggest that VZV infection of T cells appears to require efficient virion formation and egress for transfer to uninfected cells because VZV-infected T cells do not fuse with uninfected T cells (Fig. 15.3). T cell tropism exhibited no tolerance for severe disruption of virion assembly, as occurred in the rOkaΔgI, rOka47ΔC and rOka47D-N mutants. However, T cell infection tolerated reductions in gE that compromised skin replication, and disrupting gE trafficking by the AYRV mutation decreased T cell infection only slightly, compared with a substantial effect in skin. T cell tropism was also compatible reduced gI expression along with the associated effects on the cellular localization of gE trafficking, e.g. the gI promoter dual Sp1/USF mutant replicated in T cells but not skin. Interestingly, blocking ORF66 protein expression had no effect on VZ virion assembly or cell–cell spread in skin, whereas virion production was reduced dramatically in T cells. This difference suggests that the retention of IE62 in the nucleus, which occurs in the absence of ORF66 protein (Kinchington, 2001) may have a more substantial impact on limiting virion formation in T cells than skin, or that cellular homologs compensate for the ORF66 deletion in skin but in T cells. It is also possible that the capacities of ORF66 protein to inhibit apoptosis and downregulate the IFN pathway are more important in T cells than skin.

Investigations of VZV T cell tropism, innate immune responses, and immune evasion

Although the pathogenesis of VZV infection encompasses the neurotropism that is a defining characteristic of the alpha-herpesviruses, our experiments in T cell xenografts in the SCIDhu model demonstrate that VZV also infects T cells productively (Fig. 15.2). The clinical evidence is that the virus must be transported from respiratory sites of inoculation to the skin. We propose that VZV T cell tropism is the mechanism of the cell-associated viremia that is essential for the progression of primary VZV infection in the natural host (Ku *et al.*, 2004). In addition, VZV T cell tropism may be responsible for the dissemination of the virus from cutaneous sites of replication to the lungs, liver and other organs, which causes life-threatening infection in immunocompromised patients with varicella and herpes zoster. VZV replication results in cell lysis within two days in cultured cells *in vitro*. In contrast, VZV infection of differentiated cells within human skin xenografts evolves much more slowly, producing visible cytopathic changes only after ten days or longer. SCID mice lack the capacity to develop an adaptive, antigen-specific immune response. Therefore, the skin xenograft system has allowed an evaluation of the role of innate immunity in controlling progression of VZV infection within the cutaneous microenvironment *in vivo*. VZV, like the other herpesviruses, also appears to have immunomodulatory defenses that transiently facilitate viral replication in target cells that are important for pathogenesis.

VZV tropism for human tonsil CD4 T cells

In the human host, the upper respiratory lymphoid tissues form Waldeyer's ring, which consists of the pharyngeal, palatine and lingual tonsils. Respiratory epithelial cells cover and penetrate into tonsils, forming "crypts" that are adjacent to foci of mononuclear cells containing migratory T cells. Therefore, VZV is likely to have the opportunity to move from sites of respiratory inoculation into adjacent tonsil T cell populations. To better define the T cell tropism of VZV, we examined levels of infection, the activation state and the expression of surface markers on human tonsil T cells that were infected by co-cultivation with infected fibroblasts (Ku *et al.*, 2002). Tonsil T cells were highly susceptible to VZV infection, with 15–25% expressing VZV proteins within 48 hours. Active infection was documented by late gene transcription and was associated with the presence of large numbers of VZ virions on the surfaces of infected T cells by immune scanning EM. The tonsil CD4 T cells most likely to be infected were those that were both activated, shown by CD69 expression, and had surface markers of memory T cells. The relationship between the T cell activation state and VZV T cell tropism was further suggested by the fact that the percentage of VZV positive T cells increased after exposure to phorbol ester, indicating that viral DNA was present within these cells and that VZV gene expression was amplified by T cell activation. Skin homing markers, cutaneous leukocyte antigen (CLA) and chemokine receptor (CCR) 4, are expressed by memory T cells that carry out immune surveillance in skin (Campbell and Butcher, 2000). Whereas our hypothesis was that VZV might upregulate these skin homing markers on T cells to facilitate viral transfer to skin sites of replication, our experiments showed that VZV preferentially infected the memory CD4 T cells that already expressed these markers. VZV-infected CLA and CCR4 positive T cells continued to migrate to SDF-1 in a chemotaxis assay, demonstrating that skin migratory functions remained intact despite viral infection. Thus, tonsil T cells may be early targets of VZV following initial replication in respiratory epithelial cells and the preferential infection of tonsil T cells that express activation, memory and skin homing markers may enhance cell-associated viremia and the transport of infectious virus to skin.

VZV transport to skin by infected CD4 T cells and modulation of viral replication in skin by epidermal cell IFN-α

In order to pursue the hypothesis that VZV-infected T cells transfer the virus to skin, we inoculated SCIDhu mice that had skin xenografts with VZV-infected tonsil T cells via the tail vein (Ku *et al.*, 2004). These human T cells appeared in skin xenografts within 24 hours, with an early distribution around the hair follicles. A comparison of the phenotypes of T cells in the inoculum and those that were extracted from skin xenografts after transfer showed an enrichment of memory CD4 T cells in skin. However, this pattern was observed whether the inoculum consisted of VZV-infected or uninfected T cells. Thus, VZV appears to take advantage of programmed immune surveillance by preferentially infecting memory T cells that have this function. Importantly, transfer of VZV by infected T cells was followed by an interval of 10–21 days before cell–cell spread of the virus resulted in a lesion that penetrated through the surface of skin xenografts.

Further analysis of the VZV-infected skin xenografts demonstrated that viral gene products downregulated IFN-α in foci of cells that expressed VZV proteins. At the same time, the adjacent, uninfected epidermal cells exhibited upregulation of the innate IFN pathway, associated with prominent phosphorylation and nuclear translocation of Stat1. This evidence of innate immune control was confirmed by inoculating skin xenografts with VZV and treating the mice with neutralizing antibody against the human IFN-α/β receptor, which blocks type I IFN activity. Inhibiting IFN signaling by receptor blocking resulted in a ten-fold higher titers of infectious virus and much larger cutaneous lesions in skin xenografts from treated animals. In the intact host, the resolution of VZV lesions is associated with the appearance of T cells at the site of viral replication (Arvin, 2001). The influx of these inflammatory T cells into skin is regulated by interactions between endothelial cell adhesion molecules and chemoattractant receptors (Campbell and Butcher, 2000). A comparison of VZV-infected skin xenografts with biopsies of varicella and zoster lesions demonstrated that VZV replication alone was not sufficient to upregulate the adhesion molecules, E-selectin, ICAM-1 or VCAM-1 on endothelial cells, whereas these proteins were abundantly expressed in VZV lesion biopsy specimens. These experiments showed that VZV replication and extensive viral protein synthesis, along with marked increases in the expression of IL-1-α and IFN-α by epidermal cells, did not cause changes on capillary endothelial cells that would enhance entry of skin homing T cells into VZV-infected skin. Under these circumstances, initial VZV replication in skin should occur without an immediate signaling of host immune responses. Nevertheless, the normal trafficking of T cells through skin that occurs as part of immune surveillance should continue, which would permit the uptake of infectious VZV by these migrating T cells. Their re-entry into the circulation may be a mechanism by which VZV cell-associated viremia is amplified to assure effective delivery of VZV to more skin sites of replication. Taken together, these experiments suggest that the 10–21 day incubation period of varicella in the healthy host represent the time that is necessary for VZV to counteract IFN defenses of skin cells and produce cutaneous lesions that penetrate the skin surface

Immunomodulatory effects of VZV on T cell expression of major histocompatibility complex class I (MHC I) and class II proteins

Herpesviruses have evolved many mechanisms to modulate recognition by CD4 and CD8 T cells through downregulation of MHC proteins. If VZV-infected T cells must traffic from upper respiratory lymphoid tissue to skin, it is necessary for the infected cells to evade immune surveillance. When T cell xenografts were infected with VZV, the cell surface expression of MHC I was downregulated on the subpopulation of CD3 T cells that expressed VZV proteins but not on those that remained uninfected (Abendroth *et al.*, 2001). VZV infection had no effect on transferin receptor expression, indicating the specificity of MHC I downregulation. Since VZV infection also reduced surface MHC I expression in fibroblasts, the stage in MHC I protein trafficking that was altered by VZV was examined in these cells using pulse-chase and immunoprecipitation with and without endo H. These experiments showed that MHC I complexes were assembled in VZV infected cells and were not retained in the ER. However, VZV infection caused MHC I protein to accumulate in the Golgi compartment. Cell surface expression of MHC I was not restored by inhibiting late VZV gene expression with phosphonoacetic acid (PAA), indicating that these viral proteins did not cause MHC I mislocalization. While transfecting fibroblasts with plasmids expressing IE proteins had no effect, transient expression of ORF66 protein, which is an early gene product, reduced cell surface MHC I, suggesting that the ORF66 protein may have immunomodulatory effects.

MHC class II protein expression is limited on most somatic cells but is triggered by IFN-γ. When fibroblasts were treated with IFN-γ, cells that had been infected with VZV were protected from upregulation of MHC class II expression (Abendroth *et al.*, 2000). The inhibition of MHC class II occurred as a result of blocking transcription of the gene, based on *in situ* and northern blot analysis of infected fibroblasts as well as skin biopsies of varicella lesions. Failure to upregulate MHC class II protein was associated with inhibition of interferon regulatory factor (IRF-1) transcription and reduced synthesis of Stat1α and in particular, of Jak2. As has been suggested for other herpesviruses, this capacity of VZV to interfere with MHC class II expression is predicted to give transient protection against VZV-specific CD4 T cells which would otherwise block lesion formation and limit the transmission of VZV to other susceptibles in the population.

Summary: modeling events in the pathogenesis of primary VZV infection

Since the analysis of early events in VZV pathogenesis could not be evaluated in non-human models, the course of primary VZV infection has been compared with mousepox (Grose, 1981). These experiments in the SCIDhu mouse model suggest that infected T cells can transport VZV to skin immediately after entering the circulation during a primary viremia that follows shortly after inoculation of respiratory epithelium (Fig. 15.4). This new paradigm contrasts with the mousepox model in which VZV is presumed to infect mononuclear cells in regional lymph nodes, causing a primary viremia that carries the virus to reticuloendothelial organs, such as the liver, where viral amplification occurs; amplification has been considered to be followed by a secondary viremia in the late incubation period that results in VZV transport

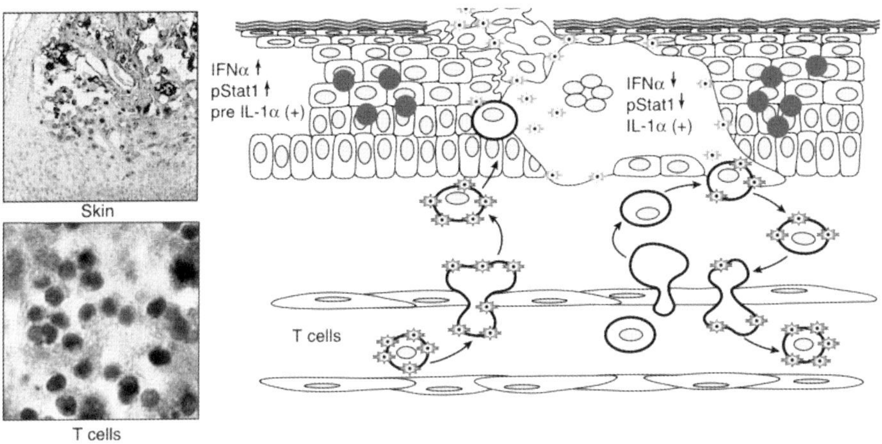

Figure 15.4 A model of the pathogenesis of primary VZV infection. This diagram depicts new concepts about events in the pathogenesis of VZV infection of skin. According to this model, T cells within the tonsil lymphoid tissues become infected by VZV transfer into these migratory cells of the immune system following the initial inoculation of respiratory epithelial cells with the virus. Infected T cells (lower left panel) enter the circulation and transport the virus to the skin shortly thereafter, exiting through capillary endothelium by the usual mechanisms for trafficking of migratory T cells. The infected T cells then release infectious VZV at skin sites of replication. The remainder of the 10–21 day incubation period is the interval required for VZV to overcome the innate IFN-α response in enough epidermal cells to create the typical vesicular lesions containing VZV at the skin surface, as shown in the upper left panel. Signaling of enhanced IFN-α production in adjacent skin cells prevents a rapid, uncontrolled cell–cell spread of VZV. Viremia may be amplified and sustained when T cells that traffic through early stage cutaneous lesions become infected. (Ku *et al.*, 2002; Ku *et al.*, 2005).

to skin and a rapid appearance of skin lesions. Instead, our data suggest that the prolonged interval between VZV exposure and the appearance of varicella skin lesions reflects the time required for VZV to overcome previously unrecognized innate immune responses of epidermal cells. It is unlikely that VZV can produce lesions at the skin surface within 3–5 days, as required in the mousepox model. While some replication may occur in reticuloendothelial organs, this phase is not necessary based on evidence from the SCIDhu model. Further amplification may occur as T cells continue to traffic through skin sites of replication as lesion formation progresses.

VZV effects on cellular gene transcription

As for other herpesviruses, human cDNA microarrays have provided a new tool to investigate changes in cellular gene transcription after VZV infection of T cells and fibroblasts *in vitro* and of skin xenografts *in vivo* (Jones and Arvin, 2003). Our comparative analyses using 40,000-spot microarray have demonstrated that pOka and vOka infection produced similar alterations in cellular gene regulation in all three cell types. These parallel effects of pOka and vOka are consistent with our observations that pOka and vOka have indistinguishable patterns of growth in T cells and in fibroblasts. While vOka exhibits delayed growth in skin xenografts, infection progresses in the absence of an adaptive immune response so that the differences between vOka and pOka become diminished at later times when VZV-infected skin xenografts were tested. However, VZV infection of T cells, fibroblasts and skin had some differential consequences depending on the target cell. For example, transcription of caspase-8, which is a major regulator of ligand-mediated as well as mitochondrial-mediated apoptosis, was inhibited in VZV-infected T cells, a finding that was confirmed by RT-PCR, whereas caspase-8 was not altered in skin or fibroblasts. As suggested from the evaluation of the pOka ORF66 stop codon mutant, blocking virus-induced apoptosis may be particularly important for preserving the VZV-infected T cell as it traffics through the circulation to skin during the pathogenesis of primary VZV infection. As a second example, transcription of the gene that encodes desmoplakin was inhibited specifically in VZV-infected skin xenografts. This protein stabilize desmosomes, which function as the main adhesive junctions in epithelial tissues. Its downregulation may make epidermal cells more susceptible to cell fusion and enhance VZV cell–cell spread in skin. Altogether, 35 host cell genes were differentially regulated in T cells and skin, using the strictest criteria for validating microarray data. Further analyses of these genes may reveal new insights about VZV interactions with differentiated human cells important for pathogenesis.

VZV neurotropism

Our most recent experiments in SCID mice have demonstrated human dorsal root ganglia (DRG) can be maintained as xenografts for prolonged periods (Zerboni *et al.*, 2005b). These DRG xenografts contain differentiated neurons, satellite cells and stromal cells that have typical cell morphologies, express human neural cell markers and are arranged in the expected cellular architecture. When the course of VZV infection of DRG was examined using two low passage clinical isolates, VZV-S and pOka, inoculation was followed by an initial phase of productive replication associated with high viral genome copy numbers (7.1×10^7 to 8.0×10^8 copies/10^5 cells) and recovery of infectious virus by infectious focus assay at 14 days. At this stage, many VZ virions were produced. However, virions were detected only in the nuclei and cytoplasm of neurons, not in satellite cells, and no completely enveloped virions were observed (Fig. 15.5). By 4 weeks, active viral replication had stopped, VZV genome copy numbers declined by two logs (3.7×10^5 to $4.7 \times 10^6/10^5$ cells), and virions were no longer detected in neurons. In this transition from active replication to persistence of the VZV genome,

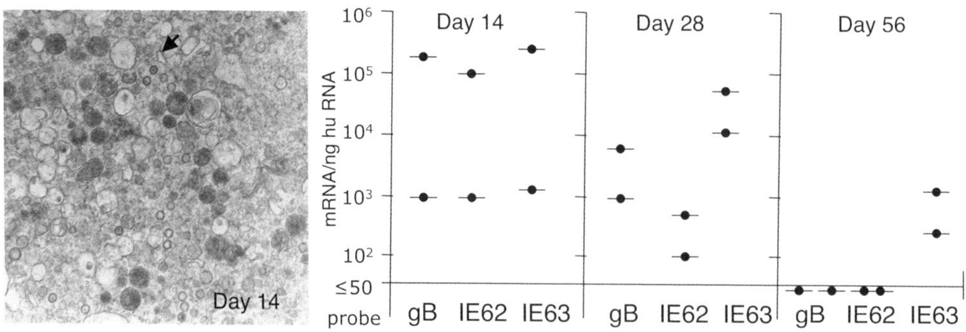

Figure 15.5 VZV infection of dorsal root ganglia xenografts in SCIDhu mice. Left: DRG sections were examined 14 days after inoculation with VZV-S; VZ virions were detected in neuronal cell nuclei and cytoplasm, magnification 5000X. Right: mRNA transcripts corresponding to gB, IE62 and IE63 were measured in DRG infected with VZV-S at day 14, 28 and 56 after infection (Zerboni *et al.*, 2005).

transcription of the gene encoding the glycoprotein B ceased while ORF63 transcription continued, albeit at significantly decreased levels, by quantitative RT-PCR for mRNA (Fig. 15.5). At 8 weeks, ORF63 mRNA was detected consistently and some xenografts also had low levels of ORF62 mRNA but gB transcripts were not detected.

The most important observation from these experiments was that VZV persistence in DRG was achieved solely as a result of VZV interactions with differentiated human neurons, without any requirement for control of viral pathogenesis by VZV-specific adaptive immune responses. This DRG xenograft model permitted the immediate fixation of infected tissues, which eliminates possible effects of the interval from death that may influence studies of VZV latency in autopsied human ganglia and avoids differences that may occur in rodent models of latency because VZV is highly species-specific and does not replicate in rodent sensory ganglia. Given these issues, experiments in DRG xenografts provide important confirmation that VZV latency appears to be characterized by prolonged ORF63 transcription and down-regulation of late glycoprotein gene expression (Mahalingam *et al.*, 1996, Lungu *et al.*, 1998, Cohen *et al.*, 2005). The inhibition of gB was consistent with our recent investigations in a chimeric NOD-SCID model which showed restricted synthesis of gE in human neural cells *in vivo* (Baiker *et al.*, 2004). Importantly, comparative evaluations of pOka and vOka showed no attenuation of the vaccine virus for infection of DRG. Finally, the DRG xenograft model allowed an assessment of the potential role of VZV T cell tropism in infection of neurons *in vivo*. In experiments similar to those in which VZV-infected T cells were shown to transfer VZV from the circulation into skin, we documented that VZV could also reached DRG via infected T cells. While VZV is likely to reach neuronal sites of latency by retrograde transport along nerve axons from varicella skin lesions, these observations suggest that the characteristic VZV lymphotropism may provide a second mechanism for viral access to DRG during VZV pathogenesis. These observations in the SCIDhu DRG model support the development of a unique pattern of VZV gene expression in human neurons associated with persistence of the viral genome.

Summary

A well-regulated balance between viral genetic mechanisms that support each of the phases of VZV pathogenesis, mediating its tropisms for T cells, skin and DRG, and the host cell responses that modulate VZV replication is necessary to avoid a clinical illness that incapacitates the host. This balance ensures that the infected individual will transmit VZV to other susceptible hosts and that VZV will persist in the human population.

Acknowledgments

This work was supported by AI053846, AI20459 and CA49605 (A.M.A.), a Stanford Graduate Fellowship from the Susan P. Markey Foundation (A.C.S.) and a National Science Foundation Graduate Fellowship (J.J.).

References

Abendroth, A., Slobedman, B., Lee, E., Mellins, E., Wallace, M., and Arvin AM. (2000). Modulation of major histocompatibility class II expression by varicella zoster virus. J. Virol. *74*, 1900–07.

Abendroth, A., Lin, I., Slobedman, B., Ploegh, H., and Arvin A.M. (2001). Varicella zoster virus retains major histocompatibility complex class I proteins in the Golgi of infected cells. J. Virol. *75*, 4878–4888.

Arvin, A.M. (1999). Varicella-zoster Virus. Persistent Viral Infections of Humans. Rafi Ahmed, Irvin S. Y. Chen, Editors. Persistent Viral Infections. John Wiley & Sons Ltd., West Sussex, UK, p. 183–208.

Arvin, A.M. (2001a). Varicella vaccine, genesis, attenuation and efficacy. Virology *284*, 153–8.

Arvin, A.M. (2001b). Varicella-zoster virus. In Fields' Virology, Knipe DM, Howley P. 4th Edition. Lippincott-Williams & Wilkins, Philadelphia, pp. 2731–2768.

Arvin, A.M., Baiker, A., Bagowski, C., Ito, H., Sommer, M., Zerboni, L., Fabell, K., Hay, J., Ruyechan, W., and Arvin, A.M. (2004). The immediate early 63 protein of varicella-zoster virus, analysis of functional domains required for replication *in vitro* and for T cell and skin tropism in the SCID hu model *in vivo*. J. Virol. *78*, 1181–1194.

Baiker, A., Fabell, K., Cozzio, A., Zerboni, L., Fabell, K., Sommer, M. Uchida, N., He, D., Weissman, I., and Arvin, A.M. (2004). Varicella-zoster virus infection of human neural cells *in vivo*. Proc. Natl. Acad. Sci. USA *101*, 10792–10797.

Baudoux, L., Defechereux, P., Rentier, B., and Piette, J. (2000). Gene activation by varixella-zoster virus IE4 protein requires its dimerization and involves both the arginine rich sequence, central part and the carboxyl-terminal cysteine-rich region. J. Biol. Chem. *275*, 32822–32831.

Besser, J., Sommer, M.H,, Zerboni, L., Bagowski, C., Ito, H., Moffat, J., Ku, C/-C., and Arvin, A.M. (2003). Differentiation of varicella-zoster virus ORF47 protein kinase and IE62 protein binding domains and their contributions to replication in human skin xenografts in the SCID-hu mouse. J. Virol. *77*, 5964–74.

Besser, J., Ikoma, M., Fabel, K., Sommer, M.H., Zerboni, L., Grose, C., and Arvin, A.M. (2004). Differential requirement for cell fusion and virion formation in varicella-zoster virus infection of skin and T cells. J. Virol. *78*, 13293–305.

Campbell, J.J. and Butcher, E.C. (2000). Chemokines in tissue-specific and microenvironment specific lymphocyte homing. Curr. Opin. Immunol. 12, 336–41.

Cohen, J.I., and Seidel, K.E. (1993). Generation of VZV and viral mutants from cosmid DNAs, VZV thymidine kinase is not essential for replication *in vitro*. Proc. Natl. Acad. Sci. USA *90*, 7376–7380.

Cohen, J. and Straus, S. (2001). Varicella-zoster virus and its replication. In Fields' Virol. Knipe DM, Howley P. 4th Edition. Lippincott-Williams & Wilkins, Philadelphia, p. 2547–86.

Cohen, J.I., Krogmann, T., Bontems, S., Sadzot-Delvaux, C., and Pesnicak, L. (2005). Regions of the varicella-zoster virus open reading frame 63 latency-associated protein important for replication in vitro are also critical for efficient establishment of latency. J. Virol. 79, 5069–77.

Cole, N.L., and Grose, C. (2003). Membrane fusion mediated by herpesvirus glycoproteins, the paradigm of varicella zoster virus. Rev. Med. Virol. *13*, 207–22.

Gershon, A.A. (2001). Live attenuated varicella vaccine. Infect. Dis. Clin. N. Am. *15*, 65–81 (2001).

Gilden, D.H., Kleinschmidt-DeMasters, B.K., LaGuardia, J.J., Mahalingam, R., and Cohrs, R.J. (2000). Neurologic complications of the reactivation of varicella-zoster virus. N. Engl. J. Med. *342*, 635–645.

Grose, C. (1981). Variation on a theme by Fenner, the pathogenesis of chickenpox. Pediatrics *68*, 735–7.

He, H., Boucaud, D., Hay. J., and Ruyechan, W.T. (2001). Cis and trans elements regulating expression of the varicella zoster virus gI gene. Arch. Virol. Suppl *17*, 57–70.

Heineman, T.C. and Cohen, J.I. (1995). The varicella-zoster virus (VZV) open reading frame 47 (ORF47) protein kinase is dispensable for viral replication and is not required for phosphorylation of ORF63 protein, the VZV homolog of herpes simplex virus ICP22. J. Virol. *69*, 7367–7370.

Heineman, T.C., Seidel, K, and Cohen, J.I. (1996). The varicella-zoster virus ORF66 protein induces kinase activity and is dispensable for viral replication. J. Virol. *70*, 7312–7.

Ito, H., Sommer, M.H., Zerboni, L., He, H., Hay, J., Ruyechan, W., and Arvin, A.M. (2003). Promoter sequences of varicella-zoster virus glycoprotein I targeted by cellular transcription factors, Sp1 and USF determine virulence for human skin and T cells in the SCID hu mouse *in vivo*. J. Virol. *77*, 489–98.

Jones, J.O., and Arvin, A.M. (2003). Microarray analysis of host cell gene transcription in response to varicella-zoster virus infection of human T cells and fibroblasts *in vitro* and SCID hu skin xenografts *in vivo*. J. Virol. *77*, 1268–80.

Kemble, G.W., Annunziato, P., Lungu, O., Winter, R.E., Cha, T.A., Silverstein, S.J., and Spaete, R.R. (2000). Open reading frame S/L of varicella-zoster virus encodes a cytoplasmic protein expressed in infected cells. J. Virol. *74*, 11311–11321.

Kenyon, T.K., Cohen, J.I., and Grose, C. (2002). Phosphorylation by the varicella-zoster virus ORF47 protein serine kinase determines whether endocytosed viral gE traffics to the trans-Golgi network or recycles to the cell membrane. J. Virol. *76*, 10980–10993.

Kinchington, P.R., Hougland, J.K., Arvin, A.M., Ruyechan, W.T., and Hay, J. (1992). The varicella-zoster virus immediate-early protein IE62 is a major component of virus particles. J. Virol. *66*, 359–366.

Kinchington, P.R., Bookey, D., and Turse, S.E. (1995). The transcriptional regulatory proteins encoded by varicella-zoster virus open reading frames (ORFs) 4 and 63, but not ORF 61, are associated with purified virus particles. J. Virol. *69*, 4274–82

Kinchington, P.R., Fite, K., and Turse, S.E. (2000). Nuclear accumulation of IE62, the varicella-zoster virus (VZV) major transcriptional regulatory protein, is inhibited by phosphorylation mediated by the VZV open reading frame 66 protein kinase. J. Virol. *74*, 2265–2277.

Ku, C.-C., Padilla, J., Grose, C., Butcher, E.C., and Arvin, A.M. (2002). Tropism of varicella-zoster virus for human tonsillar CD4+ T lymphocytes that express activation, memory and skin homing markers. J. Virol. *76*, 11425–11433.

Ku, C.-C., Zerboni, L., Ito, H., Wallace, M., Graham, B., and Arvin, A.M. (2004). Transport of varicella-zoster virus to skin by infected CD4 T cells and modulation of viral replication by epidermal cell interferon-α. J. Exp. Med. *200*, 917–925.

Ku, C.-C., Besser, J., Abendroth, A., Grose, C., and Arvin, A.M. (2005). Varicella-zoster virus pathogenesis and immunobiology, new concepts emerging from investigations in the SCIDhu mouse model. J. Virol. *79*, 5, 2651–2658.

Lungu, O., Panagiotidis, C.A., Annunziato, P.W., Gershon, A.A., and Silverstein, S.J. (1998). Aberrant intracellular localization of varicella-zoster virus regulatory proteins during latency. Proc. Natl. Acad. Sci. USA 95, 780–785.

Mahalingam, R., Wellish, M., Cohrs, R., Debrus, S., Piette, J., Rentier, B., and Gilden, D.H. (1996). Proc. Natl. Acad. Sci. USA 93, 2122–2124.

Mallory, S., Sommer, M., and Arvin, A.M. (1997). Mutational analysis of the role of glycoprotein I in varicella-zoster virus replication and its effects on glycoprotein conformation and trafficking. J. Virol. *71*, 8279–8288.

Mo, C., Suen, J., Sommer, M., and Arvin, A.M. (1999). Glycoprotein K (ORF5) of varicella-zoster virus is essential for virus growth in melanoma cells. J. Virol. 73;4197–4207.

Mo, C., Schneeberger, E.E., and Arvin A.M. (2000). Glycoprotein E of varicella-zoster virus enhances cell–cell contact in polarized epithelial cells. J. Virol. *74*, 11377–87.

Mo, C., Lee, J., Sommer, M., Grose, C., and Arvin, A.M. (2002). The requirement of varicella-zoster virus glycoprotein E for viral replication and the effects of glycoprotein I on gE in melanoma cells. Virology *304*, 176–86.

Moffat, J.F., Stein, M.D., Kaneshima, H., and Arvin, A.M. (1995). Tropism of varicella-zoster virus for human $CD4^+$ and $CD8^+$ T-lymphocytes and epidermal cells in SCID-hu mice. J. Virol. *69*, 5236–42.

Moffat, J., Zerboni, L., Stein, M., Grose, C., Kaneshima, H., and Arvin, A. (1998a). The attenuation of the vaccine Oka strain of varicella-zoster virus and the role of glycoprotein C in alphaherpesvirus virulence demonstrated in the SCID-hu mouse. J. Virol. *72*, 965–974.

Moffat, J.F., Zerboni, L., Sommer, M.H., Heineman, T.C., Cohen, J.I., Kaneshima, H., and Arvin, A.M. (1998b). The ORF47 and ORF66 putative protein kinases of varicella-zoster virus determine tropism for human T cells and skin in the SCID-hu mouse. Proc. Natl. Acad. Sci. USA 95, 11969–119.

Moffat, J., Sommer, M., Taylor, S., Mallory, S., and Arvin, A.M. (2002). VZV glycoprotein I is required for viral replication in skin and T cells. J. Virol. *76*, 8468–71.

Moffat, J., Cheng, J.J., Sommer, M., Mo, C., Zerboni, L., Stamatis, S., and Arvin, A.M. (2004). Functions of the C-terminal domain of varicella-zoster virus glycoprotein E in viral replication *in vitro* and skin and Tcell tropism *in vivo*. J. Virol. *78*, 12406–15.

Niizuma, T., Sommer, M.H., Ito, H., Hinchliffe, S., Zerboni, L., and Arvin, A.M. (2003). Mutational analysis of varicella-zoster virus ORF65 protein and its role in infection of human skin and T cell xenografts in the SCIDhu mouse model. J. Virol. *77*, 6062–5.

Olson, J., Grose, C. (1998) Complex formation facilitates endocytosis of the varicella-zoster virus gE, gI Fc receptor. J. Virol. *72*, 1542–1551.

Sato, B., Ito, H., Hinchliffe, S., Sommer, M.H., Zerboni, L., and Arvin, A.M. (2003a). Effects of mutations in open reading frames 62 and 71, encoding the immediate early transactivating protein, IE62, of varicella-zoster virus on replication *in vitro* and in skin xenografts in the SCID-hu mouse. J. Virol. *77*, 5607–20.

Sato, B., Sommer, M.H., Ito, H., and Arvin, A.M. (2003b) The requirement of varicella-zoster virus immediate early 4 protein for viral replication. J. Virol. *77*, 12369–72.

Santos, R.A., Hatfield, C.C., Faga, B.P., Padilla, J.A., Cole, N.L., Moffat, J.F., Arvin, A.M., Ruyechan, W.T., Hay, J., and Grose, C. (2000). Varicella-zoster virus gE escape mutant VZV-MSP exhibits an accelarated cell-to-cell spread phenotype in both infected cell cultures and SCID-hu mice. Virology *275*, 306–317.

Schaap, A., Fortin, J.-F., Sommer, M., Zerboni, L., Stamatis, S., Ku, C.-C., and Arvin, A.M. (2005). T cell tropism and the role of ORF66 protein in the pathogenesis of varicella-zoster virus infection. Submitted.

Sommer, M.H., Ku, C.C., Grose, C., Ruyechan, W., Hay, J., and Arvin, A.M. (2001). Mutational analysis of varicella-zoster virus ORF63/70 and ORF64/69. J. Virol. 75, 8224–39.

Spengler, M., Niesen, N., Grose, C., Ruyechan, W.T., and Hay, J. (2001). Interactions among structural proteins of varicella zoster virus. Arch. Virol. Suppl. *17*, 71–79.

Wang, Z.H., Gershon, M.D., Lungu, O., Zhu, Z., Mallory, S., Arvin, A.M. and Gershon, A.A. (2001). Essential role played by the C-terminal domain of glycoprotein I in envelopment of varicella-zoster virus in the trans-Golgi network, interactions of glycoproteins with tegument. J. Virol. *75*, 323–40.

Yao, Z., Jackson, W., and Grose, C. (1993) Identification of the phosphorylation sequence in the cytoplasmic tail of the varicella-zoster virus Fc receptor glycoprotein gpI. J. Virol. *67*, 4464–4473.

Zerboni, L., Hinchliffe, S., Sommer, M.H., Ito, H., Besser, J., Stamatis, S., Cheng, J., DiStefano, D., Kraiouchkine, N., Shaw, A., and Arvin, A.M. (2005a). Analysis of varicella-zoster virus attenuation by evaluation of chimeric parent Oka/vaccine Oka recombinant viruses in skin xenografts in the SCIDhu mouse model. Virology 332, 337–346.

Zerboni, L., Ku, C-C., Jones, C., Zehnder, J., and Arvin, A.M. (2005b). Varicella-zoster virus infection of human dorsal root ganglia *in vivo*. Proc. Natl. Acad. Sci. USA In press.

Zhu, Z., Hao, Y., Gershon, M.D., Ambron, R.T., and Gershon, A.A. (1996). Targeting of glycoprotein I(gE) of varicella-zoster virus to the trans-Golgi network by AYRV sequence and an acidic amino acid-rich patch in the cytosolic domain of the molecule. J. Virol. 70, 6563–6575.

Varicella Zoster Virus Neuropathogenesis and Latency

16

Donald H. Gilden, Ravi Mahalingam, Steven Deitch, and Randall J. Cohrs

Abstract

Varicella zoster virus (VZV) is a highly neurotropic alpha-herpesvirus. VZV causes chickenpox in 4 million children in the United States annually, after which virus becomes latent in cranial nerve, dorsal root and autonomic nervous system ganglia along the entire neuraxis. Decades later, a declining VZV host immunity allows virus to reactivate spontaneously, resulting in zoster, the primary neurologic complication of VZV reactivation, characterized by pain and rash restricted to 1–3 dermatomes. This chapter reviews the clinical features of zoster, including the two most serious neurologic complications: postherpetic neuralgia (PHN) and VZV vasculopathy. Evidence is presented to support the notion that VZV vasculopathy and PHN as well, may be caused by persistent virus infection. The chapter also provides an update on the physical state of VZV during latency in human ganglia, including the distribution and prevalence of virus in ganglia, virus abundance and configuration, cell type infected, and extent of virus transcription and translation. Finally, because VZV is an exclusively human virus, no animal model of latency and reactivation has been achieved. However, simian varicella virus (SVV), the primate counterpart of human VZV, does produce chickenpox and reactivate in monkeys. The chapter provides an update of studies, which have used SVV in primates as a model to study varicella latency and persistence.

Introduction

Varicella zoster virus (VZV) is a highly neurotropic alpha-herpesvirus. VZV causes chickenpox in 4 million children in the United States annually, after which virus becomes latent in cranial nerve, dorsal root and autonomic nervous system ganglia along the entire neuraxis. Decades later, a declining VZV-specific host immunity (Miller, 1980; Berger *et al.*, 1981) allows virus to reactivate spontaneously, resulting in zoster, characterized by pain and rash restricted to 1–3 dermatomes. Every year, zoster affects 600,000 to 1 million Americans, mostly over age 60 (NIH Shingles Prevention Study: http://www.niaid.nih. gov/shingles/cq.htm). The incidence of zoster is expected to increase. By the year 2030, 22% of Americans (65 million people) will be over age 65, and the over-85 population will triple to > 8 million. The incidence and severity of zoster can be viewed as a continuum in immunodeficient individuals, ranging from a natural decline in VZV-specific immunity with age, to more serious host immune deficits encountered in cancer patients, transplant recipients and in patients with AIDS. Furthermore, postherpetic neuralgia (PHN) develops in > 40% zoster patients (200,000–300,000 Americans a year), and zoster also precedes myelitis and vasculopathy. The latter two disorders, and possibly PHN as well, are caused by persistent VZV infection, thus emphasizing the need to know how VZV reactivates. An in-depth understanding of the mechanism(s) of VZV gene expression in latently infected human ganglia will lead to experiments designed to prevent the cascade of events leading to virus reactivation, a cause of serious neurologic disease, particularly in the rapidly increasing elderly and immunocompromised populations.

The major neurological complications of VZV reactivation: zoster, postherpetic neuralgia and vasculopathy

Zoster (shingles)

Zoster generally affects the elderly, but is also seen in individuals who are immunocompromised due to HIV infection, malignancy, chemotherapy or chronic corticosteroid use. The incidence among people over age 50 is double that of people under 50 years (Harnisch, 1984), translating into an 8- to 10-fold increased frequency in people over age 60 compared with those under 60. Varicella in infancy may predispose to zoster earlier in life (Guess *et al.*, 1985). The incidence of recurrent zoster is less than 5% (Hope-Simpson, 1965). Although varicella outbreaks occur most often in the Spring, zoster may develop at any time of the year. The risk of zoster in vaccinated individuals compared with those who developed naturally occurring chickenpox will not be known for decades. Meanwhile, some investigators predict an increased incidence of zoster with widespread use of the live attenuated varicella vaccine (Garnett and Grenfell, 1992; Wharton, 1996).

Herpes zoster is characterized by severe sharp, lancinating, radicular pain and vesicular eruption on an erythematous base in 1–3 dermatomes. Pain is often associated with itching and dysesthesias in the distribution of the affected nerve root. In affected dermatomes, sensation is decreased, yet the skin is exquisitely sensitive to touch (allodynia). Any level of the neuraxis may be involved in zoster. Thoracic zoster is most common, followed by lesions on the face, most often in the ophthalmic division of the trigeminal nerve. The latter is frequently accompanied by zoster keratitis, a potential cause of blindness if unrecognized and not treated promptly. Thus, if visual symptoms are present in patients with ophthalmic distribution zoster, they should have immediate slit-lamp examination by an ophthalmologist, especially if skin lesions extend to the medial side of the nose (Hutchinson's sign). Maxillary and mandibular trigeminal distribution zoster with osteonecrosis and spontaneous tooth exfoliation has also been described in adults (Manz *et al.*, 1986) and children (Garty *et al.*, 1985). Zoster may also be accompanied by ophthalmoplegia, most commonly affecting the third cranial nerve (Thomas and Howard, 1972), optic neuritis (Miller *et al.*, 1986), or both (Carroll and Mastaglia, 1979), and less often, lower cranial nerve palsies (Crabtree, 1968; Steffen and Selby, 1972). Zoster-associated cranial neuropathy often occurs weeks after acute VZV infection. One explanation for late-onset zoster cranial neuropathy is that after reactivation from ganglia, VZV spreads transaxonally along trigeminal and other cranial nerve ganglionic afferent fibers to the vasa vasorum of cranial nerves, results in small-vessel mediated infarction (Mayberg *et al.*, 1984). All cranial nerves receive their blood supply from the carotid circulation via small branches that supply groups of two or three cranial nerves (Lapresle and Lasjuanias, 1986). This parallels the spread of VZV from trigeminal afferent fibers to larger arteries of the anterior and posterior circulation that may lead to stroke.

The seventh cranial nerve is also commonly involved in zoster. Weakness of all facial muscles of one side develops in conjunction with rash of the external auditory canal and tympanic membrane (zoster oticus) or on the ipsilateral anterior two-thirds of the tongue or hard palate. Vesicles in either site are often overlooked. The combination of zoster oticus and peripheral facial weakness constitutes the Ramsay Hunt syndrome (RHS) (reviewed in Sweeney and Gilden, 2001). Compared with Bell's palsy (facial paralysis without rash), patients with RHS often have more severe paralysis at onset and are less likely to recover completely (Robillard *et al.*, 1986). In the only prospective study of RHS patients, 14% developed vesicles after the onset of facial weakness (Murakami *et al.*, 1998). Thus, RHS may initially be indistinguishable from Bell's palsy. Further, Bell's palsy is significantly associated with herpes simplex virus (HSV) infection (Murakami *et al.*, 1996). In light of the known safety and effectiveness of antiviral drugs against VZV or HSV, consideration should be given to early treatment of all RHS or Bell's palsy patients with a 7–10 day course of famciclovir (500 mg, 3 times daily) or acyclovir (800 mg, 5 times daily), as well as oral prednisone (60 mg daily for 3–5 days). Finally, some patients develop peripheral facial paralysis without ear or mouth rash, associated with either a 4-fold rise in antibody to VZV or the presence of VZV DNA in auricular skin, blood mononuclear cells, middle ear fluid or saliva. This indicates that a proportion of "Bell's palsy" patients have RHS zoster sine herpete. Treatment of such patients with acyclovir and prednisone within 7 days of onset has been shown to improve the outcome of recovery from facial palsy (Furuta *et al.*, 2000), although a prospective randomized treatment trial remains to be undertaken.

Pathologic changes in ganglia corresponding to the segmental distribution of rash were first noted by von Barensprung (1863), and more extensively detailed by Head and Campbell (1900) and Denny-Brown *et al.* (1944). The older literature accurately reflects the true pathology of zoster, since the lesions described were those of localized zoster in immunocompetent individuals, except perhaps for occasional zoster that developed in syphilis patients treated with arsenic. The cardinal pathologic features were inflammation and hemorrhagic necrosis, often associated with neuritis, localized leptomeningitis, unilateral segmental poliomyelitis, and degeneration of related motor and sensory roots. Demyelination was also seen in areas of mononuclear cell infiltration and microglial proliferation. Later, intranuclear inclusions (Cheatham *et al.*, 1956; Ghatak and Zimmerman, 1973), and viral antigen and herpesvirus particles (Esiri and Tomlinson, 1972; Nagashima *et al.*, 1975) were detected in ganglia, and VZV was isolated from ganglia (Bastian *et al.*, 1974). However, those studies were performed on ganglia from patients with underlying malignancies or other disorders of immune function who developed disseminated zoster just before death. There is a single report in which VZV antigen was detected and virus isolated from ganglia of a fatal case of bacterial pneumonitis on which acute thoracic zoster was superimposed (Shibuta *et al.*, 1974).

Despite the ubiquitous nature and frequency of VZV infection, the pathogenesis of zoster remains largely unknown. Our current understanding of virus pathogenesis is based on correlation of the presence of VZV in human tissues with pathologic changes in different clinical situations or *in vitro* studies confined mostly to experimental infection of human or primate cells in tissue culture. Zoster is presumed to reflect reactivation and retrograde transport of virus from ganglia to skin in a host partially immune to VZV. The dermatomal distribution of zoster lesions combined with peak VZV titers in vesicles of patients with varicella or zoster suggest a primary role for virus spread from peripheral nervous system ganglia to skin by intraaxonal transport. Intraneuronal and transsynaptic spread of VZV have also been demonstrated in the CNS (Rostad *et al.*, 1989). Alternatively, the presence of VZV in multiple organs is most consistent with hematogenous dissemination. VZV-associated viremia and lymphotropism have been documented during varicella, zoster, PHN, zoster sine herpete and even in normal healthy adults.

During varicella, MNCs are productively infected, probably contributing to spread of virus, mostly T cell-facilitated. VZV-specific DNA, transcripts and protein have been detected in MNCs of zoster and PHN patients. However, with one exception (Sato *et al.*, 1979), attempts to recover infectious VZV from MNCs of immunocompetent zoster patients have failed (Vonsover *et al.*, 1987; Gilden *et al.*, 1988). In immunocompromised patients with localized and disseminated zoster, VZV can be isolated from blood (Feldman and Epp, 1976; Feldman *et al.*, 1977; Gershon *et al.*, 1978; Myers, 1979; Gold, 1996), suggesting a role for hematogenous spread in the pathogenesis of zoster in such individuals. VZV in blood is cell-associated (Myers, 1979) and has been detected by electron microscopy in monocytes (Twomey *et al.*, 1974). The detection of VZV in macrophages (Arbeit *et al.*, 1982), B cells (Leventon-Kriss *et al.*, 1979; Cauda *et al.*, 1986), and T cells (Gilden *et al.*, 1987), particularly in activated T lymphocytes (Koropchak *et al.*, 1989), provided indirect evidence that blood MNCs are a potential site for VZV persistence. Nucleic acid hybridization studies revealed that VZV DNA did not replicate in human MNCs (Gilden *et al.*, 1987), a finding confirmed by Koropchak *et al.* (1989). In zoster and PHN patients, lack of detectable infectious virus may rest in the cell-associated nature of the virus, low-level or abortive infection of MNCs, or in a decreased frequency of MNC cell types that support productive infection compared with primary infection. Although the significance of viremia in zoster patients remains to be determined, VZV DNA has been detected by *in situ* hybridization (ISH) in blood mononuclear cells (MNCs) of four uncomplicated zoster patients for 3–7 weeks after rash (Gilden *et al.*, 1988), coinciding with the period of pain in these patients. Despite lack of recoverable virus from blood, detection of VZV DNA in MNCs of PHN and zoster sine herpete patients supports the notion that productive infection in ganglia contributes to these conditions. The ability to detect VZV DNA in MNCs of some of these patients is also a useful tool for diagnosis, leading to prompt treatment of the syndromes.

Postherpetic neuralgia (PHN)

More than 40% of zoster patients over age 60 develop PHN, pain that persists for months and sometimes years (Fig. 16.1). While not life threatening, PHN is difficult to manage. The cause and pathogenesis of PHN are unknown. Two nonmutually exclusive theories are: (1) that the excitability of ganglionic or even spinal cord neurons is altered, or (2) that there is persistent or low-grade productive virus infection in ganglia. There have been precious few instances in which ganglia from patients who had PHN before death have been analyzed pathologically. Smith's (1978) analysis of ganglia from an early case of PHN of 2.5 months duration revealed diffuse and focal infiltration by chronic inflammatory cells, an observation confirmed by Watson *et al.* (1991) who found prominent collections of lymphocytes in ganglia from a patient with PHN of 2 years' duration. The inflammatory response in ganglia of these subjects raised the possibility of prolonged viral infection. Although virologic analyses of ganglia from PHN patients are wanting, VZV DNA has been detected in blood mononuclear cells

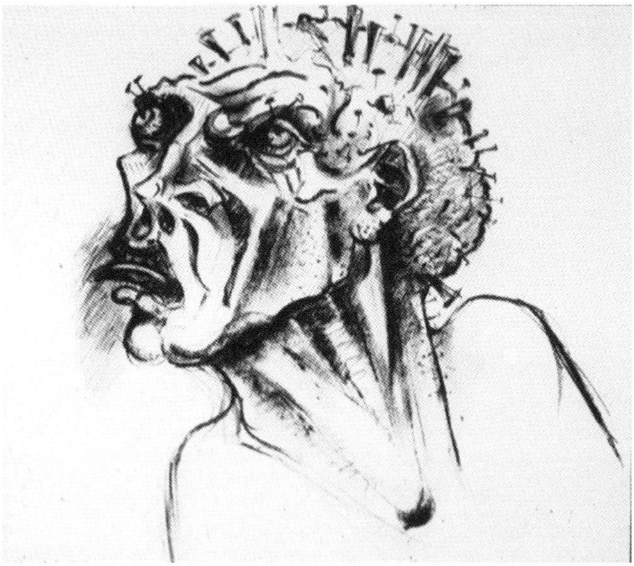

Figure 16.1 Cartoon by a professional artist depicting the painful agony of his postherpetic neuralgia.

(MNCs) of PHN patients 1–8 years after zoster (Devlin *et al.*, 1992; Mahalingam *et al.*, 1995). In contrast, in zoster patients who did not develop PHN, VZV DNA was found in MNCs only up to 38 days after zoster and not at all after disappearance of zoster pain (Gilden *et al.*, 1988). Further, a more extensive study detected VZV DNA in MNCs up to 8 years after zoster in 11/51 patients with PHN, but not in MNCs of 19 zoster patients without PHN who were analyzed 1–31 years after zoster, or in any of 11 elderly age- and gender-matched subjects with no history of zoster (Mahalingam *et al.*, 1995). Most recently, clinical and molecular virological findings were conducted over an 11-year period in an immunocompetent elderly woman with PHN (Gilden *et al.*, 2003). Initially, blood MNCs contained VZV DNA on two consecutive occasions. Random testing after treatment with famciclovir to relieve pain did not detect VZV DNA. However, the patient voluntarily stopped drug treatment five times; pain always recurred within one week, and blood MNCs contained many, but not all, regions of the VZV genome on all five occasions. Chronic VZV ganglionitis-induced PHN best explains both a gratifying clinical response to famciclovir and the recurrence of VZV DNA in MNCs whenever famciclovir was discontinued.

If viral persistence or a low-level productive infection is occurring in ganglia of patients with PHN, it is nevertheless unlikely that the same virus–cell relationship exists in MNCs of these patients. More probable is that MNCs, particularly antigen presenting cells, acquire virus while trafficking through ganglia, and ultimately digest virus. This would account for the detection of some, but not all regions of the VZV genome in circulating MNCs (Mahalingam *et al.*, 1990), compared with the presence of the entire virus genome in latently infected ganglia. Digestion of viral DNA by MNCs also helps to explain why VZV DNA is found in MNCs from only 20% of patients with PHN since there would only be a chance occurrence that fragments of viral DNA will be amplified by PCR in blood samples.

More evidence that longstanding radicular pain could be due to a chronic ganglionitis comes from studies of patients with zoster sine herpete (pain without rash). Lewis (1958) originally suggested zoster sine herpete in patients with zoster who had dermatomal distribution pain in areas distinct from zoster. A clinical link between PHN and zoster sine herpete was later provided by a report of four patients who, years after the resolution of pain from trigeminal zoster, developed zoster sine herpete in the same distribution of the trigeminal nerve. Facial surgery in three patients and a tooth abscess in one patient precipitated zoster sine herpete; unfortunately, no one was studied virologically (Schott, 1998).

Of the first two patients with zoster sine herpete to be studied virologically, both had VZV DNA in both blood MNCs and CSF, and were cured of pain after treatment with intravenous acyclovir (Gilden *et al.*,1994; Terada *et al.*, 1998). Virological analysis of another patient with lumbar distribution zoster that was followed by chronic radicular sacral distribution zoster sine herpete revealed that the persistent pain was produced by active VZV infection (Morita *et al.*, *et al.*, 2003). Another patient with ipsilateral truncal sensory deficit and ophthalmic zoster sine herpete who had reduced serum/CSF ratios of VZV antibody indicative of active virus infection, was treated with intravenous acyclovir; within seven days, the sensory deficit progression stopped, although mild neuralgia on the left face persisted (Yamada *et al.*, 2003). The most compelling evidence that persistent radicular pain without rash can be caused by a chronic active VZV ganglionitis came from analysis of a trigeminal ganglionic mass removed from an immunocompetent adult who had experienced relentless trigeminal distribution pain for more than a year. Pathological and virological studies revealed that the patient's zoster sine herpete was caused by an active VZV ganglionitis (Hevner *et al.*, 2003). Finally, a recent report described a patient with chronic active VZV ganglionitis whose MRI revealed inflammation in ganglia and nerve roots corresponding to persistent pain (Blumenthal *et al.*, 2004).

VZV vasculopathy

Vasculopathy is a serious complication of VZV reactivation. Studies of patients with CNS disease caused by VZV have shown that exclusive brain parenchymal involvement (encephalitis) is rare, and that the most common CNS disorder caused by VZV is a uni- or multifocal vasculopathy (Gilden *et al.*, 2002b), a condition produced by chronic active virus infection of arteries (Gilden *et al.*, 1996).

Two clinical presentations are apparent. The first is a unifocal vasculopathy, formerly called granulomatous arteritis, which predominates in elderly immunocompetent adults and is characterized by acute focal deficit that develops weeks to months after contralateral trigeminal distribution zoster. Vasculopathy is usually restricted to 1–3 large cerebral arteries of either the anterior (Gilden *et al.*, 2003) or posterior (Russman *et al.*, 2003) circulation. The second is VZV multifocal vasculopathy, most often seen in immunocompromised individuals, which usually produces headache, fever, mental status changes and focal deficit. A CSF mononuclear pleocytosis is usually present in both unifocal and multifocal vasculopathy. Brain MRI scanning reveals a single large infarct in unifocal vasculopathy, whereas in multifocal vasculopathy, ischemic and hemorrhagic infarcts are seen in both cortex and subcortical areas (Fig. 16.2). Cerebral angiography reveals focal arterial stenosis (Fig. 16.3). Pathological changes in affected arteries include multinucleated giant cells, Cowdry A inclusion bodies and herpesvirus particles (hallmarks of human herpesvirus infection), and virological analysis reveals both VZV DNA and antigen in affected

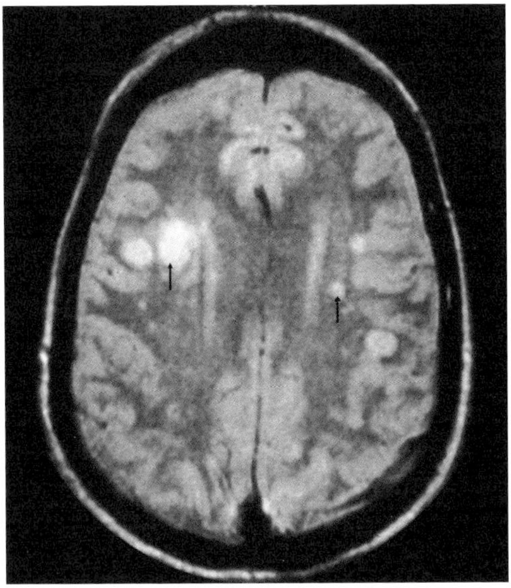

Figure 16.2 Magnetic resonance image of the brain of a patient with multifocal varicella zoster virus vasculopathy. Arrows show two of many ovoid lesions in white matter close to the gray-white matter junction.

vessels (Fig. 16.4). Recent clinical virological correlations have revealed that the manifestations of VZV vasculopathy are protean, and that both unifocal and multifocal vasculopathy may occur without rash. Vasculopathy may be recurrent and present with transient ischemic attacks remote from, and months after, acute zoster, including posterior ischemic optic neuropathy (Gilden *et al.*, 2002a).

Unifocal large vessel infarcts follow trigeminal-distribution zoster and are presumed to result from transaxonal transport of virus

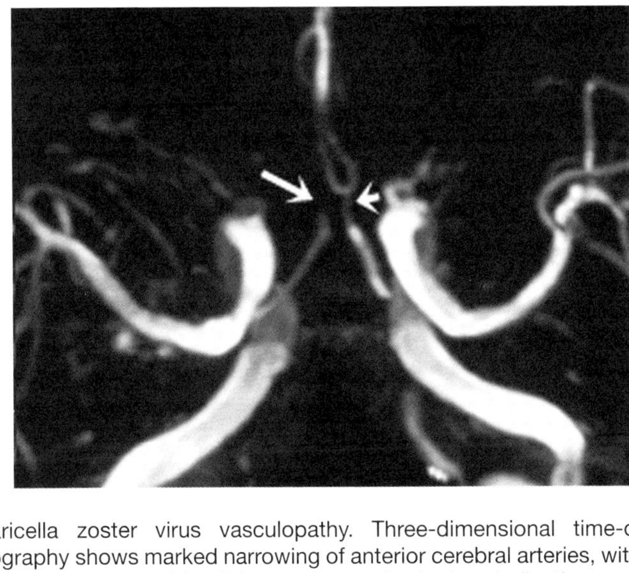

Figure 16.3 Varicella zoster virus vasculopathy. Three-dimensional time-of-flight magnetic resonance angiography shows marked narrowing of anterior cerebral arteries, with a flow gap at the junction of A1 and A2 segments of the right anterior cerebral artery, indicating occlusion (long arrow) and marked stenosis (short arrow).

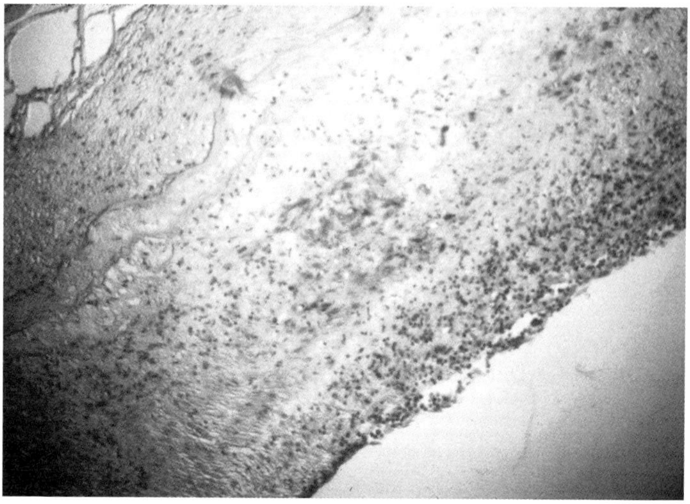

Figure 16.4 Immunohistochemical analysis of cerebral artery from a patient with VZV vasculopathy. VZV antigen (red staining) is detected after incubation of artery with rabbit antiserum again the VZV open reading frame 63 protein.

from trigeminal afferent fibers that innervate anterior circulation vessels (Mayberg *et al.*, 1981). Similarly, smaller infarcts in deep white and gray matter may reflect transport of VZV from trigeminal or cervical afferent fibers to smaller branches of posterior circulation vessels (Mayberg *et al.*, 1981; Saito *et al.*, 1989).

During the time from zoster rash until the development of vasculopathy, virus, which has reactivated from ganglia, is presumed to be actively traveling along transaxonal pathways to arteries in which virus infection remains productive. Because infection is active and often protracted, anti-VZV IgM or IgG antibody in CSF with reduced serum/CSF ratios of VZV antibody compared with total IgG or albumin is found (Gilden *et al.*, 1998). In VZV vasculopathy, the diagnostic value of detecting anti-VZV IgG antibody in CSF is greater than that of detecting VZV DNA by PCR. Although a positive PCR for VZV DNA in CSF is helpful, a negative PCR does not exclude the diagnosis of VZV vasculopathy. Only when both VZV PCR and anti-VZV IgG antibody tests in CSF are negative can the diagnosis of VZV vasculopathy be excluded. Recognition of the wide spectrum of VZV vasculopathy and its proclivity to recur is essential, since effective antiviral therapy can be curative. Finally, oral antiviral treatment is insufficient to treat VZV vasculopathy; intravenous acyclovir is required and treatment may need to be repeated if ischemia recurs.

Overall, the direct demonstration of productive VZV infection in cerebral vessels, and possibly in ganglia of PHN patients, underlines the importance of determining the mechanism(s) responsible for VZV reactivation and its resultant neurologic complications. A necessary step toward this goal is an understanding of the mechanism(s) of viral gene expression in latently infected human ganglia.

The VZV genome

Analysis of serially propagated VZV (Ecker *et al.*, 1984; Hayakawa *et al.*, 1986) and multiple isolates (Hawrami *et al.*, 1996; Muir *et al.*, 2002) has revealed little geographic variation of VZV DNA. The 124,884-bp VZV genome has been sequenced; computer-assisted analysis has identified 71 ORFs numbered consecutively from the leftward end (Accession #s X04370, M14891 and M16612; Davison and Scott, 1986). ORFs 42 and 45 are believed to be exons from the same processed transcript, and ORFs 62, 63 and 64 are repeated (ORFs 71, 70, and 69); thus, there are 68 predicted VZV genes. However, experimental analysis has identified 3 additional transcripts. The first is VZV gene 33.5, which is 3′-coterminal with ORF 33 and encodes the virion assembly protein processed by the ORF 33 proteinase (Preston *et al.*, 1997; McMillan *et al.*, 1997). A second novel VZV gene is ORF 9A, which maps within ORF 8 (Ross *et al.*, 1997). Mutation of both ORF8 and ORF9A yields virus with impaired syncytia formation and reduced growth in tissue culture. A third is ORF (S/L), the first demonstrated spliced VZV transcript, which maps to the leftward end of the virus genome (Kemble *et al.*, 2000), and is expressed during lytic virus growth *in vitro* and *in vivo*. Overall, 71 unique VZV genes have been identified by computer or experimental analysis.

VZV latency

The study of VZV latency is fraught with obstacles. VZV is exclusively a human pathogen; no rodent model exists to study reactivation. VZV is latent only in ganglia, tissue not accessible during life, so analysis of latent VZV has been restricted to human ganglia obtained at autopsy. Unlike HSV-1, VZV cannot be isolated from human ganglia, but nucleic acid hybridization (Hyman *et al.*, 1983; Gilden *et al.*, 1983) and later PCR (Mahalingam *et al.*, 1990) proved that VZV is latent in ganglia. In the past 25 years, our laboratory has analyzed thousands of human ganglia. The application of molecular virologic strategies and techniques to latently infected human ganglia has revealed important differences in VZV gene expression compared with the prototype HSV-1.

Distribution and prevalence of VZV DNA in latently infected human ganglia

While HSV-1 latency is mostly restricted to cranial nerve ganglia, VZV is latent in those ganglia (Gilden *et al.*, 1983; Hyman *et al.*, 1983) as well as in dorsal root (Mahalingam *et al.*, 1990) and autonomic nervous system ganglia (Gilden *et al.*, 2001). VZV DNA is present in ganglia of > 90% of normal adults (Mahalingam *et al.*, 1992).

Abundance of VZV DNA in latently infected human ganglia

Initially, quantitative PCR detected 6–51 copies of VZV DNA per 100 ng of total ganglionic DNA (Mahalingam *et al.*, 1993). Later, real-time PCR detected 258 copies of VZV DNA per 10^5 ganglionic cells (Pevenstein *et al.*, 1999), and recent real-time PCR analysis of trigeminal ganglia from 17 individuals revealed no difference between the left and right trigeminal ganglia of the same individual in VZV DNA copy number but a wide variation of 19–3145 copies of virus DNA per 10^3 latently infected neurons (Cohrs *et al.*, 2000). Such a large range is likely to reflect the uneven amount of VZV encountered during primary infection. For example, during acute varicella, the virus load in blood varies substantially from 200–5000 copies per 150,000 MNCs, 100–1000 copies per ml whole blood, and 100 to > 10,000 copies per ml serum (Mainka *et al.*, 1998; de Jong *et al.*, 2000). The differing amounts of latent VZV DNA among individuals is not surprising, since humans are outbred, and decades of uncontrolled stimuli separate primary infection from analysis of ganglia after death.

Configuration of VZV DNA in latently infected ganglia

The configuration of latent VZV was analyzed by quantitative PCR of human ganglionic DNA using 2 different primer sets, which exploits the rare (5%) isomerization of the unique long segment of virus DNA. The first primer set corresponded to the virus termini and would yield a PCR product if viral DNA molecule were circular, concatameric (end-to end), or if the unique long region was inverted. The second primer set, located internally, would amplify VZV DNA of any configuration. After PCR amplification, the latent VZV DNA copy number with both primer sets was the same (ratio 1:1), compared with a 15:1 ratio in DNA extracted from VZV virions. The simplest interpretation of the data is that latent VZV DNA exists as a circular episome (Clarke *et al.*, 1995), like the structure of latent HSV-1 DNA (Rock and Fraser, 1985; Efstathiou *et al.*, 1986).

VZV transcription in latently infected human ganglia

In situ hybridization (ISH) studies as well as sequencing of cDNA clones identified 4 transcripts corresponding to VZV genes 21, 29, 62 and 63 (Meier *et al.*, 1993; Cohrs *et al.*, 1994, 1995). Interestingly, although the earliest studies of VZV transcription during latency did not detect VZV gene 63 transcripts, it may be the most abundantly transcribed VZV gene identified to date (Cohrs *et al.*, 2000). Most recently, VZV gene 66 expression in latently infected human ganglia was detected by immunohistochemistry and verified by sequencing cDNA clones (Cohrs *et al.*, 2003a). ISH studies of human ganglia (Kennedy *et al.*, 1999, 2000) confirmed the presence of transcripts corresponding to VZV genes 21, 29, 62 and 63, and also suggested transcription of VZV genes 4 and 18, although the latter findings await confirmation by sequencing. Essentially, 5 VZV genes corresponding to ORFs 21, 29, 62, 63 and 66 are transcribed. VZV genes 4 and 18 may also be transcribed, and possibly others. Our recent detection of VZV gene expression from every ORF by array analysis in productively infected cells in tissue culture (Cohrs *et al.*, 2003b) is being applied to latently infected human ganglia to determine the global extent of VZV gene transcription.

Of the different VZV transcripts detected during latency, VZV ORF 63 appears to be the most abundant (Cohrs *et al.*, 2000). One study indicated that VZV ORF 63 is essential for VZV replication in culture (Sommer *et al.*, 2001), while Cohen *et al.* (2004) showed that a VZV ORF-63 deletion mutant virus exhibited impaired growth in cell culture. The exact role of VZV ORF 63 in pathogenesis and latency is unclear. The interaction of VZV ORF 63 with the viral transactivator VZV ORF 62 has been studied (Sommer *et al.*, 2001; Bontems *et al.*, 2002; Lynch *et al.*, 2002; Baiker *et al.*, 2004,). The region of VZV ORF 63 required for the interaction with VZV ORF 62 is dependent on ionic strength, and localized to the 142 amino acids at the amino terminus. Lynch *et al.* (2002) reported that: (1) VZV ORF 63 upregulates transactivation of the VZV gI promoter by VZV ORF 62 in T cells; (2) VZV ORF 62 facilitates recruitment of ORF 63 to the gI promoter; and (3) VZV ORF 63 associates with cellular RNA polymerase II. We have preliminary results showing that simian varicella virus (SVV) ORF 63 (SVV infection of primates is the counterpart of VZV infection of humans, as detailed at the end of this chapter) downregulates activation of the SVV 21 promoter by SVV ORF 62 in non-neuronal cells, but not in neuronal cells in culture. Taken together, the data suggest that the influence of ORF 63 on ORF 62 function might not only be cell-type-specific, but also promoter-specific.

VZV proteins expressed in latently infected human ganglia

Although immunohistochemical staining of normal human ganglia with polyclonal antiserum directed against VZ virions did not reveal any VZV-specific proteins, use of a monospecific polyclonal antiserum directed against VZV ORF 63 protein did identify this protein in the cytoplasm of neurons (Mahalingam *et al.*, 1996). Another study also detected proteins in the cytoplasm of neurons corresponding not only to VZV ORF 63, but also to VZV ORFs 4, 21, 29 and 62 during latency (Lungu *et al.*, 1998). In the same report, these VZV proteins were shown to be both nuclear and cytoplasmic during reactivation, and the investigators speculated that the cytoplasmic location might provide a mechanism to maintain VZV in a latent state, namely, by restricting regulatory proteins from the nucleus. This intriguing study, if confirmed, would suggest a mechanism by which latency is maintained in the presence of an efficient transactivator (e.g. IE62) of virus gene transcription.

Cell type infected in latently infected human ganglia

The first ISH studies detected VZV nucleic acid in neurons of human ganglia (Hyman *et al.*, 1983; Gilden *et al.*, 1987; Fig. 16.5). Later ISH studies by Croen *et al.* (1988) suggested that VZV was latent exclusively in non-neuronal satellite cells, although a review of Croen's ISH figures by two consultant neuropathologists at the University of Colorado and the University of Glasgow indicated that virus was only in neurons (Gilden and Kennedy, personal communication). Using ISH, Lungu *et al.* (1995) reported the presence of VZV DNA in neurons and multiple types of non-neuronal cells, an unconfirmed finding. The use of PCR combined with ISH to study VZV latency revealed VZV DNA exclusively in cytoplasm of neurons (Dueland *et al.*, 1995), perhaps due to egress of viral DNA from the nucleus during the rigors of PCR amplification *in situ*. In two further carefully controlled studies, various groups investigating VZV latency submitted human ganglia to a single laboratory for analysis by ISH and PCR-ISH; findings indicated that VZV was latent predominantly, if not exclusively, in neurons (Kennedy *et al.*, 1998, 1999). To circumvent problems associated with ISH (reviewed in Mahalingam *et al.*, 1999), modified contextual analysis technology (Sawtell, 1997) was applied to dissociated human trigeminal ganglia. VZV DNA was found primarily in cells > 20 μm in diameter (neurons); some VZV DNA was also found in cells < 10 μm in diameter (either small neurons or non-neuronal cells), again indicating that VZV is latent primarily, if

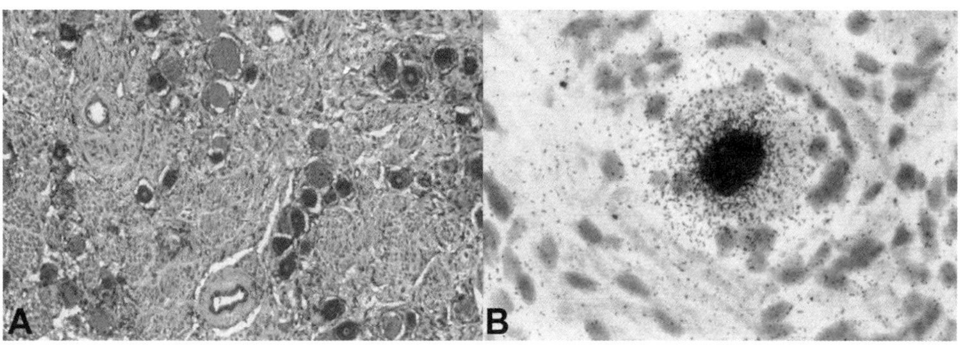

Figure 16.5 Detection of varicella zoster virus (VZV) nucleic acids in a normal human thoracic ganglion hybridized *in situ* to a ^{32}P-labeled RNA probe transcribed from one strand of the VZV SalI-P DNA fragment. Note VZV sequences in the cytoplasm of multiple neurons (A) compared with the presence of herpes simplex virus (HSV) sequences in the nucleus of a neuron after *in situ* hybridization of a normal human trigeminal ganglion to a radiolabeled probe from the HSV-1 LAT gene (B).

not exclusively, in neurons (LaGuardia *et al.*, 1999). Further analysis of neurons and non-neuronal cells sorted by size and morphology revealed VZV DNA exclusively in neurons (Levin *et al.*, 2003). Overall, neurons are the site of latent VZV DNA, as for other alpha-herpesviruses including HSV-1 (Stroop *et al.*, 1984; Stevens *et al.*, 1987), HSV-2 (Mitchell *et al.*, 1990; Croen *et al.*, 1991), bovine herpesvirus-1 (Bratanich *et al.*, 1992; Kutish *et al.*, 1990), equine herpesvirus (Borchers *et al.*, 1997) and pseudorabies virus (Priola *et al.*, 1990).

Rodent models of varicella latency

Unlike HSV-1, VZV does not reactivate from ganglia after experimental infection of primates or multiple rodent species. Experimental inoculation leads only to seroconversion without clinical signs (Takahashi *et al.*, 1975; Myers *et al.*, 1980, 1985; Matsunaga *et al.*, 1982; Wroblewska *et al.*, 1982; Walz-Cicconi *et al.*, 1986). Nevertheless, important information has been acquired from analysis of ganglia after experimental infection. In adult rats inoculated subcutaneously with VZV along the spine, analysis of dissociated dorsal root ganglia at multiple intervals for 9 months revealed ganglionic infection (Sadzot-Delvaux *et al.*, 1990). A subsequent ISH study of the same rat model detected VZV gene 63 DNA in 5–10% of neurons and VZV RNA in neurons and non-neuronal cells at a 3:1 ratio; immunocytochemistry revealed VZV gene 63 protein exclusively in neurons, both in the nuclei and cytoplasm of latently infected cells (Kennedy *et al.*, 2001). The presence of VZV primarily in neurons, and the abundant expression of VZV gene 63 nucleic acid and protein, parallels findings in human ganglia. However, VZV did not reactivate, and none of the above studies analyzed nonganglionic tissue for virus. Thus, while ganglionic infection can be established after peripheral inoculation of VZV, the rat model cannot be considered to be one of latency because: (1) VZV does not reactivate; (2) non-ganglionic tissues have not been shown to be free of VZV; and (3) VZV is not restricted to ganglionic neurons. The same holds for guinea pigs in which ocular (Tenser and Hyman, 1987) and subcutaneous (Lowry *et al.*, 1993) inoculation of VZV produced ganglionic infection; unfortunately, nonganglionic tissues were not analyzed for VZV, and virus did not reactivate. Subcutaneous inoculation of Oka VZV (vaccine strain) into the breast of a chimpanzee produced viremia and mild rash restricted to the inoculation site; however, ganglia and other tissues were not studied (Cohen *et al.*, 1996). In another study, human fetal thymus and liver were implanted under the kidney capsule or subcutaneously in the skin of severe combined immunodeficient (SCID-hu) mice, and VZV was injected into implants 3–5 weeks later (Moffat *et al.*, 1995). Human lymphocytes circulating in these mice were shown to harbor infectious VZV up to 21 days after infection, and VZV proteins were detected in CD4+ and CD8+ T cells. The disadvantage of this model is that the implants were infected externally rather than by viremia. In contrast, the closely related simian varicella virus (SVV) does reactivate from naturally infected primates (discussed below).

Simian varicella virus (SVV)

SVV is a herpesvirus that causes varicella in primates. The clinical (Padovan and Cantrell, 1986; Myers and Connelly, 1992), pathological (Wenner *et al.*, 1977; Padovan and Cantrell, 1986; Dueland *et al.*, 1992) and immunological (Felsenfeld and Schmidt, 1975, 1977) features of SVV resemble those of human varicella. Like human chickenpox, varicella in primates is preceded by an incubation period of 1 to 2 weeks, followed by fever and a papulovesicular rash of skin and mucous membranes (Gray, 2003). Primary SVV infection produces viremia, and infectious virus can be recovered from blood MNCs (Clarkson *et al.*, 1967; Wolf *et al.*, 1974; Soike *et al.*, 1984ab). Disseminated SVV infection is frequent in monkeys after intratracheal inoculation. Lung and liver are the most severely affected organs, much like disseminated varicella in immunosuppressed patients (Roberts *et al.*, 1984; Gray, 2003). Histological examination of skin and viscera reveals foci of hemorrhagic necrosis, inflammation and eosinophilic intranuclear inclusions (Clarkson *et al.*, 1967; Wolf *et al.*, 1974). Like VZV after recovery from acute disease, SVV becomes latent in sensory ganglia at all levels of the neuraxis (Mahalingam *et al.*, 2002). SVV reactivates in virus-infected monkeys exposed to social and environmental stress, and virus can be isolated from skin vesicles of these monkeys (Soike *et al.*, 1984a,b). Complement fixation, viral neutralization, immunofluorescence and immunoprecipitation assays have shown that VZV and SVV are antigenically related (Blakely *et al.*, 1973; Felsenfeld and Schmidt, 1975, 1977; Soike *et al.*, 1987). Like VZV, SVV is an enveloped, double-stranded DNA virus. The SVV genome is colinear with that of VZV (Pumphrey and Gray, 1992; White *et al.*, 1997), similar in size, and shares 70–75% DNA homology with VZV (Davison and Scott, 1986). Based on the complete sequence of the SVV genome, the virus DNA has been found to be 124,784 bp in size, 100 bp shorter than VZV DNA, with a G+C content of 40.4% compared with 46% for VZV (Fig. 16.6) (accession number AF275348; Gray *et al.*, 2001). Our comparison of the SVV DNA sequence to the previously published VZV sequence (Davison and Scott, 1985) reveals several major differences at the leftward end of the SVV genome (Mahalingam *et al.*, 2000 and unpublished observations), which is currently under study with regard to species specificity. Unlike VZV, SVV reactivates from naturally infected primates. Consequently, two models (described below) were established to study varicella pathogenesis and latency, one by intratracheal inoculation of monkeys with SVV, and the other by natural exposure of SVV-seronegative monkeys to SVV-infected monkeys.

Models of SVV pathogenesis and latency

Two models of varicella infection exist. In the first model, intratracheal inoculation of monkeys with 10^3–10^4 plaque-forming units of SVV produces varicella. Infectious virus is recovered from blood MNCs 2–11 days later (Wenner *et al.*, 1977; Iltis *et al.*, 1982; Soike *et al.*, 1984ab; Dueland *et al.*, 1995). Monkey ganglia become infected before rash appears (Mahalingam *et al.*, 2001). Histopathological analysis reveals necrosis and intranuclear inclusions in lung, liver and spleen (Gray *et al.*, 2002). Immunohistochemical analysis reveals SVV-specific antigens in multiple organs, including ganglia at all levels of the neuraxis (Dueland *et al.*, 1992; Gray *et al.*, 2002). SVV-specific immediate early, early and late transcripts are present in skin, lung, liver and ganglia of acutely infected (11–12 days postinfection) monkeys (Gray *et al.*, 2002). Months to years after intratracheal inoculation, SVV DNA and RNA are still detected in ganglia and liver, and virus DNA is detected in blood MNCs, primarily in CD4+ and CD8+ cells, but not in CD14+ or CD20+ cells (White *et al.*, 2002). The significance of continuous MNC infection is unknown. This model of persistent SVV infection can be used to study the ability of antiviral agents to drive virus into a latent state, of particular importance for patients taking immunosuppressive drugs.

In the second model, adult SVV-seronegative monkeys are exposed to monkeys previously inoculated intratracheally with SVV. After 10–14 days, mild varicella rash develops in monkeys caged with intratracheally inoculated monkeys, and PCR amplification

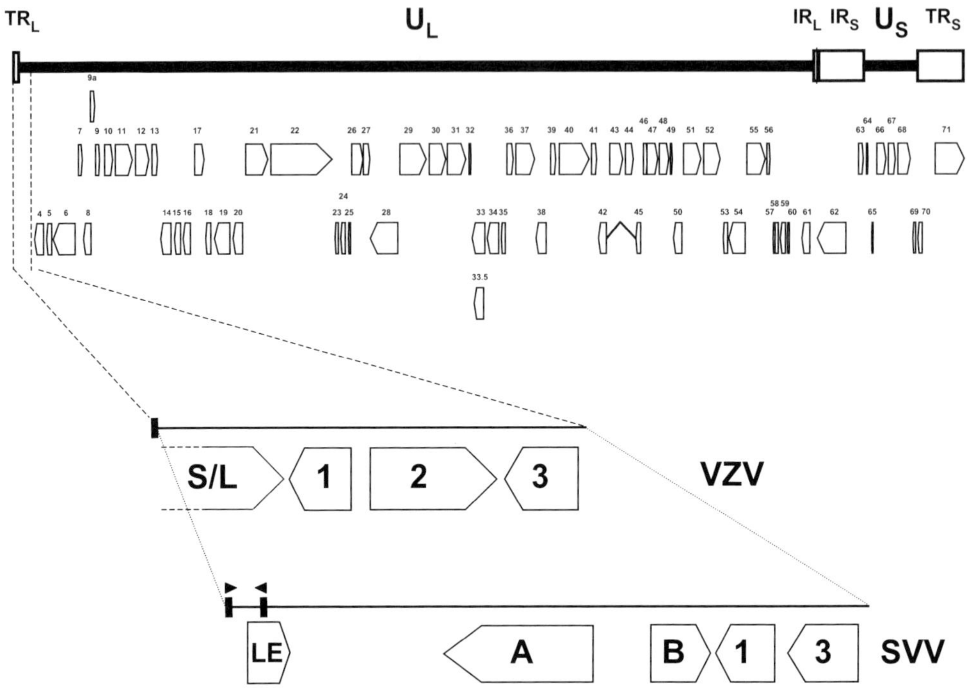

Figure 16.6. Comparison of the VZV and simian varicella virus (SVV) genomes: the double stranded DNA genome of both varicella viruses consists of unique long (UL) and short (US) segments, both of which are flanked by terminal (TRL, TRS) and internal (IRL, IRS) repeat sequences. Sequence analysis of the VZV (Davison and Scott, 1986) and SVV genomes (unpublished observations) has revealed the size of VZV and SVV to be 124 884 and 124 784 bp, and the total number of methionine-initiated open reading frames (ORFs) to be 71 and 73 respectively. The map locations of the predicted ORFs are shown. Comparative analysis reveals several important differences at the leftward end of the two varicella genomes: (1) the UL segment of the VZV genome is flanked by an 88-bp repeat, whereas the left end of the SVV genome contains 507 bp of unique sequence flanked on either side by 79 bp inverted repeats (indicated by filled triangles); part of this inverted repeat sequence (64 bp) is present at the junction of the long and short segments (IRL) of the virus genome; (2) a 672-bp ORF (S/L) initiated within the TRS extending into the UL segment was identified at the leftward end of the VZV genome; this ORF is homologous to a 342-bp ORF (B) located within the UL segment of SVV; (3) an 879-bp ORF (A) and a 420-bp ORF (LE) in SVV were absent in VZV; and (4) no homologue of VZV ORF 2 was detected on the SVV genome.

of SVV DNA in skin scrapings of naturally infected monkeys confirm that SVV caused disease. SVV DNA is detected only occasionally in blood MNCs of naturally infected monkeys at the time of varicella. At 6–8 weeks after rash, SVV DNA was present in multiple ganglia along the neuraxis, but not in lung or liver (Mahalingam *et al.*, 2002), indicating latent SVV infection. Like human VZV, latent SVV is found exclusively in neurons (Kennedy *et al.*, 2004). Thus, along with the observation that SVV reactivates in virus-infected monkeys exposed to social and environmental stress (Soike, 1984a), SVV fulfills the above-mentioned criteria of latency. One particular advantage of this model is that infected monkey ganglia can be removed from anesthetized monkeys before death to study latency, thus obviating concerns about virus reactivation after death. This not only allows a comparison of SVV transcription during latency and cell types harboring latent SVV with that found for VZV, but also serves in dissecting the molecular events in the cascade of cellular and immune factors involved in virus reactivation, studies not possible in humans.

Acknowledgments

This work was supported in part by Public Health Service grants AG 06127 and NS 32623 from the National Institutes of Health. Steven Deitch is supported by Public Health Service grant NS 07321 from the National Institutes of Health. We thank Marina Hoffman for editorial assistance and Cathy Allen for manuscript preparation.

References

Arbeit, R.D., Zaia, J.A., Valerio, M.A., and Levin, M.J. (1982). Infection of human peripheral blood mononuclear cells by varicella-zoster virus. Intervirology *18*, 56–65.

Baiker, A., Bagowski, C., Ito, H., Sommer, M., Zerboni, L., Fabel, K., Hay, J., Ruyechan, W., and Arvin, A.M. (2004). The immediate-early 63 protein of varicella-zoster virus: analysis of functional domains required for replication *in vitro* and for T-cell and skin tropism in the SCIDhu model *in vivo*. J. Virol. *78*, 1181–1194.

Bastian, F.O., Rabson, A.S., Yee, C.L., and Tralka, T.S. (1974). Herpesvirus varicellae: Isolated from human dorsal root ganglia. Arch. Pathol. 97, 331–332.

Berger, R., Florent, G., and Just, M. (1981). Decrease of the lymphoproliferative response to varicella-zoster virus antigen in the aged. Infect. Immun. 32, 24–27.

Blakely, G.A., Lourie, B., Morton, W.G., Evans, H.H., and Kaufmann, A.F. (1973). A varicella-like disease in macaque monkeys. J. Infect. Dis. *127*, 617–625.

Blumenthal, D.T., Salzman, K.L., Baringer, J.R., Forghani, B., and Gilden, D.H. (2004). MRI abnormalities in chronic active varicella zoster virus infection. Neurology *63*, 1538–1539.

Bontems, S., Di Valentin, E., Baudoux, L., Rentier, B., Sadzot-Delvaux, C., and Piette, J. (2002). Phosphorylation of varicella-zoster virus IE63 protein by casein kinase influences its cellular localization and gene regulation activity. J. Biol. Chem. *277*, 21050–21060.

Borchers, K., Wolfinger, U., Lawrenz, B., Schellenbach, A., and Ludwig, H. (1997). Equine herpesvirus 4 DNA in trigeminal ganglia of naturally infected horses detected by direct *in situ* PCR. J. Gen. Virol. *78*,1109–1114.

Bratanich, A.C., Hanson, N.D., and Jones, C.J. (1992). The latency-related gene of bovine herpesvirus 1 inhibits the activity of immediate-early transcription unit 1. Virology *191*, 988–991.

Carroll, W.M., and Mastaglia, F.L. (1979). Optic neuropathy and ophthalmoplegia in herpes zoster oticus. Neurology *29*, 726–729.

Cauda, R., Chatterjee, S., Tiden, A.B., Grossi, C.E., and Whitley, R.J. (1986). Replication of varicella zoster virus in Raji cells. Virus Res. *4*, 337–342.

Cheatham, W.J., Weller, T.H., Dolan, T.F., and Dower, J.C. (1956). Varicella: Report on two fatal cases with necropsy, virus isolation, and serologic studies. Am. J. Pathol. 32, 1015–1035.

Clarke, P., Beer, T., Cohrs, R., and Gilden, D.H. (1995). Configuration of latent varicella-zoster virus DNA. J. Virol. *69*, 8151–8154.

Clarkson, M.J., Thorpe, E., and McCarthy, K. (1967). A virus disease of captive vervet monkeys (*Cercopithecus aethiops*) caused by a new herpesvirus. Arch. Gesamte Virusforsch *22*, 219–234.

Cohen, J.I., Moskal, T., Shapiro, M., and Purcell, R.H. (1996). Varicella in chimpanzees. J. Med. Virol. *50*, 289–292.

Cohen, J.I., Cox, E., Pesnicak, L., Srinivas, S., and Krogmann, T. (2004). The varicella-zoster virus open reading frame 63 latency-associated protein is critical for establishment of latency. J. Virol. *78*, 11833–11840.

Cohrs, R.J., Srock, K., Barbour, M.B., Owens, G., Mahalingam, R., Devlin, M.E., Wellish, M., and Gilden, D, H (1994). Varicella zoster virus (VZV) transcription during latency in human ganglia: Construction of a cDNA library from latently infected human trigeminal ganglia and detection of a VZV transcript. J. Virol. *68*, 7900–7908.

Cohrs, R.J., Barbour, M.B., Mahalingam, R., Wellish, M., and Gilden, D.H. (1995). Varicella zoster virus (VZV) transcription during latency in human ganglia: Prevalence of VZV gene 21 transcripts in latently infected human ganglia. J. Virol. *69*, 2674–2678.

Cohrs, R.J., Randall, J., Smith, J., Gilden, D.H., Dabrowski, C., van der Keyl, H., and Tal-Singer, R. (2000). Analysis of individual human trigeminal ganglia for latent HSV-1 and varicella zoster virus nucleic acids using real-time PCR. J. Virol. *74*, 11464–11471.

Cohrs, R.J., Gilden, D.H., Kinchington, P.R., Grinfeld, E., and Kennedy, P.G.E. (2003a). Varicella zoster virus gene 66 transcription and translation in latently infected human ganglia. J. Virol. *77*, 6660–6665.

Cohrs, R.J., Hurley, M.P., and Gilden, D.H. (2003b). Array analysis of viral gene transcription during

lytic infection of cells in tissue culture with varicella zoster virus. J. Virol. *77*, 11718–11732.
Crabtree, J.A. (1968). Herpes zoster oticus. Laryngoscope *78*, 1853–1879.
Croen, K.D., Ostrove, J.M., Dragovic, L.J., and Strauş, S.E. (1988). Patterns of gene expression and sites of latency in human nerve ganglia are different for varicella-zoster and herpes simplex viruses. Proc. Natl. Acad. Sci. USA *85*, 9773–9777.
Croen, K.D., Ostrove, J.M., Dragovic, L., and Straus, S.E. (1991). Characterization of herpes simplex virus type 2 latency-associated transcription in human sacral ganglia and in cell culture. J. Infect. Dis. *163*, 23–28.
Davison, A.J., and Scott, J.E. (1985). DNA sequence of the major inverted repeat in the varicella-zoster virus genome. J. Gen. Virol. *66*, 207–220.
Davison, A.J., and Scott, J.E. (1986). The complete DNA sequence of varicella-zoster virus. J. Gen. Virol. *67*, 1759–1816.
de Jong, M.D., Weel, J.F., Schuurman, T., Wertheim-van Dillen, P.M., and Boom, R. (2000). Quantitation of varicella-zoster virus DNA in whole blood, plasma, and serum by PCR and electrochemiluminescence. J. Clin. Microbiol. *38*, 2568–2573.
Denny-Brown, D., Adams, R.D., and Fitzgerald, P.J. (1944). Pathologic features of herpes zoster: A note on "geniculate herpes." Arch. Neurol. Psychiatry *51*, 216–231.
Devlin, M.E., Gilden, D.H., Mahalingam, R., Dueland, A.N., and Cohrs, R. (1992). Peripheral blood mononuclear cells of the elderly contain varicella-zoster virus DNA. J. Infect. Dis. *165*, 619–622.
Dueland, A.N., Martin, J.R., Devlin, M.E., Wellish, M., Mahalingam, R., Cohrs, R., Soike, K.F., and Gilden, D.H. (1992). Acute simian varicella infection: clinical, laboratory, pathologic, and virologic features. Lab. Invest. *66*, 762–773.
Dueland, A.N., Ranneberg-Nilsen, T., and Degre, M. (1995). Detection of latent varicella zoster virus DNA and human gene sequences in human trigeminal ganglia by *in situ* amplification combined with *in situ* hyridization. Arch. Virol. *140*, 2055–2066.
Ecker, J.R., Kudler, L., and Hyman, R.W. (1984). Variation in the structure of varicella-zoster virus DNA. Intervirology *21*, 25–37.
Efstathiou, S., Minson, A.C., Field, H.J., Anderson, J.R., and Wildy, P. (1986). Detection of herpes simplex virus-specific DNA sequences in latently infected mice and in humans. J. Virol. *57*, 446–455.
Esiri, M.M., and Tomlinson, A.H. (1972). Herpes zoster: Demonstration of virus in trigeminal nerve and ganglion by immunofluorescence and electron microscopy. J. Neurol. Sci. *15*, 35–48.
Feldman, S., and Epp, E. (1976). Isolation of varicella-zoster virus from blood. J. Pediatr. *88*, 265–267.
Feldman, S., Chaudary, S., Ossi, M., and Epp, E. (1977). A viremic phase for herpes zoster in children with cancer. J. Pediatr. *91*, 597–600.
Felsenfeld, A.D., and Schmidt, N.J. (1975). Immunological relationship between delta herpesvirus of patas monkeys and varicella-zoster virus of humans. Infect. Immun. *12*, 261–266.
Felsenfeld, A.D., and Schmidt, N.J. (1977). Antigenic relationship among several simian varicella-like viruses and varicella-zoster virus. Infect. Immun. *15*, 807–812.
Furuta, Y., Ohtani, F., Mesuda, Y., Fukuda, S., and Inuyama, Y. (2000). Early diagnosis of zoster sine herpete and antiviral therapy for the treatment of facial palsy. Neurology 55, 708–710.
Garnett, G.P., and Grenfell, B.T. (1992). The epidemiology of varicella-zoster virus infections: The influence of varicella on the prevalence of herpes zoster. Epidemiol. Infect. *108*, 513–528.
Garty, B.-Z., Dinari, G., Sarnat, H., Cohen, S., and Nitzan, M. (1985). Tooth exfoliation and osteonecrosis of the maxilla after trigeminal herpes zoster. J. Pediatr. *106*, 71–73.
Gershon, A.A., Steinberg, S., and Silber, R. (1978). Varicella-zoster viremia. J. Pediatr. *92*, 1033–1034.
Ghatak, N.R., and Zimmerman, H.M. (1973). Spinal ganglion in herpes zoster. Arch. Pathol. 95, 411–415.
Gilden, D.H., Vafai, A., Shtram, Y., Becker, Y., Devlin, M., and Wellish, M. (1983). Varicella zoster virus DNA in human sensory ganglia. Nature *306*, 478-480.
Gilden, D.H., Hayward, A.R., Krupp, J., Hunter-Laszlo, M., Huff, J.C., and Vafai, A. (1987). Varicella zoster virus infection of human mononuclear cells. Virus Res. *7*, 117–129.
Gilden, D.H., Devlin, M., Wellish, M., Mahalingam, R., Huff, C., Hayward, A., and Vafai, A. (1988). Persistence of varicella zoster virus DNA in blood MNCs of patients with varicella or zoster. Virus Genes 2, 299–305.
Gilden, D.H., Wright, R.R., Schneck, S.A., Gwaltney, J.M., Jr, and Mahalingam, R. (1994). Zoster sine herpete, a clinical variant. Ann. Neurol. *35*, 530–533.
Gilden, D.H., Kleinschmidt-DeMasters, B.K., Wellish, M., Hedley-Whyte, E.T., Rentier, B., and Mahalingam, R. (1996). Varicella zoster virus, a cause of waxing and waning vasculitis. NEJM case 5–1995 revisited. Neurology *47*, 1441–1446.
Gilden, D.H., Bennett, J.L., Kleinschmidt-DeMasters, B.K., Song, D.D., Yee, A.S., and Steiner, I. (1998).

The value of cerebrospinal fluid antiviral antibody in the diagnosis of neurologic disease produced by varicella zoster virus. J. Neurol. Sci. *149*, 140–144.

Gilden, D.H., Gesser, R., Smith, J., Wellish, M., LaGuardia, J.J., Cohrs, R.J., and Mahalingam, R. (2001). VZV and HSV-1 DNA in human nodose and celiac ganglia. Virus Genes *23*, 145–147.

Gilden, D.H., Lipton, H.L., Wolf, J.S., Akenbrandt, W., Smith, J.E., Mahalingam, R., and Forghani, B. (2002a). Two patients with unusual forms of varicella zoster virus vasculopathy. N. Engl. J. Med. *347*, 1500–1503.

Gilden, D.H., Mahalingam, R., Cohrs, R.J., Kleinschmidt-DeMasters, B.K., and Forghani, B. (2002b). The protean manifestations of varicella zoster virus vasculopathy. J. Neurovirol. *8(suppl 2)*, 75–79.

Gilden, D.H., Cohrs, R.J., Hayward, A.R., Wellish, M., and Mahalingam, R. (2003). Chronic varicella zoster virus ganglionitis: a possible cause of postherpetic neuralgia. J. Neurovirol. *9*, 404–407.

Gold, E. (1996). Serologic and virus-isolation studies of patients with varicella or herpes-zoster infection. N. Engl. J. Med. *274*, 181–185.

Gray, W.L. (2003). Pathogenesis of simian varicella virus. J. Med. Virol. *70*, S4-S8.

Gray, W.L., Starnes, B., White, M.W., and Mahalingam, R. (2001). The DNA sequence of the simian varicella virus genome. Virology *284*, 123–130.

Gray, W.L., Mullis, L., and Soike, K.F. (2002). Viral gene expression during acute simian varicella virus infection. J. Gen. Virol. *83*, 841–846.

Guess, H.A., Broughton, D.D., Melton, L.J., and Kurland, L.T. (1985). Epidemiology of herpes zoster in children and adolescents: A population-based study. Pediatrics *76*, 512–517.

Harnisch, J.P. (1984). Zoster in the elderly: Clinical, immunologic and therapeutic considerations. J. Am. Geriatr. Soc. *32*, 789–793.

Hawrami, K., Harper, D., and Breuer, J. (1996). Typing of varicella zoster virus by amplification of DNA polymorphisms. J. Virol. Methods *57*, 169–174.

Hayakawa, Y., Yamamoto, T., Yamanishi, K., and Takahashi, M. (1986). Analysis of varicella-zoster virus DNAs of clinical isolates by endonuclease HpaI. J. Gen. Virol. *67*, 1817–1829.

Head, H. and Campbell, A.W. (1900). The pathology of herpes zoster and its bearing on sensory localization. Brain *23*, 353–523.

Hevner, R.F., Vilela, M.D., Rostomily, R.C., Cohrs, R.J., Mahalingam, R., Wellish, M., and Gilden, D.H. (2003). An unusual cause of trigeminal distribution pain and tumour. Lancet Neurol. *2*, 567–572.

Hope-Simpson, R.E. (1965). The nature of herpes zoster: A long-term study and a new hypothesis. Proc. R. Soc. Med. *58*, 9–20.

Hyman, R.W., Ecker, J.R., and Tenser, R.B. (1983). Varicella-zoster virus RNA in human trigeminal ganglia. Lancet *2*, 814–816.

Iltis, J.P., Arrons, M.C., Castellano, G.A., Madden, D.L., Sever, J.L., Curfman, B.L., and London, W.T. (1982). Simian varicella virus (delta herpesvirus) infection of patas monkeys leading to pneumonia and encephalitis. Proc. Soc. Exp. Biol. Med. *169*, 266–279.

Kemble, G.W., Annunziato, P., Lungu, O., Winter, R.E., Cha, T.A., Silverstein, S.J., and Spaete, R.R. (2000). Open reading frame S/L of varicella-zoster virus encodes a cytoplasmic protein expressed in infected cells. J. Virol. *74*, 11311–11321.

Kennedy, P.G.E., Grinfeld, E., and Gow, J.W. (1998). Latent varicella-zoster virus is located predominantly in neurons in human trigeminal ganglia. Proc. Natl. Acad. Sci. USA *95*, 4658–4662.

Kennedy, P.G.E., Grinfeld, E., and Gow, J.W. (1999). Latent varicella-zoster virus in human dorsal root ganglia. Virology *258*, 451–454.

Kennedy, P.G.E., Grinfeld, E., and Bell, J.E. (2000). Varicella-zoster virus gene expression in latently infected and explanted human ganglia. J. Virol. *74*, 11893–11898.

Kennedy, P.G.E., Grinfeld, E., Bontems, S., and Sadzot-Delvaux, C. (2001). Varicella-zoster virus gene expression in latently infected rat dorsal root ganglia. Virology *289*, 218–223.

Kennedy, P.G.E., Grinfeld, E., Traina-Dorge, V., Gilden, D.H., Mahalingam, R. (2004). Neuronal localization of simian varicella virus DNA in ganglia of naturally infected African green monkeys. Virus Genes *28*, 273–276.

Koropchak, C.M., Solem, S.D., Diaz, P.S., and Arvin, A.M. (1989). Investigation of varicella-zoster virus infection of lymphocytes by *in situ* hybridization. J. Virol. *63*, 2392–2395.

Kutish, G., Mainprize, T., and Rock, D. (1990). Characterization of the latency-related transcriptionally active region of bovine herpesvirus 1 genome. J. Virol. *64*, 5730–5737.

LaGuardia, J.J., Cohrs, R.J., and Gilden, D.H. (1999). Prevalence of varicella-zoster virus DNA in dissociated human trigeminal ganglia neurons and non-neuronal cells. J. Virol. 73, 8571–8577.

Lapresle, J., and Lasjuanias, P. (1986). Cranial nerve ischemic arterial syndromes. Brain *109*, 207–215.

Leventon-Kriss, S., Gotlieb-Stematsky, T., Vonsover, A., and Smetana, Z. (1979). Infection and persistence of varicella-zoster virus in lymphoblastoid Raji cell line. Med. Microbiol. Immunol. *167*, 275–283.

Levin, M.J., Cai, G.Y., Manchak, M.D., and Pizer, L.I. (2003). Varicella-zoster virus DNA in cells isolated from human trigeminal ganglia. J. Virol. *77*, 6979–6987.

Lewis, G.W. (1958). Zoster sine herpete. Br. Med. J. *2*, 418–421.

Lowry, P.W., Sabella, C., Koropchak, C.M., Watson, B.N., Thackray, H.M., Abbruzzi, G.M., and Arvin, A.M. (1993). Investigation of the pathogenesis of varicella-zoster virus infection in guinea pigs by using polymerase chain reaction. J. Infect. Dis. *167*, 78–83.

Lungu, O., Annunziato, P.W., Gershon, A., Staugaitis, S.M., Josefson, D., LaRussa, P., Silverstein, S.J. (1995). Reactivated and latent varicella-zoster virus in human dorsal root ganglia. Proc. Natl. Acad. Sci. USA *92*, 10980–10984.

Lungu, O., Panagiotidis, C.A., Annunziato, P.W., Gershon, A.A., and Silverstein, S.J. (1998). Aberrant intracellular localization of varicella-zoster virus regulatory proteins during latency. Proc. Natl. Acad. Sci. USA *95*, 7080–7085.

Lynch, J.M., Kenyon, T.K., Grose, C., Hay, J., and Ruyechan, W.T. (2002). Physical and functional interaction between the varicella zoster virus IE63 and IE62 proteins. Virology *302*, 71–82.

Mahalingam, R., Wellish, M., Wolf, W., Dueland, A.N., Cohrs, R., Vafai, A., and Gilden, D.H. (1990). Latent varicella zoster virus DNA in human trigeminal and thoracic ganglia. N. Engl. J. Med. *323*, 627–631.

Mahalingam, R., Wellish, M.C., Dueland, A.N., Cohrs, R.J., and Gilden, D.H. (1992). Localization of herpes simplex virus and varicella zoster virus DNA in human ganglia. Ann. Neurol. *31*, 444–448.

Mahalingam, R., Wellish, M., Lederer, D., Forghani, B., Cohrs, R., and Gilden, D.H. (1993). Quantitation of latent varicella zoster virus DNA in human trigeminal ganglia by PCR. J. Virol. *67*, 2381–2384.

Mahalingam, R., Wellish, M., Brucklier, J., and Gilden, D.H. (1995). Persistence of varicella-zoster virus DNA in elderly patients with postherpetic neuralgia. J. Neurovirol. *1*, 130–133.

Mahalingam, R.,.Wellish, M., Cohrs, R., Debrus, S., Piette, J., Rentier, B., and Gilden, D.H. (1996). Expression of protein encoded by varicella-zoster virus open reading frame 63 in latently infected human ganglionic neurons. Proc. Natl. Acad. Sci. USA *93*, 2122–2124.

Mahalingam, R., Kennedy, P.G.E., and Gilden, D.H. (1999). The problems of latent varicella zoster virus in human ganglia: Precise cell location and viral content. J. Neurovirol. *5*, 445–448.

Mahalingam, R., White, T., Wellish, M., Gilden, D.H., Soike, K., and Gray, W.L. (2000). Sequence analysis of the leftward end of simian varicella virus (EcoRI-I fragment) reveals the presence of an 8-bp repeat flanking the unique long segment and an 881-bp open-reading frame that is absent in the varicella zoster virus genome. Virology *274*, 420–428.

Mahalingam, R., Wellish, M., Soike, K., White, T., Kleinschmidt-DeMasters, B.K., and Gilden, D.H. (2001). Simian varicella virus infects ganglia before rash in experimentally infected monkeys. Virology *279*, 339–342.

Mahalingam, R., Traina-Dorge, V., Wellish, M., Smith, J., and Gilden, D.H. (2002). Naturally acquired simian varicella virus infection in African green monkeys. J. Virol. *76*, 8548–8550.

Mainka, C., Fuss, B., Geiger, H., Hofelmayr, H., and Wolff, M.H. (1998). Characterization of viremia at different stages of varicella-zoster virus infection. J. Med. Virol. *56*, 91–98.

Manz, H.J., Canter, H.G., and Melton, J. (1986). Trigeminal herpes zoster causing mandibular osteonecrosis and spontaneous tooth exfoliation. South. Med. J. *79*, 1026–1028.

Matsunaga, Y., Yamanishi, K., and Takahashi, M. (1982). Experimental infection and immune response of guinea pigs with varicella-zoster virus. Infect. Immun. *37*, 407–412.

Mayberg, M., Langer, R.S., Zervas, N.T., Moskowitz, M.A. (1981). Perivascular meningeal projections from the cat trigeminal ganglia: possible pathway for vascular headaches in man. Science *213*, 228–230.

Mayberg, M.R., Zervas, N.T., and Moskowitz, M.A. (1984). Trigeminal projections to supratentorial pial and dural blood vessels in cats demonstrated by horseradish peroxidase histochemistry. J. Comp. Neurol. *223*, 46–56.

McMillan, D.J., Kay, J., and Mills, J.S. (1997). Characterization of the proteinase specified by varicella-zoster virus gene 33. J. Gen. Virol. *78*, 2153–2157.

Meier, J.L., Holman, R.P., Croen, K.D., Smialek, J.E., and Straus, S.E. (1993). Varicella-zoster virus transcription in human trigeminal ganglia. Virology *193*, 193–200.

Miller, A.E. (1980). Selective decline in cellular immune response to varicella-zoster in the elderly. Neurology *30*, 582–587.

Miller, D.H., Kay, R., Schon, F., McDonald, W.I., Haas, L.F., and Hughes, R.A. (1986). Optic neuritis following chickenpox in adults. J. Neurol. *233*, 182–184.

Mitchell, W.J., Deshmane, S.L., Dolan, A., McGeoch, D.J., and Fraser, N.W. (1990). Characterization of herpes simplex virus type 2 transcription during latent infection of mouse trigeminal ganglia. J. Virol. *64*, 5342–5348.

Moffat, J.F., Stein, M.D., Kaneshima, H., and Arvin, A.M. (1995). Tropism of varicella-zoster virus for human CD4+ and CD8+ T lymphocytes and epidermal cells in SCID-hu mice. J. Virol. *69*, 5236–5242.

Morita, Y., Osaki, Y., Doi, Y., Forghani, B., and Gilden, D.H. (2003). Chronic active varicella zoster virus infection manifesting as zoster sine herpete, zoster paresis and myelopathy. J. Neurol. Sci. *212*, 7–9.

Muir, W.B., Nichols, R., and Breuer, J. (2002). Phylogenetic analysis of varicella zoster virus: evidence of intercontinental spread of genotypes and recombination. J. Virol. *76*, 1971–1979.

Murakami, S., Mizobuchi, M., Nakashiro, Y., Doi, T., Hato, N., and Yanagihara, N. (1996). Bell palsy and herpes simplex virus: identification of viral DNA in endoneural fluid and muscle. Ann. Intern. Med. *124*, 27–30.

Murakami, S., Honda, N., Mizobuchi, M., Nakashiro, Y., Hato, N., and Gyo, K. (1998). Rapid diagnosis of varicella zoster virus infection in acute facial palsy. Neurology *51*, 1202–1205.

Myers, M.G. (1979). Viremia caused by varicella-zoster virus: Association with malignant progressive varicella. J. Infect. Dis. *140*, 229–233.

Myers, M.G., and Connelly, B.L. (1992). Animal models of varicella. J. Infect. Dis. *166*, S48-S50.

Myers, M.G., Huer, H.L., and Hausler, C.K. (1980). Experimental infection of guinea pigs with varicella-zoster virus. J. Infect. Dis. *142*, 414–420.

Myers, M.G., Stanberry, L.R., and Edmond, B.J. (1985). Varicella-zoster virus infection of strain 2 guinea pigs. J. Infect. Dis. *151*, 106–113.

Nagashima, K., Nakazawa, M., and Endo, H. (1975). Pathology of the human spinal ganglia in varicella-zoster virus infection. Acta Neuropathol. *33*, 105–117.

Padovan, D., and Cantrell, C.A. (1986). Varicella-like herpesvirus infections of nonhuman primates. Lab. Anim. Sci. *36*, 7–13.

Pevenstein, S.R., Williams, R.K., McChesney, D., Mont, E.K., Smialek, J.E., and Straus, S.E. (1999). Quantitation of latent varicella-zoster virus and herpes simplex virus genomes in human trigeminal ganglia. J. Virol. *73*, 10514–10518.

Preston, V.G., Kennard, J., Rixon, F.J., Logan, A.J., Mansfield, R.W., and McDougall, I.M. (1997). Efficient herpes simplex virus type 1 (HSV-1) capsid formation directed by the varicella-zoster virus scaffolding protein requires the carboxy-terminal sequences from the HSV-1 homologue. J. Gen. Virol. 78, 1633–1646.

Priola, S.A., Gustafson, D.P., Wagner, E.K., and Stevens, J.C. (1990). A major portion of the latent pseudorabies virus genome is transcribed in trigeminal ganglia of pigs. J. Virol. *64*, 4755–4760.

Pumphrey, C.Y., and Gray, W.L. (1992). The genomes of SVV and VZV are colinear. Virus Res. *26*, 255–266.

Roberts, E.D., Baskin, G.B., Soike, K., and Gibson, S.V. (1984). Pathologic changes of experimental simian varicella (Delta herpesvirus) infection in African green monkeys (Cercopithecus aethiops). Am. J. Vet. Res. *45*, 523–530.

Robillard, R.B., Hilsinger, R.L., Jr., and Adour, K.K. (1986). Ramsay Hunt facial paralysis: clinical analyses of 185 patients. Otolaryngol. Head Neck Surg. *95*, 292–297.

Rock, D.L., and Fraser, N.W. (1985). Latent herpes simplex virus type 1 DNA contains two copies of the virion DNA joint region. J. Virol. *55*, 849–852.

Ross, J., Williams, M., and Cohen, J.I. (1997). Disruption of the varicella-zoster virus dUTPase and the adjacent ORF9A gene results in impaired growth and reduced syncytia formation *in vitro*. Virology *234*, 186–195.

Rostad, S.W., Olson, K., McDougall, J., Shaw, C.M., and Alvord, E.C., Jr. (1989). Transsynaptic spread of varicella zoster virus through the visual system: a mechanism of viral dissemination in the central nervous system. Hum. Pathol. *20*, 174–179.

Russman, A.N., Lederman, R.J., Calabrese, L.H., Embi, P.J., Forghani, B., and Gilden, D.H. (2003). Multifocal varicella zoster virus vasculopathy without rash. Arch. Neurol. *60*, 1607–1609.

Sadzot-Delvaux, C., Merville-Louis, M.P., Delree, P., Marc, P., Piette, J., Moonen, G., and Rentier, B. (1990). An *in vivo* model of varicella-zoster virus latent infection of dorsal root ganglia. J. Neurosci. Res. *26*, 83–89.

Saito, K., and Moskowitz, M.A. (1989). Contributions from the upper cervical dorsal roots and trigeminal ganglia to the feline circle of Willis. Stroke *20*, 524–526.

Sato, M., Urade, M., Shirasuan, K., Yura, Y., Yoshida, H., Yanagawa, T., Morimoto, M., Kobayashi, A., Hamamura, Y., Kubo, K., and Miyazaki, T. (1979). Herpes zoster of the maxillary branch of the trigeminus nerve. Int. J. Oral Surg. *8*, 149–154.

Sawtell, N.M. (1997). Comprehensive quantification of herpes simplex virus latency at the single-cell level. J. Virol. *71*, 5423–5431.

Schott, G.D. (1998). Triggering of delayed-onset postherpetic neuralgia. Lancet *351*, 419–420.

Shibuta, H., Ishikawa, T., Hondo, R., Aoyama, Y., Kurata, K., and Matumoto, M. (1974). Varicella virus isolation from spinal ganglion. Arch. Virusforsch. *45*, 382–385.

Smith, F.P. (1978). Pathological studies of spinal nerve ganglia in relation to intractable intercostal pain. Surg. Neurol. *10*, 50–53.
Soike, K.F., Baskin, G., Cantrell, C., and Gerone, P. (1984a). Investigation of antiviral activity of 1-beta-D-arabinofuranosylthymine (ara-T) and 1-beta-D-arabinofuranosyl-E-5-(2-bromovinyl)uracil (BV-ara-U) in monkeys infected with simian varicella virus. Antiviral Res. *4*, 245–257.
Soike, K.F., Rangan, S.R., and Gerone, P.J. (1984b). Viral disease models in primates. Adv. Vet. Sci. Comp. Med. *28*, 151–199.
Soike, K.F., Keller, P.M., and Ellis, R.W. (1987). Immunization of monkeys with varicella-zoster virus glycoprotein antigens and their response to challenge with simian varicella virus. J. Med. Virol. *22*, 307–313.
Sommer, M.H., Zagha, E., Serrano, O., Ku, C.C., Zerboni, L., Baiker, A., Santos, R., Spengler, M., Lynch, J., Grose, C., Ruyechan, W., Hay, J., and Arvin, A.M. (2001). Mutational analysis of the repeated open reading frames, ORFs 63 and 70 and ORFs 64 and 69, of varicella-zoster virus. J. Virol. *75*, 8224–8239.
Steffen, R., and Selby, G. (1972). "Atypical" Ramsay Hunt syndrome. Med. J. Aust. *1*, 227–230.
Stevens, J.G., Wagner, E.K., Devi-Rao, G.B., Cook, M.L., and Feldman, L.T. (1987). RNA complementary to a herpesvirus alpha gene mRNA is prominent in latently infected neurons. Science *235*, 1056–1059.
Stroop, W.G., Rock, D.L., and Fraser, N.W. (1984). Localization of herpes simplex virus in the trigeminal and olfactory systems of the mouse central nervous system during acute and latent infections by *in situ* hybridization. Lab. Invest. *51*, 27–38.
Sweeney, C.J., and Gilden, D.H. (2001). Ramsay Hunt syndrome. J. Neurol. Neurosurg. Psychiatry *71*, 149–154.
Takahashi, M., Okuno, Y., Otsuka, T., Osame, J., and Takamizawa, A. (1975). Development of a live attenuated varicella vaccine. Biken J. *18*, 25–33.
Tenser, R.B., and Hyman, R.W. (1987). Latent herpesvirus infections of neurons in guinea pigs and humans. Yale J. Biol. Med. *60*, 159–167.
Terada, K., Niizuma, T., Kawano, S., Kataoka, N., Akisada, T., and Orita, Y. (1998). Detection of varicella-zoster virus DNA in peripheral mononuclear cells from patients with Ramsay Hunt syndrome or zoster sine herpete. J. Med. Virol. *56*, 359–363.
Thomas, E.J., and Howard, F.M. (1972). Segmental zoster paresis: a disease profile. Neurology *22*, 459–466.
Twomey, J.J., Gyorkey, F., and Norris, S.M. (1974). The monocyte disorder with herpes zoster. J. Lab. Clin. Med. *83*, 768–777.
von Barensprung, F.G.F. (1987). Beitr'ge zur Kenntnis des zoster. Ann. Chir. Krankenh. *11*, 96–104.
Vonsover, A., Leventon-Kriss, S., Langer, A., Smetana, Z., Zaizov, R., Potaznick, D., Cohen, I.J., and Gotlieb-Stematsky, T. (1987). Detection of varicella-zoster virus in lymphocytes of DNA hybridization. J. Med. Virol. *21*, 57–66.
Walz-Cicconi, M.A., Rose, R.M., Dammin, G.J., and Weller, T.H. (1986). Inoculation of guinea pigs with varicella-zoster virus via the respiratory route. Arch. Virol. *88*, 265–277.
Watson, C.P.N., Deck, J.H., Morshead, C., Van der Kooy, D., and Evans, R.J. (1991). Post-herpetic neuralgia: further post-mortem studies of cases with and without pain. Pain *44*, 105–117.
Wenner, H.A., Abel, D., Barrick, S., and Seshumurty, P. (1977). Clinical and pathogenetic studies of Medical Lake macaque virus infections in cynomolgus monkeys (simian varicella). J. Infect. Dis. *135*, 611–622.
Wharton, M. (1996). The epidemiology of varicella-zoster virus infections. Infect. Dis. Clin. North Am. *10*, 571–581.
White, T.M., Mahalingam, R., Kolhatkar, G., and Gilden, D.H. (1997). Identification of simian varicella virus homologues of varicella zoster virus genes. Virus Genes *15*, 265–269.
White, T.M., Mahalingam, R., Traina-Dorge, V., and Gilden, D.H. (2002). Persistence of simian varicella virus DNA in CD4(+) and CD8(+) blood mononuclear cells for years after intratracheal inoculation of African green monkeys. Virology *303*,192–198.
Wolf, R.H., Smetana, H.F., Allen, W.P., and Felsenfeld, A.D. (1974). Pathology and clinical history of Delta herpesvirus infection in patas monkeys. Lab. Anim. Sci. *24*, 218–221.
Wroblewska, Z., Devlin, M., Reilly, K., van Trieste, H., Wellish, M., and Gilden, D.H. (1982). The production of varicella zoster virus antiserum in laboratory animals. Arch. Virol. *74*, 233–238.
Yamada, S., Atsuta, N., Tokunaga, S., and Motegi, Y. (2003). Ipsilateral truncal sensory deficit in a patient with ophthalmic zoster sine herpete. Neurology *60*, 1049–1050.

HSV-1 Latency and the Roles of the LATs

17

David C. Bloom

Abstract

Herpes simplex virus type 1 (HSV-1) latency is characterized by the persistence of viral genome as episomes in the nuclei of sensory neurons. During this period only one region of the genome is abundantly transcribed: the region encoding the latency-associated transcripts (LATs). The LAT domain is transcriptionally complex, and while the predominant species that accumulates during latency is a 2.0 kb stable intron, other RNA species are transcribed from this region of the genome, including a number of lytic or acute-phase transcripts. HSV-1 recombinant viruses with deletions within the LAT region exhibit reactivation deficits in a number of animal models, however there is evidence that some LAT deletion mutants also possess altered establishment and virulence properties. The phenotypic complexity associated with this region, including some evidence that the LATs may play a role in suppressing latent gene expression, suggests the that LAT locus may function as a regulator that modulates the transcription of key lytic and latent functions. The goal of this chapter is to provide an overview of our current understanding of the role(s) of the HSV-1 latency associated transcripts (LATs) in the pathobiology of HSV-1 infections *in vivo*.

Introduction to HSV-1 latency

HSV-1 establishes a life-long latent infection within sensory ganglia. During latency, viral genomes are maintained in the nucleus of each cell as multiple episomal copies that do not integrate into the host genome. These circular episomes are associated with histones and are organized into nucleosomes (Efstathiou *et al.*, 1986; Deshmane and Fraser, 1989). During latency, transcription of viral productive-phase (lytic) RNAs is repressed, and abundant transcription is only seen from one region of the viral genome: that region encoding the latency associated transcripts (LATs) (Spivack and Fraser, 1987; Stevens *et al.*, 1987).

HSV-1 latency is primarily studied in one of two well-established animal models, the rabbit and the mouse (for a review see (Wagner and Bloom, 1997). Latent infection in the rabbit is typically established in the trigeminal ganglia (TG) following ocular infection, whereas in the mouse either the TG or dorsal root ganglia (DRG) are the major sites of latency following ocular or footpad infections, respectively. Following such infections, as many as 10–30% of the sensory neurons in a ganglion can be demonstrated to contain latent genomes. As in the case of human infections, latent HSV-1 genomes in some latently infected neurons in mice and rabbits periodically reactivate, presumably in response to stress (for a review see Mitchell *et al.*, 2003). In experimental models of reactivation, the production of infectious virus can be induced by thermal stress and UV light (mouse model), and adrenergic agents such as epinephrine (rabbit model).

Ever since their initial discovery, the fact that the LATs are the only gene products that can be abundantly and consistently detected during latency has made these RNAs a focus of intense study. Early on it was shown that viruses with deletions of the LAT promoter were capable of establishing latent infections, but are severely restricted in their ability to be reactivated in models of induced reactivation (Leib *et al.*, 1989; Hill *et al.*, 1990). Interestingly, LAT promoter mutants can be induced to reactivate to near wild type levels *ex vivo* following explant cultivation (Devi-Rao *et al.*, 1994). Therefore, it is important to highlight that while the LATs play a key role in induced reactivation (such as epinephrine-induced reactivation in the rabbit), not all models of reactivation (such as explant) show the same degree of LAT-dependence. In a similar vein, certain LAT mutants, particularly ones possessing large deletions as described below, exhibit 2–4 fold reductions in the number of latently infected neurons (Perng *et al.*, 2000b; Thompson and Sawtell, 2001). This effect may be dependent on both the region of LAT that is deleted and the animal model (O'Neil *et al.*, 2004). The model-dependence of LAT phenotypes has made the definition of specific mechanisms for the LAT's role in these processes difficult to define. The following sections will attempt to summarize recent findings that have broadened our view of the range of phenotypes that LAT mutants possess. These findings suggest that the diverse functions attributed to the LATs may be due to regulatory functions contained within the LAT region that may modulate the transcription of other genes both during latency and reactivation and likely during the establishment of latency itself.

The LATs and their expression during HSV-1 latency

The LATs, or what are sometimes referred to singularly as the "LAT," are produced in neurons during latency as a primary 8.3–8.5 kb Poly A transcript (Fig. 17.1). This primary

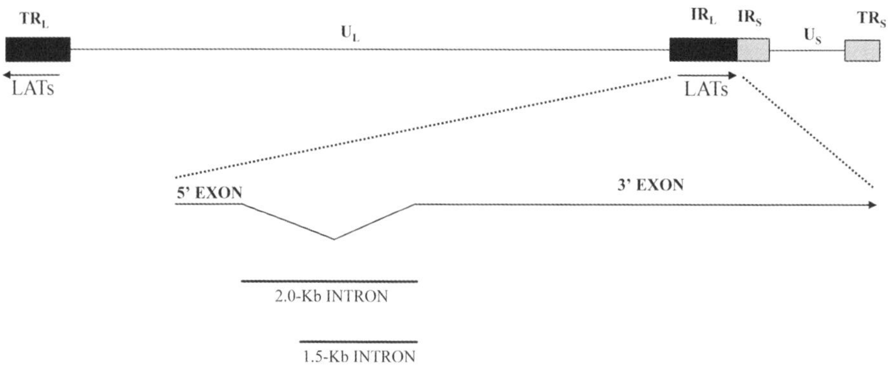

Figure 17.1 Diagram of the region encoding the HSV-1 LATs. Top: Depiction of the HSV-1 genome showing the unique long (U_L) and short (U_S) regions that are flanked by the terminal repeat long (TR_L), internal repeat long (IR_L) and the internal repeat short (IR_S) and terminal repeat short (TR_S) respectively. The location of the LATs within the R_L is shown. The primary LAT RNA is an 8.5 kb transcript that is spliced to yield a stable intron of 2.0 kb, as well as a further spliced species of 1.5 kb that is seen in a subset of neurons.

transcript is spliced to yield a stable 2.0 kb intron (also referred to as the major LAT) which is further spliced to yield a 1.5 kb intron in a subset of the neurons that produce the 2.0 kb intron (Farrell *et al.*, 1991). The 2.0 kb intron accumulates abundantly in the nuclei of latently infected neurons (Fig. 17.2), due largely to the fact it is an extremely stable intron with a half-life on the order of 24 hours (Thomas *et al.*, 2002). This impressive stability results from inefficient debranching during splicing, resulting in the intron persisting as a lariat (Wu *et al.*, 1996; Rodahl and Haarr, 1997).

While limited LAT transcription can be detected during an acute infection of sensory ganglia, the use of recombinant HSV-1 viruses engineered to contain the LAT promoter expressing a reporter have revealed that the LAT promoter is expressed almost exclusively in cells that are not expressing acute antigen (Margolis *et al.*, 1992b). These are likely cells where the virus is entering latency and strongly suggests that LAT expression and lytic gene expression are not compatible. While the major LAT accumulates abundantly in the nuclei of many latently infected neurons, it is now clear that not all neurons that harbor HSV-1 DNA during latency produce detectable levels of LAT. Carefully controlled *in situ*/PCR analyses have revealed that only 1/3 of HSV DNA-containing neurons contain detectable LAT (Gressens and Martin, 1994; Mehta *et al.*, 1995). Moreover, the LAT seems to be expressed predominantly in certain phenotypic classes of sensory neurons, and not others (Margolis *et al.*, 1992a; Yang *et al.*, 2000). Whether this tropism of expression is mediated by the virus, the neuron, or a combination of the two is not known.

Regulation of LAT transcription during latency

The LAT promoter is relatively complex compared with most other HSV promoters, and contains a number of known cellular

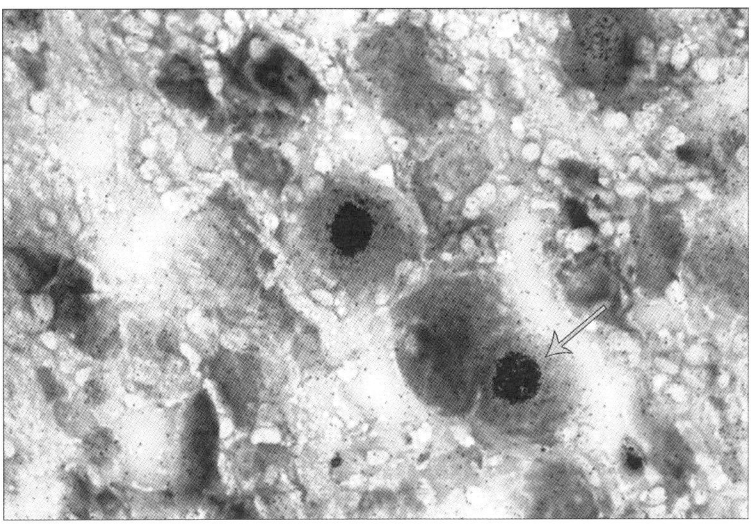

Figure 17.2 *In situ* hybridization of a latently infected rabbit TG showing the accumulation of the LAT 2.0 kb intron in the nuclei of two large sensory neurons. 5 μM section of a rabbit TG latently infected with HSV-1 strain 17*syn*+ probed with a tritium-labeled probe for the LAT 2.0 kb intron, and subsequently counterstained with Giemsa. The arrow is pointing to the nucleus of one of the two LAT positive neurons illustrating the abundance of silver grains indicating positive hybridization for LAT.

transcription factor binding sites (Fig. 17.3). While the core promoter is contained within an approximately 200 bp fragment, additional *cis*-elements which confer neuronal-specific transcription extend some 650 bp further upstream (Fig. 17.3) (Dobson *et al.*, 1995). In addition, an enhancer element that has been shown to be critical for the expression of LAT during latency resides several hundred bp downstream of the transcriptional start site (Fig. 17.4A) (Lokensgard *et al.*, 1997; Berthomme *et al.*, 2000).

Despite its relative complexity, the LAT promoter itself is not sufficient to maintain long-term latent expression of LAT or that of heterologous transgenes during HSV-1 latency (Margolis *et al.*, 1993). Expression during latency has been shown to be due, at least in part, to enhancer activity located in the region encoding the 5′ exon of LAT. The Feldman laboratory demonstrated that this 5′ exon region was able to enhance the activities of the LAT and heterologous promoters (Berthomme *et al.*, 2000). In addition, this element plays a role in maintaining LAT promoter activity during latency, and for this reason was named the long-term expression (LTE) element (Margolis *et al.*, 1993; Lokensgard *et al.*, 1994; Lokensgard *et al.*, 1997) (Fig. 17.4).

While the precise mechanism by which the LAT enhancer maintains expression of the LATs during latency is not known, there is growing data that suggests possible epigenetic regulation at the level of chromatin. Chromatin immunoprecipitation (ChIP) analyses of chromatin isolated from latently infected murine ganglia have revealed that, in contrast to HSV-1 lytic genes, the region encoding the LATs is significantly enriched in

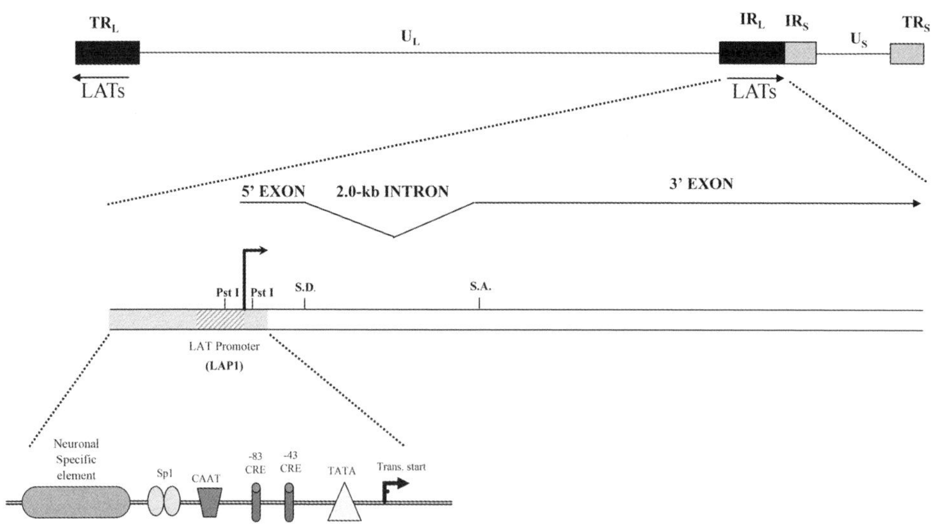

Figure 17.3 Diagram of the HSV-1 LAT promoter. Expanded view of the region encoding the 8.5 kb primary LAT transcript showing the LAT promoter (grey shaded region representing approximately 875 bp), location of the transcriptional start site (arrow) and the location of the splice donor (S.D.) and splice acceptor (S.A.) sites for the 2.0 kb intron. The core LAT promoter, containing elements necessary for basal transcription, is indicated by the cross-hatching. The expanded view of the promoter shows some of the transcription factor binding sites that have been identified, including a TATA box (Zwaagstra *et al.*, 1991), and two CRE sites (Leib *et al.*, 1991; Kenny *et al.*, 1994), as well as an upstream region shown to possess elements required for high-level expression of LAT in neurons (Zwaagstra *et al.*, 1990; Dobson *et al.*, 1995).

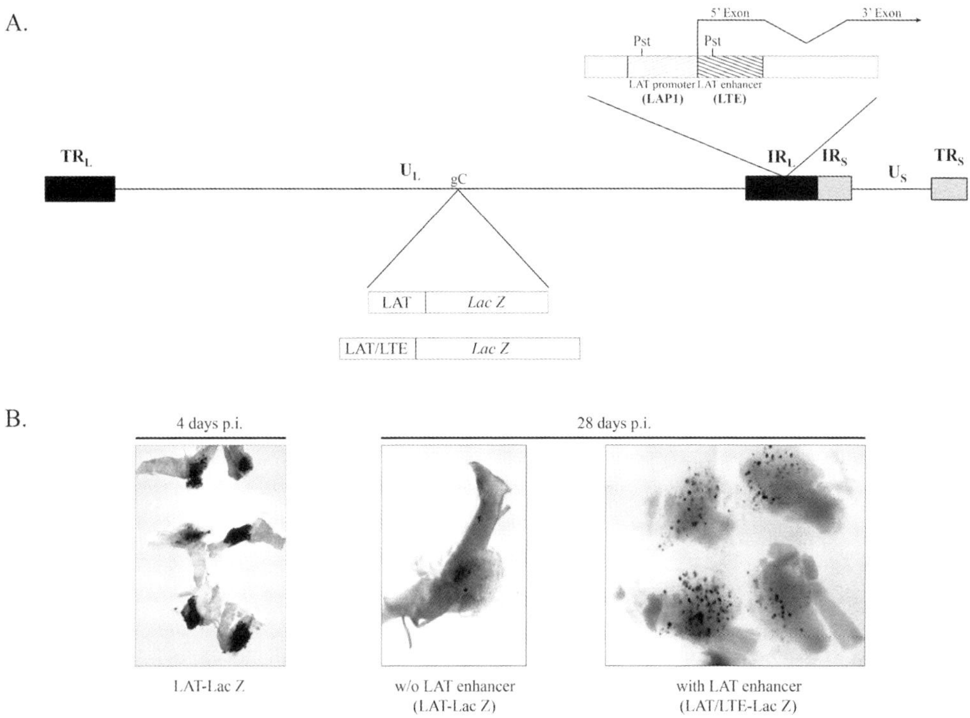

Figure 17.4 The LAT enhancer is required to maintain latent expression of a transgene from the LAT promoter. A. Above the genome: An expanded view of the IR_L shows the relative location of the LAT promoter (LAP1) and the LAT enhancer (LTE). Below the genome: the insertion of two *lac Z* reporter cassettes into the glycoprotein C (gC) gene. One cassette contains the ~875 bp LAT promoter without the LAT enhancer (LAT) and the other contains the LAT promoter and the enhancer (LAT/LTE). Mice were infected with these two recombinants by the footpad route and DRG (L4 + L5) were removed and stained with x-gal to visualize the reporter gene activity. B. Four days p.i. both recombinants exhibited strong x-gal staining (only the LAT-*Lac Z* is shown). In contrast at 28 days p.i., only a few x-gal stained neurons can be observed in the LAT-*LacZ* recombinant, whereas the LAT enhancer recombinant demonstrates sustained expression.

the modified histone acetyl H3(K9, 14), which is generally associated with transcriptionally permissive (euchromatin) regions (Kubat *et al.*, 2004b)(Fig. 17.5). Most significantly, the LAT enhancer is also hyperacetylated. This hyperacetylation remains in LAT promoter mutants (Kubat *et al.*, 2004a), suggesting that this hyperacetylation is not simply due to Pol II activity, but instead may be a property of the enhancer itself. One possibility is that the enhancer recruits histone-modifying activities that establish a transcriptionally permissive chromatin environment in this region of the genome. This is evidence that this region may play an important transcriptional control function of surrounding genes.

The transcriptional complexity of the LAT locus

The entire LAT locus is arguably one of the most transcriptionally complex regions in the HSV genome. In addition to the LATs, there are a number of other transcripts that have been mapped to this region. One is co-linear with the 2 kb intron and is actually produced from a promoter, termed the LAP-2 pro-

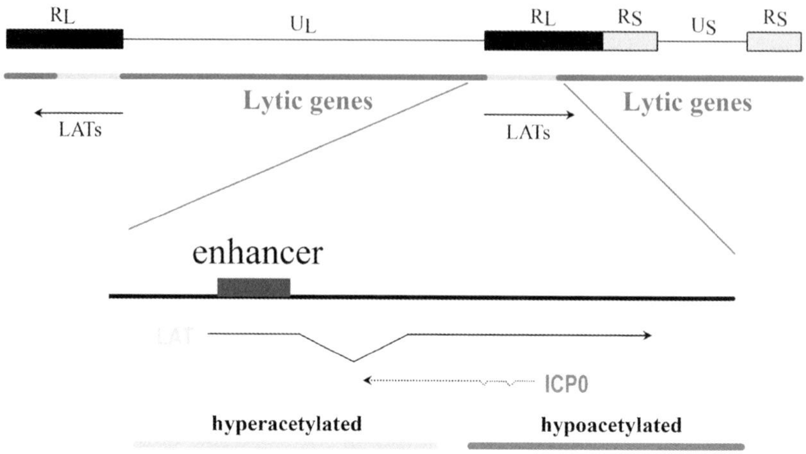

Figure 17.5 The HSV-1 latent genome is organized into transcriptionally permissive and nonpermissive domains. ChIP analysis of the HSV-1 genome during a latent infection of murine DRG has shown that all lytic genes analyzed are underenriched in the modified histone H3 acetyl (K9, 14) consistent with a transcriptionally repressed state (dark shaded line). In contrast, the LAT promoter and 5′ exon (LAT enhancer) region are enriched in this acetylated histone, consistent with a transcriptionally active chromatin state. As indicated in the expanded view, this region of hyperacetylation does not extend to the ICP0 promoter. Taken together, this suggests that the HSV-1 genome is organized into distinct chromatin domains during latency that differ in their transcriptional permissivity.

moter, which is located within the region corresponding to the 5′exon of the 8.5 kb LAT (Fig. 17.6) (Nicosia *et al.*, 1993; Goins *et al.*, 1994; Maggioncalda *et al.*, 1994). Another transcript transcribed from this region is the "antisense to LAT" (AL-LAT), described by Perng and Wechsler (Perng *et al.*, 2002). The AL-LAT promoter overlaps a portion of the LAP-2 promoter, and the transcript is transcribed antisense to the LAT promoter (Fig. 17.6). Finally, a 0.7 kb transcript is also located in this region and transcribes across the LAT promoter (Zhu *et al.*, 1999). If we also look along the full length of the 8.5 kb LAT, we see that there are some very important and well characterized lytic transcripts that are transcribed from this region including ICP4, γ34.5, and ICP0. Clearly, the opportunities for complex regulation within this region are abundant. It is important to note that that the LATs (transcribed from the LAP1 promoter) are the only transcripts that are transcribed abundantly during latency, whereas these other transcripts templated in this region are transcribed primarily during the acute infection (Fig. 17.6).

Phenotypes associated with LAT deletion mutants

Given the transcriptional complexity of this region, it is probably not surprising that there have been many functions attributed to the LATs, including reactivation, establishment of latency, virulence antiapoptosis and apoptosis. *A critical consideration is that it is very difficult to determine which of these phenotypes are directly encoded genetically within the LAT locus, and which of these functions might be secondary effects of regulatory properties of the LAT locus.* For example, if there were direct effects of the LAT region on the establishment of latency, it would be difficult to separate the impact of these establishment effects on reactivation from direct effects of the LAT region on reactivation. Another example is that if there were basic LAT functions responsible for

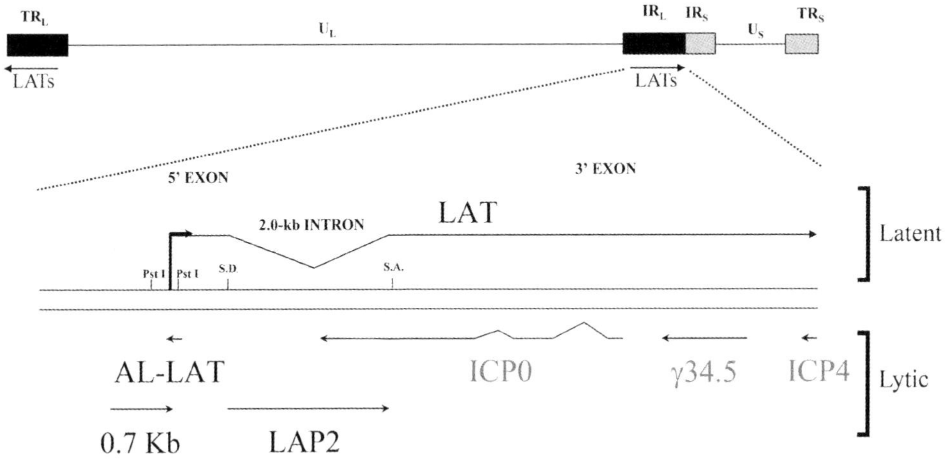

Figure 17.6 The transcriptional complexity of the HSV-1 LAT region. Expanded view of the LAT region shows transcripts that have been mapped to this region. These include transcripts that are transcribed primarily during latency: the 8.5 kb LAT, and the 2.0 and 1.5 kb intron species; and transcripts that are transcribed during the acute or lytic infection: AL-LAT, LAP2, ICP0, gamma-34.5 and ICP4.

transcriptional regulation of some lytic genes (such as ICP0) during latency, it would be difficult to separate the loss of that function and the resulting indirect induction of apoptosis that might result from the downstream effects of leaky transcription from a direct role of the LAT in suppressing cellular apoptotic effectors (Bloom, 2004). Clearly, sorting out direct functions of the LATs has been a very difficult aspect that we have all wrestled with in trying to dissect the mechanistic basis of the phenotypes in this region. With this difficulty in mind, Fig. 17.7 is an attempt to summarize the relative map positions of the phenotypic functions that have been mapped to the LAT locus. In constructing this functional genetic map, one has to consider that a range of LAT mutants that do not necessarily overlap in their deletions were used in the individual analyses, and that these studies were performed in different viral backgrounds and characterized for their phenotypes in different animal models. If you consider these caveats and realize that the boundaries for each phenotype are an approximation, the map is useful in pointing out a key point: most of the phenotypes described for LAT map to a region that extends from roughly the 5′ end of the LAT promoter through the region encoding the first 1 kb of the LAT intron, just 3′ of the ICP0 coding region. This is clearly a very important region from a regulatory point of view for the virus. Below is a summary of the data supporting each of the phenotypes attributed to the LATs shown in Fig. 17.7.

Reactivation

The first phenotype attributed to HSV-1 recombinants containing deletions in the LAT region was decreased reactivation in both mouse and rabbit models (Leib *et al.*, 1989; Bloom *et al.*, 1994; Hill *et al.*, 1990; Perng *et al.*, 1994). This restriction is most evident in models of *in vivo* reactivation, such as hyperthermia induction in the mouse (Sawtell and Thompson, 1992) and epinephrine induction in the rabbit (Bloom *et al.*, 1994), and produces a less dramatic reduction in models of explant cultivation (Block *et al.*, 1993; Devi-Rao *et al.*, 1994). Reactivation of HSV-2 in

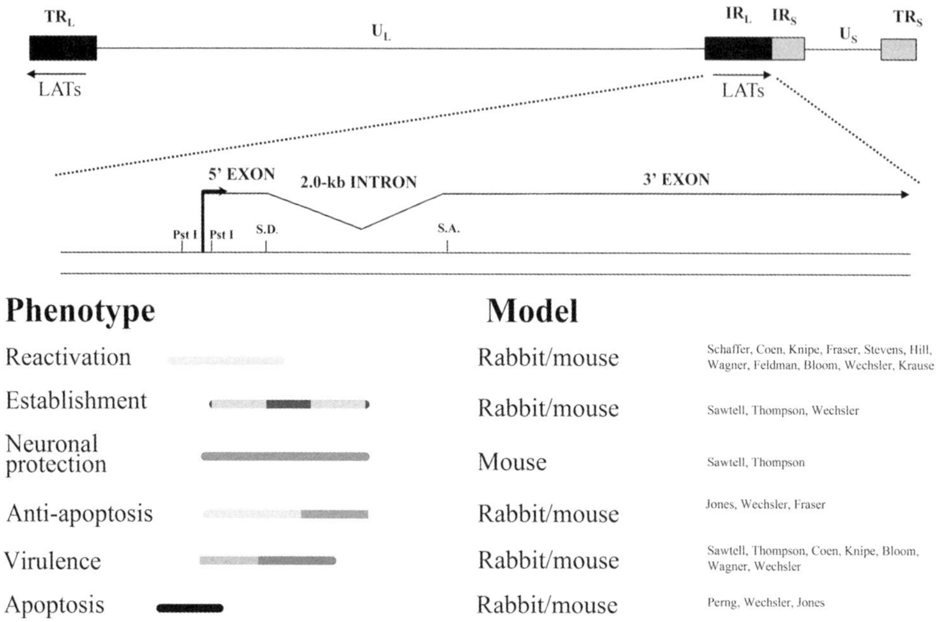

Figure 17.7 Phenotypes associated with the HSV-1 LAT region. Diagram illustrating the rough map positions for the various phenotypes that have been detected in LAT mutants. The model system (mouse or rabbit) in which these effects were noted is indicated, as well as the contributing laboratories. See the text for literature citations.

the guinea pig vaginal model is also greatly reduced for HSV-2 mutants with LAT promoter (LAP1) deletions (Krause *et al.*, 1995). Deletions within the LAT region remain the only deletion/mutation in the viruses that specifically affect reactivation efficiencies while not influencing replicative yields of the virus either *in vivo* or *in vitro*.

The region of the LAT that is linked to decreased reactivation varies slightly from model system to model system, and is largely based on the resolution afforded by the particular deletion mutants used in each case to map the location. In all cases the reactivation phenotype resides within the region encompassing the LAT promoter and the first 1.5 kb of the primary LAT transcript. In the case of induced reactivation in the rabbit, the phenotype resides within the first 663 bp of the LAT transcript, or the region corresponding to roughly the 5′ exon (Jarman *et al.*, 2002). Significantly, the presence of the 2.0 kb intron, the most prominent LAT specie, is not required for reactivation. However, deletions of ~350 nt or greater within the 5′ exon region reduce reactivation, whereas smaller deletions have no effect (Bloom *et al.*, 1996). This region, termed the reactivation critical region (*rcr*), includes the LAT enhancer, and may suggest that the enhancer element, and not the LAT RNA itself, is important in reactivation.

Establishment

Initial analysis of LAT deletion mutants and their rescues suggested that they established equivalent latent infections, as assessed by total HSV-1 DNA present in ganglia during latent times by dot blot hybridization and semi-quantitative PCR analysis (Sedarati *et al.*, 1989; Steiner *et al.*, 1989). Recently, several lines of evidence have been put forth suggesting that certain LAT mutants exhibit decreased establishment. Using contextual analysis, where individual cells from latently infected ganglia are assayed for the presence of latent HSV DNA by PCR, LAT deletion mutants were shown to be present in up to half as many cells in mouse TG as wild type virus (Sawtell, 1997). This observation was extended in studies using HSV-1 recombinants engineered to express reporter genes, either in a wild type or LAT-deletion viral backbone. Ganglia infected with the LAT deletion viruses exhibited roughly 1/3 to 1/2 as many reporter-positive cells as their wild type (or rescued) counterparts (Perng *et al.*, 2000b; Thompson and Sawtell, 2001). Finally, in a well-controlled QT-PCR analysis of a large number of murine TG demonstrated a 2–4 fold reduction in the amount of HSV-1 DNA present within the latent ganglia infected with the LAT deletion viruses, compared with wild type and rescue infected ganglia (Thompson and Sawtell, 2001). Collectively, these data make a strong case that the LATs play some role in enhancing the establishment of latency, both in terms of total HSV DNA present, and in the total number of cells harboring latent DNA. The mutants analyzed in these studies contained relatively large deletions encompassing the LAT promoter and extending 3′ through a portion of the region encoding the LAT intron, leaving some debate as to whether the decreased establishment also would occur following infection with LAT mutants possessing smaller deletions. In fact, extensive analyses have been performed with recombinants possessing smaller LAT deletions and establishment defects have not been detected, though it must be noted that the sample sizes employed in these studies would not have allowed a statistically meaningful detection of just a 2–4 fold difference (O'Neil *et al.*, 2004). None-the-less, it seems clear that some genetic function that reduces establishment by 2–4 fold resides within at least the first 1.9 kb of the region encoding the LATs. Ultimately, the most important question is whether this degree of decreased establishment directly translates into decreased reactivation. The strongest argument against a 2–4 fold decrease in establishment being responsible for the reduced reactivation levels that are observed for LAT mutants comes from studies in the rabbit model. Following an ocular infection with wild type HSV-1, there is approximately a 10-fold range in the amount of HSV-1 DNA present within the rabbit TG after latency has been established. Even given this variation, 80–100% of the ganglia can be induced to reactivate the latent virus following epinephrine induction, and there is no observed correlation between amount of DNA and the ability to reactivate. When LAT promoter mutants are examined, there is no observed difference in the overall range of establishment by LAT promoter mutants, yet the LAT promoter mutants reactivate with 0–10% the efficiency of wild type (O'Neil *et al.*, 2004). Therefore, a 2–4 fold difference in the amount of establishment between a LAT mutant would likely be lost in the larger (10-fold) variation of establishment that is inherent in the system. So while it seems likely that the LATs do exert some influence on establishment, there is strong evidence that reactivation and establishment functions of the LAT are genetically separable.

Neuronal protection

A proposed mechanism for the LAT's role in establishment of latency is in protecting neurons from being killed during the resolution of the acute infection and the initial establishment of latency. This model proposes that LAT modulates replicative functions, perhaps by regulating ICP0 expression, and helps push HSV-1 towards a latent infection. Data supporting this model comes in part from experiments where fewer neurons containing latent genomes were detected in mice infected with LAT mutants, and from data in the mouse that suggest that infections with some LAT mutants actually result in the loss of twice as many neurons as is seen in a comparable infection with wild type or rescue viruses (Thompson and Sawtell, 2001). While the data seem to strongly suggest that some function contained within the LAT region may help spare neurons and push the virus towards latency, the mutants still do establish a latent infection. Therefore, neuronal protection is clearly not an essential function for HSV-1 to establish latency, but instead may be important in protecting certain specific subtypes of sensory neurons, or perhaps be important in a multiplicity dependent manner.

Antiapoptosis

The first indication that a function encoded within the LAT region promotes cell survival was put forth by Perng *et al.* (Perng *et al.*, 2000a) using a recombinant of HSV-1 McKrae (dLAT2903) which possesses a deletion spanning −161 to +1667, relative to the transcriptional start of the primary 8.5 kb LAT (see Fig. 17.2). This study demonstrated that large numbers of TUNEL positive neurons could be observed in rabbit TG acutely infected (day 7 p.i.) with dLAT2903, compared with only a few TUNEL positive cells in TGs from rabbits infected with the rescue of dLAT2903 or the parent strain, HSV-1 McKrae. TUNEL results were further confirmed by the demonstration that p85 cleavage products of poly-ADP-ribose polymerase (PARP) were detectable in > 25% of TG neurons infected with the dLAT2903, whereas no PARP positive cells were detected in TGs infected with wild type McKrae. This provided strong evidence that a function encoded within the LAT region ultimately resulted in the inhibition of caspase 3-induced PARP cleavage, initiated during an HSV-1 infection.

While questions were initially raised concerning the specificity of the anti-PARP p85 antisera, and the detection of cytoplasmic staining (Thompson and Sawtell, 2000), the observation that IMR-90 and CV-1 cells transiently transfected with a plasmid containing nt 301–2659 of the LAT primary transcript was able to protect > 75% of the cells from ceramide or fumonisin B_1 induced apoptosis (relative to transfection with a control plasmid), strongly suggests that this region of the LAT protects against induction of apoptosis (Perng *et al.*, 2000a). The level of antiapoptotic activity afforded by the LAT plasmid was comparable to a plasmid containing the baculovirus inhibitor of apoptosis (CpIAP). Finally, it was demonstrated the LAT region also protected Neuro-2a cells (mouse neuroblastoma) from etoposide-induced apoptosis. Since ceramide and fumonisin B_1 (FB_1) induce apoptosis through different pathways (protein kinase C and TNF/FAS respectively), transient transfection studies suggested that LAT's antiapoptosis function acts through a common downstream effector.

Initial studies indicated that ocular infection of mice with the large LAT-deletion virus 17ΔAH (−1136 to +828), resulted in depletion of over half the TG neurons relative to uninfected mice, mice infected with wild type 17*syn*+, or the 17AH rescue (Thompson and Sawtell, 2001). Here the authors were unable to detect any increase in TUNEL positive cells or evidence of the cleaved form of caspase 3, leading them to conclude that the loss of neurons was not due to an apoptotic mechanism. The possibility exists that the apoptosis observed with the LAT mutants might represent a species-specific effect, however two lines of evidence argue convincingly against this possibility. The first study employed a LAT deletion mutant (17N/H) whose deletion (−359 to +1491) overlaps with that of 17ΔAH, but extends further into the intron. 17N/H showed a significant increase in TUNEL positive TG neurons compared with mice infected with 17*syn*+ or a smaller LAT deletion (17ΔSty) (Ahmed *et al.*, 2002). A second study using a transgenic mouse containing the LAT promoter and first 3 kb of the primary HSV-2 LAT transcript demonstrated diminished apoptosis, both in cell culture and *in vivo* (Wang *et al.*, 2002). In this study, hepatocytes and fibroblasts from transgenic or nontransgenic mice were treated with TNF or anti-Fas antibody. Transgenic mice showed an approximate 50% reduction in numbers of cells exhibiting annexin V staining vs. nontransgenic controls. In addition, when transgenic mice were treated with anti-Fas antibody, they demonstrated 10-fold lower serum ALT levels than the nontransgenic group. Only 37% of the transgenic mice died, compared with 100% of the nontransgenic mice. Collectively, these data suggest that the HSV LAT contains elements that are capable of blocking apoptosis both in rabbits and mice *in vivo*. One possible explanation for the authors' inability to detect evidence of apoptosis induced by 17AH is the smaller size of this recombinant's deletion.

Virulence

Several studies have mapped regions within the LAT domain, that when deleted, alter the virulence of HSV-1. Most of these effects are relatively subtle and include phenotypes that differ in the mouse and rabbit models (Perng *et al.*, 1999; Perng *et al.*, 2001). In addition, some deletions have been shown to cause increases in virulence, as measured by viral replication in target tissues and death (Perng *et al.*, 1999), whereas other deletions have been shown to cause slight decreases in overall lethality and the production of infectious virus (Garber *et al.*, 1997). It is important to note that no LAT deletions to date have altered replication or viral yields in cell culture, so these effects seem to be true "*in vivo*" phenomena.

Apoptosis

Finally, the identification of another transcript termed "antisense to LAT" or AL has been attributed to actually *inducing* apoptosis in the context of the viral infection (Perng *et al.*, 2002). This transcript is transcribed across the LAT promoter, and overlaps with just a small portion of the very 5′ end of the 8.5 kb LAT. The fact that this transcript might actually induce apoptosis initially seems paradoxical given the data suggesting that other regions of LAT inhibit apoptosis, however, the potential of this transcript to regulate other LAT functions in an antisense manner may suggest that the balance between inducing apoptosis and blocking apoptosis may be regulated. Since AL LAT is presumably expressed only during the lytic infection, its role in inducing apoptosis may be involved in viral spread.

How does LAT do all these things?

While over the past several years there have been a number of significant new developments regarding phenotypes associated with the LATs, how this region mediates these functions remains to be elucidated. Moreover, determining which of the observed phenotypes are dominant has been equally difficult. Nonetheless, there have been a number of important recent findings that have contributed to eliminating some previous hypothetical mechanisms.

Antisense regulator

Ever since its initial discovery, the potential for the 2.0 kb LAT to regulate ICP0 through an antisense mechanism has been an obvious possibility (Stevens *et al.*, 1987). Work from the Steiner lab, and others, has demonstrated that expression of the 2.0 kb intron has the potential to reduce the expression of ICP0 and viral yields when one engineers LAT to be expressed constitutively during the acute infection; however, a direct demonstration of antisense regulation in the context of the natural infection has not been made (Mador *et al.*, 2003). In fact, the Glorioso lab performed a rather convincing analysis that failed to detect functional downregulation of ICP0 in the context of the viral genome (Burton *et al.*, 2003). It should be pointed out, however, that the door is not completely closed to the possibility that the LATs could mediate some (or all) of their effects by antisense regulation. One must not forget that the 8.5 kb LAT has the *potential* to regulate a number of very important lytic genes by antisense including ICP0, γ34.5, and even ICP4. A very carefully performed study by Knipe and Coen showed that LAT mutants exhibit leakier expression of several lytic genes during latency than is seen with wild type HSV or rescue viruses (Chen *et al.*, 1997; Garber *et al.*, 1997; Kramer *et al.*, 1998). This study provided key data suggesting that the LATs have some function in helping to suppress lytic transcription during latency, either directly or indirectly.

LAT-encoded protein

While latently expressed proteins are well-described among the gamma-herpesviruses, and while the LATs contain a number of potential ORFs that could be expressed, to date no LAT-encoded protein has been directly detected by antisera against that protein during a latent infection. It remains possible that future analyses will demonstrate that low levels of protein are made from some of these ORFs. However, it is important to note that if one analyses the region of the LATs where the majority of the phenotypes reside (Fig. 17.7), potential open reading frames are not well-conserved between different strains of HSV-1. This makes the hypothesis of a LAT-encoded protein less plausible.

LAT as a noncoding RNA

Given the current data, it seems most likely that the LATs act either as noncoding RNAs or as a DNA regulatory element. As mentioned above, the possibility that LAT acts as an antisense regulator remains. Another plausible hypothesis is that LAT could act through an RNA interference mechanism, either as an siRNA or micro RNA. To date there is no evidence for those types of interactions though, there are a number of labs currently investigating this possibility.

A final RNA hypothesis is that LAT could function as a noncoding structural RNA similar to the mammalian *XIST* or *Rox* of *Drosphila* that regulate expression of the sex chromosomes, through a mechanism that has yet to be worked out (Bloom *et al.*, 1996). The *XIST* or *Rox* RNAs have been shown to associate in *cis* with the chromosome from which they are transcribed and "paint it" (Clemson *et al.*, 1996). This association has been postulated to help recruit specific histone modifying activities which ultimately alter the transcriptional status of the chromatin on that chromosome. There is no data, other than distant sequence similarities, to support this model, and association between the latent genome and LAT has been technically challenging due to difficulties detecting latent HSV-1 genomes by *in situ* hybridization.

The LAT region as a *cis*-acting regulatory region

Recent data suggests that a hypothesis that might be worthy of considering is that the 5′ exon-coding region of the LAT domain is a functionally important transcriptional regulator that regulates not only the expression of the LATs, but also has the potential to regulate or enhance the expression of surrounding genes (Berthomme *et al.*, 2000). This hypothesis is grounded in the fact that a number of phenotypes ascribed to the LATs, including reactivation, establishment and apoptosis, were mapped to the region encompassing the LAT promoter and the 5′ exon region (Fig. 17.7). Fine mapping of the reactivation critical region (*rcr*) has shown that deletions within the 5′ exon region, while still producing LAT, are sufficient to reduce reactivation (Bloom *et al.*, 1996; Loutsch *et al.*, 1999). Since the 5′ exon region possesses enhancer activity, and it appears that this enhancer is able to maintain the 5′ LAT region in a transcriptionally permissive chromatin state during latency, it is possible that this enhancer element is a key component of a regulatory system contained within the LAT region that could influence the regulation of a number of other viral genes (Kubat *et al.*, 2004a). Indeed, this might explain why LAT deletions encompass a number of phenotypes.

A key component of this model would include a means of insulating the other neighboring genes from the effects of the LAT enhancer during latency. It has been shown that the effects of the enhancer do not normally extend to the nearby ICP0 promoter, since this promoter is associated with hypo-acetylated histone H3 during latency (Kubat *et al.*, 2004a), suggesting that a chromatin boundary with enhancer-blocking properties exists between the LAT enhancer and the ICP0 promoter. The recent identification of conserved binding sites for the cellular boundary protein CTCF, and demonstration by ChIP that these CTCF sites are occupied during a latent infection suggests that these CTCF binding sites might form the core of a chromatin boundary that is functional during latency (Amelio *et al.*, 2006).

In this model, two chromatin boundaries would flank the LAT enhancer and establish a "transcriptionally permissive zone" as indicated by the enrichment of this region in the specifically modified histone H3 acetyl (K9, 14). This hyperacetylated region encompasses the LAT promoter and the 5′ LAT exon, but does not extend to the ICP0 promoter (Fig. 17.8). A reactivation stimulus (such as heat stress or epinephrine induction) could cause a collapse of the boundary separating the LAT enhancer from ICP0, and allow the enhancer to facilitate transcription of ICP0 (Fig. 17.8). ICP0 has long been proposed to be a key player in the initial stages of HSV-1 reactivation (Cai *et al.*, 1993; Halford and Schaffer, 2001). Aside from the role of ICP0 as a promiscuous transactivator, work by Everett has implicated ICP0 in chromatin remodeling activities (Everett *et al.*, 1999). Expression of ICP0 could therefore play an important role in altering the transcriptionally nonpermissive histones that are associated with the other HSV-1 immediate early and lytic genes and potentially facilitate a productive reactivation event. While the details of this model remain to be proven, it is easy to envision how mutations in various components of such a multicomponent LAT regulatory switch could ultimately modulate a large number of key viral functions and result in the multiple phenotypes that have been described for this region.

Conclusions and future directions

The LAT region and its encoded LATs have been implicated in a number of functions including establishment of latency, neuronal protection and antiapoptosis, and reactivation from latency. The mechanism by which LAT modulates these processes is unclear, but focus has shifted to regulatory functions either at the noncoding RNA or DNA level. Given the location of the LAT locus in a region of the genome that is rich in key regulatory proteins, the idea of LAT acting as a regulatory component is an attractive model. Thinking about the LAT domain as a multicomponent regulatory region, composed of the LAT enhancer and specific regulatable boundaries with the potential to regulate other transcripts in this region initially takes one away from the actual LAT RNAs that have been the focus of so much attention. These LATs may still play important roles as noncoding

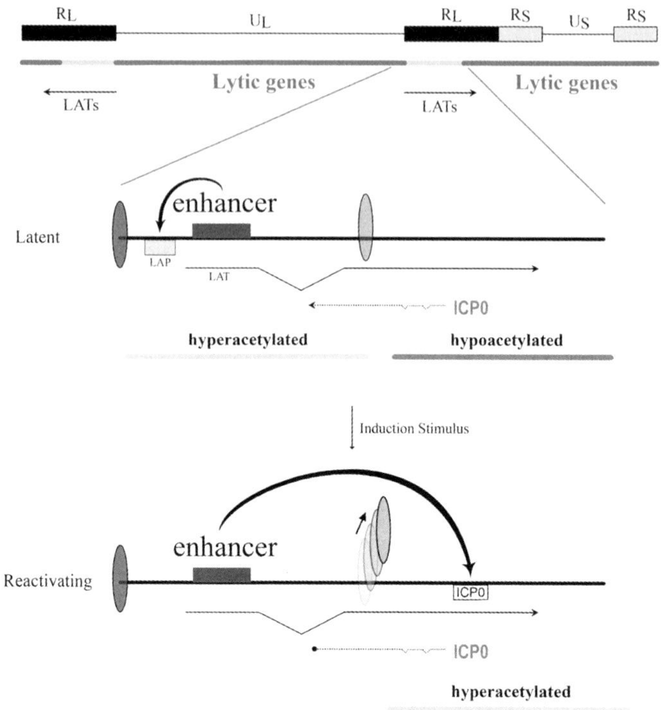

Figure 17.8 Model of the LAT region as a multicomponent regulator of transcription. The top depiction of the HSV-1 genome illustrates the organization of the latent genome into transcriptionally permissive (light grey) and transcriptionally nonpermissive (dark grey) chromatin domains. The expanded view (latent) shows the LAT region and the relative locations of the LAT enhancer and the chromatin boundaries (ovals) that would be proposed to insulate the other genome regions (and ICP0) from the effects of the LAT enhancer. During latency the enhancer would act on the LAT promoter. The bottom expanded view (reactivating) depicts the induction stimulus resulting in a dissolution of the chromatin boundary, allowing the LAT enhancer to hyperacetylate the ICP0 promoter and make it permissive for transcriptional activation.

RNAs; elements which have been shown to be transcribed from chromosomal boundary regions, are believed to mediate recruitment of histone modifying activities. Whatever the role of the LATs, if this region is shown to function as a dominant regulator of transcription, it could well explain why it has been so difficult to tease apart the functions that have been attributed to the LAT region. Further characterization of the LAT enhancers and the boundary regions from a functional point of view should shed some additional light on how this region may function in modulating the balance between the lytic infection, latency and reactivation.

Acknowledgments

Thanks to Antonio Amelio, Nicole Giordani and Zane Zeier for helpful comments on the preparation of the manuscript, and to Peterjon McAnany for help with the figures.

References

Ahmed, M., Lock, M., Miller, C.G., and Fraser, N.W. (2002). Regions of the herpes simplex virus type 1 latency-associated transcript that protect cells from apoptosis *in vitro* and protect neural cells *in vivo*. J. Virol. *76*, 717–729.

Amelio, A.L., McAnany, P.J., and Bloom, D.C. (2006). A chromatin insulator-like element in the herpes simplex virus type 1 latency-associated transcript region binds CCCTC-binding factor and displays enhancer-blocking and silencing activities. J. Virol. *80*, 2358–2368.

Berthomme, H., Lokensgard, J., Yang, L., Margolis, T., and Feldman, L.T. (2000). Evidence for a bidirectional element located downstream from the herpes simplex virus type 1 latency-associated promoter that increases its activity during latency. J. Virol. *74*, 3613–3622.

Block, T.M., Deshmane, S., Masonis, J., Maggioncalda, J., Valyi, N.T., and Fraser, N.W. (1993). An HSV LAT null mutant reactivates slowly from latent infection and makes small plaques on CV-1 monolayers. Virology *192*, 618–630.

Bloom, D.C. (2004). HSV LAT and neuronal survival. Int. Rev. Immunol. *23*, 187–198.

Bloom, D.C., Devi-Rao, G.B., Hill, J.M., Stevens, J.G., and Wagner, E.K. (1994). Molecular analysis of herpes simplex virus type 1 during epinephrine induced reactivation of latently infected rabbits *in vivo*. J. Virol. *68*, 1283–1292.

Bloom, D.C., Hill, J.T., Wagner, E.K., Feldman, L.F., and Stevens, J.G. (1996). A 348-bp region in the latency associated transcript facilitates herpes simplex virus type 1 reactivation. J. Virol. *70*, 2449–2459.

Burton, E.A., Hong, C.S., and Glorioso, J.C. (2003). The stable 2.0-kilobase intron of the herpes simplex virus type 1 latency-associated transcript does not function as an antisense repressor of ICP0 in nonneuronal cells. J. Virol. *77*, 3516–3530.

Cai, W., Astor, T.L., Liptak, L.M., Cho, C., Coen, D.M., and Schaffer, P.A. (1993). The herpes simplex virus type 1 regulatory protein ICP0 enhances virus replication during acute infection and reactivation from latency. J. Virol. *67*, 7501–7512.

Chen, S.H., Kramer, M.F., Schaffer, P.A., and Coen, D.M. (1997). A viral function represses accumulation of transcripts from productive-cycle genes in mouse ganglia latently infected with herps simplex virus. J. Virol. *71*, 5878–5884.

Clemson, C.M., McNeil, J.A., Willard, H.F., and Lawrence, J.B. (1996). XIST RNA paints the inactive X chromosome at interphase: evidence for a novel RNA involved in nuclear/chromosomal structure. J. Cell Biol. *132*, 259–275.

Deshmane, S.L., and Fraser, N.W. (1989). During latency, herpes simplex virus type 1 DNA is associated with nucleosomes in a chromatin structure. J. Virol. *63*, 943–947.

Devi-Rao, G.B., Bloom, D.C., Stevens, J.G., and Wagner, E.K. (1994). Herpes simplex virus type 1 DNA replication and gene expression during explant induced reactivation of latently infected murine sensory ganglia. J. Virol. *68*, 1271–1282.

Dobson, A.T., Margolis, T.P., Gomes, W.A., and Feldman, L.T. (1995). *In vivo* deletion analysis of the herpes simplex virus type 1 latency associated transcript promoter. J. Virol. *69*, 2264–2270.

Efstathiou, S., Minson, A.C., Field, H.J., Anderson, J.R., and Wildy, P. (1986). Detection of herpes simplex virus-specific DNA sequences in latently infected mice and in humans. J. Virol. *57*, 446–455.

Everett, R.D., Earnshaw, W.C., Findlay, J., and Lomonte, P. (1999). Specific destruction of kinetochore protein CENP-C and disruption of cell division by herpes simplex virus immediate-early protein Vmw110. EMBO J. *18*, 1526–1538.

Farrell, M.J., Dobson, A.T., and Feldman, L.T. (1991). Herpes simplex virus latency-associated transcript is a stable intron. Proc. Natl. Acad. Sci. USA *88*, 790–794.

Garber, D.A., Schaffer, P.A., and Knipe, D.M. (1997). A LAT-associated function reduces productive-cycle gene expression during acute infection of murine sensory neurons with herpes simplex virus type 1. J. Virol. *71*, 5885–5893.

Goins, W.F., Sternberg, L.R., Croen, K.D., Krause, P.R., Hendricks, R.L., Fink, D.J., Straus, S.E., Levine, M., and Glorioso, J.C. (1994). A novel latency-active promoter is contained within the herpes simplex virus type 1 UL flanking repeats. J. Virol. *68*, 2239–2252.

Gressens, P., and Martin, J.R. (1994). *In situ* polymerase chain reaction: localization of HSV-2 DNA sequences in infections of the nervous system. J. Virol. Methods *46*, 61–83.

Halford, W.P., and Schaffer, P.A. (2001). ICP0 is required for efficient reactivation of herpes simplex virus type 1 from neuronal latency. J. Virol. *75*, 3240–3249.

Hill, J.M., Sedarati, F., Javier, R.T., Wagner, E.K., and Stevens, J.G. (1990). Herpes simplex virus latent phase transcription facilitates *in vivo* reactivation. Virology *174*, 117–125.

Jarman, R.G., Loutsch, J.M., Devi-Rao, G.B., Marquart, M.E., Banaszak, M.P., Zheng, X., Hill, J.M., Wagner, E.K., and Bloom, D.C. (2002). The region of the HSV-1 latency-associated transcript required for epinephrine-induced reactivation in the rabbit does not include the 2.0 kb intron. Virology *292*, 59–69.

Kenny, J.J., Krebs, F.C., Hartle, H.T., Gartner, A.E., Chatton, B., Leiden, J.M., Hoeffler, J.P., Weber, P.C., and Wigdahl, B. (1994). Identification of a second ATF/CREB-like element in the herpes simplex virus type 1 (HSV-1) latency-associated transcript (LAT) promoter. Virology *200*, 220–235.

Kramer, M.F., Chen, S.H., Knipe, D.M., and Coen, D.M. (1998). Accumulation of viral transcripts and DNA during establishment of latency by herpes simplex virus. J. Virol. *72*, 1177–1185.

Krause, P.R., Stanberry, L.R., Bourne, N., Connelly, B., Kurawadwala, J.F., Patel, A., and Straus, S.E.

(1995). Expression of the herpes simplex virus type 2 latency-associated transcript enhances spontaneous reactivation of genital herpes in latently infected guinea pigs. J. Exp. Med. *181*, 297–306.
Kubat, N.J., Amelio, A.L., Giordani, N.V., and Bloom, D.C. (2004a). The herpes simplex virus type 1 latency-associated transcript (LAT) enhancer/rcr is hyperacetylated during latency independently of LAT transcription. J. Virol. *78*, 12508–12518.
Kubat, N.J., Tran, R.K., McAnany, P., and Bloom, D.C. (2004b). Specific histone tail modification and not DNA methylation is a determinant of herpes simplex virus type 1 latent gene expression. J. Virol. *78*, 1139–1149.
Leib, D.A., Bogard, C.L., Kosz, V.M., Hicks, K.A., Coen, D.M., Knipe, D.M., and Schaffer, P.A. (1989). A deletion mutant of the latency-associated transcript of herpes simplex virus type 1 reactivates from the latent state with reduced frequency. J. Virol. *63*, 2893–2900.
Leib, D.A., Nadeau, K.C., Rundle, S.A., and Schaffer, P.A. (1991). The promoter of the latency-associated transcripts of herpes simplex virus type 1 contains a functional cAMP-reponse element: Role of the latency-associated transcripts and cAMP in reactivation and viral latency. Proc. Natl. Acad. Sci. USA *88*, 48–52.
Lokensgard, J.R., Bloom, D.C., Dobson, A.T., and Feldman, L.T. (1994). Long-term promoter activity during herpes simplex virus latency. J. Virol. *68*, 7148–7158.
Lokensgard, J.R., Berthomme, H., and Feldman, L.T. (1997). The latency-associated promoter of herpes simplex virus type 1 requires a region downstream of the transcription start site for long-term expression during latency. J. Virol. *71*, 6714–6719.
Loutsch, J.M., Perng, G.C., Hill, J.M., Zheng, X., Marquart, M.E., Block, T.M., Nesburn, A.B., and Wechsler, S.L. (1999). Identical 371-base-pair deletion mutations in the LAT genes of herpes simplex virus type 1 McKrae and 17syn+ results in different *in vivo* reactivation phenotypes. J. Virol. *73*, 767–771.
Mador, N., Braun, E., Haim, H., Ariel, I., Panet, A., and Steiner, I. (2003). Transgenic mouse with the herpes simplex virus type 1 latency-associated gene: expression and function of the transgene. J. Virol. *77*, 12421–12429.
Maggioncalda, J., Mehta, A., Fraser, N.W., and Block, T.M. (1994). Analysis of a herpes simplex virus type 1 LAT mutant with a deletion between the putative promoter and the 5′ end of the 2.0 kilobase transcript. J. Virol. *68*, 7816–7824.
Margolis, T.P., Dawson, C.R., and LaVail, J.H. (1992a). Herpes simplex viral infection of the mouse trigeminal ganglion. Immunohistochemical analysis of cell populations. Invest. Ophthalmol. Vis. Sci. *33*, 259–267.
Margolis, T.P., Sedarati, F., Dobson, A.T., Feldman, L.T., and Stevens, J.G. (1992b). Pathways of viral gene expression during acute neuronal infection with HSV-1. Virology *189*, 150–160.
Margolis, T.P., Bloom, D.C., Dobson, A.T., Feldman, L.T., and Stevens, J.G. (1993). Decreased reporter gene expression during latent infection with HSV LAT promoter constructs. Virology *197*, 585–592.
Mehta, A., Maggioncalda, J., Bagasra, O., Thikkavarapu, S., Saikumari, P., Valyi-Nagy, T., Fraser, N.W., and Block, T.M. (1995). *In situ* DNA PCR and RNA hybridization of herpes simplex virus sequences in trigeminal ganglia of latently infected mice. Virology *206*, 633–640.
Mitchell, B.M., Bloom, D.C., Cohrs, R.J., Gilden, D.H., and Kennedy, P.G. (2003). Herpes simplex virus-1 and varicella-zoster virus latency in ganglia. J. Neurovirol. 9, 194–204.
Nicosia, M., Deshmane, S.L., Zabolotny, J.M., Valyi-Nagy, T., and Fraser, N.W. (1993). Herpes simplex virus type 1 latency-associated transcript (LAT) promoter deletion mutants can express a 2-kilobase transcript mapping to the LAT region. J. Virol. *67*, 7276–7283.
O'Neil, J.E., Loutsch, J.M., Aguilar, J.S., Hill, J.M., Wagner, E.K., and Bloom, D.C. (2004). Wide variations in herpes simplex virus type 1 inoculum dose and latency-associated transcript expression phenotype do not alter the establishment of latency in the rabbit eye model. J. Virol. *78*, 5038–5044.
Perng, G.-C., Dunkel, E.C., Geary, P.A., Slanina, S.M., Ghiasi, H., Kaiwar, R., Nesburn, A.B., and Wechsler, S.L. (1994). The latency-associated transcript gene of herpes simplex virus type 1 (HSV-1) is required for efficient *in vivo* spontaneous reactivation of HSV-1 from latency. J. Virol. *68*, 8045–8055.
Perng, G.C., Slanina, S.M., Yukht, A., Drolet, B.S., Keleher, W., Jr., Ghiasi, H., Nesburn, A.B., and Wechsler, S.L. (1999). A herpes simplex virus type 1 latency-associated transcript mutant with increased virulence and reduced spontaneous reactivation. J. Virol. *73*, 920–929.
Perng, G.C., Jones, C., Ciacci-Zanella, J., Stone, M., Henderson, G., Yukht, A., Slanina, S.M., Hofman, F.M., Ghiasi, H., Nesburn, A.B., and Wechsler, S.L. (2000a). Virus-induced neuronal apoptosis blocked by the herpes simplex virus latency-associated transcript. Science *287*, 1500–1503.
Perng, G.C., Slanina, S.M., Yukht, A., Ghiasi, H., Nesburn, A.B., and Wechsler, S.L. (2000b). The latency-associated transcript gene enhances es-

tablishment of herpes simplex virus type 1 latency in rabbits. J. Virol. *74*, 1885–1891.

Perng, G.C., Esmaili, D., Slanina, S.M., Yukht, A., Ghiasi, H., Osorio, N., Mott, K.R., Maguen, B., Jin, L., Nesburn, A.B., and Wechsler, S.L. (2001). Three herpes simplex virus type 1 latency-associated transcript mutants with distinct and asymmetric effects on virulence in mice compared with rabbits. J. Virol. *75*, 9018–9028.

Perng, G.C., Maguen, B., Jin, L., Mott, K.R., Kurylo, J., BenMohamed, L., Yukht, A., Osorio, N., Nesburn, A.B., Henderson, G. *et al.* (2002). A novel herpes simplex virus type 1 transcript (AL-RNA) antisense to the 5′ end of the latency associated transcript produces a protein in infected rabbits. J. Virol. *76*, 8003–8010.

Rodahl, E., and Haarr, L. (1997). Analysis of the 2-kilobase latency-associated transcript expressed in PC12 cells productively infected with herpes simplex virus type 1: evidence for a stable, nonlinear structure. J. Virol. *71*, 1703–1707.

Sawtell, N.M. (1997). Comprehensive quantification of herpes simplex virus latency at the single-cell level. J. Virol. *71*, 5423–5431.

Sawtell, N.M., and Thompson, R.L. (1992). Herpes simplex virus type 1 latency associated transcription unit promotes anatomical site-dependent establishment and reactivation from latency. J. Virol. *66*, 2157–2169.

Sedarati, F., Izumi, K.M., Wagner, E.K., and Stevens, J.G. (1989). Herpes simplex virus type 1 latency-associated transcription plays no role in establishment or maintenance of a latent infection in murine sensory neurons. J. Virol. *63*, 4455–4458.

Spivack, J.G., and Fraser, N.W. (1987). Detection of herpes simplex virus type 1 transcripts during latent infection in mice. J. Virol. *61*, 3841–3847.

Steiner, I., Spivack, J.G., Lirette, R.P., Brown, S.M., MacLean, A.R., Subak, S.J. H., and Fraser, N.W. (1989). Herpes simplex virus type 1 latency-associated transcripts are evidently not essential for latent infection. EMBO J. *8*, 505–511.

Stevens, J.G., Wagner, E.K., Devi, R.G. B., Cook, M.L., and Feldman, L.T. (1987). RNA complementary to a herpesvirus alpha gene mRNA is prominent in latently infected neurons. Science *235*, 1056–1059.

Thomas, D.L., Lock, M., Zabolotny, J.M., Mohan, B.R., and Fraser, N.W. (2002). The 2-kilobase intron of the herpes simplex virus type 1 latency-associated transcript has a half-life of approximately 24 hours in SY5Y and COS-1 cells. J. Virol. *76*, 532–540.

Thompson, R.L., and Sawtell, N.M. (2000). HSV latency-associated transcript and neuronal apoptosis. Science *289*, 1651.

Thompson, R.L., and Sawtell, N.M. (2001). Herpes simplex virus type 1 latency-associated transcript gene promotes neuronal survival. J. Virol. *75*, 6660–6675.

Wagner, E.K., and Bloom, D.C. (1997). Experimental investigation of herpes simplex virus latency. Clin. Microbiol. Rev. *10*, 419–443.

Wang, K., Prikhodko, E., Pesnicak, L., Cohen, J.I., and Straus, S.E. (2002). Mice transgenic for the HSV-2 LAT resist Fas-mediated apoptosis and hepatic necrosis. Paper presented at: 27th International Herpesvirus Workshop (Cairns, Australia).

Wu, T.-T., Su, Y.-H., Block, T.M., and Taylor, J.M. (1996). Evidence that two latency associated transcripts of herpes simplex type 1 are non-linear. J. Virol. *70*, 5962–5967.

Yang, L., Voytek, C.C., and Margolis, T.P. (2000). Immunohistochemical analysis of primary sensory neurons latently infected with herpes simplex virus type 1. J. Virol. *74*, 209–217.

Zhu, J., Kang, W., Marquart, M.E., Hill, J.M., Zheng, X., Block, T.M., and Fraser, N.W. (1999). Identification of a novel 0.7 kb polyadenylated transcript in the LAT promoter region of HSV-1 that is strain specific and may contribute to virulence. Virology *265*, 296–307.

Zwaagstra, J.C., Ghiasi, J.C., Slanina, S.M., Nesburn, A.B., Wheatley, S.C., Lillycrop, K., Wood, J., Latchman, D.S., Patel, K., and Wechsler, S.L. (1990). Activity of herpes simplex virus type 1 latency-associated transcript (LAT) promoter in neuron-derived cells: Evidence for neuron specificity and for a large LAT transcript. J. Virol. *64*, 5019–5028.

Zwaagstra, J.C., Ghiasi, H., Nesburn, A.B., and Wechsler, S.L. (1991). Identification of a major regulatory sequence in the latency associated transcript (LAT) promoter of herpes simplex virus type 1 (HSV-1). Virology *182*, 287–297.

Immunity to Herpes Simplex Virus: Present but Not Perfect

18

Christopher D. Pack and Barry T. Rouse

Abstract

Herpes simplex virus types 1 and 2 both represent significant human pathogens that infect a majority of the population. They should be controllable by vaccines, but, despite notable effort, satisfactory vaccines remain unavailable. One hopes that fundamental studies of both the virus and how it interacts with the host will uncover clues useful for the design of effective prophylactic and therapeutic vaccines. This chapter focuses on host responses to the virus revealed in both animal models and human patient studies.

Initial immune events and their consequences

Initial invasion by pathogens is met by numerous components of innate defense. These may serve to curtail infection, but this outcome is probably uncommon with HSV unless the infecting dose is minimal. The caveat of such failure is that virus can gain access to sensory nerve endings, is transported to sensory ganglia and establishes latency or perhaps persistence. These events occur well before effector T cells or antibodies are induced. Hence, if sterile immunity is to be achieved against primary infection it must rely solely on innate immune defenses. Many components participate in innate immunity, but their respective roles against HSV have received little attention. This may be because, until recently, we had few if any therapeutic options to manipulate them. However, there is now an expanding array of approaches that can activate aspects of innate immunity which could affect the course of infection and the pattern of immunity. Some of these have been tested against HSV infections in humans and animal models (Bernstein, 2001). We also realize that manipulating innate immune function may impact on the subsequent magnitude and quality of the adaptive immune response. This chapter discusses the major components of innate immunity that can perhaps be targeted for future modulation.

Upon entry into the host infectious virus is exposed to numerous innate defenses. These include several humoral molecules as well as a number of cellular defenses. Recently, most mechanistic studies have focused on cellular components of defense, especially dendritic cells (DC) and to a lesser extent natural killer (NK) cells, macrophages, and neutrophils (Thomas *et al.*, 1998; Biron and Brossay, 2001; Kobelt *et al.*, 2003). However, humoral factors may well be pivotal and likely curtail infection with low doses. The humoral factors involved include complement factors, lactoferrin, natural antibody, and of course interferons. Most studies involve the latter and are discussed later. A few key studies have evaluated the role of complement, of which the most compelling are the observations of the Friedman group (Lubinski *et al.*, 2002). They showed that the major glycoprotein gC of HSV bound to C3b, an important molecule in the complement cascade. In addition, gC binding to C3b prevented the otherwise destructive effects of complement activation. In support of this idea, gC minus mutants were only pathogenic in C3 depleted animals (Lubinski *et al.*, 2002). Nevertheless, whether or not variations in C3 levels, which can occur during certain chronic diseases, are associated with increased susceptibility to HSV infection is not clear. However, it is known that mice unable to make C3 are notably more susceptible to infection than control animals (Verschoor *et al.*, 2001).

Natural antibody is another humoral innate component that acts as a hindrance to certain infections which include HSV (Ochsenbein *et al.*, 1999). This antibody either acts to neutralize and so prevent infection or perhaps mediate antibody dependent cell mediated cytotoxicity (ADCC) (Ochsenbein and Zinkernagel, 2000).

Support for a possible role for natural antibody mediated defense was reported by Deshpande *et al.* (Deshpande *et al.*, 2000). They showed that $\mu M^{-/-}$ mice, which lack B cell-mediated immunity, were considerably more susceptible to HSV, but their resistance could be partially reconstituted by transfer of naïve serum containing natural antibody.

With regard to interferons, the story is complex as is the interferon system itself (Biron, 1998). Herpes simplex virus infection does induce various interferon species and viral replication is inhibited by interferons. Plasmacytoid DC, upon exposure to HSV, produce an abundance of type I interferons, and this may be a major source of IFN-α (Kadowaki and Liu, 2002). Moreover, knockout mice lacking either the IFN-α or γ receptor are extremely susceptible to infection (Leib *et al.*, 1999; Vollstedt *et al.*, 2001). Interferons are further addressed in a subsequent section.

Dendritic cells and HSV

Dendritic cells (DC) are one of the first immune cells to interact with pathogens, including HSV-1. These specialized antigen presenting cells (APC) play a critical role in acquiring antigen, processing it, and presenting foreign epitopes on MHC class I and II molecules in order to efficiently prime T cells. In addition, they also must serve as "sentinels" of the body, determining what type of immune response, if any, is needed to combat a potential pathogen. These "danger" signals are primarily orchestrated through Toll-like receptors, which possess the ability to recognize pathogen-associated molecular patterns. Examples of such patterns (and the TLR that they bind) include peptidoglycan (TLR2), dsRNA (TLR3), LPS (TLR4), flagellin (TLR5), ssRNA (TLR7), and CpG methylated dsDNA (TLR9) (Kopp and Medzhitov, 2003). Recent studies have emphasized that herpesviruses may provide ligands for certain TLRs and implied that interaction of such ligands with TLRs on DC and perhaps other cells represents a pivotal event in viral pathogenesis. However, most observations have been made *in vitro* with purified and likely unphysiological concentrations of ligands. Studies have shown that gB protein of human CMV binds TLR2 (Compton *et al.*, 2003), as does an unknown protein of HSV (Kurt-Jones *et al.*, 2004). HSV DNA is also known to bind TLR9 (Hochrein *et al.*, 2004), although whether or not this is a significant event *in vivo* remains in doubt. Recently, plasmacytoid DC were shown *in vitro* to react to HSV-2 DNA and secrete IFN-α, but *in vivo* studies were not performed (Lund *et al.*, 2003). Similarly, HSV-1 DNA stimulated plasmacytoid DC to produce IFN-α upon exposure to HSV-1 via TLR9, but, surprisingly, the majority of IFN-α came from other cell types, such as macrophages, in a TLR9 independent manner (Hochrein *et al.*, 2004).

The outcome of initial interaction between viruses and DC influences the nature, magnitude, and quality of the adaptive immune response. If DC are impaired in their ability to mature or properly prime T cells, a chronic infection or death of the host may ensue with some viruses (Steinman *et al.*, 2003). We do not know, however, if HSV interaction with DC represents a pivotal event, particularly *in vivo*. Some have suggested that HSV infection impairs DC function by preventing maturation, thus serving to retard or even prevent immune induction (Salio *et al.*, 1999). The Lanzavecchia group also showed that infected DCs failed to upregulate costimulatory molecules, produce cytokines, and were not responsive to chemokines, a process crucial for migration to secondary lymphoid organs. Similarly, mature DCs infected *in vitro* with HSV-1 changed both their phenotype and their ability to stimulate T cells (Kruse *et al.*, 2000). However, these *in vitro* data may have been overinterpreted, since DC themselves seem not to be a major target for HSV infection *in vivo*, and in the mouse may avoid direct infection (Sato and Iwasaki, 2004). Nevertheless, the type of interaction made between HSV and DC may impact on the quality of the adaptive immune response made, but if and how this occurs *in vivo* needs to be proven.

Some recent studies in the mouse have focused on DC and shed light on early events in T cell immune induction. For example, it was assumed that Langerhans cells, a predominant subtype of DC in the skin, are one of the first cells to contact HSV-1. This is followed by transport of virus by a now mature DC to the draining lymph node where T cell priming occurs (Kobelt *et al.*, 2003). However, the Carbone group, using footpad infection of mice, demonstrated that neither Langerhans cells ($CD11c^{+}CD205^{high}CD8\alpha^{low}$) nor plasmacytoid DC ($CD11b^{+}CD11c^{+}CD8\alpha^{low}$ were responsible for priming of $CD8^{+}$ T cells. Instead, a different and uncommon DC subpopulation was responsible, the $CD11c^{+}CD8\alpha^{+}$ DC (Allan *et al.*, 2003; Belz *et al.*, 2004). It appears that these cells may not be directly infected by HSV, but instead acquire viral antigens from other cell types by a process referred to as cross-priming (Heath *et al.*, 2004). Cross-priming currently enjoys popularity among mouse immunologists, but whether or not the process is a major event in human T cell immune induction remains to be established. Moreover, whereas $CD8\alpha^{+}$ DC may be the most effective T cell priming cells, it is likely that other cell types, especially abundant ones, can also participate. The situation could also be different at mucosal entry sites from that reported in the skin. For instance, the Iwasaki lab reported that vaginal submucosal $CD11b^{+}CD8\alpha^{-}$ DC were responsible for the generation of protective Th1 responses to HSV-2 infection (Zhao *et al.*, 2003). Intravaginal infection led to rapid recruitment of these submucosal DC to the infected epithelium. Careful analysis of various DC populations from the draining lymph nodes revealed that only the $CD11b^{+}$ submucosal DCs, but not Langerhans cell-derived or $CD8\alpha^{+}$ DCs, presented viral antigens to $CD4^{+}$ T cells and induced IFN-γ secretion, an unanticipated role for this DC subset (Zhao *et al.*, 2003). Moreover, no evidence of direct infection of the DC could be found (Sato and Iwasaki, 2004).

In contrast, a recent study from the Katz group indicated that human myeloid DC in the skin do become infected by HSV (Pollara *et al.*, 2003). Curiously, this was followed by both infected and uninfected cells maturing and presumably migrating to the draining lymph node. It also seemed that the viral component from infected cells responsible for the bystander activation of uninfected DC was gD. This also resulted in the release of type I interferon. They suggested that the stage is set for a race between the maturation and priming capacity of the DC and the ability of HSV to disrupt such events (Pollara *et al.*, 2003).

We may conclude that one or more subsets of DC do indeed play an important role in immunity to HSV. DC seem responsible for mediating some innate defense but are perhaps even more important for the influence they exert on the subsequent adaptive immune response. Manipulating the function of DC does represent a logical target for vaccine design. Some initial studies have shown promise in this regard. Thus, the Spraunce group and others demonstrated that the TLR 7/8 agonists imiquimod and resiquimod showed some efficacy against the severity and frequency of herpetic disease in humans (Bernstein *et al.*, 1993; Hengge *et al.*, 2001; Spruance *et al.*, 2001). In mice as well, manipulating DC function with TLR ligands was shown to raise the barrier to infection. Thus, at least two groups have shown that the mucosal delivery of CpG motifs provides temporary protection against vaginal challenge with HSV (Gallichan *et al.*, 2001; Holmgren *et al.*, 2003). These TLR9 activating compounds may also act as excellent adjuvants permitting mice at least to respond and make $CD8^+$ T cell responses to peptide antigens (Gierynska *et al.*, 2002).

Assuming HSV under *in vivo* conditions does indeed infect DC, there are additional effects which the virus can exert on DC function. These relate to effects on the antigen-presenting machinery, the expression of costimulatory molecules involved in antigen presentation, as well as the production of immunomodulatory cytokines (Iwasaki, 2003). Indeed, the DC has become almost overexposed by its avid followers. Comprehensive reviews on DC in relation to their influence on infectious disease have been published and should be consulted for the full story (Iwasaki, 2003; Kobelt *et al.*, 2003).

Role of NK and NKT cells in HSV infection

These cell types are both considered to participate in the initial response to an infection. They can both act to recognize infected cells and kill them or release numerous cytokines that participate directly or indirectly in immune defense (Biron and Brossay, 2001). We have known since the pioneering studies of Lopez that NK cells influence susceptibility to HSV infection (Bloomfield and Lopez, 1980). Most notably, susceptibility to severe herpes infection occurs in humans with genetic defects in the NK cell response (Biron *et al.*, 1989), and in mice genetic susceptibility differences may be explained by NK cell functional differences (Pereira *et al.*, 2001), but details of this genetic control remain unclear (Lundberg *et al.*, 2003). We know that the NK cell response, and perhaps also NKT cells, helps prevent initial infection, but the NKT response for HSV it is not clear if the virus–NK cell interaction affects the pattern of the subsequent adaptive immune response, as it clearly does in mouse CMV infection (Biron, 1999). It has become quite evident that the NK cell system is diverse and that several phenotypic and some functional subsets exist (Raulet, 2004). The basis of the phenotypic diversity is explained by the multiple receptors expressed on NK cells, some of which are expressed on all NK cells and others limited to different subsets (Raulet, 2004). The NK cell receptors fall into two major functional groups: inhibitory receptors and activating. The inhibitory receptors mainly recognize MHC proteins, and when such ligands are diminished, such as can occur on a virus infected or tumor cell, the inferior signal stimulation results in NK cell activation (Moser *et al.*, 2002; Raulet, 2004). This "missing self" hypothesis could explain why HSV infected cells, which do down regulate MHC molecules, result in NK cell activation (Jennings *et al.*, 1985). NK cells additionally express many activating receptors although little is currently known about the ligands for such receptors. One exciting story, however, involves the herpesvirus MCMV. Here it was shown that the CMV M157 protein bound to the activating receptor Ly-49H, and mice strains encoding Ly-49H were resistant to infection (Arase *et al.*, 2002). It is not known if other herpesviruses such as HSV encode proteins that engage NK activating receptors, but doubtless the search for such molecules is ongoing. Conceivably, HSV could also interact with NK cells by acting as a ligand for TLR2 or TLR9, as both TLRs are present on NK cells (Sivori *et al.*, 2004). Thus, HSV is reported to present ligands for these TLRs (Kurt-Jones *et al.*, 2004; Lund *et al.*, 2003).

It seems likely that the major function that NK cells play during the initial response is to act as a source of cytokines that impact antiviral defense and immune induction. Thus, NK cells act as a major source of IFN-γ and a minor source of IL-12, and abundant production of each are an early event in HSV infection (Kanangat *et al.*, 1996). The release of IFN-γ activates other cells, such as macrophages, and possesses potent antiviral properties. IL-12 is a critical cytokine involved in shaping adaptive immunity, tipping the response to Th1 (Trinchieri *et al.*, 2003). NK cells express other molecules involved in antiviral defense and immune induction, but they do not directly participate in antigen presentation.

Although the relevance of NK cells in HSV immunobiology has been somewhat neglected, recent studies addressed the contribution of NKT cells. These cells share identity markers (NK1.1+) and function much like NK cells, but they additionally express a highly limited T cell receptor repertoire. Most cells in the mouse recognize the nonpeptide antigen α-galactosylceramide presented by the nonclassical MHC molecule CD1d. Mice deficient in CD1d have few if any NKT cells and show delayed clearance and increased mortality upon infection with HSV (Grubor-Bauk *et al.*, 2003). Other data also indicate that NKT cells contribute to defense against HSV. Rosenthal showed that mice deficient in IL-15, a critical factor for NKT cells, were highly susceptible to infection (Ashkar and Rosenthal, 2003).

We can conclude that NK cells are important cells in HSV immunity, but it is not yet evident how we can influence the function of these cells with ligands that target them. However, we anticipate that the topic will be receiving attention in the future, likely coming primarily from observations in the MCMV field.

Neutrophils and macrophages in HSV infection

Both these cell types are prominent early participants in inflammatory reactions to HSV infection (Eloranta and Alm, 1999). Older studies focused heavily on macrophages and showed that depleting such cells or impeding their function enhanced susceptibility to infection (Kodukula *et al.*, 1999). The macrophage has become a somewhat neglected cell in immunology and many of them have been renamed as a subset of DC. In the mouse, if not in man, stimulated macrophages can exert powerful antiviral effects by producing reactive radicals such as NO (Reiss and Komatsu, 1998). Accordingly, iNOS knockout mice showed enhanced susceptibility to infection with HSV-1 (MacLean *et al.*, 1998).

Neutrophils also exert antiviral innate defense, and several groups have shown this in mouse systems (Thomas *et al.*, 1998; Milligan, 1999). Accordingly, if neutrophils are depleted by specific antibody mice show significantly enhanced susceptibility to infection (Tumpey *et al.*, 1996; Milligan *et al.*, 2001). How neutrophils exert their defense is not fully understood, but reactive radical production and the secretion of antiviral cytokines are suspected mechanisms.

The role of adaptive immunity

Observations in humans and studies in mouse model systems leave little doubt that acquired immunity is crucial to contain infection (Cunningham and Mikloska, 2001). Thus, as mentioned, innate immunity can handicap the chance of infection, but if it proves unsuccessful removing the virus becomes primarily the task of the adaptive immune response. With regard to resolving primary infection, most would agree that T cells are crucial for this to occur. In fact, T cell defects, either acquired (as occurs in AIDS), congenital, or constructed (as in some mouse models), result in prolonged and severe infections (Whitley, 2001). In T cell deficient mice the outcome is usually lethal encephalitis (Adler *et al.*, 1999). The role of antibody in the control of primary infection is usually considered as minor, especially since this response occurs later in comparison to T cell induction and antibody-mediated immunity is largely directed at extracellular virus. However, older studies did emphasize that perhaps ADCC was an important participant in antiviral defense (Fujimiya *et al.*, 1978). More recently, however, the Friedman group has questioned this, since HSV, via gE and gI proteins (which act as viral Fcγ receptors), seems to have an evasive maneuver against antibody-mediated immunity (Lubinski *et al.*, 2002).

Antibody, however, is an important component to prevent infection. This was best illustrated in humans by the observation that vaginally delivered infants exposed to HSV-2 were protected if they received maternal antibody (Ashley *et al.*, 1992). Furthermore, there is good evidence that the presence of type common antibody induced by HSV-2 protects against HSV-1 infection, although the reverse situation seems not to have a prominent effect (Koelle and Corey, 2003). It is also well known that once infected with an HSV-1 strain, patients generally do not become infected with additional strains of the same type. This situation is most likely mediated by antibody. Protection of susceptible mucosal surfaces from reinfection appears to be performed most efficiently by antibody as was shown by mouse studies (Kuklin *et al.*, 1998; Parr and Parr, 2003). However, mucosal memory tends to be of less duration than systemic, so mucosal surfaces tend to lose antibody-mediated protection at a faster rate in comparison to systemic immunity (Kuklin *et al.*, 1998). Nevertheless, also in support of a role for antibody is the evidence that the currently used GlaxoSmithKline (GSK) vaccine seems to protect females from infection, but not males (Stanberry *et al.*, 2002). Any vaccine designed to protect against HSV must be capable of inducing neutralizing antibody, preferably including mucosal antibody.

Most of the focus on immunity mechanisms against HSV has been directed at T cell immunity. Since it has considerable implication vaccine design, effort over the years has gone into defining what aspects of T cell immunity dominate antiviral defense and how protection proceeds mechanistically. Several studies have been done in mice, a mixed blessing since the mouse can often convey a misleading message, particularly since this non-mouse pathogen shows several differences in pathogenesis. Mice are more likely to suffer from herpetic encephalitis upon infection, and curiously almost any form of immunization can be contrived to protect them. In contrast, humans, fortunately, seldom succumb to encephalitis, but so far no vaccine seems able to protect them. Our understanding of human adaptive immunity to HSV comes largely from groups led by Corey in Seattle and Cunningham in Sydney. Both have written recent valuable reviews which cover the topic more comprehensively than the following summary saga (Koelle and Corey, 2003; Jones and Cunningham, 2004).

Primary infection by HSV, if not successfully dampened by innate defenses, becomes clinical with lesions persisting up to 3 weeks and viral shedding lasting for 10 days or so. During this time an adaptive immune response is being generated, but it may be retarded by HSV infecting DC, making such cells inferior APC (Jones and Cunningham, 2004). Lesion cessation appears to be a consequence of both $CD4^+$ and $CD8^+$ T cell function. In the recurrent situation, which is better investigated since it is easier to study, things happen more quickly. Lesions last up to 10 days and viral shedding for around 4 days (Corey, 2001). The issue has been why clinical lesions occur at all, since, invariably, latently infected patients retain circulating antibody and have memory $CD4^+$ and $CD8^+$ T cell immunity. Moreover, the extent of these immune parameters do not show significant differences in magnitude nor pattern from latently infected patients who never suffer recurrences (Corey, 2001). However, when following the sequence of immunological events studied by serial biopsies of recurrent lesions a pattern emerges. It seems that $CD4^+$ T cells, along with neutralizing antibody through leaky vessels, are the first to invade (Cunningham and Mikloska, 2001). $CD4^+$ T cells can be induced to produce an abundance of IFN-γ and other cytokines.

$CD8^+$ T cells invade later, but their arrival seems to coincide with infection control (Cunningham and Mikloska, 2001). This has given rise to the notion that perhaps $CD8^+$ T cells are principally responsible for resolution, an idea supported by the observation that the severity of genital herpes in AIDS patients seemed to correlate best with the magnitude of the $CD8^+$ T cell response (Posavad *et al.*, 1997). However, it is by no means certain how the $CD8^+$ T cells subserve their protective function. Thus, HSV infection of cells may cause them to downregulate MHC restricting molecules by a variety of mechanisms (Jennings *et al.*, 1985). Moreover, the Jerome laboratory just described another hazard for $CD8^+$ T cell-mediated immunity. They showed that human HSV infected cells somehow deliver an off function signal to $CD8^+$ CTL when they come in contact (Sloan *et al.*, 2003). Hence, the $CD8^+$ T cells theoretically should not be able to kill the virus infected epithelial cells. Presumably this negative interaction is counteracted by other molecules in the milieu, most notably IFN-γ. The latter is present in high concentration in fever blisters as are a number of other cytokines and chemokines. Most notably these include TNF-α, other interferons, and IL-12. These molecules help orchestrate the inflammatory response with its arsenal of antiviral defenders. Amongst the protective molecules antibody must also be considered since this can neutralize extracellular virus even if retarded by the mechanisms described by the Friedman group (Lin *et al.*, 2004).

Defense against HSV is usually advocated to largely result from the action of cytotoxic $CD8^+$ and $CD4^+$ T cells and the products of Th1 $CD4^+$ T cells (Koelle and Corey, 2003; Jones and Cunningham, 2004). However, formal proof for this opinion is lacking. Mouse studies clearly show that IFN-γ and perhaps TNF-α play a major role in antiviral defense (Yu *et al.*, 1996). Curiously, IFN-γ may be directed at infected cells and cause them to purge themselves of viral products. Some elegant studies from the Hendrick's lab showed that antigen-specific $CD8^+$ T cells in the infected mouse trigeminal ganglion appeared to purge cells of HSV and prevent productive infection (Khanna *et al.*, 2003). This process was even advocated as the true means by which HSV sustains latency—a hard pill to swallow for the countless souls who have worked on molecular mechanisms of HSV latency!

What the mouse tells us about HSV infection

Many groups have evaluated the role of the adaptive immune components against HSV in the mouse. Such studies more or less confirm many of the concepts that have emerged from human investigations, but some discrepancies do exist. The biggest problem with mouse studies is that almost all of them use primary infection models, while the main problem in humans is recrudescence. Moreover, virus can readily spread to the CNS and cause lethal encephalitis in the mouse. Protection against such encephalitis appears to be the primary domain of $CD8^+$ T cell mediated immunity (Simmons and Nash, 1984). This was shown more than twenty years ago, and studies done since that time have merely confirmed and extended these initial observations. A problem has always been to explain how $CD8^+$ T cells act to protect the CNS, since neurons have little or no MHC class I molecules and HSV infection is supposed to downregulate whatever they have (Pereira and Simmons, 1999). However, adoptive transfer experiments comparing the protective efficacy of $CD8^+$ T cells with other cell types, along with observations in models which $CD8^+$ T cell function was compromised, all point to them as neuroprotectors (Newell *et al.*, 1989; Simmons and Tscharke, 1992).

It might be that the neuroprotection proceeds other than by cytotoxicity, certainly an unwanted effect on neurons since such cells are not replaced. Thus, as mentioned previously, the Hendricks group demonstrated that in the mouse PNS neuron protection appeared to be mediated by IFN-γ secreted from HSV immune $CD8^+$ T cells (Liu *et al.*, 2001). The IFN-γ was proposed to purge the neurons of viral products, a concept first described by Frank Chisari and colleagues for the protective effects of $CD8^+$ T cells against Hepatitis B infected liver cells (Guidotti *et al.*, 1999).

Control of infection at peripheral sites in the mouse can be accomplished by both $CD4^+$ and $CD8^+$ T cells as well as by antibody. On a cell to cell basis $CD4^+$ T cells appear to be more effective, but this may vary between mouse strains (Manickan *et al.*, 1995). Moreover, the protective effects of both $CD4^+$ and $CD8^+$ T cells appears to depend on the IFN-γ response (Jennings *et al.*, 1991). In accord with this, $STAT1^{-/-}$ and IFN-γ $receptor^{-/-}$ mice are more susceptible to infection, as are IFN-$\gamma^{-/-}$ mice (Cantin *et al.*, 1999; Leib *et al.*, 1999). Curiously, the effect of IFN-γ may act via NO induction, primarily released by IFN-γ activated macrophages (Kodukula *et al.*, 1999).

Another aspect of HSV immunity that the mouse has helped us understand is how various aspects of immunity are induced and regulated. Focus has been directed at a variety of dendritic cells and their role as APC as mentioned previously. Surprisingly, of the large number of DC subtypes, $CD8\alpha^+$ lymphoid DC appear to be the most successful at inducing T cell immunity (Allan *et al.*, 2003; Belz *et al.*, 2004). Moreover, such DC appear to receive their antigen from other cell types by a process termed cross-priming (Heath *et al.*, 2004). Currently, it is not clear if cross-priming forms part of a significant immune induction event in humans.

The other series of events which the mouse has helped elucidate is the influence of regulatory cells on immune induction and effector function. Suvas *et al.* recently demonstrated that both the magnitude and function of $CD4^+$ and $CD8^+$ T cell responses following either infection or vaccination was hampered by a concomitant regulatory T cell response (Suvas *et al.*, 2003). Curiously, mice were more protected from HSV challenge if their Treg response was abrogated. Subsequent studies also showed that recall vaccination responses as well as humoral responses at mucosal sites were similarly compromised by the Treg response (Toka *et al.*, 2004b). The relevance of T regulation to HSV pathogenesis in the mouse as well as in human systems continues to be investigated. One might wonder why such a process exists since it hampers the efficacy of host defense against the virus. The answer to this seems to lie in the fact that Tregs also serve a beneficial role and act to minimize collateral tissue damage associated with the effector function of T cell mediated immunity (Suvas *et al.*, 2004). Such concepts were recently reviewed (Rouse and Suvas, 2004). It could be that vaccines that fail to activate the Treg response along with effector function might be more efficacious. Such issues are under investigation.

The memory T cell response to HSV: why is it unable to prevent reactivation?

A major issue for vaccinology is how to maximally induce and retain memory responses to pathogens so that re-infection, or in the case of HSV, reactivation, is kept at a subclinical level. Recently, a number of fundamental studies have established ground rules governing the generation and maintenance of $CD8^+$ T cell memory (Welsh *et al.*, 2004; Wherry and Ahmed, 2004). These studies were performed primarily using acute infection models (LCMV) and model antigen systems (ovalbumin). Far less is known about $CD4^+$ T cell memory, although it appears that this component is usually reduced in magnitude and less sustained than $CD8^+$ T cell memory. Moreover, alteration in the functional subsets of $CD4^+$ Th memory cells has been observed over time in some chronic viral infections. It appears that Th cells lose the ability to produce critical cytokines over the course of the disease, starting with a decline in IL-2 and then followed with reduced levels of TNF-α. Surprisingly, IFN-γ levels appear to remain constant (Seder and Ahmed, 2003).

The focus of memory studies has been on $CD8^+$ T cell memory, a component that may be most critical for clearance of HSV infection, at least in the natural host. In consequence, the application of a fundamental understanding of $CD8^+$ T cell memory will likely have application to vaccine design, especially the long sought for therapeutic vaccines. It is evident that short-term stimulation of naïve antigen-specific $CD8^+$ T cells is sufficient to trigger a program of proliferation and differentiation into effector cells that can react directly against antigen expressing cells (Kaech and Ahmed, 2001). These cells are generally both cytotoxic and able to generate a range of cytokines critical for antiviral control, especially IFN-γ. After the peak of the effector response, which occurs at approximately 1 week after infection, the vast majority of these effectors die of apoptosis. A small percentage of this population, about 5–10%, survive the contraction phase (days 10–15) and go on to form the long-lived memory pool. These cells produce elevated levels of antiapoptotic proteins, such as bcl-2, and express (or re-express) important receptors for survival, such as IL-7 receptor (Wherry and Ahmed, 2004). It has been noted that cells which are destined for memory also express the homodimeric CD8αα molecule on their surface, but the role for this surface protein is not clear at present (Madakamutil *et al.*, 2004). Some of these cells remain in the lymphoid tissues (so-called central memory cells), due to re-expression of CCR7. However, the effector memory subset loses CCR7 on their surface and migrates to peripheral sites, retaining the capacity to make cytokines and kill infected cells upon contact. These effector memory cells then patrol the peripheral tissues indefinitely.

The central memory cells remain more or less static in number over prolonged periods, but the small fraction that do die are replaced by a slow turnover of such cells. This so-called homeostatic division is not driven by antigen, but is believed to be a response to endogenous cytokines such as IL-15 and IL-7 (Schluns and Lefrancois, 2003). Antigen re-exposure results in a rapid recall of the central memory cells into proliferation and differentiation into effector cells. Such cells also tend to move into tissues, the usual site of new antigen introduction. The details of such events are thought to differ depending on the different types of infection and perhaps other factors. For instance, in an infection model of LCMV clone 13, which results in a persistent viral infection, functional exhaustion of at least some antigen specificities appears to occur. It appears that constant availability of antigen tends to prevent the development of a stable central effector memory pool (Wherry and Ahmed, 2004). A similar event may exist in HSV infection in the sensory ganglia, but the details of events in the memory response to HSV infection have yet to be worked out.

Curiously, whereas the initial expansion and effector response by $CD8^+$ T cells appears to occur in response to a variety of antigen preparations and can occur without signals from other lymphocytes, the long-term memory response appears to require signals from other helper T cells at the time of immune induction (Bevan, 2004). It was shown, for example, that when $CD4^+$ helper T cells are absent or not stimulated during the time of $CD8^+$ induction, a defective long-term memory response ensues (Janssen *et al.*, 2003; Shedlock and Shen, 2003; Sun and Bevan, 2003). In the case of murine responses to HSV, we observed that antigen formulations that engage naïve $CD8^+$ specific T cells, but not antigen specific $CD4^+$ T helper cells, resulted in normal acute responses of the $CD8^+$ T cells, but a poor memory response (Kumaraguru *et al.*, 2002). In addition to the size of the memory pool, the function of these cells was also diminished. Upon antigen re-exposure the recalled population demonstrated lower overall avidity in comparison to $CD8^+$ T cells that had received help during priming (Kumaraguru *et al.*, 2002). Currently, the actual mechanism by which helper T cells communicate signals resulting in optimal memory responses is not fully understood. One idea, however, is that signals from the helper T cells influence the maturation status and function of the DC that is presenting antigen. Most likely this "signal" is linked to CD40 on the DC binding to CD40L present on $CD4^+$ T cells. This interaction, termed licensing by the Matzinger group, results in a number of effects on DC, including enhanced costimulatory molecule expression (some of which may be critical for memory development) as well as the secretion of key cytokines such as IL-12 (Alpan *et al.*, 2004).

There seems to be general agreement that helper cell costimulation is generally required for optimal memory $CD8^+$ T cell responses, although with some agents the pathogens themselves promote a range of nonspecific signals, primarily via the strong stimulation of various TLRs that may substitute for helper effects. Less is known about the necessity for helper cell stimulation during other phases of the $CD8^+$ T cell response profile, such as the contraction, maintenance, and recall phases. Furthermore, it is important to understand if in conditions of defective $CD8^+$ T cell memory some form of rescue allowing normal function can be achieved. Thus, in HSV immunity, it could be that those subject to frequent or severe recurrences may have some deficiency in central or tissue residing effector memory $CD8^+$ T cells. In mouse systems with a defective $CD8^+$ T cell memory response we have achieved some level of rescue in various ways. Toka *et al.* co-administered IL-15 plasmid DNA at the time of priming with plasmid DNA encoding gB as well as during the memory phase. This treatment led to enhanced memory pool numbers as well as increased functional capacity of the responding CD8$^+$ T cells (Toka *et al.*, 2004a). Efforts might also be directed at increasing IL-7 receptor levels on either effector cells of existing memory cells in order to enhance or maintain adequate memory responses.

It will be important to develop and test vaccines in humans that express IL-15 or other such molecules that may serve to "top up" the efficacy of CD8$^+$ T cell memory responses against HSV. Other forms of activation may also be successful including immunization with persistent cross-reacting vehicles, use of TLR agonists, and possibly the use of cytokine activators.

The immune response to HSV and immunopathology

Effector T cell recognition of virus infected or viral antigen expressing cells can result in target cell destruction, but usually an inflammatory reaction is also elicited that causes collateral tissue damage. This outcome, which is usually set off by CD4$^+$ effector T cells recognizing their cognate antigens, is mediated in large part by inflammatory cytokines (Streilein *et al.*, 1997). Such inflammatory reactions of varying intensity invariably occur at tissue sites of virus infection. Although often painful and tissue damaging, they are a small price to pay for control of infection. Moreover, the tissue at most sites is usually repaired and thus functions normally. Alas, there are exceptions to this generality. At some sites inflammatory reactions can cause profound and often permanent damage to function. Sites such as the eye and brain represent examples. With respect to HSV, we understand best the consequences of chronic inflammatory reactions in the eye (Streilein *et al.*, 1997; Streilein, 2003). Thus, HSV infection commonly affects the eye, but typically only the epithelium is affected. The painful event resolves, especially if treated with antiviral drugs. However, in about 20% of cases the corneal stroma becomes involved and a chronic inflammatory reaction occurs whose management requires long-term treatment with powerful anti-inflammatory drugs. This stromal keratitis lesion (SK) is mainly a CD4$^+$ T cell-mediated immunopathological response to virus. In support of this, CD4$^+$ T cells with antiviral specificity can be recovered from corneal buttons long after primary infection (Verjans *et al.*, 1998).

Understanding of the details of SK pathogenesis comes mainly from experimental infections in mice models. In this model, SK results from primary ocular infection with HSV and, as in humans, lesions are immunopathological and orchestrated principally by CD4$^+$ T cells (Streilein *et al.*, 1997; Banerjee *et al.*, 2004). Multiple other events are involved that include additional cell types, cytokines, chemokines, angiogenic factors, and stress proteins. The pathogenesis of SK has been reviewed elsewhere (Streilein *et al.*, 1997; Deshpande *et al.*, 2004). Curiously, the virus which initiates the lesion is only briefly present in the eye and then usually not as infectious virions in the inflamed stromal tissues. Despite this, the lesion is usually progressive and certainly increases in severity for some time after virus has been controlled by the immune response. This observation, which parallels what usually occurs in human SK, supports the notion that the lesion may be an immunopathogical reaction against viral antigens, but subsequently becomes sustained by an autoinflammatory process (Deshpande *et al.*, 2004).

An additional site where HSV sets off an immunopathological reaction is in the sensory ganglia where HSV sets up latency. In the mouse, and perhaps in humans too, a long-term inflammatory reaction in the sensory ganglia occurs after virus is transmitted from peripheral sites of infection (Khanna *et al.*, 2003). This reaction is dominated by CD8$^+$ T cells, many of which, at least in the C57BL/6 mouse, are viral antigen-specific. Some suggest that these $CD8^+$ T cells somehow repress virus in neurons from replicating abundantly, which would result in the destruction of these irreplaceable cells (Khanna *et al.*, 2004). This process, which has been called viral purging, may be independent of true HSV latency (Khanna *et al.*, 2003). This fascinating topic was recently reviewed (Khanna *et al.*, 2004).

Finally, HSV is an occasional and usually tragic cause of encephalitis. Most would consider the human syndrome to largely represent a viral-driven inflammatory event with little or no role for immunopathology (Whitley, 2001). However, recent studies in a mouse model have indicated that HSV-induced encephalitis may represent largely an immunopathogical reaction reminiscent of the situation in the cornea (Kurt-Jones *et al.*, 2004). The brain, like the eye, cannot sustain long-term immunopathological reactions without compromising its function.

HSV vaccines

HSV 1 and 2 stand out as two of mankind's more prevalent and significant viral pathogens that so far have not been satisfactorily controlled by vaccines. It is not for lack of trying, and many candidate preparations have been put on the market place. Many, in fact, have appeared efficacious when tested by their developer, but none have stood up to long-term evaluation by disinterested parties. Currently, a vaccine is in the final phases of testing and may soon be available in the USA. This GSK vaccine is an adjuvanted protein vaccine that upon initial testing by its developers was shown to provide significant protection of seropositive women against HSV-2 infection from their seropositive partners (Stanberry *et al.*, 2002). The obverse situation provided no such protection. This GSK vaccine appears to be showing protection, even though the approach differs little from a previous vaccine discarded because of discouraging results. However, the adjuvant being used for the apparently successful GSK vaccine (alum and deacylated monophoshorylipid A) is different and may be able to better emphasize the $CD4^+$ type one response pattern considered important for protection (Stanberry *et al.*, 2002). Another possibility is that the vaccine may be conferring some protection to women since it induces a good protective antibody response in the genital tract. These issues require investigation, as do the reasons for the female-only protection.

We would, from our previous discussions, speculate that vaccines need to that additionally induce notable $CD8^+$ T cell responses to be optimally protective. Although not formally tested, the GSK vaccine is unlikely to provide epitopes for all MHC types, and the adjuvant may not represent an ideal one to induce $CD8^+$ T cell responses. The induction of optimal HSV specific CD8 T cell response will likely to require different technology than the GSK vaccine approach. For example, attenuated viral vaccines would be more likely to induce $CD8^+$ T cells. Such vaccines have been tested in the past and none have been adopted for widespread use. One candidate looked particularly appealing at least from an armchair perspective. This was the discontinuous replicating vaccine (DISC) that was a glycoprotein H deletion mutant that, except in complementary cells, would only undergo a single cycle of replication (McLean *et al.*, 1996). This DISC vaccine did well in animal testing and had at least a phase II trial in humans. Its discard by the industry probably means it is not sufficiently immunogenic or too complex to develop commercially.

HSV vaccinologists are left therefore in a state of frustration, since currently there is no ideal solution on the horizon. It means that more fundamental research on HSV immunobiology is required in model systems, but more particularly in humans themselves. We emphasize the latter, since animal models tend to be the bearers of good news, at least when it comes to prophylactic vaccine candidates. Were we rodents, in fact, we would already have a pharmacy full of valuable vaccines. Thus, killed vaccines, attenuated mutants, recombinant vaccines of numerous types, DNA vaccines, adjuvanted proteins, and peptides may all provide useful immunity, at least when used prophylactically. Alas we are primates, but fortunately with brains and imagination. The challenge remains to use our creative skills and overcome the vaccine road block. New clues from those optimists who think outside the box are encouraged to take up the challenge.

Conclusions

Additional work is clearly needed to further understand how HSV accommodates itself so well to the perils of both innate and adaptive immunity. We need to devise prophylactic vaccines which, even if inadequate to preclude infection, can dampen such to subclinical levels. These future vaccines will likely need to take advantage of our ever expanding knowledge of adjuvants and the mechanisms by which they shape and enhance defenses. We anticipate that successful prophylactic vaccines may be only just over the horizon. Therapeutic vaccines, which in the Western world may provide more rewards to the marketplace, could prove more problematic to develop. We shall be challenged to improve upon nature's vaccine since the virus itself, forever resident in those subject to recurrent disease, appears unable to confer adequate protective immunity. Modern medicine has solved many of mankind's problems. One hopes therapeutic anti-HSV vaccines become added to the list.

References

Adler, H., Beland, J.L., Del-Pan, N.C., Kobzik, L., Sobel, R.A., and Rimm, I.J. (1999). In the absence of T cells, natural killer cells protect from mortality due to HSV-1 encephalitis. J. Neuroimmunol. 93, 208–213.

Allan, R.S., Smith, C.M., Belz, G.T., van Lint, A.L., Wakim, L.M., Heath, W.R., and Carbone, F.R. (2003). Epidermal viral immunity induced by CD8alpha+ dendritic cells but not by Langerhans cells. Science *301*, 1925–1928.

Alpan, O., Bachelder, E., Isil, E., Arnheiter, H., and Matzinger, P. (2004). "Educated" dendritic cells act as messengers from memory to naive T helper cells. Nat. Immunol. 5, 615–622.

Arase, H., Mocarski, E.S., Campbell, A.E., Hill, A.B., and Lanier, L.L. (2002). Direct recognition of cytomegalovirus by activating and inhibitory NK cell receptors. Science *296*, 1323–1326.

Ashkar, A.A., and Rosenthal, K.L. (2003). Interleukin-15 and natural killer and NKT cells play a critical role in innate protection against genital herpes simplex virus type 2 infection. J. Virol. *77*, 10168–10171.

Ashley, R.L., Dalessio, J., Burchett, S., Brown, Z., Berry, S., Mohan, K., and Corey, L. (1992). Herpes simplex virus-2 (HSV-2) type-specific antibody correlates of protection in infants exposed to HSV-2 at birth. J. Clin. Invest. *90*, 511–514.

Banerjee, K., Biswas, P.S., and Rouse, B.T. (2004). Elucidating the protective and pathologic T cell species in the virus-induced corneal immunoinflammatory condition herpetic stromal keratitis. J. Leukoc. Biol.

Belz, G.T., Smith, C.M., Eichner, D., Shortman, K., Karupiah, G., Carbone, F.R., and Heath, W.R. (2004). Cutting edge: conventional CD8 alpha+ dendritic cells are generally involved in priming CTL immunity to viruses. J. Immunol. *172*, 1996–2000.

Bernstein, D.I. (2001). Potential for immunotherapy in the treatment of herpesvirus infections. Herpes *8*, 8–11.

Bernstein, D.I., Miller, R.L., and Harrison, C.J. (1993). Effects of therapy with an immunomodulator (imiquimod, R-837) alone and with acyclovir on genital HSV-2 infection in guinea-pigs when begun after lesion development. Antiviral Res. *20*, 45–55.

Bevan, M.J. (2004). Helping the CD8(+) T-cell response. Nat. Rev. Immunol. *4*, 595–602.

Biron, C.A. (1998). Role of early cytokines, including alpha and beta interferons (IFN-alpha/beta), in innate and adaptive immune responses to viral infections. Semin. Immunol. *10*, 383–390.

Biron, C.A. (1999). Initial and innate responses to viral infections--pattern setting in immunity or disease. Curr. Opin. Microbiol. *2*, 374–381.

Biron, C.A., and Brossay, L. (2001). NK cells and NKT cells in innate defense against viral infections. Curr. Opin. Immunol. *13*, 458–464.

Biron, C.A., Byron, K.S., and Sullivan, J.L. (1989). Severe herpesvirus infections in an adolescent without natural killer cells. N. Engl. J. Med. *320*, 1731–1735.

Bloomfield, S.E., and Lopez, C. (1980). Herpes infections in the immunosuppressed host. Ophthalmology *87*, 1226–1235.

Cantin, E., Tanamachi, B., Openshaw, H., Mann, J., and Clarke, K. (1999). Gamma interferon (IFN-gamma) receptor null-mutant mice are more susceptible to herpes simplex virus type 1 infection than IFN-gamma ligand null-mutant mice. J. Virol. *73*, 5196–5200.

Compton, T., Kurt-Jones, E.A., Boehme, K.W., Belko, J., Latz, E., Golenbock, D.T., and Finberg, R.W. (2003). Human cytomegalovirus activates inflammatory cytokine responses via CD14 and Toll-like receptor 2. J. Virol. *77*, 4588–4596.

Corey, L. (2001). Immunobiology of herpes simplex virus (HSV) infections. Herpes *8 Suppl 1*, 1A.

Cunningham, A.L., and Mikloska, Z. (2001). The Holy Grail: immune control of human herpes simplex virus infection and disease. Herpes *8 Suppl 1*, 6A-10A.

Deshpande, S.P., Zheng, M., Daheshia, M., and Rouse, B.T. (2000). Pathogenesis of herpes simplex virus-induced ocular immunoinflammatory lesions in B-cell-deficient mice. J. Virol. *74*, 3517–3524.

Deshpande, S., Banerjee, K., Biswas, P.S., and Rouse, B.T. (2004). Herpetic eye disease: immunopathogenesis and therapeutic measures. Expert Rev. Mol. Med. *2004*, 1–14.

Eloranta, M.L., and Alm, G.V. (1999). Splenic marginal metallophilic macrophages and marginal zone macrophages are the major interferon-alpha/beta producers in mice upon intravenous challenge with herpes simplex virus. Scand. J. Immunol. *49*, 391–394.

Fujimiya, Y., Rouse, B.T., and Babiuk, L.A. (1978). Human neutrophil-mediated destruction of antibody sensitized herpes simplex virus type I infected cells. Can. J. Microbiol. *24*, 182–186.

Gallichan, W.S., Woolstencroft, R.N., Guarasci, T., McCluskie, M.J., Davis, H.L., and Rosenthal, K.L. (2001). Intranasal immunization with CpG oligodeoxynucleotides as an adjuvant dramatically increases IgA and protection against herpes simplex virus-2 in the genital tract. J. Immunol. *166*, 3451–3457.

Gierynska, M., Kumaraguru, U., Eo, S.K., Lee, S., Krieg, A., and Rouse, B.T. (2002). Induction of CD8 T-cell-specific systemic and mucosal immunity against herpes simplex virus with CpG-peptide complexes. J. Virol. *76*, 6568–6576.

Grubor-Bauk, B., Simmons, A., Mayrhofer, G., and Speck, P.G. (2003). Impaired clearance of herpes simplex virus type 1 from mice lacking CD1d or NKT cells expressing the semivariant V alpha 14-J alpha 281 TCR. J. Immunol. *170*, 1430–1434.

Guidotti, L.G., Rochford, R., Chung, J., Shapiro, M., Purcell, R., and Chisari, F.V. (1999). Viral clearance without destruction of infected cells during acute HBV infection. Science *284*, 825–829.

Heath, W.R., Belz, G.T., Behrens, G.M., Smith, C.M., Forehan, S.P., Parish, I.A., Davey, G.M., Wilson, N.S., Carbone, F.R., and Villadangos, J.A. (2004). Cross-presentation, dendritic cell subsets, and the generation of immunity to cellular antigens. Immunol. Rev. *199*, 9–26.

Hengge, U.R., Benninghoff, B., Ruzicka, T., and Goos, M. (2001). Topical immunomodulators--progress towards treating inflammation, infection, and cancer. Lancet Infect. Dis. *1*, 189–198.

Hochrein, H., Schlatter, B., O'Keeffe, M., Wagner, C., Schmitz, F., Schiemann, M., Bauer, S., Suter, M., and Wagner, H. (2004). Herpes simplex virus type-1 induces IFN-alpha production via Toll-like receptor 9-dependent and -independent pathways. Proc. Natl. Acad. Sci. USA *101*, 11416–11421.

Holmgren, J., Czerkinsky, C., Eriksson, K., and Mharandi, A. (2003). Mucosal immunisation and adjuvants: a brief overview of recent advances and challenges. Vaccine *21 Suppl 2*, S89–95.

Iwasaki, A. (2003). The role of dendritic cells in immune responses against vaginal infection by herpes simplex virus type 2. Microbes Infect. *5*, 1221–1230.

Janssen, E.M., Lemmens, E.E., Wolfe, T., Christen, U., von Herrath, M.G., and Schoenberger, S.P. (2003). CD4+ T cells are required for secondary expansion and memory in CD8+ T lymphocytes. Nature *421*, 852–856.

Jennings, S.R., Rice, P.L., Kloszewski, E.D., Anderson, R.W., Thompson, D.L., and Tevethia, S.S. (1985). Effect of herpes simplex virus types 1 and 2 on surface expression of class I major histocompatibility complex antigens on infected cells. J. Virol. *56*, 757–766.

Jennings, S.R., Bonneau, R.H., Smith, P.M., Wolcott, R.M., and Chervenak, R. (1991). CD4-positive T lymphocytes are required for the generation of the primary but not the secondary CD8-positive cytolytic T lymphocyte response to herpes simplex virus in C57BL/6 mice. Cell Immunol. *133*, 234–252.

Jones, C.A., and Cunningham, A.L. (2004). Vaccination strategies to prevent genital herpes and neonatal herpes simplex virus (HSV) disease. Herpes *11*, 12–17.

Kadowaki, N., and Liu, Y.J. (2002). Natural type I interferon-producing cells as a link between innate and adaptive immunity. Hum. Immunol. *63*, 1126–1132.

Kaech, S.M., and Ahmed, R. (2001). Memory CD8+ T cell differentiation: initial antigen encounter triggers a developmental program in naive cells. Nat. Immunol. *2*, 415–422.

Kanangat, S., Thomas, J., Gangappa, S., Babu, J.S., and Rouse, B.T. (1996). Herpes simplex virus type 1-mediated up-regulation of IL-12 (p40) mRNA expression. Implications in immunopathogenesis and protection. J. Immunol. *156*, 1110–1116.

Khanna, K.M., Bonneau, R.H., Kinchington, P.R., and Hendricks, R.L. (2003). Herpes simplex virus-specific memory CD8+ T cells are selectively activated and retained in latently infected sensory ganglia. Immunity *18*, 593–603.

Khanna, K.M., Lepisto, A.J., Decman, V., and Hendricks, R.L. (2004). Immune control of herpes simplex virus during latency. Curr. Opin. Immunol. *16*, 463–469.

Kobelt, D., Lechmann, M., and Steinkasserer, A. (2003). The interaction between dendritic cells and herpes simplex virus-1. Curr. Topics Microbiol. Immunol. *276*, 145–161.

Kodukula, P., Liu, T., Rooijen, N.V., Jager, M.J., and Hendricks, R.L. (1999). Macrophage control of herpes simplex virus type 1 replication in the peripheral nervous system. J. Immunol. *162*, 2895–2905.

Koelle, D.M., and Corey, L. (2003). Recent progress in herpes simplex virus immunobiology and vaccine research. Clin. Microbiol. Rev. *16*, 96–113.

Kopp, E., and Medzhitov, R. (2003). Recognition of microbial infection by Toll-like receptors. Curr. Opin. Immunol. *15*, 396–401.

Kruse, M., Rosorius, O., Kratzer, F., Stelz, G., Kuhnt, C., Schuler, G., Hauber, J., and Steinkasserer, A. (2000). Mature dendritic cells infected with herpes simplex virus type 1 exhibit inhibited T-cell stimulatory capacity. J. Virol. *74*, 7127–7136.

Kuklin, N.A., Daheshia, M., Chun, S., and Rouse, B.T. (1998). Role of mucosal immunity in herpes simplex virus infection. J. Immunol. *160*, 5998–6003.

Kumaraguru, U., Gierynska, M., Norman, S., Bruce, B.D., and Rouse, B.T. (2002). Immunization with chaperone-peptide complex induces low-avidity cytotoxic T lymphocytes providing transient protection against herpes simplex virus infection. J. Virol. *76*, 136–141.

Kurt-Jones, E.A., Chan, M., Zhou, S., Wang, J., Reed, G., Bronson, R., Arnold, M.M., Knipe, D.M., and Finberg, R.W. (2004). Herpes simplex virus 1 interaction with Toll-like receptor 2 contributes to lethal encephalitis. Proc. Natl. Acad. Sci. USA *101*, 1315–1320.

Leib, D.A., Harrison, T.E., Laslo, K.M., Machalek, M.A., Moorman, N.J., and Virgin, H.W. (1999). Interferons regulate the phenotype of wild-type and mutant herpes simplex viruses *in vivo*. J. Exp. Med. *189*, 663–672.

Lin, X., Lubinski, J.M., and Friedman, H.M. (2004). Immunization strategies to block the herpes simplex virus type 1 immunoglobulin G Fc receptor. J. Virol. *78*, 2562–2571.

Liu, T., Khanna, K.M., Carriere, B.N., and Hendricks, R.L. (2001). Gamma interferon can prevent herpes simplex virus type 1 reactivation from latency in sensory neurons. J. Virol. *75*, 11178–11184.

Lubinski, J.M., Jiang, M., Hook, L., Chang, Y., Sarver, C., Mastellos, D., Lambris, J.D., Cohen, G.H., Eisenberg, R.J., and Friedman, H.M. (2002). Herpes simplex virus type 1 evades the effects of antibody and complement *in vivo*. J. Virol. *76*, 9232–9241.

Lund, J., Sato, A., Akira, S., Medzhitov, R., and Iwasaki, A. (2003). Toll-like receptor 9-mediated recognition of Herpes simplex virus-2 by plasmacytoid dendritic cells. J. Exp. Med. *198*, 513–520.

Lundberg, P., Welander, P., Openshaw, H., Nalbandian, C., Edwards, C., Moldawer, L., and Cantin, E. (2003). A locus on mouse chromosome 6 that determines resistance to herpes simplex virus also influences reactivation, while an unlinked locus augments resistance of female mice. J. Virol. *77*, 11661–11673.

MacLean, A., Wei, X.Q., Huang, F.P., Al-Alem, U.A., Chan, W.L., and Liew, F.Y. (1998). Mice lacking inducible nitric-oxide synthase are more susceptible to herpes simplex virus infection despite enhanced Th1 cell responses. J. Gen. Virol. *79 (Pt 4)*, 825–830.

Madakamutil, L.T., Christen, U., Lena, C.J., Wang-Zhu, Y., Attinger, A., Sundarrajan, M., Ellmeier, W., von Herrath, M.G., Jensen, P., Littman, D.R., and Cheroutre, H. (2004). CD8alphaalpha-mediated survival and differentiation of CD8 memory T cell precursors. Science *304*, 590–593.

Manickan, E., Rouse, R.J., Yu, Z., Wire, W.S., and Rouse, B.T. (1995). Genetic immunization against herpes simplex virus. Protection is mediated by CD4+ T lymphocytes. J. Immunol. *155*, 259–265.

McLean, C.S., Ni Challanain, D., Duncan, I., Boursnell, M.E., Jennings, R., and Inglis, S.C. (1996). Induction of a protective immune response by mucosal vaccination with a DISC HSV-1 vaccine. Vaccine *14*, 987–992.

Milligan, G.N. (1999). Neutrophils aid in protection of the vaginal mucosae of immune mice against challenge with herpes simplex virus type 2. J. Virol. *73*, 6380–6386.

Milligan, G.N., Bourne, N., and Dudley, K.L. (2001). Role of polymorphonuclear leukocytes in resolution of HSV-2 infection of the mouse vagina. J. Reprod. Immunol. *49*, 49–65.

Moser, J.M., Byers, A.M., and Lukacher, A.E. (2002). NK cell receptors in antiviral immunity. Curr. Opin. Immunol. *14*, 509–516.

Newell, C.K., Sendele, D., and Rouse, B.T. (1989). Effects of CD4+ and CD8+ T-lymphocyte depletion on the induction and expression of herpes simplex stromal keratitis. Reg. Immunol. *2*, 366–369.

Ochsenbein, A.F., and Zinkernagel, R.M. (2000). Natural antibodies and complement link innate and acquired immunity. Immunol. Today *21*, 624–630.

Ochsenbein, A.F., Fehr, T., Lutz, C., Suter, M., Brombacher, F., Hengartner, H., and Zinkernagel, R.M. (1999). Control of early viral and bacterial distribution and disease by natural antibodies. Science *286*, 2156–2159.

Parr, M.B., and Parr, E.L. (2003). Vaginal immunity in the HSV-2 mouse model. Int. Rev. Immunol. *22*, 43–63.

Pereira, R.A., and Simmons, A. (1999). Cell surface expression of H2 antigens on primary sensory neurons in response to acute but not latent herpes simplex virus infection *in vivo*. J. Virol. *73*, 6484–6489.

Pereira, R.A., Scalzo, A., and Simmons, A. (2001). Cutting edge: a NK complex-linked locus governs acute versus latent herpes simplex virus infection of neurons. J. Immunol. *166*, 5869–5873.

Pollara, G., Speidel, K., Samady, L., Rajpopat, M., McGrath, Y., Ledermann, J., Coffin, R.S., Katz, D.R., and Chain, B. (2003). Herpes simplex virus infection of dendritic cells: balance among activation, inhibition, and immunity. J. Infect. Dis. *187*, 165–178.

Posavad, C.M., Koelle, D.M., Shaughnessy, M.F., and Corey, L. (1997). Severe genital herpes infections in HIV-infected individuals with impaired herpes simplex virus-specific CD8+ cytotoxic T lymphocyte responses. Proc. Natl. Acad. Sci. USA *94*, 10289–10294.

Raulet, D.H. (2004). Interplay of natural killer cells and their receptors with the adaptive immune response. Nat. Immunol. *5*, 996–1002.

Reiss, C.S., and Komatsu, T. (1998). Does nitric oxide play a critical role in viral infections? J. Virol. *72*, 4547–4551.

Rouse, B.T., and Suvas, S. (2004). Regulatory cells and infectious agents: detentes cordiale and contraire. J. Immunol. *173*, 2211–2215.

Salio, M., Cella, M., Suter, M., and Lanzavecchia, A. (1999). Inhibition of dendritic cell maturation by herpes simplex virus. Eur J. Immunol. *29*, 3245–3253.

Sato, A., and Iwasaki, A. (2004). Induction of antiviral immunity requires Toll-like receptor signaling in both stromal and dendritic cell compartments. Proc. Natl. Acad. Sci. USA.

Schluns, K.S., and Lefrancois, L. (2003). Cytokine control of memory T-cell development and survival. Nat. Rev. Immunol. *3*, 269–279.

Seder, R.A., and Ahmed, R. (2003). Similarities and differences in CD4+ and CD8+ effector and memory T cell generation. Nat. Immunol. *4*, 835–842.

Shedlock, D.J., and Shen, H. (2003). Requirement for CD4 T cell help in generating functional CD8 T cell memory. Science *300*, 337–339.

Simmons, A., and Nash, A.A. (1984). Zosteriform spread of herpes simplex virus as a model of recrudescence and its use to investigate the role of immune cells in prevention of recurrent disease. J. Virol. *52*, 816–821.

Simmons, A., and Tscharke, D.C. (1992). Anti-CD8 impairs clearance of herpes simplex virus from the nervous system: implications for the fate of virally infected neurons. J. Exp. Med. *175*, 1337–1344.

Sivori, S., Falco, M., Della Chiesa, M., Carlomagno, S., Vitale, M., Moretta, L., and Moretta, A. (2004). CpG and double-stranded RNA trigger human NK cells by Toll-like receptors: induction of cytokine release and cytotoxicity against tumors and dendritic cells. Proc. Natl. Acad. Sci. USA *101*, 10116–10121.

Sloan, D.D., Zahariadis, G., Posavad, C.M., Pate, N.T., Kussick, S.J., and Jerome, K.R. (2003). CTL are inactivated by herpes simplex virus-infected cells expressing a viral protein kinase. J. Immunol. *171*, 6733–6741.

Spruance, S.L., Tyring, S.K., Smith, M.H., and Meng, T.C. (2001). Application of a topical immune response modifier, resiquimod gel, to modify the recurrence rate of recurrent genital herpes: a pilot study. J. Infect. Dis. *184*, 196–200.

Stanberry, L.R., Spruance, S.L., Cunningham, A.L., Bernstein, D.I., Mindel, A., Sacks, S., Tyring, S., Aoki, F.Y., Slaoui, M., Denis, M. *et al.* (2002). Glycoprotein-D-adjuvant vaccine to prevent genital herpes. N. Engl. J. Med. *347*, 1652–1661.

Steinman, R.M., Granelli-Piperno, A., Pope, M., Trumpfheller, C., Ignatius, R., Arrode, G., Racz, P., and Tenner-Racz, K. (2003). The interaction of immunodeficiency viruses with dendritic cells. Curr. Topics Microbiol. Immunol. *276*, 1–30.

Streilein, J.W. (2003). Ocular immune privilege: the eye takes a dim but practical view of immunity and inflammation. J. Leukoc. Biol. *74*, 179–185.

Streilein, J.W., Dana, M.R., and Ksander, B.R. (1997). Immunity causing blindness: five different paths to herpes stromal keratitis. Immunol. Today *18*, 443–449.

Sun, J.C., and Bevan, M.J. (2003). Defective CD8 T cell memory following acute infection without CD4 T cell help. Science *300*, 339–342.

Suvas, S., Kumaraguru, U., Pack, C.D., Lee, S., and Rouse, B.T. (2003). CD4+CD25+ T cells regulate virus-specific primary and memory CD8+ T cell responses. J. Exp. Med. *198*, 889–901.

Suvas, S., Azkur, A.K., Kim, B.S., Kumaraguru, U., and Rouse, B.T. (2004). CD4+CD25+ regulatory T cells control the severity of viral immunoinflammatory lesions. J. Immunol. *172*, 4123–4132.

Thomas, J., Kanangat, S., and Rouse, B.T. (1998). Herpes simplex virus replication-induced expression of chemokines and proinflammatory cytokines in the eye: implications in herpetic stromal keratitis. J. Interferon Cytokine Res. *18*, 681–690.

Toka, F.N., Pack, C.D., and Rouse, B.T. (2004a). Molecular adjuvants for mucosal immunity. Immunol. Rev. *199*, 100–112.

Toka, F.N., Suvas, S., and Rouse, B.T. (2004b). CD4+ CD25+ T cells regulate vaccine-generated primary and memory CD8+ T-cell responses against herpes simplex virus type 1. J. Virol. *78*, 13082–13089.

Trinchieri, G., Pflanz, S., and Kastelein, R.A. (2003). The IL-12 family of heterodimeric cytokines: new players in the regulation of T cell responses. Immunity *19*, 641–644.

Tumpey, T.M., Chen, S.H., Oakes, J.E., and Lausch, R.N. (1996). Neutrophil-mediated suppression of virus replication after herpes simplex virus type 1 infection of the murine cornea. J. Virol. *70*, 898–904.

Verjans, G.M., Remeijer, L., van Binnendijk, R.S., Cornelissen, J.G., Volker-Dieben, H.J., Baarsma, S.G., and Osterhaus, A.D. (1998). Identification and characterization of herpes simplex virus-specific CD4+ T cells in corneas of herpetic stromal keratitis patients. J. Infect. Dis. *177*, 484–488.

Verschoor, A., Brockman, M.A., Knipe, D.M., and Carroll, M.C. (2001). Cutting edge: myeloid complement C3 enhances the humoral response to peripheral viral infection. J. Immunol. *167*, 2446–2451.

Vollstedt, S., Franchini, M., Alber, G., Ackermann, M., and Suter, M. (2001). Interleukin-12- and gamma interferon-dependent innate immunity are essential and sufficient for long-term survival of passively immunized mice infected with herpes simplex virus type 1. J. Virol. *75*, 9596–9600.

Welsh, R.M., Selin, L.K., and Szomolanyi-Tsuda, E. (2004). Immunological memory to viral infections. Annu. Rev. Immunol. *22*, 711–743.

Wherry, E.J., and Ahmed, R. (2004). Memory CD8 T-cell differentiation during viral infection. J. Virol. *78*, 5535–5545.

Whitley, R.J. (2001). Herpes simplex viruses. In Fields Virology, D. Knipe, ed. (Philadelphia, Lippincott Williams and Wilkins), pp. 2461–2509.

Yu, Z., Manickan, E., and Rouse, B.T. (1996). Role of interferon-gamma in immunity to herpes simplex virus. J. Leukoc. Biol. *60*, 528–532.

Zhao, X., Deak, E., Soderberg, K., Linehan, M., Spezzano, D., Zhu, J., Knipe, D.M., and Iwasaki, A. (2003). Vaginal submucosal dendritic cells, but not Langerhans cells, induce protective Th1 responses to herpes simplex virus-2. J. Exp. Med. *197*, 153–162.

Anti-alpha-herpesvirus Drugs

19

Donald M. Coen

Abstract

The human alpha-herpesviruses, herpes simplex virus 1 and 2 and varicella zoster virus, cause significant disease in patients. This has led to the development of drugs active against these viruses, in particular, acyclovir, which is one of the most successful antiviral agents. Acyclovir and its relatives act via a two step mechanism, entailing selective phosphorylation by viral thymidine kinase and selective inhibition of viral DNA polymerase. Molecular genetic, biochemical, and crystallographic studies have illuminated the details of the mechanisms of these drugs, as well as how the virus can mutate to evade the drug, yet retain pathogenicity. Similar studies of how alpha-herpesviruses replicate in cells have identified interesting new drugs and drug targets. However, much more work will be needed to pinpoint molecular targets and specific strategies that could cure latent alpha-herpesvirus infections.

Introduction

The human alpha-herpesviruses cause diseases that are not usually life threatening. However, ocular and genital herpes, caused by herpes simplex virus (HSV) 1 and 2, and chickenpox and shingles, caused by varicella zoster virus (VZV), are serious diseases that affect large numbers of patients. In patients with reduced immunity, which include patients with AIDS, transplant patients, and newborns, infections by these viruses can be devastating and even fatal.

The medical need resulting from alpha-herpesviruses, and the recurrent nature of some of these diseases in the immunocompetent have led to the development by drug companies of antiviral drugs active against HSV and VZV. These agents have also provided excellent laboratory tools for investigating viral biology and biochemistry. One of these drugs, acyclovir (ACV), which was first approved for clinical use in the early 1980s, has been remarkably successful. Since that time, we have gained an excellent understanding of how it and other approved drugs act against HSV and VZV, and how the virus can mutate to become drug-resistant. This chapter will review these mechanisms. The problem of resistance has been one of several spurs to efforts to discover new anti-alpha-herpesvirus drugs, often starting with basic research on virus replication such as that covered by other chapters in this volume. This chapter will review some of the efforts in this direction. Finally, as discussed in Chapters 16 and 17, HSV and VZV form latent infections in which, viral genomes reside inside cells of the peripheral nervous system while expressing abundantly at most a few genes, thus avoiding immune surveillance. The viruses can then reactivate long after primary infection and cause disease. No available antiviral drug attacks viruses during latency; they act only on actively replicating virus. The chapter will close with a brief discussion of the prospects for therapeutic strategies that target latent virus.

The replication cycle of alpha-herpesviruses

An understanding of the mechanisms of antiherpesvirus drugs requires an understanding of the virus replication cycle (see (Cohen and Straus, 2001; Roizman and Knipe, 2001) and the other chapters of this volume for more details). The HSV cycle and several of the proteins involved are described below. The VZV replication cycle and VZV proteins, which are similar to those of HSV, will not be described separately for the sake of brevity (although by no means are these viruses identical).

Like all herpesviruses, HSV is a large, enveloped virus that contains double stranded DNA encoding scores of proteins. At the start of infection, the virus attaches to the host cell. Attachment is mediated by viral envelope proteins that bind specifically to particular host membrane components. For HSV, glycoproteins gB, gC, and gD have been implicated in such binding. Next, the virion enters the cell (entry). This process is also mediated by viral glycoproteins, in this case, gB, gD, gH, and gL. The result is an de-enveloped virion in the cytoplasm. This particle must then be transported to the nucleus, a process that requires the cell microtubule machinery, and, the viral DNA released into the nucleus (uncoating). The details of these processes from attachment to uncoating are poorly understood.

Once in the nucleus, the viral DNA becomes available for transcription by host RNA polymerase II into mRNAs that then undergo transport to the cytoplasm and translation by cellular ribosomes. These steps of gene expression occur in three phases—immediate early (the first phase), early (which requires the immediate early protein ICP4), and late (which requires the immediate early proteins ICP4 and ICP27 and viral DNA replication for maximal expression). Many of the early proteins are involved in the step of the cycle that follows gene expression, viral DNA replication. Some of these early proteins, such as thymidine kinase and the two subunit ribonucleotide reductase, are involved indirectly by increasing the pools of deoxynucleoside triphosphates, while others, such as the two subunit DNA polymerase and the three subunit helicase-primase, participate directly in DNA synthesis.

Following DNA replication in the nucleus, the completed viral genomes are packaged by preformed capsids to yield nucleocapsids. This assembly process entails several proteins that comprise the DNA encapsidation/cleavage machinery. Assembly is accompanied by a major change in capsid structure (maturation), which entails the action of a viral protease. This is followed by a very complicated process of egress. According to the currently favored model, the nucleocapsid first buds through the inner nuclear membrane, acquiring an envelope. It then enters the cytoplasm by fusing with the outer nuclear membrane, thereby losing its initial envelope. Subsequently, it becomes re-enveloped by budding into intracellular membranous compartments. It then proceeds through the cytoplasm and out of the cell. The final result is production of many new enveloped virions for each infecting virion.

It should be evident that specific host and/or viral proteins are involved in each of these steps. In principle, differences between viral and host proteins at any of the stages of the replication cycle can be targeted for antiviral therapy. The viral proteins that have thus far served as the best targets for antiviral drugs are the thymidine kinase (TK) and DNA polymerase enzymes involved in DNA replication. However, as will be reviewed below, many other steps in the viral life cycle are being evaluated as targets for antiviral agents.

Approved alpha-herpesvirus agents

Drugs active against alpha-herpesviruses were among the first antiviral agents approved for clinical use in the U.S. The first of these approved drugs were nucleoside analogs such as vidarabine (arabinosyadenine) and trifluridine (trifluorothymidine), but their use was eventually supplanted by a less natural-looking nucleoside analog, the highly successful drug, acyclovir. In this chapter, we will not discuss the older analogs further, as they have greater toxicity than acyclovir, and so are not widely used. We will instead focus on the drugs approved for clinical use in the US. These are acyclovir and its prodrug, valacyclovir; and the related analog, penciclovir; and its prodrug, famciclovir, and foscarnet (Fig. 19.1). For the sake of brevity, we will refer mainly to the HSV literature, as the mechanisms of action of and resistance to these drugs are similar for HSV and VZV.

Acyclovir

Acyclovir is the paradigmatic antiviral nucleoside analog. It consists of a guanine base attached to an acyclic sugar-like molecule (Fig. 19.1). It was synthesized at Burroughs Wellcome as part of a program to discover adenosine deaminase inhibitors, and its antiviral activity was discovered, in effect, serendipitously. Biochemical studies of the mechanism of acyclovir against HSV and VZV were undertaken by Gertrude Elion and colleagues (Elion *et al.*, 1977) and the results from these investigations were subsequently confirmed

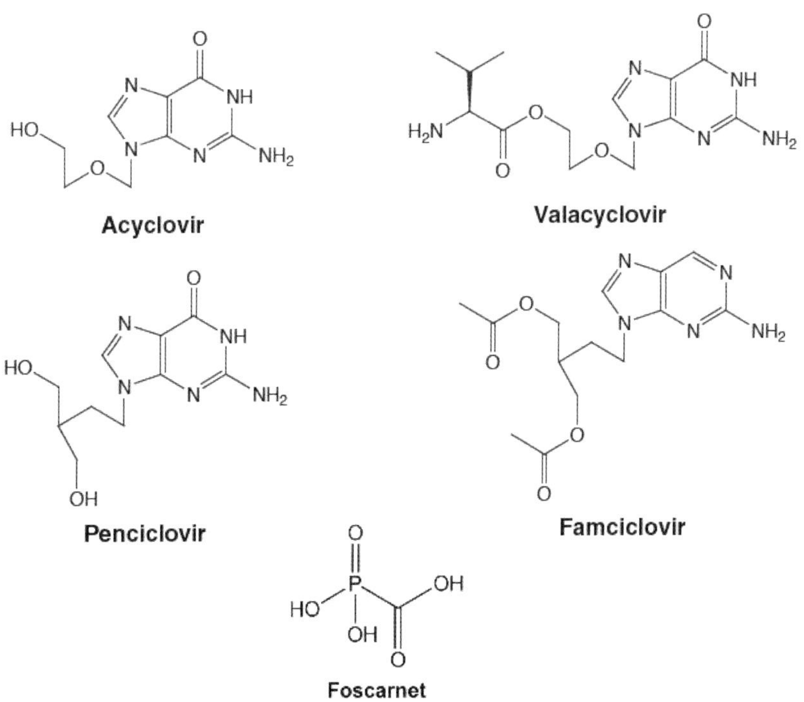

Figure 19.1 Structures of drugs approved in the US for treatment of HSV and VZV.

by studies of drug-resistant mutants. These latter genetic studies reveal drug mechanism because mapping a drug-resistance mutation to a gene identifies that gene's product as being essential for at least some portion of selective drug action. Indeed, drug-resistance mutations almost always identify the targets of antiviral drugs.

The mechanism of acyclovir action (Fig. 19.2) begins with the TK encoded by HSV and VZV. These enzymes, which are not essential for viral replication in cell culture, are able to phosphorylate not only thymidine (dT) but other pyrimidines such as deoxyuridine, deoxycytidine (dC), and a variety of nucleoside analogs, including some—like acyclovir—which contain a purine base. Initially, the strikingly broad substrate specificity of these TKs was mysterious. Subsequently, sequence alignments (Harrison *et al.*, 1991) revealed similarities between the viral TKs and human dC kinase, an enzyme that can phosphorylate deoxyguanosine and deoxyadenosine (Durham and Ives, 1970; Sarup and Fridland, 1987; Bohman and Eriksson, 1988; Datta *et al.*, 1989) and with nucleoside monophosphate kinases (NMPKs) such as adenylate kinase. Subsequent crystallographic studies showed that the HSV TK exhibits high structural homology with NMPKs (Brown *et al.*, 1995; Wild *et al.*, 1995). Crystal structures of HSV TK in complex with dT, acyclovir, and other ligands (Brown *et al.*, 1995; Champness *et al.*, 1998; Wild *et al.*, 1997) show that the guanine moiety of acyclovir interacts with same residues as does dT, albeit with differences in water-mediated binding and bond angles. Interestingly, single amino acid changes can convert HSV TK into an enzyme that can mainly phosphorylate guanosine analogs (Balzarini *et al.*, 2002). However, the function of HSV TK appears to be generation of thymine based nucleotides, as human TK and human thymidylate kinase, which do not detectably phosphorylate guanine-based nucleosides, can substitute for HSV TK for replication in and reactivation from sensory ganglia following peripheral inoculation of mice (Chen *et al.*, 1998). Moreover, human dC kinase, which can phosphorylate guanine-based nucleosides, does not substitute for HSV TK.

Acyclovir is not a particularly good substrate for HSV or VZV TK. This appears to be due, at least in part, to the paucity of bonding to the acyclic moiety; other guanine analogs with larger acyclic moieties (e.g. penciclovir, Fig. 19.1) are much better substrates and form more hydrogen bonds with the enzyme (Champness *et al.*, 1998). Despite this, no mammalian enzyme phosphorylates acyclovir nearly as efficiently as the HSV and VZV TKs. Accordingly, HSV- and VZV-infected cells contain much more (30- to 100-fold) phosphorylated acyclovir than do uninfected cells; this finding accounts for much of acyclovir's antiviral selectivity. This biochemical evidence for selectivity is confirmed by genetic studies showing that mutations that ablate (TK-negative) or reduce (TK-partial) TK activity or alter TK so that it fails to phosphorylate acyclovir, but continues to phosphorylate dT (TK-altered) confer resistance to acyclovir (Fig. 19.2; reviewed in Coen, 1994).

Phosphorylation of acyclovir produces the compound acyclovir-monophosphate. This compound is converted to acyclovir-diphosphate and then to acyclovir-triphosphate by cellular enzymes, with biochemical studies suggesting that GMP kinase and any of several kinases, respectively, carry out these conversions (Miller and Miller, 1980; Miller and Miler, 1982). Acyclovir-triphosphate then inhibits HSV and VZV DNA polymerases. As these polymerases are essential for viral replication, their inhibition results in antiviral activity. The mapping of acyclovir-resistance mutations in the HSV and VZV *pol* genes

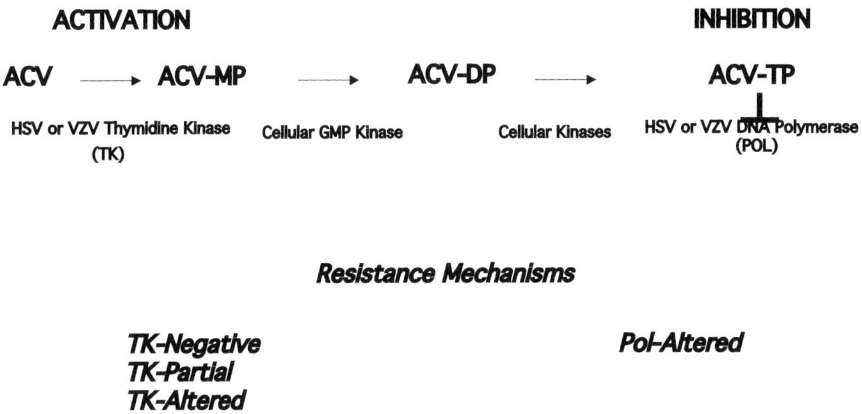

Figure 19.2 Mechanisms of acyclovir (ACV) action and resistance. MP, monophosphate; DP, diphosphate; TP, triphosphate.

(Fig. 19.2; all point lesions that result in altered polymerase) demonstrates that these enzymes mediate selectivity of the drug.

Inhibition of HSV DNA polymerase *in vitro* is a three-step process (Reardon and Spector, 1989; Fig. 19.3). In the first step, acyclovir-triphosphate competitively inhibits dGTP incorporation (high concentrations of dGTP can reverse inhibition at this early step). Next, acyclovir-triphosphate acts as a substrate and is incorporated into the growing DNA chain opposite a C residue. Finally, the polymerase translocates to the next position on the template, but cannot add a new deoxynucleoside triphosphate because there is no 3′-hydroxyl on acyclovir-triphosphate; hence acyclovir-triphosphate is an obligate chain terminator. Provided that the next deoxynucleoside triphosphate is present, the viral polymerase freezes at this final step in a "dead-end complex," leading to apparent inactivation of the enzyme. There is biochemical evidence for selectivity at each of these steps (Reardon, 1989; Martin *et al.*, 1994; Ilsley *et al.*, 1995). For example, acyclovir-triphosphate appears to bind less avidly to cellular DNA polymerase α than to HSV DNA polymerase, and is not detectably incorporated by cellular DNA polymerase β (Reardon, 1989). Nevertheless, cellular DNA polymerases α, δ, and ε can incorporate acyclovir triphosphate into DNA (Reardon, 1989; Ilsley *et al.*, 1995). However, when the third step entailing, in effect, potency of inhibition by the next deoxynucleoside triphosphate, is considered, then considerable selectivity relative to cellular DNA polymerase α is observed (Reardon, 1989).

The importance of these steps for inhibition of virus replication in infected cells has been addressed by enzymological studies of polymerase from acyclovir-resistant mutants in which the incorporation of acyclovir triphosphate and dGTP were compared (Huang *et al.*, 1999). Certain mutants exhibited a specific defect in K_m for acyclovir triphosphate, suggesting a defect in acyclovir triphosphate binding, others a specific defect for k_{cat}, suggesting a defect in acyclovir triphosphate incorporation, and a third class exhibited defects in both. The locations of the residues altered in these mutants on a model of a polymerase active site (derived from the crystal structure of a related polymerase) could explain the effects of the mutations on enzyme kinetics (Fig. 19.4). These results argue that acyclovir-

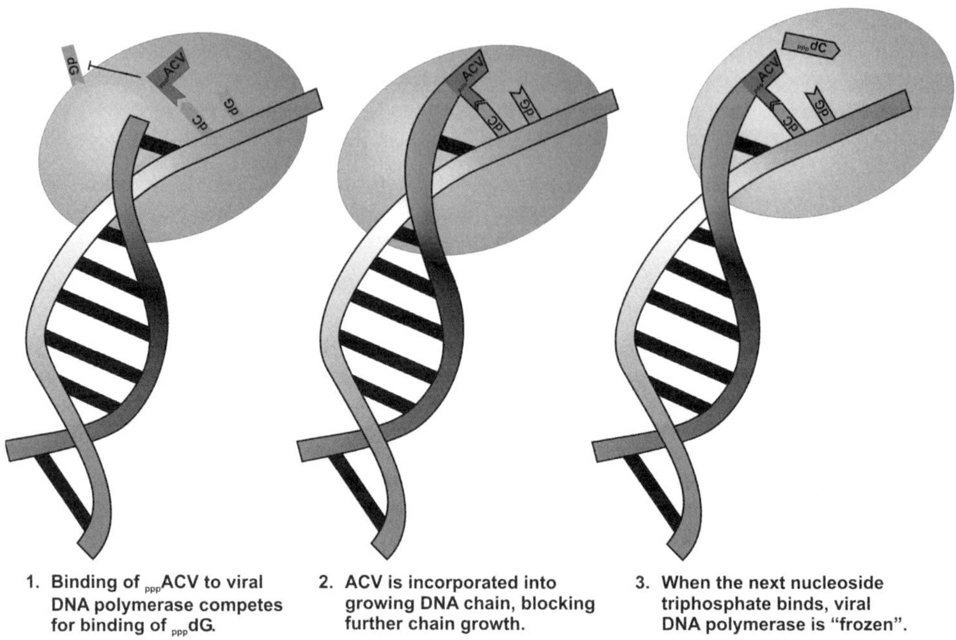

Figure 19.3 Three-step model for inhibition of HSV DNA polymerase by acyclovir triphosphate. (1) Acyclovir triphosphate (pppACV) competes with dGTP (pppG) for binding to the active site of the polymerase, thus, acting as a competitive inhibitor. (2) Acyclovir triphosphate acts as a substrate for the polymerase and is incorporated into DNA across from a templating dC residue. This terminates DNA replication, because acyclovir lacks a 3′-OH group. (3) The incorporated acyclovir leads to inactivation of the polymerase when the next dNTP binds (in this case dCTP, pppC). Reprinted with permission from Yeh and Coen (2004).

triphosphate is not simply a competitive inhibitor of viral DNA polymerase, but rather that its incorporation is crucial, consistent with the three step model (Fig. 19.3). Further work is required to determine whether the third step is, in fact, as crucial for selectivity in infected cells as it is *in vitro*.

Valacyclovir

The oral bioavailability of acyclovir is limited (15–20%). This limits its effectiveness in several clinical settings, including shingles. This led to the development of more orally available anti-alpha-herpesvirus agents including famciclovir (see below) and a prodrug of acylcovir, valacyclovir (Fig. 19.1). This compound consists of an acyclovir structure covalently attached to a valine moiety. The improved bioavailability of valacyclovir appears to be due to its improved intestinal absorption, which although this prodrug is not a peptide, is mediated by a peptide transporter (Balimane *et al.*, 1998; Ganapathy *et al.*, 1998; Han *et al.*, 1998). The prodrug is then rapidly converted to acyclovir, mediated by a hydrolase. One human enzyme that can perform this latter reaction is a biphenylhydrolase-like protein (Kim *et al.*, 2003). The end result is a prodrug with 3 to 5 times the oral bioavailability of acyclovir.

Penciclovir and famciclovir

Following the success of acyclovir, several pharmaceutical companies sought to discover related molecules that might improve upon acyclovir and capture some of its market.

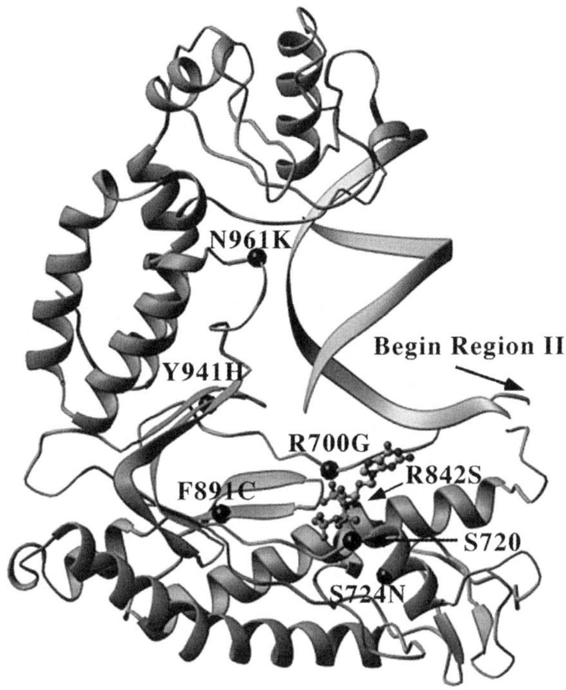

Figure 19.4 Model of locations of drug-resistance mutations on HSV polymerase active site. The figure shows a cutaway diagram of an α-like DNA polymerase active site containing primer-template and acyclovir triphosphate. The α-carbon backbone of the protein chain derived from the structure of RB69 DNA polymerase (Wang *et al.*, 1997) is presented as a gray ribbon diagram. The location of the N-terminus of conserved region II is indicated with an arrow. The positions of amino acids altered by HSV drug resistance mutations (e.g. R700G) as well as a conserved serine (S720) relative to the RB69 sequence are indicated with black dots and were aligned as described (Wang *et al.*, 1997). Primer-template is presented as two ribbons in the upper right quadrant of the figure and acyclovir triphosphate is presented as gray dots connected by gray lines. Reprinted from (Huang *et al.*, 1999).

One of these is penciclovir, which replaces the ether oxygen of the acyclic moiety with a CH_2. The acyclic moiety is also extended with a CHOH, providing, in effect, a 3′ hydoxyl group (Fig. 19.1). Penciclovir's mechanism of action is very similar to that of acyclovir, with only quantitative differences detected by both biochemical assays and analyses of resistant mutants. Penciclovir is more efficiently activated by HSV and VZV TK than is acyclovir (Vere Hodge and Cheng, 1993). As mentioned above, this appears to be due to better binding mediated by the extended acyclic moiety (Champness *et al.*, 1998). Accordingly, a mutant that expresses low levels of active TK remains sensitive to penciclovir while exhibiting resistance to acyclovir (Pelosi *et al.*, 1998a). On the other hand, TK-negative mutants are less resistant to penciclovir than they are to acyclovir (Boyd *et al.*, 1993; Pelosi *et al.*, 1998a), indicating less selectivity in the activation of penciclovir vs. acyclovir. This may be due to greater phosphorylation of penciclovir than acyclovir in uninfected cells, perhaps due to mitochondrial dG kinase (Lowe *et al.*, 1995). Penciclovir triphosphate is a less selective inhibitor of the viral DNA polymerases than is acyclovir triphosphate (Hannah *et al.*, 1989). This correlates with HSV *pol* mutants exhibiting less resistance to penciclovir than to acyclovir (Chiou *et al.*, 1995; Ertl *et al.*, 1995; Pelosi *et al.*, 1998a).

Penciclovir is not very orally available, but is used topically in an ointment for treating cold sores caused by HSV. To overcome the poor oral availability of penciclovir, famciclovir (Fig. 19.1) was developed. Famciclovir, which is a diacetyl ester of 6-deoxypenciclovir, can be converted to penciclovir by human intestinal tissue extract, suggesting conversion during adsorption, although oxidation of the 6 position may also occur in the liver (Vere Hodge and Perkins, 1989). The net result is oral bioavailability similar to that of valacyclovir (Pue and Benet, 1993). Famciclovir is used in the treatment of HSV infections and shingles.

Foscarnet

Foscarnet (phosphonoformic acid, PFA; structure in Fig. 19.1) is an analog of pyrophosphate, which is a product of polymerization of nucleic acids. Unlike the other drugs described above, it does not require activation by either cell or viral enzymes but rather inhibits Pol directly. Inhibition is *not* competitive with deoxynucleoside triphosphates. Rather, it appears that foscarnet acts as a product analog, evidently by binding to the site normally occupied by pyrophosphate and preventing normal pyrophosphate release so the polymerase cannot complete the catalytic cycle (Ostrander and Cheng, 1980; Erriksson *et al.*, 1982). Selectivity arises from the viral Pol being more sensitive than cellular enzymes, which is confirmed by the existence of foscarnet-resistant *pol* mutants (Erikkson and Oberg, 1979). Resistance is due to an altered polymerase which is less susceptible to foscarnet inhibition. Foscarnet is not as selective as ACV; correspondingly, the highest degree of resistance observed is about 20-fold. Clinically, foscarnet is approved for treatment of serious HSV (and cytomegalovirus) infections that are resistant to other drugs (e.g. acyclovir).

Resistance to anti-alpha-herpesvirus drugs in cell culture

In cell culture, mutations conferring resistance to the approved anti-alpha-herpesvirus drugs arise fairly frequently. For HSV-1, most laboratory and clinical isolates contain acyclovir-resistant mutants at a level of about 0.01–0.1% (reviewed in Coen, 1991). For at least certain HSV-2 strains, mutation frequencies are an order of magnitude higher (Sarisky *et al.*, 2000). As discussed above, there are four classes of these mutants: mutants with no TK activity (TK-negative), mutants with reduced TK activity (TK-partial), mutants with altered TK substrate specificity (TK-altered), and mutants with altered DNA polymerase (all *pol* mutants). Acyclovir and penciclovir select mainly for *tk* mutants, although some *pol* and double *tk-pol* mutants are also found (Coen and Schaffer, 1980; Sarisky *et al.*, 2001; Suzutani *et al.*, 2003). Foscarnet thus far has only selected for *pol* mutants (e.g. Derse *et al.*, 1982; Gibbs *et al.*, 1988).

Because TK is not essential for virus replication in cells or in certain locations in mammalian hosts, nonsense, frameshift, and missense mutations that confer drug-resistance can conceivably arise nearly anywhere in the HSV or VZV *tk* gene. In HSV, Hwang and Chen were the first to observe "hot spots" of frameshift mutations in homopolymeric runs of Cs or Gs (Hwang and Chen, 1995). There is some controversy regarding the relative frequency of these mutations following selection by passage in increasing concentrations of acyclovir or penciclovir. One group found that acyclovir selected mainly frameshift mutations, while penciclovir selected roughly equal numbers of frameshift and point mutations (Sarisky *et al.*, 2001). The second group found the opposite (Suzutani *et al.*, 2003). The different results may arise from when the mutations arose (prior to passage in drug vs. during passage in drug) as well as differences in the numbers of passages and the drug concentrations used. Differences in the kinds of mutations selected could reflect differences in how much TK activity is sufficient to permit sensitivity to the drug. As mentioned above, much less TK is needed for full sensitivity to penciclovir relative to that needed for full sensitivity to acyclovir (Pelosi *et al.*, 1998a). As certain frameshift mutants express low levels of TK (due to unusual translational mechanisms, as discussed below) (Hwang *et al.*, 1994; Griffiths and Coen, 2003; Griffiths and Coen, 2005), it is possible that these were not selected by the concentrations of penciclovir used by one of the groups.

Interestingly, as just alluded to, nonsense and frameshift mutations that might be expected to inactivate TK do not necessarily do so. For example, a nonsense mutation between the first and second methionine codons in the HSV *tk* gene permits the synthesis of a truncated TK at about 10% the level found for the wild type polypeptide (Irmiere *et al.*, 1989). This is evidently due to reinitiation of translation at the second methionine codon. The truncated polypeptide retains TK activity with a specific activity similar to wild type resulting in a TK-partial phenotype. A second example in HSV involves a single base insertion within a run of Gs that shifts the translational reading frame upstream of residues critical for TK activity (Hwang *et al.*, 1994). This results in the synthesis not only of the expected truncated polypeptide, but also low levels of full-length TK polypeptide and some TK enzyme activity. This can be ascribed to frameshifting wherein the translational machinery compensates for the insertion mutation about 1% of the time (Hwang *et al.*, 1994; Horsburgh *et al.*, 1996). A similar mechanism may operate when two Gs are inserted into the run of Gs (Griffiths and Coen, 2003). Another example is a deletion of a single base from a run of Cs. In this case, it appears that the normal TK reading frame downstream of the mutation is synthesized via an internal ribosome entry site (IRES) (Griffiths and Coen, 2005). It has been proposed that this C-terminal TK fragment can complement the out-of-frame TK polypeptide.

TK-altered mutations would be expected to exhibit a much narrower spectrum of locations than TK-negative and TK-partial mutation. Indeed, a large proportion of TK-altered mutations (e.g. Biron *et al.*, 1982; Darby *et al.*, 1986), lie within a small portion of the *tk* gene which encodes a conserved segment (roughly residues 162–188 in HSV-1; 163–189 in HSV-2; and 129–155 in VZV). The proposal that this segment forms the nucleoside binding site was based in part on these mutations. This proposal was nicely confirmed by X-tray crystallographic studies of HSV-1 TK and nucleoside ligands (Brown *et al.*, 1995; Wild *et al.*, 1997; Champness *et al.*, 1998).

Because Pol is essential for viral replication, all drug-resistant *pol* mutants express an enzyme that is less inhibited by drug than is the wild type enzyme (Pol-altered). Although these mutations occupy numerous sites in the *pol* gene, the vast majority lie within about 1/4 of the gene stretching from its midpoint towards its 3′ end. Within this region, the mutations cluster in specific segments that encode motifs that are conserved among the large family of α-like DNA polymerases (reviewed in Coen, 1996). This family includes the eukaryotic replicative polymerases α, δ, and ε. These conserved motifs include segments found in nearly every α-like polymerase termed regions I-III, V, and VII. An additional motif containing mutations was originally termed region A (Gibbs *et al.*, 1988), but following the sequencing of mammalian DNA polymerase δ was recognized to be a segment termed δ-region C (Zhang *et al.*, 1991), which is shared among cellular polymerases δ and ε and certain viral enzymes. Because drug-resistance mutations alter these motifs, and because the drugs mimic and/or compete for binding with dNTPs and pyrophosphate, it was hypothesized that at least some of these motifs help form the binding sites for the natural ligands (Gibbs *et al.*, 1985; Larder *et al.*, 1987; Gibbs *et al.*, 1988). This hypothesis has been borne out by X-ray crystallographic analysis of a related polymerase bound to DNA and dNTP (Franklin *et al.*, 2001). Furthermore, as reviewed above, enzymological analysis of mutant polymerases has identified mutations whose enzymological effects are consistent with their affecting specifically ACV-TP binding or incorporation (Huang *et al.*, 1999). Some of these mutants are also affected (but not as much) for dNTP binding or incorporation. The mutations that mainly affect binding are in regions II, III, and δ-region C, and those that mainly affect incorporation are in regions I and δ-region C. Although the related polymerase whose structure has been solved does not contain δ-region C, the enzymological analysis is consistent with the crystallographic analysis, and with mutational and enzymological analyses of other α-like DNA polymerases (reviewed in Wang, 1996).

At least two other genes can affect susceptibility to the approved anti-alpha-herpesvirus drugs. The first gene encodes the large subunit of the ribonucleotide reductase. HSV or VZV mutations that knock this gene out confer increased susceptibility (hypersensitivity) to polymerase inhibitors including acyclovir and foscarnet (Coen *et al.*, 1989a; Heineman and Cohen, 1994; Mineta *et al.*, 1994). The mechanism responsible for this is likely to be the depletion of the pools of deoxynucleoside triphosphates normally generated by the reductase. For drugs that mimic deoxynucleoside triphosphates, this would then lead to less competition for binding. For drugs like foscarnet that do not inhibit competitively with deoxynucleoside triphosphates, depletion of these pools nevertheless is expected to increase potency, albeit to a lesser extent. This is in fact what is observed (Coen *et al.*, 1989a). Although not yet tested, mutations that inactivate the small subunit of the reductase would also be expected to confer drug-hypersensitivity. Conversely, one can imagine that mutations that lead to greater reductase activity might confer drug-resistance. This has not been reported for herpesviruses, but a reductase overexpressing virus has been described for vaccinia virus (Slabaugh *et al.*, 1988).

A second locus for drug hypersensitivity is the HSV gene encoding the single-stranded DNA binding protein, ICP8. Mutations in this gene can confer hypersensitivity to certain polymerase inhibitors (Chiou *et al.*, 1985). The mechanism for this is not certain, but an appealing possibility is that the mutations alter the ability of ICP8 to interact with polymerase, leading to its becoming more sensitive to the drugs. In the absence of a clearcut mechanism, it is difficult to know whether mutations in the *ICP8* gene could confer drug-resistance.

Pathogenicities of drug-resistant mutants

As discussed above, studies of drug resistance can be invaluable in establishing drug mechanisms. Such studies can also shed light on interesting biochemical and biological mechanisms. However, resistance to anti-alphaherpesvirus drugs is not merely an academic exercise. Drug-resistant HSV and VZV constitute an important clinical problem, especially in immunocompromised patients (reviewed in Gilbert *et al.*, 2002; Safrin, 1996).

In the case of HSV, 5–10% of immunocompromised patients with HSV disease develop drug-resistant infections.

For a virus to cause disease that is resistant to an antiviral drug, it must mutate not only to evade drug action, but also to retain pathogenicity. How this occurs has been addressed by characterizing drug-resistant mutants for pathogenicity in animal models. Although there is always the question of whether animal models are predictive for human disease, as will be discussed below, the kinds of viruses isolated from patients suffering drug-resistant HSV disease can be understood in terms of the results of animal studies.

In the human host, productive replication by HSV and VZV can be manifested by lesions at various sites, most typically the skin or eyes. However, like all herpesviruses, HSV and VZV share the property of latency; that is, after the symptoms of primary infection are gone, viral genomes persist in a relatively quiescent state in cellular reservoirs protected from immune surveillance. From this state of latency, HSV and VZV can reactivate to cause recurrent disease. The reservoirs for HSV and VZV are ganglia of the peripheral nervous system, which is a common characteristic of alpha-herpesviruses.

Considerable effort has gone into developing animal models for both the disease manifestations of these viruses and for latency. Animal models for VZV disease and latency have become available relatively recently (see Chapters 16 and 17). As yet, there have been few studies of drug-resistant VZV mutants in these models. However, there are numerous and longstanding models of HSV disease and latency. This has led to much study of the effects of HSV mutations, including those that confer drug resistance, on pathogenesis. For the purpose of brevity only two kinds of studies using immunocompetent mice will be reviewed here. The first involves lethal infection of the central nervous system (CNS), which is known as neurovirulence. The second, which mimics the natural course of most infections of humans more, involves inoculation at a peripheral site such as the footpad, snout, or cornea. The virus replicates at the peripheral site, gains access to nerve terminals, and is transported back to neuronal cell bodies in sensory ganglia. In animal models, at least, the virus replicates acutely in ganglia. Over several days time, acute replication ceases and the virus establishes a latent infection in neurons, where no infectious virus can be detected in ganglion homogenates. The latent virus can, however, be reactivated, most commonly by explant of the ganglia. The relevance of either assay for HSV disease and latency in humans is not certain. It is possible, however, to correlate the behavior of drug-resistant clinical isolates in such animal models and the clinical courses associated with the isolates. As will be discussed, the correlations are positive.

Behaviors of drug-resistant mutants in animal models

Nearly all drug-resistant HSV mutants exhibit some degree of attenuation in these assays of pathogenesis (reviewed in Coen, 1994; Larder and Darby, 1984). TK-negative mutants are generally the most attenuated. They are somewhat deficient for lethal infection of the CNS and for replication at certain peripheral sites. They can establish latent infections following peripheral inoculation, albeit with lower efficiency. Most strikingly, there is little or no replication of virus in ganglia acutely and no reactivation from latency following normal explant procedures when common laboratory strains have been used. These ganglionic defects can definitely be ascribed to a *tk* mutation (Jacobson *et al.*, 1993). The failure to reactivate from latency is largely due to a requirement for TK during the reactivation process, rather than during establishment of latency (Jacobson *et al.*, 1993; Chen *et al.*, 2004).

The attenuation of TK-partial mutants depends on how little TK they express. As little as 5–10% of wt levels of TK activity permit normal replication and reactivation from latency and full neurovirulence (Coen *et al.*, 1989b; Chen *et al.*, 1998; Pelosi *et al.*, 1998a). Below that amount, generally speaking, the less TK-activity expressed by the mutant, the less pathogenesis. TK-altered and *pol* mutants are variably attenuated for lethal CNS infection, but most replicate robustly at the periphery and in the ganglia and reactivate from latency. Thus, these two classes of mutants include the most pathogenic drug-resistant mutants. The *pol* mutants are interesting in part because of their variability, with some much more attenuated than others. One *pol* mutant has the interesting property of being attenuated for neurovirulence, but not exhibiting any defect for viral replication in the CNS (Pelosi *et al.*, 1998b).

Only a few double mutants have been assayed for pathogenicity. So far, combining TK-negative or TK-partial and *pol* mutations yields considerably more attenuation than either mutation alone (Darby *et al.*, 1984; Jacobson *et al.*, 1995). However, not all combinations have been assayed. It may be that certain combinations of *pol* and TK-altered mutations could permit considerable pathogenicity.

This picture becomes even more complicated when mixtures of drug-resistant mutants with each other or with drug-sensitive virus is considered. Experimentally constructed mixtures of such viruses can complement each other for pathogenicity or drug-resistance or both in mice (Tenser *et al.*, 1981; Field, 1982; Field and Lay, 1984; Tenser and Edris, 1987; Sedarati *et al.*, 1988; Ellis *et al.*, 1989). Selection for drug-resistance in mice yields even more complex mixtures that are especially resistant to therapy (Field, 1982; Field and Lay, 1984). It is possible that *pol* mutations might play an especially important role in such mixtures because such mutations are co-dominant; i.e. they allow both resistant and sensitive viruses to replicate in the same cell during drug treatment (Coen *et al.*, 1980). At least one heterogeneous mixture from a patient retained full neurovirulence and latency-competence in mice (Sacks *et al.*, 1989; Pelosi *et al.*, 1992; Pelosi and Coen, unpublished observations). Thus, it is important to consider heterogeneity as a factor influencing drug-resistance and pathogenicity in the clinic.

Drug-resistant HSV isolated from patients

Most of the properties of drug-resistant HSV mutants isolated from patients who suffer disease that does not respond to antiviral therapy can be understood by considering the findings just reviewed. In immunocompetent patients, there have been only a few reports of drug-resistant HSV infection (Kost *et al.*, 1993; Mouly *et al.*, 1995; Swetter *et al.*, 1998; Kriesel *et al.*, 2005). This can be attributed to the limited nature of viral disease in such patients, which results in less replication of the virus and less exposure to and selection by drug. The mutation rate of the virus, which is substantial but limited when compared with those of RNA viruses and retroviruses, is also a factor.

The first report of drug-resistant HSV from an immunocompetent patient described mixed populations of virus that were predominantly TK-altered with a minority of TK-positive virus (Kost *et al.*, 1993). As outlined above, TK-altered and mixed populations of HSV are both pathogenic and drug-resistant in animal models. Thus, the properties of this isolate are much as would be expected from animal studies. The characterization of subsequent drug-resistant isolates from immunocompetent patients, however, has revealed frameshift mutants, including those with insertions in a run of 7 Gs and a deletion in a run of 6 Cs (Swetter *et al.*, 1998; Kriesel *et al.*,

2005). Given the above discussion regarding the importance of TK in pathogenesis, this could be surprising given that such mutations would be expected to result in a TK-negative phenotype. However as discussed above (and below), such mutants are not actually TK-negative. Moreover, in at least one case (Swetter *et al.*, 1998), some of the isolates contained TK-positive virus as well as *tk* mutants.

Most cases of drug-resistant HSV (and all cases of drug-resistant VZV) have arisen in immunocompromised individuals. In this setting, without the immune system to control virus replication, the virus replicates more, higher virus loads are achieved, and there is more exposure to and selection by drug. Additionally, there is more opportunity for drug-resistant mutants to replicate in cells that are more permissive for such mutants than are, for example, neurons.

Despite this last point, it has nevertheless been surprising that most of the viruses isolated from immunocompromised patients have mutations (genotype) predictive of the phenotype that is most attenuated in animal models --TK-negative (Safrin, 1996; Gilbert *et al.*, 2002). Three explanations for this paradox have emerged. The first is that many of these viruses are not, in fact, TK-negative. Indeed, the most common *tk* mutations found in clinical isolates are insertions of one or two bases on a run of Gs and a deletion of one base from a run of Cs. As reviewed above, these mutations do not totally abolish TK activity, because they are compensated by ribosomal frameshifting (Hwang *et al.*, 1994; Horsburgh *et al.*, 1996; Griffiths and Coen, 2003) or, evidently, internal ribosome entry (Griffiths and Coen, 2005). The amounts of TK made are low, but sufficient to permit reactivation from latency in explanted mouse ganglia (Griffiths *et al.*, 2003). Other mutants (e.g. missense) reported to be TK-negative may also be TK-partial instead, as the assays of TK activity that have been used in many studies are not very sensitive.

A second explanation is that many clinical isolates are not pure populations, but rather contain TK-positive subpopulations (e.g. Sacks *et al.*, 1989; Horsburgh *et al.*, 1998). Indeed, it is common for frameshift mutants, particularly those on homopolymeric runs (e.g. runs of Gs) to revert to a TK-positive phenotype, resulting in mixed populations. This often occurs during studies of such mutants in animal models (Sasadeusz and Sacks, 1996; Griffiths *et al.*, 2003). In one case, that of a mutant containing a two base insertion in a run of Gs, phenotypic reversion (due to addition of a G, restoring reading frame) occurs so frequently that stocks of virus contain ~3% TK-positive virus (Grey *et al.*, 2003; Griffiths and Coen, 2003). The majority population (97%) of these mutant stocks also expresses low levels of TK, evidently due to ribosomal frameshifting, but the mixed population replicates in ganglia and reactivates from latency much more efficiently than other viruses that express similar levels of TK.

A third explanation, which was not anticipated in earlier studies using laboratory strains, is that clinical isolates can contain alleles that compensate for loss of TK. This phenomenon has now been observed with two different clinical isolates (Horsburgh *et al.*, 1998; Griffiths *et al.*, 2003). The first report was of an isolate from a bone marrow transplant patient who developed acyclovir-resistant disease, which then subsided as immune function returned. However, several years later, the patient needed a second bone marrow transplant. Despite high dose acyclovir prophylaxis, the patient reactivated HSV (Horsburgh *et al.*, 1998). Although the reactivating isolate was a mixed population, other drug-resistant isolates appeared to be homogenous populations with either no or very, very low TK activity, due to a missense mutation. However, these viruses could reactivate from latency in mouse ganglia in explanted mouse ganglia, as could an engineered *tk* deletion mutant, albeit inefficiently. The second example was also from a bone marrow transplant recipient (Sacks *et al.*, 1989). Again, it was found that a *tk* deletion mutant engineered from this isolate could reactivate from latency in mouse ganglia, even more efficiently than the *tk* null mutant from the initial example (Griffiths *et al.*, 2003). It will be interesting to learn what strain differences account for this ability to reactivate from latency in the absence of TK. Regardless, such differences may be important for the pathogenicity of drug-resistant mutants in human patients.

Drugs other than nucleoside analogs and foscarnet

In part because of the problem of resistance to approved antiviral drugs, considerable effort has gone into discovering new anti-alpha-herpesvirus drugs. Although some efforts have focused on other nucleoside and pyrophosphate analogs, others have looked for different chemical entities. Some of these compounds target other stages in the virus replication cycle. For example, several drugs that inhibit attachment have been identified. Most of these have been relatively large, charged compounds exemplified by heparin, which is known to bind to certain glycoproteins that mediate the initial attachment of the virus to heparan sulfate on the surface of cells (Pertel and Spear, 1996). At least one of these compounds, an oligomer derived from a plant, has been evaluated in clinical trials as a topical treatment for HSV lesions, and showed a trend towards efficacy (Orozco-Topete *et al.*, 1997). Drawbacks of these molecules include unfavorable pharmacokinetic features, including a lack of oral availability.

Pointing towards new targets for this stage of infection, recent research has provided much information about the interaction of specific glycoproteins and cell surface receptors (Spear *et al.*, 2000), including crystal structures of interacting partners (Carfi *et al.*, 2001). Although there are concerns regarding multiple receptors and the sizes of the interaction interfaces, these protein–protein interactions might make attractive drug targets.

In principle, gene expression is a desirable stage at which to inhibit viral replication. Multiple gene products of alpha-herpesviruses are required for viral gene expression and are sufficiently different from their host-encoded counterparts to be targeted selectively. Moreover, as for drugs that target attachment and entry, drugs that inhibit viral gene expression would limit the synthesis of downstream, potentially toxic viral proteins. Two classes of small molecules have been reported to inhibit immediate early gene expression of HSV. One class, SCH 43478 and analogs, was reported to inhibit expression of immediate early proteins without affecting the kinetics of virus entry (Albin *et al.*, 1997). One of these compounds showed efficacy in a guinea pig model of HSV disease following subcutaneous administration. The second compound, a benzothiophene, also inhibited immediate early gene expression, without evident effects on entry of viral DNA into the nucleus (Boulware *et al.*, 2001). In both cases, pharmacokinetic issues (e.g. poor oral availability) prevented further development. Intriguingly, in each case, the actual target of the drug was not identified, although the investigators hypothesized an effect on either viral or cellular targets involved in the induction of immediate early gene expression. Thus, these drugs might shed light on this process. Regardless, these are the first of what hopefully will be many small molecules that target herpesvirus gene expression.

Given the success of acyclovir and its emulators, it is not surprising that new drugs that target herpesvirus DNA replication and, specifically, DNA polymerase are still being developed. One goal has been to seek drugs that would target sites on polymerase other than those targeted by nucleoside analogs. Perhaps the most interesting of the new drugs are the 4-oxo-dihydroquinolines (Brideau *et al.*, 2002; Oien *et al.*, 2002; Thomsen *et al.*, 2003). Like the anti-HIV non-nucleoside reverse transcriptase inhibitors (NNRTIs), these drugs were discovered via high-throughput screening of large libraries of compounds for inhibition of DNA polymerase activity. Remarkably, and in contrast to the highly specific NNRTIs, these drugs are active against a broad range of herpesviruses and their DNA polymerases. They are much less potent against human DNA polymerases α, δ, and γ and one of these compounds, PNU-183792, exhibits no cytotoxicity in various cell lines at concentrations up to 100 μM. The drugs appear to inhibit polymerase competitively with dNTPs, and studies of drug resistant mutants together with modeling exercises suggest a binding site where the drug could inhibit binding of polymerase to both dNTPs and to the DNA template strand (Thomsen *et al.*, 2003) PNU-183792 is orally available and shows efficacy in mice.

Other DNA replication inhibitors have been discovered by targeting protein–protein interactions. Such interactions can be highly specific and this has made them attractive potential drug targets. Previous work had shown that interactions between the two subunits of HSV ribonucleotide reductase (RR), could be targeted. The first such inhibitor, which worked only *in vitro*, was a nonapeptide that forms part of the interface between two RR subunits (Cohen *et al.*, 1986; Dutia *et al.*, 1986; McClements *et al.*, 1988; Paradis *et al.*, 1988). Subsequently, highly potent peptidomimetics derived from an even smaller peptide were developed, and these have activity in infected cells, and topically in animal models (Liuzzi *et al.*, 1994; Moss *et al.*, 1996). (However, these compounds suffered from pharmacokinetic difficulties and from the RR not being essential except under certain circumstances.)

As with RR, peptides derived from the subunit interface of HSV DNA polymerase, which is a heterodimer of Pol and a processivity subunit, UL42, are active *in vitro* (Marsden *et al.*, 1994; Digard *et al.*, 1995). Delivery of one of these peptides into cultured cells using a carrier protein specifically inhibits HSV replication (Loregian *et al.*, 1998). Bolstering the case that this interaction could be targeted by small molecules are the results of mutational and crystallographic analyses that indicate that alteration of individual hydrogen-bonding residues can abolish the interaction (Zuccola *et al.*, 2000; Bridges *et al.*, 2001). Indeed, a small molecule that disrupts this interaction *in vitro* and exhibit antiviral activity has been identified following a high-throughput screen for inhibitors of the Pol-UL42 interaction (Pilger *et al.*, 2004). That compound is too toxic for clinical development. Nevertheless, this strategy has promise.

Another essential viral DNA replication enzyme that has been targeted for drug development is the three subunit helicase-primase. Three different companies developed compounds active against this enzyme and against HSV replication (Spector *et al.*, 1998; Crute *et al.*, 2002; Kleymann *et al.*, 2002). At least one of these compounds appears to act by increasing the binding of the drug-bound enzyme to DNA (Crute *et al.*, 2002). This could hinder the movement of the enzyme as it unwinds DNA and lays down RNA primers during DNA replication. Two of these companies developed these helicase-primase inhibitors through the stage of demonstration of activity in animal models, where they demonstrated good efficacy (Crute *et al.*, 2002; Kleymann *et al.*, 2002). At this writing, whether further development will occur is unclear.

The clinical efficacy of anti-HIV protease inhibitors (in combination with other anti-HIV drugs) has led to numerous efforts to exploit proteins involved in the late stages of herpesvirus replication as new drug targets. Two different classes of compounds have been identified that inhibit the encapsidation and cleavage of alpha-herpesviral DNA into genome length molecules, an activity that herpesviruses require, but cells do not perform. One, an acridone derivative, is active against HSV, but its target is not known (Akanitapichat and Bastow, 2002). This drug is not very potent. The second class, thiourea inhibitors, appears to target the HSV UL6 protein that forms the portal through which DNA enters the capsid (van Zeijl *et al.*, 2000). Evidently, this class of drug causes the failure of UL6 and at least one other protein to be stably incorporated into capsids (Newcomb and Brown, 2002). This would prevent encapsidation of the DNA. These drugs, however, exhibit variable potency against different HSV strains, and have not been developed further.

Regardless of the prospects for the various new compounds, good drug targets can be found among multiple viral proteins that function during different stages of alpha-herpesvirus replication.

Can latent infections be cured?

The ultimate goal of antiherpesvirus drug discovery should be to cure patients of latent infections, yet all commercially availabe drugs target replicating and not latent virus. Why is this so? A rational approach to curing latency would be to disrupt the ability of the virus to maintain a latent infection. In principle, this could cause the virus to reactivate from latency. Theoretically, immune mechanisms might clear the drug. Alternatively, co-administration of drugs such as acyclovir could inactivate the replicating virus. This approach is not without potential dangers, especially if reactivation and/or immune mechanisms killed the infected neurons. A more pressing practical issue is that we presently do not understand what viral and cellular mechanisms control latency. In the case of HSV, there is evidence that the locus that encodes the latency associated transcripts (LATs) helps to maintain latency inasmuch as a deletion in this locus leads to increased expression of productive cycle genes during latency in mouse trigeminal ganglia (Chen *et al.*, 1997). However, this mutation does not abrogate latency, indicating that multiple mechanisms maintain latency. Even if LATs were sufficient to maintain latency, how to target these RNAs for drug discovery is unclear.

In the case of VZV, animal models with which to study latency have been available for less time than for HSV. However, several viral proteins are expressed in latently infected human tissue and in latently infected animal models (reviewed in Chapter 16). If these proteins are shown to be important for maintenance of latency, it may be possible to use established methods for drug discovery. This could possibly provide a path to the "holy grail" of herpesvirus research—a cure.

References

Akanitapichat, P., and Bastow, K.F. (2002). The antiviral agent 5-chloro-1,3-dihydroxyacridone interferes with assembly and maturation of herpes simplex virus. Antiviral Res. 53, 113–126.

Albin, R., Chase, R., Risano, C., Lieberman, M., Ferrari, E., Skelton, A., Buontempo, P., Cox, S., DeMartino, J., Wright-Minogue, J. *et al.* (1997). SCH 43478 and analogs: *in vitro* activity and *in vivo* efficacy of novel agents for herpesvirus type 2. Antiviral Res. 35, 139–146.

Balimane, P.V., Tamai, I., Guo, A., Nakanishi, T., Kitada, H., Leibach, F.H., Tsuji, A., and Sinko, P.J. (1998). Direct evidence for peptide transporter (PepT1)-mediated uptake of a nonpeptide prodrug, valacyclovir. Biochem. Biophys. Res. Commun. *250*, 246–251.

Balzarini, J., Liekens, S., Esnouf, R., and De Clercq, E. (2002). The A167Y mutation converts the herpes simplex virus type 1 thymidine kinase into a guanosine analogue kinase. Biochemistry *41*, 6517–6524.

Biron, K.K., Fyfe, J.A., Noblin, J.E., and Elion, G.B. (1982). Selection and preliminary characterization of acyclovir-resistant mutants of varicella zoster virus. Am. J. Med. *73*, 383–386.

Bohman, C., and Eriksson, S. (1988). Deoxycytidine kinase from human leukemic spleen: preparation and characterization of the homogeneous enzyme. Biochemistry *27*, 4258–4265.

Boulware, S.L., Bronstein, J.C., Nordby, E.C., and Weber, P.C. (2001). Identification and characterization of a benzothiophene inhibitor of herpes simplex virus type 1 replication which acts at the immediate early stage of infection. Antiviral Res. *51*, 111–125.

Boyd, M.R., Safrin, S., and Kern, E.R. (1993). Penciclovir: a review of its spectrum of activity, selectivity, and cross-resistance pattern. Antiviral Chem. Chemother. *4 Suppl.1*, 3–11.

Brideau, R.J., Knechtel, M.L., Huang, A., Vaillancourt, V.A., Vera, E.V., Oien, N.L., Hopkins, T.A., Wieber, J.L., K.F., W., Rush, R.D. *et al.* (2002). Broad-spectrum antiviral activity of PNU-183792, a 4-oxo-dihydroquinoline, against human and animal herpesviruses. Antiviral Res. *54*, 19–28.

Bridges, K.G., Chow, C.S., and Coen, D.M. (2001). Identification of crucial hydrogen-bonding residues for the interaction of herpes simplex virus DNA polymerase subunits via peptide display, mutational, and calorimetric approaches. J. Virol. *75*, 4990–4998.

Brown, D.G., Visse, R., Sandhu, G., Davies, A., Rizkallah, P.J., Melitz, C., Summers, W.C., and Sanderson, M.R. (1995). Crystal structures of the thymidine kinase from herpes simplex virus type-1 in complex with deoxythymidine and ganciclovir. Nature Struct Biol 2, 876–880.

Carfi, A., Willis, S.H., Whitbeck, J.C., Krummenacher, C., Cohen, G.H., Eisenberg, R.J., and Wiley, D.C. (2001). Human herpes virus glycoprotein D bound to the human receptor HveA. Mol. Cell. *8*, 169–179.

Champness, J.N., Bennett, M.S., Wien, F., Visse, R., Summers, W.C., Herdewijn, P., De Clercq, E., Ostrowski, T., Jarvest, R.L., and Sanderson, M.R. (1998). Exploring the active site of herpes simplex virus type-1 thymidine kinase by X-ray crystallography of complexes with aciclovir and other ligands. Proteins 32, 350–361.

Chen, S.-H., Kramer, M.F., Schaffer, P.A., and Coen, D.M. (1997). A viral function represses accumulation of transcripts from productive-cycle genes in mouse ganglia latently infected with herpes simplex virus. J. Virol. *71*, 5878–5884.

Chen, S.-H., Cook, W.J., Grove, K.L., and Coen, D.M. (1998). Human thymidine kinase can functionally replace herpes simplex virus type 1 thymidine kinase for viral replication in mouse sensory ganglia and reactivation from latency upon explant. J. Virol. *72*, 6710–6715.

Chen, S.-H., Pearson, A., Coen, D.M., and Chen, S.-H. (2004). Failure of thymidine kinase-negative herpes simplex virus to reactivate from latency following efficient establishment. J. Virol. *78*, 520–523.

Chiou, H., Weller, S.K., and Coen, D.M. (1985). Mutations in the herpes simplex virus major DNA-binding protein gene leading to altered sensitivity to DNA polymerase inhibitors. Virology *145*, 213–226.

Chiou, H.C., Kumura, K., Hu, A., Kerns, K.M., and Coen, D.M. (1995). Penciclovir-resistance mutations in the herpes simplex virus DNA polymerase gene. Antiviral Chem. Chemother. 6, 281–288.

Coen, D.M. (1991). The implications of resistance to antiviral agents for herpesvirus drug targets and drug therapy. Antiviral Res. *15*, 287–300.

Coen, D.M. (1994). Acyclovir-resistant, pathogenic herpesviruses. Trends Microbiol. *2*, 481–485.

Coen, D.M. (1996). Viral DNA polymerases. In DNA Replication in Eukaryotic Cells, M. DePamphilis, ed. (Cold Spring Harbor, Cold Spring Harbor Laboratory Press), pp. 495–523.

Coen, D.M. and Schaffer, P.A. (1980). Two distinct loci confer resistance to acycloguanosine in herpes simplex virus type 1. Proc. Natl. Acad. Sci. USA. *77*, 2265–2269.

Coen, D.M., Goldstein, D.J., and Weller, S.K. (1989a). Herpes simplex virus ribonucleotide reductase mutants are hypersensitive to acyclovir. Antimicrob. Agents Chemother. 33, 1395–1399.

Coen, D.M., Irmiere, A.F., Jacobson, J.G., and Kerns, K.M. (1989b). Low levels of herpes simplex virus thymidine- thymidylate kinase are not limiting for sensitivity to certain antiviral drugs or for latency in a mouse model. Virology *168*, 221–231.

Cohen, J.I., and Straus, S.E. (2001). Varicella-zoster virus and its replication. In Fields Virology, D.M. Knipe, P.M. Howley, D.E. Griffin, R.A. Lamb, M.A. Martin, B. Roizman, and S. E. Straus, eds. (Philadelphia, Lippincott Williams & Wilkins), pp. 2707–2730.

Cohen, E.A., Gaudreau, P., Brazeau, P., and Langelier, Y. (1986). Specific inhibition of herpesvirus ribonucleotide reductase by a nonapeptide derived from the carboxy terminus of subunit 2. Nature *321*, 441–443.

Crute, J.J., Grygon, C.A., Hargrave, K.D., Simoneau, B., Faucher, A.-M., Bolger, G., Kibler, P., Liuzzi, M., and Cordingley, M.G. (2002). Herpes simplex virus helicase-primase inhibitors are active in animal models of human disease. Nature Med. 8, 386–391.

Darby, G., Churcher, M.J., and Larder, B.A. (1984). Cooperative effects between two acyclovir re-

sistance loci in herpes simplex virus. J. Virol. *50*, 838–846.
Darby, G., Larder, B.A., and Inglis, M.M. (1986). Evidence that the "active centre" of the herpes simplex virus thymidine kinase involves an interaction between three distinct regions of the polypeptide. J. Gen. Virol. *67*, 753–758.
Datta, N.S., Shewach, D.S., Hurley, M.C., Mitchell, B.S., and Fox, I.H. (1989). Human T-lymphoblast deoxycytidine kinase: purification and properties. Biochemistry *28*, 114–123.
Derse, D., Bastow, K.F., and Cheng, Y. (1982). Characterization of the DNA polymerases induced by a group of herpes simplex virus type I variants selected for growth in the presence of phosphonoformic acid. J. Biol. Chem. *257*, 10251–10260.
Digard, P., Williams, K.P., Hensley, P., Brooks, I.S., Dahl, C.E., and Coen, D.M. (1995). Specific inhibition of herpes simplex virus DNA polymerase by helical peptides corresponding to the subunit interface. Proc. Natl. Acad. Sci. USA *92*, 1456–1460.
Durham, J.P., and Ives, D.H. (1970). Deoxycytidine kinase. II. Purification and general properties of the calf thymus enzyme. J. Biol. Chem. *245*, 2276–2284.
Dutia, B.M., Frame, M.C., Subak-Sharpe, J.H., Clark, W.N., and Marsden, H.S. (1986). Specific inhibition of herpesvirus ribonucleotide reductase by synthetic peptides. Nature *321*, 439–441.
Elion, G.B., Furman, P.A., Fyfe, J.A., de Miranda, P., Beauchamp, L., and Schaeffer, H.J. (1977). Selectivity of action of an antiherpetic agent, 9-(2-hydroxyethoxymethyl)guanine. Proc. Natl. Acad. Sci. USA *74*, 5716–5720.
Ellis, M.N., Waters, R., Hill, E.L., Lobe, D.C., Selleseth, D.W., and Barry, D.W. (1989). Orofacial infection of athymic mice with defined mixtures of acyclovir-susceptible and acyclovir-resistant herpes simplex virus type 1. Antimicrob. Agents Chemother. *33*, 304–310.
Erikkson, B., and Öberg, B. (1979). Characteristics of herpes virus mutants resistant to phosphonoformate and phosphonacetate. Antimicrob. Agents Chemother. *15*, 758–762.
Erriksson, B., Öberg, B., and Wahren, B. (1982). Pyrophosphate analogs as inhibitors of DNA polymerases of cytomegalovirus, herpes simplex virus and cellular origin. Biochim. Biophys. Acta *696*, 115–123.
Ertl, P., Snowden, W., Lowe, D., Miller, W., and Littler, E. (1995). A comparative study of the *in vitro* and *in vivo* antiviral activities of aciclovir and penciclovir. Antiviral Chem. Chemother. *6*, 89–97.
Field, H.J. (1982). Development of clinical resistance to acyclovir in herpes simplex virus-infected mice receiving oral therapy. Antimicrob. Agents Chemother. *21*, 744–752.
Field, H.J., and Lay, E. (1984). Characterization of latent infections in mice inoculated with herpes simplex virus which is clinically resistant to acyclovir. Antiviral Res. *4*, 43–52.
Franklin, M.C., Wang, J., and Steitz, T.A. (2001). Structure of the replicating complex of a pol α family DNA polymerase. Cell *105*, 657–667.
Ganipathy, M.E., Huang, W., Wang, H., Ganapathy, V., and Leibach, F.H. (1998). Valacyclovir: a substrate for the intestinal and renal peptide transporters, PEPT1 and PEPT2. Biochem. Biophys. Res. Commun. *246*, 470–475.
Gibbs, J.S., Chiou, H.C., Hall, J.D., Mount, D.W., Retondo, M.J., Weller, S.K., and Coen, D.M. (1985). Sequence and mapping analyses of the herpes simplex virus DNA polymerase gene predict a C-terminal substrate binding domain. Proc. Natl. Acad. Sci. USA *82*, 7969–7973.
Gibbs, J.S., Chiou, H.C., Bastow, K.F., Cheng, Y.C., and Coen, D.M. (1988). Identification of amino acids in herpes simplex virus DNA polymerase involved in substrate and drug recognition. Proc. Natl. Acad. Sci. USA *85*, 6672–6676.
Gilbert, C., Bestman-Smith, J., and Boivin, G. (2002). Resistance of herpesviruses to antiviral drugs: clinical impacts and molecular mechanisms. Drug Resistance Updates 5, 88–114.
Grey, F., Sowa, M., Collins, P., Fenton, R.J., Harris, W., Snowden, W., Efstathiou, S., and Darby, G. (2003). Characterization of a neurovirulent aciclovir-resistant variant of herpes simplex virus. J. Gen. Virol. *84*, 1403–1410.
Griffiths, A., and Coen, D.M. (2003). High-frequency phenotypic reversion and pathogenicity of an acyclovir-resistant herpes simplex virus mutant. J. Virol. *77*, 2282–2286.
Griffiths, A., and Coen, D.M. (2005). An unusual internal ribosome entry sequence in the herpes simplex virus thymidine kinase gene. Proc. Natl. Acad. Sci. USA *102*, 9667–9772.
Griffiths, A., Chen, S.-H., Horsburgh, B.C., and Coen, D.M. (2003). Translational compensation of a frameshift mutation affecting herpes simplex virus thymidine kinase is sufficient to permit reactivation from latency. J. Virol. *77*, 4703–4709.
Han, H.-k., de Vrueh, R.L.A., Rhie, J.K., Covitz, K.-M., Smith, P.L, Lee, C.-P., Oh, D.-M., Sadée, and Amidon, G.L. (1998). 5′-amino acid esters of antiviral nucleosides, acyclovir, and AZT are absorbed by the intestinal PEPT1 peptide transporter. Pharmaceutical Res. *15*, 1154–1159.
Hannah, J., Tolman, R.L., Karkas, J.D., Liou, R., Perry, H.C., and Field, A.K. (1989). Carba-acy-

clonucleoside antiherpetic agents. J. Heterocyclic Chem. *26*, 1261–1271.

Harrison, P.T., Thompson, R., and Davison, A.J. (1991). Evolution of herpesvirus thymidine kinases from cellular deoxycytidine kinase. J. Gen. Virol. *72*, 2583–2586.

Heineman, T.C., and Cohen, J.I. (1994). Deletion of the varicella-zoster virus large subunit of ribonucleotide reductase impairs growth of virus *in vitro*. J. Virol. *68*, 3317–3323.

Horsburgh, B.C., Kollmus, H., Hauser, H., and Coen, D.M. (1996). Translational recoding induced by G-rich mRNA sequences that form unusual structures. Cell *86*, 949–959.

Horsburgh, B.C., Chen, S.-H., Hu, A., Mulamba, G.B., Burns, W.H., and Coen, D.M. (1998). Recurrent acyclovir-resistant herpes simplex in an immunocompromised patient: can strain differences compensate for loss of thymidine kinase in pathogenesis? J. Infect. Dis. *178*, 618–625.

Huang, L., Ishii, K.K., Zuccola, H., Gehring, A.M., Hwang, C.B.C., Hogle, J., and Coen, D.M. (1999). The enzymological basis for resistance of herpesvirus DNA polymerase mutants to acyclovir: relationship to the structure of α-like DNA polymerases. Proc. Natl. Acad. Sci. USA *96*, 447–452.

Hwang, C.B., and Chen, H.J. (1995). An altered spectrum of herpes simplex virus mutations mediated by an antimutator DNA polymerase. Gene *152*, 191–193.

Hwang, C.B.C., Horsburgh, B., Pelosi, E., Roberts, S., Digard, P., and Coen, D.M. (1994). A net +1 frameshift permits synthesis of thymidine kinase from a drug-resistant herpes simplex virus mutant. Proc. Natl. Acad. Sci. USA *91*, 5461–5465.

Ilsley, D.D., Lee, S.-H., Miller, W.H., and Kuchta, R.D. (1995). Acyclic guanosine analogues inhibit DNA polymerases α, δ, and ε with very different potencies and have unique mechanisms of action. Biochemistry *34*, 2504–2510.

Irmiere, A.F., Manos, M.M., Jacobson, J.G., Gibbs, J.S., and Coen, D.M. (1989). Effect of an amber mutation in the herpes simplex virus thymidine kinase gene on polypeptide synthesis and stability. Virology *168*, 210–120.

Jacobson, J.G., Ruffner, K.L., Kosz-Vnenchak, M., Hwang, C.B. C., Wobbe, K.K., Knipe, D.M., and Coen, D.M. (1993). Herpes simplex virus thymidine kinase and specific stages of latency in murine trigeminal ganglia. J. Virol. *67*, 6903–6908.

Jacobson, J., Kramer, M., Rozenberg, F., Hu, A., and Coen, D.M. (1995). Synergistic effects on ganglionic herpes simplex virus infections by mutations or drugs that inhibit the viral polymerase and thymidine kinase. Virology *206*, 263–268.

Kim, I., Chu, X.-y., Kim, S., Provoda, C.J., Lee, K.-D., and Amidon, G.L. (2003). Identification of a human valacylovirase: biphenyl hydrolase-like protein as valacyclovir hydrolase. J. Biol. Chem. *278*, 25348–25356.

Kleymann, G., Fischer, R., Betz, U.A. K., Hendrix, M., Bender, W., Schneider, U., Handke, G., Eckenberg, P., Hewlett, G., Pevzner, V. *et al.* (2002). New helicase-primase inhibitors as drug candidates for the treatment of herpes simplex disease. Nature Med. *8*, 392–398.

Kost, R.G., Hill, E.L., Tigges, M., and Straus, S.E. (1993). Brief report: recurrent acyclovir-resistant genital herpes in an immunocompetent patient. N. Engl. J. Med. *329*, 1777–1782.

Kriesel, J.D., Spruance, S.L., Prichard, M., Parker, J.N., and Kern, E.R. (2005). Recurrent antiviral-resistant genital herpes in an immunocompetent patient. J. Infect. Dis. *192*, 156–161.

Larder, B.A., and Darby, G. (1984). Virus drug-resistance: mechanisms and consequences. Antiviral Res. *4*, 1–42.

Larder, B.A., Kemp, S.D., and Darby, G. (1987). Related functional domains in virus DNA polymerases. EMBO J. *6*, 169–175.

Liuzzi, M., Déziel, R., Moss, N., Beaulieu, P., Bonneau, A.M., Bousquet, C., Chafouleas, J.G., Garneau, M., Jaramillo, J., Krogsrud, R.L. *et al.* (1994). A potent peptidomimetic inhibitor of HSV ribonucleotide reductase with antiviral activity *in vivo*. Nature *372*, 695–698.

Loregian, A., Papini, E., Satin, B.Marsden, H.S., Hirst, T.R., and Palù, G. (1999). Intranuclear delivery of an antiviral peptide mediated by the B subunit of Escherichia coli heat-labile enterotoxin. Proc. Natl. Acad. Sci. USA *96*, 5221–5226.

Lowe, D.M., Alderton, W.K., Ellis, M.R., Parmar, V., Miller, W.H., Roberts, G.B., Fyfe, J.A., Gaillard, R., Ertl, P., Snowden, W., and Littler, E. (1995). Mode of action of (*R*)-9-[4-hydroxy-2-(hydroxymethyl)butyl]guanine against herpesviruses. Antimicrob. Agents Chemother. *39*, 1802–1808.

Marsden, H.S., Murphy, M., McVey, G.L., MacEachran, K.A., Owsianka, A.M., and Stow, N.D. (1994). Role of the carboxy terminus of herpes simplex virus type 1 DNA polymerase in its interaction with UL42. J. Gen. Virol. *75*, 3127–3135.

Martin, J.L., Brown, C.E., Matthews-Davis, N., and Reardon, J.E. (1994). Effects of antiviral nucleoside analogs on human DNA polymerases and mitochondrial DNA synthesis. Antimicrob. Agents Chemother. *38*, 2743–2749.

McClements, W., Yamanaka, G., Garsky, V., Perry, H., Bacchetti, S., Colonno, R., and Stein, R.B. (1988). Oligopeptides inhibit the ribonucleotide reductase of herpes simplex virus by causing subunit separation. Virology *162*, 270–273.

Miller, W.H., and Miller, R.L. (1980). Phosphorylation of acyclovir monophosphate by GMP kinase. J. Biol. Chem. *255*, 7204–7207.
Miller, W.H., and Miler, R.L. (1982). Phosphorylation of acyclovir diphosphate by cellular enzymes. Biochem. Pharmacol. *31*, 3879–3884.
Mineta, T., Rabkin, S.D., and Martuza, R.L. (1994). Treatment of malignant gliomas using ganciclovir-hypersensitive, ribonucleotide reductase-deficient herpes simplex viral mutant. Cancer Res. *54*, 3963–3966.
Moss, N., Beaulieu, P., Duceppe, J.-S., Ferland, J.-M., Garneau, M., Gauthier, J., Ghiro, E., Goulet, S., Guse, I., Jaramillo, J. *et al.* (1996). Peptidomimetic inhibitors of herpes simplex virus ribonucleotide reductase with improved *in vivo* antiviral activity. J. Med. Chem. *39*, 4173–4180.
Mouly, F., Baccard, M., Scieux, C., Schnell, L., Locq-Ebner, S., Morinet, F., and Morel, P. (1995). Chronic recurrent acyclovir-resistsant genital herpes in an immunocompetent patient. Dermatology *190*, 177.
Newcomb, W.W., and Brown, J.C. (2002). Inhibition of herpes simplex virus replication by WAY-150138: assembly of capsids depleted of the portal and terminase proteins involved in DNA encapsidation. J. Virol. *76*, 10084–10088.
Oien, N.L., Brideau, R.J., Hopkins, T.A., Wieber, J.L., Knechtel, M.L., Shelly, J.A., Anstadt, R.A., Wells, P.A., Poorman, R.A., Huang, A. *et al.* (2002). Broad-spectrum antiherpesvirus activities of 4-hydroxyquinoline carboxamides, a novel class of herpesvirus polymerase inhibitors. Antimicrob. Agents Chemother. *46*, 724–730.
Orozco-Topete, R., Sierra-Madero, J., Cano-Dominguez, C., Kershenovich, J., Ortiz-Pedroza, G., Vazquez-Valls, E., Garcia-Cosio, C., Soria-Cordoba, A., Armendariz, A.M., Teran-Toledo, X. *et al.* (1997). Safety and efficacy of 'Virend® for topical treatment of genital and anal herpes simplex lesions in patients with AIDS. Antiviral Res. *35*, 91–103.
Ostrander, M., and Cheng, Y.-C. (1980). Properties of herpes simplex virus type 1 and type 2 DNA polymerase. Biochim. Biophys. Acta *609*, 232–245.
Paradis, H., Gaudreau, P., Brazeau, P., and Langelier, Y. (1988). Mechanism of inhibition of herpes simplex virus (HSV) ribonucleotide reductase by a nonapeptide corresponding to the carboxy-terminus of its subunit 2. J. Biol. Chem. *263*, 16045–16050.
Pelosi, E., Hicks, K.A., Sacks, S.L., and Coen, D.M. (1992). Heterogeneity of a herpes simplex virus clinical isolate exhibiting resistance to acyclovir and foscarnet. Adv. Exp. Med. Biol. *312*, 151–158.
Pelosi, E., Mulamba, G.B., and Coen, D.M. (1998a). Penciclovir and pathogenesis phenotypes of drug-resistant herpes simplex virus mutants. Antiviral Res. *37*, 17–28.
Pelosi, E., Rozenberg, F., Coen, D.M., and Tyler, K.L. (1998b). A herpes simplex virus DNA polymerase mutation that specifically attenuates neurovirulence in mice. Virology *252*, 364–372.
Pertel, P.E., and Spear, P.G. (1996). Modified entry and syncytium formation by herpes simplex virus type 1 mutants selected for resistance to heparin inhibition. Virology *226*, 22–33.
Pilger, B.D., Cui, C., and Coen, D.M. (2004). Identification of a small molecule that inhibits herpes simplex virus DNA polymerase subunit interactions and viral replication. Chem. Biol. *11*, 647–654.
Pue, M.A., and Benet, L.Z. (1993). Pharmacokinetics of famciclovir in man. Antiviral Chem. Chemother. *4*, 47–55.
Reardon, J.E. (1989). Herpes simplex virus type 1 and human DNA polymerase interactions with 2′-deoxyguanosine-5′-triphosphate analogues. J. Biol. Chem. *264*, 19039–19044.
Reardon, J.E., and Spector, T. (1989). Herpes simplex virus type 1 DNA polymerase. Mechanism of inhibition by acyclovir triphosphate. J. Biol. Chem. *264*, 7405–7411.
Roizman, B., and Knipe, D.M. (2001). Herpes simplex viruses and their replication. In Fields Virology, D.M. Knipe, P.M. Howley, D.E. Griffin, R.A. Lamb, M.A. Martin, B. Roizman, and S. E. Straus, eds. (Philadelphia, Lippincott Williams & Wilkins), pp. 2399–2459.
Sacks, S.L., Wanklin, R.J., Reece, D.E., Hicks, K.A., Tyler, K.L., and Coen, D.M. (1989). Progressive esophagitis from acyclovir-resistant herpes simplex. Clinical roles for DNA polymerase mutants and viral heterogeneity? Ann. Internal Med. *111*, 893–899.
Safrin, S. (1996). Herpes simplex virus and varicella zoster virus: nucleosides and foscarnet: clinical aspects. In Antiviral Drug Resistance, D.D. Richman, ed. (Chichester, John Wiley & Sons), pp. 103–122.
Sarisky, R.T., Nguyen, T.T., Duffy, K.E., Wittrock, R.J., and Leary, J.J. (2000). Difference in incidence of spontaneous mutations between herpes simplex virus types 1 and 2. Antimicrob. Agents Chemother. *44*, 1524–1529.
Sarisky, R.T., Quail, M.R., Clark, P.E., Nguyen, T.T., Halsey, W.S., Wittrock, R.J., Bartus, J.O., Van Horn, M.M., Sathe, G.M., Van Horn, S. *et al.* (2001). Characterization of herpes simplex viruses selected in culture for resistance to penciclovir or acyclovir. J. Virol. *75*, 1761–1769.

Sarup, J.C., and Fridland, A. (1987). Identification of purine deoxyribonucleoside kinasees from human leukemia cells: substrate activation by purine and pyrimidine deoxyribonucleosides. Biochemistry *26*, 590–597.

Sasadeusz, J.J., and Sacks, S.L. (1996). Spontaneous reactivation of thymidine kinase-deficient, acyclovir-resistant type 2 herpes simplex virus: masked heterogeneity or reversion? J. Infect. Dis. *174*, 476–482.

Sedarati, F., Javier, R.T., and Stevens, J.G. (1988). Pathogenesis of a lethal mixed infection in mice with two nonneuroninvasive herpes simplex virus strains. J. Virol. *62*, 3037–3039.

Slabaugh, M., Roseman, N., Davis, R., and Mathews, C. (1988). Vaccinia virus-encoded ribonucleotide reductase: sequence conservation of the gene for the small subunit and its amplification in hydroxyurea-resistant mutants. J. Virol. *62*, 519–527.

Spear, P.G., Eisenberg, R.J., and Cohen, G.H. (2000). Three classes of cell surface receptors for alphaherpesvirus entry. Virology *275*, 1–8.

Spector, F.C., Liang, L., Giordano, H., Sivaraja, M., and Peterson, M.G. (1998). Inhibition of herpes simplex virus replication by a 2-amino thiazole via interactions with the helicase component of the UL5-UL8-UL52 complex. J. Virol. *72*, 6979–6987.

Suzutani, T., Ishioka, K., De Clercq, E., Ishibashi, K., Kaneko, H., Kira, T., Hashimoto, K.-I., Ogasawara, M., Ohtani, K., Wakamiya, N., and Saijo, M. (2003). Differential mutation patterns in thymidine kinase and DNA polymerase genes of herpes simplex virus type 1 clones passaged in the presence of acyclovir and penciclovir. Antimicrob. Agents Chemother. *47*, 1707–1713.

Swetter, S.M., Hill, E.L., Kern, E.R., Koelle, D.M., Posavad, C.M., Lawrence, W., and Safrin, S. (1998). Chronic vulvar ulceration in an immunocompetent woman due to acyclovir-resistant, thymidine kinase-deficient herpes simplex virus. J. Infect. Dis. *177*, 543–550.

Tenser, R.B., and Edris, W.A. (1987). Trigeminal ganglion infection by thymidine kinase-negative mutants of herpes simplex virus after *in vivo* complementation. J. Virol. *61*, 2171–2174.

Tenser, R.B., Ressel, S., and Dunstan, M.E. (1981). Herpes simplex virus thymidine kinase expression in trigeminal ganglion infection: correlation of enzyme activity with ganglion virus titer and evidence of *in vivo* complementation. Virology *112*, 328–341.

Thomsen, D.R., Oien, N.L., Hopkins, T.A., Knechtel, M.L., Brideau, R.J., Wathen, M.W., and Homa, F.L. (2003). Amino acid changes within conserved region III of the herpes simplex virus and human cytomegalovirus DNA polymerases confer resistance to 4-oxo-dihyroquinolines, a novel class of herpesvirus antivirals. J. Virol. *77*, 1868–1876.

van Zeijl, M., Fairhurst, J., Jones, T.R., Vernon, S.K., Morin, J., LaRocque, J., Feld, B., O'Hara, B., Bloom, J.D., and Johann, S.V. (2000). Novel class of thiourea compounds that inhibit herpes simplex virus type 1 DNA cleavage and encapsidation: resistance maps to the UL6 gene. J. Virol. *74*, 9054–9061.

Vere Hodge, R.A., and Cheng, Y.-C. (1993). The mode of action of penciclovir. Antiviral Chem. Chemother. *4 Suppl. 1*, 13–24.

Vere Hodge, R.A., and Perkins, R.M. (1989). Mode of action of 9-(4-hydroxy-3-methylbut-1-yl)guanine (BRL 39123) against herpes simplex virus in MRC-5 cells. Antiviral Chem. Chemother. *33*, 223–229.

Wang, T.S. F. (1996). Cellular DNA polymerases. In DNA Replication in Eukaryotic Cells, M.L. DePamphilis, ed. (Cold Spring Harbor, Cold Spring Harbor Laboratory Press), pp. 461–493.

Wang, J., Sattar, A.K.M.A., Wang, C.C., Karam, J.D., Konigsberg, W.H., and Steitz, T.A. (1997). Crystal structure of a pol α family replication DNA polymerase from bacteriophage RB69. Cell *89*, 1087–1099.

Wild, K., Bohner, T., Aubry, A., Folkers, G., and Schulz, G.E. (1995). The three-dimensional structure of theymidine kinase from herpes simplex virus type 1. FEBS Lett. *368*, 289–292.

Wild, K., Bohner, T., Aubry, A., Folkers, G., and Schulz, G.E. (1997). The structures of thymidine kinase from herpes simplex virus type 1 complexed with substrates and a substrate analogue. Protein Sci. 6, 2097–2106.

Yeh, R., Coen, D.M. (2004). Pharmacology of viral replication. In: Golan DE, Tashjian AH Jr, Armstrong EJ, Galanter JM, Armstrong AW, Arnaout RA, Rose HS, eds. Principles of Pharmacology: The pathophysiologic basis of drug therapy. Baltimore: Lippincott, Williams & Wilkins., 545–564.

Zhang, J., Chung, D.W., Tan, C.-K., Downey, K.M., Davie, E.W., and So, A.G. (1991). Primary structure of the catalytic subunit of calf thymus DNA polymerase δ: Sequence similarities with other DNA polymerases. Biochemistry *30*, 11742–11750.

Zuccola, H.Z., Filman, D.J., Coen, D.M., and Hogle, J.M. (2000). The crystal structure of an unusual processivity factor, herpes simplex virus UL42, bound to the C-terminus of its cognate polymerase. Mol. Cell. 5, 267–278.

Imaging HSV-1 in Living Animals

20

Gary D. Luker

Abstract

Studies of viral and host factors that influence pathogenesis largely have used experimental mouse models that rely upon sacrifice of infected mice to determine distribution and titers of virus. While this experimental paradigm has provided important data, it precludes real-time investigations of the same animal over the entire course of disease progression. Therefore, conventional assays may miss unexpected sites or patterns of viral dissemination because appropriate tissues were not collected or analyzed properly. Our laboratory and others have begun to exploit recent advances in technology for imaging small animals for studies of viral-host pathogenesis. In particular, we have shown that bioluminescence imaging is a sensitive, reproducible method for monitoring recombinant a recombinant strain KOS HSV-1 virus that expresses firefly luciferase. Infection with luciferase-expressing viruses can be detected at multiple anatomic sites, including brain, and quantitative differences in emitted light correlate with relative differences in viral titer measured by standard plaque assays. In this chapter we summarize current technologies for small animal imaging and their use in studies pathogenesis of HSV-1 and other microbial pathogens. We also highlight future directions in which new imaging technologies may provide innovative approaches to interrogate viral-host pathogenesis in living animals.

Introduction

Herpes simplex virus type 1 (HSV-1) is a widespread human pathogen that infects approximately 80% of all persons by adulthood (Whitley, 1996). HSV-1 infection occurs at epithelial and mucosal surfaces, where viral replication and lysis of epithelial and stromal cells permit infection of sensory nerve termini. Virus is transported retrogradely in axons to sensory nerve ganglia. During acute infection, HSV-1 undergoes limited replication in ganglia, and virus may be transported in an anterograde fashion to cause disease via zosteriform spread to the same sensory dermatome as the initial site of infection. Host immunity ultimately controls acute replication of HSV-1, leading to establishment of latent infection in sensory nerve ganglia. Latent virus remains within neurons for the life of the host, and HSV-1 may reactivate from latency in response to poorly characterized stimuli to produce new infectious virions. Primary infection with HSV-1 and episodes of reactivation cause morbidity and some mortality from diseases including stromal keratitis and encephalitis.

HSV-1 infection also can spread widely to organs including lung, liver, and adrenals, particularly in neonates and patients with

AIDS, cancer, organ transplantation, or other immunocompromising diseases (Pillay *et al.*, 1993; Irie *et al.*, 1998; Kimberlin *et al.*, 2001). Diagnosis of disseminated herpes infection frequently is delayed in patients (Kimberlin *et al.*, 2001), particularly those without skin lesions (Arvin *et al.*, 1982). Because optimal therapy with antiviral drugs must be initiated before widespread dissemination of virus (Whitley *et al.*, 1991), systemic HSV-1 infection produces significant morbidity and mortality (Kimura *et al.*, 2002), and only modest improvements in patient prognosis have occurred in the past 20 years (Kimberlin *et al.*, 2001; Kimberlin, 2004). Twenty-five percent of neonates with herpes infection develop disseminated infection, and approximately 30% of these patients die from the disease (Kimberlin, 2004). In the neonatal patient population, severity of herpes hepatitis, as quantified by elevations of liver function enzymes, is used to determine prognosis, and liver dysfunction due to hepatitis is a common cause of death in these patients (Kimberlin *et al.*, 2001; Kimberlin, 2004).

Mouse models typically are used to investigate HSV-1 viral-host pathogenesis. These studies typically quantify disease progression based on observations of cutaneous disease, cultures of virus obtained from accessible epithelial surfaces, and sacrifice of animals at defined time points to measure distribution and titer of virus. While these methods have allowed investigators to define viral and host genes that regulate replication and virulence of HSV-1 *in vivo* (Pyles *et al.*, 1992; Leib *et al.*, 1999; Leib *et al.*, 2000; Taus and Mitchell, 2001; Feldman *et al.*, 2002), these assay techniques preclude real-time monitoring of the kinetics and extent of disease progression in the same animal. Significant insights from animal-to-animal variations in host–pathogen relationships may be missed with conventional protocols, particularly for experiments that rely upon sacrifice of mice to localize and quantify viral infection and response to therapy. Furthermore, spread of HSV-1 to unexpected anatomic sites beyond the site of infection and innervating neurons also may be missed because the infected tissue is not assayed for virus.

In this chapter, I describe the current technologies for small animal imaging that are available to study HSV-1 and other pathogens in mouse models of disease. In particular, I focus on bioluminescence imaging and its applications to monitor spatial and temporal progression of disease *in vivo*. Bioluminescence and other noninvasive imaging techniques are reproducible assays for noninvasively detecting and measuring viral replication throughout an animal. These imaging technologies enable effects of host immunity or therapeutic interventions on HSV-1 pathogenesis to be investigated in living mice. I expect that noninvasive imaging techniques will continue to develop as important new tools to study HSV-1 and other pathogens.

Small animal imaging technologies

Recent advances in small animal imaging technologies now enable us to interrogate cellular and molecular processes in the context of a living mouse (Weissleder, 2002). Specific molecular events regulating processes such as cell migration, signal transduction, proliferation, and apoptosis all can be investigated with noninvasive imaging, allowing studies to be performed in the integrated physiology of an intact animal. Small animal imaging provides the opportunity to study disease progression in the same cohort of animals over time. This strategy enables detection of animal-to-animal variations in infection and host immunity. Noninvasive imaging also reduces the number of animals needed to obtain statistically significant data in an experiment. Described below are key instruments used for noninvasive imaging in small animal models.

Bioluminescence imaging (BLI)

BLI detects light produced by the reaction of luciferase enzymes, such as those from firefly (*Photinus pyralis*) or sea pansy (*Renilla reniliformis*) with defined substrate molecules (luciferin for firefly and coelenterazine for *Renilla*). Pathogens and mammalian cells can be engineered to express one or more luciferase enzymes as reporters for *in vivo* imaging. Mammalian tissues do not produce the substrates for these enzymes, so either luciferin or coelenterazine must be administered to a mouse before each imaging study. Luciferin rapidly distributes throughout a mouse and penetrates intact blood-tissue barriers, such as the placenta and brain, after intraperitoneal injection. Bioavailability of coelenterazine, the substrate for *Renilla* luciferase, is limited *in vivo*, at least in part by the multidrug resistance P-glycoprotein (MDR1) (Pichler *et al.*, 2004). Because firefly and *Renilla* luciferase use different substrates, it is possible to use these reporters to image two different biological processes in the same mouse. This imaging strategy can be limited for imaging HSV-1 infection because of limited access of coelenterazine into the nervous system (Luker *et al.*, 2002a). By comparison, bacteria, such as *Photorhabdus luminescens* or *Xenorhabdus luminescens*, contain a lux operon that encodes genes for synthesis of luciferase and luciferin. Other bacteria have been engineered to express the entire lux operon from a plasmid or chromosomal site, making these organisms constitutively bioluminescent. However, the lux operon has not been successfully transferred to other species.

BLI uses a sensitive charge-coupled device camera to detect bioluminescence *in vivo*. Because BLI has very low background noise, it is very sensitive for detection of bioluminescence. We have shown that ocular infection with as few as 1×10^2 PFU of strain KOS engineered with a firefly luciferase reporter (KOS/dlux) can be detected in a living mouse (Luker *et al.*, 2002a). BLI is a rapid and relatively inexpensive imaging modality that allows large numbers of mice to be examined repetitively. In our experience, approximately 20 mice can be imaged in approximately 1 hour. Bioluminescence can be measured by computer analysis of emitted photons, allowing relative quantification of data. Assuming that expression of the luciferase enzyme is maintained by the engineered pathogen or cell, increases numbers of viruses or cells correlate directly with greater bioluminescence. However, absolute values for bioluminescence produced by a defined number of viruses will differ in various anatomic sites because of absorption and scattering of light by overlying tissues.

Although BLI is the most commonly used imaging technique for studying pathogenesis, there are important limitations to this modality. In addition to attenuation by overlying tissues, bioluminescence is decreased by pigmentation of organs, such as liver and spleen, and fur. This affects studies of viral-host pathogenesis because many genetically engineered mice are developed in SV129 or C57BL/6 backgrounds. Shaving mice with electric clippers and removing fur with a depiliatory agent (such as Nair) can minimize this problem. Alternatively, mice can be bred into an albino C57BL/6 background. BLI currently is a two-dimensional imaging technique with a spatial resolution of 2–3 mm. To localize sites of bioluminescence in three-dimensional space, it is necessary to acquire images in multiple projections. There is ongoing research to develop BLI instruments with capabilities for three-dimensional imaging. Even with advanced instrumentation, it still may be difficult to separate photons produced by viruses in two immediately adjacent sites, such as eye and periocular tissues. Finally, BLI is unlikely to be used in humans because of the limitations in detecting light produced in deep tissues.

Optical imaging of fluorescent probes

Fluorescence imaging is another relatively inexpensive, high-throughput technique for small animal studies. Viruses and other pathogens can be engineered to express fluorescent proteins, such as green fluorescent protein (GFP) and its derivatives. Although these reporters are commonly used for *in vitro* studies, sensitivity of detection *in vivo* is limited by tissue autofluorescence and absorption and scattering of the excitation and emitted light from these fluorescent proteins. Therefore, imaging of fluorescent proteins in mice typically is best in superficial sites or surgically exposed tissues.

Absorption of light by tissues and autofluorescence are dependent highly on the wavelength of light. Imaging with near infrared fluorescence (700–900 nm) (NIRF) maximizes penetration into tissues and minimizes autofluorescence (Weissleder, 2002). Several NIRF fluorescent molecules, such as Cy5.5, have been conjugated to peptides, antibodies, or enzyme substrates and used for *in vivo* imaging (Ntziachristos *et al.*, 2003). For example, a NIRF probe for HSV-1 UL26 protease has been used for *in vivo* imaging in mice (see below).

Most whole-body fluorescence imaging has been performed with two-dimensional detection systems. Even with the improved tissue penetration of NIRF probes, two-dimensional fluorescence imaging lacks sensitivity for detecting fluorescence from deep tissues and internal organs. Similar to BLI, relative fluorescence can be quantified in various anatomic sites, although absolute levels of fluorescence are affected by properties including pigmentation and amounts of overlying tissue. Recently, a system has been developed for fluorescence mediated tomography, a three-dimensional imaging technique (Graves *et al.*, 2003). This system enables sensitive, quantitative detection of NIRF probes, even in deep tissues. In living animals, fluorescence mediated tomography has 1–2 mm spatial resolution. The primary limitation of this technique is that the instrumentation is not available commercially.

Positron emission tomography (PET)

PET is an imaging technique designed to detect positron-emitting radiotracers such as ^{18}F and ^{124}I. Within tissues, positrons emitted from these radionuclides interact with electrons and release energy in the form of γ-rays. These γ-rays are detected by a ring of detectors that allow the site and amount of released energy to be quantified by computer algorithms. PET isotopes typically are used as radiotracer tags for molecules that bind to cell surface receptors or are substrates for specific enzymes, such as HSV-1 thymidine kinase.

One of the primary advantages of PET is that amounts of radionuclide in a tissue can be quantified precisely. Therefore, PET can be used to repetitively quantify kinetics of ligand-receptor binding or enzyme activity, at least in humans. PET images can be acquired as tomographic "slices" through animals or patients, enabling three-dimensional localization of radiotracer accumulation with 1–2 mm spatial resolution. PET also is used in clinical medicine, so molecular probes validated in animal studies potentially can be translated directly into patients.

PET is limited by the high cost of instrumentation for both animal and human imaging. Because the commonly used isotopes for PET have half-lives ranging from minutes to a days, imaging probes must be synthesized very shortly before use. In most cases, this means that a cyclotron and dedicated radiochemistry laboratory must be on site to efficiently use this imaging technology. The short half-lives of PET isotopes also limit the numbers of animals that can be studied in an experiment because 2–4 animals typically can be imaged per day.

Magnetic resonance imaging (MRI)

MRI is based on the principle that unpaired nuclear spins, such as hydrogen atoms in water, align with an externally applied magnetic field. This alignment is disturbed with a rapid radiofrequency pulse. After the radiofrequency pulse, electromagnetic energy is released as nuclei re-align with the magnetic field. The timing of the radiofrequency pulse and recording of the released energy can be varied to produce images with differing magnetic contrast. Variations in the magnetic properties of tissues and administered probes for imaging allow tissues to be distinguished.

MRI images have high tissue contrast and excellent spatial resolution (10–100 μm), making MRI the optimal imaging technique for defining anatomy. MRI is a three-dimensional imaging technique, so tissues and imaging probes can be localized in all parts of an animal or patient. Because of the ability to load cells *ex vivo* with large amounts of magnetic nanoparticles, MRI frequently is used for studies of cell trafficking. For example, as few as 10–100 labeled cells have been detected *in vivo* with MRI (Lewin *et al.*, 2000). Because MRI is the primary imaging modality used for studies of the nervous system in patients, results from small animal models of HSV-1 pathogenesis potentially can be translated directly to clinical medicine.

Despite the success of MRI for tracking cell populations *in vivo*, this imaging technique is limited by the difficulty in concentrating sufficient amounts of a contrast agent in a given anatomic site to produce a detectable change in signal. Typically, micromolar amounts of a magnetic contrast agent are required, as compared with the nanomolar or picomolar amounts of imaging probe needed for NIRF or PET. MRI studies are relatively long (30–60 minutes), so it is not a high throughput imaging modality. Finally, instrumentation for small-animal MRI is not widely available because of its cost.

Imaging HSV-1

The noninvasive imaging technologies described in section II have been used to study HSV-1 infection and gene therapy applications using amplicon vectors based on HSV-1. In both of these paradigms, noninvasive imaging provides the unique opportunity to detect and quantify spatial and temporal progression of viral spread in a living animal. These imaging techniques can be performed repetitively in the same mouse, thereby enabling investigators to control for animal-to-animal variations in disease outcome. As described below, the use of imaging to provide nearly real time monitoring of viral infection has provided important new insights into HSV-1 biology *in vivo*.

Imaging HSV-1 pathogenesis with reporter viruses

To image HSV-1 infection in mouse models of disease, we utilized a recombinant strain KOS reporter virus that expresses firefly (FL) and *Renilla* (RL) luciferases from the promoters of early genes UL29 and UL30, respectively (Kos/Dlux) (Summers *et al.*, 2001; Summers and Leib, 2002). The reporter construct was inserted into a site that interrupts the C-terminal 12 amino acids of UL49.5 (Fig. 20.1). This insertion likely did not affect synthesis or targeting of the UL49.5 protein. Unlike many HSV-1 reporter viruses, replication and spread of Kos/Dlux was comparable to wild type strain Kos *in vitro* or *in vivo*. This important feature of Kos/Dlux allowed us to probe the pathogenesis of HSV-1 infection *in vivo* without confounding effects of disrupting important determinants of viral virulence.

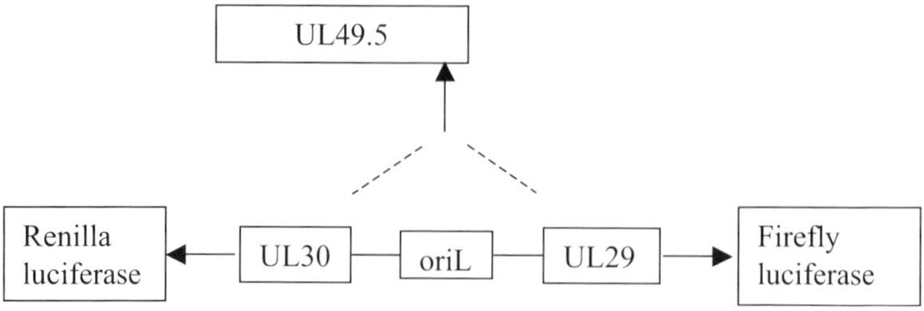

Figure 20.1 Kos/Dlux. The reporter cassette was inserted into a unique site at the C-terminus of UL49.5. Firefly and *Renilla* luciferase enzymes are expressed from the promoters of UL29 and UL30, respectively.

We used BLI of Kos/Dlux to investigate HSV-1 replication in living mice. Robust expression of FL from the reporter virus allowed viral replication to be detected and quantified following multiple routes of infection, such as footpad, intraperitoneal, corneal, and intracranial sites (Fig. 20.2) (Luker *et al.*, 2002a). By comparison, the RL reporter had limited efficacy *in vivo*. Bioluminescence from RL could not be detected after systemic administration of the coelenterazine substrate, despite evidence of RL activity in tissue ho-

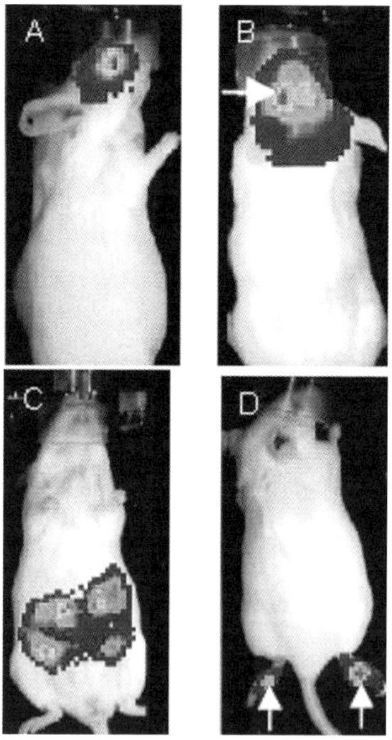

Figure 20.2 Imaging infection in multiple anatomic sites. Bioluminescence images 3 days after Kos/Dlux infection via corneal (A),intracranial (B), intraperitoneal (C),and footpad (D) routes. 2 x 10^6 PFU Kos/Dlux was used for corneal, intraperitoneal, and footpad infections, while intracranial infections were performed with 2 x 10^4 PFU. Images were acquired for 5 minutes with a 10 cm field of view (FOV) and 256 x 256 imaging matrix. Arrows denote foci of increased bioluminescence from Kos/Dlux at sites of injection for intracranial and footpad infections. Relative bioluminescence is depicted with a pseudocolor scale where the highest and lowest amounts of photon flux are shown in red and blue, respectively.

mogenates. RL activity only could be detected with BLI when the substrate was applied topically to the site of infection. The limited biodistribution of coelenterazine to sites of HSV-1 infection may be in part due to effects of P-glycoprotein as described above. Because of the limitations of RL, all subsequent studies of HSV-1 pathogenesis with Kos/Dlux were performed with BLI of the FL reporter.

Replication and temporal progression of Kos/Dlux as measured with BLI reproduced the pathogenesis of HSV-1 infection established by conventional assays. In the corneal model of infection, FL activity quantified by region-of-interest analysis increased in direct proportion to increasing titers of virus (Fig. 20.3). Bioluminescence from Kos/Dlux peaked at days 5–6 postinfection, corresponding to the known time course of HSV-1 spread retrograde from trigeminal ganglia to periocular tissues (Summers *et al.*, 2001). Decreases in FL activity quantified by BLI directly correlated with reductions in viral titers measured in tear film material following systemic therapy with the established antiviral drug valacyclovir. Collectively, these data validated BLI of Kos/Dlux as a reporter for HSV-1 infection and showed the value of this technique for monitoring effects of therapy directed against HSV-1.

BLI imaging of Kos/Dlux also has been used to investigate effects of interferon signaling on HSV-1 pathogenesis. Interferons are the major component of innate immunity to viral infection because they induce key antiviral genes in infected cells, control antiviral activity in uninfected cells, and regulate adaptive immunity (Stark *et al.*, 1998). In particular, type I interferons have been established as key mediators of host immunity to acute HSV-1 infection. However, functions of interferons or other components of host immunity in restricting HSV-1 replication to epithelial surfaces and innervating neurons

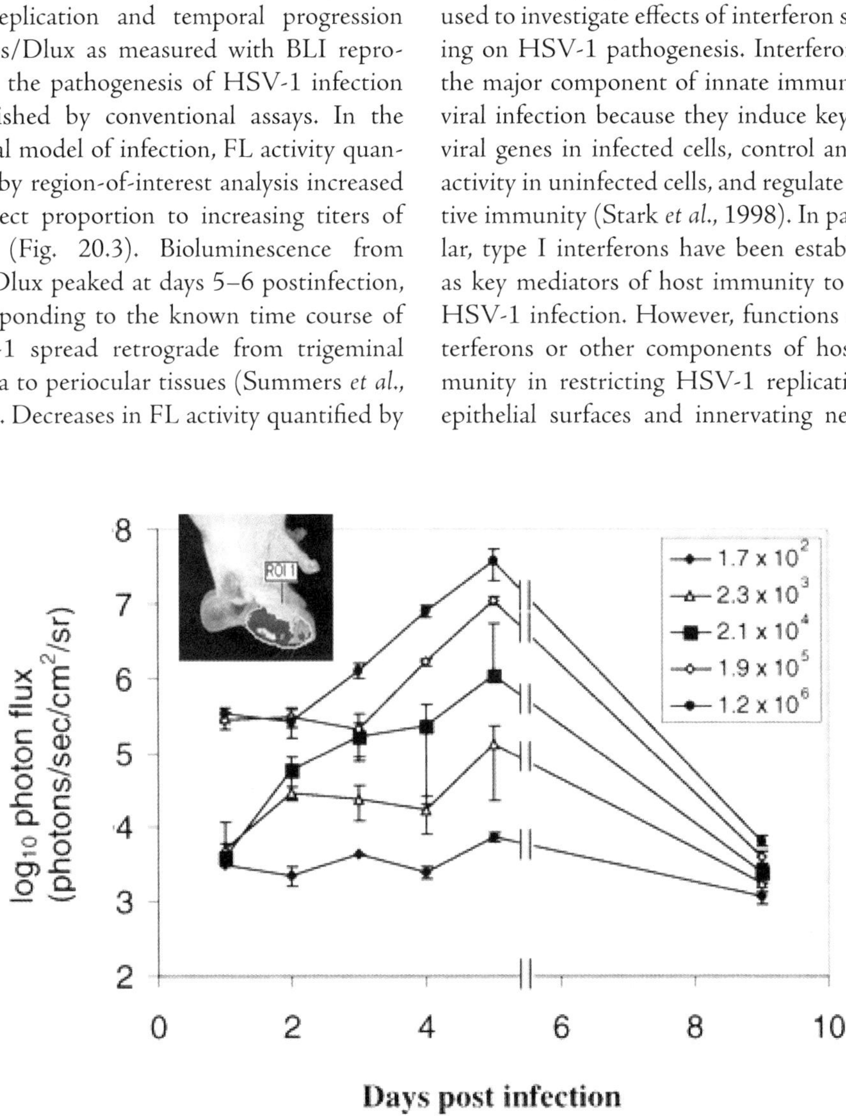

Figure 20.3 Increasing viral titers produce greater bioluminescence. Mice were infected with various doses of Kos/Dlux ranging from 1.7 x 10^5 to 1.2 x 10^6. Bioluminescence was over the 9-day course of acute infection quantified by region of interest analysis as shown on the inset figure of a mouse.

had not been established. Using BLI imaging, we confirmed that HSV-1 replication at the site of infection increased in mice with genetic deletion of type I or type I and II interferon receptors (Luker *et al.*, 2003b). Unexpectedly, systemic spread of HSV-1 occurred in mice lacking type I interferon receptors (Fig. 20.4). Systemic dissemination of virus occurred after footpad or corneal infection, showing that it was independent of the route of infection. Mice lacking receptors for both type I and II interferon had significantly greater spread of HSV-1 to multiple parenchymal organs. In particular, these mice had very high levels of HSV-1 in the liver, as evidenced by viral titers of approximately 10^7 PFU/ml/g of tissue. While mice lacking type I interferon receptors survived acute infection, mice lacking both type I and II interferon receptors had systemic spread of Kos/Dlux and rapidly died from disseminated disease. By comparison, mice lacking only type II interferon receptors did not differ from wild type mice. Collectively, these studies showed the power of noninvasive imaging to identify unexpected sites of viral infection and established that interferons are key regulators of HSV-1 tropism.

Imaging HSV-1 amplicons

HSV-1 amplicon vectors have been used commonly for gene therapy applications because they infect a wide-range of dividing and nondividing cell types, have a large capacity for insertion of transgenes, and are easy to manipulate. As investigators continue to develop these and other vectors for gene therapy, it has become apparent that monitoring the distribution of vectors and levels of gene expression are essential for interpreting experimental outcomes. A variety of noninvasive imaging techniques have been used to interrogate these nonreplicating derivatives of HSV-1 in living mice.

Effective delivery of sufficient HSV-1 amplicons to the target site is one important limitation on successful therapy with these agents. It has been difficult to analyze effective delivery of HSV-1 amplicons using conventional techniques. To quantify relative numbers of amplicons delivered *in vivo*, Schellingerhout *et al.* labeled amplicons with a radiotracer prior to use in an animals bearing experimental brain tumors (Schellingerhout *et al.*, 2000). Distribution and clearance of radiolabeled amplicons could be detected with

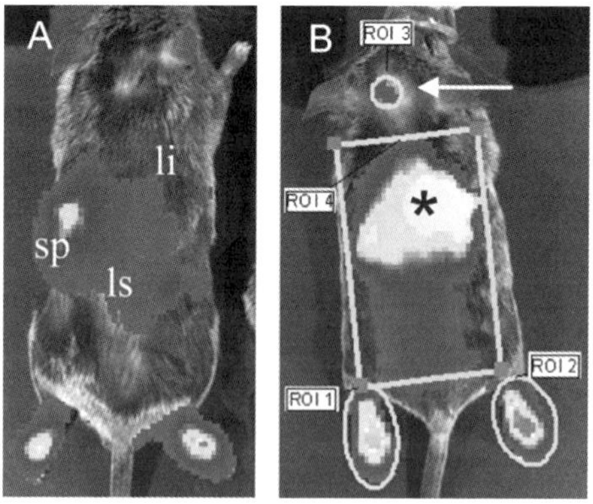

Figure 20.4 Bioluminescence images of mice lacking type I (A) or types I and II interferon receptors (B) 3 days after footpad infection with 2 x 10^6 PFU Kos/Dlux. Images show dissemination of virus to liver (li), spleen (sp), and lumbar spine (ls). The asterisk in B shows markedly greater bioluminescence in the liver of the mouse lacking type I and II interferon receptors, while the white arrow shows focal HSV-1 spread to the brain. Manually defined regions of interest for quantifying data are shown in B.

radiotracer imaging, allowing researchers to analyze various routes of administration for targeting and persistence in brain tumors. These mass distribution studies of HSV-1 allow studies of organ and vector features that affect successful targeting *in vivo*.

Imaging also has been used successfully to monitor therapeutic effects of transgenes expressed by HSV-1 amplicons in living mice. For example, HSV-1 vectors have been used to target a biologically active form of caspase-1 to tumors *in vitro* and in tumor xenografts *in vivo* (Messerli *et al.*, 2004). Activity of caspase-1 expressed from the viral vector was assayed with NIRF, using a probe that was activated to produce fluorescence after specific cleavage by caspase 1. Functional caspase 1 and the subsequent induction of apoptosis in tumor cells could be detected with fluorescence imaging. This system provided an effective method to quantify biologic effects of gene therapy with HSV-1 amplicons in living mice. This study provides an important proof-of-principle that molecular imaging can be used to detect and quantify specific effects of HSV-1 amplicon gene therapy.

In vivo imaging of other host–pathogen interactions

In addition to studies of HSV-1 pathogenesis, noninvasive imaging has been used to study other viral and bacterial pathogens. Similar to published work with HSV-1, most of these studies have been performed with BLI, primarily due to the ability to engineer viral and bacterial pathogens to express luciferase enzymes. For viral studies, investigators have used recombinant viruses that express luciferase enzymes such as firefly luciferase (*Photinus pyralis*). As described above, firefly luciferase requires systemic administration of the D-luciferin substrate to produce bioluminescence. By comparison, many bacterial pathogens have been engineered to express the lux operon from *Photorhabdus luminescens* or *Xenorhabdus luminescens*. The lux operon encodes all proteins necessary to produce the enzyme and substrate for luciferase activity, so bacteria expressing the lux operon are constitutively bioluminescent. Noninvasive imaging also has been used to localize and monitor HSV-1 amplicon vectors used for gene therapy or treatment of cancer. Examples of these applications of noninvasive imaging in host–pathogen interactions are described below. In these examples, we will highlight the important insights and discoveries about host–pathogen interactions that have been identified with molecular imaging.

BLI of Sindbis virus infection

Alphaviruses are positive-stranded RNA viruses that cause encephalitis in animals and humans (Irusta and Hardwick, 2004). Sindbis virus causes encephalitis and paralysis in mice, making this virus an important model for studying pathogenesis of alphaviruses. Cook and Griffin engineered Sindbis viruses to express firefly luciferase and used BLI to investigate mechanisms of viral spread from the periphery to the central nervous system (Cook and Griffin, 2003). This experimental question had been difficult to study with conventional techniques because free virus in blood interferes with determinations of viral titers in organs and tissues. BLI detects only virus that is replicating within cells because firefly luciferase requires ATP to produce bioluminescence. Therefore, this important characteristic of BLI allowed these researchers to distinguish intracellular virus from extracellular virus free in plasma.

Using BLI imaging to monitor viral replication over time, these investigators determined that there are two routes for recombinant Sindbis viruses to enter the central nervous system. Viruses replicated in noses of infected mice before entering the brain, indicating that the nasal neuroepithelium provided access to the central nervous system.

Viruses also replicated in the lower spinal cord prior to entry into the brain, suggesting that Sindbis virus is transported retrogradely in axons from a peripheral site of infection to the spine and brain. These conclusions were based on BLI data showing that viral replication occurred in the nose or lower spinal cord, respectively, before viral bioluminescence was detected in brain.

This study also emphasized important caveats of BLI for studying viral–host pathogenesis. Sindbis viruses expressing firefly luciferase were attenuated compared with the parental viruses. Because the degree of attenuation was similar among all tested strains, useful comparisons among various viruses were possible. BLI was approximately 10-fold less sensitive than plaque assay for detecting virus in tissues. Reduced sensitivity of BLI is due to attenuation of light by fur and pigmentation of skin and organs. Loss of firefly luciferase gene expression also could reduce detection of virus *in vivo*, although the authors found that this occurred very rarely with their recombinant Sindbis viruses. These potential pitfalls of reporter viruses and BLI are necessary to consider for all imaging studies of microbial pathogenesis.

PET imaging of HSV-1 thymidine kinase (TK)

TK from HSV-1 has been used extensively in a variety of preclinical models of suicide gene therapy for cancer. Compared with endogenous mammalian thymidine kinase, HSV-1 TK preferentially phosphorylates nucleoside analogs such as ganciclovir or penciclovir. Phosphorylated nucleoside analogs are retained within cells and incorporated into newly synthesized DNA, arresting cells in S phase of the cell cycle and ultimately triggering apoptosis. Because of the clinical interest in monitoring gene therapy *in vivo*, imaging strategies have been developed for detecting and quantifying activity of HSV-1 TK in living animals and humans. Imaging of HSV-1 TK has been accomplished with PET or single-photon (SPECT) radiotracer imaging, using various radiolabeled substrates of HSV-1 TK as imaging probes.

Bennett *et al.* used [^{124}I]-5-iodo-2′-fluoro-1-beta-d-arabinofuranosyl-uracil (FIAU) as a PET reporter probe for detecting TK in recombinant HSV-1 in a model on oncolytic viral therapy (Bennett *et al.*, 2001). Based on accumulation of ^{124}I-FIAU, TK activity could be detected by PET 8 hours after injection of virus into subcutaneous tumor xenografts in nude mice. Amounts of radiotracer in tumors increased progressively through 48 hours, showing that PET could detect viral replication *in vivo*. PET also could detect 0.5 log increments in amounts of virus injected into tumors, demonstrating that this imaging modality can be used to quantify amounts of virus *in vivo*.

^{124}I FIAU and 9-[4-[^{18}F]fluoro-3-(hy droxymethyl)butyl]guanine ([^{18}F]FHBG), another PET radiotracer for HSV-1 TK, have been administered to humans in preliminary imaging studies. In normal volunteers, [^{18}F]FHBG showed desirable properties of stability, rapid clearance from blood, low background, and acceptable radiation dosimetry in humans, making this tracer a potential probe for HSV-1 TK in patients (Yaghoubi *et al.*, 2001). A phase I/II clinical trial of gene therapy for the brain tumor glioblastoma multiforme showed the feasibility of using PET to image HSV-1 TK in humans (Jacobs *et al.*, 2001). Localization and magnitude of expression of HSV-1 TK could be determined based on accumulation of ^{124}I FIAU in tumors. In principle, these studies also demonstrate the potential for ultimately using molecular imaging to specifically detect HSV-1 infection in patients.

There also are important limitations of PET for *in vivo* studies of HSV-1 pathogenesis. The sensitivity of PET for detecting HSV-1 is significantly less than optical techniques such as BLI. While PET did not detect less than 1×10^7 PFU of recombinant HSV-1 in a tumor (Bennett *et al.*, 2001), as few as 10^2–10^3 PFU of KOS/dlux could be detected in an ocular model of infection (Luker *et al.*, 2002a). Nucleoside substrates for HSV-1 TK currently require synthesis and radiotracer labeling in a dedicated radiochemistry laboratory. For compounds labeled with ^{18}F, the short half-life of the PET isotope (approximately 2 hours) limits the number of studies that can be performed per day. The radiolabeled nucleoside analogs also do not cross the intact blood–brain barrier, potentially limiting early detection of HSV-1 infection. These examples highlight the strengths and weaknesses of PET versus BLI or other optical techniques that must be considered in imaging HSV-1 viral–host pathogenesis.

NIRF imaging of HSV-1 protease activity

The UL26 gene of HSV-1 encodes a protease that cleaves peptides with a recognition motif of LVLA*SSSFGYC (* = cleavage site) (Liu and Roizman, 1991). Function of the UL26 protease has been imaged *in vivo* by exploiting the cleavage specificity of this enzyme. Shah *et al.* designed a NIRF probe that included the LVLASSSFGYC peptide sequence between two Cy5.5 fluorochromes (Shah *et al.*, 2004). In the intact molecule, fluorescence is self-quenched because of the close proximity of fluorophores. Cleavage of the imaging substrate by UL26 separates the Cy5.5 molecules and produces fluorescence in the near infrared spectrum, which has favorable transmission through tissues.

The UL26 protease was expressed from a HSV-1 amplicon vector in subcutaneous tumor xenografts in mice. After systemic administration of the NIRF probe, fluorescence was detected in tumors that were transduced with UL26, while no fluorescence was observed in control tumors. These results demonstrate the stability of the UL26 reporter probe in the absence of specific viral protease activity, thereby producing a favorable signal-to-noise ratio for imaging. Although these data were obtained in the context of HSV-1 amplicon vectors for gene therapy of cancer, the results show that HSV-1 protease activity can be detected *in vivo*. This same imaging strategy should be applicable to imaging studies of HSV-1 infection, providing a means to detect and quantify virus in mouse models of disease. As technologies for imaging near infrared fluorescence are developed for humans, this NIRF probe or molecules similar to it may enable specific diagnosis of HSV-1 infection *in vivo*.

BLI of bacterial infection

Noninvasive imaging has been used more extensively to study pathogenesis of bacterial infections, and most of these studies have been performed with BLI. As described above, bacteria typically are engineered with a lux operon, thereby making the organism constitutively bioluminescent. While use of the lux operon eliminates the need to administer luciferin to mice for imaging, loss of luciferase expression is a common problem unless the operon is integrated into the bacterial genome (Hardy *et al.*, 2004; Yu *et al.*, 2004). Selected studies of bacterial pathogenesis using BLI are described below, highlighting the use of noninvasive imaging to provide new insights into host–pathogen interactions.

Hardy *et al.* used bioluminescent *Listeria monocytogenes* to monitor progression of infection in living mice (Hardy *et al.*, 2004). Based on BLI data for localization of bacteria during the course of infection, these researchers made the new observation that extracellular *L. monocytogenes* was present in the lumen of the gall bladder. This localization occurred in both diseased and asymptomatic mice, suggesting that listerosis could be spread asymptomatically through a fecal–oral route of dissemination.

BLI also has been used to identify specific populations of immune cells that are needed for the host response to infection. Bitmansour *et al.* used a bioluminescent strain of *Pseudomonas aeruginosa* to study the localization and kinetics of clearance of the pathogen following stem cell transplantation (Bitmansour *et al.*, 2002). By following the progression of infection over time, these authors determined that common myeloid and granulocyte-monocyte progenitor cells were needed to reduce the bacterial load early after transplantation. This study emphasizes the use of BLI to monitor microbial-host interactions in the same mouse, thereby limiting the number of animals needed to define the kinetics of bacterial infection and clearance.

Future directions

While initial studies have demonstrated that noninvasive imaging can be used to investigate HSV-1 and other pathogens in living animals, imaging of host–pathogen interactions is in the early stages of development. Essentially all research in this field has used reporter viruses to detect, localize, and quantify sites of infection. In this regard, imaging has been used as a tool for cell tracking, using a labeled pathogen. However, noninvasive imaging technologies offer the potential to interrogate both viral and host components of pathogenesis on a cellular and molecular level.

T lymphocytes have been established as key components of the host immune response to both acute and latent HSV-1 infection (Khanna *et al.*, 2003; Smith *et al.*, 2004). Characterizing the spatial and temporal responses of these cells to HSV-1 is likely to provide important new insights into progression and control of infection. Imaging has been used to investigate kinetics and localization of T lymphocyte trafficking *in vivo*. For example, T lymphocytes have been labeled with iron oxide particles (Kircher *et al.*, 2003) or transduced ex vivo with viral vectors for firefly luciferase (Nakajima *et al.*, 2001) and/or HSV-1 thymidine kinase (Dubey *et al.*, 2003) for adoptive transfer studies. Trafficking of T cells to tumors or sites of inflammatory arthritis have been studies in living mice with these labeling methods. Similar approaches can be used to interrogate T lymphocytes in the context of HSV-1 infection. Studies of T lymphocyte trafficking may have particular importance in investigating the viral and host factors that regulate latency and reactivation from latency. This is an important potential direction for imaging studies of HSV-1 because of the difficulty in studying dynamic processes in latency and reactivation in the *in vivo* setting.

In addition to tracking cellular events, techniques are available and are under continued development to monitor specific molecular events *in vivo*. Included in these are strategies to monitor protein–protein interactions in signaling pathways (Luker *et al.*, 2002b; Paulmurugan *et al.*, 2002; Ray *et al.*, 2002; Luker *et al.*, 2004), protein degradation (Luker *et al.*, 2003a; Zhang *et al.*, 2004), apoptosis (Laxman *et al.*, 2002), and regulation of gene expression (Doubrovin *et al.*, 2001). Progress in design and delivery of chemical and pharmacologic imaging reporters, such as the probe for HSV-1 UL26 described above, also will contribute to detecting molecular signals and events *in vivo*. Using a combination of selective reporter molecules with genetically modified viruses and host animals, it should be feasible to monitor defined signaling pathways in host immunity and the extent to which HSV-1 alters these responses during infection.

Imaging studies of pathogenesis of HSV-1 and other infectious agents have been performed on a whole-body level. However, recent studies have shown that two-photon microscopy can be used to investigate normal physiology and disease processes *in vivo* (Alexandrakis *et al.*, 2004; Sumen *et al.*, 2004). For example, intravital microscopy has been used to characterize phases of antigen presentation by dendritic cells to T lymphocytes in intact lymph nodes (Mempel *et al.*, 2004). These techniques have not been applied to studies of microbial pathogenesis. Using a combination of reporter genes, such as luciferase and fluorescent proteins, it should be feasible to track HSV-1-host interactions on the macroscopic scale and then focus a microscopic analysis on a specific tissue of interest, such as the local site of infection or a regional lymph node.

Collectively, published data demonstrate that noninvasive imaging methods are powerful new tools for investigating pathogenesis of HSV-1 and other viruses *in vivo*. In particular, *in vivo* imaging has enabled investigators to detect unexpected sites of infection that potentially would have been missed with conventional assays of excised organs and tissues. We expect that macroscopic and microscopic *in vivo* imaging of HSV-1 pathogenesis will become increasingly important for defining key viral and host mediators of disease.

Acknowledgments

I thank David Leib and Kathryn Luker for critical review of this chapter and their ongoing support and encouragement.

References

Alexandrakis, G., Brown, E., Tong, R., McKee, T., Campbell, R., Boucher, Y., and Jain, R. (2004). Two-photon fluorescence correlation microscopy reveals the two-phase nature of transport in tumors. Nat. Med. *10*, 203–207.

Arvin, A., Yeager, A., Bruhn, F., and Grossman, M. (1982). Neonatal herpes simplex virus infection in the absence of mucocutaneous lesions. J. Pediatr. *100*, 715–721.

Bennett, J., Tjuvajev, J., Johnson, P., Doubrovin, M., Akhurst, T., Malholtra, S., Hackman, T., Balatoni, J., Finn, R., Larson, S. *et al.* (2001). Positron emission tomography imaging for herpes virus infection: Implications for oncolytic viral treatments of cancer. Nat. Med. *7*, 859–863.

Bitmansour, A., Burns, S., Traver, D., Akashi, K., Contag, C., Weissman, I., and Brown, J. (2002). Myeloid progenitors protect against invasive aspergillosis and Pseudomonas aeruginosa infection following hematopoietic stem cell transplantation. Blood *100*, 4660–4667.

Cook, S., and Griffin, D. (2003). Luciferase imaging of a neurotropic viral infection in intact animals. J. Virol. *77*, 5333–5338.

Doubrovin, M., Ponomarev, V., Beresten, T., Balatoni, J., Bommann, W., Finn, R., Humm, J., Larson, S., Sadelain, M., Blasberg, R., and Gelovani Tjuvajev, J. (2001). Imaging transcriptional regulation of p53-dependent genes with positron emission tomography *in vivo*. Proc. Natl. Acad. Sci. USA *98*, 9300–9305.

Dubey, P., Su, H., Adonai, N., Du, S., Rosato, A., Braun, J., Gambhir, S., and Witte, O. (2003). Quantitative imaging of the T cell antitumor response by positron-emission tomography. Proc. Natl. Acad. Sci. USA *100*, 1232–1237.

Feldman, L., Ellison, A., Voytek, C., Yang, L., Krause, P., and Margolis, T. (2002). Spontaneous molecular reactivation of herpes simplex virus type 1 latency in mice. Proc. Natl. Acad. Sci. USA *99*, 978–983.

Graves, E., Ripoll, J., Weissleder, R., and Ntziachristos, V. (2003). A submillimeter resolution fluorescence molecular imaging system for small animal imaging. Med. Phys. *30*, 901–911.

Hardy, J., Francis, K., DeBoer, M., Chu, P., Gibbs, K., and Contag, C. (2004). Extracellular replication of Listeria monocytogenes in the murine gall bladder. Science *303*, 851–853.

Irie, H., Koyama, H., Kubo, H., Fukuda, A., Aita, K., Koike, A., Yoshimura, A., Yoshida, T., Shiga, J., and Hill, T. (1998). Herpes simplex virus hepatitis in macrophage-depleted mice: the role of massive, apoptotic cell death in pathogenesis. J. Gen. Virol. *79*, 1225–1231.

Irusta, P., and Hardwick, J. (2004). Neuronal apoptosis pathways in Sindbis virus encephalitis. Prog. Mol. Subcell. Biol. *36*, 71–93.

Jacobs, A., Voges, J., Reszka, R., Lercher, M., Grossman, A., Kracht, L., Kaestle, C., Wagner, R., Wienhard, K., and Heiss, W. (2001). Positron-emission tomography of vector-mediated gene expression in gene therapy for gliomas. Lancet *358*, 727–729.

Khanna, K., Bonneau, R., Kinchington, P., and Hendricks, R. (2003). Herpes simplex virus-specific memory CD8+ T cells are selectively activated and retained in latently infected sensory ganglia. Immunity 25, 593–603.

Kimberlin, D. (2004). Neonatal herpes simplex infection. Clin. Microbiol. Rev. *17*, 1–13.

Kimberlin, D., Lin, C., Jacobs, R., Powell, D., Frenkel, L., Gruber, W., Rathore, M., Bradley, J., Diaz, P., Kumar, M. *et al.* (2001). Natural history of neonatal herpes simplex virus infections in the acyclovir era. Pediatrics *108*, 223–229.

Kimura, H., Ito, Y., Futamura, M., Ando, Y., Yabuta, Y., Hoshino, Y., Nishiyama, Y., and Morishima, T. (2002). Quantitation of viral load in neonatal herpes simplex virus infection and comparison between type 1 and type 2. J. Med. Virol. *67*, 349–353.

Kircher, M., Allport, J., Graves, E., Love, V., Josephson, L., Lichtman, A., and Weissleder, R. (2003). *In vivo* high resolution three-dimensional imaging of antigen-specific cytotoxic T-lymphocyte trafficking to tumors. Cancer Res. *63*, 6838–6846.

Laxman, B., Hall, D., Bhojani, M., Hamstra, D., Chenevert, T., Ross, B., and Rehemtulla, A. (2002). Noninvasive real-time imaging of apoptosis. Proc. Natl. Acad. Sci. USA *99*, 16551–16555.

Leib, D., Harrison, T., Laslo, K., Machalak, M., Moorman, N., and Virgin, H. (1999). Interferons regulate the phenotype of wild-type and mutant herpes simplex viruses *in vivo*. J. Exp. Med. *189*, 663–672.

Leib, D., Machalak, M., Williams, B., Silverman, R., and Virgin, H. (2000). Specific phenotypic restoration of an attenuated virus by knockout of a host resistance gene. Proc. Natl. Acad. Sci. USA *97*, 6097–6101.

Lewin, M., Carlesso, N., Tung, C., Tang, X., Cory, D., Scadden, D., and Weissleder, R. (2000). Tat peptide-derivatized magnetic nanoparticles allow *in vivo* tracking and recovery of progenitor cells. Nat. Biotechnol. *18*, 410–414.

Liu, F., and Roizman, B. (1991). The herpes simplex virus 1 gene encoding a protease also contains within its coding domain the gene encoding the more abundant substrate. J. Virol. *65*, 5149–5156.

Luker, G., Bardill, J., Prior, J., Pica, C., Piwnica-Worms, D., and Leib, D. (2002a). Noninvasive bioluminescence imaging of herpes simplex virus type 1 infection and therapy in living mice. J. Virol. *76*, 12149–12161.

Luker, G., Sharma, V., Pica, C., Dahlheimer, J., Li, W., Ochesky, J., Ryan, C., Piwnica-Worms, H., and Piwnica-Worms, D. (2002b). Noninvasive imaging of protein–protein interactions in living animals. Proc. Natl. Acad. Sci. USA *99*, 6961–6966.

Luker, G., Pica, C., Song, J., Luker, K., and Piwnica-Worms, D. (2003a). Imaging 26S proteasome activity and inhibition in living mice. Nat. Med. *9*, 969–973.

Luker, G., Prior, J., Song, J., Pica, C., and Leib, D. (2003b). Bioluminescence imaging reveals systemic dissemination of HSV-1 in the absence of interferon receptors. J. Virol. *77*, 11082–11093.

Luker, K., Smith, M., Luker, G., Gammon, S., Piwnica-Worms, H., and Piwnica-Worms, D. (2004). Kinetics of regulated protein–protein interactions revealed with firefly luciferase complementation imaging in cells and living animals. Proc. Natl. Acad. Sci. USA *101*, 12288–12293.

Mempel, T., Hendrickson, S., and von Andrian, U. (2004). T-cell priming by dendritic cells in lymph nodes occurs in three distinct phases. Nature *427*, 154–159.

Messerli, S., Prabhakar, S., Tang, Y., Shah, K., Cortes, M., Murthy, V., Weissleder, R., Breakefield, X., and Tung, C. (2004). A novel method for imaging apoptosis using a caspase-1 near-infrared fluorescent probe. Neoplasia *6*, 95–105.

Nakajima, A., Seroogy, C., Sandora, M., Tarner, I., Costa, G., Taylor-Edwards, C., Bachman, M., Contag, C., and Fathman, C. (2001). Antigen-specific T cell-mediated gene therapy in collagen-induced arthritis. J. Clin. Invest. *107*, 1293–1301.

Ntziachristos, V., Bremer, C., and Weissleder, R. (2003). Fluorescence imaging with near-infrared light: new technological advances that enable *in vivo* molecular imaging. Eur. Radiol. *13*, 195–208.

Paulmurugan, R., Umezawa, Y., and Gambhir, S. (2002). Noninvasive imaging of protein–protein interactions in living subjects by using reporter protein complementation and reconstitution strategies. Proc. Natl. Acad. Sci. USA *99*, 15608–15613.

Pichler, A., Prior, J., and Piwnica-Worms, D. (2004). Imaging reversal of multidrug resistance in living mice with bioluminescence: MDR1 P-glycoprotein transports coelenterazine. Proc. Natl. Acad. Sci. USA *101*, 1702–1707.

Pillay, D., Lipman, M., Lee, C., Johnson, M., Griffiths, P., and McLaughlin, J. (1993). A clinico-pathological audit of opportunistic viral infections in HIV-infected patients. AIDS *7*, 969–974.

Pyles, R., Sawtell, N., and Thompson, R. (1992). Herpes simplex virus type 1 dUTPase mutants are attenuated for neurovirulence, neuroinvasiveness, and reactivation from latency. J. Virol. *66*, 6706–6713.

Ray, P., Pimenta, H., Paulmurugan, R., Berger, F., Phelps, M., Iyer, M., and Gambhir, S. (2002). Noninvasive quantitative imaging of protein–protein interactions in living subjects. Proc. Natl. Acad. Sci. USA *99*, 3105–3110.

Schellingerhout, D., Rainov, N., Breakefield, X., and Weissleder, R. (2000). Quantitation of HSV mass distribution in a rodent brain tumor model. Gene Ther. *7*, 1648–1655.

Shah, K., Tung, C., Yang, K., Weissleder, R., and Breakefield, X. (2004). Inducible release of TRAIL fusion proteins from a proapoptotic form for tumor therapy. Cancer Res. *64*, 3236–3242.

Smith, C., Wilson, N., Waithman, J., Villadangos, J., Carbone, F., Heath, W., and Belz, G. (2004). Cognate CD4(+) T cell licensing of dendritic cells in CD8(+) T cell immunity. Nat. Immunol. *Oct 10 [Epub ahead of print]*.

Stark, G., Kerr, I., Williams, B., Silverman, R., and Schreiber, R. (1998). How cells respond to interferons. Ann Rev Biochem *67*, 227–264.

Sumen, C., Mempel, T., Mazo, I., and von Andrian, U. (2004). Intravital microscopy: visualizing immunity in context. Immunity *21*, 315–329.

Summers, B., and Leib, D. (2002). Herpes simplex virus type 1 origins of DNA replication play no role in the regulation of flanking promoters. J. Virol. *76*, 7020–7029.

Summers, B., Margolis, T., and Leib, D. (2001). Herpes simplex virus type 1 corneal infection results in periocular disease by zosteriform spread. J. Virol. *75*, 5069–5075.

Taus, N., and Mitchell, W. (2001). The transgenic ICP4 promoter is activated in schwann cells in trigeminal ganglia of mice latently infected with herpes simplex virus type. J. Virol. *75*, 10401–10408.

Weissleder, R. (2002). Scaling down imaging: molecular mapping of cancer in mice. Nat. Rev. Cancer *2*, 11–18.

Whitley, R. (1996). Herpes simplex viruses. In B Fields, D Knipe, PM Howley *et al.* (ed) Fields Virology, 3rd edn Lippincott-Raven Publishers, Philadelphia, PA, pp. 2297–2342.

Whitley, R., Arvin, A., Prober, C., Corey, L., Burchett, S., Plotkin, S., Starr, S., Jacobs, R., Powell, D., Nahmias, A., *et al.* (1991). Predictors of morbidity and mortality in neonates with herpes simplex virus infections. The National Institute of Allergy and Infectious Diseases Collaborative Antiviral Study Group. N. Engl. J. Med. *324*, 450–454.

Yaghoubi, S., Barrio, J., Dahlbom, M., Iyer, M., Namavari, M., Satyamurthy, N., Goldman, R., Herschman, H., Phelps, M., and Gambhir, S. (2001). Human pharmacokinetic and dosimetry studies of [(18)F]FHBG: a reporter probe for imaging herpes simplex virus type-1 thymidine kinase reporter gene expression. J. Nuclear Med. *42*, 1225–1234.

Yu, Y., Shabahang, S., Timiryasova, T., Zhang, Q., Beltz, R., Gentschev, I., Goebel, W., and Szalay, A. (2004). Visualization of tumors and metastases in live animals with bacteria and vaccinia virus encoding light-emitting proteins. Nat. Biotechnol. *22*, 313–320.

Zhang, G., Safran, M., Wei, W., Sorenson, E., Lassota, P., Zhelev, N., Neuberg, D., Sharpiro, G., and Kaelin, W.J. (2004). Bioluminescent imaging of Cdk2 inhibition *in vivo*. Nat. Med. *10*, 643–648.

Index

The index is not exhaustive. Page numbers listed correspond to the first appearance of each entry.